HANDBOOK OF SYNFUELS TECHNOLOGY

OTHER McGRAW-HILL HANDBOOKS OF INTEREST

American Institute of Physics • American Institute of Physics Handbook
Baumeister • Marks' Standard Handbook for Mechanical Engineers
Brater and King • Handbook of Hydraulics
Chow • Handbook of Applied Hydrology
Condon and Odishaw • Handbook of Physics
Considine • Process Instruments and Controls Handbook
Considine • Energy Technology Handbook
Crocker and King • Piping Handbook
Davis and Sorensen • Handbook of Applied Hydraulics
Dean • Lange's Handbook of Chemistry
Fink and Beaty • Standard Handbook for Electrical Engineers
Fink and Christiansen • Electronics Engineers' Handbook
Harris • Handbook of Noise Control
Harris and Crede • Shock and Vibration Handbook
Hicks • Standard Handbook of Engineering Calculations
Juran • Quality Control Handbook
Karassik, Krutzsch, Fraser, and Messina • Pump Handbook
Lyman • Handbook of Chemical Property Estimation Methods
Mantell • Engineering Materials Handbook
Maynard • Handbook of Business Administration
Maynard • Industrial Engineering Handbook
Morrow • Maintenance Engineering Handbook
O'Brien • Scheduling Handbook
Perry • Engineering Manual
Perry and Chilton • Chemical Engineers' Handbook
Raznjevic • Handbook of Thermodynamic Tables and Charts
Rosaler and Rice • Standard Handbook of Plant Engineering
Rothbart • Mechanical Design and Systems Handbook
Schweitzer • Handbook of Separation Techniques for Chemical Engineers
Shugar • Chemical Technicians' Ready Reference Handbook
Smeaton • Switchgear and Control Handbook
Streeter • Handbook of Fluid Dynamics
Tuma • Engineering Mathematics Handbook
Tuma • Handbook of Physical Calculations
Tuma • Technology Mathematics Handbook
Turner and Malloy • Thermal Insulation Handbook
Turner and Malloy • Handbook of Thermal Insulation Design Economics

HANDBOOK OF SYNFUELS TECHNOLOGY

Robert A. Meyers, Editor in Chief
TRW Chemical Technology Operations
Redondo Beach, California

McGRAW-HILL BOOK COMPANY

New York	Johannesburg	Panama
St. Louis	London	Paris
San Francisco	Madrid	São Paulo
Auckland	Mexico	Singapore
Bogotá	Montreal	Sydney
Hamburg	New Delhi	Tokyo

7227-5637

CHEMISTRY

Library of Congress Cataloging in Publication Data
Main entry under title:

Handbook of synfuels technology.

    Includes index.
    1. Synthetic fuels—Handbooks, manuals, etc.
I. Meyers, R. A. (Robert Allen), date.
TP360.H36   1984   662'.66   83-17505
ISBN 0-07-041762-8

1 2 3 4 5 6 7 8 9 0   KGP KGP   8 9 8 7 6 5 4 3

ISBN 0-07-041762-8

The editors for this book were Harold B. Crawford and Ruth L. Weine, the
production supervisor was Thomas G. Kowalczyk. It was set in Garamond by
University Graphics, Inc.

Printed and bound by The Kingsport Press

CONTENTS

v

ADVISORY BOARD

About the Editor

ROBERT A. MEYERS, Ph.D., is presently Manager, Energy and Natural Resources, TRW Energy Development Group, Redondo Beach, California. He was graduated from California State University (San Diego) with distinction in chemistry, received his Ph.D. in chemistry from UCLA, and was a post-doctoral fellow and faculty member at California Institute of Technology. He is the author or editor of four previous books, has published 50 scientific papers, and holds 12 U.S. patents. He is a member of the American Chemical Society and a Senior Member of the American Institute of Chemical Engineers.

CONTRIBUTORS

Joseph H. Absil, Roberts & Schaefer Resource Service Inc., Chap. 4-6

Friedrich Bögner, Davy McKee Aktiengesellschaft, Köln, FRG, Chap. 3-6

Clement W. Bowman, Ph.D., Alberta Oil Sands Technology and Research Authority, Edmonton, Alberta, Canada, Chap. 5-2

R. L. Braddock, Tosco Corporation, Chap. 4-2

Robert G. Brand, Foster Wheeler Synfuels Corporation, Chap. 3-8

Dillason F. Bress, Foster Wheeler SPEC, Inc., Chap. 3-8

James C. Bronfenbrenner, International Coal Refining Company, Chap. 1-3

C. Terry Brooks, British Gas Corporation, Westfield Development Centre, Scotland, Chap. 3-4

T. A. Cavanaugh, Exxon Research and Engineering Company, Chap. 1-1

Richard S. Chillingworth, Cities Service Research and Development Co., Chap. 6-3

Scott L. Darling, Combustion Engineering, Inc., Chap. 3-3

H. Z. Dokuzoguz, KBW Gasification Systems, Inc., Chap. 3-5

David P. Domenicis, Westinghouse Electric Corporation, Chap. 3-11

J. M. Dufallo, Air Products and Chemicals, Inc., Chap. 6-4

W. R. Epperly, Exxon Research and Engineering Company, Chap. 1-1

Harold L. Erskine, Erskine Associates, Chap. 5-1

G. Escher, Vebe Oel Entwicklungsgesellschaft mbH, Gelsenkirchen-Scholven, FRG, Chap. 1-4

Joseph M. Glassett, P.E., Paraho Development Corporation, Chap. 4-3

Kenneth E. Hastings, Cities Service Research and Development Co., Chap. 6-3

Charles K. Holt, Westinghouse Electric Corporation, Chap. 3-11

D. E. Jones, The Babcock & Wilcox Company, Chap. 3-5

David M. Jones, Babcock Woodall-Duckham, Ltd., Crawley, England, Chap. 3-9

John B. Jones, Jr., Paraho Development Corporation, Chap. 4-3

Anthony Y. Kam, Mobil Research and Development Corporation, Chap. 2-3

J. F. Kamody, Koppers Company, Inc., Chap. 3-5

Allen R. Kuhns, International Coal Refining Company, Chap. 1-3

J. Langhoff, Ruhrkohle Oel und Gas GmbH, Bottrop, FRG, Chap. 1-4

Mitchell A. Lekas, Geokinetics Inc., Chap. 4-5

Peter F. Mako, Fluor Engineering, Inc., Advanced Technology Division, Chap. 2-1

S. F. Massenzio, Exxon Research and Engineering Company, Chap. 6-1

H. J. Michaels, Koppers Company, Inc., Chap. 3-5

Dr. Rer. Nat. Reiner Mueller, Saarberg + Dr. C. Otto Gesellschaft für Kohledruckvergasung mbH, Saarbruecken, FRG, Chap. 3-10

James R. O'Leary, International Coal Refining Company, Chap. 1-3

Joseph E. Papso (Deceased), Hydrocarbon Research, Inc. (Dynalectron Corporation), Chaps. 1-2 and 6-2

Marianne T. Phillips, International Coal Refining Company, Chap. 1-3

Raymond S. Phillips, Alberta Oil Sands Technology and Research Authority, Edmonton, Alberta, Canada, Chap. 5-2

Dipl. Math. Hartmut Pitz, Saarberg + Dr. C. Otto Gesellschaft für Kohledruckvergasung mbH, Saarbruecken, FRG, Chap. 3-10

John D. Potts, Cities Service Research and Development Co., Chap. 6-3

P. B. Probert, The Babcock & Wilcox Company, Chap. 3-5

Dr. Ing. Rochus F. Quinkler, Lurgi Kohle und Mineralöltechnik GmbH, Frankfort am Main, FRG, Chap. 2-4

Dr. Ing. Roland W. Rammler, Lurgi Kohle und Mineralöltechnik GmbH, Frankfurt am Main, FRG, Chap. 4-1

Philip C. Reeves, Roberts & Schaefer Resource Service Inc., Chap. 4-6

P. L. Rogerson, Imperial Chemical Industries, Ltd., Cleveland, England, Chap. 2-2

Paul F. H. Rudolph, D.Sc., Lurgi Kohle und Mineralöltechnik GmbH, Frankfurt am Main, FRG, Chap. 3-7

William A. Samuel, P.E., Fluor Engineering, Inc., Advanced Technology Division, Chap. 2-1

W. D. Schlinger, Texaco Inc., Chap. 3-1

Max Schreiner, Mobil Research and Development Corporation, Chap. 2-3

W. A. Schwartz, Air Products and Chemicals, Inc. Chap. 6-4

Charles E. Scott, Cities Service Research and Development Co., Chap. 6-3

Thomas E. Siebert, Tosco Corporation, Chap. 4-2.

John Ward Smith, Consultant, Chap. 4-7

Gregory D. Snyder, International Coal Refining Company, Chap. 1-3

Antonio V. Soriano, Allis-Chalmers Coal Gas Corporation, Chap. 3-12

D. C. Spence, Air Products and Chemicals, Inc., Chap. 6-4

Aldred L. Stevens, Occidental Oil Shale, Inc., Chap. 4-4.

Henry J. F. Stroud, D.Sc., British Gas Corporation, Research and Development Division, Midlands Research Station, England, Chap. 3-4

Dipl. Ing. Emil Supp, Lurgi Kohle und Mineralöltechnik GmbH, Frankfurt am Main, FRG, Chap. 2-4

Marney D. Talbert, Occidental Oil Shale, Inc., Chap. 4-4.

Keith R. Tart, British Gas Corporation, Research and Development Division, Midlands Research Station, England, Chap. 3-4

Calvin H. Taylor, Babcock Woodall-Duckham Ltd., Crawley, England, Chap. 3-13

L. Robert Turner, Alberta Oil Sands Technology and Research Authority, Edmonton, Alberta, Canada, Chap. 5-2

Maarten J. Vanderburgt, Shell Internationale Petroleum, Maatschaapij B.V., The Hague, Netherlands, Chap. 3-2

Erich V. Vogt, Shell Internationale Petroleum Maatschaappij B.V., The Hague, Netherlands, Chap. 3-2

D. T. Wade, Exxon Research and Engineering Company, Chap. 1-1

C. S. Waitman, Tosco Corporation, Chap. 4-2

Dr. Ing. Hans-Juergen Weiss, Lurgi Kohle und Mineralöltechnik GmbH, Frankfurt am Main, FRG, Chap. 4-1

Paul J. Weller, Shell Internationale Petroleum, Maatschaapij B.V., The Hague, Netherlands, Chap. 3-2

Karl Wintrup, Davy McKee Aktiengesellschaft, Köln, FRG, Chap. 3-6

Sergei Yurchak, Mobil Research and Development Corporation, Chap. 2-3

Samuel Znaimer, International Coal Refining Company, Chap. 1-3

PREFACE

Within the next twenty years, world petroleum and natural gas production will level off and begin a slow decline, even though there will be both new discoveries and application of advanced recovery methods. However, increasing world population and aspirations toward improved quality of life require increased utilization of liquid fuels for transportation and gaseous fuels for heating and chemicals production.

Synthetic fuels from coal, oil shale and oil sands are needed to make up the shortfall in liquid and gaseous fuels. Plants producing these synthetic fuels are expected to provide on the order of five million barrels/day ($8 \times 10^5 m^3$ or $8 \times 10^8 L/day$) of liquid fuels by the year 2000 and an even larger energy equivalent of synthesis, medium-Btu and synthetic (substitute) natural gases.

This Handbook is international in scope and presents chapters on each of the synthetic fuels technologies which are either presently on stream or at pilot plant or demonstration plant scale. Most of the technologies are presently licensable and will be heavily utilized in meeting future global fuel needs.

Technologies developed in the Federal Republic of Germany, England, the Netherlands, Canada, and South Africa are presented along with the latest systems developed in the United States.

The Handbook is intended to become the major basic source of synfuels technology information. Engineers and scientists in government and industry who are performing research or planning projects will use the Handbook to provide comparative data on synfuels technology and economics. Managers and legal, advocate, and media personnel will use the Handbook as a reference for evaluating proposals, performing planning functions, assessing information, and for preparing presentations to higher management or to the public.

All of the oil, coal and natural resource companies should have access to this Handbook at a number of levels: central engineering, corporate research, technology and licensing and commercial development. Engineering, design and construction companies as well as consultants must also have copies of this book. Finally, the book will serve as a resource text for college courses on synthetic fuels production.

Insofar as is appropriate for each technology, the technology chapters were prepared by the process developers or the licensors in accordance with the following chapter format.

1. General process description and block flow diagram

2. Process chemistry and thermodynamics for each major process unit

3. Pilot plant description and data (This section of each chapter is applicable to processes not yet proven on a commercial scale.)

4. **Process Perspective.** Developers, sponsors, location, and specifications of all test and commercial plants; near-term and long-term plans.

5. **Detailed process description.** Process flow diagram(s) with mass and energy balances for major process variations and feeds; details of unique or key equipment; illustrations of each major unit operation and the internals of special equipment with legends and annotations; and photographs of total plant, if constructed.

6. **Product and By-Product Specifications.** Detailed analyses of all process synfuels and by-products as a function of processing variations and feeds, product stability, and inter-changeability with conventional fuels.

7. **Upgrading technology.** Conversion of raw product to refinery feed grade, as appropriate.

8. **Wastes and emissions.** Analyses of process solid, liquid, and gaseous emissions as a function of processing variations and feeds.

9. **Process economics.** Installed capital cost by major section for one or more developer-selected plant sizes; total capital investment, with the basis of each factor stated; operating costs by category, with utility and feedstock costs defined; annualized capital costs with basis; price for each synfuel and by-product on a stated basis.

10. **Summation.** Process cost per unit of product; energy efficiency; cost of water, electricity, and other utilities per unit of product.

Advanced technologies for the production of liquid fuels from coal and a chapter on the important Sasol synfuels complex in South Africa are presented in Parts 1 and 2 of the Handbook. These processes produce, variously, synthetic petroleum crudes, refined synthetic petroleum liquids, methanol, and gasoline.

Part 3 contains twelve gasification technologies; of these, several have been in operation at many sites for a number of years and others are at major pilot plant scale. In addition, there is a concluding chapter on synthetic gas treatment. Part 4 of the Handbook contains separate chapters on five shale oil processes, a chapter on preretorting beneficiation, and a survey chapter comprising the history of shale oil production and descriptions of additional technologies.

Oil from oil sands (Part 5) is presented in two chapters: one on the single technology now in commercial application and a survey chapter on the large number of emerging technologies in this field. The final section of the book deals with the upgrading and refining of the synthetic fuels obtained in the various processes. The four chapters of this section present major licensable upgrading technologies.

ACKNOWLEDGMENTS

A distinguished group of more than sixty scientists and engineers from twenty-seven firms and a research institute prepared the thirty-three chapters of this Handbook. These contributors are listed on pages ix to xi, and we wish to acknowledge the support of their sponsoring organizations.

An Advisory Board was established whose members contributed to the guidance documents which were supplied to each of the chapter authors. The members of this board are listed on page vii.

Our special thanks to Verna Malough who provided executive secretarial support and manuscript preparation liaison with the chapter authors.

I thank my wife, Ilene, for her constant encouragement and advice during the three years of organizing, writing, and assembling the Handbook.

PART 1

COAL LIQUEFACTION

Four plant-proven technologies for the direct conversion of coal to liquid hydrocarbon fuels are presented in this section. These are the Exxon Donor Solvent (EDS) Coal Liquefaction process, the H-Coal® process, the SRC-1 Coal Liquefaction process, and the process used at the coal-hydrogenation plant at Bottrop.

The technology of the latter process is important to an understanding of the history of liquefaction and is also an important technology for the future. Direct coal liquefaction began in the first quarter of this century with the work of Bergius, who produced coal liquids by reacting slurries or pastes of pulverized coal in oil with hydrogen at high temperatures and pressures. Bergius received the Nobel Prize in Chemistry for this work. The technology utilized at the coal-hydrogenation plant at Bottrop is based on Bergius' work and the subsequent Pier (IG-Farben or IG process) technology. The operational pressure was reduced from near 10,000 lb/in² (70 MPa) to approximately 4500 lb/in² (30 MPa) and other improvements were also introduced.

The three U.S. technologies presented in this section were developed within the past 20 years in response to government and industrial efforts to supplement potentially dwindling supplies of petroleum liquids with synthetic liquids derived from coal.

All the processes involve the addition of hydrogen to coal (either from the gas phase or via an externally hydrogenated coal-derived oil). Dissolver operating temperatures range from 425 to 500 °C and pressures from 2000 to 4500 lb/in² (14 to 30 MPa). Major differences arise in the mode of hydrogenation. The Bottrop technology utilizes a disposable "red mud" catalyst to hydrogenate coal directly, while the H-Coal® process, which is also a direct hydrogenation process, utilizes a reusable catalyst system and a novel reactor unit. Catalysts are not added, other than those naturally present in coal mineral matter, in the SRC-1 process during the primary hydrogenation of the coal slurry. The EDS process utilizes a hydrogenated solvent as the hydrogen transfer agent. The solvent is a coal-derived recycle oil that is catalytically hydrogenated externally.

The primary products obtained from the first stage of direct liquefaction are noncondensable gases, naphtha, middle distillates, vacuum gas oils, and undistillable high-pour-point "bottoms" that have an aromatic carbon content about double that of petroleum crude oils. The hydrogen-to-carbon ratio of the typical bituminous coal feed to a liquefaction plant is compared to that of synthetic coal liquids and petroleum in the table that follows. It can be seen that liquefaction produces a wide range of hydrogen

Hydrogen-to-carbon ratio for fuels

Fuel	H/C atomic ratio	Oxygen content, wt %
Coal	0.7–0.8	4–8
Synthetic coal liquids	0.8–1.9	0.5–4
No. 6 fuel oil	1.5	Nil
Arabian light crude	1.8	0.2

content in the liquid products. The lower hydrogen values correspond to the undistillable bottoms and the higher values are secondary "cracked" products produced from these initial liquids. The secondary products approach the hydrogen-to-carbon ratio of crude oil and are similar to the various cracked liquids obtained in a petroleum refinery, with the exception of somewhat higher contents of oxygen as well as nitrogen and sulfur.

The major upgrading technologies for producing the secondary synfuel liquid products are presented briefly in this section and again in detail in Part 5 of this Handbook.

EDS
COAL-LIQUEFACTION PROCESS

T. A. CAVANAUGH
W. R. EPPERLY
D. T. WADE

Exxon Research and Engineering Company
Clinton Township, New Jersey

GENERAL

Introduction

This chapter describes the EDS Coal Liquefaction process. Included are an overview of this government-and-industry cost-shared project and a description of Exxon's integrated approach to the development of the process. The status of the laboratory and engineering research and development studies and the status of the 250-ton/day (230-Mg/day) pilot plant program are also presented. The process description includes discussions of coal-feed flexibility, process flexibility, and potential product utilization schemes.

The EDS process is a hydrogenation process for the direct conversion of a broad range of coals to liquid hydrocarbons. The chemistry of the EDS process involves several steps. First, the coal dissolves in a hydrocarbon solvent and thermal cracking occurs under reaction conditions that convert the large coal molecules into unstable smaller segments called *free radicals*. Second, these unstable free radicals react with the hydrogenated solvent and gaseous hydrogen to form lower-molecular-weight compounds which are liquids at room temperature. The hydrogenated solvent is called a *donor solvent* because it "donates" hydrogen to the free radicals. Gaseous hydrogen is used to replenish the hydrogen content of the solvent in a separate reactor sequence. A distinctive feature of the EDS process is the controlled hydrogenation of the recycle solvent in a separate reactor. This procedure extends the life of the hydrogenation catalyst, improves the control of hydrogen consumption, and is accomplished by means of conventional petroleum technology.

The timing of a coal-liquefaction industry will depend upon economic and political conditions and the supply and demand for crude oil and other sources of energy. However, in view of the long lead time involved, coal-liquefaction technology has been developed to be in a position to form the technical base for an industry. Recent EDS project activities have demonstrated commercial readiness for the EDS process. Operation of the 250-ton/day (230-Mg/day) pilot plant has been completed, incorporating process and design improvements intended to reduce the cost of synthetic liquids. EDS coal-liquefaction technology is now at the point that a pioneer commercial-size plant could be designed and built.

Process Perspective

The EDS process has been developed as a joint undertaking between the U.S. government and private industry. This development project is a 10-year effort running from 1976 to 1985, at a total cost of about $341 million. Since the project began in 1976, sponsorship has increased and is now international in scope. The U.S. Department of Energy (DOE) funds a little less than one-half of the project. The second largest contributor is Exxon, which contributes about one-quarter of the total funding. Other sponsors include the Electric Power Research Institute (EPRI), the Japan Coal Liquefaction Development Company (JCLD, a consortium of 12 Japanese companies), Phillips Coal Company, Anaconda Minerals Company, Ruhrkohle AG (a coal producer from the Federal Republic of Germany), and ENI (the Italian National Oil Company).

The EDS project is administered according to a unique "cooperative agreement." Technical work on this project is carried out by Exxon Research and Engineering Company and other Exxon organizations as though it were an activity totally funded by Exxon. Responsibility for the management of the project also lies with Exxon Research and Engineering Company. Responsibility for policy guidance and the rights to technology and licensing royalties are shared among the sponsors. Also, the sponsors have provided technical guidance and background information to assist in process, economic, and engineering studies.

In order to best achieve commercial readiness, the EDS program integrates all phases of process development. Continuing bench-scale research, operation of small pilot units, and studies of engineering design and technology support the results of the operation of the 250-ton/day (230 Mg/day) coal-liquefaction pilot plant (ECLP). The design data for the development areas, e.g., slurry drying, liquefaction, distillation, solvent hydrogenation, bottoms processing, product quality, and environmental control, have been obtained in the most appropriate project area at the minimum development cost. An ancillary program has been developed within the EDS project to evaluate bottoms processing alternatives such as (1) a hybrid boiler for direct combustion to provide process heat and plant steam, and (2) partial oxidation to provide hydrogen and fuel gas. Other alternatives for bottoms processing such as FLEXICOKING* and other gasification techniques continue to be studied but were not part of the pilot plant program to demonstrate the commercial readiness of the EDS process.

General Process Description

A generalized configuration of the EDS process is shown in Fig. 1-1. Feed coal is crushed and then dried by mixing with hot recycle donor solvent and vacuum tower bottoms. The slurry of coal, solvent, and bottoms is fed along with gaseous hydrogen to the liquefaction reactor. The design of the liquefaction reactor is relatively simple, consisting of an upward plug-flow reactor with design conditions of 425 to 500 °C and about 2000 to 3000 lb/in² (14 to 21 MPa). The reactor product is separated via conventional atmospheric and vacuum distillation into light gases, C_5-to-177 °C naphtha, 177-to-427 °C distillate, 427-

*Exxon service mark.

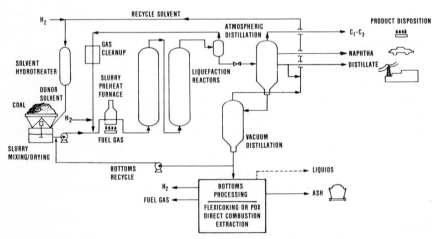

FIG. 1-1   Simplified flow diagram of EDS coal-liquefaction process.

to-538 °C vacuum gas oil (VGO), and vacuum bottoms containing 538 °C+ liquids, unconverted coal, and coal mineral matter. Conversion and liquid yields can be subsequently increased by recycling liquefaction VGO and/or vacuum bottoms to the liquefaction reactor. Part of the distillate stream is taken as the recycle solvent. This spent (dehydrogenated) solvent stream is hydrogenated in a conventional fixed-bed hydrotreater using a commercially available catalyst.

Utilization of the bottoms stream, which contains one-quarter to one-half of the available carbon in the feed coal, is necessary to achieve hydrogen and plant-fuel balances for the overall process, good carbon utilization, and minimum cost. The primary hydrocarbon sources for hydrogen and plant fuel are vacuum bottoms, coal, and light hydrocarbon gas. Bottoms can be processed in a FLEXICOKING unit to produce additional liquids and plant fuel and/or in a partial-oxidation unit to produce plant fuel or hydrogen, or it can be burned directly for process heat in a hybrid boiler. Coal is an alternative feed for the partial-oxidation unit and/or the hybrid boiler. Light hydrocarbon gas can be steam reformed to make hydrogen, burned directly as plant fuel, or sold as a product.

As mentioned, three alternatives are being considered for bottoms processing. The vacuum-bottoms stream can be fed to the FLEXICOKING process to produce additional liquid products and low-Btu fuel gas while concentrating the coal mineral matter for disposal. FLEXICOKING is a commercial petroleum process that employs integrated coking and gasification reactions in circulating fluidized beds. The process is a low-pressure [<50 lb/in^2 (7 kPa)] and intermediate-temperature (482 to 469 °C in the coker and 815 to 982 °C in the gasifier) operation. Up to 80 percent of the feed carbon can be converted to product liquid or fuel gas in the FLEXICOKING process. The remaining carbon is used for process fuel or is purged from the unit with the coal mineral matter.

Alternatively, partial oxidation of the vacuum bottoms can produce either process hydrogen or intermediate-Btu fuel gas. Using vacuum bottoms in this manner results in freeing the light-gas stream for sales or furnace fuel. In high-conversion coal-liquefaction operating modes, process hydrogen can also be generated by partial oxidation of coal. Partial oxidation is a commercial process employing oxygen to gasify petroleum fractions. Currently the option of feeding coal is being developed. This process does not recover additional liquid product but has the potential to consume effectively all the feed carbon in the production of hydrogen or fuel gas. Partial-oxidation units typically operate in the range of 1371 to 1538 °C and at 400 to 1000 lb/in^2 (3 to 7 MPa).

In another process option, the liquefaction bottoms, possibly with supplemental coal, can be fed to a hybrid boiler which supplies both the process heat and steam required for the plant. The hybrid boiler combines coal-fired boiler and process furnace technology. Steam is generated in the radiant section, and process heat is supplied from the convection section to the feed slurry of the liquefaction reactor. The hybrid boiler replaces the gas-fired preheat furnace for the liquefaction of slurry as well as much of the offsite steam-generation capacity.

Feed for the hybrid boiler can come from either liquefaction bottoms or coal. The boiler optionally incorporates recirculation of flue gas to vary the ratio of steam to process heat produced, depending on plant balances of steam and bottoms. Conventional flue-gas cleanup facilities are provided, including a flue-gas desulfurization unit and either electrostatic precipitators or baghouse filters.

Other options for providing plant hydrogen and/or fuel include a large number of different coal-gasification technologies.

Process Options

Slurry drying

The first step in the liquefaction of coal is the simultaneous slurrying and drying of crushed coal with a hot recycle solvent (and vacuum bottoms). Crushed coal is introduced to the slurry dryer vessel by means of rotary-valve or screw feeders. In the slurry dryer, the coal is thoroughly blended with the recycle streams and the resultant slurry is dried to a moisture content of less than 4 percent by weight of the dry coal feed. Figure 1-2 shows a simplified flow plan of the slurry drying system.

For high-moisture coals, such as Wyoming coal and Texas lignite, a series of mixed vessels may be utilized to efficiently dry the slurry to the desired moisture level. In all cases, an unmixed vapor disengagement vessel (not shown) is situated between the mixed vessel(s) and the slurry pumparound pumps to remove any entrained water vapor. The heat required for drying the coal is provided by the hot recycle streams and by heat exchangers in the slurry pumparound loop. These vertical downflow exchangers use steam to raise the slurry temperature. The recycle slurry is mixed with the hot recycle solvent (and vacuum bottoms) and returned to the slurry dryer. In a staged system, however, the hot recycle streams would be mixed with the coal in the first stage to provide heat for the initial drying. Heat for the final stage(s) could be provided by the slurry pumparound heat exchangers.

The overhead vapors from the slurry dryer(s) and the vapor disengagement drum are condensed and separated into sour water, distillate liquid, and off-gas. The distillate liquid is recycled to the suction of the high-pressure liquefaction feed pumps to prevent it from continually being stripped out of the slurry and possibly accumulating in the pumparound circuit. The off-gas is compressed and sent to offsite boilers as fuel.

Liquefaction

The dry feed slurry is pumped to high pressure, preheated, mixed with treat gas, and fed to a combined slurry and treat-gas furnace (see Fig. 1-3). The three-phase mixture is heated

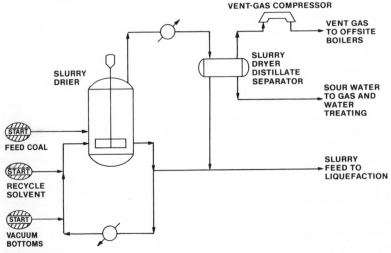

FIG. 1-2 EDS process scheme for slurry drying.

to the temperature of the liquefaction reactor inlet. The furnaces are vertical-tube box designs with vertical convection sections. To improve efficiency, each furnace preheats combustion air and fuel gas with flue gas.

The liquefaction reactors were designed for the appropriate residence time at reaction conditions. Two or three separate vessels in series are provided for each of the plant's parallel liquefaction lines. Flow is upward, with each reactor containing reaction zones of equal volume that are separated by internal distributors which allow an approach to plug-flow conditions. Solids withdrawal nozzles are provided to allow periodic removal of over-size coal or ash particles.

The total effluent from the reactor is separated into a hydrogen-rich off-gas stream and a coal-slurry stream in the liquefaction-reactor effluent separator. This vessel is designed to minimize entrainment of solids in the overhead vapor by using oil-washed sheds. Wash oil is recycled liquid from the downstream hot-separator drum. The coal-slurry stream from the reactor effluent separator is sent as feed to the flash zone of the atmospheric fractionator. The vapor from the reactor effluent separator is partially condensed in a conventional manner. Part of the hot liquid from the separator is recycled to the reactor effluent separator as wash oil, while the remainder becomes feed for the atmospheric fractionator. Vapor from the hot separator drum is quenched with recycled cold water from the separator to control deposition of ammonium salt. The quenched stream is cooled and separated in the cold separator drum into recycle gas, hydrocarbon liquid, and water. The hydrocarbon liquid is sent to the atmospheric fractionator. Part of the water is used as quench water, the remainder being sent for treating. The hydrogen-rich cold vapor from the separator is scrubbed with an amine solution in order to remove CO_2 and H_2S. Hydrogen makeup is combined with the recycle gas after a portion of this hydrogen gas is purged to control the purity of the treat gas. The resulting treat gas is compressed, preheated, and returned upstream of the slurry preheat furnace.

Distillation of liquefaction products

The liquefaction products are distilled in facilities consisting of an atmospheric fractionator, a vacuum fractionator and vacuum system, and the associated drums, pumps, and heat exchangers. All of these facilities employ conventional refining equipment, although modifications have been required because of the characteristics of the materials being handled.

The atmospheric fractionator separates the liquefaction products into off-gas, naphtha, a side stream of fuel oil and spent solvent, and a stream of bottoms slurry. The bottoms slurry from the atmospheric fractionator, which contains all the mineral matter, is fed to the vacuum fractionator. Products from the vacuum fractionator include off-gas, liquid distillate, a side stream of fuel oil and spent solvent, a side stream of VGO, and the vacuum-bottoms slurry, which is fed to downstream bottoms processing or recycled partially to the liquefaction reactor. As an alternative the stream of VGO can be recycled to extinction to the liquefaction reactor, by blending it with spent solvent.

Solvent hydrogenation

The primary purpose of solvent hydrogenation is to restore hydrogen to the spent donor solvent before it is recycled to the slurry dryer in the liquefaction section. The processing sequence for solvent hydrogenation closely resembles that of conventional petroleum-hydrotreating systems.

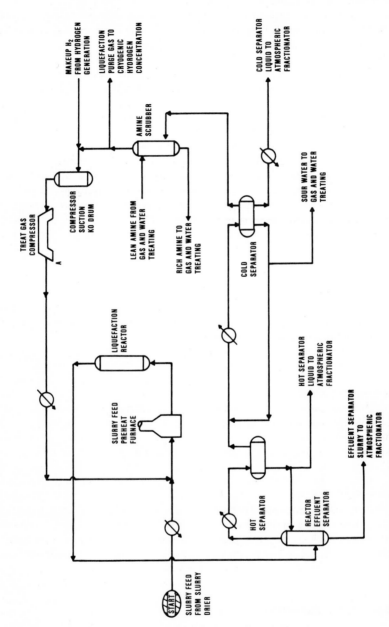

FIG. 1-3 EDS process scheme for liquefaction.

1-10

The spent solvent is hydrotreated in a fixed-bed catalytic reactor. Effluent from the solvent hydrogenation reactors is cooled and separated into a hydrogen-rich gas and a hydrotreated liquid stream. The gas is scrubbed to remove CO_2 and H_2S, a purge is taken to control the impurity level, and the remaining gas, along with makeup hydrogen, is recycled. The hydrotreated liquid is fed to the solvent fractionator.

The EDS process is unique in the use of an externally hydrogenated donor solvent. The external hydrogenation of the recycle solvent has two major impacts on the process: (1) the hydrogen donating property of the solvent improves the level of coal conversion and increases the flexibility of the process so that it can handle a wide range of coals and (2) the product slate and properties are affected by the multiple passes across the catalyst that the solvent boiling range coal liquids experience. The external and separate hydrogenation step allows precise tailoring of the hydrogenation of the donor solvent so as to achieve maximum conversion. The impact of solvent hydrogenation on the EDS product slate and properties is achieved through production of saturated and partially saturated cyclic structures along with heteroatom removal by virtue of multiple passes through the solvent hydrogenation reactor. The saturated cyclics are more amenable to thermal cracking in the liquefaction reactor, thus allowing partial control of the product slate through solvent hydrogenation. Heteroatom removal, which is necessary for downstream utilization, improves the quality of the products.

Hydrogen generation

The generation of hydrogen via partial oxidation is carried out in facilities consisting of partial-oxidation gasifiers, oxygen plants, and synthetic-gas upgrading units. In the partial-oxidation gasifiers, vacuum bottoms, oxygen, and steam are reacted to form synthesis gas (a mixture of hydrogen and carbon monoxide). The hot reactor effluent syngas is cooled and scrubbed to remove the unreacted carbon and ash. The syngas leaves the scrubber saturated with water and is fed to multiple stages of shift reactors. Most of the CO_2 and H_2S in the shifted syngas is removed in the acid-gas removal system and is sent to the H_2S removal facilities, where sulfur is recovered as a product and CO_2 is vented to the atmosphere. The product stream of treated hydrogen is blended with the hydrogen from the hydrogen-recovery section prior to the final compression step. Oxygen is supplied to partial oxidation by air-separation oxygen plants.

Hybrid boiler

The hybrid boiler replaces both the gas-fired liquefaction slurry preheat furnace and most of the offsite steam boilers employed in alternative EDS configurations. (See Fig. 1-4.) Solidified vacuum bottoms or coal is used as the fuel for the hybrid boiler. Heat recovered in the radiant section is used to generate steam in conventional waterwall tubes and a radiant steam superheater. The flue gas produced in the radiant section of the boiler enters the convection section. In the convection section, coal-slurry feed in admixture with hydrogen treat gas is preheated to the liquefaction temperature. The remaining low-level heat in the flue gas is recovered in an air preheater. Particulate matter is removed and flue gas is desulfurized before it is released to the atmosphere. Recovery of process heat can be maximized if desired by recirculation of flue gas.

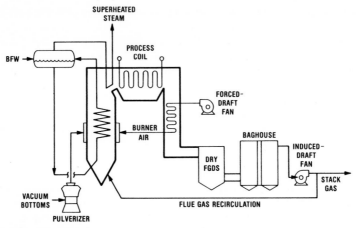

FIG. 1-4   Simplified hybrid boiler process scheme.

FLEXICOKING

The FLEXICOKING process unit utilizes three heat-integrated fluid-bed vessels: reactor, heater, and gasifier. Vacuum bottoms are fed to the reactor vessel, where liquid and gas products are recovered and coke is formed. The heat for this endothermic reaction is supplied by a circulating coke stream which transports heat from the higher-temperature heater vessel. The heater vessel is in turn kept hot by circulation of gas and solids from the gasifier vessel, where coke formed in the reactor is gasified with steam and air to form low-Btu gas. This fuel gas, after passing through the heater and waste-heat boilers, is cleaned of particulates and sulfur components. The clean low-Btu gas enters the fuel-gas system and supplies virtually all of the on-site fuel requirements and part of the fuel for the offsite boilers.

Liquid and gas products from the reactor are cooled and scrubbed in the scrubber section of the reactor, where a solids-laden stream is condensed and withdrawn. This solids-laden stream is recycled to the FLEXICOKING reactor. The level of recycle is set by the solid specifications in the fuel oil and by fuel-balance considerations. The overhead effluent from the scrubber is sent to the coker fractionator, which separates the coker naphtha from heavy fuel oil. The olefinic coker gas, after compression, cooling, and incremental naphtha recovery, can be sent to the cryogenic hydrogen recovery unit.

PROCESS DEVELOPMENT

Process Chemistry

Coal is a generic term covering a broad group of carbonaceous materials which differ in origin, geological age, properties, composition, and structure. Coals are classified (or ranked) as lignite, subbituminous, bituminous, and anthracite. This ranking is the order in which their aromaticity increases and their volatile matter and moisture content decreases. It is also roughly the order of increasing heat of combustion.

The conversion of coal to clean liquid fuels requires a reduction of molecular weight, the removal of heteroatoms and ash, and a substantial increase in the hydrogen-to-carbon ratio. The hydrogen-to-carbon (H/C) atomic ratio in coal is of the order of 0.8, while the hydrogen-to-carbon atomic ratio of petroleum crude is typically 1.8. Thus, considerable hydrogen is needed to improve the hydrogen-to-carbon atomic ratio of coal. This, of course, can be achieved either by removal of carbon as coke or as hydrogen-deficient char (as in pyrolysis) or by the addition of hydrogen (as in various solvent-assisted processes). Therefore, in its simplest form, the chemistry of coal liquefaction may be considered as the conversion of low-hydrogen coal to high-hydrogen liquid fuels with simultaneous removal of "impurities" such as nitrogen, oxygen, sulfur, and ash.

The investigation of widely differing coal types as potential liquefaction feedstocks has been pursued in the EDS program. By investigating a wide variety of coals as potential liquefaction feedstocks, the process response to differing petrographic composition, mineral matter content, and coal composition may be approximated. Because of the heterogeneity of coal, it is difficult to define the response that a given coal will exhibit under liquefaction process conditions. Nevertheless, the organic portion of coal can be classified into different chemical substituents referred to as *macerals,* which relate to the conversion potential of the coal.

The behavior of a coal under liquefaction conditions depends on the extent that the plant-derived macerals have undergone the coalification process. This coalification process defines the rank of a given coal and is primarily reflected in the extent to which the chemical functionalities of the original plant have become altered. This alteration results in a condensation of the original chemical structures and is mainly seen as an increase in the aromatic character of the coal substituents and a decrease in the coal's oxygen content. A simplified representation of the results of the coalification process is shown in Fig. 1-5. The representative structures for three coal types are depicted: lignite (the lowest in rank of these coals), subbituminous, and high-volatile bituminous. The lignite is characterized by a rather "loose" structure with high concentrations of oxygen bonded in various ways. As coalification or the aging process continues, many of the chain bridges between the aromatic clusters shown for the lignite give way to the partially aromatic linkages that are shown as characteristic of the subbituminous coal. In addition, a marked reduction of the oxygen functional groups occurs during this increase in rank. Still further coalification results in increased aromaticity, or condensation, and a reduction in oxygen content.

Along with the representative structures of different coals, Fig. 1-5 also shows the stage these coals have reached in the pilot plants in their evaluation as feedstocks in the EDS process.

Typical yields of C_4-to-537 °C fractions from EDS liquefaction are shown in Fig. 1-6 for the bituminous, subbituminous, and lignite coals that have been investigated. The liquid yields on a dry, ash-free (daf) basis that result from once-through and bottoms-recycle liquefaction are shown. Studies with bituminous coals, subbituminous coals, and lignite using once-through operations produce about 37 percent by weight, 31 to 35 percent by weight, and about 32 percent by weight, respectively, of liquids. The liquefaction conversion can be increased substantially by recycling vacuum bottoms to liquefaction. As shown, yields for bituminous coals are increased to 43 to 44 percent by weight liquids, subbitu-

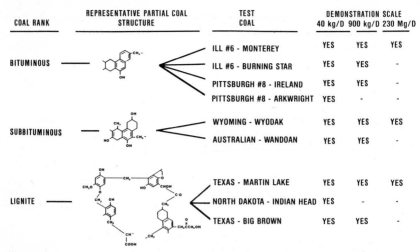

FIG. 1-5 Flexibility of EDS process to handle various types of coal. "Yes," work already completed.

minous coals to 38 to 41 percent by weight liquids, and lignite to 39 percent by weight liquids.

Although not shown in Fig. 1-6, additional heavy-distillate liquids could be recovered by FLEXICOKING the vacuum-bottoms stream.

The process chemistry involved in the remaining steps of the EDS process is well known in the petroleum industry and will not be discussed in detail. It includes gasification in

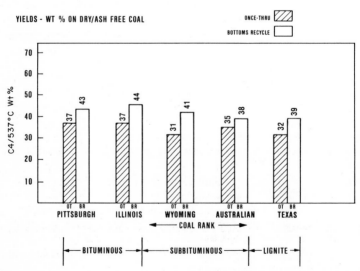

FIG. 1-6 Typical yields of C_4-to-537 °C fractions from the EDS liquefaction process using various types of coals.

bottoms processing (FLEXICOKING and partial oxidation), hydrogen manufacture from either gas re-forming or partial oxidation, donor-solvent hydrogenation via conventional hydrodesulfurization technology, and product upgrading via conventional petroleum processes.

Pilot Plant Operation

The EDS program has integrated all phases of process development. This includes bench-scale research, operation of small pilot units, operation of the 250-ton/day (230-Mg/day) coal-liquefaction pilot plant, and studies of engineering design and technology.

Before operation of the large pilot plant, development of the EDS process relied heavily on bench-scale experiments and on the small, integrated coal-liquefaction pilot units. Figure 1-7 illustrates the relative size and role of each pilot unit which was used in the program. The batch autoclave fed only 3 g of coal per test; it provided relative coal-conversion data and was used to screen new liquefaction concepts. The recycle coal liquefaction unit (RCLU) fed approximately 75 lb (40 kg) of coal per day; this unit recycled solvent and vacuum bottoms in a semibatch fashion for feed-slurry preparation and was used primarily to study process variables and to obtain yield data. The coal liquefaction pilot plant (CLPP) fed about 1 ton (900 kg) of coal per day; it incorporated continuous recycle of solvent and bottoms and used an integrated vacuum tower for product distillation. The CLPP also provided product samples for further analysis and testing.

EDS coal-liquefaction pilot plant (ECLP)

The EDS coal-liquefaction pilot plant (ECLP) was built at a cost of $120 MM and was operated in the United States by personnel of the Exxon Company, U.S.A. to gather data

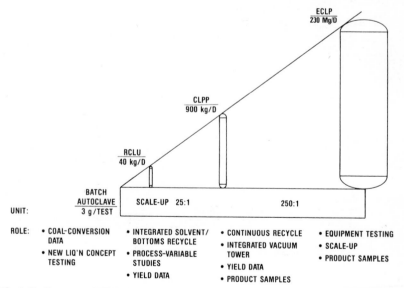

FIG. 1-7   Summary of EDS pilot plant capabilities.

on engineering design, scale-up, and operability for the EDS process. It was a 250-ton/day (230-Mg/day) (of coal) unit located near the Baytown, Texas, refinery of Exxon Company, U.S.A. (see Fig. 1-8 for an overview of the plant).

In conventional developments of petroleum processes, the developers can safely assume that equipment needed to commercialize their process, such as pumps and valves, can be purchased from vendors when needed. This is not the case with coal liquefaction. The EDS process must handle viscous, abrasive slurries of up to about 45 percent by weight solids, at up to 482 °C and about 2500 to 3000 lb/in² (18 to 21 MPa). Pumps, pressure-letdown valves, and check valves that reliably handle such streams may not exist commercially at the present time. The main purpose of the ECLP was to provide information on equipment design and equipment performance needed for commercial application. In operating the ECLP, scale-up data were obtained, correlations and guidelines were defined, and an understanding of the limitations of process equipment and process configuration were gained. Although the ECLP was used to obtain scale-up data which have a direct impact on process yields, the ECLP was not used as the primary tool to study process variables and process effects. This could best be done in the smaller units.

The ECLP was specifically planned and sized in order to obtain the design information required to engineer a commercial-size EDS liquefaction plant. A coal-liquefaction plant is actually the sum of several interrelated processes. It was necessary to demonstrate that these several processes can operate together in an integrated fashion over a sustained time period. In addition, it was necessary to characterize the wastes produced in an EDS plant under likely commercial operation conditions and to generate sufficient quantities of material, such as wastewater, to select the necessary waste-treating schemes. Hence, the ECLP was built with a process configuration similar to that in the liquefaction portion of a commercial-scale EDS plant. It included coal grinding, slurry preparation, liquefaction, fractiona-

FIG. 1-8 EDS coal-liquefaction pilot plant. Key: A, feed preparation; B, coal storage silo; C, liquefaction reactors, D, product fractionation; E, wastewater treatment; F, cooling towers; the large building at center right is the vacuum bottoms storage building.

tion, and solvent hydrogenation. These processes operated in an integrated fashion typical of a commercial plant.

The 250-ton/day (230-Mg/day) coal feed rate (and associated solvent and gas rates) was selected to provide a commercially representative tube size in the slurry-preheat furnace and to achieve the liquid and gas velocities which calculations and correlations predicted would give good contacting in the reactor. The refined correlations developed as a result of tests in the ECLP will be used to size the reactors for a commercial plant at approximately 30 to 60 times the ECLP coal rate (7500 to 15,000 tons/day of coal feed per reactor line).

The materials selected for the ECLP were those which might be selected in a commercial plant rather than more exotic materials which might be used to ensure a high service factor in a pilot unit. The purpose was to gather corrosion and erosion experience under commercial process and fluid flow conditions. Lines were sized such that flow velocities were generally in the range of expected commercial operation. The objective was to confirm the adequacy of particle saltation correlations and to gather erosion experience under commercial process and fluid dynamic conditions. The intention was to operate the ECLP under sufficiently severe operating conditions so that potential problems could be uncovered and solutions addressed prior to the design and construction of a commercial-size facility.

ECLP Test Program   A comprehensive test program was developed for the ECLP. The original test program consisted of 24 months of operation on three coals: 12½ months on an Illinois No. 6 bituminous coal from the Exxon Monterey mine, 6½ months on a subbituminous Wyoming coal from the Wyodak mine, and 5 months on a Texas lignite. Each coal operation was to consist of a start-up and shakedown period followed by a period of testing during which operations under steady and variable conditions would be conducted. The testing period would be followed by a turnaround during which equipment would be opened and inspected. The inspection would provide a basis upon which to judge the effect(s) of various coals and conditions on the plant internals.

Early operations on Illinois No. 6 coal followed the planned program quite closely, but persistent problems encountered during the early runs, together with a shift in emphasis toward bottoms recycle operation, prompted modification of the program. The Wyoming bottoms recycle run was ended 2 months earlier than originally planned to make changes aimed at solving the problems of poor solvent quality and slurry-furnace coking. After this turnaround, Illinois No. 6 coal was run again, but in the bottoms recycle mode. Coal feed was switched from Illinois to Wyoming coal after 3 months (without a turnaround), and ECLP completed the 2 months of Wyoming coal operation which had been interrupted by the early turnaround. Finally, the test program was extended 3 months to enable a complete testing period on Texas lignite. Figure 1-9 presents the actual ECLP operating history and demonstrates how knowledge gained and modifications made in the earlier runs contributed to improved service factors in the later runs. Details of these runs are described in Table 1-1 and Figs. 1-10 through 1-15.

Major ECLP Learnings

Early operations at ECLP encountered three persistent process-related problems: localized coking in the slurry preheat furnace, lower-than-anticipated coal conversion, and low sol-

TABLE 1-1 ECLP operating history—June 24, 1980 through August 20, 1982

Run	Hours in	Hours out	Service factor, %	Prime source of shutdown
1	128	242	35	Plugging—Heat tracing
2	509	233*	69	Erosion of vacuum-tower transfer line
3	32	187	15	Plugging—Heat tracing
4	134	223	38	Feed conveyor and plugging—Dowtherm
5	77	112	41	Plugging—Heat tracing
6	77	109	41	Plugging—Heat tracing
7	173	129	57	Hydrogen compressor, feed system
8	115	T/A†	——	Slurry carryover into overhead separators
Shakedown summary	1245	1235	51	
9	29	58	33	Solidification belt stoppage and damage
10	226	128	64	Block valves around slurry feed pumps leaking
11	317	67	83	Preheat furnace coking
12	49	254	19	Preheat furnace coking
13	6	92	6	Refractory in atmospheric-tower bottoms pumps
14	738	244	75	Solidification belt replacement and furnace coking
15	398	161	71	Atmospheric tower corrosion
16	39	43	48	Refractory in atmospheric-tower bottoms pumps
17	856	2	T/A†	Coal-out voluntarily
Illinois No. 6 coal testing summary	2658	1049	72	
18	1389	325	81	Plugged sheds in vacuum tower during solids withdrawal
19	453	T/A†	——	Backflow of reactor solids during preheat furnace swing
First Wyoming coal test summary	1842	325	85	
20	1031	121	90	Freezing (15 °F)
21	20	22	48	Voluntary due to cold weather
22	281	49	85	Plugged feed line to atmospheric tower
23	692	——	100	Switched to Wyoming coal feed
Illinois No. 6 coal bottoms recycle test summary	2024	192	91	

TABLE 1-1   ECLP operating history—June 24, 1980 through August 20, 1982 (*Continued*)

Run	Hours in	Hours out	Service factor, %	Prime source of shutdown
24	98	83	54	Plug valves at slurry feed pumps failed
25	42	28	60	Vacuum bottoms pump packing
26	291	70	81	Seal weld leak on hot separator letdown valve bypass
27	317	47	87	Plugged barometric leg of vacuum tower
28	9	96	9	Slurry feed-pump valve plug and crankcase failure
29	105	14	T/A†	Eroded hole in atmospheric-feed preheat exchanger
Second Wyoming coal test summary	862	338	72	
30	471	216	69	Cable failure to Substation 4
31	53	43	55	Prepared coal feeder failure
32	37	96	28	Solids carryover due to excessive recycle gas rate
33	925	26	97	Explosion at adjacent plant
34	309	188	62	Solidification belt failure
35	266	T/A†	——	Voluntary end of operations
Texas lignite test summary	2061	569	78	

*Excludes 120 hours for plant shutdown during hurricane.

†Turnaround.

vent quality. Although at first these problems appeared to be distinct, it later became apparent that they were closely related to one another. A concerted effort by laboratory, engineering, and operations personnel determined that process interactions between separate areas of the plant were negatively affecting ECLP performance.

Briefly, the interactions which caused problems in the early ECLP runs may be summarized in the following manner. To begin with, conversion was low, which meant that there was a low yield of solvent-range material as well as a reduced process exotherm (i.e., low liquefaction temperature). With less net solvent production, the recycle solvent was gradually depleted of high-quality solvent components, resulting in solvent of generally poor quality. As discussed below, the poor solvent quality led to accelerated coking in the slurry preheat furnace. This coking, in the form of localized "hot spots" on the furnace tubes, forced operation at a lower coil-outlet temperature which caused even lower liquefaction-reactor temperatures. This had two negative effects. First, it tended to further reduce conversion. Second, at these low liquefaction temperatures saturated compounds in the recycle solvent did not crack to lighter components as they should have. Hence, an undesirable buildup of saturates occurred in the solvent which limited hydrotreating severity (to prevent excessive saturates production) and, consequently, exacerbated the existing solvent-quality problem.

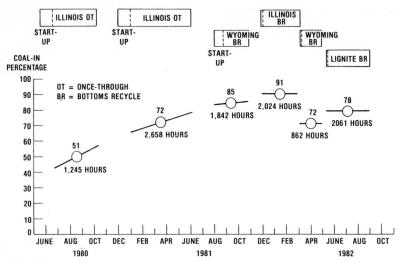

FIG. 1-9   ECLP operating history and service factors.

Major problems were mechanical rather than process related

   Breakdown of solids conveyors and gate-lock coal feeder

   Plugging of slurry heat exchangers

   Short packing life of slurry reciprocating pumps

   Erosion of elbow in vacuum-tower transfer line

Steady operation was hampered by plugging from solids and solidification of heavy streams

   Undersized jacketed heating system aggravated problem

   Time was required to unplug equipment largely responsible for low service factor

Excellent performance of slurry letdown valves and centrifugal pumps

Turnaround revealed little corrosion and isolated areas of severe erosion

   Small area corroded in upper section of atmospheric tower

   Bottom heads of reactors—slurry bypassed and dislodged plugged distributors

   Significant erosion in tangential vacuum-tower inlet

   Gas-swept mill cyclones and atmospheric-tower feed mix chamber also eroded

Significant equipment deficiencies were corrected during turnaround

   Replaced gate-lock with rotary-valve coal feeder

   Improved jacketed heating system

   Installed refractory-lined vacuum-tower transfer line

   Repaired eroded areas and added protection

FIG. 1-10   Summary of ECLP start-up and shakedown operations (June 24 to December 30, 1980).

Most major problems were process-related

   Low coal conversion gave little excess solvent production

   Solvent quality was poor

   Localized coking of the slurry preheat furnaces in form of hot spots

   Upper section of atmospheric tower corroded through

Rotary-valve coal feeder gave much better service than gate-lock feeder

Excellent performance of slurry letdown valves continued

Turnaround exposed several deposits in process equipment

   Hard, cokelike deposits in liquefaction-reactor effluent separator

   Coke deposits in atmospheric-tower feed preheat exchanger

   Fouling of stripping baffles in vacuum tower

Severe erosion was discovered in atmospheric-feed mix chamber and vacuum-tower inlet

All significant equipment problems were corrected

   Corroded atmospheric-tower shell was replaced with a stainless-steel-lined section

   Vacuum-tower inlet was lined with alumina bricks for erosion resistance

Process and equipment changes were made for the test program

   Redesigned distributor trays were installed in first two liquefaction reactors

   Second two liquefaction reactors were placed in operating reactor train

   Redesigned stuffing boxes were installed on atmospheric-tower bottoms pump

   Completed facilities to recycle vacuum-tower bottoms to liquefaction

FIG. 1-11   Summary of Illinois coal once-through operations (December 30, 1980 to July 31, 1981).

During most of 1981, ECLP operations were limited by the problems just described. However, process changes and equipment modifications made toward the end of 1981 overcame the detrimental effects of these process interactions. Detailed descriptions of the changes which were made may be found in the EDS project literature. The analysis which follows highlights the major changes and principal knowledge gained as a result of the ECLP experience.

Furnace coking

The tendency toward coking in an EDS slurry-preheat furnace depends on a number of factors. After considerable effort, it was determined that these variables included the slurry-film temperature in the furnace coil, the solids concentration of the slurry, the fraction of solvent that is vaporized (which is determined primarily by hydrogen gas rate and solvent boiling range), and the composition of the solvent as specified by the content of hydroaromatics (donatable hydrogen) and saturates.

Laboratory studies showed that coking does not occur at appreciable rates below about 425 °C (800 °F) and that the time required for coking to occur decreases rapidly as temperature is raised. Thus, it was deduced that the higher the slurry-film temperature, the

No major mechanical failures occurred

    New stuffing boxes improved packing life in atmospheric-bottoms pumps

    Successfully commissioned vacuum-bottoms recycle facilities

    Completed longest continuous run during ECLP operations (57.9 days)

Low coal conversion, low solvent quality, and slurry-furnace coking persisted

    Radioactive-tracer tests conducted to determine gas holdup in reactors

Process changes improved solvent quality from poor to fair

    Reduced solvent-hydrogenation-catalyst volume to limit saturates

    Removed additional low-boiling saturates from solvent by distillation

Turnaround revealed minimal corrosion/erosion

    No corrosion/erosion detected in atmospheric tower

    Minor erosion in vacuum-tower transfer line

    Significant thinning of chimneys on lower trays of vacuum tower

Few deposits were discovered in process equipment

    Small accumulation of loose solids in liquefaction reactors

    Minimal plugging of liquefaction-reactor distributor trays

    Large amount of soft solids in liquefaction-reactor effluent separator

Process equipment was modified for smoother operations

    Slurry preheat furnace was converted into two-zone (high-flux/low-flux) design to lower slurry-film temperatures

    Redesigned stuffing boxes were installed on slurry feed pumps

FIG. 1-12   Summary of First Wyoming coal bottoms recycle operations (July 31 to November 28, 1981).

more likely it was to form hot spots due to localized coking on the tubes. The furnace was modified during the October 1981 turnaround to contain two cells—the inlet low-temperature cell having a high heat flux and the outlet high-temperature cell having a relatively low heat flux. This modification reduced the slurry-film temperature at the coil outlet by 45 °F at equivalent coil-outlet temperatures, thereby reducing the likelihood of coking.

Other important variables include the slurry solids concentration and the general area of solvent quality, which includes solvent boiling range and composition. Research studies indicate that the large planar aromatic molecules in the coal slurry may be rearranged during heating to form a liquid crystal mesophase. The role of the mesophase in preheater coking has been postulated to be that of an adhesive. Spherules of the liquid crystal behave as colloids which may stick to the preheater wall or precipitate from the bulk liquid phase due to incompatibility with, or vaporization of, the solvent. This mesophase deposition and subsequent pyrolysis could form a low-ash coke layer on the tube wall similar to that observed at ECLP. In addition, the mesophase spherules may act as a binding agent which agglomerates coal particles, allows them to stick to the thin coke layer on the preheater wall, and finally causes rapid formation of a coke deposit. The presence of hydroaromatics may arrest this sequence by preventing formation of the mesophase. Furthermore, if less solvent vaporizes, the concentration of slurry solids remains low and any liquid crystal formed may be less likely to precipitate. As a result of these hypotheses, more of the light fraction of the solvent was removed in the solvent fractionator and the heavy vacuum gas oil was added to the solvent stream before hydrogenation.

Increasing solvent quality decreased the coking tendency at a given set of furnace conditions for Illinois and Wyoming coals. The coking tendency for Texas lignite, on the other hand, was apparently independent of solvent quality over the relatively narrow range of conditions tested. This fact emphasizes that different coals exhibit different responses to the coking variables. In other words, the operating conditions necessary to avoid coking with one coal may not prevent coking with a second coal. The existing data base and operating experience, however, can be interpolated to other coals using the predictive tools and coal-specific laboratory tests developed for ECLP. These are fully described in the EDS project literature (in particular, July–September 1982 and October–December 1982 *EDS Coal Liquefaction Project Technical Progress Reports*). In essence, the extrapolation would involve taking measurements of the relative coking tendency for a coal in a special bench-scale unit and applying ECLP-derived correlations to yield the anticipated coking behavior of that coal in a large-scale plant. The relative coking tendency for several coals in a bottoms-recycle liquefaction mode has been established as follows: Wyodak > Wandoan > Martin Lake > Ireland ≃ Monterey > Burning Star.

Coal conversion

During the Illinois No. 6 coal once-through operations in 1980 and 1981, it was observed that conversions were approximately 5 to 10 percent by weight lower at ECLP than in the smaller pilot units when compared at equivalent nominal residence times. Material balances that were conducted during the Wyoming coal bottoms recycle operations in 1981 showed

---

Achieved excellent coal conversion and solvent quality without furnace coking

   Demonstrated conversion of 60 percent 538 °C— (1000 °F—) on daf coal

   Consistently good-to-excellent solvent quality

   Trouble-free operation of modified slurry-preheat furnace

Attained commercial design points at solvent-to-coal-to-bottoms (S/C/B) ratios of 2.3:1:0.5 and 1.6:1:0.5

   Smooth operations gave 91 percent coal-in service factor

New or redesigned process equipment installed and operated successfully

   Screw feeders to inject coal below slurry dryer liquid level

   Vertical, one-pass slurry heat exchangers to replace horizontal, two-pass units

   Streamlined slurry letdown valve to improve wear life

Coal feed switched to Wyoming coal without a turnaround

FIG. 1-13 Summary of Illinois coal bottoms recycle operations (November 28, 1981 to February 28, 1982).

Excellent process operability continued

Demonstrated conversion of 66 percent 538 °C— (1000 °F—) on daf coal

High-quality donor solvent produced

Troublefree operation of modified slurry-preheat furnace

Attained commercial design point at S/C/B ratio of 2.0:1:0.5

Mechanical problems limited coal-in service factor to 72 percent

Installed new equipment for test program

Slurry dryer vapor disengagement drum to knock out entrained water vapor

Smaller diameter tubes in outlet row of slurry preheat furnace to increase slurry mass velocity

Erosion of atmospheric-tower feed preheat exchanger forced turnaround

Extensive cokelike deposits in the exchanger shell

Few significant problems elsewhere in the plant

Large amount of cokelike material in liquefaction-reactor effluent separator

Several misplaced chimney hats found in vacuum tower

Moderate erosion of vacuum-tower transfer line

Serious equipment deficiencies corrected in relatively short turnaround

Atmospheric-tower feed preheat exchanger shell replaced

Chimney hats secured in vacuum tower

Deposits cleared from effluent separator

FIG. 1-14 Summary of second Wyoming coal bottoms recycle operations (February 28 to May 2, 1982).

that conversions were approximately 10 to 15 percent by weight lower than expected, based on small pilot plants results. After evaluating the material balance data from ECLP and the smaller pilot units, it was suspected that at equivalent nominal residence times, the actual residence times differed among the various units. As a result, a sophisticated series of radioactive tracer studies were conducted at ECLP and the smaller units. The results of these studies confirmed that actual liquid residence time difference was the major factor in explaining the lower-than-expected conversions at ECLP.

All of the reactors used in the EDS process are upflow. As the superficial gas velocity increases in this type of reactor, the fraction of liquid (or liquid holdup) in the reactor decreases. This in turn reduces the slurry holdup or slurry residence time. Previously, nominal slurry residence times were calculated for all pilot units based on volumetric feed rate and volume of reactor. Figure 1-16 (p. 1-27) illustrates the effect of superficial gas velocity on gas holdup or, by difference, liquid holdup. These data were collected from the 900 kg/day pilot plant (CLPP) using solvent oil and hydrogen only. The superficial gas velocity in the ECLP reactors was about 6 cm/s, while that in CLPP was about 3 cm/s. Thus, the actual slurry residence time in ECLP was lower than in CLPP, which, in turn, resulted in reduced conversion of the coal.

The other factor that may have affected conversion of Wyoming coal, but not of Illinois coal, was the method by which it was prepared in the gas-swept mill at ECLP. In this mill, the coal was contacted with flue gas at 230 to 315 °C (450 to 600 °F) that contains up to 13 percent O_2. To determine the effect of this hot oxidizing atmosphere on conversion, raw coal fed to the gas-swept mill and dried coal from the mill were liquefied in small pilot units. The results of these experiments showed about a 5-percent reduction in conversion with the dried coal versus the raw feed coal. When the same set of experiments was conducted with Illinois coal, the conversion was about the same with both the dried and raw coals. A conversion debit similar to Wyoming coal was observed with the Texas lignite but, interestingly, the difference in conversion vanished for bottoms recycle operations. Thus, drying some lower rank coals in an oxygen atmosphere at elevated temperatures may reduce liquefaction conversion. This problem can be avoided by designing the

---

Coal conversion and solvent quality remained high

    Demonstrated conversion of 66% 538 °C— (1000 °F—) on daf coal

    Donor solvent somewhat lighter but same quality as previous operations

Attained commercial design point for S/C/B ratio of 1.6:1:0.5

    Lower S/C/B ratios tested (down to 1.2:1:0.5)

    Higher slurry film temperatures demonstrated with smaller furnace tubes

Several successful tests conducted

    Demonstrated operation with coal ash content on 9–20+ percent by weight

    Demonstrated slurry drying of high-moisture lignite

    Controlled reactor solids build-up with on-line solids withdrawal

Turnaround revealed only limited areas of severe and moderate erosion

    Gas-swept mill venturi scrubber circulation and blowdown pumps

    Main slurry letdown valve plug tip and seat

    Moderate erosion of atmospheric-tower inlet transfer line

    Moderate wear of vacuum-tower transfer line, ceramic bricks at tangential inlet, and carbon-steel chimneys of lower trays

Distillation columns exhibited minor corrosion

    Carbon-steel shell of atmospheric tower below stainless-steel lined section

    Cracking of some stainless-steel hardware in atmospheric tower

    Carbon-steel trays and internals in upper section of solvent fractionator

Hard, cokelike deposits in all liquefaction reactors and effluent separator

    Severe fouling of liquefaction-reactor distributor trays

    Large (8–10 ft) plug in fourth liquefaction reactor

    Large deposit and loose solids in bottom head of effluent separator

---

FIG. 1-15   Summary of Texas lignite bottoms recycle operations (May 2 to September 1, 1982).

gas-swept mill for lower temperature operation or by using an impact mill and slurry drying the coal.

Solvent quality

Solvent quality is a term commonly used to denote the relative suitability of hydrogenated solvent for coal conversion under EDS process conditions. Attributes of good quality solvent are a relatively high concentration of hydrogen donor species (hydroaromatics) and a low concentration of saturated compounds. The actual measure of hydrogen donor species in recycle solvent is proprietary. Nevertheless, the interaction between liquefaction and solvent hydrogenation with respect to solvent quality may be discussed on a relative basis.

Figure 1-17 illustrates the relationship between relative solvent quality and the operating conditions in both the solvent hydrotreater and liquefaction. It is based on predictions made for Illinois coal using an integrated process model developed from laboratory data. As mentioned before, the objective of good solvent quality is high donor hydrogen and low saturates. However, increasing the solvent hydrogenation severity (along the solid lines in the figure) increases the amount of both these compounds in the solvent.

The key to increasing solvent donor hydrogen level without also raising the saturates level lies in operating liquefaction at a higher temperature. The dashed lines on the figure represent constant solvent hydrogenation severity at various average liquefaction temperatures. Therefore, it is possible to maintain high solvent quality (a high level of solvent donor hydrogen) at a relatively low saturates level by operating at a sufficiently high liquefaction temperature. This is possible because the saturates crack into lighter boiling compounds at high liquefaction temperatures. Once these relationships were quantified, ECLP personnel were able to modify the solvent-hydrogenation operating philosophy to achieve and maintain good solvent quality, in spite of coal changes and process upsets.

Equipment

The primary purpose of ECLP in the EDS program was to test equipment designs which would be part of a commercial-sized EDS plant. Considerable progress was made during ECLP operations towards understanding equipment behavior and that knowledge was successfully applied to improve several critical pieces of equipment. Advances were made in the materials and piping designs to be used in highly erosive services. Various valve body and trim designs were used and evaluated. Particularly good service was given by the slurry block valves and the high-pressure slurry letdown valves. In the machinery area, the coal feed system to the slurry dryer was improved significantly by replacing the unwieldy gate-lock system with the rotary-valve feeder and, finally, with flushed screw feeders. Centrifugal pump service throughout the plant was good, and reciprocal pump packing life was lengthened markedly by stuffing box and packing material changes.

In the area of process equipment, the heat exchangers in the slurry dryer pumparound loop were redesigned into a vertical downflow configuration to eliminate saltation problems experienced in the original horizontal unit. Furthermore, a vapor disengagement drum was placed between the slurry dryer and the centrifugal slurry booster pumps to remove water vapor entrained from the mixed vessel during slurry drying operations. This vessel eliminated the cavitation problems which had plagued earlier tests of the slurry dryer. Pump cavitation did occur with this vessel at high feed rates of high-moisture coals, but data gathered during those tests have led to further improvements of the slurry dryer design. These changes include staging of mixed vessels for high-moisture coals to improve drying

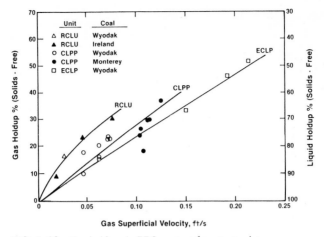

FIG. 1-16   Gas holdups in EDS reactors from tracer data.

efficiency and minimizing slurry residence time between the vapor disengagement vessel and the centrifugal pumps to avoid vapor generation in the pump suction line.

Liquefaction reactor distributor trays were also redesigned to avoid the dislodging and bypassing which occurred during the shakedown run in 1980. The trays were removed for the Illinois coal once-through testing period and new trays were fitted to the two lead reactors in the July 1981 turnaround. The new trays were capable of withstanding a 300-lb/in² pressure drop (versus 15 lb/in² for the original trays) and were sealed better to prevent bypassing. The redesigned trays revealed no bypassing but were found to be severely plugged during the final turnaround in September 1982. Analysis of the process data indicates that the plugging probably occurred during a severe upset and flow stoppage. Design of the commercial-scale reactors has consequently been modified to anticipate a potential flow stoppage. Finally, the reactor solids withdrawal system was redesigned and control of

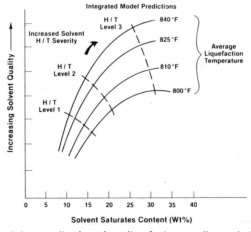

FIG. 1-17   Solvent quality depends on liquefaction as well as on hydrotreating.

TABLE 1-2 EDS bottoms processing alternatives—highlights

	Case A	Case B	Case C
Illinois coal feed rate, kt/sd*			
Process coal	25.0	25.0	25.0
Boiler coal	2.0	0.7	3.1
Hydrogen source	Bottoms POX†	Bottoms POX	Bottoms POX
Fuel source	FLEXICOKING	High-Btu gas	Hybrid boiler (Coal and bottoms)
Product rates, kFOEB/SD‡			
High-Btu gas	12.6	0.0	9.0
C_3-C_4 LPG	4.0	9.1	9.3
Naphtha	19.7	27.6	27.6
Distillate/fuel oil	35.6	29.7	29.7
Total product	71.9	66.4	75.6

* 10^3 tons of dry coal per stream day.

†Partial oxidation.

‡Thousands of fuel oil equivalent barrels (FOEB) per stream day. One FOEB equals 6.415 MBtu (HHV).

pressure drop across the liquefaction reactors by on-line withdrawal of inert reactor solids was demonstrated.

PROCESS CHARACTERISTICS

Processing Alternatives

There are several process options available in the EDS process to meet a wide range of process objectives. Three cases have been chosen to represent the typical range of process options and serve as a basis for describing the unit operating principles. The main characteristics of these cases are outlined in Table 1-2 and described in Fig. 1-18. These cases will also be used in subsequent discussions on yield and investments. The cases vary primarily on the disposition of the liquefaction bottoms as described below:

Case A: About 50 percent of the vacuum bottoms is fed to the FLEXICOKING process for additional liquids recovery and production of plant fuel. The remaining vacuum bottoms are fed to a partial-oxidation gasifier to produce hydrogen for use in the plant. In this case vacuum bottoms are not recycled to the liquefaction reactor.

Case B: A portion of the vacuum bottoms is recycled directly to the liquefaction reactors for additional conversion. The remaining vacuum bottoms are fed to both a partial-oxidation gasifier for production of hydrogen and to offsite boilers for production of steam. Process-derived high-Btu gas supplies the fuel requirements of the plant.

Case C: A portion of the vacuum bottoms is recycled to the liquefaction reactors for additional conversion. Vacuum bottoms are fed to a partial-oxidation gasifier to satisfy the

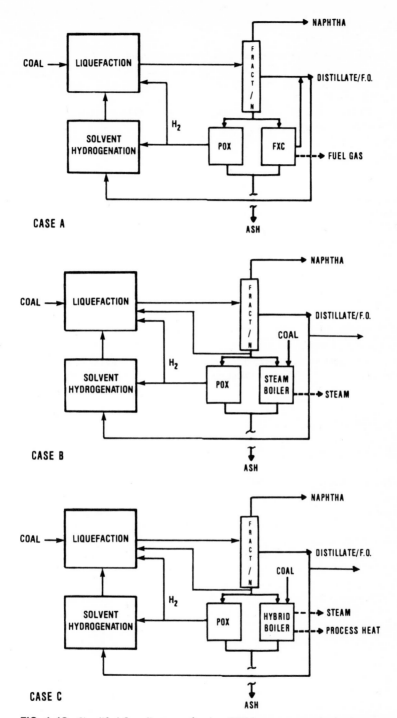

FIG. 1-18 Simplified flow diagrams of various EDS bottoms-processing options.

hydrogen requirements of the process. The remaining vacuum bottoms, with supplemental coal, fire a hybrid boiler which supplies both the heat requirements of the liquefaction process and most of the steam requirements of the plant.

The overall energy balance for a plant with a feed rate of 25,000 tons (dry basis) per stream day (23,000 Mg per stream day) is given in Table 1-3.

Product Yields

The EDS process offers a wide range of product slates depending on the process configuration, process conditions, and possible upgrading options. The yields presented in this section illustrate typical ranges, and no attempt was made to represent the maximum achievable or the optimum, since this will depend on the specific plant objectives.

In Fig. 1-19, typical product yields from liquefaction are shown as percentages by weight of dry, ash-free (daf) coal. The bottoms recycle operations provide an additional 7 to 12 percent by weight of liquid products while maintaining the flexibility to yield different product mixes. Variations in products were achieved by different combinations of reactor temperature, residence time, and pressure. Product slates similar to those shown in Fig. 1-19 for Illinois coal have also been observed for subbituminous and lignitic coals.

Figure 1-20 presents possible plant yields in another fashion. Simulated balanced product yields for a commercial plant using Illinois coal are presented for different EDS bottoms-recycle operating configurations. The first column shows the net liquid product yields in the maximum fuel oil mode of operating (once-through liquefaction). The yield is about 0.2 bbl of C_3, 1.0 bbl of naphtha, and 1.5 bbl of fuel oil, for a total of 2.7 bbl. In a mixed clean products mode, the EDS process produces about 0.4 bbl of C_3, 1.6 bbl of naphtha,

TABLE 1-3 Energy balance of EDS process using Illinois coal

Energy (GBtu/sd)*	Case A GBtu	%	Case B GBtu	%	Case C GBtu	%
Input						
Process coal	628.2	84.8	628.2	88.7	628.2	81.7
Off-site coal	51.4	6.9	17.8	2.5	75.9	9.9
Electric power	61.1	8.3	62.0	8.8	64.8	8.4
Total input	740.7	100.0	708.0	100.0	768.9	100.0
Output						
Products						
High-Btu gas	79.2	10.5	——	——	57.5	7.5
C_3-C_4 LPG	27.2	3.7	58.2	8.2	59.9	7.8
Naphtha	126.8	17.1	177.1	25.0	177.1	23.0
Distillate/fuel oil	227.4	30.7	190.2	26.9	190.2	24.7
Total products	460.6	62.0	425.5	60.1	484.7	63.0
By-products						
Ammonia	2.4	0.3	3.3	0.5	3.3	0.4
Sulfur	8.5	1.1	8.3	1.1	8.3	1.1
Total by-products	10.9	1.4	11.6	1.6	11.6	1.5

*Energy given in 10^9 GBtu per standard day based on higher heating value.

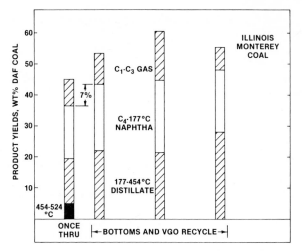

FIG. 1-19 Demonstration of flexibility of EDS product yield.

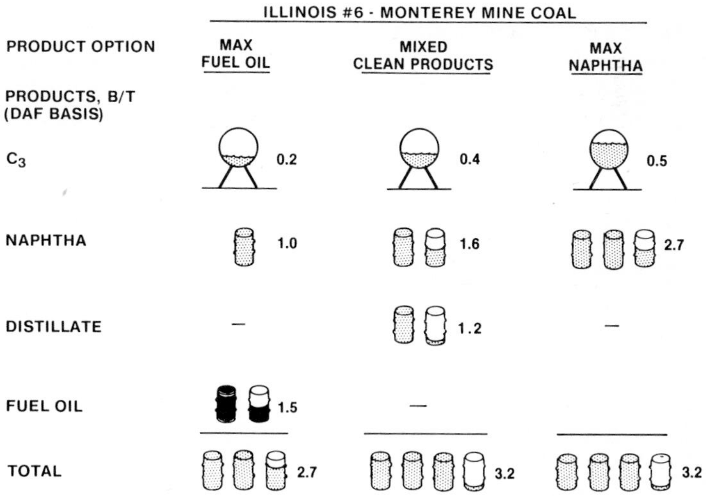

ILLINOIS #6 - MONTEREY MINE COAL

PRODUCT OPTION	MAX FUEL OIL	MIXED CLEAN PRODUCTS	MAX NAPHTHA
PRODUCTS, B/T (DAF BASIS)			
C_3	0.2	0.4	0.5
NAPHTHA	1.0	1.6	2.7
DISTILLATE	—	1.2	—
FUEL OIL	1.5	—	
TOTAL	2.7	3.2	3.2

FIG. 1-20 Range of product yield patterns available from EDS process.

and 1.2 bbl of distillate, for a total of 3.2 bbl. Finally, in the maximum naphtha case, the EDS process produces 0.5 bbl of C_3 and 2.7 bbl of naphtha, for a total of 3.2 bbl. In the last case, conventional extinction recycle hydrocracking of distillate (discussed below) was used to produce the naphtha and lighter products. As indicated, these results are for Illinois coal and are on a daf coal basis.

The production of the high clean-product slates described in Figs. 1-19 and 1-20 was assisted by recycle of the VGO fraction to near extinction in the liquefaction reactor. Comparison of operations with and without VGO recycle shows relatively small differences in coal conversion or total liquid yields. However, VGO recycle allows a product slate that does not contain fuel oil. This minimizes potential concerns associated with the production and disposition of heavy coal liquids.

Table 1-4 presents the yields corresponding to the plant configurations (cases A, B, and C) previously outlined under "Processing Options." The yields are given in kilobarrels of product and by-products per stream day from an EDS plant operating on 25,000 tons (dry basis) per stream day of Illinois No. 6 coal to liquefaction.

Product Properties

The specific objectives of the EDS product quality program were to conduct studies of product quality, define potential end-use applications for coal liquids, and evaluate the properties of these coal liquids relative to the requirements of these potential applications. The EDS coal-liquefaction process exhibits a broad flexibility in terms of the feed coal and product slate. It handles a wide range of coals (bituminous, subbituminous, and lignitic), and can produce a product slate varying from predominantly heavy fuel oil, through a clean product only slate (consisting of naphtha and distillate) to a high naphtha yield slate. Options were also explored to recover a "jet fuel" fraction from the stream of EDS distillate product. Table 1-5 outlines the disposition and/or primary processing options for the various streams.

The C_4-to-71 °C EDS naphtha fraction will likely be utilized after hydrotreating as motor gasoline blending stock. The 71-to-177 °C EDS naphtha, after hydrotreating and catalytic re-forming, will also be used in motor gasoline and represents a premium motor gasoline pool component (105 + RON). Alternatively, or in conjunction with disposition to motor gasoline, EDS naphtha can serve as a source of valuable chemicals. For example, raw naphtha (135-to-177 °C fraction) is an abundant source of phenol and cresols, while the reformate contains high concentrations of BTX (benzene, toluene, and xylene). Conventional

TABLE 1-4 Comparisons of product and by-product yield*

	Case A	Case B	Case C
Products, kbbl/SD			
C_3 LPG	4.2	9.8	10.2
C_4 LPG	2.5	4.7	4.7
C_5-to-177 °C naphtha†	23.4	32.6	32.6
177-to-343 °C distillate	21.1	27.3	27.3
343 °C+ heavy distillate	14.5	2.6	2.6
Total liquid products	65.7	77.0	77.4
HBG product, kFOEB/SD‡	12.6	———	9.0
By-products			
Ammonia, (st/SD)	126	172	172
Sulfur (Lt/SD)¶	948	934	934

*All cases feed 25000 standard tons per stream day (st/SD) of (dry) coal to the plant.

†Includes recovered phenols.

‡High-Btu gas (HBG), in 10^3 fuel oil equivalent barrels (kFOEB/SD) per stream day at higher heating values (HHV). One FOEB equals 6.415 MBtu (HHV).

¶Long tons per stream day.

TABLE 1-5   EDS product disposition and/or primary processing options

Stream	Disposition	Primary processing
C$_1$ to C$_2$	HBG	
C$_3$ to C$_4$	LPG	
C$_4$-to-177 °C naphtha	Motor gasoline blendstock	Hydrotreating and re-
	Chemicals	forming
		Extraction
177-to-399 °C distillate*	Turbine fuel	Direct use
	Home heating oil blendstock	Direct use
	Diesel fuel blendstock	Direct use
	Fuel oil blendstock	Direct use
	Jet fuel	Extraction (177 to 288 °C)
	Jet fuel/auto diesel fuel/home heating oil	Severe hydrotreating
399-to-538 °C VGO†	Fuel oil blendstock	Direct use or hydrotreating
Coker gas oil† and/or scrubber liquids	Fuel oil blendstock	Direct use or hydrotreating

*Potential for use of 177-to-204 °C portion as motor gasoline blendstock being investigated.

†Potential conversion feeds (catalytic cracking, hydrocracking).

extractive processing is needed to recover these and perhaps other desirable chemicals. The 177-to-204 °C hydrotreated heavy naphtha may potentially find application as a direct motor gasoline blending stock or possibly as a blending component in the distillate pool.

With additional upgrading, the 177-to-399 °C EDS distillate might be utilized as a stationary turbine fuel, or as a blendstock for home heating oil, auto diesel fuel, or fuel oil. Initial test work in this area has been promising. While meeting most of the performance specifications for home heating oil, the 177-to-399 °C distillate typically contains 75 percent by weight aromatics and does not meet the 30° API specific gravity specification for ASTM no. 2 fuel oil. Combustion tests have shown that raw and hydrotreated EDS mid-distillates perform well in a domestic heating-oil burner relative to a reference no. 2 petroleum fuel oil at comparable combustion conditions. However, some adjustments may be required to optimize burner operation. The coal distillates burned cleanly, giving essentially the same smoke number as the petroleum fuel at a given level of excess air. Alternatively, the hydrotreated solvent can be extracted with an appropriate solvent to segregate a saturates-rich raffinate containing about 20 percent by weight aromatics. On the basis of available data, current specifications for jet fuel can be met with the 177-to-288 °C fraction of the raffinate stream. However, certain critical tests (e.g., thermal stability) have not been performed. The aromatics-rich extract, which contains only about 10 percent by weight nonaromatics, would necessarily become fuel-oil blending stock or, if proved acceptable, stationary turbine fuel. While disposition of this saturates-rich stream to jet and/or diesel fuel is considered feasible, additional testing is required to augment present test data.

Additional hydrotreating of the EDS distillate coupled with fractionation can be used to produce no. 2 fuel oil, auto diesel, or jet fuel from the total distillate product. Saturation of aromatics to achieve a 30° API specific gravity or a 40 cetane number for auto diesel is possible by using conventional hydrotreating. However, if the nearly acceptable performance of the starting EDS mid-distillate is considered, the need for upgrading solely for the purpose of meeting an existing specification should be reexamined. In the case of auto diesel fuel, the use of cetane improvers may also be appropriate.

Unless upgraded, the VGO and/or coker gas oil is not compatible with petroleum-derived stocks. In this context, incompatibility refers to sediment formation under simulated handling conditions for a comingled petroleum- and coal-derived stream. Raw EDS VGO would therefore require a dedicated use and is classified as a specialty fuel oil. Compatibility can be achieved via conventional hydrotreating with or without EDS mid-distillate in admixture. Alternatively, the VGO stream can be converted to motor gasoline and distillate products. On the basis of previous government-sponsored studies with a blend of an EDS solvent and VGO (204-to-538 °C fraction), it has been shown that hydrotreating plus catalytic cracking and/or hydrocracking coupled with re-forming are feasible conversion options.

Typical properties of streams produced directly from the EDS process are shown in Table 1-6. They are presented in such a way that a determination can be made concerning the further testing and/or upgrading needed to produce final petroleum products. The properties given are for Illinois No. 6 coal-derived products and there will, of course, be variations according to the type of coal and process sequence employed in the EDS process.

Environmental Considerations

The first step in formulating an environmental program for the EDS project is defining those areas expected to be different from experience in the petroleum industry. Three general areas have been identified. (1) The coal feed is expected to have an impact on the following environmental areas: Air as a result of fugitive dust emissions generated during coal handling and crushing and during disposal of coal fines, noise generated during coal crushing, and worker health as a consequence of potential dust emissions and noise levels. (2) The products are anticipated to pose a potential health hazard to workers because of the products' high content of aromatics. (3) Plant discharges are expected to have an enviromental impact as a consequence of fugitive dust and hydrocarbon emissions to the air, discharge of aromatic hydrocarbons and phenols from process and runoff water, and leaching of solid waste in landfills.

Strategy

Figure 1-21 shows the development strategy used in addressing these concerns. The general approach was to define the problem using ECLP operating data and then to follow up with engineering solutions based on existing control technology from petroleum refining and the electric power industry. The EDS processing steps are extensions of existing technology for coal and petroleum refining, and thus emissions are expected to be similar to those encountered in these industries. Programs to protect workers from both coal- and product-based emissions are based on more than 14 years of coal-liquefaction research at Exxon and on Exxon's general industrial experience.

TABLE 1-6 Typical properties of EDS products based on Illinois No. 6 coal

Property	Stream						176-to-	
	Unhydrotreated naphtha	Unhydrotreated distillate	Hydrotreated distillate	Vacuum gas oil	Vacuum bottoms	Coker gas oil	538 °C fuel oil	176 °C+ fuel oil
Composition, wt%								
C	85	89	89	87	70	86.0	88.5	87.7
H	12	9	10	7	4	7.4	8.7	8.0
N	0.2	0.3	0.2	1	2	1.1	0.5	0.8
S	0.5	0.2	0.01	1	2	1.0	0.4	0.6
O	2.3	1.3	0.4	4	2	4.5	1.9	2.5
Ash	0.0	0.0	0.0	0.15	20	0.0	0.03	0.4
Distillation, °C at								
5	48	203	190	332	538 at 4%	188	200	221
50 } wt % off	121	277	268	482		371	271	368
95	168	455	415	554		515	463	543
Specific gravity								
16/16 °C	0.82	0.99	0.96	1.1	1.4	1.1	1.01	1.08
°API	41	11	16	-4	-30	-4	9	-1
Viscosity, cS								
32 °C	0.5	5.0	3.7	18,000	100,000	393	5.5	—
99 °C	0.3	1.5	1.3	342	1,000	16	2.0	4.
Heating value, Btu/lb	19,600	18,200	18,700	16,400	12,500	17,200	17,500	16,700
Flash point, °C	<21	70	27	>213	——	>213	92	92

Air

    Quantify emissions from ECLP coal crushing and handling
    Adapt emission controls for coal crushing and handling from electric power industry
    Burn clean fuels in process furnaces
    Adapt emissions controls for fugitive hydrocarbons from petroleum refining

Water

    Characterize untreated waters from large pilot plants
    Simulate treating scheme using water samples from the ECLP

Solids

    Perform studies of leaching and properties of solids

Noise

    Identify and quantify sources in large pilot plants
    Adapt controls from electric power and petroleum refining industries

Occupational health

    Monitor workplace in large pilot plants
    Assess adequacy of health programs
    Develop toxicology program

FIG. 1-21   Development strategy for environmental controls.

Atmospheric emissions

The expected atmospheric emissions for a commercial EDS plant that is fed 30,000 ton/day of an Illinois No. 6 coal containing 4.4 percent sulfur are shown in Table 1-7. These values are based on the 1978 EDS Study Design by the Exxon Research and Engineering Co. and show annual rates of about 6150 tons of sulfur oxide, 740 tons of carbon monoxide, 3200 tons of nitrogen oxides, 525 tons of particulate matter, and 1575 tons of hydrocarbons. When these numbers are stated on an annual basis, they appear large, but it should be noted that the emission rate of sulfur oxides is less than 20 ton/day. Calculations using a dispersion model have shown that the ground-level concentration (GLC) for major pollutants from such a continuous source is below the allowable GLC for prevention of significant deterioration (PSD) for a Class II area. Thus an EDS plant could probably be located in a Class II attainment area that has the meteorological conditions typical of southern Illinois.

Wastewater emissions

Table 1-8 gives an analysis of the wastewater stream from an EDS plant that feeds 30,000 ton/day of an Illinois No. 6 bituminous coal. The stream has been stripped, and extracted to remove phenol, but has not been treated. The wastewater treating scheme included in the study design is shown in Fig. 1-22. The flow plan is based on the "best available technology" to control the water pollutants as found in the water samples from the small pilot plant. On the basis of laboratory tests, the proposed treatment scheme is predicted to meet existing Illinois effluent standards. Sour process waters are stripped to remove H_2S

TABLE 1-7 Potential air emissions from a
30-kt/d commercial EDS plant*

Pollutant	t/yr
SO_2	6150
CO	740
NO_x	3200
Particulates	525
Hydrocarbons	1575

*Illinois No. 6 coal feed containing 4.4% sulfur;
feed rate on a dry-coal basis is 25 kt/d.

TABLE 1-8 Potential wastewater
contaminants from a 30-kt/day EDS plant
(25 kt/day dry)*

Contaminant	ppm by wt.
NH_3	80
H_2S	1
CO_2	80
Free CN^-	Nil
SCN^-	10
Cl^-	90
Na^+	500
Ca^{2+} (as $CaCO_3$), Mg^{2+}	16
Fe^{2+}, Fe^{3+}	4
Phenolics	50
Organic acids	1500
Oil and grease	30
BOD_5	1410
Total suspended solids	30
Total dissolved solids	2400

*After stripping and phenolics extraction but
before treatment of wastewater stream; feed rate of
Illinois No. 6 coal containing 4.4% sulfur is 25 kt/d
on a dry-coal basis; maximum rate of wastewater
flow is 2550 gal/min.

and NH_3. Wastewater from the slurry dryer is then added, and the mixture is fed to the phenol extraction process. The nonprocess wastewater and oily rainwater which have passed through the API separator for oil removal are then added and the mixture is subjected to dissolved air flotation, activated sludge biological oxidation, dual media filtration, and activated carbon treatment. Sludge from the activated-sludge unit is processed in dissolved-air flotation thickener, an aerobic digester, and a filtration unit. The resulting solids are disposed of in a suitable landfill. Approximately 60 percent of the treated water is reused as cooling tower makeup, with the remainder being discharged.

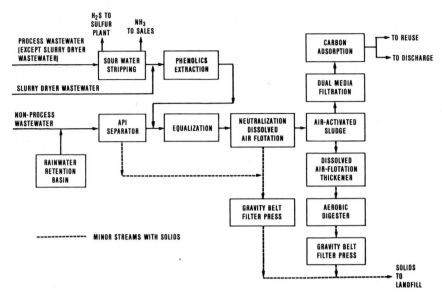

FIG. 1-22 Wastewater-treating scheme for a commercial EDS plant based on Illinois No. 6 coal.

Occupational health

A broad occupational health program was conducted at the ECLP. The program was patterned after one utilized for the past 14 years in the Baytown Laboratories and by other operators of coal-liquefaction pilot plants. The program contained seven specific areas that combine engineering control of emissions, monitoring of industiral hygiene, medical surveillance, personal hygiene, operating and laboratory work practices, and health education. An overall testing program to study the toxicity of EDS products, by-products, intermediates, and wastes is underway.

An extensive program of industrial hygiene monitored all pilot plant operations and determined the degree of employee exposure to toxic substances and physical agents. In-depth studies were carried out during the operation of the plant on Illinois No. 6 coal. Extensive monitoring of the air in the work area with both personal and area sampling was conducted routinely to assure a healthy environment; this provided an expanded data base for future production facilities. Contaminants of concern were coal dust, hydrocarbons, phenols, particulates, hydrogen sulfide, and noise.

A program of personal hygiene at the ECLP prevented and/or minimized contact with or ingestion of any coal-derived material and the breathing of respirable particles or mists and vapors. It was also designed to prevent any contaminated person, clothing, or equipment from leaving the plant site. A dirty/clean locker room with an interconnecting shower was provided.

Specific work practices were required to prevent exposure to contaminants during equipment repairs and handling of samples and to prevent the transmission of contaminants to areas outside the plant. All work permits defined the type of work to be performed and were required before the job began. Guidelines were established for the type of safety equipment needed for each job.

Health education of all employees was a fundamental part of the program. The entire occupational health program was presented to employees in detail through group meetings and informational booklets. Employees were then tested to determine if they absorbed the meaning of the materials presented. In addition, a program was formulated to supply updated information to all supervisory personnel (both operational and contractor) on an as-needed basis, but at least once per year.

The medical surveillance program was designed to be particularly sensitive to the effect of polynuclear aromatics. Target organs were skin, lungs, and the urinary tract. The program consisted of two parts—routine periodic examinations, similar to those provided for all Exxon employees, and special examinations directed at the target organs.

Health problems were not anticipated, but preparations were made to detect problems early. The program outlined above did this, and ensured that any hazards of coal liquefaction would not cause significant occupational disease.

Process Economics

For the purpose of calculating plant investments, it is assumed that the EDS coal-liquefaction plant is located in western Illinois, 50 to 60 mi from a major metropolitan area. The feed rate of coal for liquefaction is 25,000 tons per stream day of cleaned, dry Illinois No. 6 coal (30,000 tons per stream day "as received"). There is also some coal used in each case for steam generation. The plant is composed of two EDS liquefaction trains, each one-half of the total plant capacity. The trains are constructed in stages to reduce peak construction personnel. The two trains start up in the first quarter of 1985 and the first quarter of 1986, respectively.

Table 1-9 shows the predicted total erected cost for the plant just described based on a 1985 start-up. Table 1-10 shows other project information that is needed to analyze process economics. Figure 1-18 shows the plant configurations.

The economic viability of a coal-liquefaction plant depends on the cost of the product produced by that plant and the value which potential customers assign to the product. The actual market revenues derived from EDS products will depend on market factors, such as availability of alternative supplies, over which the plant owner has no control. However, the plant operator can strive to produce a product of acceptable quality at the lowest possible production cost. Calculating the cost of coal liquids from this conceptual plant requires making numerous assumptions about the course of future events. For example, certain economic factors must be specified: the unit costs of coal, labor, and electricity; annual escalation rates for these unit costs and for the selling price of the product; and interest rates required on borrowed monies. Technical factors, such as the service factor of the mature plant and the rate at which the plant reaches this on-stream performance, must also be selected. There are numerous assumptions which must be specified in order to calculate a selling price for coal liquids. A different set of assumptions will result in a different calculated product cost.

Major project items to be specified include the following:

- Project life
- Tax rate
- Debt and equity

TABLE 1-9   EDS coal-liquefaction plant investment

	Case A	Case B	Case C
Onsites			
Liquefaction and fractionation	786	1107	731
Solvent hydrogenation	177	275	275
Cryogenic H_2 recovery	102	115	115
Partial oxidation	397	511	511
Oxygen plant	109	166	166
FLEXICOKING	325	——	——
Hybrid boiler	——	——	680
Product upgrading	——	32	32
Other onsites	221	176	195
Total onsites	2117	2382	2705
Offsites			
Coal preparation	118	116	120
Offsite boilers	341	311	163
Wastewater treating	170	123	121
Utilities	357	419	419
Tankage	171	240	237
Other off-sites	571	745	838
Total off-sites	1728	1954	1898
Total on-sites plus off-sites	3845	4336	4603
Process contingency	279	416	433
Project contingency	1153	1301	1381
Total erected cost (2d qtr, 1985)	5277	6053	6417

Total erected cost (MM$) for 1985 start-up

- Depreciation
- Investment tax credit
- Working capital
- Nonrecurring expenditures
- Capital expenditure schedule
- Production schedule
- Recovery factor for plant product
- Coal price and escalation of coal price
- Product price and escalation of product price
- Plant investment (including escalation)
- Utility costs
- By-product prices to be considered

In general, project-specific economic factors exert a greater impact on project product cost than the technical factors. Among the most important economic factors are assumed inflation rate, tax treatment of investment and products, and desired rate of return.

TABLE 1-10   EDS coal-liquefaction project basis (1985 start-up)

Specification	Case A	Case B	Case C
Location	Western Illinois (50 miles from major city)		
Coal Supply			
Type	Illinois No. 6		
Number of mines	3–(6 miles apart)		
Feed rate, kT/sd			
As received	32.4	30.8	33.7
Dried	27.0	25.7	28.1
Plant configuration			
Number of trains	2	2	2
H_2 production	POX*	POX*	POX*
Onsite fuel	FLEXICOKING	HBG	Hybrid boiler
Products, kbbl/sd			
C_3 to C_4 LPG	6.9	14.5	14.9
C_5–177 °C naphtha	23.4	32.6	32.6
177 °C+ distillate and fuel oil	21.1	27.3	27.3
High-Btu gas	12.6	——	9.0
Product recovery, %	90	90	84
Thermal efficiency, %	63.6	61.7	64.5
Utilities			
Purchased power, MW	266	304	317
Other operating-cost factors			
Personnel	3700	3700	3900
Repair material, % TEC	1.6	1.6	1.6
Catalyst and chemicals, MM$/yr	29	26	29
Land requirements, acres	1300	1300	1300
Start-up year	1985	1985	1985

*Partial oxidation

Major technical factors consist of the stage of process development, the stage of project development, and the plant reliability. The stage of process development refers to the current status of the technology with respect to commercialization. Since the EDS process is a developing technology, some allowances must be made for identification of unknown factors which will have an impact on the cost. The stage of project development refers to the depth of engineering detail provided at any particular point. In doing rough project economics, a project contingency of approximately 30 percent and a process contingency of 8 percent can be added to reflect these uncertainties. This level of contingency allowance would increase the product cost by about 15 percent. Obviously, these contingencies remain a matter of judgment until a commercial plant has actually been constructed.

An in-depth analysis of scheduled and possible unscheduled outages at a plant, based on EDS study designs, indicates that early commercial plants can expect to produce from 84 to 90 percent of the theoretical yearly production for an operation with zero equipment outages, depending on the plant configuration. The analysis takes into account service factors for individual units, the level of intermediate tankage, and sparing philosophy. The result, in economic terms, is to cause an increase in the product costs in the range of $+10$ percent.

The principal conclusion is that the cost of coal liquid can be accurately described only within the context of a consistent evaluation of specific supply alternatives. That is, the principal question facing those interested in building liquefaction plants is what are the most economically attractive ways to provide an additional fuel supply. Realistically, this decision can only be made by evaluating the actual investment alternatives for specific technologies at specific locations using specific feedstocks and making specific products within the framework of a consistent set of economic factors.

Outlook

EDS coal liquefaction has been successfully developed for liquefaction of a broad range of coals, but the prospects and timing for an economically attractive coal-liquids industry remain uncertain. A continuing effort on process improvements will be made in an effort to reduce the cost of synthetic liquids. The project will also continue to focus attention on the selection of alternative bottoms processing.

The operation of the ECLP has met or exceeded targets. Process innovations such as recycle of liquefaction bottoms were tested along with characterization of various types of coal feed. Valuable experience in the operation and maintenance of conventional equipment under the severe conditions of liquefaction was obtained. Of equal importance was the feedback from the environmental, plant, and personal hygiene programs, which will lead to a safe and environmentally acceptable commercial design.

BIBLIOGRAPHY

Ansell, L. L., et al.: "Bottoms Recycle Studies in the EDS Process Development," *Second Chemical Congress of North America, American Chemical Society, Las Vegas, Aug. 27, 1980.*

Becker, P. W., et al., "Conversion of Lignite to Liquid Fuels Using the Exxon Donor Solvent (EDS) Process," *Proc. Gulf Coast Lignite Conference, Houston, Nov. 4, 1982.*

Blaser, D. E., and A. M. Edelman: "FLEXICOKING For Improved Utilization of Hydrocarbon Resources," *Proc. API 43rd Midyear Meeting, Toronto, May 8-11, 1978, American Petroleum Institute, Washington, DC, 1978, pp. 216-223.*

Boyer, G. T., et al.: "Water Pollution Control in the Exxon Donor Solvent Coal Liquefaction Process," *AIChE 87th National Meeting, Boston, Aug. 21, 1979.*

Brackett, R. H., T. J. Clunie and A. M. Goldstein: "Exxon Donor Solvent Liquefaction Process: Exxon Coal Liquefaction Plant Operations," *6th Annual Contractors Conference of Coal Liquefaction, May 13-14, 1981.*

———, et al: "Exxon Donor Solvent Coal Liquefaction Process: ECLP Operating Experience III," *API 47th Midyear Refining Meeting, New York, May 12, 1982.*

Cohen, S. J., and R. E. Payne: "Exxon Donor Solvent Coal Liquefaction Process: ECLP Operating Experience II," *Proc. 8th Annual Conference of Coal Liquefaction, Gasification and Conversion to Electricity, Pittsburgh, Aug. 5, 1981, pp. 57-66.*

Corning, H. F., et al.: "Liquid Fuels from Coal Utilizing the EDS Process," *Energy Technology Conference and Exposition, Washington, D.C., Feb. 28–March 2, 1983.*

EDS Coal Liquefaction Project Technical Progress Reports (prepared for the U.S. Department of Energy). Annual Reports: January–December 1976, FE-2353-9; July 1, 1977–June 30, 1978, FE-2893-17; July 1, 1978–June 30, 1979, FE-2893-35; July 1, 1979–June 30, 1980, FE-2893-53; July 1, 1980–June 30, 1981, FE-2893-74; July 1, 1981–June 30, 1982, FE-2893-93; July 1, 1982–June 30, 1983, FE-2893-116. Interim Reports: "Summary of EDS Predevelopment," FE-2894-16; "EDS Commercial Plant Study Design Update, Illinois Coal," FE-2893-61; "EDS Enviromental Program," FE-2893-79; "EDS Product Quality Final Report," FE-2893-97; "EDS Wyoming Coal B/R Study Design-Summary," FE-2893-108; and "Final Report for Phase IIIA," FE-2353-20.

Epperly, W. R.: "Cooperative Agreement—A New Mechanism for Joint Government/Industry Projects," *Contract Manage., July 1979*, p. 4.

Epperly, W. R., K. W. Plumlee, and D. T. Wade "Exxon Donor Solvent Coal Liquefaction Process, Development Program Status V," *Proc. 8th Energy Technology Conference and Exposition—New Fuels Era, Washington, D.C., March 9–11, 1981*, pp. 726–749.

———, ———, and ———: "Exxon Donor Solvent Coal Liquefaction Process: Development Program Status," *Intersociety Energy Conversion Engineering Conference, Aug. 18–22, 1980.*

———, ———, and ———: "Exxon Donor Solvent Coal Liquefaction Process: Development Program Status," *Seventh Annual International Conference on Coal Gasification, Liquefaction, and Conversion to Electricity, Aug. 5–7, 1980,* Seattle, Wash.

———, ———, and ———: "Exxon Donor Solvent Coal Liquefaction Process: Development Program Status," *Proc. American Mining Congress International Coal Show, Chicago, May 5–9, 1980.*

———, ———, and ———: "Exxon Donor Solvent Coal Liquefaction Process: Development Program Status," *Proc. 1980 NPRA Annual Meeting, New Orleans, Mar. 23–25, 1980.*

Epperly, W. R., and J. W. Taunton: "Progress in Development of Coal Liquefaction Process," *Proc. 72nd AICHE Annual Meeting, San Francisco, Nov. 28, 1979.*

———, and ———: "Development of the Exxon Donor Solvent Coal Liquefaction Process," *The Annual Fall Conference of the Canadian Society of Chemical Engineering, Sarnia, Ontario, Sept. 30–Oct. 3, 1979.*

———, and ———: "Exxon Donor Solvent Coal Liquefaction Process Development," *Coal Conversion Technology,* ACS Symp. Ser. No. 110, 1979, p. 71.

———, and ———: "Donor Solvent Coal Liquefaction Process," *Coal Process. Technol.,* vol. 5, 1979, p. 78.

———, and ———: "Exxon Coal Liquefaction Process Development," *Proc. of the Thirteenth Intersociety Energy Conversion Engineering Conference, Aug. 20–25, 1978, San Diego.*

———, and ———: "Status and Outlook of the Exxon Donor Solvent Coal Liquefaction Process Development," *Proc. of the Fifth Energy Technology Conference, Feb. 27–Mar. 1, 1978, Washington, D.C.*

———, and ———: "Development of the Exxon Donor Solvent Coal Liquefaction Process," *85th National AIChE Meeting, Philadelphia, June 7, 1978.*

———, and ———: "Exxon Donor Solvent Coal Liquefaction Process Development," *COAL DILEMMA II—American Chemical Society, Industrial and Engineering Chemistry Division, Colorado Springs, Feb. 12–13, 1979.*

———, and ———: "Status of Exxon Donor Solvent Coal Liquefaction Process Development," *Sixth Annual International Conference on Coal Gasification, Liquefaction, and Conversion to Electricity, Pittsburgh, Aug. 2, 1979.*

————, et al.: "Exxon Donor Solvent Coal Liquefaction Process: Development Program Status II," *Fifth Annual Conference on Materials for Coal Conversion and Utilization, Gaithersburg, Md., Oct. 7-9, 1980.*

————, et al.: "Exxon Donor Solvent Coal Liquefaction Process: Development Program Status III," *EPRI Conference on Synthetic Fuels: Status and Directions (sponsored by EPRI and Kernforschungsanlage Julich), San Francisco, Oct. 13-16, 1980.*

————, et al.: "Donor Solvent Coal Liquefaction," *Chem. Eng. Prog.,* May 1981.

Fant, B. T., and W. J. Barton: "Refining of Coal Liquids," *Proc. API 43rd Midyear Meeting, Toronto, May 8-11, 1978.*

Furlong, L. E., E. Effron, L. W. Vernon, and E. L. Wilson: "The Exxon Donor Solvent Process," *Chem. Eng. Prog.,* vol. 72, August 1976, p. 69.

Given, P. H., Walker, P. L., W. Spackman, A. Davies, and R. G. Jenkins: "Characterization of Mineral Matter in Coals and Coal Liquefaction Residues," Electric Power Research Institute Annual Report, AF-832, Research Project 3361, Pennsylvania State University, December 1978.

Goldstein, A. M.: "The Exxon Donor Solvent Process—Equipment and Scale-up Issues," *Second World Congress of Chemical Engineering, Montreal, Oct. 4-9, 1981.*

————, et al.: "Exxon Donor Solvent Plant: Design Update," *Chemical Engineering Progress,* April 1982, pp. 76–80.

Green, R. C.: "Environmental Controls for the Exxon Donor Solvent Coal Liquefaction Process," *Proc. Second DOE Environmental Control Symposium, Mar. 19, 1980, Reston, Va.*

Hu, Y. A., et al.: "Application of Semi-Empirical Model for the EDS Coal Liquefaction Process," *AIChE National Meeting, New Orleans, Nov. 8-12, 1981.*

Lendvai-Lintner, E., and G. Sorell: *Mater. Perform.,* vol. 19, no. 4, 1980, p. 19–25.

————, and ————: "Materials Evaluation Program for the EDS Coal Liquefaction Process," *COR-ROSION/79, National Association of Corrosion Engineering, Atlanta, March 12-16, 1979.*

Levasseur, A. A., et al.: "Pilot Scale Combustion Evaluation of EDS Coal Liquefaction Bottoms," *ASME Energy Sources Technology Conference, Houston, Jan. 30-Feb. 3, 1983.*

Maa, P. S., et al.: "Solvent Effects in EDS Coal Liquefaction," *American Chemical Society, Fuels Division Meeting, New York, August 1981.*

Montgomery, C. H.: "Coal Liquefaction Occupational Health Program," *65th Annual Meeting of the American Occupational Medical Association, Detroit, April 21-25, 1980.*

National Energy Plan II, National Technical Information Service, 1978, U.S. Department of Energy, Washington, D.C.

Pan, G., and A. J. DeRosset: "Hydrotreating and Reforming EDS Process Naphtha and Fuel Oil," UOP Report, FE-2566-25, February 1979.

Payne, R. E., R. P. Souther, and W. J. York: "Exxon Donor Solvent Coal Liquefaction Process: ECLP Operation Experience I," *Proc. API 40th Mid Year Refining Meeting, Chicago, May 13, 1981.*

Pepper, M. W., et al.: "Combustion of Fuel Oils Derived from the Exxon Donor Solvent Process," *American Chemical Society Meeting—Combustion of Synthetic Fuels Symposium, Las Vegas, April 2, 1982.*

Platt, R. J.: "High Pressure Slurry Letdown Valve Designs for Exxon Coal Liquefaction Pilot Plant," *DOE/Argonne National Laboratory Symposium on Instrumentation and Control for Fossil Energy Processes, Virginia Beach, Va., June 9-11, 1980.*

————: "High Pressure Slurry Letdown Valve Designs for Exxon Coal Liquefaction Pilot Plant," *Scientific Apparatus Manufacturers Association Meeting, Washington, D.C., March 19, 1981.*

————: "High Pressure Slurry Letdown Valve Performance for Exxon Coal Liquefaction Pilot Plant," *Annual Conference of Instrument Society of America, Anaheim, Calif., Oct. 6-8, 1981.*

————: "Severe Service Instrumentation," *Oil Gas J.,* January 1983.

Quinlan, C. W., and C. W. Siegmund: "Combustion Properties of Coal Liquids from the Exxon Donor Solvent Process," *ACS National Meeting, Anaheim, Calif., March 14, 1978.*

Reidl, F. J., and A. J. DeRosset: "Hydrocracking of EDS Process Derived Gas Oils," UOP Report, FE-2566-33, November 1979.

Robin, A. M.: "Hydrogen Production from Coal Liquefaction Residues," Electric Power Research Institute Final Report, AF-233, Research Project 714-1, Texaco, Inc., December 1976.

Ryan, D. F., and T. Aczel: "EDS Coal Liquefaction Products as Petrochemical Feedstocks," *AIChE National Meeting, New Orleans, Nov. 8-12, 1981.*

————, and S. K. Poddar: "Upgrading of Exxon Donor Solvent Liquefaction Products," *NPRA 80th Annual Meeting, San Antonio, Mar. 23, 1982.*

Scarborough, C. E., et al.: "Development of a Commercial Coal Slurry Preheater Furnace," *DOE Advanced Two-Stage Liquefaction Contractor Review Meeting, Albuquerque, N. Mex., Feb. 26-27, 1980.*

Sexton, R. J.: "The Hazards to Health in the Hydrogenation of Coal," *Arch. Environ. Health,* vol. 1, 1960, pp. 181–233.

Sorell, G., et al.: "Materials Performance in the EDS Coal Liquefaction Pilot Plant: Illinois No. 6 Coal," *CORROSION/82, Houston, Mar. 22-26, 1982.*

Stober, B. K., et al.: "Experience With Heat Transfer Equipment in the EDS Process," *ASME Winter Meeting, Washington, D.C., Nov. 16-20, 1981.*

Stone, J. B., et al.: "Calcium Carbonate Deposit Formation During the Liquefaction of Low Rank Coals," *ACS/Chemical Society of Japan Congress, Honolulu, April 1-6, 1979.*

Stretzoff, S.: "Partial Oxidation for Syngas and Fuel," *Hydrocarbon Process.,* vol. 53, December 1974, pp. 79–88.

Sullivan, R. F.: "Refining and Upgrading of Synthetic Fuels from Coal and Oil Shale by Advanced Catalytic Processes," Chevron January–March 1979 Quarterly Report, FE-2315-37, 1979.

Swabb, L. E., Jr.: "Liquid Fuels from Coal: From R&D to an Industry," *Science,* vol. 199, 1978, p. 619–622.

————: "Prospects for Coal Liquefaction," National Academy of Engineering Annual Meeting, Washington, D.C., Nov. 2, 1978.

————, G. K. Vick, and T. Aczel: "The Liquefaction of Solid Carbonaceous Materials," *World Conference on Future Sources of Organic Raw Materials (CHEMRAWN), Toronto, July 10-13, 1978.*

Tao, F. F., and R. Billimoria: "The Interrelationship of Chemical Characterization and Rheological Behavior of Coal Liquefaction Bottoms," *AIChE Annual Meeting, Los Angeles, Nov. 14-18, 1982.*

————, et al.: "Sintering Phenomena of Coal Ash in Fluidized Bed Gasification," *AIChE Annual Meeting, Los Angeles, Nov. 14-18, 1982.*

Taunton, J. W., K. L. Trachte, and R. D. Williams: "Coal Feed Flexibility in the Coal Liquefaction Process," *Fuel,* vol. 60, no. 9, September 1981, pp. 788–793.

Thomas, R. L.: "Environmental Program and Plans for the EDS Coal Liquefaction Project," *EPA Symposium on Environmental Aspects of Fuel Conversion Technology, Denver, Oct. 26-30, 1981.*

van der Burgt, M. J., and H. F. Kraayveld: "Technical and Economic Prospect of the Shell-Koppers Coal Gasification Process," *Proc. ACS National Meeting, Anaheim, Calif., Mar. 16, 1978.*

Vernon, L. W.: "Free Radical Chemistry for Coal Liquefaction", *ACS/Chemical Society of Japan Congress, Honolulu, April 1-6, 1979*. See also *Fuel,* vol. 59, 1980, p. 45.

Vick, G. K. and W. R. Epperly: "Status of the Development of EDS Coal Liquefaction," *Science,* vol. 217, July 23, 1982, pp. 311–316.

Wade, D. T., et al.: "Exxon Donor Solvent Coal Liquefaction Process: Development Program Status VII," *51st Congress of ANZAAS, U. of Queensland, Brisbane, Queensland, Australia, May 14, 1981.*

————, et al.: "Coal Liquefaction," *Chemtech,* vol. 12, April 1982, pp. 242–250.

Wilson, G. M., and R. H. Johnston: "Volatility of Coal Liquids at High Temperatures and Pressures," *I&EC Process Design and Development,* vol. 20, 1981, p. 94.

Winegartner, E. C. and D. K. Sondhi: "Combustion Properties of EDS Liquefaction Bottoms, *ASME Energy Sources Technology Conference, Houston, Jan. 30-Feb. 3, 1983.*

Zaczepinski, S., et al.: "The Exxon Donor Solvent (EDS) Process—Technical Status of the Development," *Pan-Pacific Synfuels Conference Tokyo '82, Tokyo, Nov. 17-19, 1982.*

————, P. W. Kamienski, H. J. Toups, R. S. Smith, and D. W. Turner: "Upgrading of Coal Liquids," *Proc. API MidYear Meeting, May 14-17, 1979, San Francisco.*

THE H-COAL® PROCESS

JOSEPH E. PAPSO

Hydrocarbon Research, Inc. (Dynalectron Corporation)
Lawrenceville, New Jersey

BACKGROUND

H-Coal® is a direct catalytic hydroliquefaction process developed by Hydrocarbon Research, Inc. (HRI), for conversion of coal to high-quality clean liquids.[1] The process can be modified to produce a variety of liquid fuels ranging from all-distillate syncrude to heavy fuel oil. The novelty of H-Coal® (U.S. Patent 3,321,393) resides in its use of the commercially proven ebullating-bed reactor in combination with other process steps to achieve C_4-to-524 °C distillate yields in the range of 40 to 50 percent by weight of dry coal (Table 2-1).

The H-Coal® process has been under development for more than 18 years and has been used in bench-scale units that process up to 25 lb/day of coal and in a process-development unit (PDU) that handles 3.5 tons/day of coal. The bench-scale units are utilized for studying process improvements, evaluating catalysts, and testing new coals, while the PDU studies have concentrated on confirming the design basis, operating conditions, and modes of operation for the pilot plant and commercial plant projects.

The feasibility of the process is currently being demonstrated on a large scale at the 200–600 ton/day H-Coal® pilot plant in Catlettsburg, Kentucky. The projected 2-year evaluation at Catlettsburg will provide the experience necessary for scale-up to commercial plants. Performance and product yields will be confirmed during the plant operations, and a firm foundation will be established for designing and constructing full-scale commercial H-Coal® facilities. The design of a commercial H-Coal® plant has been initiated under the sponsorship of the U.S. Department of Energy (DOE). Other commercial plant designs are in the feasibility stage.

Continuing research and development on the H-Coal® process have led to the discovery of better catalysts and improvements in modes of operation and have demonstrated the versatility of the ebullating-bed reactor in processing various coals.[2] The current H-Coal® development program consists of laboratory R&D studies and PDU operations; engineering process-development and economic studies; product testing, upgrading, and end-use studies; and pilot plant construction and operation. The pilot plant ran through the end of 1982 and cost a total of 320 million dollars (Table 2-2).

Initially, Dynalectron Corporation, HRI's parent company, supported the development program and as the process advanced, funding became available through other sources. Currently, the sponsors are DOE, the Electric Power Research Institute, Ashland Oil, Inc., Standard Oil Company of Indiana, Conoco Coal Development Company, Mobil Oil Corporation, the Commonwealth of Kentucky, and Ruhrkohle AG.

TABLE 2-1 Background of the H-Coal® process

- Patented catalytic hydroliquefaction process developed by HRI
- Produces C_4–524 °C distillates in the range of 40 to 50 percent by weight of feed coal
- More than 18 years of development and improvement, from bench unit process feasibility study to commercial design
- Over 54,000 hours of operation in bench-scale and process-development units
- Bench-scale operations have optimized process and evaluated catalysts and 18 types of coal
- Process development unit operations have confirmed design basis
- Process confirmation in the 600-t/d pilot plant at Catlettsburg, Ky

TABLE 2-2 Catlettsburg project

Type of project	Large H-Coal® pilot plant
Operator	Ashland Synthetic Fuels, Inc.
Participants	U.S. Department of Energy (DOE)
	Kentucky Department of Energy
	Ashland Oil, Inc.
	Electric Power Research Institute (EPRI)
	Mobil Oil Company
	Standard Oil Company of Indiana
	Conoco Coal Development Company
	Ruhrkohle AG
Location	Catlettsburg, Ky
Schedule	Phase I: Design (completed 12/77)
	Phase II: Construction (1/77 to 5/80)
	Phase III: Shakedown (12/79 to 5/80);
	Operation (5/80 to 9/82)
Status	Complete break-in operation on Kentucky coal in 11/80
	Complete 45-day run with Illinois coal in syncrude mode
	in 4/81
	Almost 12,100 tons coal processed
Resource requirements:	
Coal	200 tons/day (syncrude mode)
Personnel	600 tons/day (fuel oil mode)
	Construction: 830 workers
	Operation: 240 workers
Capital	Total project cost: $320 MM
Product:	
Oil	Plant design capacity is approximately 600 bbl/day of
	synthetic distillates in the syncrude mode

H-COAL® PROCESS DESCRIPTION

In the H-Coal® process, coal is crushed to smaller than 20 mesh, dried and slurried with a process-derived oil, pumped to reactor pressure, mixed with hydrogen, heated, and fed to the reactor. There, the coal, recycled oil, and hydrogen react in the presence of a synthetic catalyst. The reactor typically operates at a temperature of about 454 °C and a pressure of 3000 psig. Depending on the process severity selected, the net product yield can be all-

TABLE 2-3 Main features of the H-Coal® process

- High yields of distilled low-sulfur liquids from bituminous and subbituminous coals and lignites—typically 2.8 to 3.5 bbl per ton of dry coal
- Catalytic ebullating bed combines coal liquefaction, solvent hydrogenation, and product upgrading in a single reactor
- Hydroclones to recover low-solids-residuum oil stream to slurry coal feed
- Liquefaction effluent separated by distillation
- Process hydrogen requirements can be met by partial oxidation of liquefaction bottoms and/or steam re-forming of light hydrocarbons

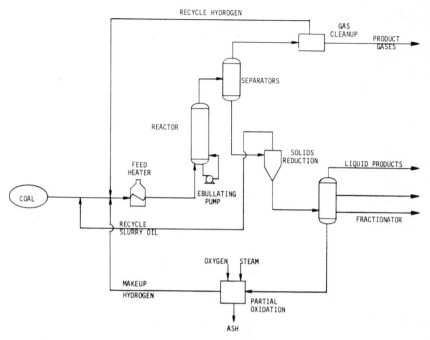

FIG. 2-1   H-Coal® process.

distillate material or, at low severities, distillates and heavy fuel oil. The reactor effluent slurry is processed through hydroclones to reduce its solids content. Low-solid-content oil is recycled as a slurry oil for the feed coal. The balance of the liquid is fractionated to produce an all-distillate product. The vacuum residuum, which contains nondistillate oils, unconverted coal, and ash, can be fed to a partial-oxidation unit to produce the hydrogen for the process or can be used for plant fuel. Figure 2-1 is a schematic of the H-Coal® process.

Table 2-3 lists some of the main features of the H-Coal® process. High yields of distilled low-sulfur liquids have been achieved with bituminous and subbituminous coals and lignites. The presence of the catalyst in the coal-liquefaction reactor significantly improves conversion of heavy coal liquids to products within the distillate boiling range. Typically, 2.8 to 3.5 bbl of C_3–524 °C oil are produced per ton of dry coal fed to liquefaction. The catalytic ebullating-bed reactor combines coal liquefaction, solvent hydrogenation, and product upgrading in a single reactor. This results in fewer process steps than are necessary in other coal-liquefaction technologies. H-Coal's simplified flow scheme helps to reduce plant costs, increases process efficiency, and improves the plant service factor.

H-Coal® Reactor

Figure 2-2 is a sketch of the ebullating-bed reactor which is the heart of the H-Coal® reactor design.

The reactor feed and recycle stream from the ebullating pump enter the bottom of the reactor. The liquid flow causes the catalyst bed to expand and fluidize. The catalyst, which

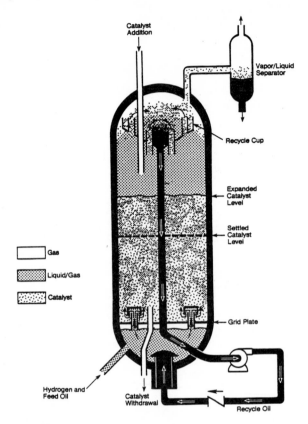

Catalyst Addition

Vapor/Liquid Separator

Recycle Cup

Expanded Catalyst Level

Settled Catalyst Level

Gas

Liquid/Gas

Catalyst

Grid Plate

Hydrogen and Feed Oil

Catalyst Withdrawal

Recycle Oil

FIG. 2-2 Ebullating-bed reactor.

is a $\frac{1}{16}$-in extrudate, remains in the bed. The reactor products, including the unconverted coal and ash solids, leave the bed and are separated in a vapor-liquid separator so that they can be further processed. Because the catalyst is constantly in motion, a portion of it can be withdrawn and replaced with fresh catalyst to maintain high catalyst activity. On a daily basis, about 1 or 2 percent of the catalyst inventory is removed for this purpose. The ebullating-bed reactor system has over 27 unit-years of commercial operations in the H-Oil® petroleum residuum hydroconversion process. The current H-Coal® catalyst has also been demonstrated commercially in H-Oil® operations.

The ebullating-bed reactor allows intimate contact between catalyst particles, hydrogen, and the coal-oil slurry. Isothermal reaction conditions are essentially achieved by the use of low, constant-differential pressure in the reactor through the use of the ebullating pump. This pump is a high-volume low-differential-pressure pump constructed with sufficient impeller clearance to pass any catalyst that might burp over into the suction line. Other major advantages of the H-Coal® reactor system are listed below.

- High liquid yields of high quality are achieved in the presence of a synthetic catalyst and are not dependent on the catalytic effect of coal ash.

• Continuous catalyst replacement controls deactivation, provides constant product quality, allows the possibility of continuous catalyst regeneration, and provides for high unit service factors.

• Operating conditions can be varied to meet flexible product slate requirements.

• Direct catalytic hydrogenation of coal offers the potential for use of different and improved catalysts in the future as product requirements change.

• The ebullating bed assures good temperature control throughout the reactor by using the energy of the reaction to heat the feed slurry to reaction temperatures. The continuous liquid phase in this well-mixed system provides an excellent heat sink to assure reactor stability and a high degree of operability.

The fluid dynamics of the H-Coal® reactor system is a critical element of scale-up and has been studied extensively in the PDU and in cold-model simulations of both the PDU and the pilot plant.[3] Significant improvements have been made to the design of reactor internals, and transient responses of the system under upset conditions have been quantified.

Development Plan for Commercialization

The development path for commercialization of H-Coal® is similar to that used by HRI for scale-up of the commercial H-Oil® residuum and heavy crude oil hydroconversion process.[4] The H-Oil® reactor system was scaled up from the bench, through the PDU and a large pilot plant demonstration unit, and finally to the commercial-scale plant. The reactor diameters are shown below.

	H-Oil® reactor diameter	H-Coal® reactor diameter
Bench unit	¾ in	¾ in
Process-development unit (PDU)	8½ in	6 in, 8½ in
Large pilot plant	4½ ft	5 ft
Commercial plant	13 ft	10–13 ft

The H-Coal® commercialization steps follow the same reactor scale-up criteria. The 5-ft-diameter H-Coal® reactor is in operation at Catlettsburg, while commercial-scale reactors are being designed as part of the Phase Zero H-Coal® commercial plant project. An H-Coal® commercial plant would have several reactors in parallel, depending on the economy of scale desired by the operator and the availability of capital. In terms of the individual reactor train, the commercial-scale reactor would have about 10 times the throughput of the pilot plant, with a diameter scale-up of 2 to 3 times.

DOE has authorized the design work for a commercial-scale H-Coal® liquefaction plant. This plant is to be located in Breckinridge County, Kentucky, and will feed about 23,000 tons/day of run-of-mine Illinois No. 6 coal, to produce approximately 50,000 bbl/day of hydrocarbon liquid products and about 30MM stdft³/day of SNG. The Phase Zero program includes:

TABLE 2-4 Breckinridge project

Type of project	H-Coal® commercial plant
Participants	U.S. Department of Energy (DOE)
	Ashland Synthetic Fuels, Inc. (ASFI)
	Airco Energy Corp., Inc. (AECI)
	Kentucky Department of Energy
Location	Breckinridge County, Ky
Schedule (as of 10/81)	Phase Zero:
	Detailed process design
	Preliminary engineering cost estimates
	Economic analysis and environmental
	assessment (4/80 to 10/82)
	Phase I:
	Detailed engineering design (10/81 to 10/85)
	Phase II:
	Construction (10/83 to 10/87)
	Phase III:
	Initial production (1988)
Resource requirements:	
Coal	22,500 t/d (approximate)
Personnel	Construction: 5000 workers (average)
	Operation: 1500 workers
Capital	Total project cost (approximate): $3 billion
	(1981$)
Products	51,500 barrels/day of synthetic refinery feedstocks
	and fuel oils

- Design of the commercial plant
- Cost estimate and economic evaluation
- Detailed plans for construction and operation

Phase Zero was a 9-million-dollar cooperative effort, involving DOE, Ashland Oil, Inc., and Airco, Inc., which extended through April 1981. That schedule called for follow-on phases of detailed engineering, procurement, and construction, leading to start-up of the commercial plant about mid-1986 (Table 2-4).

HRI is currently involved in feasibility studies for other commercial H-Coal® liquefaction facilities. These include a major program for an overseas client, involving several coals, which will extend the H-Coal® data base significantly. Several runs with bench-scale units have been undertaken to optimize process parameters and depth of coal cleaning and to establish the variability of the client's coal resource. A PDU program to confirm the engineering design basis has been completed.

H-Coal® Performance in Laboratory Testing

HRI has experience with a large number of coal feeds over a wide range of operating conditions. Table 2-5 is a summary of some of the coals run in the H-Coal® process. The

TABLE 2-5   Coals run in H-Coal® process

Eastern U.S. coals
Illinois No. 6
Indiana No. 5
Kentucky Nos. 9, 11, and 14
Pittsburgh seam (Consol No. 8)
Western U.S. coals
Wyodak
Utah D seam
Big Horn
Colorado
Black Mesa
U.S. lignites
Texas
North Dakota
Foreign coals
Australian brown
German "Steinkohle"
Others

eastern U.S. coals processed include Illinois No. 6; Indiana No. 5; Kentucky Nos. 9, 11, and 14; and Pittsburgh seam coal—all of which are bituminous coals. The western U.S. coals processed include both bituminous and subbituminous.

Our most extensive experience with western subbitumiunous coal has been with Wyodak coal. This coal has been difficult to process in other direct coal-liquefaction systems because of the formation of calcium carbonate deposits in the liquefaction reactor. This has not been a problem in our well-mixed catalytic H-Coal® reactor system. A major PDU run of 45 days was completed on Wyodak coal. This run used an advanced catalyst and demonstrated high yields and maintenance of activity while operating free of calcium carbonate deposition.

Lignites also have been successfully processed, as have Australian brown coal and German "Steinkohle."

Table 2-6 summarizes some typical H-Coal® yields on the basis of pounds per 100 pounds of dry coal. The first two columns compare yields from an Illinois No. 6 coal for two different modes of operation—syncrude and fuel-oil. In the syncrude mode, high yields of distillate liquids are achieved, in this case 47.8 percent by weight of C_4–524 °C liquid product. The yield of bottoms material is adequate to meet hydrogen requirements if the bottoms are processed via partial oxidation to produce hydrogen.

In the fuel-oil mode, operating conditions are less severe and produce a heavier product slate. The heavy fuel oil is recovered using a solids-liquid separation technique such as Lummus antisolvent de-ashing or Kerr-McGee critical solvent de-ashing. Hydrogen consumption is also much lower than in the syncrude mode. Other product slates intermediate to these may be produced to meet particular market needs.

The third column shows the yields achieved when processing Wyoming subbituminous coal in the syncrude mode. Hydrogen consumption is higher as a result of the increased yield of water from this high-oxygen-content coal. Yields of distillate liquid (C_4–524 °C

TABLE 2-6 Typical H-Coal® yields (lb/100 lb of dry coal)*

	Coal Used		
	Illinois Bituminous (Burning Star)		Wyoming Subbituminous (Wyodak),
Product	Syncrude mode	Fuel-oil mode	Syncrude mode
H_2	(5.3)†	(3.4)†	(6.2)†
H_2O, CO, CO_2	7.1	6.5	20.0
H_2S, NH_3	3.6	2.2	1.6
C_1 to C_3 fraction	11.2	6.8	12.3
C_4–204 °C naphtha	18.7	13.4	25.8
204–524 °C fuel oil	29.1	20.8	18.6
524 °C+ bottoms (including ash)	35.6	53.7	27.9
Total	100.0	100.0	100.0

*Based on lined-out operation in the 3-t/d PDU at HRI's R&D center.

†Hydrogen consumption.

fraction) of 44.4 percent by weight were achieved. Less severe conditions could again be utilized to obtain a heavier product slate and to lower hydrogen requirements.

Some typical properties and compositions of H-Coal® products are presented in Table 2-7. The analyses are given for coal liquids produced from Illinois and Wyoming coals used in the syncrude operating mode. These qualities were achieved at lined-out operating conditions in HRI's 3.5-t/day PDU. These H-Coal® liquids are very low in sulfur when compared with typical petroleum fractions; the oxygen and nitrogen contents, however, are higher. Unlike petroleum crude oils and products from some other direct coal-lique-faction technologies, no residual-oil products (compounds with boiling points greater than 524 °C) are produced.

H-Coal® PDU Experience

The H-Coal® PDU has been operated regularly over the last 14 years to demonstrate scale-up of yield data and equipment operability and to obtain products for downstream testing. Nine PDU runs, typically of about 30 days' duration were carried out in the most recent H-Coal® development program. Some of the major accomplishments are summarized below.

- Illinois No. 6, Kentucky No. 11, and Wyodak coals were successfully processed.
- Equilibrium catalyst conditions were simulated by means of continuous catalyst addition and withdrawal.
- Syncrude, fuel-oil, and intermediate modes of operation were demonstrated.
- Emergency operating procedures for the pilot plant were tested, and at the same time operator training in these procedures was provided.
- Critical operating limits such as maximum gas velocity were evaluated.

TABLE 2-7   Typical properties and compositions of H-Coal® liquid products*

	Coal	
	Illinois bituminous (Burning Star)	Wyoming subbituminous (Wyodak)
Naphtha (IBP† to 177 °C)		
Specific gravity, °API	52.3	55.8
Elemental analysis, wt %		
C	85.3	84.7
H	13.8	14.0
O	0.56	1.25
N	0.24	0.10
S	0.07	0.02
Mid-distillate (177 to 343 °C)		
Specific gravity, °API	18.5	27.8
Elemental analysis, wt %		
C	88.4	87.0
H	10.1	11.4
O	1.0	1.3
N	0.47	0.22
S	0.08	0.03
Distillate boiler fuel (260 to 427 °C)		
Specific gravity, °API	4.9	10.9
Elemental analysis, wt %		
C	89.4	88.4
H	8.6	9.4
O	1.3	1.7
N	0.63	0.46
S	0.08	0.03

*Syncrude operating mode used.

†Initial boiling point.

- A two-stage slurry letdown system that was designed for the pilot plant was demonstrated.
- An irradiated catalyst was used to test ebullating-bed mixing and catalyst deactivation.
- The demonstration run on Illinois No. 6 coal was used as the basis for the H-Coal® commercial plant.

PILOT PLANT PROJECT

The H-Coal® process has been thoroughly tested in bench-scale and PDU-size equipment and is now being demonstrated in large-scale equipment at the Catlettsburg pilot plant.

This plant is the largest coal-liquefaction pilot plant built in the United States to date. The plant is designed to feed up to 600 tons/day of coal and to produce up to 1800 bbl/day of liquid product, depending on the process mode. Ashland Oil, Inc. is responsible for pilot plant operations. The pilot plant has several major objectives beyond those attained in laboratory-scale equipment.

These objectives include:

- Demonstrating the mechanical operability and reliability of commercial-scale equipment
- Providing products for commercial testing at rates of 100 to 300 t/day
- Verifying yields in commercial-size equipment
- Collecting scale-up and engineering data
- Determining appropriate materials for construction
- Establishing maintenance requirements for key items of equipment

Two operating configurations have been designed into the plant, and a 2-year demonstration program is planned that encompasses operation in the syncrude and boiler-fuel modes and uses three different coals. Plans for the first year include syncrude operations with Kentucky No. 11 and Illinois No. 6 coals. The schedule for the second year calls for boiler-fuel operations with those two coals and a return to the syncrude mode using Wyodak coal.

During the initial start-up of the equipment, oil was first fed to the reactor system on February 26, 1980. A series of scheduled oil operations were then carried out, processing first a light gas oil, then a heavy gas oil, and finally a residual fuel oil to eliminate any deficiencies in the operating equipment and to provide operator training.

The plant is now operating. Coal was first fed to the H-Coal® reactor on May 29, 1980. Initial break-in operations used Kentucky No. 11 coal at a targeted feed rate of 200 Mg/day (220 t/day) in the syncrude operating mode. Coal conversions as high as 95 percent have been achieved. On-stream days, to date, have been somewhat limited because of maintenance requirements associated with the commissioning of a new process in commercial-size operating equipment. Table 2-8 gives a description of the problems met so far; most of these were mechanical in nature and all of them have been resolved successfully.

Following a major turnaround for equipment inspection, operations were initiated in February 1981 using Illinois No. 6 coal in the syncrude mode. Continuous smooth oper-

TABLE 2-8   H-Coal® process—major problems during initial coal operations

- Willis letdown valves
- EPG ball block valves
- Instrumentation—especially level controls
- Oil-seal system unreliable—pump seal failed
- Insufficient isolation valves, blocks, and bypasses
- High-pressure absorber system inoperable
- Inadequate line tracing—plugging of line and pump
- Flaker unreliable
- Weigh-feeder operation unreliable
- Ebullating-pump vane and diffuser assembly failed

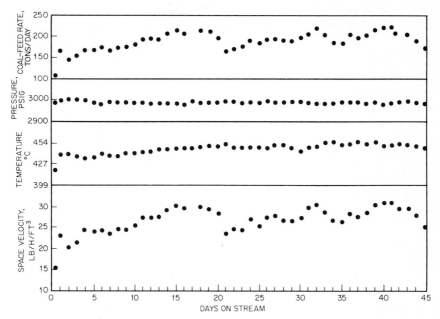

FIG. 2-3 H-Coal® pilot plant operating data—run with No. 6, Illinois coal.

ations in excess of 45 days have been achieved. This successful pilot plant performance is illustrated in Fig. 2-3, which presents plotted operating data and feed rate during the entire run.

Table 2-9 presents a summary of the H-Coal® pilot plant design basis yields for the different coal feeds and different modes of operation that are scheduled for demonstration in actual operations. Table 2-10 represents a typical day's operation in the 45-day run following the first major turnaround of the unit. Included are both the actual operating conditions compared to the target values and the product yields obtained compared to those obtained in the PDU at the same conditions. These data were supplied by Jack C. Swan of Ashland Synthetic Fuels, Inc., in a paper presented at the Sixth Annual Contractor's Conference on Coal Liquefaction.[5] This conference was sponsored by EPRI (the Electric Power Research Institute) in Palo Alto, California, on May 13, 1981. The consensus of most conference attendees was that coal-conversion technology has an optimistic future, based on H-Coal® achievements.

FUTURE PROSPECTS FOR H-COAL®

The cost of coal liquids produced by direct hydroliquefaction is generally considered to fall in the range of $30 to $60 per barrel (1981 dollars).[6] This wide range of cost estimates derives from the variations in basic assumptions made and the level of detail incorporated in the calculations. These costs are presently about equal to the average cost of products from imported oil at OPEC prices. In addition, the balance-of-payments and security-of-supply issues have led the U.S. government to act further to stimulate commercialization of a coal-liquids industry.

TABLE 2-9 H-Coal® pilot plant design yields

	Product		
	Synthetic crude*	**Low-sulfur fuel oil***	**Synthetic crude†**
Yield content (dry basis), wt %			
$H_2S + NH_3$	4.8	4.2	1.0
$H_2O + CO_2$	10.0	10.1	19.8
C_1 to C_4 fraction	10.7	8.7	12.3
C_5 to 204 °C fraction	15.9	9.9	22.8
204 to 260 °C fraction	7.9	6.0	5.8
260 to 524 °C fraction	23.9	21.0	15.1
524 °C+ fraction	17.0	27.5	11.6
Solids	14.3	15.8	17.3
Total	104.5	103.2	105.7
C_5 + liquid yield (solids-free basis), b/t dry coal	3.7	3.5	3.5
Chemical H_2 consumption, scf/t dry coal	16,900	12,000	21,400

*Coal feedstock is Illinois Burning Star.

†Coal feedstock is Wyoming Wyodak.

TABLE 2-10 Example of typical operating day in H-Coal® pilot plant: March 27–28, 1981

Conditions	Actual	Target
Total pressure, psi	3000	3000
Reactor temperature, °F	845	850
Coal feed, t/d	222	219
Oil-to-solid ratio	1.65	1.75
Space velocity, lb dry coal/(h)(ft³)	31.6	31.2
Hydroclone recycle flow, % of slurry oil	47	66

	Actual		Target	
Yields	**wt %**	**bbl/t**	**wt %**	**bbl/t**
C_1 to C_3 fraction	10.96	——	10.68	——
Naphtha (C_4 to 204 °C)	22.71	1.66	18.74	1.40
Distillate (204 to 524 °C)	23.89	1.46	28.33	1.63
Residuum oil (524 °C+)	21.62	0.94	19.00	0.86
Unconverted coal	3.47	——	5.78	——
Ash	11.22	——	11.51	——
Total		4.06		3.89

In part because of the wide variation in product costs calculated for coal liquefaction, comparisons of the various processes are difficult to make and infrequently reported. One such comparison, though, was made in July 1979 by the Engineering Societies Commission on Energy (ESCOE), under DOE Contract no. EF-77-C-01-2468. Product costs estimated by ESCOE (Table 2-11) are summarized for various coal-liquefaction processes and are calculated by two alternate methods. The first column lists costs of producing coal liquids for the various technologies on an energy basis in terms of dollars per million Btu of energy produced. Since different products of different compositions result from each process, it is necessary to adjust the product costs to reflect the value of the products in the marketplace. In the second column, the individual products are assigned value factors, based on current market-price relationships. These factors provide a basis for determining an effective cost for a multi-product slate and to simulate the cost incurred if all products were transformed to gasoline product.

The H-Coal® syncrude mode appears to produce products at the lowest estimated cost for all processes reported by ESCOE. While these data are not conclusive, H-Coal® would appear to be a front-runner in terms of lowest-cost product. Low costs for H-Coal® reflect the superior liquid yields demonstrated in the development of the process to date.

The H-Coal® process, as studied by ESCOE and as now being commercialized, represents a translation to large-scale reactor operations and process configurations designed in the early 1970s. Improvements and variations are being evaluated by means of HRI's new H-Coal® commercial plant planning (LP) model. These studies, supported by an ongoing experimental and engineering R&D program, promise even lower H-Coal® product costs. Staged operations, further catalyst improvements, and superior disposition of bottoms and

TABLE 2-11 Comparison of coal-liquefaction processes

	Energy cost, $/MM Btu	Reference price, $/MM Btu
Direct liquefaction yielding solid products:		
SRC*-I process	3.38	6.67
Direct liquefaction yielding liquid products:		
SRC*-II process	3.62	5.59
EDS† process	3.96	5.40
H-Coal® fuel-oil process	3.30	5.09
H-Coal® syncrude process	3.58	4.81
Indirect liquefaction:		
Fischer-Tropsch process	4.99	5.52
Methanol-to-gasoline process	4.89	4.91
Methanol process	4.37	4.54

*Solvent-refined coal.

†Exxon Donor Solvent.

Source: From Ref. 6 with permission.

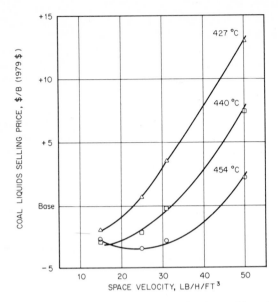

FIG. 2-4 H-Coal® commercial plant planning model: coal-liquids selling price vs. space velocity.

gas are of particular interest. Combined with the inherent flexibility of a direct-catalytic process and the proven capability to handle a full range of coal types, the ESCOE assessment suggests a bright future indeed for the H-Coal® process.

H-COAL® ECONOMICS

Coal-conversion technology is complex and highly investment-intensive. To further define the cost of coal liquids, HRI has developed an H-Coal® commercial plant planning (LP) model which uses linear programming to optimize process configuration, operating conditions, and product options for the complete coal-liquefaction plant.[7] The model simultaneously balances hydrogen, fuel gas, steam generation, and bottoms disposition in an optimal manner. A complete utilities system is designed by the model for each case, including electric power generation, cooling water, steam generation, and raw-water requirements. Several options are available for producing hydrogen and plant fuel, and product upgrading (i.e., catalytic re-forming of naphtha and product hydrotreating) may be included.

An example of results from the LP model is shown in Fig. 2-4. H-Coal® liquefaction yield slates for Illinois No. 6 coal were estimated for a range of H-Coal® process variables (temperature, coal-space velocity, and catalyst addition rate) using a reactor simulator correlated from the H-Coal® experimental data base. The range of process variables evaluated includes:

- Temperatures from 427 to 454 °C
- Coal-space velocities from 15 to 50 lb of coal per hour per cubic foot of reactor
- Catalyst replacement rates from 0.5 to 3.0 lb of catalyst per ton of coal

The H-Coal® LP model was used to estimate the impact on the overall economics of changes in the H-Coal® operating conditions and yields. Figure 2-4 shows the selling price of coal liquids required to achieve the target rate of return on investment vs. coal-space velocity. Separate curves are presented for reactor temperatures of 427, 440, and 454 °C. The high-severity operations (low coal-space velocity) are preferred.

Reductions of several dollars per barrel in the required selling price of coal liquids have been demonstrated by means of the H-Coal® LP model. The results of the LP model study are being used to set operating conditions for specific experimental studies, to provide guidance to the continuing H-Coal® R&D program and to develop planning bases for projected H-Coal® commercial plants.

REFERENCES

1. Layng, E. T., and K. C. Hellwig: "Liquid Fuels from Coal by the H-Coal Process," *Min. Congr. J.*, April 1969.

2. Merdinger, M., and A. Comolli: "Recent Developments in the Processing of Western Coal with the H-Coal® Process," *Sixth Annual EPRI Contractor's Conference on Coal Liquefaction, Palo Alto, CA, May 13, 1980.*

3. Li, A., and D. Lin: "Scaleup Performance and Thermal Stability Analysis of H-Oil® and H-Coal® Ebullated-Bed Reactors," *Second World Congress of Chemical Engineering, Montreal, Canada, October 1981.*

4. Burke, P.: "The Rough Road to a Synfuels Industry," *Chem. Week,* July 9, 1980, pp. 18–27.

5. Swan, J.: "H-Coal Pilot Plant Operations," *Sixth Annual EPRI Contractor's Conference on Coal Liquefaction, Palo Alto, CA, May 13, 1980.*

6. Roger, K. A., and R. F. Hill: *Coal Conversion Comparison,* U.S. Department of Energy Report no. FE-2468-51, July 1979.

7. Duddy, J. E., and J. B. MacArthur: "H-Coal Process Optimization via Linear Programming Model," 91st National AIChE Meeting, Detroit, Michigan, August 1981.

THE SRC-I COAL LIQUEFACTION PROCESS

MARIANNE T. PHILLIPS, *Editor and Writer*
JAMES C. BRONFENBRENNER
ALLEN R. KUHNS
JAMES R. O'LEARY
GREGORY D. SNYDER
SAMUEL ZNAIMER

International Coal Refining Company
Allentown, Pennsylvania

INTRODUCTION AND BACKGROUND

International Coal Refining Company (ICRC), a partnership of Air Products and Chemicals, Inc. (APCI) and Wheelabrator-Frye Inc., is currently completing design of the SRC-I process for the direct liquefaction of coal to be used to refine high-sulfur bituminous coal.

In the first stage of the process, coal is dissolved and hydrogenated at high temperature and pressure in a process-derived recycle solvent. Sulfur, ash, nitrogen, and oxygen are removed, yielding an environmentally acceptable solid fuel, or solvent-refined coal (SRC), as well as liquid distillate products. In the second stage, the SRC is hydrocracked and further hydrogenated to produce solid and liquid fuel products, the amounts of which can be varied, depending upon current market needs for energy.

ICRC believes this two-stage-liquefaction (TSL) technology is superior to less flexible, single-stage, direct-liquefaction processes because of its efficient use of hydrogen, its ability to treat a wide range of coals and to produce a flexible product slate, and its low yield of gas by-product.

The SRC-I process design has been supported by the operation of two pilot plants, one at Wilsonville, Alabama [sponsored by the U.S. Department of Energy (DOE) and the Electric Power Research Institute (EPRI)], and the other at Ft. Lewis, Washington (sponsored by DOE). In addition, DOE is providing major funding for the design of a 6000-ton/day coal-refining facility to demonstrate the technical, economic, and environmental feasibility of the SRC-I process.

GENERAL DESCRIPTION OF THE SRC-I PROCESS

Figure 3-1 is a block flow diagram of the major SRC-I process areas. Washed coal is delivered to the plant and then unloaded, conveyed, stored, reclaimed, dried, and ground to 200 mesh in the coal preparation area. About 93 percent of the coal is fed to the SRC process unit, and the remainder is conveyed to the gasification area.

In the SRC process unit, the coal is slurried in a process-derived solvent (boiling point range of 232 to 455 °C), pumped to a reaction pressure of roughly 2000 psig (14 MPa), mixed with a hot, hydrogen-rich gas stream, and heated in a fired heater to about 427 °C. The coal dissolves in the fired heater, and hydrocracking reactions begin. At the exit of the fired heater, additional hydrogen-rich gases are added to the coal slurry, and the mixture flows to the dissolvers. Coal-hydrocracking reactions are completed in the dissolvers, generating SRC, oils, light hydrocarbon gas, H_2S, NH_3, and CO_2. The high-pressure hydrogen-rich gas is separated from the product slurry by flashing and distilling. Oils boiling up to 455 °C from the SRC, expanded-bed hydrocracker, and coker-calciner units are combined in a common fractionator and distilled into three product cuts: naphtha (C_5–205 °C), middle distillate (205 to 345 °C), and heavy distillate (345 to 455 °C). The residual slurry of SRC, ash, and unreacted coal is sent to the Kerr-McGee critical solvent de-ashing (CSD) unit, where most of the SRC is recovered and the ash, unconverted coal, and small amounts of SRC are rejected. The rejected stream combines with the coal from the coal preparation area and is sent into the gasification unit. The carbon content of the ash concentrate and coal is partially oxidized at atmospheric pressure with oxygen and steam to produce makeup hydrogen. This partially oxidized gas is compressed and passed over a shift-conversion catalyst, where residual CO reacts with H_2O to yield CO_2 and H_2. After purification,

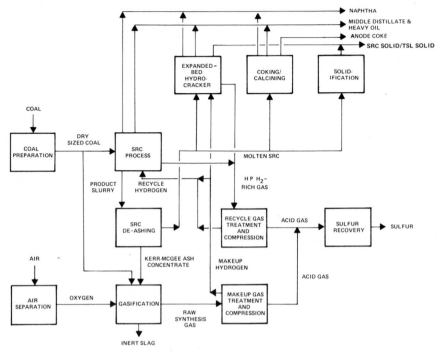

FIG. 3-1 Block flow diagram of SRC-I process.

the hydrogen stream divides; about two-thirds is sent to the SRC process unit and one-third is passed to the expanded-bed hydrocracker.

One-third of the molten SRC recovered in the Kerr-McGee unit is sent to the expanded-bed hydrocracker, one-third to the delayed coker-calciner, and one-third to the product solidifier. Expanded-bed hydrocracking catalytically hydrocracks the SRC into naphtha, middle and heavy distillates, and low-sulfur solid fuel. The coker-calciner converts the SRC to anode coke and produces naphtha and middle distillate as by-products. The solidifier essentially cools the molten SRC into solid SRC.

Gas and wastewater streams generated in the major process areas undergo effective pollutant-mitigation treatment, as discussed later under "Wastes and Emissions."

SRC-I PROCESS CHEMISTRY AND THERMODYNAMICS

Preheater dissolver kinetics and chemistry

The kinetics of coal conversion are complex and are dependent upon such factors as coal character, pressure, residence time, solvent quality, catalyst concentration (through solids accumulation), mixing, and reaction temperature. In the presence of a hydrogen-donor solvent, coal liquefaction occurs very rapidly once reaction temperatures are reached. As it dissolves, the reactable coal (feed coal less ash and insolubles) proceeds through a sequence of liquid yield reactions paralleled by gas yield reactions. The production of SRC (i.e., the

455 °C-to-end point fraction composed primarily of asphaltenes and preasphaltenes) is followed by the formation of successively lower-boiling-point liquids [i.e., heavy distillate (345 to 455 °C), middle distillate (205 to 345 °C), and naphtha (C_5–205 °C)]. Gas production includes the formation of H_2S, by which the organic sulfur content of the feed coal is depleted. Sulfur removal increases with increasing reaction temperature. Oxygen-containing species, which include CO, CO_2, and H_2O, are formed from the breakdown of preasphaltenes. Hydrocarbon (C_1 to C_4) gases are a product of dissolution and thermal-hydrocracking reactions. Little nitrogen is removed during the preheating and dissolving step.

De-ashing unit

The separation of ash and unconverted coal from SRC involves extraction of the SRC into a proprietary de-ashing solvent (DAS) at temperature and pressure conditions near the solvent's critical point. By then raising the temperature of the DAS above its critical point, the DAS density is sharply reduced, resulting in the rejection of part or all of the dissolved SRC. The DAS density is further reduced by decreasing the pressure, which results in the rejection of the remaining SRC. This rejected fraction is termed *light SRC*.

Gasifier

Gasification is an autothermal process in which the Kerr-McGee ash concentrate (K-MAC)-plus-coal feedstock is partially oxidized in an entrained bed. The small particle size of the feedstock combined with a high reaction temperature (1600 °C) results in a high degree of carbon conversion, yielding a raw synthesis gas composed mainly of CO, H_2, and CO_2. The raw synthesis gas contains only traces of CH_4 (<0.1 percent by volume) and no higher-molecular-weight organic compounds. Almost all organic or inorganic sulfur in the gasifier feedstock appears in the raw synthesis gas as H_2S and COS. Nitrogen in the feedstock and in the gaseous oxygen reacts in the gasifier to form traces of NH_3, HCN, and NO_x.[1]

Expanded-bed hydrocracker

The feed to the expanded-bed hydrocracker [i.e., de-ashed SRC (nominal 455 °C+ cut), recycle solvent (nominal 260-to-455 °C cut), and hydrogen] is passed over a heterogeneous catalyst to partially convert the SRC to distillates and reduce the sulfur and nitrogen content of the unconverted SRC. The catalyst bed is expanded by a large internal recycle flow, resulting in a well-mixed bed and nearly isothermal reaction conditions. The catalytic reactions, including hydrogenation, hydrodesulfurization, hydrodenitrogenation, and hydrocracking, occur on a dual-function metal sulfide acidic support catalyst, typically molybdenum, with either cobalt or nickel on an alumina support. It is not clear how the catalyst interacts with the SRC. However, it is hypothesized that the solvent transfers hydrogen to the high-molecular-weight species, which then thermally split into lower-molecular-weight species that directly interact with the catalyst. A high-activity catalyst will reduce the reactor temperature required, thus increasing the efficiency of hydrogen use by minimizing gas yield.

Naphtha hydrotreater

In this severe (high-temperature and high-pressure) hydrotreating process, nitrogen, oxygen, and sulfur contaminants of the raw naphtha feed react with hydrogen over a cobalt-molybdenum or nickel-molybdenum catalyst. These heteroatoms are converted to H_2O, NH_3, and H_2S, and the feed achieves re-former feedstock specifications of less than 100 ppm oxygen, 1 ppm nitrogen, and 1 ppm sulfur.

Delayed coker-calciner

Delayed coking is a severe thermal-cracking process, usually occurring between 425 and 485 °C. The carbon (coke) formed during the process contains some volatile material (i.e., light hydrocarbons) and heavy hydrocarbons. To produce an anode-grade coke product, calcination of the coke at temperatures exceeding 1100 °C is required. This high-temperature treatment drives off the light hydrocarbons, as well as moisture, and converts the heavy hydrocarbons into additional coke.

PROCESS PERSPECTIVE OF PILOT AND DEMONSTRATION PLANTS

Two pilot plants with differing coal-processing capabilities have been built and operated to demonstrate the basic feasibility of the SRC process. These plants have logged over 10 operating years and have provided the data base for design of the proposed 6000-ton/day (5442-Mg/day) SRC-I demonstration plant. Although the larger of the two, a facility processing 50 tons/day (45 Mg/day) of coal at Ft. Lewis, Washington, is no longer in operation, the 6-ton/day (5-Mg/day) pilot plant in Wilsonville, Alabama, continues to provide design and process information.

Wilsonville Pilot Plant

In 1973, Catalytic, Inc., an APCI subsidiary, engineered, built, and began operating a pilot plant in Wilsonville, Alabama, under the sponsorship of Southern Company Services, Inc. (SCS), and EPRI.[2] In 1976, DOE assumed responsibility for 65 percent of the operating expense, while EPRI retained ownership and responsibility for capital improvements. SCS, principal contractor for the work to date, is the utility host for the plant, which is located on Alabama Power Company land next to its coal-fired Gaston steam station.

The immediate goal was to produce a solid fuel that would meet environmental standards. The SRC produced at this plant from high-sulfur coal contains less than 0.16 percent ash, and when burned it yields less than 1.2 lb of SO_2 per million Btu. Once the basic feasibility of the SRC-I process was demonstrated, Wilsonville personnel began testing a wide variety of feed coals and plant operating conditions; they also began introducing process improvements. In 1978, the Kerr-McGee CSD process was introduced at Wilsonville to separate the ash and unconverted coal residue from the SRC. In 1981, an expanded-bed hydrocracker, designed and built by Hydrocarbon Research, Inc., came on stream.

A main goal of the current 5-year research program is to provide a sound design basis to ensure the operability of the 6000-ton/day demonstration plant. Research programs will focus on operation of the high-pressure separator, the ash concentrator, and the SRC solidifier and will address schemes for chloride treatment. Samples will be drawn for wastewater treatment, chloride mass balance, and thermodynamic correlation research programs. In addition, the program should provide data to optimize coal-liquefaction processes.

These include confirmation of the potential benefits of the Kerr-McGee CSD process (see detailed description under "Kerr-McGee Critical Solvent De-ashing") and the upgrading of SRC (including light SRC) by hydrogenation.

SRC-I Demonstration Plant

The proposed 6000-ton/day SRC-I demonstration plant is the culmination of nearly a decade of work in synthetic fuels development by Wheelabrator-Frye Inc. and APCI. In early 1977, Wheelabrator-Frye and the Commonwealth of Kentucky entered into an agreement for the design of a 2000-ton/day (1814-Mg/day) SRC-I demonstration plant. Part of this work led to the selection of a 1500-acre (6.073-km^2) plant site at Newman, Daviess County, Kentucky. A year later, the Air Products/Wheelabrator-Frye joint venture was formed to act as the sole prime subcontractor to SCS, the principal contractor for the project's initial design phase (Phase Zero).

During Phase Zero, the joint venture prepared the conceptual design, preliminary cost estimates, marketing assessments, economic evaluation, and environmental appraisal. Process options were evaluated, critical technology areas requiring additional data were identified, and the economics was assessed for both a 30,000-ton/day (27 211-Mg/day) commercial plant and a commercial plant of the same size that would be an expansion of the 6000-ton/day demonstration plant, producing the equivalent of 20,000 bbl/day of petroleum. The latter option was chosen. Phase Zero work was completed in July 1979, and in October 1979, DOE authorized Phase I, the detailed engineering of the project.

In March 1980, the Joint Venture became International Coal Refining Company (ICRC), and in August of that year, a cost-sharing agreement between ICRC and DOE was signed that covers the remainder of the demonstration plant program through start-up and operation. Under the terms of the agreement, ICRC will invest $90 million in the project, the Commonwealth of Kentucky will invest $30 million, and DOE will fund the balance. The contract states that ICRC can buy out the federal and state governments' interest. At that time, ICRC would plan to expand the facility up to fivefold, or to 30,000 tons/day of coal, to produce a nominal 100,000 bbl/day of petroleum product equivalent.

Under the cost-sharing agreement, ICRC became the project's prime contractor. As an important subcontractor to ICRC, SCS will continue its role in the development of SRC technology by providing broad technical reviews and studies of product use. In addition, six other subcontractors are providing engineering, procurement, and construction services for defined areas of the plant.

Design, construction, and operation of the 6000-ton/day SRC-I demonstration plant will accomplish the following objectives:

- Provide an accurate basis for determining investment and operating costs for commercial application of this technology, including equipment and operating procedures required for environmental control measures.

- Demonstrate the technical feasibility of certain process steps at commercial capacity and provide other technical information on process integration, materials of construction, equipment design, and fabrication techniques needed to optimize the economics of commercial operation.

- Provide significant technology advances that would not otherwise occur. Such advances will also serve to optimize the use of other coal-liquefaction technologies in such areas as mineral ash separation and hydrogen generation.

- Integrate the SRC-I process with a second liquefaction stage that results in a more flexible solid and liquid product slate.

- Monitor plant emissions and evaluate environmental control and mitigation measures.

- Provide adequate supplies of SRC-I fuel products for commercial-scale combustion tests necessary to demonstrate product use in existing and new electric utility and industrial applications.

- Provide adequate supplies of SRC-I liquid fuel products to test their applicability as gasoline blendstock and turbine and distillate fuel oils.

- Provide adequate supplies of SRC-I fuel products for commercial-scale emission testing necessary to demonstrate the environmental acceptability of the fuels for direct use and to characterize the control equipment or modifications necessary for compliance with applicable emission limitations.

- Provide adequate supplies of SRC-I solid and liquid fuel products for commercial-scale testing of potential upgrading techniques.

- Provide adequate supplies of upgraded products to determine their effectiveness and economic viability as feedstocks to petrochemical industries and as raw materials for aluminum and steel industries.

- Allow assessment of the marketability of the range of products obtainable from a given coal.

Once these objectives are realized, ICRC will have the basis necessary to successfully scale up the demonstration plant to full commercial size.

SRC-I PILOT PLANT OPERATION

Wilsonville Pilot Plant

As depicted in Fig. 3-2, the Wilsonville pilot plant incorporates all major process units that are to be included in the demonstration plant excepting the coal preparation unit, gasifier, delayed coker-calciner, and naphtha hydrotreater; these units have been adequately demonstrated on a commercial scale.

This pilot plant, which has been on stream since 1976, has processed a variety of coals (Table 3-1) under a wide range of operating conditions (Table 3-2). The resulting yields and hydrogen consumption rates from these test runs are provided in Table 3-3.

The runs conducted at Wilsonville have achieved the following goals:

- Demonstrated that the SRC-I process is able to produce a specification-grade (low-sulfur and low-ash) product

- Demonstrated that the SRC-I process is capable of using a variety of coals

- Demonstrated that the SRC-I process can not only achieve solvent balance but also produce a net yield of process solvent

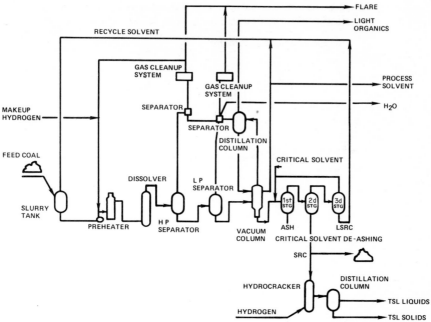

FIG. 3-2 Block flow diagram of Wilsonville pilot plant.

- Demonstrated an SRC-I solid separation step that maintains a high recovery of on-spec SRC-I product
- Determined optimal process conditions for efficient hydrogen consumption and high product yields
- Demonstrated the operability of process equipment, such as the preheater, the high-pressure separator, and letdown valves

TABLE 3-1 Feed coals processed at Wilsonville pilot plant since 1976

Type of coal	Mine
Kentucky no. 9	Fies
	Lafayette
	Dotiki
	Pyro
	Colonial
Kentucky no. 6 and 11	Pyro
Kentucky no. 14	Colonial
Indiana no. V	Old Ben 1
Utah	Emery
Amax Belle Ayr	Amax Wyodak
Illinois no. 6	Monterey 1

TABLE 3-2 Ranges of operating conditions tested in process runs conducted at Wilsonville pilot plant since 1976

Process condition parameter	Range
Slurry concentration, % mf* coal	10–38.5
Coal feed rate, lb/h (g/s)	200–900 (0.0393–0.4570)
Space velocity, lb/(h)(ft³) [g/(m³)(s)]	11–128 (0.0393–0.4570)
Gas feed rate, scf/h	1000–11,700
Hydrogen purity, mol %	85–94
Dissolver pressure, psig (MPa)	1650–2400 (11.48–16.65)
Dissolver temperature, °C	402 to 463

*Moisture-free

TABLE 3-3 Ranges of net yields from process runs conducted at Wilsonville pilot plant since 1976

Products	Yield range, wt % maf* coal
SRC (455 °C+)	45–70
Process solvent (232 to 455 °C)	10–25
Light oil (C$_5$ to 232 °C)	5–12
Gas	4–10
Hydrogen consumption	1.5–3

*Moisture- and ash-free

Ft. Lewis Pilot Plant

Owned by DOE and operated by Gulf Oil Corporation, this facility provided data on the fired heater and slurry components of the SRC-I process until its closing in August 1981.

Fired-heater tests

Fired-heater tests were conducted from June 1 to August 8, 1980 to establish a heat-transfer and pressure-drop data base in support of the design of the slurry-fired heater in the demonstration plant. The fired heater heats a complex three-phase (i.e., coal, solvent, and gas) non-Newtonian mixture to reaction temperatures, permitting dissolution of the coal and initiating its conversion to SRC.

The primary objective of the experimental program was to gain insight into the apparent gel-formation zone of the coal slurry, where low heat-transfer coefficients are experienced, thereby increasing the possibility of coke formation due to high slurry-film temperatures. In order to determine the allowable heat fluxes for the demonstration plant, it was necessary to correlate the heat-transfer characteristics of the three-phase Ft. Lewis system.

Two important results were achieved:

• The solvent-only baseline runs assessed the reliability and accuracy of the instrumentation, resulting in improved temperature measurements of the tube wall by revision of the thermocouples.

• Reliable data were obtained for the SRC-I system at demonstration plant conditions.

Slurry-mix tests

These tests, performed in August 1980, determined the feasibility of preparing a 38.5 percent by weight slurry from feed coal and process solvent at temperatures of 177 °C and higher. Demonstration plant design has specified a temperature of 177 °C in the blending tank. Blending at this temperature reduces the cooling required for the recycle process solvent, thus improving thermal efficiency.

In four experiments at temperatures ranging from 177 to 216 °C, no operability problems were encountered in spite of a viscosity increase from 30 to 100 cP. Thus, verification of design and further potential improvements in thermal efficiency were achieved.

In addition, a special study determining the effects of coal concentration and residence time expanded the matrix of data available for understanding and designing the slurry-mix system.

SRC-I PROCESS—DETAILED DESCRIPTION

The production of SRC involves three key processes not previously used at the scale of the 6000-ton/day SRC-I demonstration plant or for this specific application.

Preheating and Dissolving

In the SRC process unit (Fig. 3-3), 93 percent of the output of the coal preparation unit is slurried in the coal-derived process solvent. The slurry-mix system is designed to operate at 177 °C with 38.5 percent by weight of 200-mesh coal in the slurry. Since the economics of the process would be greatly enhanced by the ability to operate at higher temperature with a coarser coal, the flexibility of operating at a temperature up to 216 °C with 20-mesh coal is being included in the design. The slurry mixture is then pumped to a reaction pressure of about 2000 psig (14 MPa), mixed with a hot, hydrogen-rich gas stream, and heated first against hot returning process solvent in an exchanger and then in the fired heater to 427 °C before it enters the dissolver. Each slurry-fired heater has been designed to operate at a flow velocity ranging from 12 ft/s (3.7 m/s) to 30 ft/s (9.1 m/s) and to accommodate slurry flows up to three times normal flow. Within the fired heater, coal

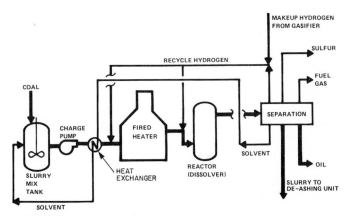

FIG. 3-3   Flow diagram of preheating and dissolving process.

dissolution is accomplished and hydrogenation begins. Additional hydrogen-enriched gas is added at the exit of the fired heater, and the three-phase mixture flows to the dissolvers, where hydrogenation and desulfurization are completed. Current design calls for the operation of two dissolvers in series with a total coal-feed residence time of 30 min. However, the demonstration plant will be designed to permit parallel operation as well. In addition, the amount and temperature of hydrogen introduced at the inlet of each dissolver will be variable.

The effluent liquid from the dissolver must be quickly cooled to below 416 °C to minimize retrograde reaction. Demonstration plant design avoids the need to manifold a three-phase slurry into multiple parallel heat exchangers by separating the vapor and liquid phases and then cooling each phase. In this separation of vapor and liquid, the high-pressure, hydrogen-rich gas separated from the product slurry is sent to the Selexol and diethanolamine (DEA) acid-gas removal units and then is combined with syngas from the gasifier unit for further treatment. After purification, the hydrogen stream is recycled to the SRC process unit. The slurry is flashed and distilled to remove process solvent and lighter components. The remaining slurry, containing ash, unconverted coal, and SRC, is sent to the de-ashing unit.

Kerr-McGee Critical Solvent De-Ashing (CSD)

A key segment of the SRC-I process is the separation of ash and unconverted coal residue from the SRC leaving the reactor. Of the filtration, solvent-extraction, and centrifugal separation techniques tested at Wilsonville, solvent extraction, specifically Kerr-McGee CSD, has proved to be the most effective. Because ash poisons the catalyst used in the hydrocracker, the significantly lower ash content of SRC from the CSD unit makes it an excellent feedstock for upgrading. In addition, the Kerr-McGee process improves the quality of the recycle solvent by producing a light SRC product that, when recycled to the first liquefaction stage, reduces the need for makeup hydrogen, increases coal conversion, and reduces process severity (i.e., temperature, pressure, and residence time) while maintaining solvent balance. The improved characteristics of the SRC recycle solvent and hydrocracker feedstock distinguish SRC-I from other coal-liquefaction technologies.[2]

CSD is a continuous de-ashing solvent-extraction process that is operated near the critical temperature and pressure of the DAS (de-ashing solvent). The SRC slurry is mixed with recycled DAS and flows to the first-stage settler (Fig. 3-4), where ash, unconverted coal solids, and insoluble SRC are separated as a heavy fluid phase. This heavy phase is withdrawn from the bottom of the settler and flashed to recover DAS. The ash, unconverted coal, and SRC form fine particulates (referred to as *Kerr-McGee ash concentrate*, or *K-MAC*) which are removed from the solvent separator as a free-flowing powder. The K-MAC is conveyed to a gasifier to generate hydrogen for coal liquefaction and to stabilize the ash.

The light phase from the first-stage settler, consisting of SRC dissolved in DAS, is removed from the top of the settler and heated to reduce the DAS density. This decrease in density causes a partial rejection of the soluble SRC as a heavy phase. The two phases flow to the second-stage settler, where the rejected de-ashed SRC is withdrawn and sent to a second solvent separator to recover DAS. The stream of ash-free SRC product is withdrawn from the separator, and one-third is sent to each of three locations: (1) the solidification section, (2) the coker-calciner area to produce anode coke, or (3) the expanded-bed hydrocracker to convert SRC material to distillate oil products.

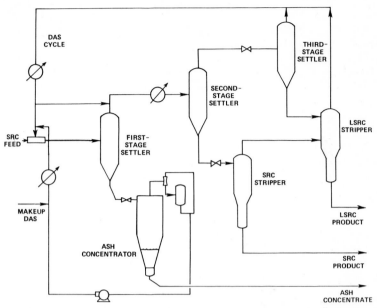

FIG. 3-4   Flow diagram of Kerr-McGee critical solvent de-ashing (CSD) process.

The density of the DAS in the light phase from the second-stage settler is further decreased in the third-stage settler, and the remaining soluble SRC is rejected as a heavy fluid phase and sent to a third solvent separator to recover the DAS. The product from the third stage is light SRC. By adjusting process conditions in the second-stage settler, the relative quantities of SRC and light SRC can be varied over a wide range.

Gasifying

Gasification will employ technology supplied by Gesellschaft für Kohle-Technologie (GKT).[1] This process generates the hydrogen necessary for coal liquefaction and SRC upgrading and converts ash from the coal to an environmentally acceptable slag.

As a first step, K-MAC (entrained in nitrogen) and supplemental coal (7 percent of that processed in the coal preparation unit) are conveyed to the dust preparation unit (DPU) (Fig. 3-5), where the K-MAC is humidified. A nominal mixture of 20 percent coal and 80 percent K-MAC is then sent to the service bunkers in the GKT gasifier area.

The mixture of K-MAC and coal is fed via the variable-speed screw feeder to the four-headed GKT gasifier, where the feed is partially oxidized with oxygen and steam to form raw gas which consists mainly of CO and H_2. The raw gas leaves the gasifier at about 1540 °C and is cooled in a waste-heat boiler, quenched with water to remove particulates, and conveyed to a common raw-gas holder. In the gasifier, part of the ash forms a slag, which exits from the bottom; another portion is entrained in the raw-gas stream.

Raw wash water containing slag, fly ash, and particulates is sent to a common enclosed clarifier. The solids separate from the water and are then removed as sludge by a sludge hopper. The sludge is pumped into a settling pond, and the clean wash water from the clarifier and the settling pond is recycled.

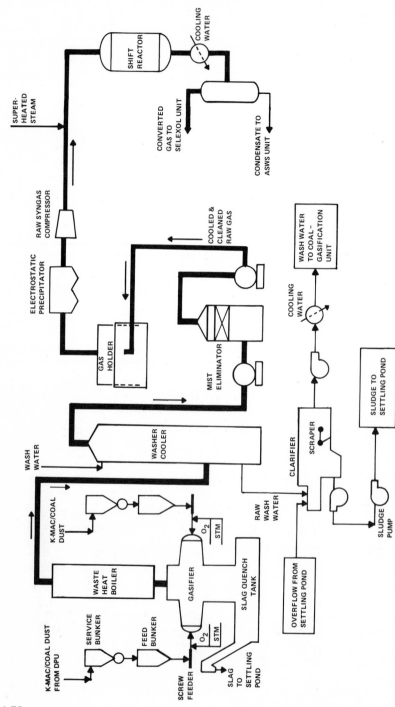

FIG. 3-5 Flow diagram of GKT gasification process.

1-78

Raw syngas from the gas holder passes through electrostatic precipitators for final particulate removal. The gas is compressed and mixed with superheated steam before it enters the shift reactor, where almost all CO is catalytically converted to H_2 and CO_2. Finally, the converted gas, together with sour off-gases from the SRC, LC-Finer, and coker-calciner process units are sent to the Selexol and/or DEA acid-gas removal units. Condensate separated from the gas is sent to the ammonium sulfide water stripper (ASWS). Acid gases stripped in the Selexol and DEA units and ammonia and hydrogen sulfide gases stripped in the ASWS pass to the Claus unit, where elemental sulfur is recovered and where NH_3 is converted to nitrogen. The tail gas from the Claus unit receives final cleaning in the Beavon-Stretford unit. The purified hydrogen stream is divided: two-thirds becomes makeup hydrogen for the SRC process unit and one-third supplies the expanded-bed hydrocracker.

UPGRADING TECHNOLOGY

For certain applications, SRC can be further upgraded to other desired products. The design of the SRC-I demonstration plant includes an expanded-bed hydrocracker or LC-Finer which will process one-third of the total SRC yield, producing additional distillate fuels and improving product quality. A hydrotreater will upgrade raw naphtha from the LC-Finer to render it suitable for refinery needs. In addition, the coker-calciner will convert one-third of the SRC to high-grade anode coke for the aluminum industry.

LC-Fining

Through a hydrocracking process, the LC-Finer hydrogenates SRC and partially converts it to distillates while removing sulfur and reducing nitrogen and oxygen contents. Since ash, which is poisonous to the catalyst, will have been removed from the SRC by the CSD process, hydrocracker operation will be more efficient than that of other coal-liquefaction processes.

A schematic flow drawing of the LC-Fining process is shown in Fig. 3-6. De-ashed SRC and recycle solvent are blended in a 70:30 weight ratio, heated, and then combined with hot hydrogen as feed to the bottom of the LC-Finer expanded-bed reactor. Since the expanded catalyst bed is very fluid, catalyst can be added and withdrawn during operation to maintain constant activity. The reactor effluent is flashed at high and low pressure to recover recycle hydrogen and process gas. The liquid products are fractionated into naphtha, middle and heavy distillates, and hydrotreated SRC. Makeup hydrogen is required to replace that consumed in the reactions.

The hydrogen partial pressure and the temperature in the reactor are approximately 2000 psig (14 MPa) and 430 °C, respectively. Typically, 40 to 50 percent by weight of the SRC feed is converted to distillate in each pass, together with the removal of 40 percent by weight of the nitrogen and 86 percent by weight of the sulfur. By recycling unconverted SRC, the conversion of SRC feed can be increased to 75 to 80 percent by weight and the removal of nitrogen and sulfur to 85 and 96 percent by weight, respectively. Yield structures for the once-through low-conversion operation and the SRC recycle, high-conversion operation are shown in Table 3-4.

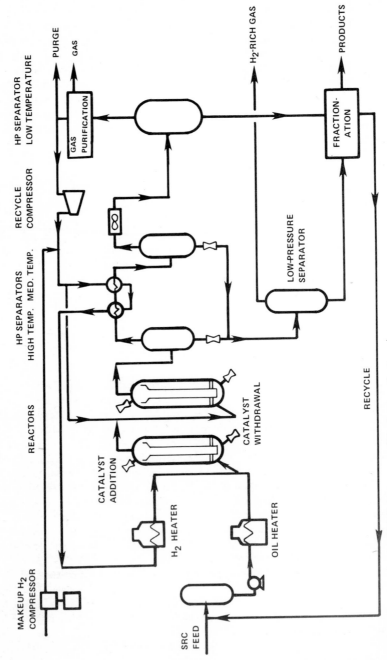

FIG. 3-6 Flow sketch of LC-Fining process. (*The Lummus Company.*)

TABLE 3-4 LC-Finer yield structures*

	Wt % of SRC feed	Sp. gravity (15.6 °C/ 15.6 °C)	Analysis, wt % of Fraction				
			H	S	N	O	Ash
SRC feed							
345-to-455 °C fraction	5.1	1.12	6.7	0.77	1.16	4.3	0
455 °C fraction	94.9	1.21	6.2	0.85	1.68	4.5	0.11
Total SRC feed	100.0	1.20	6.2	0.85	1.65	4.5	0.10
Low-Conversion Product Yields (Once-through SRC)							
Products							
H_2S, NH_3, H_2O	5.5						
C_1-C_3 fraction	8.9	——	20.3				
C_4–205 °C fraction	9.3	0.84	11.8	0.02	0.22	0.84	0
205-to-345 °C fraction	15.5	0.97	9.8	0.04	0.21	0.33	0
345-to-455 °C fraction	9.3	1.08	8.2	0.07	0.39	0.35	0
455 °C+ fraction	54.0	1.26	5.4	0.19	1.68	1.67	0.19
Total products	102.5	——	8.4	0.12	1.00	1.1	0.10
455 °C+ conversion, vol %	45.6						
H_2 consumption, std ft^3/bbl	1980						
Sulfur removal, wt %	86						
Nitrogen removal, wt %	39						
High-Conversion Product Yields (Recycle SRC)							
Products							
H_2S, NH_3, H_2O	7.4						
C_1-C_3 fraction	14.1	——	20.1				
C_4–205 °C fraction	16.7	0.84	11.8	0.01	0.10	0.23	0
205-to-345 °C fraction	31.2	0.96	10.0	0.01	0.10	0.21	0
345-to-455 °C fraction	15.3	1.09	8.1	0.02	0.18	0.22	0
455 °C+ fraction	20.3	1.24	5.9	0.10	0.77	0.69	0.49
Total products	105.0	——	10.6	0.03	0.23	0.28	0.10
455 °C+ conversion, vol %	79.2						
H_2 consumption, std ft^3/bbl	3950						
Sulfur removal, wt %	96						
Nitrogen removal, wt %	86						

*Specific gravity and analysis "total" values are weighted averages.

Naphtha Hydrotreating

Demonstration plant product liquids from the SRC, LC-Finer, and coker areas will be combined and fractionated to yield about 3700 bbl/stream day of naphtha having a boiling range of C_5–205 °C (191 °C at 95 percent point). ICRC determined that raw SRC-I naphtha was not appropriate for use as a boiler or turbine fuel or as a feedstock for synthetic natural gas, benzene-toluene-xylene extraction, or catalytic re-forming. However, naphtha can be upgraded to a feedstock for catalytic re-forming for use as a high-octane gasoline blendstock. This application will generate higher revenues compared with the other markets evaluated.

The hydrotreating process reacts raw naphtha with hydrogen over a conventional hydrotreating catalyst. Before entering the hydrotreater, makeup hydrogen from the gasifiers is methanated to remove CO and is then mixed with the stream of recycle hydrogen and the raw naphtha. The mixture is preheated and passed over the catalyst in the fixed-bed reactor at high temperature and pressure. The reactor effluent is cooled against the incoming streams and sprayed with water to prevent plugging and to facilitate the removal of ammonium hydrosulfide (NH_4HS) in a flash separator operating at about 38 °C. Except for a purged side stream, the hydrogen-rich flash vapors are recompressed and recycled to the reactor. The flash liquids are sent to a steam stripper unit to remove residual sour gases and light ends. These gases are compressed, combined with the recycle side stream, and sent to the gas cleanup unit before they are used as fuel gas. The stripped product is pumped to a storage tank.

Although the hydrotreated naphtha from the demonstration plant will be re-formed in a petroleum refinery, the addition of a re-former unit to a commercial SRC-I refinery will produce not only gasoline blendstock but also excess hydrogen for process use.

Delayed Coking and Calcining

In the first step of this proven and widely used technology, de-ashed molten SRC from the SRC-I process unit is fed to the delayed-coker unit, where it is mixed with recycle oil in the bottom of a fractionation tower. The mixture is then heated in a coking heater and fed into the bottom of one of two coke drums, in which the coke is formed. One coke drum is on stream and receives the coking heater effluent while the other is being decoked. Vapor from the top of the on-stream coke drum is sent back to the fractionation tower, in which it is brought into contact with refluxed liquids and fractionated into sour off-gas, naphtha, and middle distillate.

When a coke drum is filled, it is taken off stream, steamed, and cooled. The coke is then hydraulically cut from the drum, crushed, and dewatered.

Green coke, still containing approximately 12 percent moisture, is fed to the rotary-kiln calciner to burn off any remaining volatile material. The coke moves counter to the hot flue gas generated within the kiln. The hot, calcined coke is then transferred to a direct-contact rotary coke cooler, where it is cooled by controlled water injection. The cooled coke is then conveyed to storage. After waste-heat recovery and before it is discharged to the atmosphere, the flue gas from the calciner is scrubbed to remove SO_2 and particulates.

SRC-I PRODUCT AND BY-PRODUCT ANALYSES

Product and process flexibility

The flexibility of product types, quantities, and quality with efficient use of hydrogen and high thermal efficiency is a unique and important advantage of the SRC-I process.[3] In a commercial plant, a variety of coals could be processed to yield as much as 85 percent solids with 15 percent liquids to as little as 10 percent solids with 90 percent liquids. Specifically, the first stage of the SRC process will produce 75 percent by weight solid product and 25 percent by weight liquid product. The second, or LC-Fining, stage of this

TSL process can then be operated in either a high-severity or a low-severity hydrocracking mode, depending upon the proportions of solid and liquid desired. The TSL process, coupled with additional refining through naphtha hydrotreating or coking and calcining, will enhance the production of a varied solid and liquid product slate, as shown in Tables 3-4 and 3-5.

Product marketability

SRC-I solids   SRC-I solids are produced in two forms, SRC solids, made directly from the SRC process, and TSL solids, the residue remaining after SRC is hydrocracked in the LC-Finer.[3] The major difference between the two solid products is their sulfur and oxygen content; SRC solids are typically 0.85 percent by weight S and 4.3 percent by weight O_2, and TSL solids are usually less than 0.3 percent by weight S and about 1.5 percent by weight O_2. One-third of the SRC will be further processed in a delayed coker to produce green coke, which will be calcined to a high-quality anode coke for aluminum smelting.

SRC and TSL solids are significantly different from and offer definite advantages over coal (Table 3-6). These solids also display many similarities to no. 6 oil, especially a low ash content. The low sulfur content results in fuels that comply with stringent New Source Performance Standards (NSPS), thus eliminating the need for scrubbing devices. In addition, the higher Btu content, as compared with coal, reduces both transportation and handling costs. Finally, such unique properties as very low ash content, low melting temperature, and very high Hardgrove grindability index enable SRC to be used in three firing modes that are applicable to oil- and coal-designed boilers and furnaces: pulverized, melted and atomized like oil, and mixed in SRC liquid.

SRC-I burn tests   In 1977, SCS performed a large-scale test of the shipping, storage, handling, and combustion characteristics of SRC solids at the Plant Mitchell Generating Station (Table 3-7).[4] The combustion test, which was conducted in a 22-MW coal-designed boiler for 18 days, demonstrated that SRC solids can be burned in a utility boiler much like coal, while providing these distinct advantages:

• The test revealed that after some modifications were made, no problems were encountered in pulverizing SRC solids. In addition, the pulverizer tests showed a decrease of almost 25 percent in the power requirement for grinding SRC solids.

• Emissions measurements indicated that SRC solids complied with NSPS for SO_2 and NO_x. Particulate emissions from the furnace were one-seventh to one-tenth the levels resulting from burning coal, and precipitators of conventional design could be used to meet the emissions standards.

• Tests indicated that after the 18-day burn, virtually no ash buildup occurred in the superheater sections. In addition, there was no need to deslag the burner front at any time during the burn test.

Burn tests conducted at the DOE's Pittsburgh Energy Technology Center were carried out with SRC fuel using all three previously described firing modes for solids (Table 3-8).[5] These tests indicated that SRC solid-liquid mixtures and SRC melt can be burned at full rating in a boiler designed for oil. A combustion efficiency of 99.8 percent and a boiler efficiency of 81.3 percent for the SRC solid-liquid mixture were similar to those achieved

TABLE 3-5 SRC-I product and by-product markets

Product	Product, t/SD*	Yields,† MM Btu/h	Application(s)	Market(s)
Product				
Naphtha (C$_5$–205 °C fraction oils)	533	875	Catalytic re-forming feedstock	Transportation fuel and petrochemical industry
Middle distillate (205-to-345 °C fraction oils)	762	1155	No. 6 fuel oil substitute	Utility and industrial fuel
			No. 2 fuel oil substitute	Utility and industrial fuel
			Refinery-cracking feedstock	Transportation and industrial fuel
			Turbine fuel	Industrial fuel
			Low-speed diesel fuel	Transportation and industrial fuel
			Ethylene feedstock	Petrochemical industry
Heavy oil (345-to-455 °C fraction oils)	196	294	Boiler and furnace fuel	Utility and industrial fuel
			No. 6 fuel oil substitute	Utility and industrial fuel
			Carbon-black feedstock	Industrial fuel
			Needle-coke feedstock	Industrial fuel
			Refinery-cracking feedstock	Transportation and industrial fuel
			Component of solid-liquid mixture	
SRC‡	985	1333	Solid fuel for boilers and furnaces	Utility and industrial fuel
			No. 6 fuel oil substitute	Utility and industrial fuel
			Needle-coke feedstock	Industrial fuel
			Anode-coke feedstock	Industrial fuel
			Component of solid-liquid mixture	
Calcined anode coke	571	NA§	Aluminum-smelting anodes	Industrial uses
By-product				
Liquefied petroleum gas	55	99	Ethylene feedstock	Petrochemical industry
			Fuel	Industrial fuel
Sulfur	196	NA§	Sulfuric acid	Chemical industry
CO$_2$ (89 wt %)	—	—	Enhancer of oil recovery	Petroleum
Mineral aggregate	—	—	Cement fill	Construction
			Roadbed fill	Construction

*Stream day

†Yields based on SRC-I demonstration plant design and on higher heating value of feedstock; MM = million

‡Low sulfur, 0.3–0.8%; low ash, 0.1–0.5%

TABLE 3-6 Comparison of properties of SRC-I solids with properties of coal (maf) and fuel oil

Property	Kentucky no. 9 coal[a]	SRC solid	TSL solid[b]	No. 6 fuel oil
Higher heating value, MJ/kg	30.00	36.75	37.8	41.40
Moisture, wt %	11.5	2	2	<2
Ash, wt %	10.5	0.1	0.2	0.1
Proximate analysis, wt %				
Volatiles	46	57	62	NA[c]
Fixed carbon	54	43	38	12[d]
Ultimate analysis, wt %				
Carbon	79.3	87	91	87
Hydrogen	5.6	5.9	5.0	11
Oxygen	9.4	4.3	1.5	——[e]
Nitrogen	1.7	2.0	1.9	0.4
Sulfur	3.8	0.85[f]	0.26[f]	0.3–2.2
Bulk density, kg/m³	897	513	529	NA
Apparent density, kg/m³	1350	1230	1270	993
Hardgrove grindability index	59	176	176	NA[c]
Softening point, °C	NA[c]	150	100	NA[c]

[a]Typical coal, washed
[b]From LC-Fining of SRC solid
[c]Not applicable
[d]Carbon residue
[e]Negligible value
[f]Design basis

when burning the reference no. 6 fuel oils. Results from the tests conducted with pulverized SRC solids indicated that they too can be successfully burned in an oil-designed boiler. A combustion efficiency of 99.4 percent and a boiler efficiency of 82.4 percent were achieved. All tests indicated that SRC used in oil-designed boilers would comply with sulfur emissions standards.

TABLE 3-7 SRC fuel burn test in a 25-MW coal-designed utility boiler[*]

Fuel	Pulverizer energy usage,[†] kWh/MM Btu	Boiler efficiency,[‡] %	Combustion efficiency,[‡] %	Emissions, lb/MM Btu		
				SO₂	NOₓ	Particulates[†]
Coal	0.52	87	98.0	1.01	0.47	0.07
SRC fuel	0.37	86	98.5	0.97	0.40	0.04

[*]Performed at SCS's Plant Mitchell
[†]Fuel feed rate of 7000 lb/h
[‡]At full-load HHV basis
[§]Outlet of electrostatic precipitator

TABLE 3-8   SRC fuel burn test in fire-tube, oil-designed boiler

	Boiler combustion rating efficiency,	Boiler efficiency,	Uncontrolled Emissions, lb/MM Btu		
	hp	%	%	SO$_2$	NO$_x$	Particulates
No. 6 fuel oil	100	99.7	82.4	0.65	0.24	0.14
SRC fuel–liquid mixture	100	99.8	81.3	0.64	0.76	0.17
SRC fuel melt	100	99.2	82.2	1.01	0.71	0.31
SRC fuel finely pulverized	50*	99.4	82.4	1.15	0.98	0.35

*The physical design of the burner limited the feed rate of fuel to 50 percent of the boiler's rated capacity.

SRC-I liquids   SRC-I liquids produced by either the first or second stage of the SRC-I process can perform like those derived from petroleum. The demonstration plant will be producing a hydrotreated C,–205 °C naphtha fraction that will be catalytically re-formed in a commercial refinery to produce a high-octane unleaded gasoline blendstock, a 205 to 345 °C cut that can be used as fuel oil similar to no. 2 fuel oil, and a 345-to-455 °C heavy-oil fraction that can be used either directly as a substitute for no. 6 fuel oil or blended with SRC solids to form a mixture that can be fired as a substitute for no. 6 fuel oil.

All raw SRC-I liquids have significantly higher oxygen content than petroleum distillates (Table 3-9). Nitrogen compounds are also prevalent in SRC-I liquid streams and exceed the amount usually found in petroleum fractions.

To produce a suitable SRC-I naphtha re-former feed for conversion into a high-octane gasoline blendstock, sulfur and nitrogen content must be lowered, typically to less than 1 ppm, and oxygen to less than 100 ppm to prevent deactivation of the re-forming catalyst. Severe hydrotreating removes these heteroatoms, yielding naphtha with a high aromatic and naphthene content. In addition, research has indicated that conventional catalytic re-forming of SRC-I naphtha can result in higher octane numbers ($>$100) and higher yields than those typically obtained from re-forming petroleum naphthas under similar conditions.

Because the SRC-I middle distillate and heavy oils have a somewhat higher density than similar petroleum fuels (Table 3-9), they may handle slightly differently from no. 2 and no. 6 fuel oils. ICRC presently expects the stream of middle distillate to be a good no. 2 fuel oil substitute and the heavy-oil fraction to be used directly or blended with pulverized solids as a substitute for no. 6 fuel oil.

In anticipation of commercial-scale production, ICRC is also evaluating SRC-I naphtha as a catalytic re-forming feedstock to produce benzene, toluene, and xylene for the petrochemical industry; middle distillate as a hydrocracker feedstock to produce more gasoline, diesel fuel, and turbine fuel; and heavy oil as a feedstock for carbon products.

By-products   The SRC-I process will produce several major by-products. Liquefied petroleum gas (LPG) will be a mixed stream of propane and butane that can be fractionated

TABLE 3-9 Typical SRC-I and petroleum characterization

	Raw SRC-I naphtha	Typical petroleum naphtha	SRC-I middle distillate (205 to 345 °C)	Typical no. 2 fuel oil	SRC-I heavy distillate (345 to 455 °C)	Typical no. 6 fuel oil
Sulfur, wt %	0.55	0.05	0.25	0.05	0.25	0.3–2.2
Nitrogen, wt %	0.25	——*	0.50	0.20	0.50	0.4
Oxygen, wt %	3.2	——*	2.15	——*	1.45	——*
Specific gravity, °API	37	58	11	30	−2	12
Specific gravity 15.6 °C/15.6 °C	0.84	0.75	0.99	0.88	1.08	0.99
Viscosity (at 38 °C), cSt	1	1	5	3	N/A†	25–1,000
Pour point, °C	——‡	——‡	−26	−29	−7	16
Higher heating value, MJ/m³	——‡	——‡	41,250	39,020	42,920	40,690

*Negligible value

†Not available

‡Not applicable

to yield the two component gases. The removal and recovery of sulfur from the coal permits the production of a sulfur by-product. The gasification and gas treatment areas of the demonstration plant will produce gas consisting mainly of carbon dioxide (89 percent by weight) and nitrogen (10 percent by weight). Finally, the SRC-I process will yield large amounts of mineral aggregate comprising 70 percent fly ash and 30 percent slag.

Anticipated demand A recent market analysis indicates that SRC-I solids, in either a solid or pulverized form or dissolved in process solvent to make a residual fuel oil, are a viable fuel for the 1990s in markets such as oil-fired noncoal-capable utility and industrial boilers; oil-fired coal-capable utility boilers with technological, environmental, or economic conversion problems; new intermediate-load utility boilers; new industrial boilers; and industrial furnaces and kilns.

The potential of SRC-I solids to displace oil in utility and industrial markets is enhanced by the following key factors:

• Low sulfur content, thus eliminating the need for sulfur pollution control systems

• Low ash content, eliminating slag and bottom-ash problems and substantially reducing fly ash

• No boiler derating

• Product uniformity, which reduces equipment downtime

- More efficient use of hydrogen with two-stage liquefaction
- Use of domestic resources, labor, and refining, resulting in increased supply reliability

SRC-I solids applications within the electric utility boiler sector should offer a significant incentive for commercialization. The total size of this market for SRC-I solids by the year 2000 will be 210,000 tons/day (190,476 Mg/day), which is equivalent to 1.1×10^6 bbl of oil per day. The oil-fired non-coal-capable segment of the market provides the opportunity for SRC-I solids to displace 0.5×10^6 bbl of oil per day during the 1990s. Serving a small share of this market need would require several SRC-I coal refineries.

As previously stated, SRC-I solids can also be used to produce calcined coke for the aluminum industry. Conventional anode coke is made from selected residual material (e.g., petroleum coke) from the petroleum-refining process. However, the declining availability of low-sulfur crude oil and low-sulfur residual oil necessitates the development of alternate high-quality sources of calcined coke. Anode coke derived from SRC-I solids appears to be superior to conventional petroleum coke in several respects, including lower sulfur content and potentially lower power consumption during the aluminum-smelting process.

The United States is the largest producer of calcined coke. In addition to domestic demand, this country supplies about half of the rest of the demand of the free world. By the year 2000, total demand in the free world is projected to increase to approximately 17 million tons, with exports in excess of one billion dollars per year. Much of this increment can be supplied by nonpetroleum sources such as SRC solids.

Catalytic re-forming presents an excellent commercial application for SRC-I naphtha, once it is hydrotreated. It is estimated that, by the year 2000, available markets for gasoline and petrochemical feedstocks will be 5.1 million bbl/day of crude oil equivalent.

Through the existing refinery technologies of hydrofinishing, hydrocracking, fluidized catalytic cracking, thermal cracking, and ethylene pyrolysis, SRC-I middle distillate can be upgraded to fill the needs for turbine and jet transportation fuel. By the year 2000, available markets for these applications will total 5.3 million bbl/day of crude oil equivalent.

Because of its high concentration of complex molecules and its corresponding high boiling point, SRC-I heavy distillate is most suitable for use as a boiler fuel. By the year 2000, the available market for utility and industrial boiler fuel will be 5.5 million bbl/day of crude oil equivalent.

By-products from the SRC-I process can be used to fill existing commercial needs. Contacts with potential customers have determined that the LPG can be marketed for fractionation into butane and propane. The elemental sulfur would meet commercial purchasing specifications. Approximately 85 percent of the elemental sulfur produced in the United States is converted to sulfuric acid, which in turn is converted to products used in the agricultural, chemical, and paper industries. Elemental sulfur is projected to be in short supply, especially in northeastern markets. Several firms have shown interest in using the carbon dioxide–rich vent stream to enhance oil recovery. Carbon dioxide flooding has been proven to be a technically and economically viable means of obtaining oil that would otherwise be inaccessible. Presently ICRC is exploring markets for SRC-I fly ash and slag. This mineral-aggregate by-product has potential use as an additive or raw material for cement, as a raw material for brick manufacture, for structural fills and embankments, for soil improvement and land reclamation, for metals and minerals recovery, and for highway construction.

WASTES AND EMISSIONS

Information in this section was obtained principally from DOE.[6]

Air

Particulate control measures to be incorporated in the demonstration plant include the use of gaseous and liquid fuels in fired heaters, partial enclosure of coal-handling equipment and fabric filters to control emissions from such equipment, and the use of dust-suppression techniques to control fugitive emissions from the storage of coal and solid SRC, as well as areas for slag disposal and handling. Sulfur oxides will be controlled primarily through the recovery of elemental sulfur, scrubbing, and the use of low-sulfur gaseous fuels. Emissions of nitrogen oxides will be controlled through the use of gaseous nitrogen-free fuel and burners designed to emit low amounts of nitrogen oxides. Hydrocarbon emissions resulting from leaks in process equipment and piping will be minimized through an extensive inspection, monitoring, and maintenance program. Hydrocarbon vapors from vessels and storage tanks will be flared.

Basis for emission estimates

In many cases, the performance levels of commonly used control devices chosen for the demonstration plant are well known. For example, baghouse efficiency is known to be greater than 99 percent. In cases in which the control device is more process-specific (e.g., burners that emit low levels of NO_x), control efficiencies were obtained from vendors.

Pollutants formed during combustion were estimated using U.S. Environmental Protection Agency (EPA) emissions factors for firing of natural gas.[7] Adjustments to these factors were a NO_x reduction of 30 percent and an SO_2 emission rate corresponding to the sulfur content of the fuel gas (50 ppm H_2S by volume). Particulate emissions from baghouses in the coal-handling area were assumed to be 0.01 gr/std ft[3]. The carrier-gas flow rate is the ventilation rate of the coal-handling equipment. It is assumed that the coker-calciner scrubber will remove 85 percent of the SO_2 and 92 percent of the particulates (worst case). Although the scrubbing system has not yet been chosen, it will provide removal efficiencies at least this high.

Estimates of fugitive emissions of volatile organic compounds from standard process line fittings are based on Wetherhold, Provost, and Smith[8] and the EPA.[9] Estimates for fugitive total suspended particulates are derived from Martin and Catizone[10] and the EPA.[11]

The flow rates and composition of process waste gas, such as tail gas from sulfur recovery and waste CO_2, were calculated from heat and material balances.

Emissions summary

Table 3-10 summarizes the predicted normal operating emission rates of SO_2, NO_x, CO, total suspended particulates (TSP), and nonmethane hydrocarbons (NMHC). These rates comply with EPA standards.

Fugitive emissions

In contrast to point-source emissions, such as those from baghouses and combustor stacks for which relatively precise emission factors and design material balances are available,

TABLE 3-10 Estimated emissions to air during normal operation of SRC-I demonstration plant

Type	Rate, g/s				
	CO	SO$_2$	TSP*	NO$_x$	NMHC†
Flue gas from combustion	2.13	6.1	1.48	25.4	0.16
Baghouse emissions	——	——	2.27		
Emissions from storage tanks and bins	Trace	——	——	——	Trace
Miscellaneous emissions					
Cryogenic system	——	——	——	——	0.001
Emissions from gas systems (CO$_2$ vent)	11.7				
Coker-calciner flue gas	0.13	2.28	0.613	0.73	0.025
Emissions from gas treatment and sulfur recovery	9.7				
Fugitive emissions	——	——	0.37	——	1.77
Total	23.66	8.38	4.73	26.13	1.96

*Total suspended solids.

†Nonmethane hydrocarbons.

Source: From Ref. 6.

fugitive emissions cannot be easily quantified. Estimates of fugitive emissions from equipment leaks and from open coal piles and product-storage piles are less precise because they depend on many factors that vary from one facility to another. Estimates used in Table 3-10 are derived from literature references in which the reported values are based on performances at other installations. Many of these emissions were not controlled as thoroughly as they will be at the demonstration plant. Therefore, these estimates are believed to be conservative.

Cooling tower drift

About 60 gal/min (227 L/min) of drift will be discharged from the evaporative cooling tower. Minerals in the drift will normally consist of those in the makeup water. The composition will vary, depending upon the sources of the makeup water. For example, when treated river water is the only source and is concentrated up to seven times through evaporation, the composition will approximate that listed in Table 3-11.

Water

The design of the demonstration plant incorporates a primary, secondary, and tertiary system that will treat wastewater streams generated by all major process areas. In the "zero-discharge" mode, the treated wastewater will then enter the cooling tower as makeup water. Blowdown from the cooling tower will enter an evaporation system to concentrate the solids, which will be disposed of in a secure landfill. At times (e.g., during evaporator shutdown), treated wastewater will be discharged to the Green River.

The estimated composition of the treated wastewater and the estimated concentrations of trace organics are presented in Tables 3-12 and 3-13, respectively. The listed values have been extracted from the literature, since more definitive composition estimates cannot be determined until a wastewater characterization and treatability study has been completed. In the preparation of a National Pollution Discharge Elimination System (NPDES) permit application, the EPA and the Commonwealth of Kentucky will develop effluent limitations based on water quality in the Green River. Proposed methods for the treatment of wastewater from the demonstration plant will then be modified if necessary to meet these limitations.

Solids

Several solid-waste residues will be stored outdoors on site in secure landfills. The three principal types of solid wastes generated at the SRC-I plant will be fly ash and slag from the gasifier [700 tons/day (635 Mg/day)], metal-bearing sludges [8 tons/day (7.3 Mg/day)] and biological sludge [2 tons/day (1.8 Mg/day)] from the wastewater treatment system, and a sludge from the treatment of raw river water [8 tons/day (7.3 Mg/day)]. In the zero-discharge operation mode, approximately 35 tons/day (31.7 Mg/day) of water-soluble solid wastes will also be generated. When subjected to rain, these solids may generate leachates, which will be collected and channeled through the plant's wastewater treatment facility.

TABLE 3-11 Estimated composition of drift from the SRC-I demonstration plant

	Quantity, mg/L	
Parameters	**Average**	**Maximum**
Calcium	190	380
Magnesium	50	100
Sodium	60	120
Manganese	1	2
Phosphorus	1	2
Potassium	0	0
Zinc	13	26
Aluminum	0.5	1
Bicarbonate	144	288
Sulfate	600	1200
Nitrate	7	14
Chloride	100	200
Ammonia	0	0
Organic phosphate inhibitor*	10	10
Silica	0	0
Total dissolved solids	1100	2200
Total suspended solids	5	10
pH	7.2	6.5–7.5

*If chromate inhibitor is used, replace with 10 mg/L of chromium.

Source: From Ref. 6.

TABLE 3-12 Estimated composition and properties of treated wastewater in the projected SRC-I effluent after all treatments

Parameter	Values*	
	Average	Maximum
Temperature, °C	16	38
pH, pH units	6–9	6–9
Biological oxygen demand	20	30
Chemical oxygen demand	150	300
Total organic carbon	150	300
Oil and grease	10	20
Total suspended solids	20	35
Total dissolved solids	TBD†	TBD†
Chlorides	TBD†	TBD†
Phenolics	0.1	0.6
Ammonia (as N)	5	20
Phosphate (as P)	5	10
Nitrate	TBD†	TBD†
Sulfide	0.04	0.09
Arsenic	0.1	0.25
Barium	1	2
Cadmium	0.15	0.30
Chlorine (residual, total)	0.5	1.0
Chromium (total)	1	2
Cyanide	0.45	0.9
Fluoride	TBD†	TBD†
Lead	0.25	0.4
Selenium	0.35	0.7
Silver	0.1	0.2
Zinc	1	2
Nickel	0.5	1.6
Iron	2	4
Vanadium	1	5
Copper	1	2
Manganese	2	4
Mercury	0.001	0.002
Beryllium	0.005	0.01
Fecal coliform bacilli	TBD†	(below 400/100 mL)
Boron	TBD†	TBD†
Alkalinity	TBD†	TBD†
Thiocyanate	TBD†	TBD†
Sulfate	TBD†	TBD†

*Values are in mg/L unless otherwise indicated.

†To be determined.

Source: From Ref. 6.

TABLE 3-13 Organic compounds in the projected SRC-I effluent

Chemical compound	Concentration, mg/L		
	In untreated effluent[*]	After activated-carbon filtration and 75% reduction	After dilution by 7-day flow[†]
Methylindan	15	2.1	13
Tetralin	0.1	0.01	0.1
Dimethyltetralin	0.5	0.10	0.6
Naphthalene	5	0.20	1.2
Dimethylnaphthalene	0.3–2	0.28	1.7
2-Isopropylnaphthalene	0.7	0.10	0.6
1-Isopropylnaphthalene	2	0.28	1.7
Biphenyl	0.2	0.03	0.2
Acenaphthalene	< 0.1	0.02	0.1
Dimethylbiphenyl	0.2–0.5	0.28	1.7
Dibenzofuran	0.6	0.08	0.5
Xanthene	0.1	0.02	0.1
Dibenzothiophene	1.5	0.01	0.4
Methyldibenzothiophene	< 0.1	0.02	0.1
Thioxanthene	0.1	0.02	0.1
Fluorene	0.3	0.02	0.1
9-Methylfluorene	0.3	0.04	0.2
1-Methylfluorene	0.2	0.01	0.1
Anthracene and phenanthrene	1.1	0.07	0.4
Methylphenanthrene	0.2–0.3	0.03	0.2
Dimethylanthracene	< 0.05	0.003	0.02
Fluoranthene	0.4	0.16	1.0
Dihydropyrene	< 0.05	0.008	0.1
Pyrene	0.6	0.06	0.4
Dimethyldibenzothiophene	< 0.05	0.008	0.1

[*]Data from J. S. Fruchter, M. R. Peterson, J. C. Laul, and P. W. Ryan, "High Precision Trace Element and Organic Constituent Analysis of Oil Shale and Solvent-Refined Coal Materials," Battelle Memorial Institute Pacific Northwest Laboratories Report no. BNWL-SA-6001, 1976, Table 3.

[†]Rate of flow is lowest rate in a 10-yr period, or the worst case.

Source: From Ref. 6.

TABLE 3-14   Chemical composition of slag and fly ash from
GKT gasifier (SRC-I residue)

Element	Slag (run 179), %	Fly ash (run 179), %
Aluminum	14.78	7.16
Calcium	0.50	0.43
Iron	11.80	8.89
Potassium	0.85	0.73
Magnesium	0.38	0.36
Sodium	0.34	0.29
Titanium	0.36	0.32
Silicon	21.74	18.13
Sulfur*	0.50	3.74
Carbon	0.53	22.44

*The sulfur content in the fly ash from the prototype plant will be sig-
nificantly lower because of different operating conditions.

Source: From Ref. 6.

Gasifier slag and fly ash

Preliminary findings indicate that the materials in gasifier slag and fly ash are within the
range of nonhazardous material as defined by the Resource Conservation and Recovery
Act (RCRA). Table 3-14 provides the general chemical composition of this material and
Table 3-15 lists selected elements in both Kentucky No. 9 coal feedstock and the resulting
slag and fly ash.

The results of a leachate analysis for eight priority trace elements included in the Primary
Drinking Water Standards (PDWS) and RCRA standards appear in Table 3-16. The analysis indicates that leachates of slag and fly ash will be nonhazardous. Copper, nickel, van-

TABLE 3-15   Concentration of trace elements in gasifier coal feedstock
and gasifier slag and fly ash

Element	Kentucky no. 9 coal, ppm	Slag (run 179), ppm	Fly ash (run 179), ppm
Arsenic	8.9		
Barium	83	383	3.16
Beryllium	1.3	8.7	7.4
Chromium	21	3500	890
Copper	8.5	57	106
Manganese	42	298	159
Lead	2.9	4	287
Nickel	10.7	1860	260
Silver	0.02	2	0.7
Zinc	55	33	365

Source: From Ref. 6.

TABLE 3-16 Results of analysis of leachate from gasifier slag and fly ash using the EPA extraction procedure*

Element	Slag (run 179)	Fly ash (run 179)	RCRA standard†
Arsenic	4	4	5,000
Barium	3.3	73	100,000
Cadmium	0.098	45	1,000
Chromium	0.46	8.5	5,000
Lead	1.1	78.5	5,000
Mercury	0.022	0.047	200
Selenium	5	5	1,000
Silver	0.01	0.01	5,000
Copper	0.0013	1.53	——‡
Nickel, ppm	0.10	3.71	——‡
Zinc, ppm	0.012	7.9	——‡
Vanadium, ppm	0.27	1.6	——‡

*Values are in ppb unless otherwise indicated.

†Resource Conservation and Recovery Act standard is based on 100 times the limits set in the Primary Drinking Water Standards.

‡Elements not listed in the Interim Primary Drinking Water Standards.

Source: From Ref. 6.

adium, and zinc concentrations were also determined because of their environmental importance and the possibility that they may eventually be added to the list of hazardous elements.

Coal leachate

Results of field leaching tests performed on Kentucky no. 9 coal at Oak Ridge National Laboratory (ORNL) are presented in Table 3-17. Mean concentrations of each parameter were compared with EPA PDWS and proposed Secondary Drinking Water Standards. Because several parameters exceed the standards, special handling, including the collection and treatment of rainwater runoff, will be required to avoid contamination of surface water and ground water.

SRC-I solid product leachate

The results of leaching tests performed on two samples of SRC-I product by ORNL are reported in Table 3-18. Primary and Secondary Drinking Water Standards are included for comparison. Leachate concentrations were well below the recommended levels for all primary standards but exceeded manganese concentrations and were more acidic than the pH range recommended in the Secondary Standard. Organic constituents of the leachate are under study.

TABLE 3-17   Summary of leachate data collected on Kentucky no. 9 coal

Parameter	Observations	Mean concentration	Multiple by which mean exceeds drinking water standard
Runoff, % of rainfall	8	71	
Percent transmission before filter	20	32.5	
Percent transmission after filter	19	48.0	
pH	20	2.1	4.4×
Acidity (as $CaCO_3$), mg/L	15	33,100	
Electric conductivity, $\mu\mho$/cm*	20	10,500	
Sulfate, mg/L	20	27,300	109×
Iron, mg/L	20	9,850	32,833×
Arsenic, μg/L	8	9,050	181×
Barium, mg/L	8	<0.2	0
Cadmium, μg/L	8	166	16.6×
Chromium, μg/L	8	724	14.4×
Lead, μg/L	8	12	0
Selenium, μg/L	8	829	82.9×
Silver, μg/L	8	0.05	
Mercury, μg/L	8	0.20	0

*More properly expressed in microsiemens (μS) per centimeter.

Source: From E. C. Davis, "A Laboratory and Field Analysis of Factors Affecting Quality of Leachate from Coal Storage Piles," Ph.D. dissertation, Vanderbilt University, Nashville, 1980; Ref. 6 with permission.

PROCESS ECONOMICS

The economics presented is for a commercial TSL plant designed to process 30,000 tons per calendar day or 33,333 tons per stream day. The LC-Fining unit, operating in a low-conversion mode, will process the entire 455 °C+ fraction (SRC), converting approximately 50 percent of the SRC solids into liquid products and retaining the other 50 percent in a solid form.

Tables 3-19 and 3-20 list estimates of capital and production costs, as well as estimates of product price. Calculations were made on the following basis:

1990 on-stream year (5-year schedule for design and construction), de-escalated to 1980 on stream

65:35 debt-to-equity ratio

9 percent interest on debt

TABLE 3-18 Chemical analysis of leachates produced from SRC-I using the ASTM Type "A" leaching procedure

Constituent	Sample No. 1	Sample No. 2	Drinking water standards[*]
Arsenic, ppb	0.5	0.1	50
Barium, ppm	0.10	0.03	1
Cadmium, ppb	0.07	0.33	10
Chromium, ppb	0.22	0.18	50
Lead, ppb	17	0.7	50
Mercury, ppb	0.024	0.001	2
Selenium, ppb	3	3	10
Silver, ppb	0.02	0.02	50
Copper, ppb	4.0	1.8	1,000
Nickel, ppb	3.4	1.2	NS[†]
Zinc, ppb	5.0	0.44	5,000
Calcium, ppm	0.54	0.44	NS
Chloride, ppm	1.1	6.6	250
Iron, ppm	0.012	0.0012	0.3
Magnesium, ppm	0.12	0.06	NS
Manganese, ppm	0.1	0.1	0.05
Potassium, ppm	0.5	0.5	NS
Sodium, ppm	2.05	1.62	NS
Sulfate, ppm	1.5	1.0	250
pH (final)	5.5	5.6	6.5–8.5
Electric conductivity, $\mu\mho/cm$[‡]	1.6	2.0	NS
Alkalinity, ppm	4	3.75	NS
Total dissolved solids, ppm	64	68	500
Total filterable residue, ppm	2,440	2,520	NS
Total organic carbon, ppm	7.5	6.3	NS
Chemical oxygen demand, ppm	23	21	NS

[*]EPA Primary and Secondary Drinking Water Standards.

[†]No standard promulgated or proposed.

[‡]More properly expressed in microsiemens (μS) per centimeter.

Source: Written communication from W. J. Boegly, Environmental Sciences Division, Oak Ridge National Laboratory, June 2, 1980; Ref. 6.

TABLE 3-19 Summary of capital investment estimates for a commercial TSL plant operated in a low-conversion mode and processing 30,000 tons per calendar day

Capital investment parameter	De-escalated to 1980 on stream,* $ MM	1990 on stream, $ MM
Major process areas		
SRC liquefaction and de-ashing	710	1,270
Expanded-bed hydrocracker	365	655
Hydrogen production and gas treatment	625	1,115
Utilities, offsites, and coal preparation	450	810
Subtotal	2,150	3,850
License fees	35	65
Initial catalysts and chemicals	30	50
Land	5	10
Contingency (20%)	445	795
Working capital	165	300
Start-up	110	200
Interest during construction	290	520
Total investment	3,230	5,790

*De-escalation factor from 1990 on stream = 1/1.791; costs rounded to nearest $5 million.

15 percent discounted cash-flow return on equity

328.5 stream days per year for years 2 through 20 (182.5 stream days for year 1)

50 percent income tax rate

10 percent income tax credit

13-year tax life (double-declining-balance depreciation method)

6 percent yearly inflation average between 1980 and 1990 (based on Energy Information Administration (EIA) projections[12])

20-year plant operating life

Coal and power costs on 1980 EIA projections

Prices of liquid coproduct derived from EIA

Because EIA does not provide detailed pricing for the entire range of SRC-I products, the 1990 prices for naphtha and LPG (50 percent butane and 50 percent propane) were obtained by adjusting the EIA no. 2 fuel price by the 1990 expected price differentials for those products.

The revenues from the liquid coproduct and sulfur by-product were subtracted from the annual operating cost to arrive at a residual revenue from the TSL solids required to achieve the return on equity provided above. This method of calculation resulted in a production cost for TSL solids of $2.82 per million Btu for 1990 expressed in 1980 constant dollars. On the basis of delivery to a distance of 500 mi, the cost of TSL solids was estimated at $3.25 per million Btu.

TABLE 3-20 Summary of estimated production costs and revenues of commercial TSL plant operated in low-conversion mode and processing 30,000 tons per calendar day

Parameter	Quantity	Price, $	$ MM[a] (1980 basis)
Production costs			
Coal	33,333 tpsd[b]	1.75/MM Btu	488
Power	421,000 kWh/h	0.046/kWh	153
Critical solvent	5.24 t/h	0.15/lb	12
Chemicals and lubricants	——	——	32
Maintenance			
Material	2% of TEC[c]	2,580 MM	52
Labor	926 workers	68,800/worker	64
Operating labor	530 workers	48,100/worker	25
Insurance and taxes	1.5% of PI[d]	2,665 MM	40
General and administrative	10% of direct labor	——	4
Capital charges	15.7% of TI[e]	3,230 MM	505
Total			1,375
Revenue			
Naphtha	54,116 MMM Btu/yr[g]	8.96/MM Btu	485
Middle distillate	57,514 MMM Btu/yr	7.85/MM Btu	451
Heavy distillate	16,738 MMM Btu/yr	6.80/MM Btu	114
LPG	9,272 MMM Btu/yr	8.19/MM Btu	76
Sulfur	358 M ton/yr[h]	57.80/t	21
TSL solid[f]	80,898 MMM Btu/yr	2.82/MM Btu	228
Total			1,375

[a]Million.

[b]tpsd = tons per stream day.

[c]Total erected cost.

[d]Plant investment (TEC plus license fees, initial catalyst and chemicals, and land).

[e]Total investment (PI plus working capital, start-up cost, and interest during construction).

[f]Calculated.

[g]MMM = billion.

[h]M = thousand.

SUMMARY

Estimates of Economic Factors

Factor	Estimate
Product cost*	$37/bbl of oil equivalent ($6.20/MM Btu)†
Energy efficiency	71%
Water consumption	25 gal/MM Btu product
Electricity consumption	15 kWh/MM Btu product

*Based on 20-yr plant operating life; assumes 1990 plant start-up with prices de-escalated to 1980 on stream.

†MM = million

REFERENCES

1. Gesellschaft für Kohle-Technologie: "Process Description, Gasification Unit, Project no. H2 4613, SRC-I Demo Plant," Gesellschaft für Kohle-Technologie, Essen, FRG, 1981.

2. Electric Power Research Institute: "Refining the Process That Refines the Coal," *EPRI J.*, May 1980, pp. 21–25.

3. Tao, J. C., R. K. Malhotra, T. M. Sukel, E. P. Foster, and S. M. Morris: "Solvent-Refined Coal (SRC-I). Technology, Product Markets, and Economics," *3rd International Coal Utilization Exhibition and Conference, Houston, Texas, 1980.*

4. Southern Company Services: "Solvent Refined Coal Burn Test," Final Report, SCS, Birmingham, Alabama, 1979.

5. Pan, Y. S., G. T. Bellas, D. E. Wieczenski, R. B. Snedden, J. I. Joubert, D. R. Hart, E. P. Foster, and T. G. Ingham: "Combustion of Solvent-Refined Coal in a 100-hp Firetube Boiler," *16th Intersociety Energy Conversion Engineering Conference, Atlanta, Georgia, 1981.*

6. U.S. Department of Energy: *Final Environmental Impact Statement: Solvent-Refined Coal-I Demonstration Project, Newman, Daviess County, Kentucky,* U.S. Department of Energy Report no. DOE/EIS-0073, vol. 2:C-61-74, 1981.

7. U.S. Environmental Protection Agency: *Compilation of Air Pollutant Emission Factors,* 3d ed., U.S. Environmental Protection Agency, Office of Air and Waste Management, Office of Air Quality and Planning Standards, Research Triangle Park, N.C., Publication no. AP-42, 1977.

8. Wetherhold, R. G., L. P. Provost, and C. D. Smith: *Assessment of Atmospheric Emissions from Petroleum Refining,* U.S. Environmental Protection Agency Report no. EPA-600/2-80-075C, vol. 3, Appendix D, 1980.

9. U.S. Environmental Protection Agency: *Emission Factors and Frequency of Leak Occurrence for Fittings in Refinery Process Units,* U.S. Environmental Protection Agency Report no. EPA-600/2-79-004, 1979.

10. Martin, D. J., and P. Catizone: "Study of Fugitive Emissions Controls for Coal Handling and Storage Operations," Prepared for International Coal Refining Co. by TRC Environmental Consultants, Inc., Report no. 1628-L31-02, 1981.

11. U.S. Environmental Protection Agency: *Technical Guidance for Control of Industrial Process Fugitive Particulate Emissions,* U.S. Environmental Protection Agency Report no. EPA-450/3-77-010, 1977.

12. Energy Information Administration: "Annual Report to Congress, U.S. Department of Energy," U.S. Department of Energy Report no. DOE/EIA-0173(79)/3, 1979.

THE COAL HYDROGENATION PLANT AT BOTTROP

J. LANGHOFF

Ruhrkohle Oel und Gas GmbH,
Bottrop, Federal Republic of Germany

G. ESCHER

Veba Oel Entwicklungsgesellschaft mbH,
Gelsenkirchen-Scholven, Federal Republic of Germany

INTRODUCTION

The objective of current coal liquefaction activities in the Federal Republic of Germany (FRG) is improvement in the industrial-scale technology for producing liquid products from coal. The development work has reached the pilot plant stage. These plants were designed by extrapolation of data from laboratory-scale plants and by utilizing certain data from the IG process collected by Badische Anilin und Soda Fabrik. The capacity of these pilot plants was chosen to minimize future risks in the design of construction of large-scale industrial plants.

The largest pilot plant under construction in the FRG at the moment is the coal hydrogenation plant at Bottrop. The plant is designed to process 200 tons* of moisture- and ash-free (maf) coal per day, producing 30 tons of light oil (C_5 and those fractions distilling below 200 °C) and 70 tons of middle oil (200 to 325 °C). The total investment including commissioning is 400 MM DM. The project is sponsored by the government of North Rhine–Westphalia. Industrial partners of the project are Ruhrkohle AG and Veba Oel AG.

The basic engineering of the pilot plant at Bottrop was carried out by Ruhrkohle Oel und Gas GmbH and Veba Oel AG. The detailed engineering was done by two German engineering companies, Firma Carl Still and Didier Engineering.

GENERAL PROCESS DESCRIPTION

Background

The German technology used in the pilot plant is based on the process developed by Bergius and Pier (IG process), which has been modified by introducing new advanced process-unit operations and equipment.

The operational pressure, which was 70 MPa (10,150 lb/in²) in the IG process, has been brought down to 30 MPa (4350 lb/in²), and the same cheap iron catalyst, i.e. red mud, will be used. Instead of the thin centrifuge liquid used in the IG process, only distillates consisting of 40 percent middle oil and 60 percent heavy oil are to be used as solvents for preparing the slurry. This prevents asphaltene enrichment of the coal slurry. The residue from commercial plants will be used as feedstock for gasification units in order to produce the hydrogen necessary for the liquefaction process.

Objectives

The following objectives have been set for the pilot plant at Bottrop:

- Plant operation in order to test the expected results already achieved on a laboratory scale (PDU)
- Optimization of the entire process

*Throughout this chapter the ton used is the metric ton (1000 kg or 1 Mg).

- Testing and improving of plant components
- Collection of basic data that can be used for planning the layout of commercial plants

Process Concept

The plant has been erected close to a coking plant at Bottrop. Crude gas, wastewater, and granulated residue will be sent to the coking plant for reuse or treatment. Fresh hydrogen will be supplied by a hydrogen pipeline belonging to Chemische Werke Hüls AG.

The block flow diagram (Fig. 4-1) shows the main streams. In the coal and catalyst preparation unit, the feed coal is dried, ground, and mixed with a solvent. At this point the catalyst is added.

The reaction with hydrogen takes place in the hydrogenation unit. Part of the hydrogen is supplied as makeup hydrogen from the hydrogen pipeline; the other part is recycle gas from the process.

The coal-derived oil is distilled into light, middle, and heavy oil in the atmospheric-fractionation unit, which operates at normal pressure (1 standard atmosphere). The light oil is further treated in a stabilized column. The bottom product of the hot separator, which contains minerals, and undissolved coal, and catalyst as solids, is expanded and further treated in a combined vacuum-distillation and flash-evaporation unit. This is necessary in order to obtain as much oil as possible from the liquefaction residues. The resulting distillate is returned to the coal-slurrying unit. The residue containing minerals, unreacted coal, and catalyst as solids is cooled on a steel belt and further processed to a granular substance, which is sent to the coking plant.

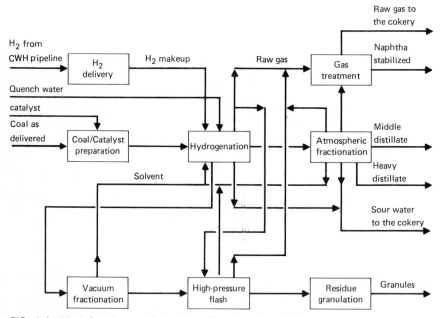

FIG. 4-1   Block flow diagram of the 200-ton/day pilot plant at Bottrop.

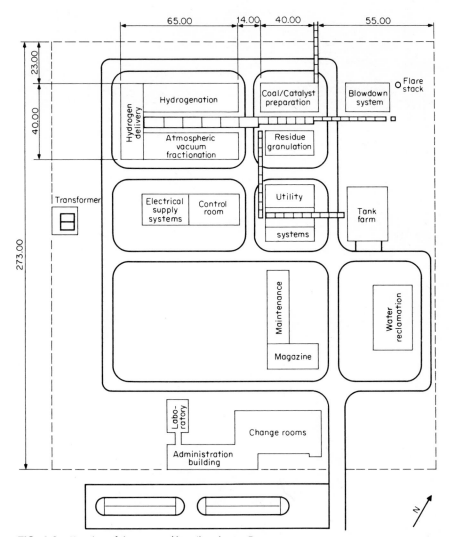

FIG. 4-2   Site plan of the 200-ton/day pilot plant at Bottrop.

PILOT PLANT DESCRIPTION

Figure 4-2 is a plot plan of the 200-ton/day pilot plant. The plant has a length of 273 m (896 ft), a width of 215 m (705 ft), and therefore a total area of 58,700 m^2 (14.5 acres). The main process facilities require about 25 percent of the total area. A flash and blow-down system and a tank farm for the products are situated next to the main process units.

The feed coal is supplied by a coking plant via a conveyor. The residue from the pilot plant is also disposed of via this conveyor.

FIG. 4-3 The 200-ton/day pilot plant at Bottrop (May 1981).

FIG. 4-4 A 1:25 scale model of the 200-ton/day pilot plant at Bottrop.

The building housing the administrative offices and the changing rooms as well as the maintenance building, the magazine, and the repair area are shown. The open area in the center may be used later for expansion of the pilot plant.

Figure 4-3 shows a bird's eye view of the plant. The hydrogen-compressor building can be seen on the right. The hydrogenation chamber and the coal preparation building are located in the foreground.

A detailed model of the main units of the plant was built at a scale of 1:25 while the engineering of the plant was being designed; the model shows details such as pipes, valves, and flanges. The plant has a very large number of pipes, and thus this model was very helpful to the engineers while they were plotting the isometrics of the plant. Figure 4-4 shows a small section of the model: On the left are two hot separators and next to them are the absorber and desorber (oil wash for the recycle gas).

CONSTRUCTION STATUS

In November 1977 Ruhrkohle AG and Veba Oel AG decided to build and operate jointly a pilot plant based on German technology. As the result of a site selection study, Bottrop, in North Rhine–Westphalia, was chosen as the site for the pilot plant. Scale-up limits determined the input capacity at 200 ton/day of maf coal.

The necessary approvals for erecting and operating the plant were obtained during 1978 and 1979, the construction site was formally opened in May 1979, and the detailed engineering was finished at the end of 1980. Construction was finished in early 1981.

Operation was initiated in 1982 and several thousand tons of coal have been processed in the unit.

DETAILED PROCESS DESCRIPTION

Coal Preparation

Figure 4-5 shows the coal preparation and slurrying unit. After the coal has been ground to a size smaller than 1 mm and dried to a moisture content lower than 1 percent by weight, it is slurried in a mix tank with a ratio of middle oil to heavy oil of 40 to 60. This slurried coal is ground to a size smaller than 0.2 mm, and the catalyst is added. The final slurry contains approximately 40 percent by weight of solids.

Hydrogenation

The hydrogenation step, consisting mainly of high-pressure pumps, preheaters, three reactors in series, separators, and an gas scrubber, is shown in Fig. 4-6. The slurry is pumped from the slurry preparation tank via high-pressure plunger pumps operating at 30 MPa (4350 lb/in^2).

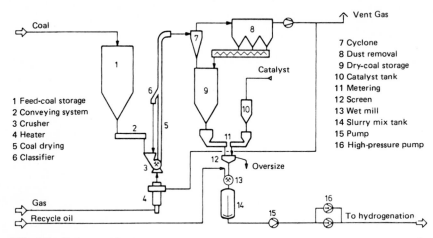

1 Feed-coal storage
2 Conveying system
3 Crusher
4 Heater
5 Coal drying
6 Classifier

7 Cyclone
8 Dust removal
9 Dry-coal storage
10 Catalyst tank
11 Metering
12 Screen
13 Wet mill
14 Slurry mix tank
15 Pump
16 High-pressure pump

FIG. 4-5   Coal preparation.

Three pumps have been installed in the Bottrop plant: Two are in operation and one is a spare. Each is equipped with three parallel pistons, which, for testing purposes, are arranged vertically in two pumps and horizontally in the third pump. Furthermore, a remote valve box is connected to each pump via a column of clean oil. As a result of the pumps' action some of the oil is lost in the valve box but is replaced by clean oil after each stroke. The expectation is that by designing the pumps with external valve boxes, the action of the pumps' pistons will be abrasion-free. In addition, inlet and outlet valves of different designs and materials have been installed in the valve boxes in order to investigate their performance with regard to corrosion and erosion.

After the slurry is pumped from the slurry preparation tank, hydrogen and recycle gas

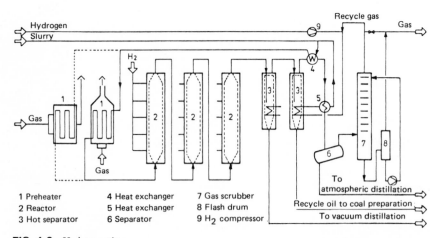

1 Preheater
2 Reactor
3 Hot separator

4 Heat exchanger
5 Heat exchanger
6 Separator

7 Gas scrubber
8 Flash drum
9 H$_2$ compressor

FIG. 4-6   Hydrogenation.

are added. The slurry passes a feed and product heat exchanger and is further heated in a furnace to 420 °C.

When the coal is heated in the solvent, it swells considerably. During this so-called gel phase, the viscosity of the slurry goes up even though theoretically it should be decreasing with rising temperatures. The viscosity reaches a maximum between 300 and 350 °C and then decreases rather steeply, assisted by the incipient chemical decomposition of the coal. Within this high-viscosity range, heat transfer deteriorates considerably as a result of laminar flow but the coking tendency increases. For these reasons special attention was given to the layout of the preheater. Two preheaters of different design—a convection heater and a radiation furnace—were installed.

Each heater has enough capacity to handle the design capacity of the plant. The first preheater is very similar to those that were used previously in Germany for hydrogenation of coal. In this type of preheater the heat transfer is achieved by circulating hot air around the tubes. The heat transfer occurs very gradually, thus preventing the slurry from coking in the hairpin tubes. The disadvantages of this heater are its size and high cost.

The second preheater that was installed has a radiation furnace of a type similar to that frequently used in oil refineries. Because it operates at a higher temperature, the risk of coking is also greater. The aim of this equipment evaluation test is to minimize this risk while utilizing the better heat-transfer characteristics of a radiation furnace as well as its smaller size and lower costs.

The preheated slurry enters the reactor from the bottom. The exothermic hydrogenation reaction occurs in the reactor and is controlled by feeding quench hydrogen into the reactor at a rate that maintains the temperature below 476 °C. The product leaves the reactor at the top.

Most reactors previously used for coal hydrogenation in Germany were made from forgings. The reactors used in the pilot plant are of a new multilayer design to which a layer of refractory bricks has been added as insulation.

A still newer type of reactor is presently in the design stage. This type will also use a multilayer design but without refractory bricks inside. The insulation will be integrated into the multilayer wall. The performance of this reactor will be tested in the pilot plant at Bottrop, and then the reactor will be removed from the plant and cut up for testing.

The hydrogenation product leaves the reactor and passes to the hot separator, where a phase separation takes place. The gases and vapors are removed from the top, while liquid products and solids are discharged from the bottom. The hot separator operates at 30 MPa (4350 lb/in²) and 450 °C.

The coal gases and vapors are cooled. The pressure of the resulting coal-oil condensates is reduced and the condensates are pumped to the atmospheric distillation unit. The gas that does not condense is piped to an absorber tower where the hydrocarbons are washed out and the concentration of hydrogen is increased. Part of the scrubbed gas is used as recycling gas; the excess is discharged as purge gas. The absorber tower operates at 30 MPa, and when the recycle gas leaves the absorber tower, it is returned to the preheater.

Atmospheric Distillation

The condensates obtained from the separator are subsequently treated by atmospheric distillation as shown in Fig. 4-7. The coal-oil condensates are distillated into gasoline and medium- and heavy-oil fractions. A gasoline stabilizer column is also provided (Fig. 4-8).

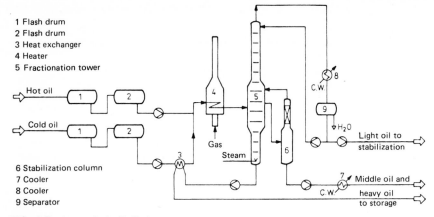

1 Flash drum
2 Flash drum
3 Heat exchanger
4 Heater
5 Fractionation tower

Hot oil

Cold oil

6 Stabilization column
7 Cooler
8 Cooler
9 Separator

Gas

Steam

C.W.

H_2O

Light oil to stabilization

Middle oil and heavy oil to storage

C.W.

FIG. 4-7 Atmospheric distillation.

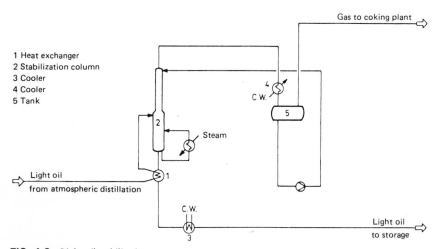

1 Heat exchanger
2 Stabilization column
3 Cooler
4 Cooler
5 Tank

Gas to coking plant

C.W.

Steam

Light oil
from atmospheric distillation

C.W.

Light oil
to storage

FIG. 4-8 Light-oil stabilization.

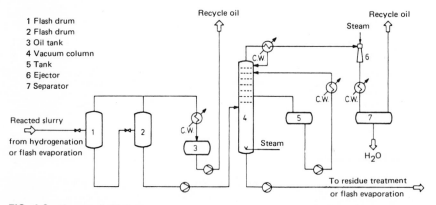

1 Flash drum
2 Flash drum
3 Oil tank
4 Vacuum column
5 Tank
6 Ejector
7 Separator

Recycle oil

Recycle oil

Steam

C.W.

C.W.

C.W.

C.W.

Reacted slurry
from hydrogenation
or flash evaporation

Steam

H_2O

To residue treatment
or flash evaporation

FIG. 4-9 Vacuum distillation.

1-10

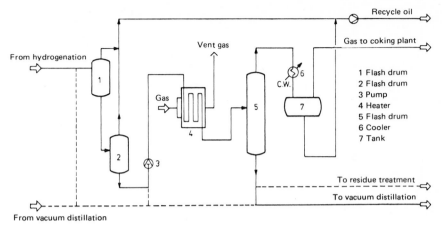

FIG. 4-10   Flash evaporation.

Vacuum Distillation

As shown in Fig. 4-9, the bottom product of the hot separator expands and is separated into distillate oil and residue by vacuum distillation. The residue, which contains minerals, undissolved coal, and catalyst as solids, is cooled on a steel belt (flaker) and further processed to a granular substance which is sent to the coking plant.

In order to attain the objective of enrichment of solids by up to approximately 50 percent by weight, different processes for treating the residue are being developed. Included are different combinations of pressure treatment, preheating, flash evaporation (Fig. 4-10) and vacuum distillation.

FEED AND PRODUCTS

Table 4-1 gives the specifications for the standard coal that is to be fed to the pilot plant at Bottrop as well as the range of flexibility of this plant. As is shown, a wide variety of coals may be used for hydrogenation in this pilot plant.

Table 4-2 summarizes the feed materials as well as the products that should normally be produced by the plant. These figures are based on tests run on standard coal by Bergbau-Forschung GmbH at Essen.

The pilot plant will be operated for special test periods. During these test runs the reaction conditions (oil-to-coal ratio, reactor pressure, hydrogen partial pressure, and recycle-gas-to-feed ratio) as well as the quality of coal feedstock will be varied. Some special programs will also be carried out, e.g., analysis of products, determination of corrosion and erosion, and pilot plant balancing. Extensive basic work on the thermodynamics and kinetics of these processes is also planned; this work will use the results from these mentioned tests.

For this reason, a most sophisticated program for measuring the material and elemental balances of the pilot plant has been installed to obtain as much information as possible

TABLE 4-1 Specifications of feed coals for the pilot plant at Bottrop

	Standard case	Range
Moisture (raw), wt %	4.30	3.8–14.4
Ash (mf), wt %	4.52	3.0–17.0
Volatiles (maf), wt %	37.9	10.0–53.8
Carbon (maf), wt %	84.30	75.0–93.0
Hydrogen (maf), wt %	4.96	3.5–7.0
Oxygen (maf), wt %	7.23	1.5–17.8
Nitrogen (maf), wt %	1.68	1.0–1.9
Sulfur (maf), wt %	1.67	0.5–5.4
Chlorine (maf), wt %	0.16	0.0–0.4
Vitrinite, vol %	64	61–77
Exinite, vol %	17	5–20
Inertinite, vol %	14	1–14
Minerals, vol %	5	1–9
SiO_2, wt %	32.92	32.11–64.8
Fe_2O_3, wt %	32.20	3.71–36.00
Al_2O_3, wt %	27.63	18.48–34.60
Grindability, °H	52	>30
Higher heating value:		
Btu/lb	14,875	13,540–15,820
kJ/kg	34,600	31,500–36,800

TABLE 4-2 Feeds and products of the pilot plant at Bottrop*

Substance	Amount
Feed:	
Coal (maf)	200 t/d
Hydrogen makeup	220,000 m³/d
Process water	41.2 t/d
Catalyst (Fe_2O_3)	4.0 t/d
Power	108,000 kWh/d
Products:	
Gas	61.3 t/d
Naphtha (stabilized) (<200 °C)	24.4 t/d
Middle oil (200 to 325°C)	69.0–74.4 t/d
Heavy oil (>325 °C)	Not significant
Residue	68.6–74.3 t/d

*Standard case coal.

from the plant while it is running. The following main units will have to be treated as separate balance areas:

- Hydrogenation
- Atmospheric distillation and light-oil stabilization
- Vacuum distillation and flash evaporation
- Residue treatment

The other units such as coal preparation, the tank farm, and water treatment will not have to be balanced. Only the input and output streams such as coal, catalyst, hydrogen, and products as well as heating gas, steam, and residue will have to be measured when they cross into a balanced area.

The following list gives a breakdown of the 221 employees required for the operating phase:

99 operating

23 technical

22 administrative

8 laboratory

59 maintenance

If necessary, an additional staff of 37 workers will be provided for extensive repairs.

UPGRADING OF OIL FROM COAL

When it leaves the hydrogenation unit, the raw oil from coal needs further upgrading in order to yield stable, marketable products. Desired products include specific fuels such as gasoline and diesel and jet fuel as well as heating fuels, which can be produced by using conventional mineral oil processing techniques and adapting them to the special needs of oil produced from coal. The oil can also serve as a source for chemical products such as BTX (benzene-toluene-xylene) aromatics, phenols, and polynuclear aromatics.

Figure 4-11 presents an upgrading scheme for the different oil fractions, which are handled separately. The main products from the light-oil (naphtha) processing are motor fuel and BTX aromatics, with phenols as by-products. The main product of middle-distillate upgrading should be heating fuel. However, diesel fuel and jet fuel can be produced by severe hydrogenation. The production of gasoline by hydrocracking of the middle distillate is discussed under "Hydrocracking." Interesting by-products are naphthalene and tetraline.

Feedstock

Coal-derived oil has characteristic properties which differ considerably from those of mineral oils. Table 4-3 contains a comparison of characteristic properties of coal-derived oil and the corresponding characteristics of Arabian light oil.

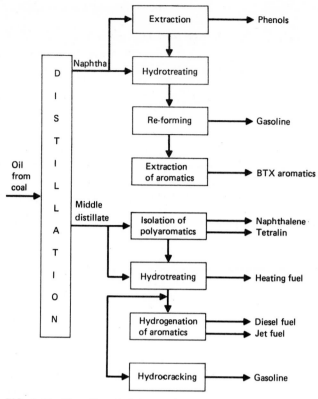

FIG. 4-11 Upgrading of oil from coal.

TABLE 4-3 Comparison of specific gravity and components of oil from coal and Arabian light oil

Characteristic	Oil from coal	Arabian light oil
Density at 15 °C, g/mL	0.950	0.856
Specific gravity, °API	17.5	34
Components		
Carbon, wt %	86.6	85.5
Hydrogen, wt %	9.05	12.6
Sulfur, wt %	0.1	1.7
Nitrogen, wt %	0.75	0.2
Oxygen, wt %	3.50	
H/C ratio	1.26	1.77
Gasoline, IBP* to 200 °C, wt %	25	23
Middle distillate, 200 to 325 °C, wt %	75	24
Vacuum gas oil, 325 to 500 °C, wt %	——	28
Vacuum residue, 500 °C, wt %	——	25

*Initial boiling point.

If one looks at the boiling range of the two oils, it can be seen that in the case of Arabian light oil the fractions are distributed rather uniformly over the whole boiling range and there exists a vacuum residue of about 25 percent by weight. The primary coal oil, however, consists of middle distillate up to 75 percent by weight. As a result of the liquefaction procedure the heavy distillate is recycled for the preparation of the coal slurry. The remaining heavy oil is removed together with the unconverted coal, ash, and catalyst. Therefore, no vacuum gas oil and no distillation residues at all occur in the net liquid hydrogenation product.

Another characteristic difference between the two oils is their content of heteroatoms. In coal oil the amount of sulfur compounds is very low but the nitrogen and oxygen contents are much higher than in Arabian light oil. Oxygen, which is practically nonexistent in natural crude oils, is bound almost completely in phenolic compounds in coal oil. Primary coal oil also has a high aromaticity, which is reflected in its high density and low hydrogen content.

Hydrotreating and Re-forming of Naphtha

The primary objective in the hydrotreating of coal naphtha is the reduction of the nitrogen, sulfur, and oxygen contents to levels acceptable for use of the product as re-former feedstock. Re-former runs have been carried out with hydrotreated naphtha.

Equipment

Continuous naphtha hydrotreating has been carried out in small bench-scale units under pressures of up to 100 bars. A simplified flow diagram of this bench-scale hydrotreating plant is shown in Fig. 4-12. Hydrogen and primary coal naphtha were mixed and were then passed in a downflow pattern over a fixed bed of a commercial hydrotreating catalyst. Figure 4-13 is a simplified flow diagram of the bench-scale re-forming plant. The re-former reactor was loaded with a commercial platinum catalyst. The capacity of the bench-scale units amounts to approximately 1 kg/h.

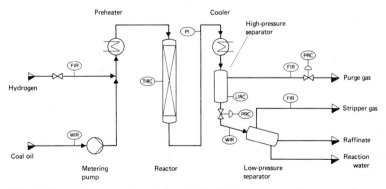

FIG. 4-12 Hydrotreater. Key: F, flow; L, liquid; T, temperature; P, pressure; C, controller; I, indicator; R, recorder; W, weight.

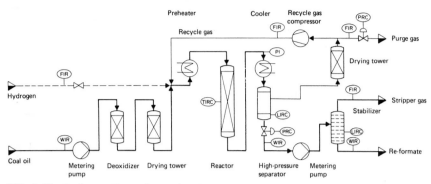

FIG. 4-13   Re-former. Key: F, flow; L, liquid; T, temperature; P, pressure; C, controller; I, indicator; R, recorder, W, weight.

Results

Raw naphtha reaches a high research octane number, or RON (clear), because of its high content of aromatics and phenols. It can be blended directly with transportation fuel up to 2 percent by weight, but further addition causes an increase in gum formation and a strong odor.

In order to reach the quality required for re-former feed, very severe hydrotreating conditions had to be used compared to those used in conventional naphtha refining. The hydrotreating was performed as a two-stage operation with a sulfided $CoO\text{-}MoO_3/Al_2O_3$ catalyst in the first stage and a sulfided $NiO\text{-}MoO_3/Al_2O_3$ catalyst in the second stage. The results of this operation as well as the feedstock analysis are shown in Table 4-4.

The temperatures shown are average bed temperatures, and the pressure is the total pressure. The hydrogen content of the gas was approximately 97 percent by volume.

As can be seen, a two-stage operation with space velocities of 2 $kgkg^{-1}h^{-1}$ in each stage was necessary in order to meet re-former feed specifications, which means reduction of the sulfur and nitrogen contents to values lower than 1 ppm. The pressures of 6 MPa and 10 MPa, respectively, as well as the reaction temperatures of 360 °C and 400 °C, are very high. In spite of these drastic conditions the production of C_1 to C_4 gases was low. The total hydrogen consumption was 190 L/kg, which is rather high and may be related to the removal of heteroatoms and to a partial saturation of aromatics. The percentage of aromatically bound carbon decreases from 48.7 to 31.6. The phenols, which have an initial value of 18.5 percent by weight are removed completely. Since the phenols contribute significantly to the octane number and some aromatics become saturated, the RON (clear) is decreased by this hydrotreatment. Therefore, the hydrotreatment products need re-forming before they can be used as blending components for gasoline. Hydrotreatment leads to a product which is still highly aromatic and naphthenic. Therefore, the hydrotreatment product has a lower specific gravity and hydrogen content than are normally found in petroleum-based naphtha fractions.

The second-stage product was fed to the re-former at a temperature of 490 °C, a pressure of 2500 MPa, and a space velocity of 1.5 $kgkg^{-1}h^{-1}$. Table 4-5 shows the composition and some properties of the re-formate.

This re-formate is a good feedstock for the production of BTX aromatics, as it consists of benzene, toluene, xylenes, and ethylbenzene, up to 37 percent by weight. On the other hand it also has properties that make it a good blending component for gasoline.

TABLE 4-4   Two-stage hydrotreatment of the coal naphtha fraction

	Feedstock	1st stage	2d stage
Reaction conditions:			
Temperature, °C		360	400
Pressure, MPa		6	10
WHSV, kg/(kg)(h)*		2.0	2.0
H₂/oil ratio, L/L		500	500
Product properties			
Density (15 °C), g/mL	0.880	0.845	0.827
Initial boiling point, °C	76	73	73
End point, °C	212	208	205
Carbon, wt %	85.05	87.60	87.80
Hydrogen, wt %	10.80	11.75	12.25
Oxygen, wt %	3.6	0.4	0.1
Nitrogen, ppm	2400	480	1
Sulfur, ppm	295	10	1
Aromatic carbon, wt %	48.7	38	31.6
Phenols, wt %	18.5		0.1
RON (clear)	96.4	79.2	75.0
Gum, mg/100 mL	45		1
C₁ to C₄ g/kg of feed	——	2	1
Hydrogen consumption, L/kg	——	120	70

*Weight hourly space velocity in kilograms oil per kilogram of catalyst per hour.

TABLE 4-5   Properties and composition of re-formate

Density (15 °C), g/mL	0.858
H₂ yield, wt %	2.1
RON (clear)	98.9
Gum, mg/100 mL	1
Aromatics, wt %	81.9
Benzene, wt %	13.6
Toluene, wt %	13.0
Ethylbenzene, wt %	3.9
p-Xylene, wt %	1.3
m-Xylene, wt %	3.1
o-Xylene, wt %	2.5

Upgrading (catalyst) of the Middle Distillate (oil)

The middle distillate must be upgraded in order to meet the specifications required for use as a heating fuel or jet or diesel fuel. The middle distillate has a sulfur content of 800 ppm, which lies far below the upper limit of 0.3 percent by weight that was set in Germany for heating and diesel fuel. But the quantities of nitrogen and oxygen (which is bound mostly in phenolic compounds), in the middle distillate are high: 0.85 and 2.6 percent by weight, respectively. This high oxygen and nitrogen content leads to corrosivity and NO_x production beyond allowable limits if the distillate is used untreated. In addition, the very high aromaticity of the oil results in its having a cetane index near zero.

Equipment

Experiments for upgrading the middle distillate were performed in apparatus similar to that shown in Fig. 4-12 for the hydrotreating of naphtha. These units can be run under pressures of up to 30 MPa and include a gas recycle with a gas-cleaning system and stripping columns.

Hydrofining

Table 4-6 gives the middle-distillate feedstock properties and the results of subjecting this feedstock to low- and high-severity hydrogenation using commercial hydrotreating catalysts ($CoO\text{-}MoO_3/Al_2O_3$ or $NiO\text{-}MoO_3/Al_2O_3$).

The product obtained from low-severity hydrofining has sufficiently decreased contents of nitrogen and oxygen. Its storability is increased considerably, and its NO_x output is decreased.

The primary middle distillate itself can be blended with normal heating fuel up to 15 percent by weight if certain chemicals are added to stabilize the mixture. However, high-severity treatment yields a product oil which satisfies the specifications for heating fuel oil. After fractionation, it is also suitable for jet fuel production.

Hydrogenation of aromatics

To be suitable for use as a diesel fuel the middle distillate must undergo severe hydrogenation in order to saturate and partially crack the aromatic rings and obtain high cetane numbers. A two-stage operation under severe conditions is convenient for the removal of heteroatoms, with a sulfided $CoO\text{-}MoO_3/Al_2O_3$ catalyst in the first stage and hydrogenation with a sulfur-sensitive nickel catalyst in the second stage. The hydrogen consumption is approximately 600 L/kg of feed.

The fraction of aromatic carbon can be reduced from 64 percent by weight in the feed to 6 percent by weight in the product; the naphthenic carbon increases from 8 to 43 percent by weight and the paraffinic carbon from 28 to 51 percent by weight. The hydrotreated middle distillate has a cetane number of 45; it behaves well at low temperatures and its freezing point is below -35 °C.

Hydrocracking

In order to obtain a high gasoline yield the middle distillates have to be hydrocracked. To reduce the high nitrogen content of this feed, it must be hydrotreated before it can be hydrocracked; in addition, aromatic compounds in the feed have to be partially hydrogenated during hydrotreatment.

A sulfided $NiO\text{-}MoO_3/Al_2O_3$ catalyst is used. After sour gases are stripped, the products from the hydrotreatment are fed to an adjacent reactor, where the hydrocracking is performed with a sulfided nickel-tungsten catalyst possessing an acid carrier.

Depending on throughput and temperature, a maximum yield of gasoline of up to 90 percent by weight is reached. The production of gases is greater than 10 percent by weight. Hydrogen consumption for both process steps is a total of 500 to 800 L/kg of feed.

TABLE 4-6 Hydrofining of primary middle distillate

Reaction Conditions:	Type of Hydrotreatment	
	Mild	Drastic
Temperature, °C	390	420
Pressure, MPa	10	28
WHSV, kg/(kg)(h)*	1.2	1.0
H_2/oil ratio, L/L	500	600

Physical and analytical data	Feedstock	Products	
		Mild	Drastic
Density (15 °C), g/mL	0.992	0.920	0.860
Carbon, wt %	88.15	88.75	87.25
Hydrogen, wt %	8.80	10.60	12.35
Oxygen, wt %	2.6	0.3	0.2
Nitrogen, wt %	0.85	0.1	0.006
Sulfur, ppm	800	16	6
Heating value, MJ/kg	38.9	41.2	42.6
C_1–C_4, g/kg of feed		2	16.5
Hydrogen consumption, L/kg		145	478

*Weight hourly space velocity in kilograms oil per kilogram of catalyst per hour.

Isolation of Phenols and Aromatics

Phenols can be extracted from the primary coal naphtha, which contains the following phenolic compounds in measurable quantities:

Compound	Wt %
Phenol	13.0
o-Cresol	3.1
m-Cresol	1.6
p-Cresol	0.7
2,6-Dimethylphenol	0.1

In a two-stage mixer-settler battery more than 90 percent of these phenols was extracted by a 12% sodium hydroxide solution at 48 °C. The ratio of the flow rates of naphtha and the sodium hydroxide solution was 4:1.

As shown in Table 4-5, the re-formate is a good feedstock for BTX extraction. It contains benzene, toluene, xylenes, and ethyl benzene, up to 37 percent by weight. Commercial processes, e.g., extraction with N-formyl morphylate, can be used to produce BTX aromatics.

ENVIRONMENTAL PROTECTION AND EMISSIONS

In order to minimize plant emissions the most recent technological advances for protecting the environment were considered when the plant was being designed. Precautions have been taken for running the different units in a way which is compatible with the environment even in the event of disturbances or accidents. A slop system, e.g., for pump leakages and discharge operations, and an emergency blowdown system for the hydrogenation process have been provided.

Gaseous Emissions

The flue gas produced by the different process furnaces during the combustion of the fuel gas is discharged through stacks into the atmosphere. Coke-oven gas of town quality (a sulfur content of 0.25 g/m^3) is used as the fuel gas so as to minimize SO_2 emissions from the plant.

During normal plant operation all hydrocarbon-containing flows circulate in a completely enclosed system. Under exceptional operating conditions any gases and vapors escaping from pressure-relief fittings and from blowdown units are fed via blowdown manifolds to the stack-gas system. In addition, steam is fed into the flare stack to obtain a smokeless combustion of the gases that are burned.

In order to avoid hydrocarbon-containing emissions during manipulation and storage of products, the relevant plant units are equipped with shuttle gas lines.

During normal operating conditions there are also hydrocarbon emissions from flange joints, from seals of fittings and safety valves, and from shaft seals of pumps and compressors as well as emissions that occur in the course of sampling, cleaning, and other routine operations. To minimize these emissions, the following precautions as to plant design and procedures have been taken:

- Use of annular joint seals
- Reduction in the number of flange joints
- Use of bellows and safety stuffing boxes for fittings
- Use of sliding seals for pumps or use of pumps without stuffing boxes

Dust Emissions

Any dust emissions from units, e.g., from the supply and storage of coal and catalysts or from slag disposal, are subject, as far as the Federal Republic of Germany is concerned, to the regulations contained in the *Technical Instructions for Maintaining the Cleanliness of the Air*.

Dust-containing waste gases discharged from coal preparation plants and catalyst storage areas are cleaned in a filter system before being discharged to the atmosphere.

The conveyor belt leading from the coking plant to the hydrogenation plant as well as the transfer points for coal and slag is enclosed to prevent dust development by wind action.

Noise

Sound insulation is provided in order to decrease noise emissions from machinery and other equipment of units, e.g., relief valves. Sound insulation is implemented partly as hoods or enclosures around single pieces of machinery or valves. In the case of compressor units, sound insulation is provided by the erection of compressor buildings.

Wastewater

Separate systems are provided for the treatment of process and surface water.
  Process water is discharged from the following units:

• Hydrogenation

• Atmospheric distillation

• Vacuum distillation

• Slop tank

The process water from these units is fed to a process effluent tank via a closed pipeline system, which is connected to the breather line. Gas emissions are burned in the flare stack. Process water is then fed to the water-recycling system of the coking plant.
  The surface water is treated in a conventional way.

ECONOMICS

The results from the operation phase of the pilot plant at Bottrop will be used to form the basis for a more accurate design of industrial plants and thus for the next step in commercializing the German technology.
  Since coal oil can substitute for mineral oil products, it has a big market potential. The market prospects, however, depend on the prices of crude oil and of the competitive mineral oil derivates.
  The production costs of coal oil are composed of the costs of capital and raw material and the plant operating costs.
  For privately financed commercial plants in the Federal Republic of Germany (FRG), the raw material costs (a coal price of approximately US\$125 per ton) amount to 40 to 50 percent of the production costs. The resulting cost of liquid products from coal hydrogenation (syncrude) thus is about double that of the corresponding mineral oil products. More favorable results can be achieved in countries with low-price coal, although the capital costs are somewhat higher.
  The economic break-even point of coal hydrogenation depends critically on the future development of crude oil prices. The competitiveness of coal liquefaction products in the FRG could be attained toward the end of this decade, provided that the disproportionate price escalation of crude oil against coal continues, which is, however, unlikely.

PART 2

2

PRODUCTION OF LIQUID FUELS FROM COAL-DERIVED SYNTHESIS GAS

The production of liquid fuels by first gasifying coal and then synthesizing liquid fuels from the resulting gas is termed *indirect liquefaction*. Gasoline, fuel oil, and methanol can be produced from medium-heat-content coal gas which has been purified and shifted to the proper ratio of hydrogen to carbon monoxide (and carbon dioxide) to form synthesis gas, as shown in Fig. I-1.

The two principal technologies for conversion of synthesis gas to methanol, the ICI low pressure methanol process and the Lurgi Low Pressure Methanol process, are presented in this section. The single test-plant-proven process for the direct conversion of methanol to gasoline, the Mobil Methanol-to-Gasoline (MTG) process, is also presented.

Fischer-Tropsch technology was developed in Germany. As early as 1913, a patent for the production of oxygenated compounds and aliphatic hydrocarbons by catalytic hydrogenation of carbon monoxide was awarded to Badische Anilin und Soda Fabrik. The procedures utilized today were developed mainly by Fischer and Tropsch in the 1920s in Germany, where liquids were produced from hydrogen and carbon monoxide at 1500 to 2200 lb/in² (10 to 15 MPa) and 400 to 450 °C utilizing alkalized iron catalysts. A number of Fischer-Tropsch plants were constructed in the 1930s, with product yield totaling approximately 570,000 tons (6×10^5 Mg) of oil and gasoline annually—all from synthesis gas derived from coal. In 1955, the South African Coal, Oil and Gas Corporation placed in operation a commercial-size facility for the production of motor fuels from coal that utilized the Fischer-Tropsch technology for liquids production and second-generation plants, known as Sasol Two and Three are on stream. The Sasol complex is described in detail in this section.

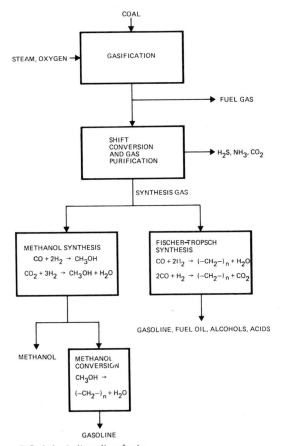

FIG. I-1   Indirect liquefaction.

Methanol has been produced industrially from coal for many years, predating the liquid-fuel plants based on Fischer-Tropsch technology. The original process normally required pressures of 3000 to 6000 lb/in^2 (20 to 40 MPa), but this pressure requirement has been reduced to 450 to 1500 lb/in^2 (3 to 10 MPa) in the processes covered in this section. Methanol synthesis quickly became well established, and by 1940 U.S. production alone had reached 130,000 gal/day (500,000 L/day). After World War II low-cost natural gas and light petroleum distillates replaced coal as the feedstock, and today little methanol is being manufactured from coal.

The properties of methanol as a transportation fuel are compared with gasoline and ethanol in the table that follows. Methanol has a lower heat content than either ethanol or gasoline. However, the octane numbers of both methanol and ethanol are quite high, a property which contributes to their desirability as either raw transportation fuels or additives to unleaded regular gasoline.

Methanol offers a major advantage over the use of ethanol or ethanol-based gasohol because of its much larger potential for production. Ethanol is produced from biomass

Properties of synthetic and petroleum-based transportation fuels

Fuel	Energy content of fuel		Octane number (research)
	Btu/gal	**MJ/L**	
Methanol	59,000	15.9	112
Ethanol	76,000	21.2	111
Unleaded regular gasoline	115,000	32.1	92
With 10% methanol	109,000	30.4	96
With 10% ethanol*	111,000	31.0	96

*Gasohol.

by fermentation or by hydration of ethylene. However, since ethylene is already a valuable petrochemical, the utilization of ethanol as a transportation fuel will most likely depend on its production from biomass rather than ethylene. In order to produce approximately one million barrels (160 million liters) of ethanol annually (5 to 6 percent of U.S. annual transportation needs), a yearly consumption of 6 to 6½ billion bushels of corn would be necessary. However, the annual total corn crop in the United States is only 4½ billion bushels (160 million cubic meters).

In contrast, the production of one million barrels of methanol daily requires about 250,000 tons (227 000 Mg) of coal daily and would supply about 5 percent of U.S. transportation needs. This is about one-tenth of the current average daily rate of production of coal in the United States. Thus, establishment of a methanol transportation fuel industry based on coal would require a 10 percent increase in coal production, which is not an unreasonable increment over a 20- to 30-year period.

THE SASOL APPROACH TO LIQUID FUELS FROM COAL VIA THE FISCHER-TROPSCH REACTION

PETER F. MAKO
WILLIAM A. SAMUEL, P. E.

Fluor Engineers, Inc.
Advanced Technology Division
Irvine, California

HISTORY OF SASOL

The Fischer-Tropsch reaction is the foundation of the Sasol plants. Known since the twenties, it is represented generally by the equations:

$$2nH_2 + nCO \rightleftharpoons (CH_2)n + nH_2O + heat$$

$$nH_2 + 2nCO \rightleftharpoons (CH_2)n + nCO_2 + heat$$

The quantity of patents and literature that has appeared on the subject in the past 60 years makes it virtually impossible to retrace stepwise the developments of the Fischer-Tropsch synthesis. The two equations above are also an oversimplification of the reactions that occur. The resulting hydrocarbons may vary from C_1 to about C_{60}, from paraffins to olefins, and products may vary from alcohols to ketones to acids; any combination may occur depending upon the catalyst used and conditions chosen.

The origin of South African coal conversion activities can be traced back as far as 1927.[1-4] At that time, a government white paper recommended the establishment of a coal-gasification and low-temperature carbonization plant. In order to implement these recommendations, the mining concern Anglovaal (Anglo Transvaal Consolidated Investment Co.) and the British Burmah Company formed a joint venture to mine shale in the Ermelo district and to extract oil from it. Because large coal resources were also available close to the shale deposits, they also acquired rights to the German Fischer-Tropsch process. However, in the pre-World War II era and during the war, no attempts were made to build a large-scale plant based on this technology in South Africa.

Toward the end of World War II, Anglovaal acquired the rights to the Hydrocol process, a modification of the Kellogg moving-catalyst-bed process used with the Fischer-Tropsch method. In November 1945, Anglovaal officially announced its desire to continue with an oil-from-coal plant. Enabling legislation was passed in 1947. In January 1948, the South African Liquid Fuel and Oil Industry Advisory Board considered Anglovaal's application for a license to manufacture oil from coal. The license was issued in 1949. Because of capital shortages, Anglovaal then asked the government for a loan guarantee for its project, which was estimated to cost about $100 million. The government was not able to grant the loan guarantee, and U.S. capital was also unavailable. Nevertheless, the project was not abandoned, but other vehicles were tried to move the venture forward.

A committee was formed that consisted of members of the government, the Industrial Development Corporation (a government agency), and Anglovaal. The committee was to form a company which would obtain Anglovaal's licenses and secure other licensing and know-how to put the project together. In the United States, the M. W. Kellogg Co. negotiated to channel its know-how and patents and to construct the plant. Other rights to the Fischer-Tropsch process were obtained from the Federal Republic of Germany (FRG) from Ruhrchemie AG and Lurgi Gesellschaft für Wärmetechnik.

With the necessary licensing and other rights available, the committee recommended the formation of a state-backed company to start the construction of the plant.

ACKNOWLEDGMENTS

The authors greatly appreciate the review and comments by J. J. Kovach, Manager of Process Engineering, of Fluor's Advanced Technology Division. Thanks are also extended to Sasol for permission to publish this material and to Mr. P. Naudé, Sasol's General Manager of Technology Transfer, for many helpful comments.

Sasol was formed in 1950; the acronym reflects the company's original name—Suid-Afrikaanse Sintetiese Olie (South African Synthetic Oil) Limited. Later the company's name was changed to Suid-Afrikaanse Steenkool, Olie en Gaskorporasie, but the acronym *Sasol* remained.

Design of Sasol's first coal-to-oil complex started in 1950, and 5 years later the first product was obtained. Although serious problems were experienced initially, these were overcome and design production was achieved in 1960. In 1964 Sasol (which today is Sasol One) started production of butadiene and styrene, ammonia, nitric acid, and ammonium nitrate—all based on coal. Also in 1964, Gascor—South African Gas Distribution Corporation—was formed to distribute gas. A 175-km pipeline was built to provide the network to the customers.

In the late 1960s, Sasol participated in forming South Africa's largest crude oil refinery, National Petroleum Refiners of South Africa (NATREF). With 52 percent of the shares of NATREF, Sasol is in not only the coal-conversion business but also in oil refining.

By the 1970s Sasol was using coal conversion to produce a small but significant share of South Africa's total needs for liquid fuels. But as long as the price of Middle East crude oil was low, the process was not economically competitive and no major expansion was justified.

Political events in the Middle East in 1973 changed the oil supply situation of the world. The South African government foresaw the continuation of the price increases for crude oil and the threat of politically motivated embargoes. On December 5, 1974, the decision to build Sasol Two, mainly to produce motor fuels, was announced.

After the overthrow of the Shah of Iran, South Africa's crude oil sources became greatly limited. As a result of this, the decision to build Sasol Three was announced in February 1979.

Until 1979 the Industrial Development Corporation was the only shareholder in Sasol on behalf of the government. On June 26, 1979, Sasol Limited was incorporated as the holding company of the Sasol group.

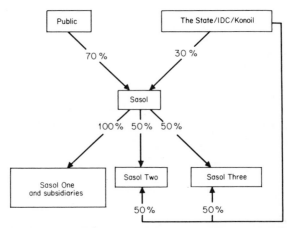

FIG. 1-1   Sasol financial structure *(From Refs. 1 and 3, by permission).*

Currently 70 percent of Sasol Limited is owned by the public and 30 percent by the state, via the Industrial Development Corporation and Konoil. Sasol Limited presently holds a 100 percent share interest in Sasol One (Pty) Ltd. and a 50 percent share interest in both Sasol Two (Pty) Ltd. and Sasol Three (Pty) Ltd. The state indirectly holds, through the Industrial Development Corporation and Konoil, the other 50 percent share interests in Sasol Two and Sasol Three. Through that arrangement the government supplies the major portion of the initial and construction financing of Sasol Two and Sasol Three. (See Fig. 1-1.) However, it is the expressed intention that once Sasol Two and Three are in production, the government's shares will be bought out by private shareholders.

SASOL ONE—PROCESS DESCRIPTION

The strategy for building Sasol plants reflects a commitment to long-range planning. Sasol One, which represented a large investment for a moderate production capacity, was built to *establish* a technology for reducing South Africa's dependence on imported oil. Accordingly, Sasol One originally was designed with only nine Mark III gasifiers (discussed under "Operating Experience with Sasol One"), but downstream of the gasifiers both the fixed- and fluid-bed Fischer-Tropsch technologies were implemented. The fluid-bed approach looked more attractive, but it represented unproven technology, while the fixed-bed route had commercial experience from the World War II days in Germany. Thus a compromise was made.

The process is illustrated in Fig. 1-2.[5,6] Coal is obtained from the Sigma mine about 3 km from the plant. The mine has a combined yearly capacity of 5.5 million tons, using room-and-pillar as well as longwall mining. The coal is crushed, and the sized coal (6 mm or about ¼ in) is sent to the gasifiers; the fines are sent to the power plant. A typical Sigma coal analysis is given in Table 1-1, and the typical size distribution for the Lurgi gasifiers is listed in Table 1-2. Coal, steam, and oxygen are fed to the Lurgi gasifiers, where they react to produce synthesis gas and other products typical of fixed-bed Lurgi gasifiers, such as gas liquor, tars, and oil. These by-products are separated from the raw gas and then distilled into solvents, creosotes, and road tars. The gas liquor contains water-soluble condensates which are recovered in the Phenosolvan plant and ammonia recovery plant.

The raw gas is purified using the Lurgi Rectisol (refrigerated methanol) system, which removes all of the H_2S and most of the carbon dioxide. The purified gas can be used in various ways. The main destinations are the Arge (fixed-bed) reactors and the Synthol (fluid-bed) reactors. The clean gas can also go to the gas-blending plant, from which it is piped to the town-gas distribution system. The products from the Arge and Synthol reactors are separated and refined into salable products in the process units associated with the plant.

The gas from the synthesis units contains some noncondensible hydrocarbons as well as unreacted CO and H_2. This gas may be blended into the industrial gas that is distributed by the Gascor system, or it can be scrubbed to remove carbon dioxide and then fed to the partial-oxidation re-formers to be converted into more Synthol feed gas. Some of the Synthol tail gas joins a stream of pure gas that has undergone a water-gas shift. After carbon dioxide and hydrocarbon removal, this stream becomes feed for the ammonia synthesis units. The nitrogen for the ammonia is a by-product of the air-separation unit.

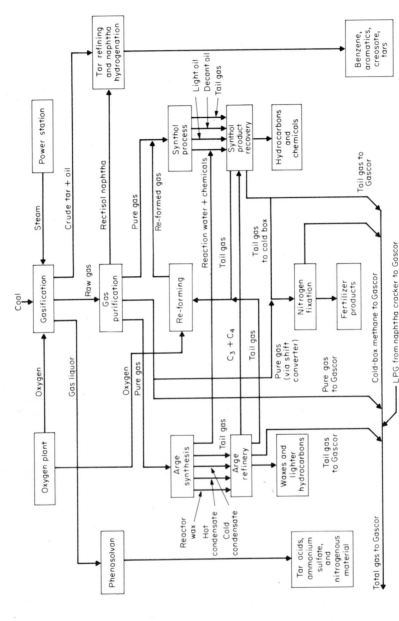

FIG. 1-2 Process flow sheet of Sasol One (*From Refs. 5 and 6, by permission*).

TABLE 1-1 Evaluation of typical Sigma bituminous coal

Parameter	Quantity
Moisture (as received), wt %	10.7
Volatiles (dry basis), wt %	22.3
Ultimate analysis (dry basis), wt %	
Ash	35.9
Sulfur	0.5
Nitrogen	1.2
Carbon	50.8
Hydrogen	2.8
Oxygen	8.8
Ash fusion temperature, °C	
Softening point	1340
Melting point	1430
Fluid point	1475
Energy content, (dry) MJ/kg	20–22
Energy content (dry), Btu/lb	8620–9480

Source: From Ref. 5, by permission.

As mentioned above, two different types of reactors are used at Sasol One to perform the Fischer-Tropsch synthesis; one uses the fixed-bed Arge process developed in Germany and the other the fluid-bed Synthol process originated by M. W. Kellog Co. and perfected by Sasol.

Fixed-Bed Arge Reactors

The designs of the fixed-bed Arge reactors were based on the original technology developed in Germany. Each reactor of the five utilized at Sasol One (see Fig. 1-3) has a diameter of 9.7 ft (2.95 m) and a vertical length of 42 ft (12.8 m). The reactors contain a vertical tube bundle containing over 2000 tubes with a nominal 2-in diameter.[5,7]

The heat of reaction is removed by generating steam on the shell side of the tubular heat exchanger. The generated steam goes to a 246-psia (1.7-MPa) steam system. The

TABLE 1-2 Typical coal size for Lurgi gasifiers

Size	%
>38 mm (1.5 in)	8
25–38 mm (1.0–1.5 in)	32
12–25 mm (0.5–1.0 in)	47
6–12 mm (0.25–0.5)	8
<6 mm (<0.25 in)	5

Source: From Ref. 5, by permission.

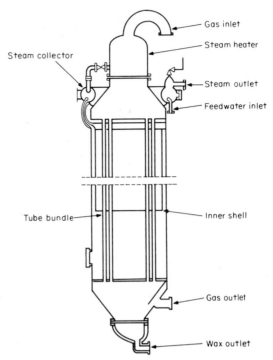

Gas inlet

Steam heater

Steam collector

Steam outlet

Feedwater inlet

Tube bundle

Inner shell

Gas outlet

Wax outlet

FIG. 1-3 Arge reactor *(From Refs. 5 and 7, by permission).*

reaction temperature at the start of the run is about 220 °C and about 245 °C at the end of the run. The duration of a cycle before the catalyst is changed is about 70 to 100 days. Reactors operate at 360 psia (2.5 MPa), with a booster compressor providing the needed gas pressure. The recycle ratio is from 1.5 to 2.5, depending upon whether the load is a maximum or minimum. A conversion of about 65 percent is obtained based on the H_2 and CO in the fresh feed. Table 1-3 gives the reactor conditions and selectivities.[5,8] For comparison, the Synthol yields are also shown and the implications are reviewed later in this chapter. As can be seen in Fig. 1-3, wax accumulates at the base of the reactor. It is transferred to the vacuum distillation unit, where it is split into various wax products. A schematic representation of this workup is given in Fig. 1-4,[5,9] and the composition of the tail gas is given in Table 1-4.[5,10]

Originally the catalyst for the reactors was obtained from German sources. Later, however, as the Sasol know-how was established, a plant was built at Sasol to manufacture the catalyst with proper modifications.

As with the Synthol catalyst, the manufacturing process for the Arge catalyst is a closely guarded secret. The literature states, however, that iron ore is the raw material and that alkali metals are used as promoters. The catalyst can be used for several hundred days, but at such extremes the quality of the products, especially of the waxes, deteriorates. The yields also decrease with catalyst age, although this can be partially offset by increasing the reactor temperature.

TABLE 1-3   Comparison of Arge and Synthol processes

Parameter	Arge	Synthol
Temperature, °C	232	330
Pressure, MPa	2.55	2.20
Pressure, psia	370	320
Conversion of fresh feed ($CO +$		
H_2) entering, %	65	85
H_2/CO ratio	1.9	2.10
Selectivities, %		
CH_4	5.0	10.0
C_2H_4	0.2	4.0
C_2H_6	2.4	6.0
C_3H_6	2.0	12.0
C_3H_8	2.8	2.0
C_4H_8	3.0	8.0
C_4H_{10}	2.2	1.0
Gasoline		
C_5 to C_{12}	22.5	39.0
Diesel fraction		
C_{13} to C_{18}	15.0	5.0
Heavy oil and wax		
C_{19} to C_{21}	6.0	1.0
C_{22} to C_{30}	17.0	3.0
C_{31}	18.0	2.0
NAC[*]	3.5	6.0
Acids	0.4	1.0
Total	100.0	100.0
Ratio of tail gas to fresh feed	0.53	0.33

[*]Nonacid chemicals.

Source: From Refs. 5 and 8, by permission.

Original production rates for the Arge reactors were 11,000 to 13,000 tons (10,000 to 12,000 Mg) yearly. With increasing know-how and catalyst improvements, the yields were doubled. Rights to the fixed-bed process are jointly owned by Sasol, Lurgi, and Ruhrchemie.

Fluid-Bed Synthol Reactor

The Synthol entrained-fluid-bed reactors produce a range of hydrocarbons concentrated in the diesel and gasoline range. The Synthol reactors for Sasol One were designed by M. W. Kellogg Co., on the basis of limited pilot plant tests. The first liquid products from the Sasol One Synthol reactors were produced in mid-1955, but it took about 5 more years of development work by Sasol to solve all problems and to ensure reliable operations. The reactor is schematically represented in Fig. 1-5.[5,7,11] The gas enters the reactor through a

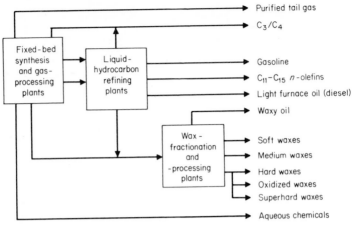

FIG. 1-4   Arge product refining *(From Refs. 5 and 9, by permission).*

horizontal line (about 3 ft (1.0 m) in diameter) and meets the catalyst that falls from the standpipe via a slide valve. The catalyst at this point acts as a heat-transfer medium by raising the temperature of the preheated gas to between 300 and 315 °C.

The feed gas consists of a stream of recycle gas and fresh feed. The latter is partly straight Rectisol effluent gas and partly re-formed gas from the methane re-formers. The reactor pressure is about 320 psig (2.3 MPa). By the time the catalyst and the gas have reached the vertical section (the riser) of the reactor, the reaction has begun. The Fischer-Tropsch reaction is exothermic, as pointed out previously, and the heat of reaction may cause carbon deposits on the catalyst. In the cooling section, which is about 7.4 ft (2.25 m) in diameter, the temperature is controlled and steam at a pressure of 174 psia (1.2 MPa) is generated. A temperature of 340 °C is maintained at the top of the reactor. In the catalyst-settling section of the reactor, cyclones separate the gaseous products from the catalyst. The catalyst particles drop into the catalyst standpipe, while the gases leave the reactor and enter the gas-cooling train.

Table 1-5 gives Synthol selectivities and indicates that as the catalyst ages, more lower hydrocarbons are formed at the expense of light oil.[5,8,10] The extent of that trend is one

TABLE 1-4   Composition of Arge tail gas

H_2, vol %	47
N_2, vol %	2
CO, vol %	22
CO_2, vol %	1
CH_4, vol %	28
C_2 to C_4, vol %	Trace
Energy content (HHV)*, MJ/m³	19.0
Energy content (HHV*, Btu/ft³	513

*Higher-heating value.

Source: From Refs. 5 and 10, by permission.

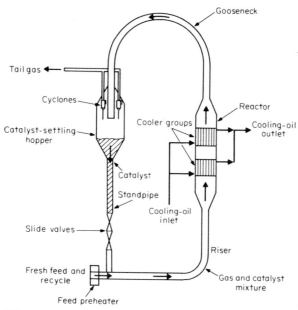

FIG. 1-5 Synthol reactor *(From Refs. 5, 7, and 11, by permission).*

TABLE 1-5 Synthol selectivities: commercial results[*]

	Start (day 0–5)	End (day 35+)	Average
CH_4	7	13	10
C_2H_4	4	3	4
C_2H_6	3	9	6
C_3H_6	10	13	12
C_3H_8	1	3	2
C_4H_8	7	9	8
C_4H_{10}	1	2	1
C_5+	6	9	8
Light oil[†]	40	30	35
Decanted oil[‡]	14	2	7
NAC[§]	6	6	6
Acids[¶]	1	1	1

[*] Selectivity is the percentage of C atoms in the CO and CO_2 in the fresh feed that are converted and end up in the indicated products.

[†] Boiling point range of 50 to 290° C (see Fig. 1-6).

[‡] Boiling point range of 160 to 490°C (see Fig. 1-6).

[§] Nonacid chemicals.

[¶] Water-soluble acids.

Source: From Refs. 5, 8, and 10, by permission.

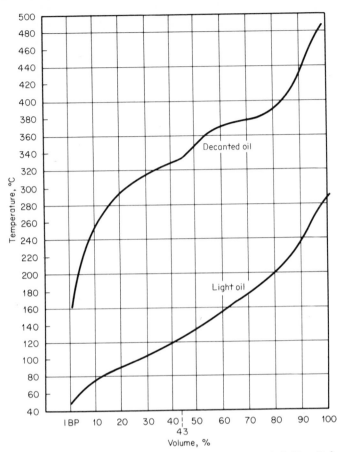

FIG. 1-6   ASTM distillation curves for light oil and decanted oil *(From Refs. 5 and 10, by permission)*.

major criterion that can be used to decide when the catalyst should be changed. Figure 1-6[5,10] shows the ASTM (American Society for Testing and Materials) distillation curves of a typical Synthol light oil and decanted oil, and Table 1-6[5,10] gives the carbon-atom distribution for Synthol light oil and decanted oil. Table 1-7 lists the composition of nonacid chemicals and organic acids, and Table 1-8[5] gives the Synthol tail-gas composition.

The catalyst for the Synthol reactors is manufactured on site. Improvements in the catalyst and the technology of its manufacture were major contributions in improving the operation of the fluid-bed technology. The precise details of the process constitute part of Sasol's proprietary know-how. It has been reported, however, that mill scale provides the iron base, promoters are added and fused together with the base, and finally the catalyst is reduced in a hydrogen atmosphere.

About 130 to 145 tons (120 to 130 Mg) of catalyst are estimated to be required for

TABLE 1-6 Carbon-atom distribution for Synthol light oil and decanted oil

	Light oil, %	Decanted oil, %
Carbon atoms		
4	3.4	
5	7.7	
6	14.3	
7	16.3	
8	14.5	
9	13.1	
10	8.2	1.9
11	6.8	1.2
12	6.5	2.3
13	5.0	2.7
14	2.1	3.2
15	2.1	3.7
16–20		23.2
21–25		28.0
26–30		19.9
31–35		9.5
>35		4.4
Mono-olefins	75	75
Paraffins	10	
Oxygenates	8	
Aromatics	7	

Source: From Refs. 5 and 10, by permission.

one load in a reactor at Sasol One. The circulating rate is about 8000 tons/h in each reactor. A batch of catalyst is used for an average of 40 to 45 stream days.[5]

Comparison of Arge and Synthol Products

Table 1-3 compares the Arge and Synthol product distributions and shows their gasoline and diesel quantities. The Synthol data refer to the average run as given in Table 1-5. These data indicate that the Arge reactor produces more diesel and heavier fractions than the Synthol unit. Because more straight-chain paraffins are produced in the Arge reactors, the gasoline quality therefrom is inferior to that from Synthol. The diesel quality, on the other hand, because of straight-chain paraffins, is very good. This is reflected in a high cetane number. Arge also produces a smaller quantity of nonacid chemicals (predominantly alcohols) and fewer acids.

The conversion is appreciably higher with the Synthol operation (85 percent vs. 65 percent), and the H_2/CO ratio in the feed is also higher. The most marked difference, however, is the high wax production with the Arge reactors. If market limitations exist on wax

TABLE 1-7 Percentage nonacid chemicals and organic acids*

Chemicals	% by weight
Nonacid chemicals	
Acetaldehyde	3.0
Propionaldehyde	1.0
Butyraldehyde	0.6
Acetone	10.6
Methylethyl ketone	3.0
Diethyl ketone–methyl propyl ketone	0.8
n-Butylketone	0.2
Methanol	1.4
Ethanol	55.6
Isopropanol	3.0
n-Propanol	12.8
2-Butanol	0.8
Isobutanol	4.2
n-Butanol	4.2
2-Pentanol	0.1
n-Pentanol	1.2
C_6+ alcohols	0.6
Acids	
Acetic acid	70.0
Propionic acid	16.0
Butyric acid	9.0
Valeric acid and higher	5.0

Source: From Refs. 5 and 10, by permission.

production, the wax fractions can readily be hydrocracked to increase the yield of high-quality diesel fuel.

OPERATING EXPERIENCE WITH SASOL ONE

As pointed out in the previous section, the initial purpose of Sasol One was to establish a technology. After the initial years, as the processes worked successfully, additions and expansions of the original plant followed. Additional boilers, gasifiers, and Rectisol, Synthol, Phenosolvan, and oxygen units were added.

Originally, nine Lurgi (Mark III) moving-bed gasifiers were installed. In 1959 a slightly modified gasifier was added, and then in 1966 three more Mark II gasifiers were commissioned. In 1971 three Mark IV gasifiers were added to give a total of sixteen gasifiers. A Mark V gasifier was designed in 1976. It was installed in 1979 and is currently operational as a test unit.

The Mark III gasifiers have an internal diameter (ID) of 12.1 ft (3.68 m), and the Mark IV gasifiers an ID of 12.63 ft (3.85 m). The Mark V gasifier is designed with a 15.42-ft

TABLE 1-8   Snythol tail-gas composition*

Component	% by volume
H_2	45
N_2	4
CO	2
CO_2	12
CH_4	35
C_2	2

Source: From Ref. 5, by permission.

(4.70-m) ID and is 41.0 ft (12.5 m) long, weighing 220 tons. The significant difference between the Mark III and Mark IV gasifiers is the method of driving the grate. In the older gasifiers, the motor is located at the top and drives the grate at the bottom of the reactor vessel via a long vertical shaft. In the Mark IV, as shown in Fig. 1-7, this shaft has been removed by placing the grate drive motor at the bottom and having it drive the outer edge

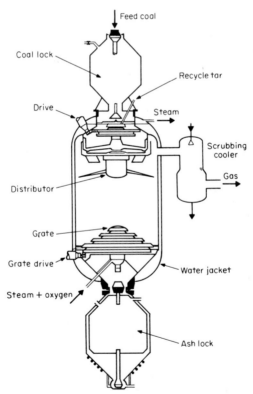

FIG. 1-7   Mark IV Lurgi Gasifier. Note: Sasol gasifiers do not have distributors *(From Ref. 12, by permission).*

of the grate.[12] The removal of the shaft provides increased reactor volume. One other change made in the Mark IV gasifier was an increase in the size of the lock-hopper to about 350 ft³ (10 m³).

Significant improvements have been made in the operation of the gasifiers, leading to about 85 percent overall availability of the gasifier plant; that is, only two gasifiers can be off-line. This availability is achieved by employing efficient servicing during operations and by scheduling preventive maintenance. With current operational know-how, the maintenance of the ash-lock system is on an average schedule of 3 months, whereas the feed lock-hopper requires only annual maintenance. The scheduling of pressure tests required by the government has been extended by experience to the point where a complete reconditioning is required only after a 3-year period.

The ash-disposal system also has been drastically improved over the years. Coarse ash is sluiced to the disposal area through special channels covered by ceramic tiles. This system has eliminated the erosion of metal linings.

Besides the catalyst development which was already mentioned, significant improvements were also made to the Synthol reactor system. One of these involved a solution to the problem of blockages which occurred in the intercoolers. The original shell-and-tube exchangers were removed, and serpentine coils were installed. Now the mixture of gas and catalyst is on the outside of the tubes, and there is no further blockage problem.[5,13]

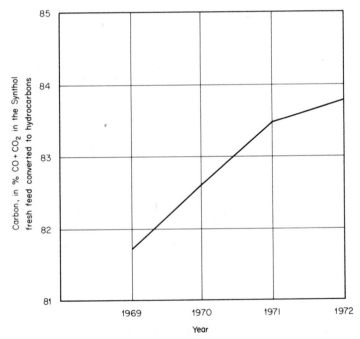

FIG. 1-8 Synthol synthesis data for CO + CO₂ (*From Refs. 5 and 14, by permission*).

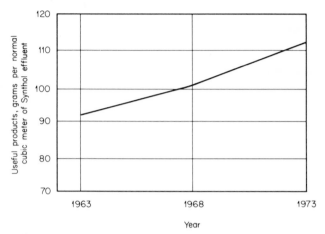

FIG. 1-9   Useful products yield from Synthol synthesis process. Useful products include all hydrocarbons and oxygenates except methane, ethane, ethylene, and organic acids *(From Refs. 5 and 14, by permission).*

Over the lifetime of the Sasol One facility, the performance of all the process units has improved to well over the original design levels, especially in the Synthol unit. Figures 1-8 and 1-9,[5,14] respectively, indicate how the CO and CO_2 conversion improved in the period from 1969 to 1972 and how useful product yield improved between 1963 and 1973. This has been further improved more recently, but data are not yet available.

HISTORY OF SASOL TWO

Reasons for Sasol Two

As long as the price of Middle East crude oil was low, Sasol-derived synthetic fuels were not competitive in the world market. As a result, major expansion could not be justified. With the 1973 oil crisis, the entire energy scene changed, as was noted in the first part of this chapter. As a result of soaring oil prices and because of national interests, the government of South Africa decided on December 5, 1974, to build Sasol Two. In early 1975, the Fluor Corporation was selected as the general contractor to design and build the complex, along with a number of engineering firms who participated in detailed engineering and construction of various units. Along with enforcing strict schedules, Sasol's philosophy was to divide major process and offsite units among contractors to assure sufficient personnel and know-how in special tasks. Fluor Corporation was responsible for the overall integration and coordination of the project as well as for a number of detailed designs.

It was well known to the decision makers that even with the OPEC price pegged at just over $11.00/bbl in the middle of 1974, Sasol Two products might not be competitive. But with increasing prices and inflation, it was felt that by the early 1980s, when Sasol Two was to come fully on stream, production costs would come close to prevailing world prices. Even with the softened oil prices of 1982–1983, we must applaud the foresight of those decision makers.

Design of Sasol Two

Long before the construction of Sasol Two was announced, a number of studies were carried out by Sasol to determine the optimum location and configuration of Sasol Two.

Like Sasol One, Sasol Two was to be designed as a mine-mouth facility. The available coal resources thus narrowed down the choice of plant location. In addition, marketing considerations, water availability, and other aspects were taken into account to find the most suitable site. Only later, after the coal rights and land had been secured, was the actual location announced—87 mi (140 km) northeast of Sasolburg and 81 mi (130 km) east of Johannesburg in the eastern Transvaal near the towns of Trichard and Evander. A new town called *Secunda* (coined from the Latin word for second) would be developed, and the plant located nearby.

The design philosophy set forth by Sasol was to use only proven technologies and to utilize the 20 years of experience gained at Sasol One. In particular, the coal preparation section was carefully planned to minimize formation of coal fines. It was decided very early that Sasol Two would utilize Lurgi Mark IV gasifiers. In order to obtain maximum efficiency with these gasifiers, sized coal must be fed, while the fines are utilized in the steam generators. Wet screening offered the opportunity to minimize fines and to provide maximum economies of scale. Two facts further supported the decision to employ wet screening: it results in sharper particle-size separation, and this approach is ecologically more attractive.

Since the region is heavily agricultural, it was also laid down in the design philosophy that "zero liquid-discharge" conditions would be required in order to protect the water balance of the area. Extensive laboratory studies were carried out to identify the trace elements in the coal, and leaching tests were undertaken on the ash originating from the coal.

When the time came to select the Fischer-Tropsch reactors, there was little doubt that the Synthol route would be the choice. The selection was governed by the desired products and the ease of scaling the design.

One more line of thinking was incorporated into the Sasol Two design that differed from the philosophy of Sasol One. With Sasol Two, the product goal was to transport fuels and to consider future gas sales. By-products were to be utilized in the pool of transport fuel or converted to products which could be utilized in that pool.

Overall Block Flow Diagram of Sasol Two[15]

Figures 1-10[16,17,18] and 1-11[5] show schematic flow diagrams of Sasol Two. The following process units will be discussed in some detail:

1. Coal mining and coal preparation
2. Gasification

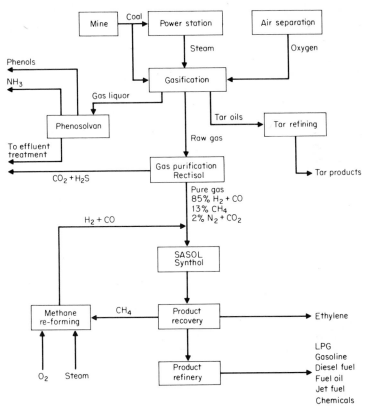

FIG. 1-10   Sasol synfuel process—simplified flow sheet for Sasol Two and Three *(From Refs. 16, 17 and 18, by permission).*

3. Rectisol

4. Phenosolvan and ammonia recovery

5. Tar separation

6. Synthol

7. Synthol tail-gas processing and methane re-forming

8. Downstream refining

9. Utilities including steam, power, oxygen plant, and water management

Coal mining and coal preparation

The Bosjesspruit mine, developed specifically for supplying coal to Sasol Two and later to Sasol Three, lies about 1 mi (1.5 km) south of the plant site.[19] Four portals will eventually supply the 30×10^6 tons (27.5×10^6 Mg) of coal needed annually for the two plants. Coal seams 7 to 16 ft (2 to 5 m) thick lie between about 400 and 600 ft (100 and 200 m)

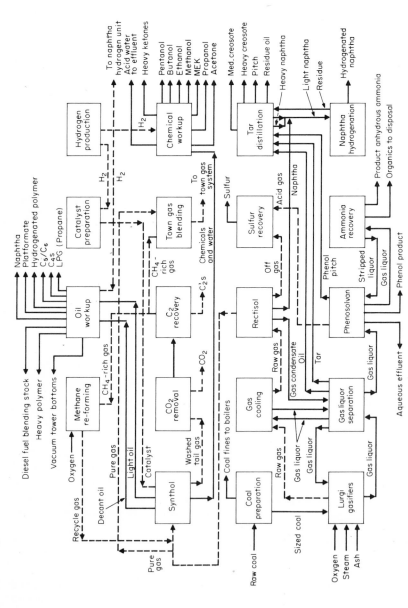

FIG. 1-11 Sasol Two flow sheet (*From Refs. 5 and 21, by permission*).

TABLE 1-9 Properties of Bosjesspruit coal (subbituminous)

Parameter	Value
Moisture (as received), wt %	5.5
Ash (moisture-free), wt %	22.5
Volatiles (moisture-free), wt %	24.8
Fixed carbon (moisture-free), wt %	52.7
Heating value, MJ/kg (Btu/lb)	23.9–24.5 (10,300–10,560)
Carbon (daf)*, wt %	79.6
Hydrogen (daf), wt %	4.3
Sulfur (daf), wt %	1.3
Nitrogen (daf), wt %	2.0
Oxygen (daf), wt %	13.6
Ash properties	
Softening point, °C	1290
Melting point, °C	1330
Fluid point, °C	1360

*Dry, ash-free.

Source: From Refs. 5, 20, and 21, by permission.

below the surface. The coal is part of the Highveld coal field, and its analysis is given in Table 1-9.[5,20,21] Currently the mine has two access systems, each with three separate shafts:

- 36-ft (11-m) diameter services access
- 33-ft (10-m) diameter ventilation
- 8.2- by 21-ft (2.5- by 6.4-m) inclined production shaft, set at a 17-degree slope

Proven mining methods such as room-and-pillar and longwalling are employed, using continuous mechanized equipment. An overall recovery rate in excess of 60 percent is expected. Primary and secondary trunk conveyors 3.9 ft (1.2 m) wide with a capacity of 1600 tons/h feed the coal from the primary source into several 4.9-ft (1.5-m) steel-cord shaft belts which have a combined capacity of 4500 tons/h (4000 Mg/h). These shaft belts feed into 12,000-ton (11,000-Mg) capacity bunkers at the portals of each mine shaft.

From the bunkers, coal is transported by overhead conveyors to an outside stockpile with a capacity of 2×10^6 tons (1.8×10^6 Mg), of which 400,000 tons (363,000 Mg) are live-storage coal. Dead-storage coal is compacted to prevent spontaneous combustion. Live storage provides a buffer between the plant and mine. Coal is mined 5½ days/week. A stocker-reclaimer system operates at the stockpiles.

Coal from the crushers and breakers is fed to the wet-screening system. Here the ⅛-in (3-mm) particle-size material is separated and fed to the gasifiers. The separation is achieved in 14 vibrating triple-deck screen systems. The wet fines (i.e., underflow) go to thickeners and subsequently to hydrocyclones before transportation to the boiler plant. The reclaimed water is recycled to the wet-screening system.

The wet-screening system has proved to give the predicted sharper particle separation. Because of the scale of operation, the use of this approach over dry screening was fully justified.

Gasification

Several Lurgi gasifiers in parallel operation are required to gasify coal in a commercial-scale plant. Sasol Two employs 36 Lurgi Mark IV gasifiers to process a nominal 31,000 tons/day (28,000 Mg/day) of sized coal. The gasifiers are arranged in two rows, each having 18 gasifiers. Furthermore, each row is composed of two banks of 9 gasifiers to constitute a total of four trains. This arrangement gives high flexibility and assures about 90 percent on-stream efficiency. Each gasifier consists of a number of sections, including a coal bunker, coal lock, gasifier, wash cooler, waste-heat exhanger, ash lock, and ash-lock expansion chamber. Figure 1-7 shows the gasifier and its associated equipment. Located near the top of the gasifier is a coal distributor that maintains a uniform flow of coal from the coal lock to the gasifier. An ash grate continuously moves ash from the gasifier into the ash lock. Both coal and ash locks are depressurized in the cycle to charge coal and to remove ash from the system.

The raw gas leaving each gasifier is quenched in the wash cooler with recycle gas liquor to water-saturation temperature. The saturated gas then flows through the waste-heat exchanger in order to recover heat in the form of low-pressure steam from the raw gas.

Oxygen required for the gasification of coal is produced in an air-separation plant and then compressed to coal-gasifier pressure. Superheated steam for the process is produced in a coal-fired steam generation plant. The ratio of steam to oxygen is controlled so as to maintain the gasifier combustion zone at temperatures below the ash softening point.

As the coal moves through the gasifier from top to bottom, it passes through several distinct zones. The first zone preheats and dries the coal by contact with hot gas leaving the reaction zone. In the reaction zone, the coal is devolatilized and the gasification reaction takes place. Combustion takes place at the bottom of the reaction zone, where the char is reacted with oxygen (producing principally CO_2) to provide the heat required in the zones above. The bottom of the gasifier is the ash zone. The ash normally contains no more than 3 percent carbon.

Operating at about 390 psia (2.7 MPa), the Sasol Two gasifiers are designed to produce raw gas at a rate of 56×10^6 ft³/h (1.65×10^6 m³/h) of the approximate composition shown in Table 1-10.[5,21] Some 27,500 to 33,000 tons/day (25,000 to 30,000 Mg/day) of coal will be reacted with 33,000 to 40,000 tons/day (30,000 to 36,000 Mg/day) of steam and 8800 to 9900 tons/day (8000 to 9000 Mg/day) of oxygen. Some 5500 tons/day (5000 Mg/day) of ash from gasification will have to be disposed of.

TABLE 1-10 Composition of raw gas

Component	% by volume
H_2	38.1
N_2	0.3
CO	19.0
CO_2	32.0
CH_4	9.4
C_2+	0.5
H_2S	0.7

Source: From Refs. 5 and 21, by permission.

In the gasifier waste-heat exchangers and the subsequent gas-cooling units, gas liquor is condensed from the raw gas. The gas liquor consists of water, oil, tar, dust, ammonia, and water-soluble hydrocarbons. The condensate is also saturated with other components contained in the raw gas.

This aqueous condensate, which is removed in each of the cool-down exchangers, is fed to the gas-liquor separation unit, where the tar, oil, and water phases in the gas liquor are separated. The tar and oil phases are combined and fed to a tar-distillation unit.

The water phase from the gas-liquor separation unit is fed to the Phenosolvan unit for extraction of phenols and then processed for ammonia recovery. Finally, the stripped gas liquor is treated for use as makeup water for the process cooling towers.

The raw gas is cooled further until it is close to the operating temperature of the Rectisol unit, where the remainder of the coal naphtha and sulfur compounds and the bulk of the carbon dioxide is removed.

Rectisol unit

The cooled gas is first washed with prewash methanol to remove the remainder of the coal naphtha (light naphtha). The rich prewash methanol is regenerated by flashing to atmospheric pressure and recycling the lean methanol. Acid gas from the flash is fed to the sulfur recovery unit.

Next, the raw gas is washed with main wash methanol. In this portion of the process, the bulk removal of sulfur compounds and carbon dioxide occurs. The rich methanol is regenerated by flashing to subatmospheric pressure, and the lean methanol is recycled. Finally, the raw gas is washed with fine wash methanol to meet the pure-gas specification for sulfur. Pure gas from the fine wash is fed to the Synthol unit for synthesis into liquid products.

The acid gas extracted in the Rectisol unit contains sulfur compounds, carbon dioxide, and some light hydrocarbons and is fed to a Stretford sulfur-recovery unit. In this unit, the sulfur is absorbed in Stretford solution and converted to elemental sulfur. The product sulfur can be sold either as a liquid or as a solid.

The treated acid gas from the Stretford unit is mostly carbon dioxide and can be vented to the atmosphere unless the hydrocarbon and residual sulfur content of the gas has an adverse environmental impact. If necessary, the gas can be incinerated to destroy the contaminants that are present.

In Sasol Two there are four Rectisol trains, each capable of producing 9.7×10^6 ft^3/h (275,000 m^3/h) of pure gas. The composition of the pure gas is given in Table 1-11.[5,21] The methanol wash towers in the Rectisol trains are the most massive single units in the plant, each weighing about 340 tons (308 Mg). Propylene is used as the refrigerant. (The Rectisol process operates near $-40°$.)

The Rectisol units are the most power-consumptive units in the plant; each 35,310 ft^3 (1000 m^3) of pure gas requires 35.7 kWh of electricity and 174.6 lb (79.2 kg) of steam, mainly for the circulating pumps and compressors used for refrigeration.

Phenosolvan unit

The aqueous gas-liquor effluent from the gasifiers contains about 2000 ppm of tar acids (phenols, cresols, etc.) and about 8000 to 9000 ppm of ammonia. This gas liquor, after gravitational separation of residual tars and oil, is extracted with diisopropyl ether (DIPE) in a countercurrent horizontal extractor to obtain the phenols. Live steam is used in a

TABLE 1-11 Composition of pure gas from Sasol Two

Component	% by volume*
H_2	56
CO	28
CO_2	1.5
CH_4	13.5
C_2H_4	0.02
C_2H_6	0.20
N_2	0.08
A	0.46
S	0.07 ppm

*Unless otherwise noted.

Source: From Refs. 5 and 21, by permission.

packed tower to strip off the organic solvent from the gas liquor without removing much of the ammonia. The rich DIPE is treated in atmospheric and then vacuum distillation columns to recover the crude phenols. The gas liquor, after the removal of the residual solvent, goes to the ammonia strippers. Ammonia recovery includes fractionation to produce a high-purity ammonia product equal to that from synthetic plants. After this unit, the gas liquor contains not over 250 ppm phenols and 250 ppm ammonia. The gas liquor is then treated in a biotreatment plant which has been described by Lurgi.[22]

Sasol Two was designed for zero effluent discharge, and the pure water from the biotreatment plant is returned to the process cooling towers as makeup water.

Tar separation and distillation

Tars and oils are separated from the gas liquor by means of decanting systems in the gas-liquor separation unit. The oil, being less dense than water, floats to the surface. The tar, being heavier than the water, sinks to the bottom. These two streams are recombined and then joined by whatever residual pitch is separated from the gas liquor in the Phenosolvan unit. The combined stream contains virtually all the fine entrained coal and ash that left the gasifier in the raw-gas stream. The solid residue is removed by filtration.

The production rate of tar, oil, and naphtha from the gasification, Rectisol, and Phenosolvan units is 20 to 25 L per ton of coal. The spearation of the combined and filtered stream occurs in the tar-distillation unit. The following products are obtained:

Heavy naphtha

Medium creosote

Heavy creosote

Residue oil

Residue

Synthol unit

The Synthol unit for Sasol One has been described in detail earlier in this chapter. Sasol Two utilized the know-how and experience gained in Sasol One. The units were scaled up

TABLE 1-12 Product selectivity for the Sasol Two Synthols

Product	Mass %
Methane	11.0
Ethane and ethylene	7.5
Propane and propylene	13.0
Butanes and butylenes	11.0
C_5-to-375 °F fraction	37.0
375-to-750 °F fraction	11.0
750-to-970 °F fraction	3.0
Heavier than 970 °F fraction	0.5
Chemicals	6.0
Total	100.0

Source: From Refs. 5 and 15, by permission.

by a factor of 2.5,[13] and a total of eight units were constructed for Sasol Two. Each of these units has the capacity to process about 10,600,000 to 12,400,000 ft³/h (300,000 to 350,000 m³/h) of fresh feed (pure gas and re-formed gas). The reactor structures are about 250 ft (75 m) high, and the actual reactor vessels are about 10 ft (3 m) in diameter. Each Synthol reaction train consists of a reactor where products are formed, a feed-gas compressor, and heat exchangers that preheat the feed gas and cool the products.

Carbon-conversion selectivity is given in Table 1-12.[15,17] Selectivity is defined as the mass percentage of carbon converted from feed CO and CO_2 to the product indicated. These selectivities are comparable with those given in Table 1-5, which represents Sasol One operations.

Heavy oil is separated out in a hot-quench tower which also oil-scrubs any catalyst which may have escaped from the reactor with the product gas. Separation of the light oil, aqueous fraction, and tail gas takes place in a separation drum after cooling. The products have to be stabilized before being sent to the product refinery.

The heavy oil containing the carry-over catalyst ("gunk") is depressurized to remove dissolved gases, cooled, and pumped into a large decanter with a rotating rake. In the gunk-removal system, catalyst solids settle and the clean oil is decanted. The concentrated catalyst slurry is pumped through a pressure filter. The filtrate of cleaned oil is then sent to the refinery for further processing.

Synthol tail-gas processing and methane re-forming

The uncondensed portion of the product from the Synthol unit, called *Synthol tail gas*, consists primarily of hydrogen, light hydrocarbon vapors, and carbon dioxide. After the tail gas is compressed, the carbon dioxide is removed using the Benfield process (activated hot potassium carbonate). The carbon dioxide absorbed in the CO_2 removal unit is vented to the atmosphere.

The hydrocarbon vapor from the CO_2 removal unit is cooled in a cryogenic process train and separated into a C_3+ stream, a mixed C_2 stream, a methane-rich stream, and a hydrogen-rich stream. The highly olefinic C_3+ stream is processed in the oil workup unit to produce transport fuels. The C_2 stream is used to produce polymer-grade ethylene. The hydrogen-rich stream is partly used as a feed to a hydrogen purification unit, which supplies

TABLE 1-13 Composition of re-formed gas*

Component	% by volume
H_2	61.6
N_2	3.7
CO	22.2
CO_2	7.2
CH_4	4.7
H_2O	0.6
Total	100.0

Source: From Refs. 5 and 23, by permission.

high-purity hydrogen to downstream hydrogenation units. The remaining part of the hydrogen-rich stream is recycled to the Synthol units.

The methane-rich stream separated in the C_2 recovery-unit train is re-formed in the methane re-forming unit to produce more feed to the Synthol unit. Lurgi's authothermal, catalytic, partial-oxidation re-formers are used for the reaction

$$CH_4 + H_2O \rightarrow 3H_2 + CO$$

In order to provide the heat for the reaction (which is highly endothermic), the feed is partially oxidized with oxygen. Also, as the preceding equation shows, steam is needed for the reaction. In addition to fulfilling the stoichometry of the reaction, excess steam is used for temperature control.

The re-former feed gas is heated from about 35 to 150 °C by heat exchange with the hot re-formed gas product. At this point, the feed is joined by high-pressure steam generated in the unit's quench boiler. The mixture cools slightly, but further heat exchange with the product streams raises the temperature of the steam and gas to about 650 °C. Partial bypassing of the heat exchanger allows for temperature control. The mixture is then introduced at the top of the reactor through mixing nozzles. Oxygen, with some sweep steam to prevent flashback, is also added here. The stainless steel nozzles are cooled with high-pressure water. The reactor is made of stainless steel; it is refractory-lined and water-jacketed.

A nickel catalyst[2] is employed, and the reaction temperature is about 1000 °C.

The re-formed gas has a composition such as that given in Table 1-13.[5,23] As can be seen, the H_2/CO ratio is 2.77. This is on the high side stoichiometrically for hydrocarbon synthesis, but it lessens carbon formation on the Synthol catalyst. After passing through the quench boiler, the feed-gas heat exchanger, a waste-heat boiler, and a scrubber, the re-formed gas enters the Synthol area.

There are eight partial-oxidation units to re-form the methane. They receive 8×10^6 ft^3/h (225,000 m^3/h) of feed gas that contains about 90 percent methane. Oxygen consumption is 120,000 ft^3/h (3400 m^3/h).

Downstream refining units

These are divided into the oil workup area and the chemical workup area. The processing units employed are for the most part quite conventional in nature. The chemical section, especially, contains nothing that is peculiar to coal processing or to synthetic fuels, and will not be discussed here.

The oil workup area, as outlined in Fig. 1-12,[5] produces the final transport-fuel products. Even though the streams are similar in composition to those in petroleum refineries,[24] this section is important enough to be reviewed briefly. In the light-ends fractionation unit, the noncondensable and water-insoluble products, namely C_3+ (from C_2 recovery), stabilized light oil, and decant oil, are fractionated to produce the following streams:

- C_3 stream that is fed to a catalytic polymerization unit
- C_4 stream that is fed to a C_4 isomerization unit for subsequent alkylation
- C_5–C_6 stream that is fed to an isomerization unit
- C_7–190 °C (375 °F) cut of naphtha that is fed to the naphtha hydrotreater and re-former unit
- Gas-oil cut that is fed to a distillate, or decant oil, hydrotreater
- Bottoms that are used as fuel oil

The coal naphtha and tar distillates are hydrotreated. In the catalytic polymerization unit, the mixed C_3–C_4 olefinic-rich feed is converted to gasoline and diesel blending components. The saturated C_3 and lighter material are fractionated from the polymerized product as LPG, and then the product is hydrotreated. The hydrotreated product is then split into gasoline and diesel components.

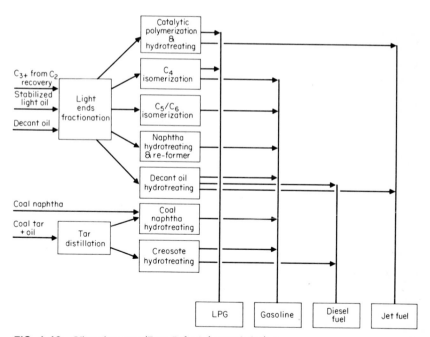

FIG. 1-12   Oil workup area *(From Ref. 26, by permission).*

The saturated C_4 material is utilized to adjust the vapor pressure of the finished gasoline. The C_5–C_6 isomerization unit improves the octane number of this gasoline blending stock. The naphtha hydrotreater and re-former unit accepts the combined C_7–190 °C (375 °F) naphtha and medium naphtha from the tar-distillation unit (creosote unit). The products are stabilized, and the re-formate added to the gasoline pool. The coal-derived naphtha from the gasifier cooling train is hydrogenated to remove sulfur compounds and diolefins. This highly aromatic cut is a valuable component in the gasoline pool.

The distillate (decant oil) hydrotreater is also a distillate selective cracker. The feed is hydrotreated, and the heavy portion of the hydrotreated material, with an end point of approximately 510 °C (950 °F), is cracked. Products from the unit are gasoline, jet fuel, and diesel fuel. Each of the fuels is separated and added to its respective pool.

Utilities: steam, power, oxygen plant, and water management

The coal fines (up to 30 percent of the input coal) go to the steam plant for process use and power generation.

The steam plant consists of six boilers, each able to deliver 1.2×10^6 lb/h (540 Mg/h) of steam at 430 °C and 615 psia (4.2 MPa). There are four large mills per boiler to pulverize the coal, and each boiler has six fans to provide combustion air.

Deutsche Babcock designed and built the boilers. Six 60-MW steam-driven turbo-generators are available for power generation, though the largest portion of Sasol Two's power requirement is purchased from the Electricity Supply Commission.

The oxygen plant was built by Air Liquide of France. The total capacity of the plant is about 250,000 stdft³/min (28,000 m³/min) of 98.5 percent purity oxygen at 493 psia (3.4 MPa). There are six identical production trains in the plant. Each train is composed of an air compressor delivering air at 105 psia (0.53 MPa), an Air Liquide cold box producing 2750 tons/day (2500 Mg/day) of oxygen and 13,100 stdft³/min (370 m³/min) of nitrogen containing 100 ppm of oxygen at 85 psia (0.395 MPa), and an oxygen compressor delivering oxygen at about 515 psia (3.4 MPa).

The air compressors (six) were supplied by GHH. They are synchronous direct-drive machines with variable-frequency start. Each is rated at about 36 MW.

The seven oxygen compressors (one is a spare) are manufactured by Sulzer and are also centrifugal induction motor–driven machines with across-the-line start. They are each rated at 14 MW.

The discharge lines of the six air compressors and the suction lines of the seven oxygen compressors are cross-connected, allowing great flexibility of operation.

Environmental regulations which must be adhered to in designing the plant require that no process-water effluents be discharged to the receiving stream. Sasol Two's design is based on zero discharge to local drainage systems. Adequate water for plant needs can be supplied, but reuse of water within the plant is mandatory.

A very simplified block flow diagram of water balance as engineered for the new plant is shown in Fig. 1-13.[21] The raw water supplied is that required to make up for losses of water, principally through evaporation in the cooling towers. Included are clean and contaminated storm waters, which are collected in ponds, and treated effluent water. The water system is divided into raw water, boiler-feed water, raw effluents, and ash-handling water. Treatment of effluents is difficult, considering that much of the effluent is water condensed from reaction products of gasifiers and Fischer-Tropsch synthesis reactors. Total water

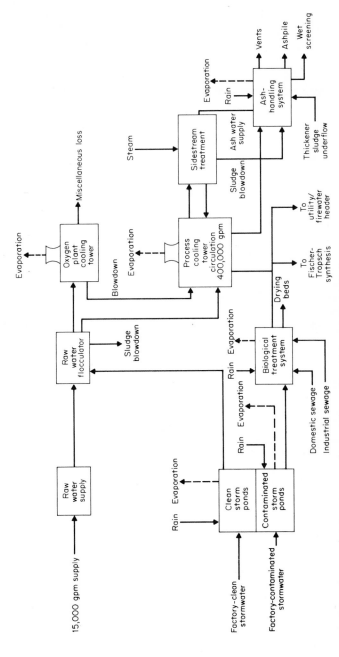

FIG. 1-13 Simplified flow diagram of water balance (*From Ref. 21, by permission*).

input to replace losses and evaporation is in the range of 14,000 to 15,000 gal/min (53 to 57 m³/min).

As noted earlier, the water-treatment system has been described in appreciable detail by Lurgi.[22]

START-UP AND MAINTENANCE EXPERIENCE

In a large complex like Sasol Two, the start-up procedure is a well-planned and lengthy program. Because the entire plant—unit by unit—consists of two or more trains, the basic philosophy of the start-up was to bring the independent portions of the plant on stream with full capacity as they were completed. As other sections were finished they were commissioned and brought to capacity.

The first string of gasifiers started break-in runs in the last months of 1979, and then gradually the other units were tested and made ready for operations.

It is appropriate to note that, except for sporadic mechanical failures which could be remedied in a short time, no major difficulties were encountered during start-up. These mechanical failures occurred mostly in areas such as coal preparation and water treatment, where extensive material handling was involved. Since only a short period has elapsed since the start-up, routine maintenance practice has yet to be established in detail.

Maintenance procedures were prescribed in advance, and the plant was designed with ample flexibility to permit these maintenance operations without interrupting regular production schedules. A well-trained maintenance crew that was trained at Sasol One is available, and it is ready to undertake the job at Sasol Two.

SASOL THREE[15-18]

Originally, Sasol Two's location was selected with the idea that coal supplies would be secure and land available when the time came for an adjacent plant to be erected. Sasol Three is practically a duplicate of Sasol Two. Except for minor improvements and two major changes, the plants are identical. These two major differences are the following:

- Sasol Three is not designed to use purchased electric power from outside.
- Creosote hydrotreating for maximizing diesel quality and quantity has been adopted.

Sasol Three was essentially mechanically finished by the end of 1981, and start-up began in 1982. Full production should be achieved in 1984 or early 1985.

ECONOMICS

Detailed information on product yields and product distribution of Sasol Two and Sasol Three constitutes proprietary data. In addition, South African law prohibits the dissemination or publication of any information concerning the production and consumption of liquid fuels. Therefore, a published study is used here to present the economics of the process. Mobil Research and Development Corporation produced the data in Table 1-14 for the U.S. Department of Energy (DOE) in 1978.[25]

TABLE 1-14 Summary data for U.S. Sasol-type plants

	Mixed output	All-liquid output
Output:		
SNG, MM stdft^3/d	173 .3	——
Gasoline, bbl/d	13,580	28,090
Other liquids fuels, bbl/d	6,011	13,140
Total liquids fuels, bbl/d	19,591	41,230
Total FOE, bbl/d	44,950	33,652
Input:		
Total coal used, 10^3 t/d	27 .8	27.8
Coal to steam plant, 10^3 lb/h	416	416
Coal to gasifier, 10^3 lb/h	1,901	1,901
Steam to gasifier, 10^3 lb/h	1,700	1,700
Oxygen to gasifier, 10^3 lb/h	458	458
Efficiency (HHV), %	57	44
Liquid fuels, bbl C$_4$+/ton dry coal	0 .92	1.94
Plant construction cost, MM $	1,186 .1	1,382.7
Capital cost, MM $	1,887	2,199
Gasoline cost, $/gal	1 .33	1.55

Fuels other than gasoline: Product	Price
SNG and C$_3$LPG	$6.17/MM Btu
C$_4$LPG‡	Gasoline — 0.30/MM Btu
Diesel fuel‡	Gasoline — 1.70/bbl
Fuel oil‡	Gasoline — 3.50/bbl
Alcohols	0.15/lb

*Fuel-oil equivalent assuming 6 MM Btu/bbl.

†All prices are October 1977 dollars.

‡Shown relative to gasoline price.

Source: From Ref. 25, by permission.

In this study, a Sasol-type technology was adapted for a U.S. location, and the economics was computed. Because at peak efficiency the combination of the dry-ash Lurgi process and the Synthol process yields over 50 percent of its energy output as methane, two product slates were considered: a mixed output and an all-liquid output. In the former, the methane was marketed as substitute natural gas (SNG) in addition to the liquid products; in the all-liquid output, the methane was re-formed and the resulting CO and H$_2$ were recycled to the Synthol reactors. As seen in Table 1-14 the all-liquid product operation mode is achieved with a considerable reduction in thermal efficiency (44 percent vs. 57 percent).

Another reference source[26] compares the maximum liquid production with SNG coproduction (Table 1-15), and it gives investment costs for various U.S. locations in mid-1979

TABLE 1-15   Maximized liquid production vs. coproduction of SNG

	Maximized liquid fuels Lurgi/ Sasol synthol	Coproduction of SNG* Lurgi/F-T synthesis	Coproduction of SNG* Lurgi/Mobil methanol conversion
Run-of-mine coal feed, t/sd†	40,000	40,000	40,000
Transport fuels, bbl/sd			
Gasoline	37,000	19,550	33,780
Jet fuel	8,120		
Diesel fuel	12,180	3,320	_____
Total	57,300	22,870	33,780
Production, bbl/ton of run-of-mine coal	1.45	0.57	0.85
Other fuels			
SNG, MM stdft^3/d	——	249.4	227.6
LPG, bbl/sd	4,000	1,800	5,760
Heavy fuel oil, bbl/sd	——	900	

*Data from cited report for DOE by Mobil R&D, factored up to 40,000 tons of coal feed per stream day.

† sd = stream day.

Sources: From Ref. 23, by permission.

dollars (Table 1-16). Note that this investment cost reflects 44,000 tons/day (40,000 Mg/day) of coal throughput. If the investment costs from Tables 1-14 and 1-16 are put on a common size basis, the results are in good agreement.

CONSTRUCTION MANAGEMENT AND INFRASTRUCTURE

The construction of Sasol Two and Sasol Three started a new era in the engineering and construction field. It signaled the birth of the multi-billion dollar project. Imagine a train

TABLE 1-16   Approximate estimates in mid-1979 dollars of Sasol-type project for U.S. location

	Installed capital cost, $ × 10^9
Sasol Two, moved to U.S.	
Gulf coast	2.5
Southern Illinois	2.75
Wyoming, Colorado, or New Mexico	3.2*
Sasol Two, modified for U.S.	
Wyoming, Colorado, or New Mexico	3.6*

*Camp and premium overtime allowance included.

Source: From Ref. 26, by permission.

with 10,000 freight cars, a train 100 miles long. That's what it would take to bring all the materials, equipment, and supplies to build Sasol Two—640,000 tons of steel, pipe, valves, electrical cable, vessels, pumps, computers, exchangers, and instruments.

As noted earlier, a new town called Secunda was created for Sasol Two. In the 1980s, Secunda will have a population of more than 30,000.

An optimistic version of the overall schedule for a typical Sasol Two–type liquefaction plant is given in Fig. 1-14.[26] The personnel requirements for home office work are given in Fig. 1-15[26] and for construction work in Fig. 1-16.[26] (Note that these figures represent typical U.S. installations and do not necessarily coincide with Sasol Two or Sasol Three.)

These figures suggest the importance of control of schedules that must be coordinated with delivery dates of critical equipments. Sophisticated management tools, such as critical path scheduling, PERT, and other techniques, were implemented at Sasol Two and Sasol Three by project teams to assure that the job was carried out on schedule and on budget.

Special emphasis was put on the careful planning of the infrastructure. *Infrastructure* can be defined as the necessary support facilities needed to accompany a "megaproject." Adverse socioeconomic effects that can result from such a plant must be minimized. The socioeconomic effects of energy conversion facilities in any one geographic area are directly related to the number of construction and operating workers employed by these complexes.

Although the exact number of employees varies according to the type of facility, a typical 250MM-stdft^3/day coal-gasification plant is used in the following discussion to demonstrate the impact of a large plant.

Labor and Population Projections

Table 1-17 shows the growth in population that a synfuels plant creates.[27] The construction labor is temporary and decreases as labor requirements for operation and maintenance increase. Secondary employment is generated by the services required by the personnel employed at the energy conversion facility. This typically ranges from 1.5 to 2.5 times the facility labor requirement.

In order to mitigate the population influx and service requirements, a dialogue must be initiated at the commencement of facility planning with the local leaders in government and the private sector. The objective of this early dialogue is to avoid a breakdown in services. The developers of the facilities, as well as local business and government officials, must develop detailed plans to manage socioeconomic impacts resulting from the growth associated with the facilities.

Services

Table 1-18 lists the key services that are immediately affected by a major population influx.[27] One of the most favorable socioeconomic impacts of an energy conversion facility is the increased tax base provided by such a capital-intensive project. The increased tax base results in additional money being available for social services to the surrounding area.

The incomes of the personnel employed are usually above average, because this type facility requires a high proportion of skilled workers. These high-income individuals further increase the tax base because of their demand for more expensive housing. Although a lag normally occurs between the influx of population and the provision of social services, the

2-38

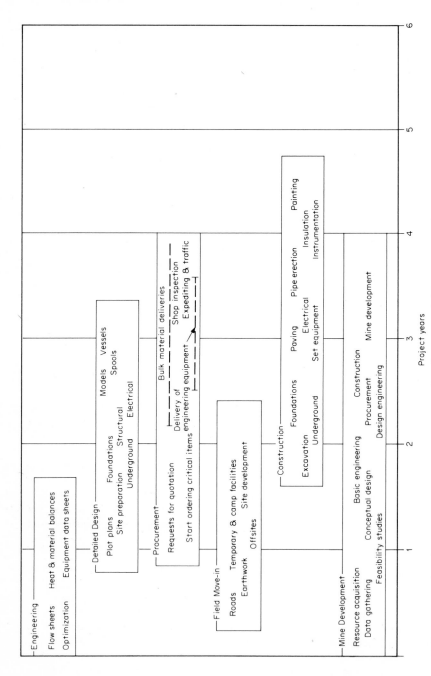

FIG. 1-14 Synfuels from coal: project planning—schedule (liquefaction plants) (*Fluor Corporation*).

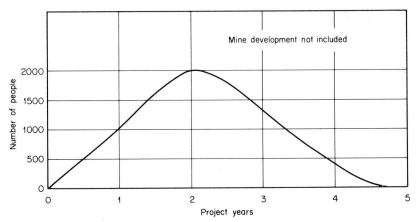

FIG. 1-15 Synfuels from coal: project planning—office workforce for one liquefaction plant *(Fluor Corporation).*

eventual result is the availability of additional money to increase social services for both the existing and newly arrived populations.

Ideally, mechanisms should be developed to provide funds for needed facilities before the people actually arrive. Proper cooperative planning, with suitable tax credits to the developers, is possible because of the long lead time between project commitment and field activity.

Employment

Increased employment is another favorable impact created by the construction of an energy conversion facility. The building of the facility employs numerous construction workers

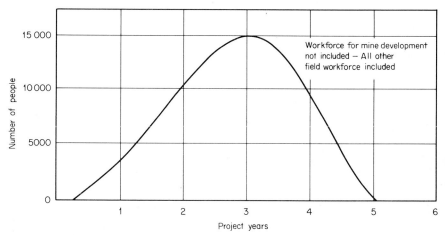

FIG. 1-16 Synfuels from coal: project planning—construction workforce for one liquefaction plant. *(Fluor Corporation).*

TABLE 1-17 Labor and population projections for a gasification facility*

Construction, mine and plant	1,000 + 11,000(temporary)
Operation and maintenance, mine and plant	500 + 650(temporary)
Secondary†	2,415
Total new jobs (operation and maintenance, mine and plant)	3,565
Population increase (approximate)‡	11,000

*Strip mining.

†Value is 2.1 times the facility's permanent labor requirement (1150 persons).

‡Value is approximately 3 times that of the total new jobs created.

Source: From Ref. 27 with permission.

from the local and surrounding areas. The construction contractor for the facility often conducts training classes to upgrade the skills of those workers employed. The local workers receiving this training will have opportunities to obtain positions in the permanent maintenance forces required by the facility. Engineering, laboratory, administration, maintenance, and operating positions become available as staff is required for operation of the plant. Additional employment is also available through the establishment of satellite industries needed to perform maintenance and to handle supplies for the energy conversion facility. Service facilities, such as stores, laundries, and automobile repair shops, are also expanded to provide further employment opportunities.

Community planning mitigates many of the impacts associated with the population expansion. Experience has shown that people are much happier in a thriving, expanding environment because of increased employment opportunities and increased spending capacity. The unfavorable impacts are usually temporary in nature, and adjustments to inconveniences are known to be temporary. Economics dictates the need to be realistic during the community development planning phase to avoid excessive expenditures on short-term impacts.

Land Use

Realistic appraisals usually result in assessments of land use that favor the installation of the energy conversion facility. Most land-use conflicts are the result of the value placed on land because of flora, fauna, or aesthetic considerations. Consideration of the productivity of usable materials per acre of land usually indicates that the installation of an energy

TABLE 1-18 Services affected

Electricity	Police
Gas	Fire protection
Water	Health care
Sewer	Education
Solid waste	Recreation
Telephone	Transportation

conversion facility provides improved land use. Land unfit for agriculture is utilized as building sites for homes, businesses, and the energy conversion facility. Restoration of the land used for coal mining may result in better agricultural land than the original overburden was, especially in areas where slopes are leveled to provide tillable land.

Housing

The most noticeable socioeconomic impact associated with an energy conversion facility is the need to provide sufficient housing accommodations for the construction workers. This impact may be mitigated effectively through the development of a temporary construction community, usually built by the owner of the facility. This construction community could be self-sufficient with regard to all of the services for personnel. Therefore, the local community would not be required to provide any extensive services for construction workers.

The permanent work force for the facility requires the addition of housing stock in the community. These houses usually are provided by local developers to meet the community housing demand.

This discussion is not meant to make light of the social and environmental impacts of the installation but rather to address the logic for obtaining local support.

SASOL TWO AND SASOL THREE—PROTOTYPES OR LAST OF A KIND?

The Sasol complex is a major milestone on the road toward the evolution of a global synthetic fuels industry. Its significance is in proving that the World War II technology utilized on a "mini" scale in Germany could be improved, scaled up, and implemented for use in our modern era. It is also clear to all involved in the development of synfuels that Sasol is just the start. As such, processes will have to be further improved and adjusted to the specific needs of various markets and to the conversion characteristics of the available feedstocks.

One should also keep in mind that coal-derived liquid fuels will never fully replace petroleum-based products, just gradually supplement the dwindling crude oil supply. Since coal deposits are so widespread, there is virtually no probability of an OPEC-like monopoly. If those who possess coal can emulate the example of South Africa with respect to foresight and determination, they can provide themselves with a measure of energy independence. Because most of the costs in synfuels ventures, in developed countries at least, are expended within the national economy, one can postulate that the net cost to the given society is but a fraction of the apparent "bookkeeping" cost. A good argument can therefore be made that the development of a synfuels industry will put a "cap" on the petroleum prices set by OPEC. Indeed, the threat to the oil marketers that is implicit in building a new facility that will produce a substitute material for 20 to 30 years could well exert downward pressure on petroleum prices.

On the other hand, the present outlook is somewhat dim for synfuels. Many energy seers forecast that the apparent excess of supply will persist till the end of the decade. Little consideration is given, however, to either the extent of our vulnerability to supply interruptions or to the fact that it would take until beyond the end of the decade to install any significant synfuels capacity.

REFERENCES

1. "Special Survey Supplement," *Financial Mail*, Nov. 16, 1979.

2. Thurston, Rodney S.: "Synfuels from Coal-Lessons from South Africa," *LASL Mini-Review*, University of California, Los Alamos Scientific Laboratory, LASL 80-13, July 1980.

3. "Sasol in a Nutshell," Sasol Ltd., Sasolburg, So. Africa, 1980.

4. Johnson, R. W.: "South Africa's Sasol Project: How to Succeed in Synfuels," *Coal Min. Process*, February 1981, pp. 42–45.

5. Pay, T. D.: "Sasol—The Commerical Experience," International Energy Technology Assessment Project, Lawrence Livermore National Laboratory Subcontract no. G2B-13837C, U.S. Department of Energy Contract no. W-7405-eng-26, November 1980.

6. De Villiers, D. P.: "Oil and Gas from Coal in South Africa," *Australian Gas Association 14th Annual Convention, Canberra, Australia, September 1975*.

7. Dry, D. M. E., and J. C. Hoogendoorn: "Technology of the Fischer-Tropsch Process," *Catalysis and Surface Science Conference, Berkeley, California, July 1980*.

8. Hoogendoorn, J. C.: "Conversion of Coal into Fuels and Chemicals in South Africa," *Third International Coal Conference, Sydney, Australia, October 1976*.

9. ———: "The Sasol Story," *American Institute of Mining, Metallurgical, and Petroleum Engineers*, Dallas, Feb. 24, 1974.

10. ———: "Experience with Fischer-Tropsch Synthesis at Sasol," *IGT* [*Institute of Gas Technology*] *Symposium Papers, Clean Fuels from Coal*, Chicago, September 1973.

11. "Coal to Cobalt," *Encyclopedia of Chemical Processing and Design, 1979*, Marcel Dekker, New York, 1979, chap. 9.

12. Rudolph, P. F. H.: "The Lurgi Process. The Route to SNG from Coal," *Fourth Synthetic Pipeline Gas Symposium, Chicago, October 1972*, pp. 175–214.

13. Brink, A.: "Fischer-Tropsch Synthesis Sasol: Past, Present and Future," *Synfuels' First Worldwide Symposium, Brussels, Belgium, October 1981*.

14. Roberts, H. L.: "Oil from Coal: The Engineering Challenge," *S. Afr. Mech. Eng.*, vol. 25, 1975, pp. 36–41.

15. Hoogendoorn, J. C., and S. B. Jackson: "Sasol Projects in South Africa," *Coal Technology '79*, Houston, November 1979.

16. Samuel, William A.: "Sasol—A Proven Prescription to Convert Tons to Barrels," *8th Technology Conference*, Washington, D.C., March 1981.

17. Samuel, William A.: "Sasol—An Update," *Coal Technology '80*, Houston, November 1980.

18. Mako, Peter F.: "Coal-Based Synfuels Project Overview, South Africa," *Coal Technology Europe '81*, Cologne, Federal Republic of Germany, June 1981.

19. "Synthetic Fuels the Sasol Way," *Eng. Min. J.*, December 1979.

20. Hoogendoorn, J. C.: "Gas from Coal for Synthesis of Hydrocarbons, Status of Sasol II," *Ninth Synthetic Pipeline Gas Symposium*, Chicago, November 1977.

21. Kronseder, J. G.: "Sasol II: South Africa's Oil-from-Coal Plant," *Hydrocarbon Process.*, July 1976, pp. 56ff.

22. Rolke, D. E.: "Treatment of Waste Water from a Lurgi Gasification Plant," *Coal Technology '80*, Houston, November 1980.

23. Hoogendoorn, J. C., and J. M. Salomon: "Sasol: World's Largest Oil-from-Coal Plant—Parts I,

II, III, and IV," *Br. Chem. Eng.*, May 1957, pp. 238–244; June 1957, pp. 308–312; July 1957, pp. 368–373; August 1957, pp. 418–419.

24. Joiner, J. R., and J. J. Kovach: "Sasol Two and Sasol Three—Versatility in a Proven Route to Synfuels," *AIChE 1981 Annual Meeting*, New Orleans, November 1981.

25. Schreiner, Max: "Research Guidance to Assess Gasoline from Coal by Methanol-to-Gasoline and Sasol-Type Fischer-Tropsch Technologies, Final Report," Mobil Research and Development Corporation, U.S. Department of Energy Report no. FE-2447-13, 1978.

26. "A Fluor Perspective on Synthetic Liquids—Their Potential and Problems," Fluor Engineers and Constructors, Inc., July 1979.

27. "Briefing Guide on Synthetic Fuels," Fluor Engineers and Constructors, Inc., 1981.

THE ICI LOW-PRESSURE METHANOL PROCESS

P. L. ROGERSON, Ph.D.

Imperial Chemical Industries, Ltd.
Cleveland, England

GENERAL PROCESS DESCRIPTION

The ICI (Imperial Chemical Industries) low-pressure methanol process takes a suitably purified mixture of hydrogen, carbon monoxide, and carbon dioxide and converts it into methanol using a copper-based catalyst operating at medium temperatures and pressures. The single-pass conversion of hydrogen, carbon monoxide, and carbon dioxide to methanol is limited by equilibrium, and therefore, a recycle system is adopted to raise the conversion efficiency to acceptable levels. The crude methanol product is purified in a distillation system to the required quality.[1]

A complete plant for producing methanol comprises four basic process units as shown in Fig. 2-1.

1. A plant for producing hydrogen, carbon monoxide, and carbon dioxide. This may be a steam re-former or partial-oxidation unit combined with an appropriate purification system. The feedstock may be natural gas, naphtha, heavy fuel oil, coal, or similar materials.

2. A compression plant. This unit is driven by process steam generated by waste heat liberated by other process units or by electricity internally generated or imported. A compression plant may not be required if high-pressure partial-oxidation gasifiers are adopted.

3. A methanol-synthesis plant.

4. A distillation plant.

PROCESS CHEMISTRY AND THERMODYNAMICS

Methanol Synthesis

The synthesis gas contains H_2, CO, and CO_2, together with inert gases such as CH_4, N_2, and Ar. The major reactions which take place in a methanol-synthesis converter are

$$2H_2 + CO \rightleftharpoons CH_3OH \qquad -90.7 \text{ kJ/mol} \qquad (1)$$

$$3H_2 + CO_2 \rightleftharpoons CH_3OH + H_2O \qquad -49.5 \text{ kJ/mol} \qquad (2)$$

$$CO_2 + H_2 \rightleftharpoons CO + H_2O \qquad +41.2 \text{ kJ/mol} \qquad (3)$$

Equilibrium data for reactions (1) and (3) are shown in Fig. 2-2.[2]
Side reactions take place; the principal ones are illustrated below.

$$2CH_3OH \rightleftharpoons CH_3OCH_3 + H_2O \qquad (4)$$

$$H_2 + CO \rightleftharpoons HCHO \qquad (5)$$

$$2nH_2 + nCO \rightleftharpoons C_nH_{2n+1}OH + (n-1)H_2O \qquad (6)$$

Ethers, formates, high alcohols, and hydrocarbons are all detected in the crude methanol product. The formation of these compounds is kinetically limited and entirely dependent on the selectivity of the catalyst.

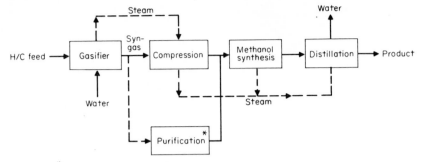

FIG. 2-1 ICI methanol process flow schematic. Purification step occurs only in partial-oxidation plants.

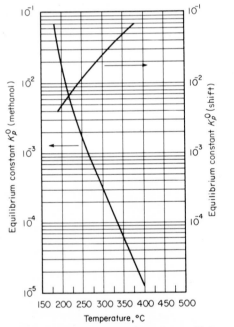

FIG. 2-2 Methanol and shift reaction equilibrium constant: K_p^0 for $CO + 2H_2 \rightleftharpoons CH_3OH$ and $H_2 + CO_2 \rightleftharpoons CO + H_2O$. [Data Sources: *A. V. Frost*, Zh. Obshch. Kim., *vol. 1, 1931, pp. 367-376; R. H. Newton and B. F. Dodge*, J. Am. Chem. Soc., *vol. 56, 1934, pp. 1287-1291; R. H. Ewell*, Ind. Eng. Chem., *vol. 32, 1940, pp. 149-152; W. J. Thomas and S. Portalski*, Ind. Eng. Chem., *vol. 50, 1958, pp. 967-970;* ICI (shift data).]

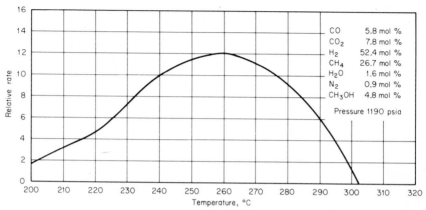

FIG. 2-3   Maximum rate curve.

The true equilibrium conversion for reaction (1) can never be achieved in practice since the reacting gases and products are not in contact with the catalyst for an infinite period of time. Fig. 2-3 shows the effect of temperature on the rate of methanol synthesis occurring over a methanol catalyst operating at the conditions shown. At low temperatures the reaction is controlled kinetically and is displaced a long way from equilibrium. As the temperature increases, the reaction rate also increases according to an Arrhenius function such that the true equilibrium state is more closely approached. As the temperature is increased further, the amount of methanol formed is controlled by the equilibrium which, as shown in Fig. 2-2, falls away with increasing temperature. The maximum rate of methanol synthesis is governed by the activity of the catalyst, and the higher the activity the lower the temperature at which the maximum product rate is achieved.

As discussed previously, the measured equilibrium conversion for a gas leaving the catalyst will be less than the true equilibrium conversion measured at infinite contact time, and this difference is expressed in terms of temperature. The measured equilibrium constant is called the *pseudo-equilibrium constant.*

The conversion of carbon monoxide and carbon dioxide to methanol through the reaction with hydrogen is an important parameter in determining operating and capital costs.[3] The so-called carbon efficiency is:

$$\frac{(\text{Moles of methanol produced}) \times 100}{\text{Moles of } (CO + CO_2) \text{ in synthesis gas}}$$

The effect of pseudo-equilibrium temperature on carbon efficiency is shown in Fig. 2-4 for different overall synthesis pressures. Figure 2-4 is based on a typical set of synthesis-loop conditions using synthesis gas derived from the steam re-forming of natural gas. A different set of curves would be obtained for synthesis gas derived from the partial oxidation of coal or heavy fuel oil where the inert gas concentration is very low and the CO/CO_2 ratio is high. In these cases the carbon efficiency would be greater than 95 percent at 750 lb/in².

The process for producing methanol from synthesis gas is essentially one of energy conversion. In order to obtain some measure of its efficiency, both the first and second laws of thermodynamics have to be considered, so as to take into account the quality and quantity of heat exchanged.

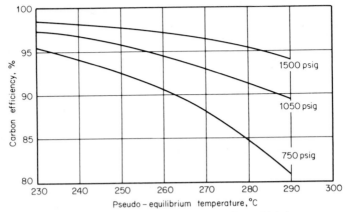

FIG. 2-4 Carbon-conversion efficiency versus pseudo-equilibrium temperature at three pressures. (*Excerpted by special permission from* Chemical Engineering, *July 4, 1977, copyright © 1977 by McGraw-Hill, Inc., New York, N.Y. 10020.*)

The synthesis reactor is quenched at various stages by the addition of cool recycle gas. Using the Carnot principle, the efficiency of this step may be expressed in the following way:

$$\frac{W_{net}}{Q} = \frac{T_1 - T_2}{T_1}$$

where Q is the heat energy supplied to the engine
W_{net} is the shaft work produced
T_1 is the hot-sink temperature
T_2 is the cold-sink temperature
The net work involved in quenching the reaction is not recoverable in practice. However this loss can be minimized by reducing the temperature difference $T_1 - T_2$. T_1 is generally fixed by the catalyst operating conditions, but T_2, or the cold-shot temperature, can be raised using low-grade heat in the synthesis loop. The thermodynamic efficiency of the conversion process therefore can be increased by using warm shot, thus enabling more heat to be recovered from the synthesis loop.

Methanol Distillation

Generally, high-purity methanol is required for chemical purposes. Therefore, the crude product must be purified, and this is generally carried out by distillation. There are two broad classes of by-products: those which have a higher relative volatility than methanol (and include dissolved gases, dimethyl ether, methyl formate, hydrocarbons, and carbonyl compounds) and those with lower relative volatilities than methanol, such as water and higher alcohols. The distillation unit, therefore, comprises two main sections: one for removing low-boiling impurities and the other for separating methanol from the high-boiling fractions. The thermodynamic principles of distillation are well known and will not be discussed here.[4]

METHANOL-SYNTHESIS CATALYST—GENERAL DESCRIPTION

The ICI methanol-synthesis catalyst is manufactured as a mixture of copper, zinc, and mixed zinc-aluminum oxides. In operation, the copper oxide is reduced to metallic copper in the form of fine crystallites. The ICI catalysts are covered by a number of patents which contain a full description of the catalysts.[5]

It has been thought generally that copper is the active constituent in low-temperature methanol-synthesis catalysts.[6] Copper metal has a relatively low melting point of 1028 °C, and the metal crystallites easily sinter at elevated temperatures. It can be shown experimentally that the activity of copper-based methanol-synthesis catalysts is inversely proportional to the copper crystallite size. Figure 2-5 illustrates the relationship between activity and copper crystallite size using x-ray diffraction line–broadening measurements of Cu_{111} and Cu_{200}. Slight curvature at the upper ends of the graph indicates the onset of internal diffusion.

The formulation of the catalyst support is extremely important in the preparation of an active and stable catalyst. Many years of experimentation have shown that the spinel $ZnAl_2O_4$ is a highly stable support material. Zinc oxide by itself is not satisfactory, but the addition of alumina or chromia considerably improves its support properties. Indeed, the early catalysts of ICI were formulated as such with no spinel, and they performed satisfactorily for many years over a limited pressure range. More recently, it has been discovered

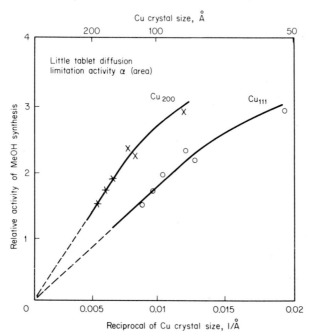

FIG. 2-5 Relation between copper crystal size and activity of low-temperature methanol-synthesis catalyst tablets prepared by differing procedures from a single Cu-Zn-Al-containing precipitate.

that the support material itself is involved in the catalytic process so that there is a catalytic interaction between copper metal and free zinc oxide.

Copper catalysts are sensitive to sulfur- and chlorine-containing compounds, both of which react with copper and zinc oxide. Fresh catalysts are better absorbents than old catalysts; this fact indicates that the structural changes which occur with time affect the absorption properties of the catalyst. Also, different formulations have different resistances to poisons. Catalyst 51-2 used in the ICI process has good resistance to poisoning, and its resistance is assisted by the presence of free zinc oxide.

Many of the by-products formed during methanol synthesis are thermodynamically more stable than methanol itself. Therefore, selectivity for methanol synthesis is gained from the catalyst itself and is not the result of thermodynamic stability. The selective properties of the catalyst can be disrupted by the presence of impurities like iron or nickel, both of which promote hydrocarbon synthesis, or by alkalies such as sodium hydroxide which promote higher-alcohol synthesis. Silica can also promote the formation of higher-molecular-weight organic compounds. It can be shown that the catalytic effects of various impurity components depend very much on where and in what form they are located on the catalyst, and therefore the influence of a given amount of impurity is very dependent on the details of the catalyst-formulation method. Figure 2-6 shows the bad effects of support components on low-temperature methanol-synthesis catalysts.

The catalyst used in the ICI methanol process has been used widely around the world. Its average life is between 3 and 5 years, and it shows good selective properties.

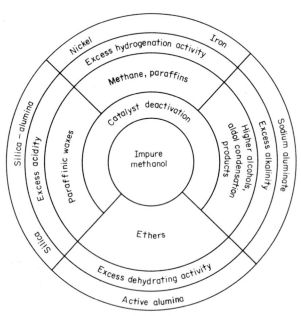

FIG. 2-6   Bad effects of support components in low-temperature synthesis catalysts.

PROCESS PERSPECTIVE

Technology Transfer

ICI's methanol technology may be acquired from a number of internationally known contractors, appointed as licensees of the technology by ICI. These contractors have been given both the means to carry out the process design of the plant and the critical engineering details of important items of equipment such as the converter. The operating and engineering experience of ICI has been made available and is always maintained up-to-date in relation to this process. The contractors have a considerable amount of their own experience which is used in the design and construction of methanol plants.

The following organizations are licensed contractors of the ICI low-pressure methanol process:

Davy McKee Ltd.

Foster Wheeler Corp.

Humphreys & Glasgow Ltd.

Lummus Industries, Inc.

Technip

Toyo Kogyo Co. Ltd.

Friedrich Uhde GmbH

Users

A list of companies who have adopted the ICI low-pressure methanol process can be found in Table 2-1.

Future Developments

There are two main areas of future development, both closely interrelated:

Process design

Much work has been carried out on the process-design development of the ICI low-pressure methanol process. This work continues for the standard designs based on the steam re-forming of hydrocarbons and on coal-based plants.

In coal-based plants the raw synthesis gas is carbon-rich and requires adjustment to achieve stoichiometry. It can be shown that in circumstances where the raw gas is carbon-rich, the maximum amount of methanol that can be made is $(H_2) + (CO)/3$ where (H_2) and (CO) signify kilogram moles per hour of H_2 and CO in the raw synthesis gas. Similarly, in hydrogen-rich gases derived from the steam re-forming of hydrocarbons, the maximum amount of methanol that can be made is $(CO) + (CO_2)$ where (CO) and (CO_2) are kilogram moles per hour of CO and CO_2 in the synthesis gas. In both cases standard designs can

TABLE 2-1   ICI low-pressure methanol process user list

Operator	Country	Size, t/d*	Operator	Country	Size, t/d*
AECI	South Africa	55	Air Products	U.S.A.	550
Alberta Gas I	Canada	600	Alberta Gas II	Canada	600
Alberta Gas III	Canada	1200	Almer	Algeria	375
Arco	U.S.A.	2000	Borden Chem.	U.S.A.	1900
Celanese	Canada	2300	Celanese	U.S.A.	1800
Cepsa	Spain	660	Chang Chun Pet.	Taiwan	140
CNTIC	China	330	CPDC	Taiwan	310
Elf	Germany	880	Georgia Pacific	U.S.A.	1000
GCFC	India	66	Gulf Pet. Ind.	Bahrain	1100
ICI I	United Kingdom	660	ICI II	United Kingdom	1430
Metanor	Brazil	165	MCN	Netherlands	1100
Methanor	Netherlands	1100	Monsanto	U.S.A.	1000
MSK	Yugoslavia	660	NEC	Trinidad	1320
NMC	Libya	1100	Nishi Nihon	Japan	1100
Ocelot Ind.	Canada	1320	Pemex	Mexico	2750
Petralgas	New Zealand	1320	Romchin	Romania	670
SABIC	Saudi Arabia	2300	SIR	Italy	500
Taesung Lumber Co.	South Korea	165	Taesung Methanol Co.	South Korea	1100
USSR I	U.S.S.R.	2750	USSR II	U S.S.R.	2750

*Short tons/day

achieve over 95 percent of the theoretical maximum; hence there seems to be little possibility of improving process-feedstock efficiency. Areas in which improvements should be sought would appear to be those requiring fuel for direct heating and steam raising, such as in the steam re-former itself. In coal-based plants, steam raising requires auxiliary boiler capacity to provide compression power and process steam from the shift reaction. Also considerable power is required for the CO_2 removal plant. It is probably in these areas that most improvements can be made.

Recovery of heat from the synthesis reaction is now commonly practiced, and most plants are designed to effectively recover all the useful heat available.

Catalysts

Better methanol-synthesis catalyst activities with longer lives would have significant effects on the overall economics of the process. Higher activities would mean smaller converters per unit output and thus would make possible the design of larger single-stream units. Lower synthesis pressures could be achieved while still maintaining high carbon or hydrogen efficiencies. Power costs could be cut as a result, a matter of particular relevance in coal-based methanol plants which use low-pressure gasification systems.

In summary, there would be important improvements in design which would lead to energy savings but perhaps more significantly to capital-cost reduction.

Methanol World Capacity and Expected Growth

The expected growth in world methanol capacity up to 1986 but excluding Comecon countries is shown in Fig. 2-7. This curve is based on known projects, many of which have

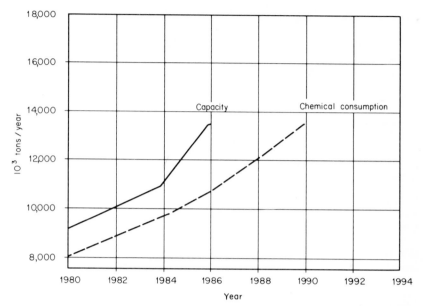

FIG. 2-7   Expected growth in consumption and capacity in the world except Comecon countries.

already been sanctioned but not constructed. About 20 percent of this growth is expected to supply the synfuel market.

The expected growth of the traditional chemical markets is also shown in Fig. 2-7. A growth rate of 5 percent has been assumed, since this has been the average experienced in the last few years. The growth in methanol markets for synfuels is not included. It is expected that substitutes for dwindling natural resources will be required, but the time frame is still a matter of conjecture.

DETAILED PROCESS DESCRIPTION

Gas Production

Steam re-forming route

Natural gas or naphtha is an ideal feedstock for manufacturing methanol. The steam re-forming of natural gas produces a synthesis gas which is rich in hydrogen. If carbon dioxide is freely available, for example, from an adjacent ammonia plant, then it is economical to add carbon dioxide to a synthesis plant to react with the excess hydrogen. If carbon dioxide is not available, then the most economical solution is to feed the synthesis gas direct to the synthesis unit and to remove the excess hydrogen as a large purge, which can be used as a fuel.[7] This arrangement is cheaper in terms of both capital and running costs than a scheme involving carbon dioxide recovery from flue gas.

The steam re-forming of naphtha produces a synthesis gas which contains approximately the correct proportion of hydrogen, carbon monoxide, and carbon dioxide. A small purge is taken from the synthesis loop to remove methane which remains unconverted by the re-former.

Partial-oxidation route

Heavy fuel oil or coal can be partially oxidized to produce synthesis gas consisting of hydrogen, carbon monoxide, and carbon dioxide. These feedstocks contain considerable quantities of sulfur and produce a synthesis gas rich in carbon. For economical operation, it is necessary to remove the excess carbon and, at the same time, purify the gas to remove sulfur compounds, which are catalyst poisons. This is achieved by directing the gas to a purification stage which, in the same process, removes sulfur compounds and the excess carbon as carbon dioxide. For this purpose, there are available several acid-gas removal processes which operate by physical or chemical absorption. In order to ensure a high-quality synthesis gas as well as to guard against the breakthrough of sulfur, a final catalytic absorption stage is recommended.

The partial oxidation of heavy fuel oil can be carried out at the methanol-synthesis pressure, and therefore no compression stage is needed. Coal-gasification processes designed to operate at the synthesis pressure are being developed.

Compression

Synthesis gas is compressed to between 30 and 100 atm before it is admitted into the synthesis loop. A centrifugal compressor, which can be used for all but the smallest plants, is driven by a direct-coupled turbine, giving a low capital cost per unit horsepower.

Steam is generated by the re-former (or by waste heat from the partial-oxidation-unit), at a pressure between 60 and 120 atm, and it is then let down through the compressor and circulator turbines. At intermediate pressures the steam is utilized in the gas-producing process, while low-pressure steam passing out of the turbines reboils the distillation columns. The overall effect is to convert heat into mechanical energy at an efficiency of around 90 percent.

Methanol Synthesis

The methanol-synthesis recycle loop contains six main plant items: circulator, converter, feed/effluent heat exchanger, heat-recovery exchanger, cooler, and separator.

The converter, containing the copper catalyst, is the principal main process item. The reaction takes place over a series of catalyst beds. Interbed cooling is achieved by the addition of quench gas or by indirect gas-heat exchange or a combination of both.

Distillation

The low-pressure methanol process produces crude methanol of considerably higher purity than the crude methanol from the old high-pressure process. Distillation-plant design employs two columns, the first of which strips volatile impurities (dimethyl ether, esters, ketones, and iron carbonyl) and the second of which removes water and higher alcohols.

By distillation of low-pressure crude methanol in this way, refined methanol can be produced easily with a purity higher than specified by U.S. federal grade AA.

A fuel-grade product may be produced by a single column which effectively removes all or part of the water as required. Organic impurities are retained in the product.

Complete Plant Flow Sheet

A flow sheet showing the main flow streams for a plant based on the steam re-forming of a hydrocarbon feedstock is shown in Fig. 2-8. Extensive heat recovery and use are achieved in the re-former, synthesis, and distillation areas. Figure 2-9 shows a similar flow diagram for a plant based on the partial oxidation of either heavy fuel oil or coal.

In order to maximize energy recovery, some of the heat generated by the methanol synthesis reaction is utilized to preheat boiler feedwater.[8,9] The heat recovered in this way is used for raising high-pressure steam, and thus high thermodynamic efficiency is achieved. The same reaction heat can also be used to heat water for saturating natural gas feedstock or unshifted synthesis gas in partial-oxidation–based plants.[10,11] With these systems more heat can be upgraded, with the effect of reducing the overall requirement for medium- and high-pressure steam.

Optimization of Mass and Energy Balance

The optimization of the ICI low-pressure methanol process has been discussed in a number of papers.[3,12]

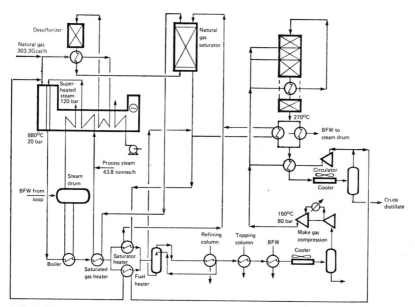

FIG. 2-8   Schematic flow sheet for 1000 tons/day of AA grade methanol.

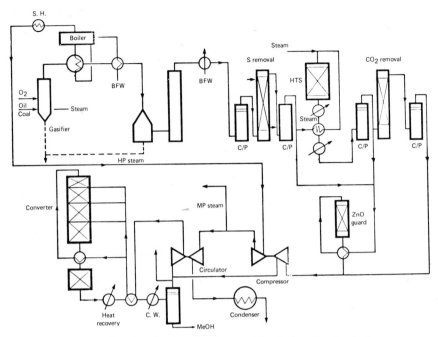

FIG. 2-9   Schematic flow sheet for plants based on partial oxidation of coal and oil.

The carbon efficiency is dependent on pressure and temperature (Fig. 2-4) and the CO/CO_2 ratio in the feedstock. Hence for a fixed converter-outlet temperature, the process feedstock requirements fall as the synthesis pressure is increased (Fig. 2-10a and b). However, fuel requirements increase, since more compression power is required. Figure 2-10a and b shows the combined effects on feed plus fuel requirements for a plant based on natural gas and naphtha. For partial-oxidation–based plants and some plants using high-temperature coal gasification, the inert gas concentration in the synthesis loop is low and the CO/CO_2 ratio is high. Thus the carbon efficiency is already in the range of 95 to 99 percent at 50 atm so that pressure has little effect on the process feedstock requirements. Therefore partial-oxidation plants and some coal-based plants tend to operate at lower synthesis pressures to save power and fuel costs.

From a thermodynamic point of view, the lower the exit temperature of the methanol converter, the higher the carbon efficiency, but the catalyst volume and the size of the reactor increase. Also the heat recovery may be significantly affected as shown in Fig. 2-11. The beneficial effects of a lower converter temperature are usually exploited during the early life of the catalyst.

The design of the synthesis loop has a significant influence on the overall capital costs of a plant. Fig. 2-12 shows the effect of synthesis pressure and temperature on differential capital cost; this effect results from changes in the loop carbon efficiency and its influence on the size of the gasification plant (a steam re-former in this case).

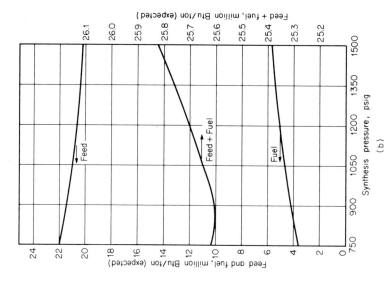

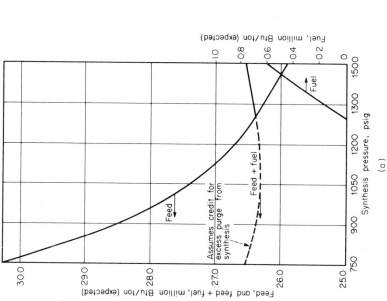

FIG. 2-10 (*a*) Synthesis pressure vs. feed and fuel requirements for natural gas. (*b*) Synthesis pressure vs. feed and fuel requirements for naphtha. (*Excerpted by special permission from Chemical Engineering, July 4, 1977, copyright © 1977 by McGraw-Hill, Inc., New York, N.Y. 10020.*)

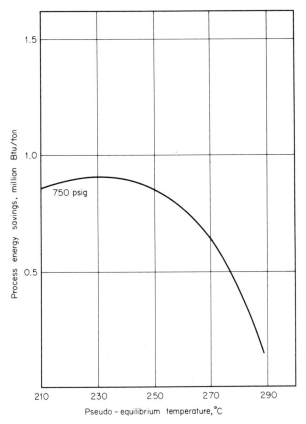

FIG. 2-11 Process energy savings vs. pseudo-equilibrium temperature. (*Excerpted by special permission from* Chemical Engineering, *July 4, 1977, copyright* © *1977 by McGraw-Hill, Inc., New York, N.Y. 10020.*)

Converter Design

Two basic converter designs are used in the ICI low-pressure methanol process. In Fig. 2-13 (p. 2-62) the converter uses distributors of ICI-patented design,[13] buried within the catalyst at different levels. These distributors allow the free passage of catalyst for easy charging and discharging but at the same time permit good mixing.

The ICI distributors, often referred to as "lozenges" because of their cross-sectional shape, have been adopted for all plants in operation or under construction up to this time. The construction of a lozenge is shown in Fig. 2-14 (p. 2-63). At a predetermined depth within the catalyst, a series of lozenges lies across the converter along a set of parallel chords. The separation of the lozenges is great enough to permit the flow of catalyst between each one, and the slope of the faces is greater than the angle of repose of the

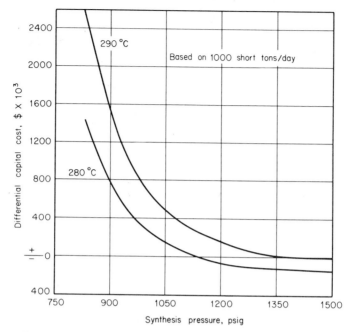

FIG. 2-12 Differential capital cost vs. synthesis pressure at two temperature levels. (*Excerpted by special permission from* Chemical Engineering, *July 4, 1977, copyright © 1977 by McGraw-Hill, Inc., New York, N.Y. 10020.*)

catalyst. This permits the catalyst to flow down the sides of the lozenges and also eliminates the possibility of voids under the lozenges which may otherwise occur when the catalyst is being loaded.

The area within a lozenge is contained by a wire mesh which excludes catalyst but allows unrestricted flow of reaction gases. Quench gas is introduced by a perforated pipe lying along the central axis of the lozenge. This pipe is connected to a supply of quench gas by manifolds, which may lie inside or outside the converter. Figure 2-15 (p. 2-63) shows the internal type of manifold.

The void within the lozenge provides a space for mixing hot and cold gas before the gases pass to the next bed. The lower pressure drop across the lozenge encourages reaction gases to pass through, and Fig. 2-16 (p. 2-64) shows the computed lines of flow across a typical set of lozenges.

The lozenges are designed as beams supported at both ends. As their length increases, the height of the lozenges must be increased to provide the strength necessary to avoid collapse. Hence for larger-diameter converters the size of the lozenges may become unacceptable, and individual beds separated by support grids and mixing chambers become a viable alternative.

A design of reactor may be employed that uses either lozenges or separate beds. The basic difference from the above designs is that the gas entering the final bed is cooled by

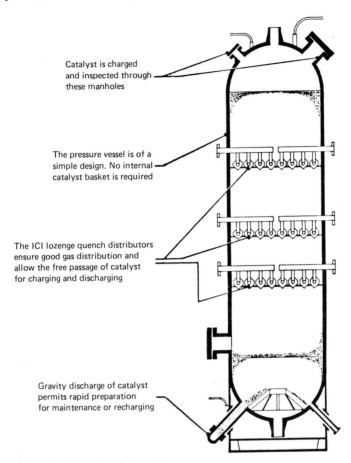

Catalyst is charged and inspected through these manholes

The pressure vessel is of a simple design. No internal catalyst basket is required

The ICI lozenge quench distributors ensure good gas distribution and allow the free passage of catalyst for charging and discharging

Gravity discharge of catalyst permits rapid preparation for maintenance or recharging

FIG. 2-13 ICI methanol converter.

indirect heat exchange with the gas entering the first bed rather than by the addition of further quench gas. The heat exchanger required in this type of reactor may be internal or external to the converter, and Fig. 2-17 (p. 2-65) shows such a design with an internal heat exchanger.

The conversion efficiency in the reactor is improved by the use of indirect cooling. Since the methanol concentration in the reaction gases is not diluted, the duty of the next catalyst bed is reduced. Thus a lower outlet temperature is possible, and the methanol content of the reaction gas will be proportionately higher. This effect is shown in Fig. 2-18 (p. 2-66).

In summary, this new type of reactor permits higher conversions to take place using a smaller number of catalyst beds and less cold shot. It may be of particular benefit in plants where the carbon efficiency is already high and an improvement in conversion efficiency does not adversely affect the carbon efficiency.

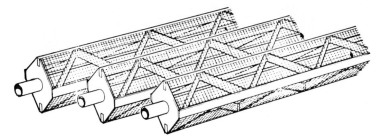

FIG. 2-14 ICI "lozenge" distributors.

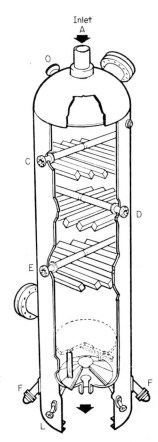

FIG. 2-15 Methanol converter showing arrangement of internal manifolds. Key: C = cold shot, D = cold shot, E = cold shot, F = catalyst dropout, L = thermocouple, O = catalyst.

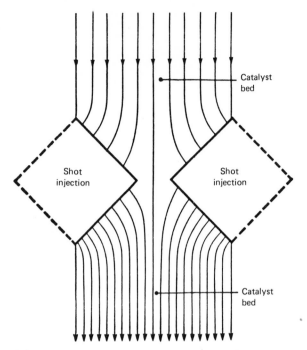

FIG. 2-16   Computed fluid flow pattern in the vicinity of the cold-shot distributors (calculated for the case of shot rate = initial flow rate).

PRODUCT SPECIFICATION AND PURIFICATION

Specification

Crude methanol usually contains between 10 and 25 percent by weight of water, depending on the feedstock, together with organic by-products classified as follows:

Dimethyl ether	100–400 ppm (w/v)
Carbonyl compounds	4–20 ppm (w/w)
Higher alcohols	1500 ppm (w/w)

Crude methanol is unsatisfactory for sale as a chemical-grade product and must therefore be purified to an acceptable specification. The most commonly used specification is that of the U.S. government. The latest edition is O-M-232F. The specifications for grades A and AA are given in Table 2-2.

Sometimes more stringent specifications are required by clients, and the most recent example of a more rigorous specification is the requirement for a fading time of 60 min.

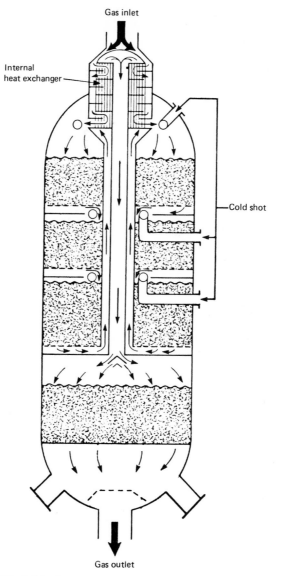

Gas inlet

Internal
heat exchanger

Cold shot

Gas outlet

FIG. 2-17 Converter showing indirect heat exchanger and separate beds.

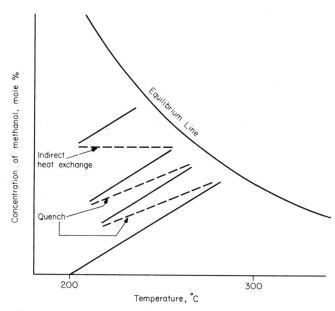

FIG. 2-18 The methanol profiles inside a methanol converter.

TABLE 2-2 Requirements for U.S. Federal Grades A and AA methanol

Characteristics	Grade A	Grade AA
Acetone and aldehydes, % max	0.003	——
Acetone, % max	——	0.002
Ethanol, % max	——	0.001
Acidity (CH_3COOH), % max	0.003	——
Appearance and hydrocarbons	Free of opalescence, suspended matter, and sediment	
Carbonizable substances	Not darker than color standard no. 30 of ASTM D1209 Pt/Co scale	
Color	Not darker than color standard no. 5 of ASTM D1209 Pt/Co scale	
Distillation range	Less than 1.0 °C and shall include 64.6 °C ± 0.1 °C at 760 mmHg	
Specific gravity, 20 °C/20 °C	0.7928	0.7928
Methanol, wt %	99.85	99.85
Nonvolatiles, g/100 mL (max)	0.001	0.001
Odor	Characteristic nonresidual	
Permanganate fading time, min	30	30
Water, % max	0.15	0.10

Purification

The purification of methanol is achieved by distillation as described earlier, and details of various systems have been published.[14,15,16]

In the ICI methanol process, three basic types of distillation schemes are offered:

Single column

Removal of light-end and higher-boiling materials is achieved in the same column. It is possible to produce Federal Grade A material, although the possibility of hydrocarbon breakthrough causing failure of the miscibility test is great. This system is shown in Fig. 2-19.

Two-column system

The most common arrangement to produce Federal Grade A material is the two-column system described earlier and shown in Fig. 2-20. This arrangement can be modified to produce Federal Grade AA material (10 ppm ethanol and acetone). Acetone is easily removed in the light-ends column without modification, but ethanol can only be effectively removed by increasing the reflux ratio and number of trays in the refining column. This has the effect of concentrating ethanol at one particular point in the tower which is coincident with the optimum position for removing higher alcohols as a purge. The loss of methanol is small and amounts to between 1 and 1.5 percent by weight of the feed.

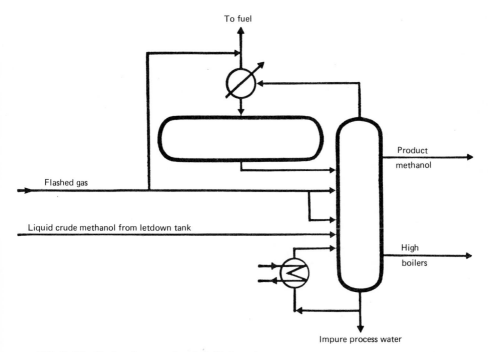

FIG. 2-19 Single-column methanol distillation scheme.

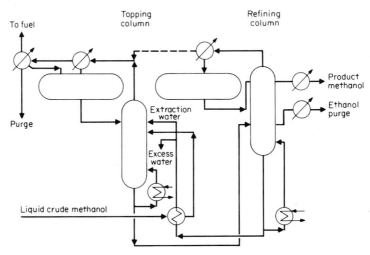

FIG. 2-20   Two-column methanol system with atmospheric refining column.

Three-column system

The three-column distillation design (Fig. 2-21) is based on the requirements for Federal Grade AA material. The first column is the standard light-ends tower, but the second column operates at slightly above atmospheric pressure and with a reflux ratio between 1 and 1.5. In order to achieve a grade AA product containing no more than 10 ppm of ethanol from this tower, it is necessary to remove about 60 percent of the methanol in the feed as a purge above the feed tray. This purge which contains 5 percent water is then

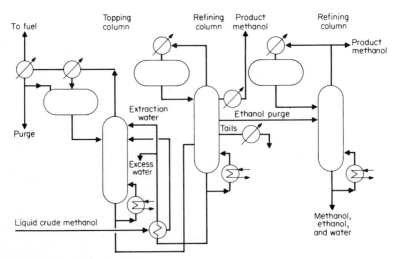

FIG. 2-21   Three-column system.

distilled in a third column under pressure to produce a grade AA product containing no more than 10 ppm of ethanol. The benefit of this system is that the reflux of the final column is at a sufficiently high temperature and contains sufficient heat to reboil the topping column and vaporize the feed to the middle column. This achieves a 30 percent saving in overall heat requirements for the distillation system, but it requires an extra column plus attendant equipment.

In many cases there is sufficient residual heat available to satisfy the requirements of a two-column system so that there is no benefit in adding a third column to improve efficiency. This is generally the case for plants based on the steam re-forming of natural gas and naphtha. However in some special cases where low-grade heat is not available for distillation, the three-column system will be of advantage.

WASTE STREAMS AND EMISSIONS

Methanol-Synthesis Section

Gaseous emissions

Gases are discharged from the methanol plant only at start-up and shutdown (Table 2-3). However, uncontrolled emissions can occur if the synthesis recycle loop relief valves are lifted.

Liquid emissions

There is no continuous liquid emission, apart from the products. During reduction, water is produced in the ratio (w/w) of 15 to 1000 STD. This water contains catalyst dust, some methanol, and the normal debris left after a shutdown (such as rust).

Methanol Distillation

Gaseous emissions

Dissolved gases in the crude methanol may be released to the atmosphere or burned. Methanol vapor may be released to the atmosphere in the very unlikely event of a main relief valve lifting.

Liquid effluents

There are two liquid effluents: a fusel oil–ethanol purge and the refining-column-bottoms purge. The fusel oil is generally burned and consists of about 50 percent alcohols and 50

TABLE 2-3   Gaseous emissions

Conditions	Approximate volume, scf/1000 STD*	Composition
Catalyst reduction	150,000	N_2, CO_2
Blowdown	350,000–700,000	H_2, CO, CO_2, CH_4, N_2 (circ. gas)

*STD means short ton day(s).

percent water. The main constituents of the refining-column-bottoms liquid are given below. This material is usually discharged to a sewage-treatment unit.

Refining-Column-Bottoms Analysis	
Total organic carbon, ppm (w/w)	80
Methanol, ppm	10
Formaldehyde, ppm	60
BOD, mg O_2/L	150
Total solids, g/m³	150
pH	7

BOD is biological oxygen demand.

PROCESS ECONOMICS

Capital Costs

The capital costs of methanol-synthesis and -distillation plants show some slight dependence on synthesis-gas composition. There is, however, a more significant effect of synthesis pressure, and this is illustrated by Fig. 2-22, which shows, in a differential form, the

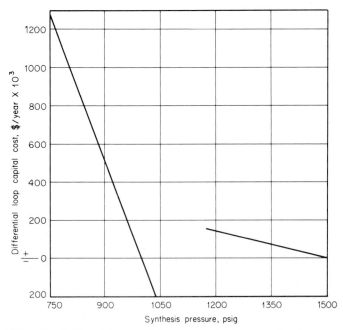

FIG. 2-22   Differential loop capital cost vs. synthesis pressure, based on 2000 ton/day.

effect of pressure on costs. The discontinuity that occurs is the result of changing design standards.

The approximate costs of a synthesis loop and distillation plant with a capacity of 3500 tons in terms of 1980 dollars are as follows:

Synthesis section	$20 million
Distillation plus storage section	$20 million

No account has been taken of the synthesis pressure or synthesis-gas composition. These numbers refer to the erected cost and factored engineering charges.

The distribution of costs for a complete plant based on steam re-forming and coal gasification is given below for each major section:

Plant section	% of total cost
Steam re-forming route	
Re-former	45
Compression	25
Synthesis	15
Distillation	15
Coal-gasification route	
Coal handling and gasification	25
Oxygen plant	16
Gas shift and purification	14
Compression	16
Methanol synthesis and purification	10
Steam distribution and offsites	19

Utility Consumption

Natural gas–based plant with no CO_2 addition

Plant Requirements per Ton of Product

Feed plus fuel, (LHV)*, MM Btu	26.2 (27.6 GJ)
Total energy†, MM Btu	25.4 (26.8 GJ)
Feedwater, tons	1.0
Cooling water, tons	60
Catalyst cost, $	1.4

*Lower-heating-value gas.

†With export power credit.

Coal-based plant using an atmospheric entrained gasifier

Plant Requirements per Ton of Product

Coal net calorific value, MM Btu/ton	27.8(32.3 MJ/kg)
Total coal (daf)* (feed plus fuel), tons	1.55
Oxygen requirement, tons	1.1
Cooling water, tons	90
Catalyst cost, $	1.4

*Dry, ash-free.

Partial oxidation of heavy fuel oil

Plant Requirements per Ton of Product

Feed plus fuel (LHV), MM Btu	35.5 (41.3 MJ/kg)
Feedwater, tons	0.75
Cooling water, tons	88
Catalyst cost, $	1.4

Energy-Conversion Efficiency

The heat of combustion of methanol (liquid) is 19.5 Btu per ton (LHV). Therefore, in the case of natural gas re-forming, the energy-conversion efficiency, taking credit for power, is about 77 percent. In coal or partial-oxidation plants, the energy-conversion efficiency is rather lower. For example, using the figures given earlier and assuming that coal has a heating value of 27.8 million Btu per short ton (LHV), the conversion efficiency is about 45 percent. Similarly for heavy fuel oil–based plants the energy conversion is about 55 percent efficient.

REFERENCES

1. "Methanol the ICI Way," *Chem. Week,* Jan. 6, 1968, p. 34.

2. Strelzoff, S.: "Methanol: Its Technology and Economics," *Chemical Engineering Progress* Symposium Series no. 98, American Institute of Chemical Engineers, New York, 1970, p. 54.

3. Pinto, A., and P. L. Rogerson: "Optimizing the ICI Low-Pressure Methanol Process," *Chem. Eng.,* July 4, 1977, p. 102.

4. Perry, R. H., and C. H. Chilton: *Chemical Engineers' Handbook,* 5th ed., McGraw-Hill, New York, 1973, sec. 13.

5. ICI, U.S. Patents 3,326,956, 1963; 3,850,850, 1973; 3,923,694, 1974.

6. Andrew, S. P. S.: "The Development of Copper-Based Catalysts for Methanol Synthesis and for Water Gas Shift," *Post Congress Symposium, Seventh International Congress on Catalysts,* Osaka, July 1980.

7. Kenard, C. J., and N. M. Nimmo: "Present Methanol Manufacturing Costs and Economics Using the ICI Process," *Chemical Engineering Progress* Symposium Series no. 98, American Institute of Chemical Engineers, New York, 1970, p. 47.

8. Pettman, M. L., and G. C. Humphreys: "Improved Designs to Save Energy," *Hydrocarbon Process.*, January 1975, p. 77.

9. ICI, U.S. Patent 4,065,483, 1975.

10. Pinto, A., and P. L. Rogerson: "Impact of High Fuel Costs on Methanol Plant Design," *Large Chemical Plants Fourth International Symposium*, Antwerp, October 1979, Elsevier, Amsterdam, p. 45.

11. ICI, U.S. Patent 4,072,625, 1976.

12. Mehta, D. D.: "Optimise Methanol Synthesis Gas," *Hydrocarbon Process.*, December 1968, p. 127.

13. ICI, U.S. Patent 3,458,289, 1965.

14. Mehta, D. D., and W. W. Pan: "Purify Methanol this Way," *Hydrocarbon Process.*, February 1971, p. 115.

15. ICI, U.S. Patent 4,210,495, 1978.

16. ICI, U.S. Patent 4,013,521, 1973.

MOBIL METHANOL-TO-GASOLINE (MTG) PROCESS

ANTHONY Y. KAM
MAX SCHREINER
SERGEI YURCHAK

Mobil Research and Development Corporation
Paulsboro, New Jersey

INTRODUCTION

The Mobil Methanol-to-Gasoline (MTG) process provides the first new route in over 40 years to synthesize high-quality high-octane gasoline from nonpetroleum sources. The process is relatively simple and highly energy-efficient and produces a readily marketable gasoline which requires minimal downstream processing and is totally compatible with conventional gasoline. More importantly, the MTG process is being commercialized.

In 1976, Mobil disclosed a novel process for converting methanol and related oxygenates to high-octane gasoline.[1] The key to this process is a highly selective zeolite catalyst discovered by Mobil. As shown in Fig. 3-1, this new process, coupled with commercially proven technologies for converting coal and natural gas to methanol, provides a novel route for synthesizing gasoline from these resources. Development work to date has produced two variants of the MTG process—fixed-bed and fluidized-bed processes. The former is being commercialized in a 14,000-bbl/day (gasoline product) plant in New Zealand, with a projected plant start-up in the mid-1980s. The latter is in the demonstration stage in a 100-bbl/day (methanol feed) plant in the Federal Republic of Germany (FRG) and is scheduled to start up in late 1982. This chapter describes the chemistry, process engineering, product characteristics, and economic perspective of the Mobil MTG process.

MTG PROCESS CHEMISTRY

ZSM-5 Catalyst

The catalyst used in the MTG process is a member of the ZSM-5 class of catalysts. Zeolite ZSM-5 was discovered in the early sixties and has found commercial application in a number of processes:

- Mobil xylene isomerization
- Ethylbenzene synthesis (Mobil-Badger)
- Mobil toluene disproportionation
- Mobil distillate dewaxing
- Mobil lubricating oil dewaxing

This zeolite posseses several important properties which are desirable not only for the above processes but also for the conversion of methanol to hydrocarbons:

- Unique molecular shape selectivity which can limit the size of both the reactants and products
- High catalytic activity
- Unusual stability in hostile environments
- Exceptionally low selectivity for conversion of many organic compounds to coke

The structure of ZSM-5 is shown in Fig. 3-2.[2] The lines represent oxygen atoms in a siliceous framework. The ZSM-5 framework forms two types of intersecting channels (Fig.

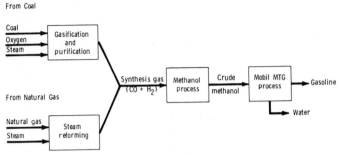

FIG. 3-1 Mobil Methanol-to-Gasoline (MTG) process route.

3-3): one with a nearly circular cross section and the other elliptical. The elliptical 10-membered rings form openings which are about 6 Å in diameter. The sizes of these channels play an important role in determining the hydrocarbon distribution from reactions occurring therein.

Chemical Reaction Paths

The detailed chemistry of the methanol conversion process is quite complex, and several paths have been suggested involving carbene or carbenium ion–type species[3,4] or oxonium-type intermediates.[5,6] The primary mechanism for forming a C—C bond from methanol remains unresolved.

The overall reaction scheme for converting methanol to gasoline is shown in Fig. 3-4. The initial step is the reversible dehydration of methanol to form dimethyl ether (DME). These two oxygenates (methanol and DME) undergo further dehydration reactions to give light olefins. These olefins then react to produce heavier olefins, which rearrange to paraffins, naphthenes, and aromatics. The formation of aromatic compounds occurs through hydrogen transfer-type reactions with the accompanying formation of paraffins. Very little molecular hydrogen is produced during methanol conversion.

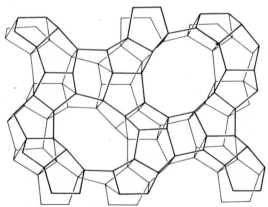

FIG. 3-2 Framework of new zeolite. *(Reprinted with permission from Ref. 2.)*

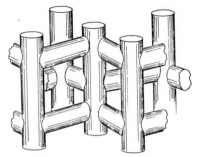

FIG. 3-3 Approximate model of pore structure. *(Reprinted with permission from Ref. 1.)*

$$2CH_3OH \rightleftharpoons CH_3OCH_3 + H_2O$$
$$\downarrow$$
$$\text{Light olefins} + H_2O$$
$$\Updownarrow$$
$$C_5^+ \text{ olefins}$$
$$\downarrow$$
$$\text{Paraffins}$$
$$\text{Cycloparaffins}$$
$$\text{Aromatics}$$

FIG. 3-4 Overall reaction path.

Almost no hydrocarbons are produced higher than C_{10} because of the shape-selective nature of the zeolite. The influence of molecular size on diffusivity is very striking. For example, the diffusivity of cyclohexane in ZSM-5 is about four orders of magnitude larger than that of 1,3,5-trimethylbenzene.[1] While bulky higher-molecular-weight aromatics could be formed, they probably undergo subsequent isomerization and other reactions to lower-molecular-weight compounds of the proper size to exit the catalyst. The conversion also produces a paraffinic product which is low in normal paraffins. This arises from the ability of the catalyst to selectively crack linear paraffins more rapidly than branched paraffins.[7] Since the ease of cracking increases with molecular weight, long-chain paraffins are cracked to lower-molecular-weight components. The selective nature of the catalyst produces MTG gasoline with high octane and an end point typical of commercial gasolines.

The MTG reaction path is shown in Fig. 3-5.[3] During the early stage of the reaction,

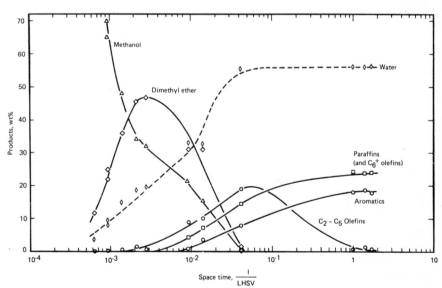

FIG. 3-5 Reaction path for methanol conversion to hydrocarbons (371 °C). LHSV stands for liquid hourly space velocity. *(Reprinted with permission from Ref. 3.)*

methanol is rapidly converted to DME and water. The initial hydrocarbons formed are rich in light olefins. As reaction time is increased, heavier olefins are formed and eventually aromatic hydrocarbons are produced. The light olefins pass through a maximum, indicating they are intermediates in the overall reaction sequence. These data indicate that by exercising proper control a product rich in light olefins could be produced.[8] The MTG process operates in the region further down the reaction path, with a product rich in paraffins and aromatics and lean in olefins.

Close examination of the data in Fig. 3-5 reveals that the initial formation rate of hydrocarbons is low. As conversion increases, the reaction rate accelerates. This behavior is typical of autocatalytic reactions as long as thermal effects do not intervene. The nature of the autocatalytic components is not resolved. One recent study suggests that these might be light olefin products.[9] On the basis of a simple kinetic scheme, the data indicate that the initial rate of formation of olefins from oxygenates (methanol and DME) is much slower than the rate at which the oxygenates react with the product olefins.

A more complex scheme involving a carbenoid intermediate has been used to fit methanol conversion and selectivity data over a wide range of pressure.[10] The following equations show this scheme on a water-free basis:

$$A \rightarrow B$$

$$A + B \rightarrow C$$

$$B + C \rightarrow C$$

$$C \rightarrow D$$

where A = oxygenates
B = (:CH$_2$)
C = olefins
D = paraffins + aromatics

Oxygenates initially react to form a reactive intermediate which can then react either with the oxygenates to form olefins (insertion) or with olefins to form higher-molecular-weight olefins (addition). The olefins then react to give paraffins and aromatics. The fit of this kinetic scheme to product selectivity data obtained at 370 °C over a pressure range of 0.6 to 735 lb/in² (4 to 5070 kPa) is shown in Fig. 3-6 and indicates the overall reasonableness of this phenomenological model. This model has also been shown to fit conversion data quite well.[10]

While the kinetic scheme just presented does not prove or disprove a particular reaction mechanism, it indicates that the conversion process occurs by a more complex route than is shown in Fig. 3-4.

Process Stoichiometry

The overall stoichiometry of converting methanol to hydrocarbons over the ZSM-5 class catalyst is shown below:

$$x\text{CH}_3\text{OH} \rightarrow (\text{CH}_2)_x + x\text{H}_2\text{O} \tag{1}$$

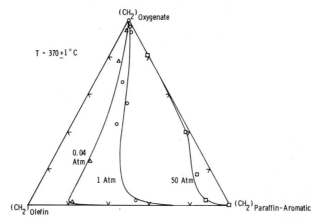

FIG. 3-6   Methanol conversion to hydrocarbons: comparison of experimental and calculated reaction path. *(Reprinted with permission from
Ref. 10.)*

For every 100 lb (kg) of methanol, about 44 lb (kg) of hydrocarbons and 56 lb (kg) of
water are produced. The yield of by-products such as CO, CO_2, H_2, and coke is very small,
typically less than 0.5 percent by weight. Only small quantities of oxygenates are present
in the aqueous-phase product. Most (85 to 90 percent by weight) of the hydrocarbon
product boils in the gasoline range. Hydrocarbons boiling above the gasoline boiling range
are not produced in significant quantities; i.e., x in Eq. 1 is ≤ 10.

Energy Efficiency

The MTG process is highly energy-efficient. The hydrocarbon product contains about 95
percent of the chemical energy content of the methanol.[11] This is on a low-heating-value
(LHV) basis, which is appropriate, since internal combustion engines do not condense
water. The remaining 5 percent of the chemical energy of methanol is liberated as heat of
reaction, which is about 750 Btu/lb (1.74 MJ/kg) of methanol. In the fluid-bed MTG
process, most of this reaction heat can be recovered as high-pressure (600 psig) steam. In
the fixed-bed process, a lesser amount may be recoverable, but this depends strongly on
the detailed design of the plant. It is estimated that with provisions for processing energy,
the overall efficiency of the MTG section of the plant will be about 90 percent.[11,12]

PROCESS DEVELOPMENT HISTORY

Development of the MTG process was initiated in the early 1970s and progressed steadily
from laboratory-scale equipment to bench-scale units, pilot plant, demonstration plant, and
commercial fixed-bed plant as shown in Fig. 3-7. Since the MTG reactions are highly
exothermic, one of the key considerations in process engineering and scale-up is removal
of the heat of reaction in order to maintain and control the reactor temperature. Other
factors to consider include methanol conversion, product yields, gasoline quality, and catalyst regeneration.

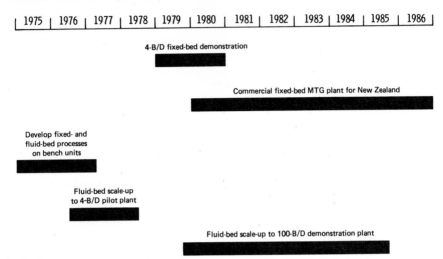

| 1975 | 1976 | 1977 | 1978 | 1979 | 1980 | 1981 | 1982 | 1983 | 1984 | 1985 | 1986 |

4-B/D fixed-bed demonstration

Commercial fixed-bed MTG plant for New Zealand

Develop fixed- and
fluid-bed processes
on bench units

Fluid-bed scale-up
to 4-B/D pilot plant

Fluid-bed scale-up to 100-B/D demonstration plant

FIG. 3-7 Development of Mobil MTG process.

After initial exploratory studies, bench-scale development of the MTG process was carried out from 1974 to 1976 in a program partially supported by the U.S. Energy Research and Development Administration (ERDA).[11] In this program, two variants of the process were developed—a fixed-bed mode and a fluid-bed mode. Studies of process variables were conducted to identify the preferred operating conditions. Product testing, including in-vehicle tests, was also carried out to evaluate the quality of the MTG gasoline product.

The major distinguishing features and advantages of the two MTG variants are summarized in Table 3-1. Detailed descriptions of the variants are given in the sections that follow. Under normal operation, single-pass methanol conversion is complete for the fixed bed and virtually complete ($>$99.5 percent by weight) for the fluid bed.

At the conclusion of the bench-scale studies, the fixed-bed MTG process was essentially ready for scale-up to commercial-size plants. In 1979, the New Zealand government announced the selection of the fixed-bed MTG process for a commercial natural gas-to-gasoline plant. In order to provide confirmational data, a 4-bbl/day fixed-bed pilot plant was built and put into operation in 1979. Design and engineering of the New Zealand plant is underway, and start-up is projected for the mid-1980s.

The bench-scale studies identified a number of potential advantages for the fluid-bed version. However, it was recognized that additional engineering development was necessary for satisfactory scale-up. Between 1976 and 1978 a program was undertaken to scale up

TABLE 3-1 Two modes of operation for MTG process

	Fixed-bed mode	Fluid-bed mode
Methanol conversion	2 steps	1 step
Heat removal	Light-gas recycle	Heat-exchanger pipes
Regeneration	Intermittent	Continuous
Operation	Cyclic	Steady-state

the fluid-bed process to a 4-bbl/day pilot plant.[13] The program was partially supported by the U.S. Department of Energy (DOE). Operation of the 4-bbl/day fluid-bed plant was extremely successful and confirmed the advantages of the fluid bed. This study provided further impetus for continuing the scale-up of the fluid bed. In the second half of 1979 another program was initiated to demonstrate the fluid-bed process in a 100-bbl/day plant. This program is an international effort between industry and government. The industrial team is made up of Mobil Research and Development Corporation (MRDC) of the United States and Union Rheinische Braunkohlen Kraftstoff (URBK) and Uhde Gmbh (Uhde); the last two are in the FRG. Financial support is partially provided by DOE and the Bundesministerium für Forschung und Technologie (BMFT) of the FRG. The plant is located at Wesseling, FRG. The start-up date was late 1982. After the completion of this demonstration phase, the fluid-bed MTG process will likely be ready for commercial-scale application.

FIXED-BED MTG PROCESS

Process Description

A major consideration in the development of the fixed-bed MTG process is managing the heat liberated during methanol conversion. The heat of reaction is about 750 Btu/lb (1.74 MJ/kg) of methanol. Uncontrolled release of this reaction heat could result in a temperature rise of about 600 °C. In the fixed-bed MTG process, removal of reaction heat is accomplished by dividing the overall reaction (and reactor system) into two steps and by light-gas recycle, as shown in Fig. 3-8. Crude methanol is vaporized and heated to reaction temperature (ca. 315 °C) by exchange with reactor effluent. It is then passed into a dehydration reactor, where it is partially dehydrated to form an equilibrium mixture of methanol, DME (dimethyl ether), and water. The dehydration catalyst is a Mobil proprietary catalyst which is different from the ZSM-5 catalyst. Because this reaction is controlled by chemical equilibrium, excessive temperatures cannot be attained and the reactor system is inherently stable. About 20 percent of the heat of reaction is liberated in this first step.

The equilibrium mixture of methanol, DME, and water is then mixed with recycle gas and passed into conversion reactors which contain the ZSM-5 catalyst. The function of the recycle gas is to limit the temperature rise in the conversion reactor to less than 60 °C. Without this heat sink, the temperature rise could be much greater. In the conversion reactors the methanol and DME are converted to gasoline-range hydrocarbons and water. In a commercial system there would be multiple conversion reactors in parallel, as shown in Fig. 3-8. The products leaving the conversion reactors are cooled by steam generation, by heat exchange with recycle gas, and finally by air and/or water cooling to a temperature of about 38 °C. The product is then separated into three phases: (1) a liquid hydrocarbon phase from which gasoline and LPG are recovered, (2) a gas phase which is mostly recycled to the conversion reactor, and (3) an aqueous phase which will be treated to remove trace oxygenates or recycled to the synthesis-gas production step. The aqueous phase contains very little methanol and DME, since essentially complete conversion of these compounds occurs over the ZSM-5 catalyst.

The liquid hydrocarbon product and the gas-phase product are processed in conventional fractionation columns to produce gasoline, LPG, and fuel gas. Additional gasoline

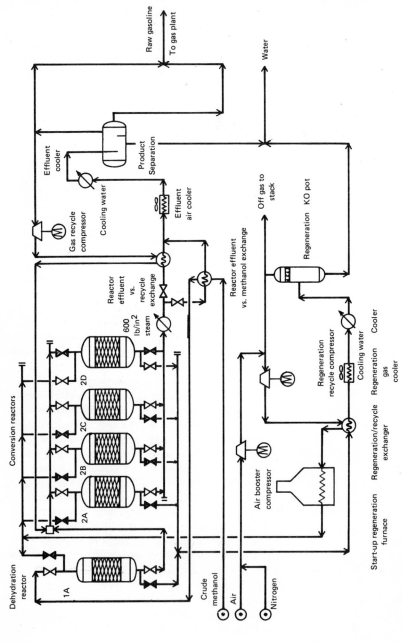

FIG. 3-8 Typical flow diagram of fixed-bed MTG reaction section.

could be produced by alkylating the propylene and butenes with isobutane produced in the process, depending on the plant size. This would, of course, require the addition of an alkylation unit.

During the conversion reaction, coke is deposited on the catalyst. This causes a temporary deactivation of the catalyst. The reactor temperature profiles shown in Fig. 3-9 indicate that the catalyst aging in a cycle occurs in a band. At the conditions of the test, about 35 percent of the catalyst in the reactor is used to complete the conversion of methanol to hydrocarbons and water. This quantity was estimated from axial temperature profiles. As the catalyst ages during a cycle, the reaction zone moves toward the reactor outlet. The significance of this band aging is that the process is being operated at an effectively variable space velocity throughout the cycle, since the quantity of catalyst actually being utilized decreases with cycle time. Catalyst upstream of the main reaction zone (which is that portion of the reactor where temperature gradients exist) is inactive for methanol conversion, as indicated by the absence of significant temperature gradients. An additional significance of this band aging is that the time, temperature, and water-pressure history of each fraction of the catalyst bed is different. Since the conversion catalyst is permanently deactivated by steam and steaming depends on these three variables, an activity gradient will be established in the catalyst bed. Catalyst located near the reactor inlet will be more active than catalyst near the outlet. Coupled with this permanent activity gradient will be a temporary activity gradient associated with coking of the catalyst. As the main reaction zone moves toward the reactor outlet, the conversion is occurring not only over less cat-

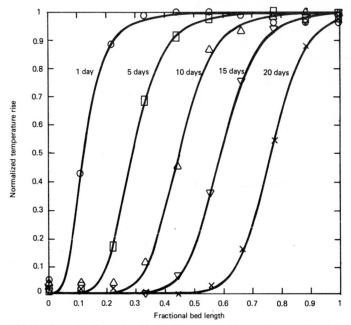

FIG. 3-9 Band aging in fixed-bed MTG process. The numbers represent days on stream.

alyst but also over less-active catalyst and product selectivities will change both within a cycle and from cycle to cycle in a manner to reflect this movement backward along the reaction path. Examples of these selectivity changes are given under "Bench-Scale Development" and "4-bbl/day Pilot Plant Scale-up Experience." Of major importance is that process conditions can be selected which give long catalyst life.[11,14]

The temporary deactivation due to coke formation is readily restored by regeneration of the catalyst. This is accomplished by burning the carbonaceous deposits off the catalyst with controlled amounts of air at relatively low tempeature (<490 °C). Regeneration would be performed every 2 to 6 weeks, depending on the age of the catalyst and the design of the regeneration facilities. By using swing reactors, most of the conversion reactors would be in the process cycle while a few would be in the regeneration cycle.[12]

Product Yields and Selectivities

Product yields and hydrocarbon selectivity representative of the fixed-bed MTG process are given in Table 3-2 along with typical operating conditions. The conversion of methanol is complete, and the yields of hydrocarbons and water are virtually stoichiometric. Only small quantities of by-products (CO, CO_2, coke, and oxygenates) are formed.

The major hydrocarbon product is gasoline. Alkylate for blending into the C_5+-synthesized gasoline can be produced by alkylating the propylene and butenes with the isobutane produced. After adding alkylate and pressurizing the gasoline to a 9-lb/in² (62-kPa) Reid vapor pressure (9 RVP) with butanes, the finished gasoline yield is 85 percent by weight of hydrocarbons. The yield of light gas (C_1 and C_2) is only 1.4 percent by weight. Since the yield of alkylate is limited by the yield of propylene and butenes, the excess isobutane is put into LPG.

The yield of LPG is 13.6 percent by weight. In the fixed-bed MTG process, the yield of alkylate is low because most of the propylene and the butenes are recycled to the reactor via the stream of recycle gas. This results in their conversion to directly synthesized gasoline.

As indicated earlier, by-product formation in the MTG process is low. A small quantity of oxygenates is produced. Because the amount is so low, they reside in the aqueous product. Our studies have demonstrated that these oxygenates can be reduced to acceptable environmental limits by employing conventional biological treatment methods.

Bench-Scale Development

After its initial discovery in 1972, the MTG process was subjected to preliminary evaluations in small-scale fixed-bed equipment (catalyst capacity of 1 to 10 cm³). The results of these studies indicated that further development of the process was warranted. Work on larger-scale units was required, since the very small units do not simulate the performance of very large units.

Development of the fixed-bed MTG process was initiated in 1974. This work was conducted on bench-scale pilot plants which have a capacity of about 1 to 2 gal/day (4 to 8 L/day).[11,14,15] Special care was taken in the design of these pilot plants so that they would simulate the performance of commercial-size units, particularly with respect to heat effects. The pilot plants were also designed to simulate the key features of the process such as use

TABLE 3-2 Yield from methanol in fixed-bed reactor system*

Temperatures, °C (°F)	
Dehydration reactor inlet	316 (600)
Dehydration reactor outlet	405 (760)
Conversion reactor inlet	360 (680)
Conversion reactor outlet	415 (780)
Pressure, psig (kPa abs)	300 (2170)
Recycle ratio, mol/mol of charge	9:1
Space velocity (WHSV)† in conversion reactor	2
Yield, wt % of methanol charged	
Methanol plus ether	0.0
Hydrocarbons	43.66
Water	56.15
CO, CO_2	0.04
Coke, oxygenates	0.15
Total	100.00
Hydrocarbon product, wt %	
Light gas	1.4
Propane	5.5
Propylene	0.2
Isobutane	8.6
n-Butane	3.3
Butenes	1.1
C_5^+ gasoline	79.9
Total	100.0
Gasoline (including alkylate)‡	85.0
LPG	13.6
Fuel gas	1.4
Total	100.0

*Charge is methanol containing 17 wt % of water.

†Weight hourly space velocity.

‡93 Unleaded Research Octane Number; 9-lb/in² (62 kPa) Reid vapor pressure (RVP).

of recycle streams and high-pressure separators in addition to operating at realistic process conditions.

A large amount of the bench-scale process development work was carried out during 1975 and 1976 under a contract jointly funded by ERDA and MRDC.[11] On the basis of this work, preferred process conditions were defined, product yields were established, the properties of the gasoline were determined, and the sensitivity of several cars to the gasoline's durene (1,2,4,5-tetramethylbenzene) content was determined. Of utmost importance, it was demonstrated that the ZSM-5 catalyst performed satisfactorily at realistic process conditions during an 8-month test during which the catalyst was regenerated 10 times. The gasoline yield during this aging test is shown in Fig. 3-10.

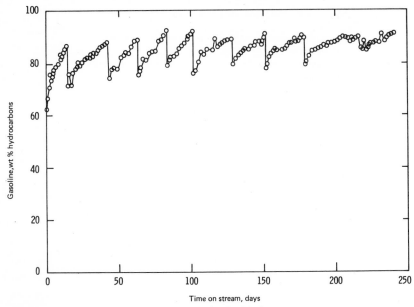

FIG. 3-10 Methanol-to-gasoline aging study for fixed-bed process.

The vertical lines indicate the completion of a process cycle and subsequent regeneration of the catalyst. Gasoline yield increases within a process cycle as a result, in part, of changes in catalyst activity caused by coke deposition. Gasoline yield is lowest at the start of the cycle as a consequence of the high production of aromatics, which results in the production of relatively large quantities of gas (mainly propane and n-butane). As the catalyst ages, gasoline yield increases and approaches 90 percent by weight of hydrocarbons (100 percent methanol conversion) near methanol breakthrough. End-of-cycle gasoline contains less aromatics than start-of-cycle gasoline, but this has relatively little effect on the octane number, since the production of high-octane olefins and isoparaffins is enhanced.[11,14] The changes in gasoline selectivity within a cycle are greatest for the first cycle and decrease substantially in subsequent cycles, resulting in increasing gasoline production as the catalyst ages. As shown, cycle lengths were about 20 days and appeared to have stabilized. Catalyst performance was still satisfactory when the aging test was arbitrarily terminated. Bench-scale work also demonstrated that the dehydration catalyst performed satisfactorily.[11,14]

Pilot Plant Scale-up Experience

Upon completion of the bench-scale work, the fixed-bed MTG process was essentially ready for scale-up to a commercial size. Extensive experience at Mobil with other fixed-bed processes indicated that such scale-up would be entirely feasible. However, to provide confirmational data and operating experience, the process was scaled up to a 4-bbl/day pilot plant.

The scale-up was carried out so that the length of the catalyst beds would be the same length as a commercial-size reactor.[16-18] In this manner, no further effects would be expected as a result of diameter scale-up.

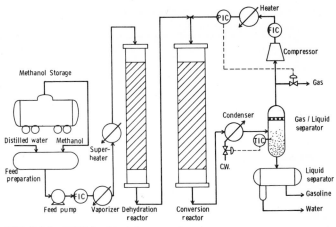

FIG. 3-11   MTG fixed-bed demonstration plant.

A schematic diagram of the 4-bbl/day demonstration plant is shown in Fig. 3-11. The dimensions of the catalyst bed for the dehydration reactor are 2 in (50 mm) in diameter by 10 ft (3 m) in length; those for the conversion reactor are 4 in (100 mm) in diameter by 8 ft (2.4 m) in length. These lengths are about 10 times longer than those in the bench-scale unit. The gas velocities in the reactors are in the range of those in commercial plants. In all other aspects, the 4-bbl/day demonstration plant is very similar to the bench-scale unit.

For the scale-up study the bench-scale unit and demonstration plant were operated at the same process conditions (Table 3-3). The same batch of catalyst and the same methanol feed were used. The methanol feed was a simulated charge of crude methanol prepared by blending 83 percent by weight of methanol with 17 percent by weight of water. This water content is typical of crude methanol which would be produced by a methanol plant based on natural gas feed.

Operation of the demonstration unit has been very satisfactory, and no detrimental scale-up effects have been observed. Product selectivity, yield, and quality measured at a given

TABLE 3-3   Fixed-bed MTG scale-up: average first-cycle process conditions

	Bench unit	4-bbl/day unit
Methanol/water ratio in charge, % w/w	83:17	83:17
Inlet temperature, °C (°F)	358 (676)	360 (680)
Outlet temperature, °C (°F)	404 (759)	407 (765)
Methanol WHSV*, lb feed/(lb catalyst)(h)	1.6	1.6
HP† separator temperature, °C (°F)	50 (121)	52 (125)
Recycle ratio, mol/mol of charge	9	9.2
Pressure, psig	299	298

*Weight hourly space velocity in conversion reactor.

†High pressure.

TABLE 3-4 Bench-scale and 4-bbl/day fixed-bed MTG units have same yield of product and hydrocarbon*

	Bench unit	4-bbl/day unit
Product, wt % methanol		
Hydrocarbons	43.73	43.75
Water	56.17	45.19
Hydrogen	0.002	0.001
Carbon monoxide	0.02	0.02
Carbon dioxide	0.08	0.04
Methanol	0.00	0.00
Dimethyl ether	0.00	0.00
Total	100.00	100.00
Hydrocarbon, wt %		
Methane	1.33	1.25
Ethane	0.82	0.86
Ethylene	0.02	0.03
Propane	8.54	8.60
Propylene	0.15	0.15
Isobutane	8.45	8.39
n-Butane	4.06	4.20
Butenes	0.71	0.74
C_5^+ gasoline	75.92	75.78
Total	100.00	100.00
9-RVP† gasoline (including alkylates)		
Yield, wt % of HC	80.2	80.2
R + O‡	95	95

*Average over first cycle.

†Reid vapor pressure.

‡R + O is unleaded research octane number.

severity or averaged over the whole cycle are similar to those of the bench unit. Typical product yields averaged over the first cycle are compared in Table 3-4. The hydrocarbon yields correspond to the stoichiometric yield. By-product formation is very low. No unreacted methanol or DME was observed until the cycle ended at methanol breakthrough. Gasoline yield and octane number are also quite similar. These results attest to the ease of scale-up of vapor-phase, fixed-bed reactors.

One difference between the performance of the 4-bbl/day unit and the bench-scale unit is that the cycle length is significantly longer in the larger unit (30 vs. 20 days). This difference is associated with the slower movement of the conversion-reactor temperature profile through the catalyst bed, as shown in Fig. 3-12. This slower movement is a result of the lower coke yield in the 4-bbl/day unit, which may be attributed to the higher linear velocity, since this is the only major difference in the operation of the two units.

The gasoline yield data for the two units are compared in Fig. 3-13. They do not coincide on a time basis because of differences in cycle lengths. This can be resolved by com-

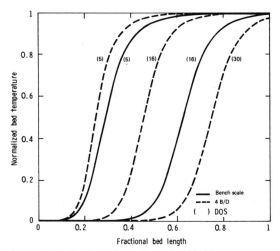

FIG. 3-12 Catalyst aging is lower in 4-bbl/day demonstration plant.

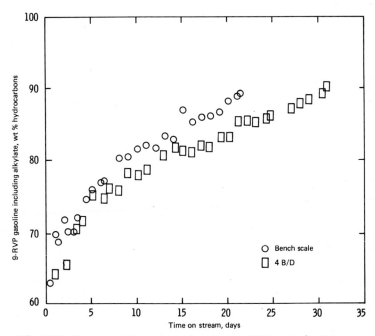

FIG. 3-13 Gasoline yields increase with cycle time. RVP stands for Reid vapor pressure.

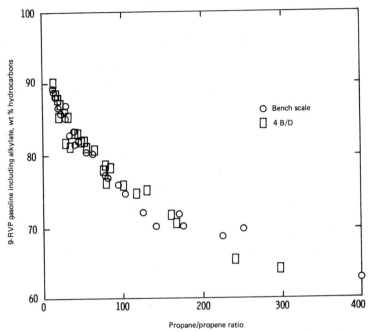

FIG. 3-14   Gasoline selectivity is the same in both units. RVP stands for Reid vapor pressure.

paring the gasoline yield for the two units at the same extent of reaction, as depicted in Fig. 3-14. Since the conversion of methanol is virtually 100 percent, the ratio of propane to propylene in the product hydrocarbon has been used to gauge the extent of reaction.[13] On this basis, it is clear that at the same catalyst state, gasoline yields in both units are the same. This is also true for individual components of the hydrocarbon product. While the propane to propylene ratio is a convenient indicator of the extent of reaction, it should not be interpreted as a direct measure of the change in catalyst activity.

In summary, these results attest to the ease of scale-up of the fixed-bed MTG process. No detrimental effects were noted on scaling up this process from a bench-scale unit with a capacity of 1 to 2 gal/day to the 4-bbl/day unit. The key scale-up, i.e., reactor length, has been completed, and the process is ready for commercialization.

Commercialization Activities

The Mobil MTG process has technical and economic advantages over the commercially established synthetic fuel processes such as Sasol's Synthol and Arge processes.[19] As demonstrated by the scale-up studies reported here, the fixed-bed MTG process has been technically ready for commercialization for several years. Actual commercialization, however, must await the establishment of the proper political and economic climate. This appears to have occurred in at least one country—New Zealand.

New Zealand natural gas-to-gasoline plant

In November 1979, the New Zealand government announced the selection of the Mobil MTG Process as the most desirable route for converting the natural gas from the Maui offshore gas field into gasoline. The gas will be converted to methanol, which will then be converted to gasoline by the fixed-bed MTG process. The MTG facility will use multiple conversion reactors in parallel. The complex will produce about 14,000 bbl/day of gasoline, which is about one-third of New Zealand's gasoline requirement. This plant, which is in the detailed design stage, will be the first commercial application of the process when it starts up in the mid-1980s.

U.S. coal-to-gasoline projects

There is substantial interest in the United States in employing the Mobil MTG process to effect the conversion of coal to gasoline. A number of projects have been proposed, and several feasibility studies are underway. Because of the political and economic climate prevailing in the United States, it is not certain that the projects outlined in this section will be brought to commercial fruition in the near future (within 10 years).

In February 1980, W. R. Grace and Co. announced plans for a coal-to-methanol-to-gasoline project. Loan and price guarantees have been requested from the government Synthetic Fuels Corp. (SFC). The initial design phase of this plant is underway and will cost $12 to $16 million. The plant will convert 29,000 tons/day of coal into 50,000 bbl/day of gasoline using the Mobil fixed-bed MTG process. It is estimated that the plant will cost more than $3 billion.[20] The plant will be located in Baskett, Kentucky.

Tennessee Synfuels Associates, a consortium of Koppers Synfuels Corp. and Citco Synfuels, Inc., has requested loan and price guarantees from the SFC for a plant to produce gasoline from coal via methanol, using the MTG process. The initial plant capacity will be about 9000 bbl/day of gasoline with possible expansion to 50,000 bbl/day of oil equivalent product.[21] The plant will be located in Oak Ridge, Tennessee. Initial design work is underway.

Hampshire Energy (Koppers Company, Inc., Kaneb Services, and Northwestern Mutual Life Insurance Company) has also requested loan and price guarantees from the SFC for a project to produce 20,000 bbl/day of gasoline from 15,000 tons/day of coal.[20,21] This project involves coal gasification, using KBW and Lurgi gasifiers, followed by methanol synthesis and then by MTG conversion of the methanol to gasoline. The complex is to be located in Gillette, Wyoming.

FLUID-BED MTG PROCESS

Process Description

The fluid-bed MTG process is shown schematically in Fig. 3-15. Crude methanol feed, which may contain variable amounts of water, is vaporized, superheated to the range of 175 to 260 °C, and charged into the dense fluid-bed reactor. Typical reactor operating conditions are shown in Table 3-5. The fluid-bed catalyst is a Mobil proprietary ZSM-5 catalyst with physical properties similar to those of commercial fluid catalytic-cracking (FCC) catalyst. In a single pass through the reactor, virtually complete methanol conversion

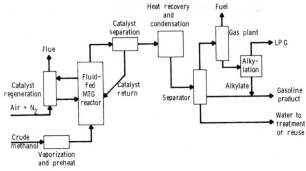

FIG. 3-15   Fluid-bed MTG process.

is achieved. After catalyst disengagement, the effluent stream is collected and separated into hydrocarbon and aqueous products. The hydrocarbons are sent through fractionation columns to yield C_5+ gasoline, C_3 and C_4 streams, and a C_2- stream. Typical yields from the reactor are also shown in Table 3-5. The C_5+-gasoline fraction is 60 percent by weight of the hydrocarbon products. However, the C_3-C_4 fraction contains significant quantities of isobutane, propene, and butenes. By means of the well-known alkylation process, these

TABLE 3-5   Yields from methanol in a 4-bbl/day fluid-bed unit

Average bed temperature, °C	413
Pressure, kPa	275
Space velocity (WHSV)*	1.0
Yields, wt % of methanol charged	
Methanol + DME	0.2
Hydrocarbons	43.5
Water	56.0
CO, CO_2	0.1
Coke, other	0.2
Total	100.0
Hydrocarbon product, wt %	
Light gas ($C_1 + C_2$)	5.6
Propane	5.9
Propylene	5.0
Isobutane	14.5
n-Butane	1.7
Butenes	7.3
C_5+ gasoline	60.0
Total	100.0
Gasoline (including alkylate)†	88.0

*Weight hourly space velocity, wt feedstock/(wt catalyst)(hour).

†Reid vapor pressure (RVP) is 9 lb/in² (62 kPa); unleaded research octane number (R + O) is 96.

TABLE 3-6 Typical properties of finished fluid-bed MTG gasoline

Components, wt %		
Butanes	2	
Alkylate	28	
C$_5$+ gasoline	70	
Total	100	
Composition, wt %		
Saturates	67	
Olefins	6	
Aromatics	27	
Total	100	

	Research	Motor
Octane no.		
Clear (unleaded)	96.8	87.4
Leaded (3 cm^3 TEL/U.S. gal)	102.8	95.1
Reid vapor pressure, kPa	76	
Specific gravity	0.730	
Sulfur and nitrogen	Nil	
Copper strip corrosion rating	1A	
ASTM distillation, °C		
10% evaporated	47	
30% evaporated	70	
50% evaporated	103	
90% evaporated	169	

components are easily converted to additional gasoline. Including the alkylate, the final overall yield for a 9-RVP (62-kPa) gasoline is 88 percent by weight of the hydrocarbons produced from methanol.

As shown in Fig. 3-15, recycle is not required for the fluid-bed MTG reactor. By taking advantage of the superior heat-transfer characteristics of a dense fluid bed, the heat of reaction can be readily removed by heat-exchange pipes immersed either internally in the reactor or externally in a catalyst cooler. With the reactor operating at 400 to 425 °C, the heat of reaction can be recovered as valuable high-quality high-pressure steam.

In the fluid-bed MTG process, catalyst regeneration is carried out continuously by circulating catalyst from the reactor through a regenerator (Fig. 3-15). Since the coke yield is extremely low, the required circulation rate is quite modest and much lower than that of, say, commercial FCC units. The MTG regenerator typically operates at around 480 °C and at pressures similar to those of the reactor.

Typical properties of the finished gasoline (including alkylate and adjusted to 11 RVP) from the fluid-bed MTG process are shown in Table 3-6. The gasoline quality is excellent, with an unleaded research octane number of 96.8 and a leaded one of 102. The boiling range is similar to that of petroleum-derived gasoline. The MTG gasoline contains no sulfur and no nitrogen.

Scale-up of Fluid-Bed MTG Process

The fluid-bed MTG process features a high-performance fluid-bed reactor which meets the following performance objectives:

- Intimate contact between catalyst and gas resulting in virtually complete methanol conversion
- Good mixing to maintain temperature uniformity
- Efficient heat removal for temperature control and stability
- Continuous regeneration to maintain optimal catalytic activity and maximize gasoline yield

An orderly, three-step scale-up has been conducted, progressing from bench-scale units, to a 4-bbl/day pilot plant, to a 100-bbl/day demonstration plant. The scale-up experience is described below.

Bench-Scale Studies

Between 1974 and 1976 initial studies of the scope of the process including effects of process variables, characterization of reaction kinetics, and catalyst aging and regeneration were carried out in the once-through, isothermal, dense fluid-bed unit shown in Fig. 3-16. The reactor is 1⅝ in (41 mm) in diameter, with a dense-bed height of 14 to 18 in (35 to 46 mm). An enlarged disengaging section serves to minimize catalyst entrainment. The reactor is also provided with internal baffles to promote contact between gas and solids. Methanol is vaporized, preheated, and then charged into the fluid bed by a "doughnut"-shaped distributor. Complete or partial methanol conversion can be achieved in this bench unit. In normal operations, the catalyst is charged batchwise and initial conversion is 100 percent. When the conversion falls below 99.5 percent by weight, the cycle is terminated. The catalyst is then regenerated in the same reactor after which a new cycle is started. Typically a cycle may last from 2 to 15 days.

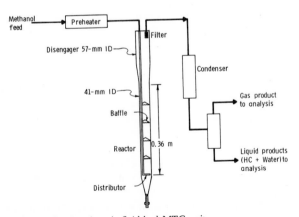

FIG. 3-16 Bench-scale fluid-bed MTG unit.

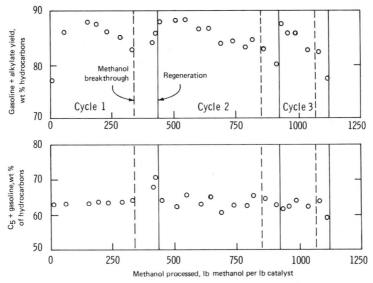

FIG. 3-17 Fluid-bed MTG gasoline yields during aging test.

The bench unit was used for extensive studies of process variables. The results show that the preferred operating region for the fluid-bed MTG process is defined by the following list.

Pressure, 25 to 50 psig (172 to 345 kPa)

Temperature, 370 to 425 °C

Weight hourly space velocity (WHSV)*, 0.5 to 3.0 wt feed/(wt catalyst)(h^{-1})

The studies also showed that the presence of water in the methanol feed is acceptable and hence crude methanol would be a satisfactory feedstock. In an integrated methanol-synthesis and MTG plant, the final distillation step for the methanol-synthesis section can thus be eliminated for substantial energy savings.

The apparent kinetics of the MTG conversion was investigated in the bench unit by running studies of partial conversion. The results showed that methanol conversion could be adequately represented by first-order kinetics.[22] A simple mathematical model which also included catalyst aging was constructed.[11]

Stability and regenerability of the fluid MTG catalyst were also confirmed in the bench unit. Three long-term aging runs were made. Data on gasoline yields from a three-cycle run are shown in Fig. 3-17. Since the bench unit was operated in a cyclic mode, product yields and selectivity varied during the cycles as expected. Note, however, that a peak in total gasoline yield can be achieved for a certain catalyst activity level. In a continuous regeneration unit, where catalyst activity can be maintained at a steady state, this point of optimal gasoline yield would be the activity target.

In summary, the bench-unit studies identified the following potential advantages of the fluid-bed MTG process:

- Optimal gasoline yield by steady-state operation
- No recycle necessary
- Ease of temperature control

The results encouraged further scale-up of the fluid-bed MTG process.

Pilot Plant (4 bbl/day)

The objectives of the 4-bbl/day pilot plant program are listed below.

- Scale-up the fluid-bed MTG reactor system by a factor of 75
- Demonstrate feasibility of steady-state operation at maximum gasoline yield and virtually complete methanol conversion
- Investigate options for removal of reaction heat

The program was carried out between 1976 and 1978 with partial support from DOE. All objectives were successfully met.[13]

A schematic of the 4-bbl/day pilot plant is shown in Fig. 3-18. The feed of crude methanol was simulated by blending commercial-grade methanol with distilled water (17 percent by weight). The feed was vaporized, superheated (to about 175 °C), and charged into the reactor. The dimensions of the adiabatic reactor were 4 in (102 mm) in diameter by 25 ft (7.6 m) in height. Topping the reactor was a disengager where most of the entrained catalyst was separated from the product gas. A cyclone and filter system completed the catalyst separation. The product was then cooled in a three-stage condenser and sent to a gravity separator from which light hydrocarbons, gasoline, and water were separately withdrawn.

As shown in Fig. 3-18, the reactor was equipped with an external catalyst circulation line through which catalyst was returned from the disengager to the bottom of the reactor.

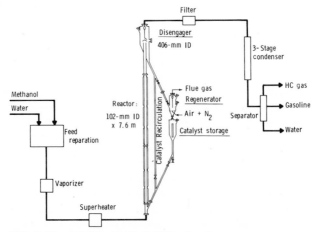

FIG. 3-18   4-bbl/day fluid-bed MTG pilot plant.

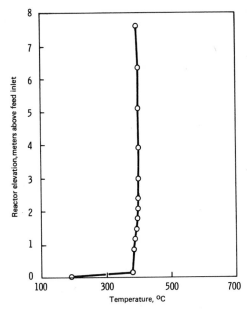

FIG. 3-19 Profile of 4-bbl/day fluid-bed reactor temperature.

This line also served as an external cooler for the catalyst. By controlling the rate of heat loss from this line, a stable and extremely uniform temperature profile was easily maintained in the reactor despite the highly exothermic nature of MTG reactions and the very large (75:1) L/D (length-to-diameter) ratio of the reactor. Figure 3-19 shows an actual profile of the reactor temperatures. The feed was introduced at 175 °C. Within a mixing zone of 2 to 3 ft, the desired reactor temperature (400 °C) was reached. From that point up, the temperature was virtually constant. These results provide an outstanding demonstration of the superior heat-transfer characteristics of a fluid bed. In addition, the system exhibited excellent thermal stability and handled upsets readily.

In order to study steady-state operation, the reactor was provided with a catalyst regenerator. Because only a small amount of catalyst needed to be regenerated per day (less than 7 kg), continuous regeneration was not practical. Instead, regeneration was performed batchwise on a daily basis. In this manner, steady-state operation could be simulated on a time-averaged basis. Figure 3-20 shows the data from a 10-day steady-state operation. Within a given day, catalytic activity declined as expected. The activity was then recovered by regeneration and some catalyst makeup. On a time-averaged basis, activity, conversion (99.5 percent by weight), and gasoline were maintained satisfactorily at the target levels.

The 4-bbl/day pilot plant was operated successfully for over 3 months. During this time, the plant afforded a demonstration of steady-state operation, accomplished virtually complete methanol conversion, and permitted studies of process variables and process concepts. The stability and regenerability of the catalyst were confirmed. The quality of the product gasoline was extensively tested in the laboratory as well as in test vehicles. Satisfactory results were obtained in both cases.

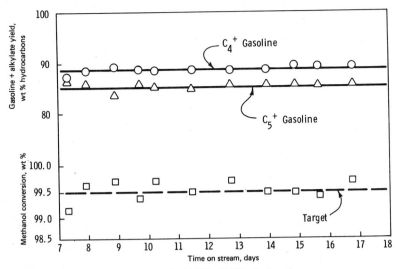

FIG. 3-20 Steady-state operation of 4-bbl/day fluid-bed pilot plant.

Demonstration Program (100-bbl/day)

A demonstration program for further scale-up of the fluid-bed MTG process was initiated in 1979. The objective of this program is to demonstrate the commercial feasibility
of the process. This 5½-year, $35 million program is an international effort of industry and
government. The industrial partners are URBK and Uhde, both of the FRG, and MRDC
of the United States. Partial financial support for the program is provided by BMFT of the
FRG and DOE.

This demonstration program encompasses the design, construction, and operation of a
100-bbl/day demonstration plant being sited at Wesseling, FRG. URBK serves as the
operating agent responsible for the day-to-day operation of the plant (by a joint URBK-
Uhde-MRDC team). Uhde is primarily responsible for the engineering and construction of
the plant, and MRDC provides the process know-how, catalyst, and technical guidance for
the program. Figure 3-21 shows the planned schedule for the program. The program has
progressed as scheduled.

A schematic of the 100-bbl/day plant is shown in Fig. 3-22. Crude methanol is simulated
by blending methanol and water. After vaporization and superheating, the feed is passed
through the reactor, which contains a 2-ft- (0.6-m-) diameter by 40-ft- (12.2-m-) high dense
fluid bed. After catalyst disengagement, the product vapor is condensed and the hydrocarbon products are separated from the aqueous product. A fractionation column splits
the hydrocarbons into C_4^- and C_5^+ fractions. The flow scheme is relatively simple and follows closely that of the 4-bbl/day plant.

The reactor system consists of three major process vessels: the reactor, the regenerator,
and the external catalyst cooler. For steady-state operation, catalyst will be continuously
circulated to the regenerator. For removal of reaction heat, two options will be tested. In
the first, catalyst will be withdrawn from the reactor, cooled in the external cooler, and

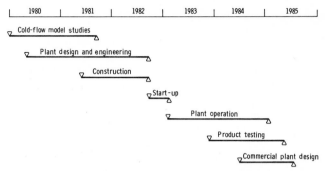

FIG. 3-21   100-bbl/day fluid-bed MTG program schedule.

then returned to the reactor. In the second, heat-exchanger pipes will be placed in the reactor fluid bed and the external cooler will not be used. Merits of each option will be evaluated.

Current plans call for an operating period of 21 months for the 100-bbl/day plant. Comprehensive product testing as well as a commercial-plant design are part of the program. Successful completion of this demonstration program will provide a firm basis for commercial application of the fluid-bed MTG process.

PROPERTIES AND PERFORMANCE OF MTG GASOLINE

Extensive testing of MTG gasoline has shown that its properties and performance are highly satisfactory. Results of tests on the fixed-bed MTG product are summarized in the following section. The fixed-bed process is being commercialized and offers more near-term interest than the developing fluid-bed process. It is expected, however, that the fluid-bed product will also show very good performance.[13]

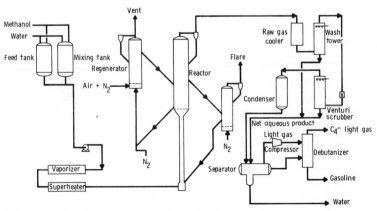

FIG. 3-22   Simplified flow diagram of 100-bbl/day MTG demonstration plant.

MTG Gasoline Composition

Fixed-bed MTG gasoline is composed primarily of directly synthesized gasoline and varying quantities of butanes and alkylate depending on volatility requirements and on the presence in the plant of an alkylation unit. MTG gasoline contains typically 60 percent by volume saturates (paraffins and naphthenes), 10 percent by volume olefins, and 30 percent by volume aromatics. Most of the saturates are branched paraffins. The gasoline contains essentially no sulfur or nitrogen compounds.

One characteristic of MTG gasoline is that the finished product contains about 2 percent by weight durene (1,2,4,5-tetramethylbenzene), whereas conventional gasoline in the United States typically contains 0.2 to 0.5 percent by weight durene. While durene has a high octane blending number [approximately 110 RON (research octane number) clear] and boils in the gasoline range (197 °C), its relatively high freezing point of 79 °C can cause automotive vehicle drivability problems if its concentration in the gasoline is excessive.[11,14] The MTG process produces about 3 to 7 percent by weight durene in the gasoline, but it can be modified to produce more or less durene. The 2 percent by weight content was selected as a target to ensure overall satisfactory performance in all aspects of distribution, marketing, and end use.[23] This level can be readily achieved by process changes or by blending with conventional gasolines.[24] Neither method will adversely affect gasoline yield or quality.

Physical Characteristics of Fixed-Bed Gasoline[23]

The antiknock quality of fixed-bed MTG gasoline meets or exceeds that of conventional unleaded regular gasolines in the United States. Leaded octane quality meets or exceeds specifications for leaded premium grades in international markets. Typical properties of the finished gasoline containing less than 2 percent by weight durene are listed in Table 3-7. This gasoline was produced from the yield of the 4-bbl/day unit. The volatility characteristics of this gasoline are similar to those of petroleum-derived gasolines. The gasoline has a low content of existent gum and is not corrosive.

TABLE 3-7 Typical fixed-bed gasoline properties

Research octane number	
Unleaded	93
Plus 3.17 g Pb/U.S. gal (0.84 g/L) as TEL	100
Motor octane number	
Unleaded	83
Plus 3.17 g Pb/U.S. gal (0.84 g/L) as TEL	91
Reid vapor pressure, lb/in^2 (101.3 kPa)	10.2
ASTM distillation, °C (°F)	
10% evaporated	46(115)
50% evaporated	99(210)
90% evaporated	166(330)
End point	204(400)
Existent gum, mg/100 mL	1
Copper strip corrosion rating	1A

Fixed-bed gasoline containing a conventional additive package also meets other quality standards established for conventional gasoline, e.g., carburetor detergency, emulsion formation, filterability, copper attack, metals retention, multimetal corrosion, and storage stability. The gasoline will not adversely affect automotive exhaust emissions.

In-Vehicle Performance of Fixed-Bed Gasoline

Drivability studies indicated that the performance of finished MTG gasoline is equivalent to that of conventional gasolines of similar volatility under a variety of ambient temperatures and driving procedures.[23]

Controlled drivability studies were carried out with six 1981-model U.S. cars which represented popular and/or expected future power-train designs in the United States. In this test, trained drivers operated and rated the vehicles as prescribed in the Coordinating Research Council (CRC) Cold Start and Driveaway Test Method. Drivability malfunctions such as stalls, hesitation, and idle roughness were noted and assigned demerits depending upon their frequency and severity. For example, one stall during acceleration would result in 42 demerits. A demerit rating of below 50 indicates the gasoline has good drivability, while a rating which exceeds 100 is considered fair to poor. The test cars were equipped with an automatic transmission and a two-barrel carburetor. Engine size ranged from 1.6- to 2.6-L displacement, and one car was equipped with a 3.6-L, V-6 engine.

The cold-start and driveaway performance of MTG and conventional gasolines is compared in Table 3-8, which also lists some of the test fuels' properties. These results indicate that the performance of MTG gasoline is excellent and is comparable to that of high-quality conventional gasolines.

Controlled drivability tests were also carried out with New Zealand–type cars in a test fleet selected to simulate recent car sales in that country. MTG gasoline containing 2 per-

TABLE 3-8   Cold-start and driveaway performance of fixed-bed MTG gasoline*

	MTG gasoline	Conventional Gasoline		
		Fuel 1	Fuel 2	Fuel 3
Fuel properties				
Reid vapor pressure, lb/in² (101.3 kPa)	11.6	13.2	12.7	11.1
ASTM distillation, °C (°F)				
10% evaporated	40 (104)	37 (98)	38 (100)	41 (106)
50% evaporated	103(218)	92 (198)	111(231)	113(236)
90% evaporated	179(354)	183(362)	166(331)	184(364)
Cold-start and driveaway performance at −4 °C (25 °F), demerits				
Six-car average	49	47	67	101
Range	16 –110	12 –96	25 –150	41 –170
Cold-start and driveaway performance at 4 °C (40 °F), demerits				
Six-car average	41	28	68	68
Range	13 –86	8 –63	10 –169	15 –127

*Average values for six 1981 cars.

cent by weight durene demonstrated as good performance as petroleum-derived gasoline of similar volatility. Vehicle ambient operating conditions ranged from −18 to 32 °C (0 to 90 °F), temperatures which bracket the ambient temperatures likely to be encountered in New Zealand.

In a consumer-type test carried out with 34 cars (23 U.S., 6 Japanese, and 5 European) over a temperature range of −12 to 32 °C (10 to 90 °F), the drivability performance of MTG gasoline compared well with high-quality petroleum-derived gasoline. Vehicle operators were unaware of which fuel was being used and completed detailed performance-evaluation forms for each operating period.

PROCESS ECONOMICS

In this section, the process economics for converting methanol to gasoline is evaluated for a commercial-sized fixed-bed MTG plant; the costs of constructing and operating such a plant are also discussed. Included in the plant description are an overall material balance and a summary of the utilities required. In addition, operating concerns such as plant effluents and emissions are addressed.

Since in the United States the primary source of methanol for the MTG process would be coal, the feed methanol for the MTG plant discussed here is typical of crude methanol derived from coal gasification. The effects of different feedstock compositions are discussed separately. The major costs of a coal-to-gasoline complex are those involved with the primary conversion of the coal into a clean synthesis gas; for such a complex the investment for the MTG portion is a small fraction of the overall investment (less than 15 percent).

Description of a Commercial MTG Plant

A block flow diagram for an MTG plant is presented in Fig. 3-23. As shown, the MTG plant can be viewed as having two sections: (1) the MTG reaction section and (2) the product distillation section. In the MTG reaction section crude methanol is converted to crude gasoline. In the product distillation section for the design discussed here, the LPG

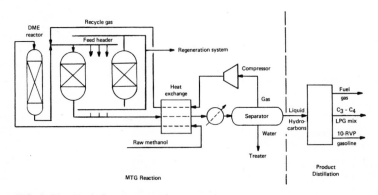

FIG. 3-23   Block flow diagram of an MTG unit.

mixture is simply separated from the MTG gasoline. It should be noted, however, that the distillation section could be more complex, depending on how the C_3 and C_4 components of the LPG mix are used. Facilities could be added, for example, to provide alkylation of the butenes and/or pentenes or separation of isobutane for sales.

Crude methanol to the MTG reaction section is fed either directly from a methanol synthesis unit or from offsite storage. The feed is vaporized and superheated by heat exchange with reactor effluent and enters into the DME reactor where the methanol is partially converted by dehydration to DME. This step is followed by complete methanol conversion in a set of four downflow, fixed-bed conversion reactors containing ZSM-5 catalyst. The effluent from the DME reactor is combined with a recycle gas to control the reaction temperature rise in the conversion reactors within a limit of about 60 °C. The heat content of the effluent from the conversion reactors is used to generate steam, reheat recycle gas, and vaporize and superheat the methanol feed.

After the reactor effluent leaves the heat-recovery train, it is condensed into three phases which are separated in a product separator. The gas phase is recycled, the water phase is pumped to a wastewater treatment unit, and the hydrocarbon phase is pumped to the product distillation section.

The operating conditions for the MTG plant are not severe and are within the range of conditions for normal petroleum processing. Typical operating conditions for the plant include inlet and outlet temperatures for the DME reactor of 316 °C and 405 °C, respectively. Inlet and outlet temperatures for the conversion reactors are approximately 360 to 415 °C, respectively, and the reactors are designed for a pressure of 2300 kPa. Approximately 9 mol of recycle gas is used per mole of crude-methanol feedstock. The weight hourly space velocities, based on neat methanol, are about 4 and 1.6 for the DME reactor and ZSM-5 reactor, respectively.

After several weeks of operation, the conversion catalyst becomes inactive because of coke buildup. Catalyst activity is restored by controlled air burning of this coke. In order to minimize throughput variations, the MTG plant design incorporates four conversion reactors operating in parallel, while a fifth is being regenerated in a separate regeneration loop.

The product distillation section in this plant consists of two simple petroleum fractionators. Crude gasoline from the MTG reaction section is first fed to a deethanizer to remove the light ends. The overhead gas from this tower is used as fuel gas, and the bottoms from this tower are fed to a gasoline stabilizer. In the stabilizer the gasoline product is separated from the LPG mixture. The recovered LPG mixture is a combination of C_3 and C_4 material. Although this mixture could be processed further, no facilities for such processing are included in the product distillation section for this plant.

Overall Material Balance and Utility Requirements

An overall material balance for a commercial MTG plant is given in Table 3-9. The MTG plant is designed to convert about 4600 Mg per stream day (sd) of methanol (computed on a neat basis) to 1690 Mg/sd (14,900 bbl/sd) of synthetic gasoline and about 280 Mg/sd (3200 bbl/sd) of a mixture of propane and butane. Although crude methanol generally contains significant amounts of dissolved gases, no dissolved gases are included in the material balance, since their composition can vary. The water content of crude methanol derived from coal typically is on the order of 4 percent by weight; as an example, this value was assumed for the MTG plant feed (see Table 3-9).

TABLE 3-9 Overall material balance for a commercial MTG plant

		Output Streams				
	Crude methanol	10-RVP* gasoline	Propane- butane mixture	MTG coke	MTG fuel gas	Process water
Mg/sd†	4792	1686	278	2	43	2783
Mg/h	199.67	70.27	11.59	0.06	1.81	115.94
kg mol/h						
Hydrogen		——	——		1.10	——
Carbon monoxide		——	——		0.80	——
Carbon dioxide		——	Trace		1.85	0.04
Methane		——	——		39.20	1.11
Ethylene		——	——		4.60	——
Ethane		——	2.6		2.60	Trace
Propylene		——	3.3		0.50	——
Propane		0.3	76.9		9.50	0.17
Isobutane		21.6	102.4		3.40	0.03
Butenes		3.9	12.0		1.20	0.02
Butane		16.7	23.2		0.40	0.02
Isopentane		140.3	0.5		0.10	Trace
Pentenes		25.0	Trace		——	0.10
n-Pentane		16.0			——	——
Cyclopentane		2.8			——	——
Hexane and heavier		521.0			0.20	0.31
Methanol	5982.0				——	0.96
Acetone	——				——	1.95
Acetic acid	——				——	1.59
Water	444.0				1.50	6414.74

*Reid vapor pressure.

†Stream day.

Estimated utility requirements for the MTG plant are listed in Table 3-10. These require-ments are based on the plant size given by the material balance (i.e., 1690 Mg/sd of gas-oline product). The power requirements for the plant compressors are shown separately from the other utilities. Because of variations in load, steam turbine drivers would normally be used for the first two services. Note that the utility requirements shown do not reflect the amount of steam and cooling water required for the compressor drivers.

Capital Investment and Manufacturing Costs

The total construction cost in 1980 dollars for a 1690-Mg/sd MTG plant is estimated to be $170 million. Process facilities for the MTG reaction section make up the largest por-tion of this investment estimate; the simple product distillation and treating of the water effluent make up the remainder of the estimate. Not included in the estimate are costs for interconnecting piping to a methanol synthesis plant, tankage, utilities, and shipping facil-ities. This plant investment estimate includes direct construction costs, field distributable costs, contractor's engineering costs and associated fees, and owner's costs. An additional 30 percent of the investment is included for contingency and estimating allowances.

TABLE 3-10   Utility requirements for a commercial MTG plant

Major compressors (utility service not included)	
Recycle-gas compressor, kW	26,110
Regeneration-gas circulator, kW	5,220
Regeneration-air compressor, kW	900
Steam	
Consumption [at 2860 kPa (400 psig)], kg/h	25,900
Production [at 4150 kPa (600 psig)], kg/h	32,750
Cooling-water circulation rate (17-°C rise), m³/h	6,750
Power consumption, kW	620
Fuel	
Average consumption (Fired, LHV), MW	1.0
Total production (LHV), MW	24.2
Nitrogen consumption*	
Maximum rate, m³(n)/h†	3,000
Maximum per year, m³(n)/h†	1,500,000

*Required for regeneration of the MTG conversion reactors; maximum frequency is 100 times per year.

†m³(n) symbolizes the normal cubic meter, that is, the volume of one cubic meter of gas measured at 0 °C and 101.3 kPa pressure.

Although the MTG plant is most likely to be part of an overall coal-to-gasoline complex, the economics is treated here simply in terms of transfer prices for methanol and for the required utilities. On this basis, the cost of manufacturing MTG gasoline is estimated to be 39¢/L (147¢/gal). A breakdown of the various costs to manufacture MTG gasoline is shown below.

	Cost, ¢/L of gasoline product
Methanol (at 15c/L)	36
Operating costs	3
LPG by-product credit (at 23c/L)	(5)
Capital charges (including 12% DCF return)*	5
Total cost at plant	39

*Discounted cash-flow return.

The largest contribution to the MTG gasoline cost is for the crude-methanol feed. To make 1 L of gasoline requires about 2.5 L of methanol; so considering the transfer price of methanol on a neat basis to be 15¢/L (55¢/gal), this contribution amounts to a cost of 36 cents per liter of gasoline, or over 90 percent of the net manufacturing cost.

The operating costs contribute about 3 cents per liter of gasoline manufactured. These costs include expenses for the operators, maintenance supplies and labor, insurance, local taxes, and miscellaneous overhead. Also included are the transfer costs for the utilities (steam, power, cooling water, etc.) required for the MTG plant, including those for the major compressors.

Credit for the LPG mixture was set at 23 cents per liter of the mixture and amounts to about 5 cents per liter of gasoline product. Further separation of this material could increase its value, thus increasing the by-product credit.

The cost of capital amounts to about 5 cents per liter of gasoline product. One of the economic asumptions made for this analysis is that an annual cost of about 21 percent of the capital investment is required to cover the costs of plant capital, income tax, and after-tax returns.

Effluents and Emissions

A significant quantity of water effluent is produced in the MTG process, as indicated in the material balance. In an integrated complex, consideration would be given to recycling part or all of this water to other plants in the complex to reduce the costs of makeup water. However, in the plant discussed here, facilities are included to provide biological treatment of the MTG wastewater.

As concerns gaseous effluents, during regeneration of the conversion catalyst an off-gas is released which during a 24-h burn period will contain from 160 to 260 kg/h of carbon monoxide. The carbon monoxide concentration will average between 2 to 3 percent by volume during this period. Regeneration of the DME catalyst is expected to produce similar emissions, but regeneration of this catalyst should not be necessary more often than once a year.

There should not be any problems with odor during regeneration, since coke does not contain any sulfur or nitrogen. In addition, both catalysts are not promoted with, nor do they contain, any volatile chemical compounds. Similarly, the release of any objectionable materials arising from heavy-metal constituents will not occur, because both catalysts do not contain such materials.

Spent catalysts from the MTG plant are suitable for disposal to a controlled landfill. Because of their potential value, however, some of the catalyst may be returned to the manufacturer for recovery. A list of spent catalysts and their annual disposal volume is shown in the following table.

Catalyst	Assumed life, yr	Annual quantity for disposal	
		m³	Mg
DME	2	37*	10*
Conversion	1	290	150

* Annual rate based on a 2-yr catalyst life.

Effects of Changes in Feedstocks

Modifications in design of the MTG plant may be required for differing compositions of crude methanol. Crude methanols may vary according to the amount of impurities they contain such as light gases, DME, higher alcohols, and water. In the case of light gases, these compounds pass unaltered through the DME and conversion reactors and exit to the fuel gas; they affect design and operation only to the extent that they affect the composition of the recycle gas. Similarly, DME and the higher alcohols, which are frequently present in crude methanol, are easily converted to gasoline and do not require design changes as long as they are not present in significant quantities. Their hydrocarbon portion (CH_2) can be considered equivalent to that of methanol in regard to forming hydrocarbon products.

Though water in crude methanol passes through the reactor system and appears in the wastewater product, increases in water content above about 7 percent by weight would require changes in the plant design that is presented here. Varying the water content will change (1) the amount of methanol dehydration in the DME reactor, (2) the operating conditions in the conversion reactors, and (3) the vaporizing or condensing sections of the heat-exchange system. All of these changes would affect the economics to some extent.

Potential for Cost Improvement

Potential cost reductions for the MTG process, like most catalytic processes, are most likely to fall into the category of catalyst improvement. If catalyst selectivity toward gasoline were increased, a lower unit cost for gasoline would result, since the LPG material usually has a reduced value compared to that of gasoline. Alternatively, if catalyst stability were improved, the conversion reactors could be operated at a reduced gas-recycle ratio. A reduced recycle ratio would lower the investment for both equipment for heat exchange between recycle gas and effluent and for the recycle-gas compressor, both of which are significant cost items.

As mentioned previously, the cost of an MTG plant is not a large percentage of the overall investment of a coal-to-gasoline complex. This is pointed out pictorially in Fig. 3-24, which shows the approximate percentages of investment associated with each of the major sections of a coal-to-gasoline plant. Offsite and support facilities for the plant are

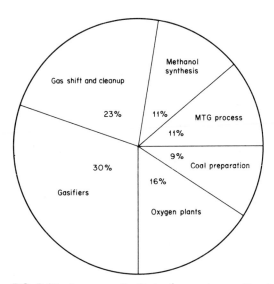

FIG. 3-24   Investment distribution for a coal-to-gasoline plant.

prorated appropriately. Note that the facilities associated with coal gasification and gas cleaning make up the largest segment of the overall investment. Any significant reductions in capital investment for the plant are more likely to come from these coal-related process facilities than from the facilities associated with methanol synthesis or methanol-to-gasoline conversion.

SUMMARY

The Mobil MTG (Methanol-to-Gasoline) process offers a novel route for synthesizing high-octane gasoline from nonpetroleum carbonaceous resources such as coal and natural gas. The process is relatively simple and highly energy-efficient and poses no environmental problems. The fixed-bed MTG process is being commercialized in a 14,000-bbl/day natural gas-to-gasoline plant in New Zealand. The fluid-bed MTG process is undergoing final stages of development.

Gasoline from the MTG process has been tested extensively, including in-vehicle evaluations, and found to perform as well as or better than conventional gasoline. The cost of MTG gasoline is strongly dependent on the cost of the methanol feed. For methanol feed at 15 cents/L, the estimated cost of the gasoline is 39 cents/L. In an integrated coal-to-gasoline plant, the MTG section would constitute only 11 percent of the required capital investment.

REFERENCES

1. Meisel, S. L., J. P. McCullough, C. H. Lechthaler, and P. B. Weisz: "Gasoline from Methanol—In One Step," *CHEMTECH*, vol. 6, 1976, p. 86.

2. Kokotailo, G. T., S. L. Lawton, D. H. Olson, and W. M. Meir: "Structure of Synthetic Zeolite ZSM-5," *Nature*, vol. 272, 1978, p. 437.

3. Chang, C. D., and A. J. Silvestri: "The Conversion of Methanol and Other O-Compounds to Hydrocarbons over Zeolite Catalysts," *J. Catal.*, vol. 47, 1977, p. 249.

4. Derouane, E. G., J. B. Nage, P. Dejaifve, J. H. C. van Hooff, B. P. Spekman, J. C. Vedrine, and C. Naccache: "Elucidation of the Mechanism of Conversion of Methanol and Ethanol on a New Type of Synthetic Zeolite," *J. Catal.*, vol. 53, 1978, p. 40.

5. Kaeding, W. W., and S. A. Butter: "Production of Chemicals from Methanol. I. Low Molecular Weight Olefins," *J. Catal.*, vol. 61, 1980, p. 155.

6. Van den Berg, J. P., J. P. Wolthuizen, and J. H. C. van Hooff: "The Conversion of Dimethyl Ether to Hydrocarbons on Zeolite H-ZSM-5—The Reaction Mechanism for Formation of Primary Olefins," *Proc. 5th International Conference on Zeolites, Naples, Italy,* L. V. C. Rees (ed.), Heyden, London, 1980, p. 649.

7. Chen, N. Y., and W. E. Garwood: "Some Catalytic Properties of ZSM-5, a New Shape Selective Zeolite," *J. Catal.*, vol. 52, 1978, p. 453.

8. Lechthaler, C. H., J. J. Wise, P. B. Weisz, and A. J. Silvestri: "Novel Technology for Conversion of Methanol and Synthesis Gas to Hydrocarbons," *13th Intersociety Energy Conversion Engineering Conference, Aug. 20-25, 1978, San Diego.*

9. Chen, N. Y., and W. J. Reagan: "Evidence of Autocatalysis in Methanol to Hydrocarbon Reactions over Zeolite Catalysts," *J. Catal.*, vol. 59, 1979, p. 123.

10. Chang, C. D.: "A Kinetic Model for Methanol Conversion to Hydrocarbons," *Chem. Eng. Sci.*, vol. 35, 1980, Pergamon Press, Ltd., p. 619.

11. Voltz, S. E., and J. J. Wise: "Development Studies on Conversion of Methanol and Related Oxygenates to Gasoline," Final Report, ERDA Contract no. E(49-18)-1773, November 1976.

12. Wise, J. J., and A. J. Silvestri: "Mobil Process Efficiently Converts Methanol to Gasoline," *Oil Gas J.*, vol. 74, 1976, pp. 140–142.

13. Kam, A. Y., and W. Lee: "Fluid-Bed Process Studies on Selective Conversion of Methanol of High Octane Gasoline," Final Report, DOE Contract no. EX-76-C-01-2490, 1978.

14. Yurchak, S., S. E. Voltz, and J. P. Warner: "Process Aging Studies in the Conversion of Methanol to Gasoline in a Fixed-Bed Reactor," *Ind. Eng. Chem. Process Des. Dev.*, vol. 18, 1979, p. 527.

15. Chang, C. D., J. C. W. Kuo, W. H. Lang, S. M. Jacob, J. J. Wise, and A. J. Silvestri: "Process Studies on the Conversion of Methanol to Gasoline," *Ind. Eng. Chem. Process Des. Dev.*, vol. 17, 1978, p. 255.

16. Satterfield, C. N.: *Mass Transfer in Heterogeneous Catalysis*, MIT, Cambridge, Massachusetts, 1970.

17. Lee, W., S. Yurchak, N. Daviduk, and J. Maziuk: "A Fixed-Bed Process for the Conversion of Methanol to Gasoline," *1980 NPRA Annual Meeting, New Orleans, Mar. 25, 1980.*

18. Liederman, D., S. Yurchak, J. C. W. Kuo, and W. Lee: "Mobil Methanol-to-Gasoline Process," *15th Intersociety Energy Conversion Engineering Conference, Seattle, Aug. 18–22, 1980.*

19. Schreiner, M.: "Research Guidance Studies to Assess Gasoline from Coal by Methanol-to-Gasoline and SASOL-type Fischer-Tropsch Technologies," Final Report, DOE Contract no. EX-76-C-01-2447, February 1978.

20. Patel, J. G.: "Low, Medium-Btu Gas from Coal Lead Conversion Routes," *Oil Gas J.*, vol. 79, no. 26, 1981, p. 90.

21. United States Synthetic Fuels Corporation: Press Release, Apr. 1, 1981.

22. Liederman, D., S. M. Jacob, S. E. Voltz, and J. J. Wise: "Process Variable Effects in the Conversion of Methanol to Gasoline in a Fluid-Bed Reactor," *Ind. Eng. Chem. Process Des. Dev.*, vol. 17, 1978, p. 340.

23. Fitch, F. B., and W. Lee: "Methanol-to-Gasoline, An Alternative Route to High Quality Gasoline," *SAE International Pacific Conference, Honolulu, Nov. 16–19, 1981.*

24. Silvestri, A. J.: "Mobil Methanol-to-Gasoline Process," *181st ACS National Meeting, Atlanta, Mar. 29–Apr. 3, 1981.*

THE LURGI LOW-PRESSURE METHANOL PROCESS

DIPL. ING. EMIL SUPP
DR. ING. ROCHUS F. QUINKLER

Lurgi Kohle und Mineralöltechnik GmbH
Frankfurt am Main, Federal Republic of Germany

INTRODUCTION

Methanol is presently one of the top 20 organic chemicals in production. It is manufactured by one of the most important high-pressure organic syntheses, and methanol production is the second largest use for synthesis gas, ammonia production being the largest.

As is discussed in the introduction to this portion of the handbook, methanol was produced industrially from coal-derived synthesis gas for many years but natural gas and petroleum distillates replaced coal as a feedstock after World War II. As the demand for synthetic transportation fuels becomes established, it can be expected that coal will once again become a major feedstock for the production of methanol from synthesis gas, which can then be blended with or converted to gasoline.

CHEMISTRY AND THERMODYNAMICS OF METHANOL SYNTHESIS

From the middle of the ninetenth century up to the nineteen twenties, methanol was produced exclusively by the distillation of wood. The first synthetic methanol was produced in 1913 by Mittasch and Schneider at Badische Anilin & Soda Fabrik AG, Ludwigshafen, Germany. In 1923 the first commercial-scale plant for production of synthetic methanol was started up by Ammoniakwerk Merseburg at Leuna, Germany. The main chemical reactions for the formation of methanol from CO or CO_2 and H_2 are

$$CO + 2H_2 \rightleftharpoons CH_3OH \qquad \Delta H = -90{,}786 \text{ kJ/mol}$$

$$CO_2 + 3H_2 \rightleftharpoons CH_3OH + H_2O \qquad \Delta H = -49{,}530 \text{ kJ/mol}$$

The reaction heat generated during methanol formation is considerable, and as the temperature increases, the equilibrium shifts toward methanol formation; however at the same time the chances for competing side reactions increase. Side reactions can lead to the formation of methane, dimethyl ether, methyl formate, higher alcohols, and acetone. Although the reaction heat must be withdrawn quickly, the temperature must also be maintained high enough to achieve acceptable reaction times and yields.

The early methods for producing synthetic methanol used a selective zinc-chromium catalyst. An improved type of this catalyst is still being used today for conventional high-pressure processes. These processes normally operate at a pressure of 4410 lb/in² (30 MPa) and at a temperature of 350 °C. Until ICI commercially applied copper catalysts which permitted methanol synthesis at low pressures, all conventional high-pressure processes were substantially identical to the process developed in 1923.

Most of the high-pressure processes operate under a CO partial pressure of 588 to 735 lb/in² (4 to 5 MPa) at the reactor inlet. Some high-pressure processes use multistage quenching with cold synthesis gas to control the reactor temperatures; others transfer the reaction heat of [about 1259 Btu/lb (2930 kJ/kg)] to cooling media via appropriate equipment in the reactor. The output of by-products, especially methane and higher alcohols, is considerable as a result of the high CO partial pressure in such high-pressure processes. Despite the use of CO-resistant steels and a copper lining in the reactor, the formation of iron carbonyl cannot be avoided. The iron carbonyl decomposes at the temperatures prevailing in the reactor, causing iron deposits on the catalyst, increasing the pressure drop, and creating the hazard of excessive methane formation.

Only one process, that of Union Rheinische Braunkohlen-Kraftstoff AG, uses a CO partial pressure below 294 lb/in^2 (2 MPa), thus avoiding the formation of iron carbonyl and methane and largely suppressing the formation of higher alcohols.

DEVELOPMENT OF THE LURGI LOW-PRESSURE METHANOL PROCESS

At the end of the fifties, Lurgi started development work on a low-pressure methanol process at its research and development center at Frankfurt am Main, Federal Republic of Germany (FRG). The first copper catalysts developed by Lurgi permitted reasonable conversion rates at temperatures below 300 °C. However, the space-time yield and the life of these catalysts proved to be unsatisfactory, and so the development work was suspended for several years.

In 1964 Lurgi resumed the research work. At that time the purification of synthesis gas was no longer a problem, since efficient chemical and physical wash processes (such as Amisol, Lurgi Rectisol, and Purisol) had become available. The Lurgi Rectisol process in particular reduces impurities—mainly sulfur compounds—in the synthesis gas to concentrations which are not detrimental to the activity of a copper-based methanol catalyst. Therefore, research and development work could be concentrated on other problems, such as

• Finding a catalyst with adequate activity and a long service life

• Developing a reactor system that assures safe, flexible operation and offers optimum conditions for the catalyst

• Finding an optimum combination of gas production, methanol synthesis, and methanol distillation, with the aim of developing an economic process that utilizes the heat from the exothermic methanol reaction

The problem of finding a suitable catalyst with specific properties, which was given top priority, was the last to be solved. Several years of development work by Lurgi were required to select, from the numerous elements available, those catalysts which give the desired efficiency and to adapt and modify the laboratory formulation for use as a commercially applicable catalyst.

The development of a suitable reactor system proceeded relatively quickly. A reactor system developed earlier by Lurgi for the Fischer-Tropsch synthesis seemed to be a promising solution to both the reactor design and heat recovery problems. The classical Fischer-Tropsch reactor in the Lurgi-Ruhrchemie process version is a tubular reactor. This Fischer-Tropsch reactor as well as the Lurgi Methanol reactor system are similar to shell-and-tube heat exchangers; however they are installed vertically.

The tubes in the Lurgi Methanol reactor contain the catalyst. The space around the tubes is filled with boiling water, maintaining a uniform catalyst temperature over the reactor cross section and over the full length of the tubes.

Comprehensive pilot plant tests were carried out to find suitable tube materials to meet the following requirements:

• Resistance to CO under the prevailing operating conditions in order to avoid iron carbonyl formation

• Resistance to vaporizing feedwater

- No formation of undesired by-products
- Same thermal expansion as normal carbon steel in order to permit a simple reactor design

The range of suitable steels was limited by the requirement for CO resistance in conjunction with the same thermal expansion as carbon steel. CO resistance was mandatory; long-term operation of a pilot plant showed that iron carbonyl formation and decomposition occur only within a small temperature range in the heat exchanger upstream of the reactor, resulting in iron deposits on the catalyst. While in most high-pressure processes the iron deposits on the catalyst lead to a rapid catalyst deactivation, the active iron particles in a low-pressure process are detrimental only insofar as they favor the formation of paraffins. Conventional steels have been found that meet these requirements.

Numerous computer calculations were made to optimize the economics of the combined scheme of gas re-forming for synthesis gas production and of methanol sysnthesis and distillation. One of the major considerations in the process development work was to achieve steam generation in the synthesis reactor at a pressure of 588 to 735 lb/in² (4 to 5 MPa).

This work resulted in a new methanol process, the Lurgi Low-Pressure Methanol process, which can be best characterized as follows:

- Tubular reactor system
- Operating pressures between 735 and 1470 lb/in² (5 to 10 MPa)
- Operating temperatures between 230 and 265 °C
- Steam production at 588 to 735 lb/in² (4 to 5 MPa)
- Proprietary catalyst

At the works of Union Rheinische Braunkohlen-Kraftstoff AG at Wesseling, FRG (UK Wesseling), two single-tube pilot units were operated for almost 3 years to study operating conditions, catalyst behavior, and other process parameters.

In April 1970, an 8,800,000-lb/yr (4000-Mg/yr) plant employing the Lurgi Low-Pressure Methanol process was started up jointly at Wesseling, FRG, by UK Wesseling and Lurgi using synthesis gas from the UK Wesseling's works system. The successful operation of this plant, which had been built for demonstration purposes and which already had large-scale dimensions, proved that the choice of process parameters was correct.

The first commercial-scale methanol plant using the Lurgi Low-Pressure Methanol process was designed and constructed by Lurgi for Veba Chemie AG of Gelsenkirchen, FRG, in 1970. By the end of 1981 more than 10 methanol plants using the Lurgi Low-Pressure Methanol process were in operation and 10 plants were in various stages of design and construction. A typical synthesis section of a Lurgi methanol plant is shown in Fig. 4-1.

THE LURGI LOW-PRESSURE METHANOL REACTOR SYSTEM

Lurgi, when dropping the high-pressure methanol technology, also dropped the quench reactor system and adopted the tubular reactor system as part of the low-pressure technology.

FIG. 4-1 Synthesis section of a Lurgi Low-Pressure Methanol plant of INA at Lendava, Yugoslavia. (*Lurgi.*)

Important features of the Lurgi reactor system are shown in Fig. 4-2. The catalyst is placed inside the tubes of the reactor. The catalyst-filled tubes are cooled by boiling water. There is a natural circulation of the boiling water between the inner shell of the reactor and a steam drum placed on top of the reactor. A uniform distribution of the synthesis gas over the multitude of catalyst-filled tubes is achieved by a pressure drop over the length of the tubes.

The steam-pressure controller at the steam drum is the only controlling device in this overall reactor–heat exchanger system. The temperature of the overall reactor system and the temperature of the catalyst in the tubes are controlled simply, accurately, and safely by this steam-pressure controller.

For starting up the plant, only a steam injector is required; it initiates the natural water circulation between the steam drum and the reactor and heats the reactor very uniformly with or without circulation gas flowing.

Because of the considerable quantities of water surrounding the tubes filled with catalyst, the Lurgi tubular methanol reactor behaves well and consistently, even under extreme operating conditions. Control of reactor temperature via the pressure of the boiling water provides simple operability.

During normal operation the pressure level of the boiling water is around 588 lb/in^2 (4 MPa). At this pressure level the temperature increase in relation to a change in the pressure is very flat; therefore pressure fluctuations in the boiling-water system scarcely affect the temperature of the catalyst in the tubes.

High or excessively low feed-gas temperatures at the reactor inlet (occurring for any reason) tend to stop the reaction in other methanol reactor systems; however, in the Lurgi

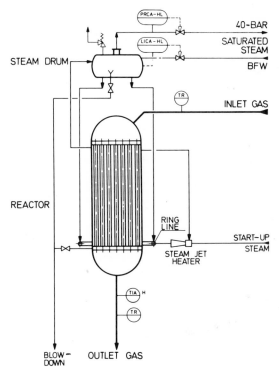

FIG. 4-2   Lurgi Boiling Water Low-Pressure Methanol reactor. (*Lurgi.*)

tubular reactor system the boiling water acts as a buffer against feed-gas temperature deviations. The top and bottom portions of the reactor tubes act as a controlling heat exchanger. Therefore, the Lurgi tubular reactor provides a more constant temperature profile over the tube length than does a quench-type reactor, and thus increases the life of the catalyst (compare in Fig. 4-3). In a quench-type methanol reactor the only buffer against changes in temperature is the catalyst; therefore the quench-type methanol reactor will cool down below the reaction start temperature about 10 times faster than the Lurgi tubular methanol reactor. These changes in temperature can occur as a result of failures in the synthesis-gas supply or in the upstream system.

The maximum temperature differential between the boiling water in the shell and the tube centerline is about 10 to 12 °C because of the intensive heat exchange between the boiling water around the tubes and the gas in the tubes. The temperature differential in a quench-type reactor is governed by gas composition, quench-gas rate, and number of quench points and is usually 30 °C or more.

Moreover, as shown in Fig. 4-3, the temperature of the catalyst in the tubes of a Lurgi Low-Pressure Methanol reactor decreases toward the outlet and thus contributes to a better equilibrium, while each stage of a quench-type reactor has an increasing temperature profile.

A particular advantage of the tubular methanol reactor is that almost the entire reaction

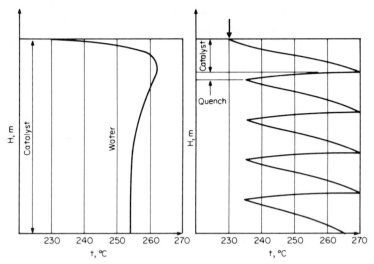

FIG. 4-3   Temperature profiles in tubular (left) and quench (right) reactors. (*Lurgi.*)

heat can be utilized for producing high-pressure steam at constant temperature levels; in a quench-type reactor the reaction heat is largely used to bring the quench gas to reaction temperature, and steam generation from the waste heat of the reactor exit gas is possible only at falling reactor-exit-gas temperatures.

The selection of the optimum design gas pressure at the inlet of the Lurgi tubular reactor [normally between 840 and 2205 lb/in² (6 and 15 MPa)] depends upon energy considerations and upon investment cost. Calculations show that 99 percent CO conversion is achieved at 735 lb/in² (5 MPa).

THE LURGI CATALYST

Type of Catalyst

The main component of the Lurgi methanol catalyst is copper. As with all other copper catalysts, this one tends to recrystallize and deactivate at temperatures above 270 °C. A catalyst of this type has to be treated gently in a quasi-isothermal system.

Use of catalyst-filled tubes in the Lurgi methanol reactor system ensures an extremely quick transfer of the reaction heat from the catalyst to a cooling medium. The medium temperature of the catalyst does not fluctuate greatly in such a quasi-isothermal system, which enables the use of highly active catalysts. Lurgi is in a position to offer a catalyst attaining, with an appropriate synthesis-gas feedstock, a methanol yield close to 1 kg/L, the magic figure for a low-pressure methanol process.

Influence of Age

Figure 4-4 shows various temperature profiles in a Lurgi tubular methanol reactor for different catalyst ages. Curve I shows the temperature profile, for a fresh catalyst, of the catalyst over the length of the tubular reactor. In this case, the gas temperature in the upper section of the tubes increases to about 10 to 12 °C above the temperature of the boiling water in the shell.

Toward the outlet of the tubes the concentrations of CO and CO_2 in the gas stream decrease, the reaction velocity decreases, and the gas temperature gradually approaches the boiling-water temperature to within 3 to 5 °C.

Curve II depicts the temperature profile for a noticeably aged catalyst and for an unchanged boiling-water temperature. Obviously, catalyst aging takes place only in the upper sections of the tubes where maximum conversion takes place. In this case the gas temperature in the upper sections of the tubes increases slowly to only about 5 to 8 °C above the boiling-water temperature; therefore, the reaction velocity in the upper sections of the tubes is less, the maximum temperature is somewhat lower than in the first case, and the center of conversion is somewhat stretched. The catalyst in the bottom sections of the tubes is aged to a lesser degree, since the temperature in the bottom sections is lower and conversion in the bottom sections occurs at a lower, more favorable equilibrium temperature.

Curve III shows the temperature profile for a catalyst after 3 to 4 years of service. The temperature profile shown in curve III has the same tendencies as that of curve II. However, this temperature profile is obtained only after the temperature of the boiling water in the shell is increased by 5 °C.

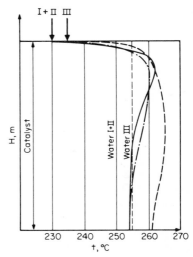

FIG. 4-4 Typical temperature profiles for Lurgi tubular methanol reactor as a function of catalyst age. (*Lurgi*.)

When the reactor is operated under the conditions shown in curves I and II, the methanol content of the gas leaving the reactor is between 6 and 8 percent by volume, depending upon the type of synthesis gas. The methanol content drops slightly when the reactor operates under the conditions shown in curve III. In order to maintain the design methanol capacity, the circulating gas rate must be appropriately increased.

Use of the Lurgi catalyst results in the formation of only small quantities of by-products. No notable rise in the formation of by-products with increasing catalyst age has been observed.

THE METHANOL PLANT

Process Description

Figure 4-5 shows the sysnthesis loop and the distillation section of a Lurgi Low-Pressure Methanol plant. The synthesis gas is compressed to synthesis pressure [normally between 735 and 1470 lb/in^2 (5 and 10 MPa)] in a turbocompressor (1) and then mixed with cir-

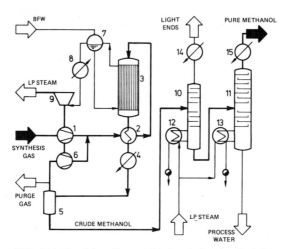

FIG. 4-5 Lurgi Low-Pressure Methanol plant (synthesis loop and distillation section). (*Lurgi.*) Key: 1, turbocompressor; 2, heat exchanger; 3, tubular reactor; 4, cooler; 5, separator; 6, recycle compressor; 7, steam drum; 8, superheater; 9, turbine; 10 and 11, distilling columns; 12 and 13, reboilers; 14 and 15, condensers.

culating gas. If the synthesis gas originates from gasification processes (such as partial oxidation) which operate at pressures above 735 lb/in² (5 MPa), compression of the synthesis gas for the Lurgi Low-Pressure Methanol process is not required. The mixture of circulating gas and fresh feed gas is heated up in a heat exchanger (2) and then enters the tubular reactor (3). The reaction mixture leaving the reactor is cooled in a heat exchanger and in a cooler (4) with air or water.

The condensed crude methanol is separated from the unreacted gas in a separator (5), and the unreacted gas is returned to the reactor inlet via a recycle compressor (6). A small amount of purge gas is exported from the loop to avoid accumulation of inert gas.

The heat from the methanol reaction is utilized for the generation of HP (high-pressure) steam. The reactor tubes are cooled with boiling water via a natural circulation system, and the resulting steam is collected in a steam drum (7).

The steam is superheated in a superheater (8), which is either associated with the methanol sysnthesis unit or integrated in the syngas production unit. The superheated steam is used to drive a turbine (9) for the syngas and recycle-gas compressors. The turbine exhaust steam is utilized for methanol distillation.

The flashed crude methanol from the separator is processed to pure methanol in a subsequent distillation unit.

The light-ends column (10) strips dimethyl ether, methyl formate, and other low-boiling impurities, while a second column (11) removes water and higher-boiling impurities. The reboilers [(12) and (13)] are heated with low-pressure steam from the turbine exhaust (9). The light ends are recovered in one condenser (14), and the pure methanol is recovered in a second condenser (15).

Lurgi's energy-saving distillation system is being applied in plants having methanol production capacities of more than 1,100,000 lb/day (500 Mg/day). In this system, the pure methanol column is divided into two sections. The first section is operated at elevated pressure, and some 50 percent of the methanol is distilled overhead. In the second section, which operates at ambient pressure, the balance of the methanol is separated from water and high-boiling impurities. While the first section is heated with LP (low-pressure) steam, the heat for the seond section is provided by condensing the methanol vapors from the first unit.

Integration of Synthesis Gas Production and Methanol Synthesis

Synthesis gas from natural gas

Figure 4-6 illustrates the production of methanol from natural gas with and without CO_2 addition. The natural gas is first desulfurized over zinc oxide and then re-formed with steam in a steam re-former. If the natural gas contains nonreactive sulfur components, these components must be hydrogenated with hydrogen-rich purge gas from the methanol synthesis unit over Nimox or Komox catalysts and then the natural gas has to be desulfurized. If external CO_2 is available, it is admixed into the cooled re-formed gas.

Up to capacities of about 2,200,000 lb/day (1000 Mg/day), Lurgi's low-pressure methanol plants are designed for pressures of about 220 lb/in² (1.5 MPa) at the steam re-former outlet and about 735 to 1030 lb/in² (5 to 7 MPa) in the synthesis loop. This permits the steam re-former to operate with a low steam-to-carbon ratio (2.4 to 2.6 depending upon the molecular weight of the natural gas). This mode of operation not only saves process steam and heat but also establishes a more favorable CO/CO_2 ratio in the synthesis gas. This design results in lower energy requirements for the synthesis gas compressor.

In larger plants, a pressure of up to 1475 psi (10 MPa) is used in the synthesis loop to allow larger single-train capacities. How-

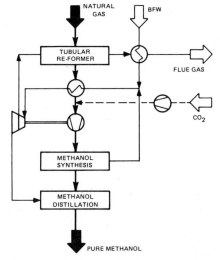

FIG. 4-6 Methanol plant based on natural gas. (*Lurgi.*)

ever, in these plants the pressure at the steam re-former outlet is raised to about 294 psi (2 MPa) to prevent the power requirement of the synthesis gas compressor from rising too high and also to permit the use of a two-casing machine in this pressure range. This, of course, requires an appropriate increase in the re-former outlet temperature in order to maintain the residual methane content at a constant rate.

Conversion of the syngas to crude methanol and its further treatment proceed as previously described.

Steam produced in the re-forming unit and in the methanol synthesis unit is superheated and used to drive turbines for the syngas and recirculating-gas compressors. The turbine exhaust steam is used as process steam for re-forming and distillation purposes.

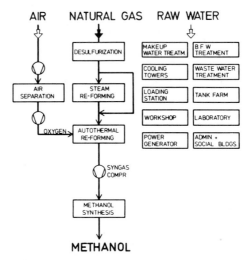

AIR   NATURAL GAS   RAW WATER

DESULFURIZATION	

MAKEUP WATER TREATM.	B F W TREATMENT
COOLING TOWERS	WASTE WATER TREATMENT
LOADING STATION	TANK FARM
WORKSHOP	LABORATORY
POWER GENERATOR	ADMIN + SOCIAL BLDGS.

AIR SEPARATION   STEAM RE-FORMING

OXYGEN   AUTOTHERMAL RE-FORMING

SYNGAS COMPR

METHANOL SYNTHESIS

METHANOL

FIG. 4-7 Methanol plant based upon combined steam and autothermal re-forming of natural gas. (*Lurgi.*)

FIG. 4-8 Steam re-former and synthesis section of a Lurgi Low-Pressure Methanol plant of PEMEX at Texmelucan, Mexico. (*Lurgi.*)

An interesting alternative to producing synthesis gas for methanol synthesis from natural gas is the combination of the conventional steam re-forming unit with a Lurgi autothermal re-forming unit, as depicted in Fig. 4-7. The residual methane content of the Syngas is adjusted in the autothermal re-former. The combining of these two different re-forming processes permits operation of the conventional steam re-forming unit at a much lower temperature and a higher pressure. Only part of the natural gas is subjected to endothermic primary re-forming and the remaining portion, together with the gas produced by steam re-forming, is treated with oxygen in a Lurgi autothermal re-forming unit. When these two processes are combined, the size of the steam re-forming unit can be reduced to about one-quarter of its size in the base scheme. If the steam re-forming unit is reduced in size, the Syngas is compressed from 470 lb/in² (3.2 MPa) to 735 to 1470 lb/in² (5 to 10 MPa) in a single-casing turbocompressor.

A typical steam re-former and synthesis section of a Lurgi methanol plant running on natural gas is shown in Fig. 4-8.

Synthesis gas from naphtha

Figure 4-9 shows a scheme for the production of methanol from naphtha. The naphtha is converted with steam to a CO- and H_2-rich synthesis gas via the Lurgi-Recatro process. The syngas is cooled and then fed to the methanol synthesis unit. Since the Lurgi-Recatro process can be operated at a low steam-to-carbon ratio, this process yields a synthesis gas which can be fed without further treatment to the Lurgi methanol reactor system.

Synthesis gas from heavy residual oil

Figure 4-10 shows a scheme for the production of methanol from heavy residual oil. The heavy residual oil is gasified in a partial-oxidation unit (using, e.g., the Shell Partial-Oxidation process) with steam and oxygen. The small amount of carbon formed during gasification is remixed with the feedstock for the partial-oxidation unit.

The crude synthesis gas has to be completely desulfurized, preferably in a two-stage absorption unit which is also used for subsequent CO_2 removal. The Lurgi Rectisol process, which uses methanol as the washing agent, is specifically suited for this purpose.

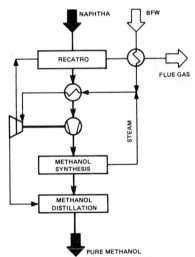

FIG. 4-9 Methanol plant based on naphtha. (*Lurgi.*)

For the methanol synthesis a stoichiometric number of about 2 expressed as $(H_2 - CO_2)/(CO + CO_2)$ is required. Since the gas produced from heavy residual oils via partial oxidation does not have this ratio, the gas composition has to be changed to this ratio via shift conversion of a partial stream of the desulfurized crude synthesis gas. This partial stream, after being freed from CO_2, is back-mixed into the main stream, which then has the correct stoichiometric ratio. Since

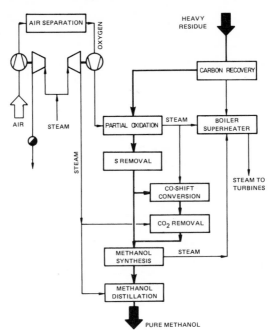

FIG. 4-10   Methanol plant based on heavy residual oil with Shell Partial-Oxidation unit. (*Lurgi.*)

the synthesis gas obtained in this way is already at the pressure level required for methanol synthesis, syngas compression can be omitted.

Synthesis gas from coal

Figure 4-11 shows a scheme for the production of methanol from coal. Coal is gasified with oxygen and steam in a Lurgi dry-bottom gasifier. The crude synthesis gas produced is rich in H_2 and CO, and for many coal types it need only be freed of sulfur and of part of the CO_2 to obtain the stoichiometrically correct methanol syngas. However, for some coal types it is necessary to convert part of the CO contained in the crude synthesis gas to CO_2, which is accomplished by shift conversion without prior removal of sulfur and condensable hydrocarbons. This partial stream is then back-mixed into the main stream.

A subsequent Lurgi Rectisol washing unit removes the higher hydrocarbons not condensed during cooling of the crude gas immediately after coal gasification, as well as sulfur and part of the CO_2. The purified synthesis gas from the Lurgi Rectisol unit has a sulfur content of less than 0.1 ppm and a CO_2 content of about 2 volume percent. The stoichiometric number is 2.05. The syngas is then compressed in a single-casing compressor from about 367 to 441 lb/in^2 (2.5 to 3.0 MPa) to 882 to 1470 lb/in^2 (6 to 10 MPa) before it is fed to the methanol synthesis unit. Methanol production and distillation proceed as described earlier in this chapter.

As the crude gas from a Lurgi dry-bottom gasifier contains 8 to 12 volume percent methane (depending on the type of coal gasified and on the gasification pressure), a substantial quantity of syngas containing up to 40 volume percent methane has to be purged

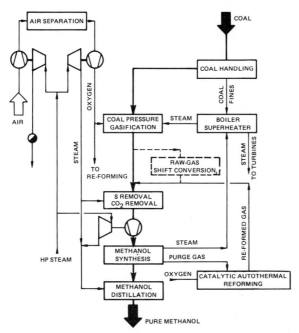

FIG. 4-11   Methanol plant based on coal with Lurgi coal pressure-gasification unit. (*Lurgi.*)

from the synthesis loop. This purge gas is treated with oxygen and steam in a Lurgi autothermal catalytic re-forming unit at about 950 °C and 514 lb/in² (3.5 MPa); the re-formed gas is back-mixed into the main gas stream before entering the Lurgi Rectisol unit.

Since the recycling of the total stream of purge gas via the Lurgi autothermal re-forming unit would result in an unwanted buildup of nitrogen and argon in the synthesis loop, a portion of this stream of purge gas is used as plant fuel, e.g., as fuel for a steam boiler.

As an alternative the total stream of purge gas can be methanated and delivered as substitute natural gas (SNG).

The by-products from the Lurgi Coal Pressure-Gasification unit, such as tars, oils, naphtha, and phenols, as well as ammonia are recovered from the gas liquor. These by-products may be marketed after adequate treatment or used as fuel within the plant.

OPERATING DATA

The consumption figures for methanol production from various feedstocks via the Lurgi Low-Pressure process that are given in Table 4-1 have been calculated for plants where the necessary infrastructure already exists and where the utilities are supplied from existing offsite units. These consumption figures cover the feedstock and utility requirements for all process units, including for example oxygen for gasification of propane-extracted asphalt or of coal. The on-site units of a complete methanol plant are synthesis gas pro-

TABLE 4-1 Consumption in the production of synthesis gas* from various feedstocks via various processes

		Feedstock and utility requirements								
Feedstock	Process	Natural gas, GJ	CO_2, m³	Naphtha (process), GJ	Heating fuel, GJ	Propane-asphalt, GJ	Bituminous coal, GJ	Electric power, kWh	Feedwater, kg	Cooling water, kg
Natural gas	Steam re-forming	31.4						0	820	50,000
	Combined re-forming	29.3						20	680	120,000
Natural gas + CO_2	Lurgi-Recatro	29.3	151					50	760	45,000
Naphtha‡				21.26	8.67			0	820	48,000
Propane-asphalt	Shell Partial Oxidation					36.8		130	840	75,000
Coal§	Lurgi Coal Pressure Gasification						40.7	0	1650	100,000

*Used to produce 1000 kg of pure methanol via the Lurgi Low-Pressure process.

†Steam and autothermal re-forming.

‡Carbon-to-hydrogen ratio is 5.5.

§Wyoming-type subbituminous coal.

duction, including gas purification where required; syngas compression; methanol synthesis; and methanol distillation. All quoted consumption figures relate to the production of 1000 kg of pure methanol and to the lower heating value (LHV).

The product specifications of methanol received from the methanol distillation unit comply with U.S. federal Grade AA standards and are as follows:

Methanol content, wt %	99.9
Specific gravity (density), $20°/4°$, kg/m^3	792
Water content (max.), wt %	0.02
Boiling interval, $°C$	0.5
Permanganate test (min.), min	60
Acid content (max.) as acetic acid, wt %	0.002
Aldehydes and ketones (max.) as acetone, wt %	0.001
Volatile iron (max.), g/m^3	0.01
Ethanol, ppm	10

REFERENCE PLANTS

Methanol plants in operation and using the Lurgi Low-Pressure Methanol process at the end of 1981 are listed in Table 4-2.

TABLE 4-2 Lurgi Low-Pressure Methanol plants (in operation at the end of 1981)

Client	Plant location	Feedstock	Methanol capacity, Mg/yr (MM lb/yr)
Veba Chemie	Gelsenkirchen, FRG	Vacuum residue	200,000 (441)
HIAG	Fischamend, Austria	Natural gas and/or naphtha	60,000 (132)
Lee Chang Yung	Taipei, Taiwan	Natural gas	50,000 (110)
Dr. Moro	Cologna Monzese, Italy	Natural gas	45,000 (99)
Celanese Chemical Co.	Bishop, Texas	Natural gas	380,000 (838)
PEMEX	Texmelucan, Mexico	Natural gas	150,000 (331)
INA	Lendava, Yugoslavia	Natural gas	185,000 (408)
Lee Chang Yung	Taipei, Taiwan	Natural gas and CO_2	60,000 (132)
E.I. Du Pont de Nemours	Deer Park, Texas	Heavy residue	605,000 (1334)
U.K. Wesseling	Wesseling, FRG	Vacuum residue	365,000 (805)
Tenneco Chemicals Inc.	Pasadena, California	Natural gas	375,000 (827)

TABLE 4-3   Lurgi Low-Pressure Methanol plants (in various stages of design, construction, and start-up at the end of 1982)

Client	Plant location	Feedstock	Methanol capacity Mg/yr (MM lb/yr)
Mitsubishi Chemical Industry	Kurosaki, Japan	Vacuum residue	100,000 (220)
CNTIC	Zibo, Shandong, People's Republic of China	Vacuum residue	100,000 (220)
Allemania Chemical Corp.	Plaquemine, Louisiana	Natural gas and CO_2	390,000 (860)
E.I. Du Pont de Nemours	Beaumont, Texas	Natural gas and CO_2	810,000 (1,786)
Getty Oil Co.	Bakersfield, California	Naphtha and/or refinery gases	300,000 (662)
Tennessee Eastman Co.	Kingsport, Tennessee	Coal	180,000 (397)
American Natural Gas	Beulah Hazen, North Dakota	Coal	6,000 (13)
Petrobras	Araucaria, Parana, Brazil	Vacuum residue	8,000 (18)
State Government of Sabah	Labuan Island, Malaysia	Natural gas	660,000 (1,455)
Pertamina	Bunyu Island, Kalimantan, Indonesia	Natural gas	330,000 (728)
VEB "Walter Ulbricht"	Leuna, German Democratic Republic	Visbreaker vacuum residue	660,000 (1,455)
Petrochemical Ind. Corp.	Seikdha, Burma	Natural Gas	150,000 (331)

Table 4-3 shows methanol plants which were in various stages of design, construction, and start-up at the end of 1982.

OUTLOOK

Methanol has been up to now one of the three primary basic chemicals (after ammonia and ethylene) and is used as chemical raw material for the manufacture of formaldehyde, dimethylterephthalate, methyl methacrylate, acetic acid, methyl halides, methylamines, solvents, and other products.

In 1980 the government of New Zealand decided to utilize part of the Kapuni and Maui natural gas reserves for the production of gasoline via methanol. The natural gas is to be transformed to methanol, which in turn is the feedstock for a Mobil MTG (methanol-to-gasoline) unit for producing gasoline.

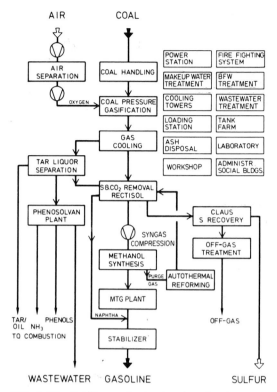

FIG. 4-12 Gasoline production based on Lurgi pressure gasification of coal. (*Lurgi.*)

The transition from this concept to that of a plant which proceeds from coal to synthesis gas to methanol and then to gasoline is only a small step and a question of time. This route for the production of gasoline from coal (see Fig. 4-12) will be competing with the more traditional routes for the production of transportation fuels from coal, the Fischer-Tropsch synthesis and direct hydrogenation of coal.

These three routes differ considerably as to their state of maturity, their thermal efficiencies, and their slates of products and by-products.

It is likely that in the future methanol will have a role as a fuel in addition to its role as a major chemical raw material. Methanol can be converted to gasoline via the Mobil MTG process; methanol can be used as a transportation fuel either by itself in methanol engines or as a gasoline-methanol mixture in conventional gasoline engines that have been slightly modified; and methanol can be used as a turbine fuel, e.g., for peak-shaving purposes in power generation in highly populated areas.

PART 3

COAL GASIFICATION

Fuel gas was first produced commercially from coal in the early nineteenth century in England. A chronology of the development of coal-gasification technology is presented in Chap. 3-9 of this section.

Coal gas is classified by heating value. High-heat-content (also termed *high-Btu*) gas has a heating value of approximately 1000 Btu/stdft³ [37 MJ/m³(n)],* which is the same as natural or pipeline-quality gas and is therefore known as *substitute* or *synthetic natural gas*, or more simply *SNG*. Medium-heat-content (also known as *medium-Btu*) gas has a heating value of 270 to 600 Btu/stdft³ [10 to 22 MJ/m³(n)].* Gas at the lower end of this range consists primarily of carbon monoxide and hydrogen, with small amounts of carbon dioxide, while gas at the higher end contains more methane. Low-heat-content (or *low-Btu*) gas has a heating value in the range of 90 to 200 Btu/stdft³ [3 to 7 MJ/m³ (n)]* and is mainly nitrogen and carbon dioxide, with the combustible components consisting of carbon monoxide, hydrogen, and methane.

The production of each of these types of gases is shown in Fig. I-1. When coal is burned with less than a stoichiometric quantity of air, with or without steam, the product is low-heat-content gas, which after purification can be used to fuel conventional or combined-cycle power plants for either new or retrofit boiler applications.

Medium-heat-content gas is produced by utilizing oxygen in place of air. It can be used without post-gasifier shift for the combustion applications just described or with shift as synthesis gas for producing liquid fuels (see Part 2 of this Handbook), SNG, or

---

*m³(n) symbolizes the normal cubic meter, that is, the volume of one cubic meter of gas measured at 0 °C and 101 kPa pressure.

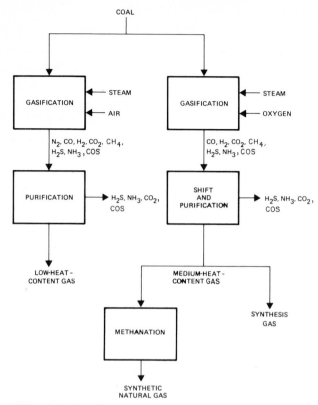

FIG. I-1 Coal-gasification operations.

ammonia. Medium-heat-content gas may not be suitable for delivery by natural-gas distribution networks because of its toxic carbon monoxide content, while SNG can be directly substituted for conventional pipeline gas.

Some of the important chemical reactions which take place during the gasification process are shown below.

Combustion

$$C + O_2 \rightarrow CO_2 \qquad \Delta H^\circ_{298} = -394 \text{ kJ/mol} \tag{1}$$

Gasification

$$C + \tfrac{1}{2}O_2 \rightarrow CO \qquad \Delta H^\circ_{298} = -111 \text{ kJ/mol} \tag{2}$$

$$C + H_2O(g) \rightarrow CO + H_2 \qquad \Delta H^\circ_{298} = +131 \text{ kJ/mol} \tag{3}$$

$$C + CO_2 \rightarrow 2CO \qquad \Delta H^\circ_{298} = +173 \text{ kJ/mol} \tag{4}$$

Shift

$$CO + H_2O(g) \rightarrow CO_2 + H_2 \qquad \Delta H^\circ_{298} = -41 \text{ kJ/mol} \tag{5}$$

Methanation

$$CO + 3H_2 \rightarrow CH_4 + H_2O(g) \qquad \Delta H^\circ_{298} = -206 \text{ kJ/mol} \qquad (6)$$

$$2CO + 2H_2 \rightarrow CH_4 + CO_2 \qquad \Delta H^\circ_{298} = -247 \text{ kJ/mol} \qquad (7)$$

The combustion portion of the gasification process is of course quite exothermic, as shown by the heats of combustion, whereas most gasification reactions are endothermic. The water gas shift reaction occurs in the gasifier and is also used to adjust the hydrogen-to-carbon monoxide ratio in the product synthesis gas. Methane formation takes place in the gasifier and is also catalytically induced in a separate methanation step to form SNG. There are many other chemical reactions which take place during the gasification process, but these are of somewhat lesser importance.

Twelve major commercial or near-commercial gasification processes are presented in this section. Some major features of each of these processes are summarized in the table that follows. There are other significant differences between the processes, such as the

Gasification processes

Process	Pressure*	Air- or oxygen-blown
Descending-bed		
Nonslagging		
Lurgi	To 1300 lb/in^2	Air or oxygen
Foster Wheeler Stoic†	Atmospheric	Air
Two-Stage Gasifier (Babcock Woodall-Duckham)	Atmospheric	Air or oxygen
Slagging		
British Gas Lurgi	To 400 lb/in^2	Oxygen
Fluidized-bed		
Nonagglomerating		
Winkler†	Atmospheric to 600 lb/in^2	Air or oxygen
Agglomerating		
Westinghouse	450 lb/in^2	Air or oxygen
Entrained-flow		
Dry coal		
KBW	Atmospheric	Oxygen
Shell	430 lb/in^2	Air or oxygen
Combustion Engineering†	Atmospheric	Air or oxygen
Slurry feed		-
Texaco	440–1200 lb/in^2	Air or oxygen
Molten slag media		
Saarberg/Otto	375 lb/in^2	Air or oxygen
Tumbling-bed		
KILnGAS	To 120 lb/in^2	Air

*kPa = lb/in^2 × 6.89.

†Two-stage.

method of feeding coal to the gasifier (e.g., lock-hopper, rotary feeder, or slurry), coal-size distribution (e.g., from 2-in to ¼-in feed to pulverized coal), gasifier design, ash-removal method, and methodology of preventing coal agglomeration. These will be discussed in detail in the various process chapters. In addition, the end of this section consists of a separate chapter on gas purification, with emphasis on removal of sulfides, carbon dioxide, and other impurities from the gas stream.

CHAPTER **3-1**

THE TEXACO COAL GASIFICATION PROCESS

W. G. SCHLINGER

Texaco Inc.
Los Angeles, California

INTRODUCTION

The Texaco Coal Gasification Process uses an entrained-bed slagging downflow gasifier capable of processing a wide range of caking and noncaking coals of bituminous or sub-bituminous origin. The concurrent flow of both coal and oxygen in this type of gasifier exposes all material leaving the gasification zone to the high temperature of the slagging environment. As a result of this exposure only components stable at the slagging temperature are present in significant quantities in the product gas. Hydrocarbon materials heavier than methane are present only at extremely small levels. This type of gasifier does not produce tars, carcinogenic compounds, or partially oxidized material which often lead to massive wastewater-treating requirements.

The Texaco Coal Gasification Process has met with wide acceptance in recent years as a result of the process's simplicity and its accompanying superior environmental qualities. A number of projects have been announced in which the Texaco process will play a key role.

The process involves feeding coal to the gasifier as a concentrated water slurry. Commercially available reciprocating pumps are used to feed the charge slurry at gasification pressures ranging from 30 to 80 atm. Hot gases leaving the gasifier are either quenched in water or passed through a waste-heat recovery section. The water-quench mode is used when hydrogen is the desired end product, in which case the quenched gases saturated with water vapor are passed over a shift-conversion catalyst where the carbon monoxide in the gas is converted to additional hydrogen. The waste-heat recovery mode is employed when downstream processing, such as methanol manufacture, oxo-synthesis processes, and fuel gas production, requires a mixture of hydrogen and carbon monoxide.

Particulate matter is readily removed from the high-pressure gasifier product by water scrubbing. Sulfur in the feed coal is converted to H_2S, and a number of proven process technologies are available for selective removal of the H_2S at process pressure.

CHEMISTRY AND THERMODYNAMICS OF COAL GASIFICATION

The chemistry and thermodynamics of coal gasification are relatively simple. Coal consists of large molecules containing carbon and hydrogen, with lesser amounts of oxygen, sulfur, and nitrogen mixed with a wide range of other elements contained in the ash of the coal. The composition of this ash material is important and determines the minimum temperature at which the gasifier can be operated under slagging conditions.

For simplicity in studying the gasification reactions coal can be represented by the chemical formula CH. When this hypothetical coal is exposed to oxygen at high temperature, it can undergo several stages of oxidation. The three basic reactions are shown in Table 1-1. Equation 1 represents the complete combustion of coal to produce carbon dioxide and water vapor. Equation 2, on the other hand, depicts the partial combustion of coal to produce a mixture of hydrogen and carbon monoxide. Equation 3 represents the "re-forming" of coal in which coal reacts with steam to produce a mixture of hydrogen plus carbon monoxide. Each of these reactions is accompanied by an enthalpy change, or heat of reaction. The heat of reaction under isothermal conditions at 20 °C is also shown in Table 1-1. In addition the enthalpy change, or heat release, of the reaction is shown under condi-

TABLE 1-1 Coal-gasification reactions

Reaction equations	Heat of reaction, Btu/lb (kJ/kg)	
	20 °C*	1400 °C*
(1) $CH + \frac{5}{4}O_2 \rightleftharpoons CO_2 + \frac{1}{2}H_2O$	−17,500 (−40,800)	−14,200 (−33,000)
(2) $CH + \frac{1}{2}O_2 \rightleftharpoons CO + \frac{1}{2}H_2$	−4,200 (−9,800)	−2,000 (−4,700)
(3) $CH + H_2O \rightleftharpoons CO + \frac{3}{2}H_2$	+3,800 (+8,800)	+7,400 (+17,100)

*Temperature of reaction products.

tions in which the reactants are allowed to rise to a temperature of 1400 °C, a typical slagging gasifier temperature.

It is immediately apparent from inspecting the enthalpy changes that a large amount of heat is released by the reaction shown in Eq. 1 and that the gaseous products will rise to a temperature well above 1400 °C unless some form of heat-extraction system is employed. By contrast Eq. 2 indicates that nearly half of the heat release associated with the partial oxidation of coal is required to raise the combustion products to 1400 °C. Under these conditions the temperature of the reaction products would still increase substantially above 1400 °C unless some type of heat extraction were included in the process. On the other hand the re-forming reaction shown in Eq. 3 is an endothermic reaction and substantial quantities of heat must be added to the system to maintain the reactants at their initial feed temperature.

Efficient coal gasification involves an optimum combination of Eqs. 1, 2, and 3. The chemical engineer must design a system to allow the desired combination of these three reactions to be realized.

In most coal-gasification processes some form of moderation of the reaction temperature is required. In the case of the Texaco Coal Gasification Process the moderator is the water contained in the feed slurry and the ash components in the coal. Other types of moderators would be satisfactory, for example, recycle synthesis gas, steam, or even nitrogen.

PILOT PLANT FACILITIES

The Texaco Coal Gasification Process was developed at Texaco's Montebello Research Laboratory near Los Angeles. Work on coal gasification at this facility was initiated in 1948. Since that time the process has evolved to its present form, which is described in detail in the next section of this chapter.

Two relatively large pilot plants, each capable of gasifying 15 to 25 tons/day of coal, have been erected at the laboratory site. These gasifiers, which are shown in Fig. 1-1, have been used for the extensive process development work that has been undertaken in recent years. One of the gasifiers has a design operating pressure of 80 atm (8100 kPa), and the other is designed for operation at 30 atm (3040 kPa). Facilities are also available at the

FIG. 1-1 Coal-gasification pilot units at the Montebello Research Laboratory *(Texaco Inc.).*

laboratory to grind coal and to prepare and store the concentrated slurries of coal and water. Equipment is also installed to remove H_2S selectively from the synthesis gas.

In addition to the facilities just described a new coal-gasification pilot plant went on stream in late 1981. This newly completed unit will also process 15 to 25 tons/day of coal at 80 atm and is equipped with additional coal-grinding and slurry-preparation equipment. A picture of the new pilot plant under construction is shown in Fig. 1-2.

PROCESS DESCRIPTION

A process flow diagram of the Texaco Coal Gasification Process is shown in Fig. 1-3. It is apparent from inspection of this figure that the process is relatively simple. The coal slurry fed to the gasifier can be prepared in several ways. Dry grinding of the coal allows for accurate classification by screening and close control of the particle-size distribution in the slurry of coal and water. Control of the particle-size distribution is important if a maximum concentration of coal in the slurry is to be achieved. Particles which are too small tend to thicken the slurry at a lower concentration of coal than intermediate-size particles. The maximum particle size is determined by the characteristics of the charge pump and by the reaction kinetics of the coal selected for the gasification process. The maximum particle size in the coal slurry may vary from plant to plant and with coal quality, but in general maximum particle diameter should be limited to 500 to 1000 μm. It has also been demonstrated that wet-grinding processes can be used to prepare satisfactory slurry feeds for the process.

Nearly every coal has its own particular slurrying properties, and it is necessary to run laboratory tests on each candidate coal. Most coals contain a small fraction of chemically bound water which is not removed by air drying the coal. Lower-rank coals (lignites) inherently contain more bound water than the higher-rank bituminous and subbituminous coals.

FIG. 1-2 Coal-gasification pilot unit under construction at the Montebello Research Laboratory *(Texaco Inc.)*.

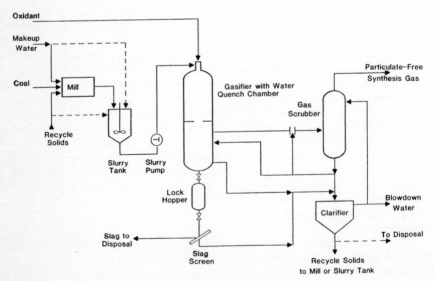

FIG. 1-3 Texaco Coal Gasification Process—water-quench mode.

This bound water, of course, decreases the maximum percentage of bone-dry coal that can be put into a given slurry. Lignites typically contain 25 to 40 percent bound water, and by contrast, subbituminous coals generally contain less than 20 percent water. Bituminous coals typically contain less than 10 percent bound water. These facts must be considered not only in selecting a coal for the process but also in preparing the slurry itself. A typical optimized slurry will contain more than 60 percent dry coal.

It is usually necessary to crush and pregrind the coal to a maximum particle size on the order of 1 cm before it is fed to the final milling system. No separation or removal of fines from this coarsely ground coal is required. In some cases it is desirable to use the fines prepared from classifying coal for other uses.

It is recommended that, after the slurry has been prepared, a small holding or run tank be employed from which the coal slurry is actually fed through the process pump to the gasifier. This holding tank tends to smooth out minor variations in slurry composition and provides a more uniform feed to the gasifier. Some agitation in this holding tank is required, and typically a circulating pump is used, which helps to maintain a uniform composition in the slurry and at the same time provides a suction head for the gasifier feed pump. The charge pump to the gasifier must be able to pump coal at a steady rate and deliver it to the slurry injector at a pressure above the level in the gasifier. Triplex positive-displacement pumps are generally used for this purpose. The pump should be equipped with either a variable-speed motor or a variable stroke so that precise changes in the coal feed rate to the gasifier can be made, since this pump also serves as a metering device for the gasifier charge.

At the injector entrance the coal and water slurry is mixed with oxygen which has been supplied by a carefully controlled flow regulating system. This oxygen should be of constant quality, although high purity is not necessarily required. The purity of the oxygen is generally dictated by the downstream use of the gasified product. In some instances oxygen might be supplied directly as air if substantial nitrogen in the product gas can be tolerated.

The coal and oxygen along with the water enter the hot gasifier, where the water is rapidly vaporized and the partial combustion and devolatilization reactions begin to occur simultaneously. The temperature in the gasifier is controlled by small changes in the ratio of oxygen to coal. An increase of only 1 percent in the oxygen-to-coal ratio will raise the gasifier temperature approximately 25 °C.

The gasifier temperature must be kept high enough to allow the ash in the coal to form a molten slag but low enough to avoid unnecessary wear and deterioration of the refractory lining in the gasifier. The refractory lining in the gasifier serves several purposes. It greatly reduces the heat loss from the reactants to the surroundings, and at the same time this refractory lining protects the steel shell of the gasifier from exposure to excessively high temperatures. The actual material selected for the lining again depends upon the properties of the coal ash and the temperature to which the lining must be exposed. It has generally been found that refractories containing a substantial amount of chromium oxide give superior performance.[1]

Figure 1-3 shows the products leaving the gasifier quenched with water. This water-quenched version is used when the next process step would be shift conversion to produce additional H_2 from the CO contained in the gas stream. Hydrogen is the desired end product in cases where the ultimate plant product is NH_3: H_2 for coal liquefaction, or even H_2 for refining low-quality crude oil.

The alternate process scheme shown in Fig. 1-4 includes recovery of some of the sensible

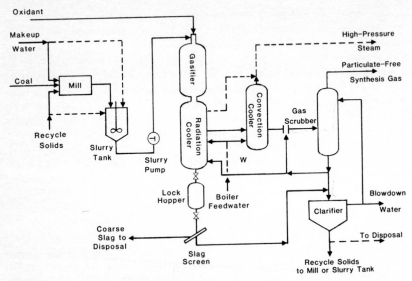

FIG. 1-4 Texaco Coal Gasification Process—gas-cooler mode.

heat in the generator exit gases by the production of high-pressure steam. Since the gas leaving the gasifier contains a significant amount of molten slag, this heat-recovery sytem is one of special design. The hot gases are first passed through a waterwalled chamber in which the gases are cooled by radiant-heat transfer to the water-cooled wall. This chamber must be designed with a diameter and length sufficient to cool the particulate matter and slag in the exit gas to a temperature below the ash solidification point. After this initial cooling by radiant transfer, the gas can be separated from most of the solid-phase material and cooled by more conventional methods.

Final gas cleaning is done by water scrubbing at full gasifier pressure. By using venturi or orifice-type scrubbers at high pressure it has been demonstrated that the particulate loading in the gas can be reduced to levels well below 1.0 mg/m³(n).*

Water enters into the process at a number of locations, as can be seen in Figs. 1-3 and 1-4. This water is extensively recirculated, and only a small fraction of the circulating water is blown down from the system. Water is reused for scrubbing and quench purposes, and it is also used for slurry preparation. Only when the dissolved solids content of the water rises to an unacceptable level is it necessary to remove water from the system. Data on composition of this blowdown water will be presented later in this chapter.

In a modified version of the Texaco Coal Gasification Process, high-ash-content liquid or fluid material can be fed to the gasifier in place of the water slurry. This technology is of particular interest to the developers of the coal-liquefaction processes described in Part 1 of this handbook. Residues from the H-Coal®, SRC-II, EDS (Exxon Donor Solvent), and the modified I.G. Farben processes are all fluids with relatively high ash contents at tem-

*m³(n) symbolizes the normal cubic meter, that is, the volume of one cubic meter of gas measured at 0 °C and 101 kPa pressure.

peratures in the range of 200 to 300 °C. These residues typically contain all of the ash present in the original coal as well as a small fraction of the most difficult portion of the coal to liquefy. It is not unusual to find coal-liquefaction residues containing as much as 35 percent ash.

The viscous residues from the H-Coal, SRC II, and EDS processes have been fed to the Texaco gasifier, using a small amount of steam as moderator to control the temperature, and have been successfully gasified in the Montebello pilot units.[2,3] Solid residues produced from processes such as SRC I have also been successfully gasified by feeding the material to the gasifier as a water slurry. Gasification of these residues in the water-quench mode followed by shift conversion produces the hydrogen required for the liquefaction process and at the same time eliminates the environmental problems associated with disposal of the residue, which in many cases has been classified as a hazardous waste, since substantial quantities of polynuclear aromatics are always present.

Gasifier Material and Energy Balance

Figure 1-5 shows a schematic representation of the gasifier system and identifies the entering and exiting streams. A material and energy balance based on pilot-plant data obtained during gasification of a western bituminous coal is shown in Table 1-2. This table contains an analysis of the coal feed as well as the product gas and slag components. The enthalpies of the streams relative to the elements at 0 K are also included in Table 1-2. The difference between the enthalpies of the entering and exiting streams is defined as the *heat loss* from the system. For purposes of this calculation the gases are assumed to be ideal and no corrections for changes in enthalpy with pressure are incorporated. The data show a heat loss from the pilot unit of just over 200,000 Btu/h (210,000 kJ/h). This figure is approx-

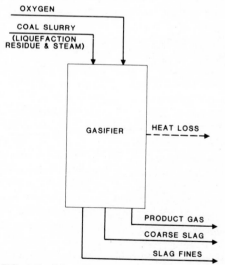

FIG. 1-5   Schematic representation of Texaco gasifier.

TABLE 1-2 Material and energy balance for western bituminous coal

	Coal	Coarse slag	Slag fines	Product gas	Oxygen
Stream					
lb/h	1030*	72.5	53.6	1943*	881
(kg/h)	(467)*	(32.9)	(24.3)	(881)*	(400)
Analysis, wt %*					
Carbon	67.93	——	27.94		
Hydrogen	4.88	——	0.62		
Nitrogen	1.11	——	——		
Sulfur	0.38	——	0.19		
Oxygen	14.97	——	——		
Ash	10.73	100.00	71.25		
Component, mol %*					
O_2				——	99.5
CO				44.69	
H_2				36.75	
CO_2				17.77	
CH_4				0.05	
A				0.15	0.5
N_2				0.45	
H_2S				0.13	
COS				0.007	
Enthalpy†					
MBtu/h	−4547	54	42	−4710	135
(MJ/h)	(−4796)	(57)	(44)	(−4966)	(142)

*Water-free basis.

†Including water.

Note: The heat loss from the pilot unit (defined as the difference between the enthalpies of the entering and exiting streams) is 202,000 Btu/h (211 MJ/h).

imately 200 Btu/lb (460 kJ/kg) of coal fed to the gasifier, or a little more than 1 percent of the heat of combustion of the coal. In a full-scale unit this heat loss would be significantly less than 1 percent.

The data presented in Table 1-2 show several interesting features. If the gasification reactions proceed according to Eq. 2 shown in Table 1-1, the atomic ratio of pure O_2 in the oxygen stream to the C in the coal feed would equal the fraction of C in the coal converted to gaseous products. If the C conversion exceeds the atomic O/C ratio, some of the gasified products are the result of reactions of the type shown in Eq. 3 of Table 1-1. Under such circumstances, O_2 contained in the slurry water or O_2 present in the coal feed has supplied a portion of the O_2 required for gasification.

The data in Table 1-2 show that western bituminous coal is gasified in the pilot unit at a carbon-conversion level of 98 percent with an atomic O/C ratio of 0.94, indicating that nearly 4 percent of the O_2 required for gasification came from the water fed with the coal or from the O_2 present in the coal. The 18 percent CO_2 present in the dry product gas is formed as a result of the shift reaction which occurs in the gas phase and in which CO is converted to H_2 and CO_2 by reaction with water vapor. This shift reaction has no bearing on the oxygen required for gasification, and the CO_2 content is not a direct indication of gasification efficiency.

Similar data obtained while processing the SRC-II residue from an Appalachian coal (Powhatan) are presented in Table 1-3. This table contains an analysis of the liquefaction residue and product gases as well as the slag components and their corresponding feed rates and enthalpies.

Inspection of the data in Table 1-3 shows that the liquefaction residue from Appalachian coal can be gasified at a conversion level in excess of 98 percent with an atomic O/C ratio

TABLE 1-3 Material and energy balance for SRC-II liquefaction residue

	Coal residue	Coarse slag	Slag fines	Product gas	Oxygen
Stream					
lb/h	709*	18	165	1246*	553
(kg/h)	(322)*	(8)	(75)	(565)*	(251)
Analysis, wt %*					
Carbon	67.06	——	4.75		
Hydrogen	4.03	——	0.08		
Nitrogen	1.22	——	——		
Sulfur	2.51	——	1.22		
Oxygen	0.50	——	——		
Ash	24.68	100.00	93.95		
Component, mol %*					
O_2				——	99.5
CO				53.46	
H_2				36.46	
CO_2				8.48	
CH_4				0.19	
A				0.03	0.5
N_2				0.49	
H_2S				0.83	
COS				0.058	
Enthalpy†					
MBtu/h	−1439	15	137	−1623	87
(MJ/h)	(−1518)	(16)	(144)	(−1712)	(92)

*Water-free basis.

†Including water.

Note: The heat loss from the pilot unit (defined as the difference between the enthalpies of the entering and exiting streams) is 119,000 Btu/h (126 MJ/h).

of 0.87. In this instance more than 10 percent of the total O_2 requirement for gasification is supplied by the water vapor, since the O_2 content of liquefaction residues is very low. Again the CO_2 found in the gasifier product is primarily a result of the reaction of the CO produced by the gasification process with excess water vapor to form additional H_2 and CO_2.

Analytical and Environmental Test Data

When making a gasification evaluation on a specific coal it is necessary to conduct extensive analytical tests to determine the distribution of the major and minor components in the coal and slag. It is also necessary during the gasification tests to obtain a large amount of detailed environmental data in order to be able to apply for and obtain the numerous permits required for plant construction. A number of coals have been subjected to this treatment, and the distribution or partition of 34 elements has been studied.

Data from tests on eastern high-sulfur and western low-sulfur coals have been reported.[4] These tests not only include a complete analysis of the discharge stream or blowdown water but also include leaching tests on the various slag components. Table 1-4 contains a detailed elemental analysis of a western bituminous coal and shows both major and minor components. Data on the quality of blowdown water are shown in Table 1-5. A complete elemental analysis of the slag components is also included in Table 1-4, and the results of leaching tests on the slag components are shown in Table 1-6. In addition, the gaseous products have been analyzed for a variety of trace components, and the results of these analyses are shown in Table 1-7. It is apparent that the gaseous product contains an insignificant quantity of any materials other than reduced sulfur species and the major components. The slag fractions were also subjected to special analyses to determine the possible presence of polynuclear aromatic hydrocarbon components on the U.S. Environmental Protection Agency (EPA) priority pollutant list. The results of these analyses are shown in Table 1-8.

It is apparent from this detailed testing that the Texaco Coal Gasification Process exhibits significant environmental advantages over conventional fossil-fuel power-generating facilities and other emerging energy-producing processes. It is not anticipated that other coals will show significant deviations from the low level of trace components reported in Tables 1-4 through 1-8.

STATUS OF COMMERCIAL AND DEMONSTRATION PROJECTS

The first installation to use the Texaco Coal Gasification technology on a demonstration scale was built in Morgantown, West Virginia, in 1957. This plant, charging 100 tons/day of coal, was operated for a substantial period during 1957 and 1958. It was, however, concluded at the end of the test period that coal gasification could not compete with the cheap natural gas and crude oil available in the United States at that time. The plant was dismantled, and it was not until the oil embargo of 1973 that major efforts to complete the process development were reinstituted.

The efforts put forth in 1973 and succeeding years resulted in a joint program involving

TABLE 1-4 Detailed elemental analysis of solid feed and product streams*

	Coal	Coarse slag	Slag fines
Element, wt %			
Carbon	68.58	0.54	12.58
Hydrogen	4.91	0.06	0.37
Nitrogen	1.14	0.01	0.09
Sulfur	0.51	0.10	0.36
Ash	9.64	99.53	82.60
Aluminum	0.58	6.4	5.4
Calcium	1.29	14.2	10.4
Iron	0.34	4.1	4.0
Magnesium	0.16	1.6	1.4
Potassium	0.03	0.37	0.34
Silicon	1.55	23.4	19.2
Sodium	0.08	2.0	2.4
Titanium	0.05	0.8	0.73
Element, ppm			
Antimony	0.2	0.2	1.6
Arsenic	0.16	3	11
Barium	262	770	705
Beryllium	0.2	0.6	0.7
Cadmium	0.3	0.2	0.2
Chromium	17	244	115
Cobalt	3.3	22	14
Copper	6	72	109
Lead	14	33	190
Lithium	3	74	64
Manganese	51	500	469
Mercury	0.05	0.5	2.6
Molybdenum	1.6	10	22
Nickel	15	55	41
Selenium	1.2	0.2	5.7
Silver	0.4	0.1	0.9
Strontium	107	835	637
Thallium	0.7	0.7	0.7
Tin	0.2	0.5	0.3
Vanadium	20	70	70
Zinc	22	26	110

*Feedstock is western bituminous coal.

TABLE 1-5 Analytical tests on blowdown water*

Parameter	Value	Parameter	Value
pH	9.0	Trace elements, ppm	
Conductivity	20,000 μ℧/cm†	Silver	0.002
Dissolved solids	682 ppm	Sodium	103
		Strontium	2
Trace elements, ppm		Thallium	<0.001
Aluminum	0.10	Tin	<0.2
Antimony	0.004	Titanium	<0.1
Arsenic	0.006	Vanadium	<0.04
Barium	0.11	Zinc	0.117
Beryllium	<0.001		
Cadmium	0.007	Anions, ppm	
Calcium	20	Bromide	2.0
Chromium	0.003	Chloride	37
Cobalt	<0.01	Fluoride	32
Copper	0.024	Cyanide	9
Iron	1	Formate	4626
Lead	0.016	Nitrate	<1.0
Lithium	0.4	Nitrite	<1.0
Magnesium	27	Phosphate	<1.0
Manganese	0.11	Sulfate	21
Mercury	<0.00001	Sulfide	61
Molybdenum	0.009	Sulfite	<1.0
Nickel	0.018		
Potassium	14	Ammonia, ppm	10,200
Selenium	0.116	Organic carbon, ppm	1,720
Silicon	26	Inorganic carbon, ppm	1,420

*Feedstock is western bituminous coal.

†Micromhos (μ℧) are more properly called microsiemens (μS).

Texaco, Ruhrkohle AG, Ruhrchemie AG, and the West German government to build a 165-ton/day demonstration unit at the petrochemical complex of Ruhrchemie in Oberhausen, FRG (Federal Republic of Germany). The unit design was based on test runs made in the Montebello pilot unit, and the plant went on stream in January 1978. In the first 4½ years of operation of this plant more than 66,000 tons of bituminous coals has been gasified during periods of operation lasting up to 30 days on stream. Some of the test results from this plant have been reported.[5,6] Table 1-9 contains a partial list of coals that have been processed in the Ruhrchemie demonstration plant.

Two additional demonstration plants are now in operation. Information relative to each of the three demonstration plants is shown in Table 1-10. Data obtained from these plants and from the pilot units at the Montebello Research Laboratory are being used to design and construct a number of large commercial-scale plants incorporating the Texaco Coal Gasification Process. The two most important projects under way are the Cool Water Coal

TABLE 1-6 Leaching tests on slag streams*

Component	Slag stream	
	Slag fines	Coarse slag
Trace elements, ppm		
Antimony	<0.5	0.5
Arsenic	<0.005	0.005
Barium	0.25	0.05
Beryllium	0.01	0.01
Cadmium	0.04	0.03
Cobalt	<0.5	0.5
Copper	<0.1	0.1
Iron	<0.3	0.3
Lead	0.3	0.3
Lithium	<0.5	0.5
Manganese	1.1	<0.05
Mercury	0.001	<0.0005
Molybdenum	<0.3	<0.3
Nickel	0.2	0.1
Selenium	0.001	0.001
Silver	0.1	<0.1
Strontium	1	<1
Thallium	<0.5	<0.5
Tin	<1	<1
Vanadium	<0.2	<0.2
Zinc	0.50	0.03
Anions, ppm		
Bromide	<0.1	<0.1
Chloride	2.0	0.14
Cyanide	<0.01	<0.01
Fluoride	3.6	0.3
Formate	112	11
Nitrate	<0.1	<0.1
Nitrite	<0.1	<0.1
Phosphate	0.5	<0.1
Sulfate	7.3	1.5
Ammonia	0.8	0.4
Inorganic carbon	6	2

*Feedstock is western bituminous coal. Leachate properties were determined using the U.S. Enviornmental Protection Agency (EPA) extraction procedure toxicity test method.

TABLE 1-7 Trace components in product gas*

Inorganic components	Content, $\mu g/m^3$	Organic components	Content, $\mu g/m^3$
Aluminum	<1	Benzene	16.5
Antimony	0.02	Toluene	3.3
Arsenic	<0.02	Ethylbenzene	2.9
Barium	0.6	o-Xylene	<0.1
Beryllium	<0.02	Naphthalene	522
Cadmium	2.0	Anthracene	8
Calcium	6	Phenanthrene	1.5
Chromium	0.5	Acenaphthene	5
Cobalt	1.4	Acenaphthylene	<125
Copper	1.3	Benz(a)anthracene	5
Iron	14	Benzo(b)fluoranthene	<0.05
Lithium	<1	Benzo(k)fluoranthene	0.1
Magnesium	5	Benzo(a)pyrene	<0.05
Mercury	3.8	Benzo(g,h,i)perylene	<0.25
Molybdenum	0.2	Chrysene	<0.6
Nickel	3.0	Dibenz(a,h)anthracene	<0.1
Potassium	1.0	Fluoranthene	1
Selenium	<0.2	Fluorene	<2.5
Silicon	20	Indeno(1,2,3-c,d,)pyrene	<0.5
Silver	<1	Pyrene	2
Sodium	25	Triphenylene	<1
Strontium	4		
Thallium	<0.02		
Tin	<2		
Titanium	<1		
Vanadium	<0.6		
Zinc	2		
Ammonia	3.5		

*Feedstock is western bituminous coal.

Gasification Program, which is under construction in the high desert of southern California with start-up scheduled for mid-1984, and the Tennessee Eastman project presently under construction at Kingsport, Tennessee, in Eastman's chemical plant.

The Cool Water Coal Gasification Program is sponsored by Texaco Inc., Southern California Edison Company, Electric Power Research Institute (EPRI), General Electric Company, Bechtel Power Corporation, a Japanese consortium, and the Empire State Electric Energy Research Company.

The program is organized to design, construct, start up, test, and operate a modular commercial coal-gasification combined-cycle power plant for a period of seven years. Approximately 1000 tons/day of western bituminous coal and other selected coals will be

TABLE 1-8 Polynuclear aromatic hydrocarbon priority pollutants detected in slag fractions

	Slag fraction	
Pollutant*, μg/kg†	Coarse slag	Fine slag
Acenaphthene	<12	<17
Acenaphthylene	<280	<420
Anthracene	0.8	1.0
Benz(a)anthracene	5	3
Benzo(b)fluoranthene	<0.1	<0.2
Benzo(k)fluoranthene	<0.1	<0.2
Benzo(a)pyrene	<0.1	<0.2
Benzo(g,h,i)perylene	<0.6	<0.8
Chrysene	<1.4	<2.4
Dibenz(a,h)anthracene	<0.2	<0.3
Fluoranthene	5	9
Fluorene	<5.7	<8.3
Indeno(1,2,3-c,d)pyrene	<1.2	<1.7
Naphthalene	<22	<32
Phenanthrene	<3.4	<4.9
Pyrene	9	10
Triphenylene	<10	<10

*Pollutant determined by methanol/benzene extraction. Feedstock is western butiminous coal.

†Most components are below detection limit as indicated by < (less than) sign.

gasified, producing a net power output of 100 MW for distribution in southern California. Details concerning the design and operation of this project, which is estimated to cost $300 million, have been reported.[7] A block flow diagram of the plant is shown in Fig. 1-6.

The Tennessee Eastman project is presently under construction. This plant which will charge 900 tons/day of Appalachian coal, will be the first commercial operation using the Texaco Coal Gasification Process to go on stream. Syngas from the plant will be used to manufacture a variety of chemical products, most of which are related to the line of photographic materials Eastman produces.[8] Tennessee Eastman also will become the first major chemical manufacturer in the United States to change its base of raw materials from petroleum and natural gas derivatives to coal.

In addition, Ube Industries, Ltd. has announced that its affiliate, Ube Ammonia Industry Co., Ltd. will construct a 1650-ton/day Texaco Coal Gasification complex to manufacture hydrogen for use in an existing ammonia plant in Ube City, Japan. The plant is scheduled to go on stream in 1984 and produce 1100 tons/day of ammonia.

In a fourth project, Ruhrchemie AG and Ruhrkohle AG are planning to construct a commercial coal gasification unit in Ruhrchemie's petrochemical complex. Known as SAR

TABLE 1-9 Coals gasified at Ruhrchemie demonstration plant

	Ruhr Coal Types					Imported Coal		
	Forge coal	Fat coal	Gas coal	Gas coal	Coal sludge	High-volatility bituminous coal	High-volatility bituminous coal	Medium-volatility bituminous coal
Composition, % wt								
C	85.4	81.9	80.8	79.4	60.0	70.9	68.6	65.8
H	4.0	4.7	5.0	4.9	3.8	5.0	4.6	3.5
N	1.6	1.5	1.5	1.6	1.1	1.2	1.2	1.5
S	0.8	1.1	1.1	1.0	1.4	3.6	0.6	0.8
O	2.6	4.0	5.1	4.6	5.8	7.1	12.4	8.6
Ash	5.6	6.8	6.5	8.5	27.9	12.2	12.6	19.8
Volatiles	16.0	21.7	32.1	29.0	24.2	38.5	37.6	23.8
Ash fusion temperature, °C	1360	—	1500	1450	—	1280	1370	1410
Hardgrove index	99	100	—	68	—	55	47	66

TABLE 1-10 Demonstration plants in operation

	Ruhrchemie/ Ruhrkohle	Dow Chemical Company	Tennessee Valley Authority
Plant location	FRG	Louisiana	Alabama
Start-up date	1978	1979	1982
Feed coal rate, t/d	165	400	190
Oxidant	Oxygen	Air	Oxygen
Mode of operation	Gas cooler	——	Quench
Pressure, psig	540	——	600
H_2 + CO production, MM scf/d	9	——	10
Ultimate product	Medium-Btu gas, oxo chemicals	Electric power (gas turbine)	Ammonia

(Synthese Anlage Ruhr), the plant is scheduled to go on stream in 1985 and gasify approximately 770 tons/day of coal to produce syngas and hydrogen. The gases will be used in Ruhrchemie's chemical facilities for the production of organic chemicals.

Table 1-11 is a summary of commercial coal gasification projects using the Texaco Coal Gasification Process which are in the design phase or under construction.

The Texaco Coal Gasification Process is one of the few process developments available today which can demonstrate significant environmental advantages over conventional fossil-fuel power-generating facilities and other emerging energy-producing processes. The environmental acceptability of the process has been proved on both pilot- and demonstration-plant scales. The process is relatively simple and can handle a wide range of high-sulfur caking and noncaking coals. This technology should make major contributions to the energy supply produced from coal in the decades ahead.

TABLE 1-11 Commercial projects using the Texaco Coal Gasification Process

	Project			
	Tennessee Eastman	Cool Water	Ube Ammonia	SAR (Ruhrchemie)
Location	Tennessee	California	Japan	FRG
Product	Chemicals	Power	Ammonia	Chemicals
Status	Construction	Construction	Construction	Engineering
Fuel	Coal	Coal	Coal	Coal
Feed rate, t/d	900	1000	1650	770
Oxidant	Oxygen	Oxygen	Oxygen	Oxygen
Gas cooling	——	Heat recovery	——	Heat recovery
Pressure, psig	——	600	——	625
H_2 + CO production, MM scf/d	——	56	77	45
Start-up	1983	1984	1984	1985

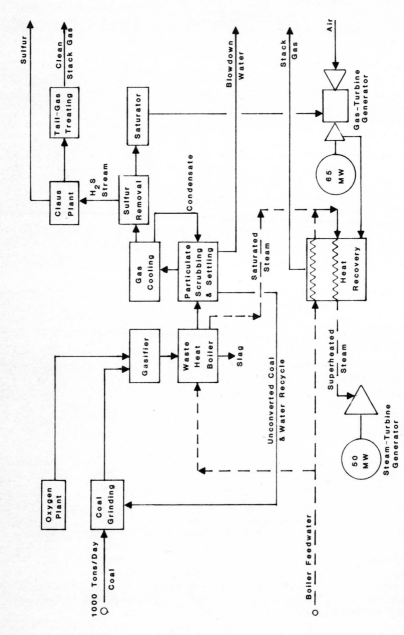

FIG. 1-6 Block flow diagram for the Cool Water program.

REFERENCES

1. Kennedy, C. R.: "Refractories for Slagging Gasifiers: Problems, Solutions and Trade-Offs," *90th National AIChE Meeting, Houston, Texas, Apr. 5–9, 1981.*

2. Robin, A. M.: "Synthesis Gas from H-Coal Liquefaction Residues," *Coal Processing Technology,* vol. IV. *AIChE Technical Manual,* AIChE, New York, 1978, pp. 35–39.

3. Robin, A. M., A. R. Catena, and E. Nour: "Gasification of Residual Materials from Coal Liquefaction," U.S. Department of Energy Report no. Fe-2247-26, May 1980.

4. Schlinger, W. G., and G. N. Richter: "Process Pollutes Very Little," *Hydrocarbon Process.,* vol. 59, October 1980, pp. 66–70.

5. Cornils, B., J. Hibbel, J. Langhoff, and P. Ruprecht: "Stand der Texaco-Kohlevergasung in der Ruhrchemie/Ruhrkohle-Variante," *Chem. Ing. Tech.,* vol. 52, no. 1, 1980, pp. 12–19.

6. Cornils, B., and R. Specks: "Experience with the Texaco Process of Coal-Dust Pressure Gasification Using the Ruhrchemie/Ruhrkohle Technical Version," *EPRI Conference on Synthetic Fuels, San Francisco, October 1980,* EPRI report WS-79-238.

7. Walter, F. B., H. C. Kaufman, and T. L. Reed: "Cool Water Coal Gasification," *Chem. Eng. Prog.,* vol. 77, no. 5, 1981, pp. 61–66.

8. Coover, H. W., and R. C. Hart: "A Turn in the Road: Eastman Chemicals from Coal," *8th International Conference on Coal Gasification, Liquefaction, and Conversion to Electricity, University of Pittsburgh, Aug. 4–6, 1981.*

THE SHELL COAL GASIFICATION PROCESS

ERICH V. VOGT
PAUL J. WELLER
MAARTEN J. VANDERBURGT

Shell Internationale Petroleum
Maatschappij B.V.
The Hague, Netherlands

INTRODUCTION

The Shell coal gasification (SCG) process for the gasification of coal under pressure is based on the principles of entrained-bed technology. The process is characterized by the following properties:

- Practically complete gasification of virtually all solid fuels
- Production of a clean gas without by-products
- High throughput
- High thermal efficiency and efficient heat recovery
- Environmental acceptability

There are numerous possible future applications for this process. The gas produced (93 to 98 percent by volume hydrogen and carbon monoxide) is suitable for the manufacture of hydrogen or reducing gas and, with further processing, substitute natural gas (SNG). Moreover, the gas can be used for the synthesis of ammonia, methanol, and liquid hydrocarbons.

Another possible application of this process is as a fuel gas supplier to a combined-cycle power station featuring both gas and steam turbines. A Shell coal gasifier with a combined-cycle power station will allow for electricity generation at more than 40 percent efficiency for a wide range of feed coals.

The development program has included the operation of a 150-ton/day (6250-kg/h) gasifier at Deutsche Shell's Harburg refinery since November 1978 (Fig. 2-1) and of a 6-ton/day (250-kg/h) pilot plant at the Shell research laboratories at Amsterdam since December 1976. Both facilities have been operated very successfully. With hard coal, a conversion of 99 percent is reached while producing a gas with only 1 percent by volume of CO_2.

PROCESS DESCRIPTION

The SCG process is based on the principle of entrained-bed gasification at elevated pressures under slagging conditions. A general flow scheme is given in Fig. 2-2.

The coal feed is ground to a size of less than 100 μm (10^{-4} m) and is normally dried to a water content of about 2 percent. The dry coal is then pressurized in a lock-hopper system and introduced into the gasification reactor together with oxygen and steam.

The reactor is basically an empty vessel, providing a residence time of a few seconds at a pressure of about 430 lb/in² (3000 kPa). Flame temperatures can be up to 2000 °C, but the reactor outlet temperature is normally of the order of 1500 °C.

Under these operating conditions the coal is virtually completely gasified without the formation of tars, phenols, or other condensate hydrocarbons. Overall carbon conversion is about 99 percent.

The high reactor temperatures also cause most of the ash to melt and to flow down the reactor wall into a water-filled compartment. The remainder of the ash leaves the reactor with the product-gas flow. In order to solidify the entrained ash droplets before the product gas enters the waste-heat boiler, the gas is cooled to about 900 °C. Depending on the

FIG. 2-1 Pilot plant with an intake of 150 tons of coal per day for the Shell Coal Gasification Process at Deutsche Shell's Harburg Refinery. *(Deutsche Shell.)*

specific product-gas application, this is done either by recycling cold gas or by quenching with water. High-quality steam, normally used for driving the oxygen-plant compressors, is raised in the waste-heat boiler.

After the solids have passed through the waste-heat boiler, they are removed from the gas via an integrated system that includes a cyclone and scrubbers. Use of this system allows the removal of solids in a dry form. If so required, these solids (ash plus some unconverted carbon) may be recycled to the reactor.

When the product gas leaves the solids-removal system, it has the following typical mole composition for a hard-coal feedstock:

Constituent	Mol %
H_2O	2
H_2	28.5
CO	65.5
CO_2	1.5
CH_4	0.1
H_2S	1.4
N_2, Ar	1.0

In addition, it contains traces of COS, HCN, and NH_3. Total gas production is about 32 stdft^3/lb (2 m^3/kg) of coal feed, and the gas has a lower heating value of about 300 Btu/stdft^3 (11.3 MJ/m^3). Depending on the ultimate use of this gas (fuel gas or synthesis gas) further treatment will be required, for which various commercially available processes can be applied.

A large pilot plant with a capacity of 150 tons/day (6250 kg/h) of solid-fuel input (Fig. 2-1) started operation at the end of 1978 at Deutsche Shell's Harburg refinery. By mid-1983 the plant had completed over 5500 running hours, with a longest uninterrupted run of 1000 h. A flow scheme of the 150-ton/day (6250-kg/h) pilot plant is given in Fig. 2-3. (The numbers in parentheses in the discussion that follows refer to different pieces of plant equipment indicated in Fig. 2-3.)

Coal is ground in a mill, dried to specification, and subsequently pneumatically transported to the atmospheric cyclone hoppers (2). In the Harburg plant, nitrogen from a liquid-nitrogen storage vessel is used as a carrier gas for the coal. To transport the coal from the cyclone hoppers (2) to the nitrogen-pressurized feed hoppers (4) a fully automatic lock-hopper system is used. The pneumatic transport of the coal from the feed hoppers to the burners is again accomplished with nitrogen.

Good control of the process mandates accurate measurement of the coal, oxygen, and steam flow to the burners. The success of the pneumatic feed system was made possible by the development of an accurate and dependable coal flowmeter.

The reactor is equipped with two diametrically opposed burners and consists essentially of a pressure shell which is protected from the hot gases by a tube wall in which saturated

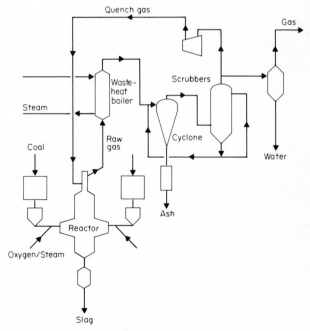

FIG. 2-2   Diagram of Shell Coal Gasification Process.

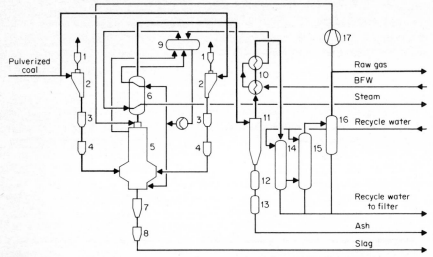

FIG. 2-3 Coal gasification process flow scheme for the 150-ton/day pilot plant at Harburg. Key: coal filter (1), cyclone hopper (2), lock-hopper (3), feed hopper (4), gasifier (5), waste-heat boiler (6), slag breaker (7), slag lock-hopper (8), steam drum (9), boiler feedwater preheater (10), cyclone (11), ash hopper (12), ash lock-hopper (13), venturi (14), scrubber (15), high-pressure separator (16), recycle gas compressor (17).

steam of 725 lb/in² (5000 kPa) is raised. The tube wall is in turn protected by a thin layer of a refractory material.

The slag which leaves the reactor via a hole in the bottom is quenched in water, crushed in a submerged mill, and then lock-hoppered out to atmospheric pressure.

The gases, which leave the reactor at about 1500 °C and 430 lb/in² (3000 kPa), are quenched with solids-free recycled synthesis gas at 100 to 900 °C in order to solidify the entrained slag particles before they enter the waste-heat boiler (6). The gases leave the waste-heat boiler at a temperature of 320 °C. In the waste-heat boiler superheated steam at 500 °C and 725 lb/in² (5000 kPa) is raised. The waste-heat boiler and the reactor tube wall have a common forced-circulation system. Some 80 percent of the solids in the gas are separated in the cyclone (11).

The remainder of the solids is washed out with water in a series of scrubbers and separators (14), (15), and (16). The gas leaving the scrubbers has a solids content of 1 mg/m³ and a temperature of 40 °C.

A summary of the operational experience with the 150-ton*/day (6250-kg/h) pilot plant is given in Table 2-1.

The main objectives of the plant are the following:

• To confirm the operational results of the 6-ton/day pilot plant on a larger scale

• To develop scale rules for scale-up to prototype and commercial plants

• To implement new technical developments and to collect component-reliability data

• To collect environmental impact data

*Throughout this chapter the ton used is the metric ton (1000 kg or 1 Mg).

TABLE 2-1   Summary of experience in operating a 150-ton/day pilot plant (status mid-1983)

Operating period, h	5500
Longest uninterrupted run, h	1000
Pressure, lb/in² (kPa)	275 (1900)
Temperature, °C	1500
Coal-feed rate, t/h	4.5
Oxygen-coal ratio (maf)*, w/w	1.05
Steam-coal ratio	0
Raw gas-coal ratio, scf/lb (m³/kg)	36 (2.1)
Carbon conversion, %	99
Gas composition, vol %:	
H_2	25.6
CO	65.1
CO_2	0.8
H_2S	0.4
N_2 + Ar†	8.1

*Moisture- and ash-free coal.

†Including N_2 used in feed system.

PROCESS PERSPECTIVE

If the total operating experience in the two plants in Amsterdam and Harburg is surveyed, it may be concluded that the basic concept of the SCG process has been demonstrated. Important in this respect are the following:

- A conversion of 99 percent is obtained.
- The CO_2 content in the product gas can be maintained between 1 and 2 percent.
- Reactor conditions (temperature) can be controlled accurately.

Apart from its direct use as a fuel gas producer, the SCG process has a whole range of further applications. Three of these are discussed below.

Use in Combination with Combined-Cycle Power Generation

A block scheme is given in Fig. 2-4. In this scheme the gasification and combined-cycle stations are kept separate.

Production of Hydrogen-Rich Synthesis Gas for Methanol, Fischer-Tropsch Type Syntheses, and SNG

Two possible flow schemes are presented in Figs. 2-5 and 2-6. The optimum route is dependent on many parameters, including the composition of the feedstock and availability and cost of utilities. Detailed design-optimization studies are at present being carried out in this field.

Production of Hydrogen in Coal-Liquefaction Schemes

This application follows the same scheme as above but includes a deeper CO shift plus a methanation step to remove the remaining CO. In direct-liquefaction processes, this scheme may be used to convert the bottom products into hydrogen, which in turn is used for liquid upgrading.

A feedstock which can be gasified for the same purpose is the fluid coke from the Canadian Athabasca tar sands. This feedstock has already been successfully gasified in Amsterdam.

DETAILED PROCESS DESCRIPTION

Additional information is presented in this section on dry- and wet-process feed systems, blast requirements, product-gas thermal efficiency, and process improvements.

SCG Process Feed Systems

One of the characteristic features of the SCG process is its dry feed system. The coal is ground and dried to a moisture content of 2 percent, pressurized in a lock-hopper system, and pneumatically fed to the reactor. The lock-hopper system is cyclic and requires the frequent opening and closing of valves in a dusty environment. An alternative would be a "wet" system. A coal-water mixture could be made pumpable and be compressed and transported directly into the reactor.

A comparison between a "dry" and a "wet" feed system for the SCG process system is presented in Tables 2-2, 2-3, and 2-4. The calculations have been made with the SIPM-developed (Shell Internationale Petroleum Maatschappij) mathematical model for entrained-bed gasification. This model has proved to be a valuable tool in various design studies. Its reliability for this type of exercise has been verified by comparing the mathematical-model results with actual plant measurements in Harburg.

All calculations are based on the standard SCG process scheme, with the only difference being the moisture-water content in the feed stream to the gasifier.

Case 1	Standard dry feed system.
Case 2	Same as case 1 but coal is not dried. (Moisture content of "as received" coal is 16.5 percent.)
Cases 3, 4, and 5	Water is added to "as received" coal: 20, 40, and 80 percent respectively.

Table 2-2 gives the main results. For increasing water content, it shows:

• Lower Syngas production ($H_2 + CO$)

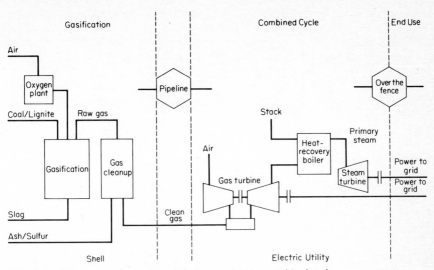

FIG. 2-4   Electricity generation via gasification used with a combined cycle.

- Higher steam production
- Higher specific oxygen consumption

In the standard SCG process scheme, high-pressure steam is generated, which is mainly used to supply power to the various consumers (the principal one is the oxygen plant). If there is not enough high-pressure steam available, a coal-fired auxiliary boiler is assumed;

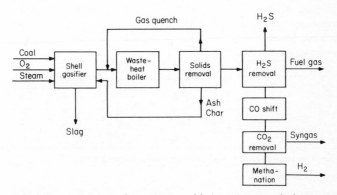

FIG. 2-5   Block scheme for production of fuel gas, Syngas, or hydrogen.

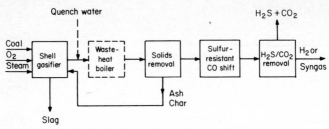

FIG. 2-6 Typical block scheme for production of hydrogen or synfuels.

TABLE 2-2 Comparison of "dry" and "wet" feed systems: gasification data

Gasifier data	Coal dried to 2% moisture	Coal as received	Water Added to Coal		
			20%	40%	80%
Coal (A.R.), lb*					
To process	2200	2200	2200	2200	2200
To auxiliaries	99	33	0	0	0
Moisture removed, lb	320	—	—	—	—
Water added, lb	—	—	440	880	1800
Oxygen to gasifier, lb	1500	1700	1800	1900	2300
Steam to gasifier, lb	33	—	—	—	—
O_2/maf† coal ratio	0.92	0.99	1.07	1.16	1.35
HP‡ steam produced, lb	2600	2900	3300	3800	4800
Raw gas produced, scf $\times 10^{-3}$	60	66	75	85	103
H_2 + CO produced, scf $\times 10^{-3}$	57	54	51	47	40
Raw gas composition, vol %					
H_2O	1.7	9.0	18.6	27.3	41.4
H_2	28.7	27.8	26.0	23.3	18.1
CO	65.6	54.0	41.7	32.5	20.5
CO_2	1.6	7.0	11.8	15.1	18.5
H_2S	1.5	1.4	1.2	1.1	0.9
N_2 + Ar	0.9	0.8	0.7	0.7	0.6

*Illinois no. 6: as received (A.R.), 16.5 percent moisture, 7.6 percent ash.

†Moisture- and ash-free coal.

‡High-pressure steam.

TABLE 2-3 Comparison of "dry" and "wet" feed systems: net plant efficiency

Feed to gasifier / Gasifier data	Coal dried to 2% moisture	Coal as received	Water Added to Coal		
			20%	40%	80%
Coal (A.R.) to process:*	100	100	100	100	100
Raw product gas	81.2	77.3	72.5	67.1	56.6
HP steam produced†	16.0	18.3	20.9	24.1	30.2
Plant energy requirement:					
Steam to gasifier	0.2				
Electric power‡	18.5	19.6	20.9	22.5	25.7
Heat for drying	1.9				
Total	20.6	19.6	20.9	22.5	25.7
Total coal in	104.6	101.3	100	100	100
Total raw gas out	81.2	77.3	72.5	67.1	56.6
Total HP steam out	0	0	0	1.6	4.5
Net efficiency	77.6	76.3	72.5	68.7	61.1

*Heat content of coal, as received (A.R.), to process is set at 100 and other values in table are calculated relative to it. Heat content is based on net heat of combustion. Raw gas is at its dew-point temperature (latent heat not included).

†Heat content of steam is based on heat content of equivalent amount of coal required to raise steam. High-pressure (HP) steam is at 1450 psi (10,000 kPa) and 520 °C.

‡Heat content of power is based on heat content of equivalent amount of coal required to raise power.

a steam surplus is considered an export product. Heat for coal drying (case 1 only) is provided by an extra amount of import coal.

Table 2-3 presents the consequences for the overall efficiency. It shows that for case 1 (dried coal), 4.6 percent extra coal is required, partly for coal drying and partly for providing extra power. For case 5 (80 percent water added), 4.5 percent of the incoming energy is exported in the form of high-pressure steam. Table 2-3 clearly illustrates the reduction of net efficiency with increasing water content.

In the standard process, the raw gas is cooled to its dew point in the solids-removal system. Subsequent cooling (if so required) is by direct air cooling. For a wet feed system, with its relatively high raw-gas dew-point temperature, it might be advantageous, depending on the process application, to use the low-level heat for process preheating or for raising low-pressure steam. This would increase capital costs but improve overall efficiency. If such heat-recovery schemes were applied, the net efficiency differences between dry and wet feed systems would be reduced to about half of those listed in Table 2-3.

Table 2-4 is indicative of the effects on plant investment costs. It gives the relative throughput of the main plant systems for plants producing the same quantity of synthesis gas.

The increase in the required oxygen-plant capacity is significant. It reflects the combined

TABLE 2-4 Comparison of "dry" and "wet" feed systems: indication of main plant system capacities

Gasifier data	Coal dried to 2% moisture*	Coal as received	Water Added to Coal		
			20%	40%	80%
Syngas production†	100	100	100	100	100
Total coal (A.R.)‡ (including coal to auxiliaries)	100	102	105	115	135
Feed to gasifier (including transport medium)	100	120	150	185	285
Oxygen	100	110	130	150	210
HP steam§	100	120	145	180	270
Raw product gas	100	115	140	170	245

*Individual system throughput for case 1 (dried coal) is set at 100 and other values in table are calculated relative to it.

†Cubic meters of H_2 + CO.

‡As received coal.

§High-pressure steam.

effect of (1) the higher coal consumption for the same quantity of synthetic gas and (2) the higher oxygen consumption for the same quantity of coal feed.

A third aspect of a dry vs. a wet system is the product-gas composition (Table 2-2). For wet systems the CO_2 content rises sharply. Even if the CO_2 itself does not have to be removed (as in most fuel gas applications), it is a disadvantage, as it significantly increases the cost of H_2S removal by requiring that an extra enrichment stage precede the sulfur-recovery unit.

For synthesis-gas applications, which require a CO shift, a somewhat higher H_2/CO ratio is beneficial. In addition, in the wet feed case, the raw gas already contains part of the steam required for the shift. If a sulfur-resistant shift is applied, this has a favorable effect on the shift economics. On the other hand, the high CO_2 content remains a disadvantage, and, more generally, introducing water into the gasifier is an expensive way of raising steam for the shift reaction.

If the above three aspects (net efficiency, plant investment, and product quality) are taken together, it is evident that a dry feed system offers remarkable advantages.

The dry feed system with lock-hoppers has performed completely satisfactorily in both Amsterdam and Harburg plants. In an effort to develop further the concept of dry pressurizing and dry feeding, SIPM is seriously studying the possibilities of a continuous dry-pressurizing system as an alternative to the cyclic lock-hopper system.

Blast Requirements, Product Gas, and Thermal Efficiency

Consumption of oxygen and steam is dependent on the coal feedstock quality. An oxygen demand of 0.9 to 1.0 ton of oxygen per ton of moisture- and ash-free (maf) coal is fairly typical of hard coals; for low-rank coals a figure of 0.8 ton of oxygen per ton of maf coal is more representative. The steam requirement is very low and, in fact, is almost zero for some brown coals and in cases where air is used as a gasifying agent. For hard coal the steam requirement is on the order of 8 percent by weight of the maf coal feed.

The raw-gas production is about 32.5 stdft³/lb (2 m³/kg) for a good-quality bituminous coal. The gas is relatively rich in CO and the CO/H_2 ratio, on a volume basis, is typically between 1.8 and 2.2 for the preferred operation at minimum steam dosage and when the recycle gas is used as a quenching medium.

The heat content of the gas is of the order of 300 Btu/stdft³ (11.3 MJ/m³). The thermal efficiency of the gasification proper is about 82 percent for oxygen-steam gasification. Apart from this percentage of the total heat in the coal which is recovered as chemical heat in the gas, about 15 percent of the heat in the coal going to the gasifier becomes available for raising steam in the waste-heat boiler via the sensible heat in the gas.

The total heat content of synthesis gas and steam represents approximately 94 to 97 percent of the heat content of the coal feed. (See Fig. 2-7). It is, however, more realistic to consider the overall thermal efficiency, taking into account the energy consumption of the oxygen plant, coal mill and dryer, coal gasifier, and auxiliary equipment. This overall thermal efficiency, calculated on a lower-heating-value basis, corresponds to approximately 78 percent for a 12 percent ash, 7 percent moisture hard coal.

To demonstrate the versatility of the SCG process *vis-à-vis* feed-coal quality, the thermal efficiency has been calculated for a range of coals, namely:

- Bituminous coal with moderate ash and moisture content (Illinois No. 6)
- Subbituminous coal (Wyodak)

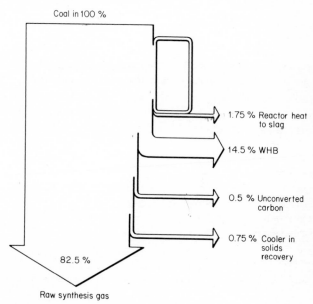

FIG. 2-7 Typical heat balance for gasifier proper.

TABLE 2-5 Shell coal gasification: coal-feed analyses

Analytical measurements	Types of Feed			
	Illinois No. 6 bituminous	Wyodak subbituminous	Coal-liquefaction vacuum bottoms	German brown coal
Elements, wt %:				
Carbon (maf)	78.1	75.6	87.1	67.5
Hydrogen	5.5	6.0	5.7	5.0
Oxygen	10.9	16.8	3.3	26.5
Sulfur	4.3	0.9	2.4	0.5
Nitrogen	1.2	0.7	1.5	0.5
Other components:				
Ash (wt %, as received)	12.0	5.9	17.6	6.4
Moisture (wt %, as received)	6.5	35.0	0	5.0
Moisture (wt % of coal to gasifier)	2.0	2.0	0	5.0
Lower heating value of coal as received				
MJ/Mg	25,800	17,160	29,410	9,990
Btu/lb	11,095	7,380	12,645	4,295

TABLE 2-6 Coal, oxygen and steam requirements for different coals*

Constant plant capacity	Illinois No. 6 bituminous	Wyodak subbituminous	Coal-liquefaction vacuum bottoms	German brown coal
Coal (A.R.) intake	573	810	494	1364
Coal to gasifier	477	489	494	626
Oxygen demand (99 vol % pure)	400	395	407	434
Steam demand	36	12	86	20
Thermal efficiency, %†				
Gasifier proper	83	83	83	79
Overall plant after subtraction of own consumption (e.g., coal drying, oxygen plant)	78	77	77	72

*Unless otherwise noted, values are in metric tons and are based on the production of 10^6 m^3 of synthesis gas ($CO + H_2$).

†Calculated on a lower-heating-value (LHV) basis.

- Coal-liquefaction vacuum bottoms
- Brown coal with a high moisture and low ash content (German brown coal)

These feedstocks cover a fairly wide range of ash and moisture contents (Table 2-5). The calorific values, correspondingly, range from 4300 to 12,700 Btu/lb (10,000 to 29,500 kJ/kg) for "as received" coal.

The data presented in Table 2-6 are indicative of the oxygen and steam consumptions and of the overall thermal efficiencies for the coal feedstocks specified. Assuming a constant production of synthesis gas ($CO + H_2$), it is evident that the amount of coal to be processed increases as the rank, and correspondingly the heating value, of the coal goes down and as moisture and ash contents go up. Variations in oxygen demand are generally well below 10 percent for the various feedstocks.

Some details of the gas composition are given in Table 2-7. Hydrogen-to-carbon monoxide ratios are not materially affected by a change in coal-feed composition.

At CO_2 concentrations not exceeding 5 percent by volume, as shown in Table 2-7, the selective removal of H_2S does not normally present a problem. Typical CO_2 concentrations are below 2 percent by volume.

Process Improvements

Apart from the recirculation of the solids from the cyclone back to the gasifier, which leads to the situation that in the end all ash entering the reactor leaves the system in the form of inert nonleaching slag, the following improvements are currently being considered:

- Pressurizing the lock-hoppers and transporting the coal with Syngas instead of nitrogen, thus avoiding dilution of the gas with nitrogen.
- Developing a noncooled, insulated, brick-lined reactor.
- Adapting the gasifier for the gasification of hydrocarbon liquids such as vacuum residues from coal-liquefaction plants. An advantage of such a process over the Shell gasification process for oil is that liquids containing unconverted coal and high percentage of ash can be gasified.

TABLE 2-7 Wet synthesis gas composition*

Component	Illinois No. 6 bituminous	Wyodak subbituminous	Coal liquefaction vacuum bottoms	German brown coal
H_2O	1.5	2.6	2.1	11.3
H_2	31.6	32.5	33.6	26.9
CO	64.0	62.8	61.8	55.0
CO_2	0.8	1.3	1.0	6.1
CH_4	—	—	0.1	—
H_2S + COS	1.4	0.3	0.7	0.2
H_2	0.5	0.3	0.5	0.3
Ar	0.2	0.2	0.2	0.2

*Percent by volume.

PROCESS ECONOMICS

The investment for a 1300×10^6 stdft³/day (36×10^6-m³/day) synthesis gas plant based on a U.S. location in mid-1980, is estimated at $1 billion. This figure applies to the processing of a hard coal containing 10 percent ash and 10 percent moisture and includes the investment of coal handling and storage, coal mill and dryer, oxygen plant, water treatment, ash-disposal facilities, and offsites.

Excluded is the investment in plants for the further processing of the dry, particulate- and sulfur-free synthesis gas.

The corresponding cost, on a heating-value basis, of the synthesis gas before and after conversion into methyl fuel and subsequently into gasoline is shown as a function of the coal-feed unit cost in Fig. 2-8.

For a low-cost coal feed ($1/MM Btu), such as would be available from opencast mining of large coal reserves, the synthesis gas cost would amount to some $3/MM Btu. For deep-mined coal ($2.50/MM Btu, as available in Europe), the cost could be as high as $5/MM Btu.

The corresponding figures for methyl fuel from a low-cost coal feed and from deep-mined coal are $6.5 and $9.50 per MM Btu and for gasoline $8 and $11 per MM Btu,

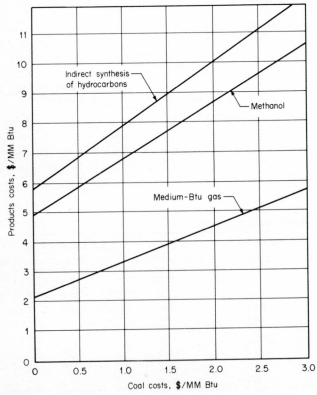

FIG. 2-8 SCG Process costs.

respectively. The economics of SCG is only marginally affected by variations in coal rank, ash content, or moisture content. Only the costs of the coal mill and dryer and of the ash-disposal facilities are significantly influenced, but these have a minor impact on the overall economics.

SUMMARY

The characteristics and advantages of the SCG process are discussed in the following text.

Complete conversion of any coal

The process is suitable for the complete gasification of a wide variety of solid fuels, such as all types of coal and petroleum coke. Fuels with a high ash content (up to 40 percent by weight) and high sulfur content (up to 8 percent by weight) can be used in the Shell gasification plant without trouble. Even a high water content in the coal does not pose a technical problem. However, on economic grounds, it is advantageous to dry the coal to a moisture content of 1 to 6 percent by weight.

Since the process in any case requires a solid fuel to be in dust form for gasification, the entire output of a mine, including fines, is acceptable as feed. Unlike fixed-bed or fluidized-bed processes, the SCG process places practically no limitation on the ash fusion behavior or caking properties of the coal.

Clean gas production without formation of by-products

The operation at very high temperatures ensures the formation of a high-quality synthesis gas essentially consisting of hydrogen and carbon monoxide (93 to 98 percent by volume). Tars, phenols, and other by-products are absent; as a rule methane concentrations in the gas do not exceed 0.2 percent by volume. In most cases, the low CO_2 concentration (typically 1 to 2 percent by volume) allows the production of a regenerator off-gas which has such a high H_2S content that it can be sent direct to a Claus unit.

Large unit capacity

Both the high temperatures (above 1400 °C) and the high pressures are responsible for the high capacities attainable. Short-term targets for coal per reactor are 50 to 100 Mg/h, which corresponds in terms of raw gas to 90MM to 180MM stdft^3/day (2.4MM to 4.8MM m^3/day).

On the basis of the application, the optimum pressure level can be selected. Apart from the beneficial effect of the elevated pressure on reactor capacity, there are spin-offs in terms of increased heat-transfer rates in the waste-heat boiler, easier gas treating, and a reduction in gas compression costs.

High thermal efficiency and efficient heat recovery

The chemically bound heat in the gas produced with oxygen gasification is equivalent to about 79 to 82 percent of the chemically bound heat contained in the coal feed. The recovery of the sensible heat from the hot gases leaving the reactor accounts for another 12 to 15 percent of the heat content of the coal feed. The steam produced by this cooling is generally sufficient to drive the compressors of the oxygen plant (see Fig. 2-9).

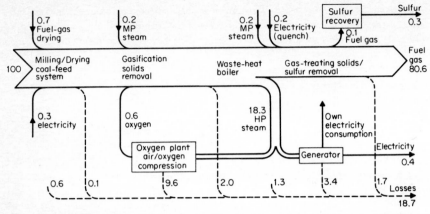

FIG. 2-9   Simplified Sanky diagram for coal gasification. Heat content of both coal and fuel gas is based on net heat of combustion. Heat content of both steam and oxygen is the sum of sensible heat and latent heat. Heat content of electricity is the direct heat equivalent of electric power.

Environmental acceptability

A negligible environmental impact can be expected from the SCG process. This is the consequence of the following:

- The clean raw gas produced
- The high thermal efficiency
- The low wastewater production
- The production of nonleachable and inert slag

THE COMBUSTION ENGINEERING COAL-GASIFICATION PROCESS

SCOTT L. DARLING

Combustion Engineering, Inc.
Windsor, Connecticut

INTRODUCTION

Combustion Engineering (C-E) is developing a process for the production of a clean low-Btu gas from coal, based on the use of an entrained-flow, air-blown, slagging-bottom, two-stage gasifier that operates at atmospheric pressure. The process is intended primarily for the production of electric power from coal in an economical and environmentally acceptable manner. In the process, coal is converted to a low-Btu gas, which is then cleaned of particulates and sulfur compounds. This gas can be used to fuel conventional or combined-cycle power plants, for both new or retrofit applications. New combined-cycle applications are especially attractive, with the potential for reduced emissions, higher thermal efficiency, and reduced generating cost. The product gas can also be used as a process fuel for industrial applications. In addition, the process can be modified to use oxygen as the oxidant in place of air, producing a medium-Btu gas with wide application for utility, industrial, and synthesis operations.

The major advantages of the C-E process are as listed:

- Ability to process all coals
- High carbon utilization
- High thermal efficiency
- Ease of operation and control
- Maximum use of proven equipment
- Environmental acceptability
- Minimum effluents
- High unit capacity
- Favorable economics

A process-development unit (PDU) with a capacity of 120 tons/day (109 Mg/day) of coal is located at C-E's Windsor, Connecticut, facility. The PDU was designed to demonstrate the capability and suitability of the process and equipment and to provide data for scale-up. This plant began operating in late 1977, and through its shutdown in mid-1981 had accumulated over 5000 h of coal-fired operation of which 3700 h were in the gas-making mode. The major process concepts and gasifier design features have been demonstrated.

PROCESS DESCRIPTION

Overall Process

The C-E coal-gasification process uses an entrained-flow, air-blown, slagging-bottom, two-stage gasifier that operates at atmospheric pressure and produces a low-Btu gas. A simplified block diagram of the process is shown in Fig. 3-1. Feed coal is pulverized and pneumatically transported with air to the gasifier. In the gasifier the coal and transport air are mixed with additional air and reacted at high temperature to produce a stream of raw product gas. The ash in the coal is fused and tapped from the bottom of the gasifier as

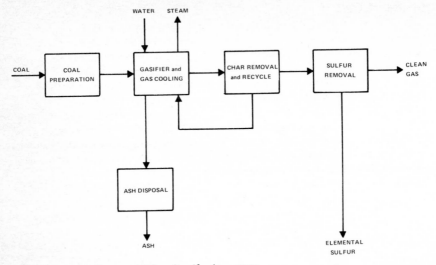

FIG. 3-1 Block diagram of C-E coal-gasification process.

molten slag. The sulfur in the coal is converted mainly to hydrogen sulfide (H_2S) with some carbonyl sulfide (COS). The product gas, containing some unconverted carbon and ash, which is called *char,* is cooled by the use of heat-recovery surface within the gasifier and ducted to a particulate-removal system, where the char is removed from the gas and recycled back to the gasifier. The cooled, char-free gas is then sent to sulfur-removal equipment, where the sulfur compounds are removed and converted to elemental sulfur. The cleaned gas stream contains CO and H_2, has a higher heating value (HHV) of approximately 110 Btu/stdft³ [4 MJ/m³(n)],* and is free of particulates and sulfur. This gas is suitable for fueling steam generators, gas turbines, or industrial process equipment without the need for stack-gas cleanup systems.

Gasifier

The process chemistry in the C-E gasifier involves two major steps:

- Combustion of a portion of the coal feed and all recycle char to provide a stream of hot gas
- Devolatilization and partial gasification of the remaining coal feed within that hot gas stream

The gasifier is designed as a two-stage reactor, or as two reactors in series, each performing one of the above steps. This allows the performance of each step to be optimized, thus optimizing overall gasifier performance.

 Functionally, the gasifier consists of three sections, a combustor and reductor connected

---

*m³(n) symbolizes the normal cubic meter, that is, the volume of one cubic meter of gas measured at 0 °C and 1.01 Pa pressure.

by a diffuser (Fig. 3-2). The functions of the combustor are (1) to provide heat for subsequent endothermic gasification reactions occurring downstream, (2) to consume char recycled from the gasifier outlet, and (3) to slag ash in the coal and recycled char. In the combustor a portion of the feed coal and all recycled char are reacted with fuel transport air and preheated secondary air. The stoichiometry is adjusted to be somewhat fuel-rich to maintain design gas temperatures within range of 1760 °C. A high degree of turbulence is provided by tangential entry of the fuel and air jets, which create a swirling vortex. This promotes rapid mixing of the reactants and good slag removal. Residence times on the order of 1.0 s are provided in order to produce a high degree of carbon conversion to gas. Principal reactions in the combustor are the evolution, cracking, and combustion of coal volatiles, followed by combustion and some gasification of the devolatilized coal and recycled char carbon with the remaining oxygen.

From the combustor, the gas enters the diffuser section through a restriction, or throat, intended to minimize the flow of gas back into the combustor. In the diffuser, reductor coal and transport air are mixed with the gas leaving the combustor. The functions of the diffuser are (1) to provide rapid and efficient mixing and devolatilization of the reductor coal in the hot combustor gases and (2) to stabilize the gas flow patterns so that the velocity

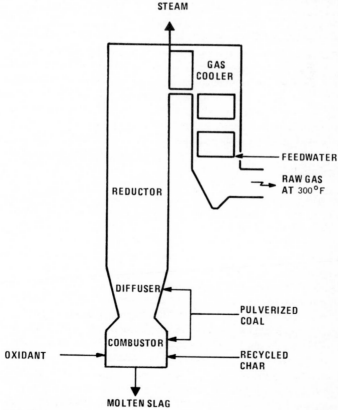

FIG. 3-2 C-E coal gasifier.

and concentration profile of the gas entering the reductor will be uniform. The principal reactions in the diffuser are the evolution and cracking of coal volatiles, along with combustion of a portion of the volatiles with the reductor-coal transport air. At the design temperature levels, devolatilization and cracking will be rapid and complete such that no tars or oils are formed, leaving only a gas mixture containing CO_2, CO, H_2O, H_2, N_2, and unreacted char.

From the diffuser the gases enter the reductor. The function of the reductor is to provide residence time for the endothermic reaction of char carbon with CO_2 and H_2O, producing additional CO and H_2. The reactions are generally allowed to proceed until the gas temperature reaches 930 to 1100 °C, at which point the reductor is terminated and recovery of sensible heat from the gas begins. This temperature range is selected because the gasification reactions have essentially stopped and, also, because the coal ash is generally below the ash softening temperature so that fouling of the downstream heat-recovery surface is avoided. This produces a stream of gas containing CO_2, CO, H_2O, H_2, N_2, and some remaining unreacted char. The char is cleaned from the gas and recycled to the combustor for conversion.

Reactions involving the sulfur in the coal produce mainly H_2S and a small amount of COS. Nitrogen-containing compounds such as NH_3 and HCN are produced in trace amounts only.

The gasifier vessel is constructed of water-cooled tubing, using the same type of construction used in conventional utility boilers. The vessel is lined with refractory in order to limit waterwall heat absorption to about 10 percent of heat input, allowing attainment of the high gas temperatures required for gasification. In commercial applications the heat absorbed by the vessel may be used to generate high-pressure steam. This steam may be added to that generated in the gas-using facility, particularly in conventional or combined-cycle power plants, and results in high overall plant efficiency. The heat absorbed by the vessel may also be used for air preheat, feedwater heating, auxiliary equipment drives, or other uses, depending on the gasifier application and customer needs.

Gasifier Auxiliaries

The major auxiliary systems within the C-E process include coal handling and feed, char removal and recycle, ash handling, and sulfur removal. Each of these systems contains equipment that has been proved commercially.

The coal-handling and feed system consists of standard C-E coal pulverizers and pneumatic transport piping. In the pulverizer the coal is dried and ground to 70 percent passing through a 200-mesh sieve. Two basic types of feed systems are possible, indirect or direct feed. In the indirect system, coal is pulverized, stored in a bin, and then pneumatically transported from the bin to the gasifier. The coal moisture is vented from the pulverizer system and does not enter the gasifier. In the direct system, the coal is pulverized and then pneumatically transported directly to the gasifier. The moisture from the coal that was dried in the pulverizer is contained in the transport air and enters the gasifier with the coal. Indirect systems generally cost more because of the added equipment, but produce a higher gas HHV as a result of the venting of the coal moisture. Both types of systems have been used on commercial utility coal-fired steam generators.

The char-removal and recycle system can consist of many different kinds of equipment, including cyclones, wet scrubbers, precipitators, and baghouses, along with feeders and pneumatic transport piping. The most attractive combination of equipment appears to be a cyclone followed by a baghouse. This combination can yield the extremely low level of dust loading required to meet the limitations of a gas turbine at reasonable capital cost and power consumption.

The ash-handling system is identical to that used on conventional slagging-bottom boilers. Molten ash flows from the slag taphole in the floor of the gasifier combustor to a water-filled quench tank, where the slag is frozen, granulated, and periodically sluiced to disposal.

For sulfur removal, many systems are commercially available for removing the H_2S and some COS from the stream of product gas.

While the C-E process was initially developed as an air-blown process, preliminary studies indicate that the gasifier can be converted to an oxygen-blown mode. As such, the gasifier would produce a synthesis gas with a heating value of approximately 300 Btu/stdft^3 [11 MJ/m^3(n)]. Such a gas has wide potential as a synfuel for industrial use, for pipeline distribution over reasonable distances, and for further ungrading via shift conversion and methanation, or other processes.

Reasons for Selection

The major reasons for the selection of an entrained-flow, air-blown, atmospheric-pressure gasifier for the C-E process are:

1. *Ability to process all coals*   All coals can be gasified without pretreatment.

2. *Ability to handle all coal sizes*   Since the coal is pulverized prior to gasification, the entire run-of-mine size range is acceptable, including fines.

3. *High carbon utilization*   All char produced is recycled and consumed, resulting in no net char production. Carbon loss in the slag is negligible. Carbon utilization approaches that of a pulverized coal-fired boiler (>99.5 percent).

4. *High thermal efficiency*   Virtually all the energy of the input coal leaves the process in the form of gas or steam.

5. *Lower water consumption*   Water consumption is minimized by use of a dry-coal feed system and the absence of steam injection for gas tempering.

6. *Ease of operation and control*   Operation, control, and maintenance of the gasifier are simplified by operation at atmospheric pressure.

7. *High reliability and availability*   High-maintenance items, such as lock-hoppers and letdown valves, are not required.

8. *Maximum use of proven equipment*   With the exception of the gasifier itself, all equipment within the process is proven, commercially available equipment. Most is identical to that used with conventional pulverized coal-fired boilers and is familiar to electric utilities.

9. *Environmentally acceptable products*   The gas can be cleaned of sulfur compounds to a high degree. The high temperature level at which gasification occurs precludes production of tars, oils, or other contaminants.

10. *Minimum effluents*   The effluents from the process consist of slagged ash and elemental sulfur, which represent the most acceptable and least costly forms of these materials for disposal.

11. *Capability to be scaled up to large sizes*   Since the gasifier is constructed much like a boiler, gasifiers can be built in sizes similar to boilers. Multiple units are not required to achieve a desired output, and economies of scale result.

12. *Reduced generating costs*   Economic studies for new plant applications show that in a combined cycle the C-E process can produce electric power from coal at rates approximately 10 percent less than conventional plants with stack-gas cleanup systems. This results from the higher thermal efficiency of the combined cycle. Studies for retrofit applications show the process is competitive for retrofitting both boilers and combined cycles. Recent work also shows several economically attractive industrial applications for the process.

DESCRIPTION OF PROCESS-DEVELOPMENT UNIT

In order to develop the C-E process, a 120-ton/day (109-Mg/day) process-development unit (PDU) was constructed at C-E's Windsor, Connecticut, site. The objectives of the PDU operation were to demonstrate the capability and suitability of the process and equipment to produce a low-Btu gas of predictable composition and cleanliness, to provide design information for scale-up to larger-size plants, and to provide information on the application of the plant equipment to the process. The PDU size and equipment were selected such that, following a comprehensive test program, the unit could be scaled directly to a demonstration-size unit.

The plant is a self-contained experimental facility and has all the major process units required in a commercial application: coal preparation and feed, gasification, char removal and recycle, ash handling, sulfur removal, and gas incineration combined with steam generation. Figure 3-3 is a schematic diagram of the plant, and Fig. 3-4 shows a view of the plant.

Gasifier

The PDU gasifier is shown in Fig. 3-5. The gasifier was designed to incorporate all the major design features of commercial gasifiers. The vessel is approximately circular in cross section, being formed of 16 flat tube panels, with an internal diameter (ID) of about 9 ft (2.7 m) in the combustor and reductor sections and an ID of 4.5 ft (1.4 m) at the throat and is approximately 90 ft (27 m) tall. The walls are formed from gas-tight, fusion-welded panels of 1½-in (38-mm) diameter, vertical water-cooled tubes on 2-in (50-mm) center lines and are insulated and cased on the outside and lined with refractory on the inside. Except for the refractory lining, this is the same construction C-E uses for conventional, industrial, or utility steam generators. Fuel nozzles are located at several elevations in both the combustor and reductor, with eight nozzles per elevation spaced evenly around the vessel perimeter.

The interior of the vessel from the combustor floor to the reductor outlet is lined with refractory to limit heat absorption by the vessel and to provide the gas temperatures required for efficient gasification. In order to limit heat absorption by the PDU gasifier vessel to the 10 percent of coal heat input that is planned for the larger commercial units, a much thicker refractory lining is required than on commercial units because of the rela-

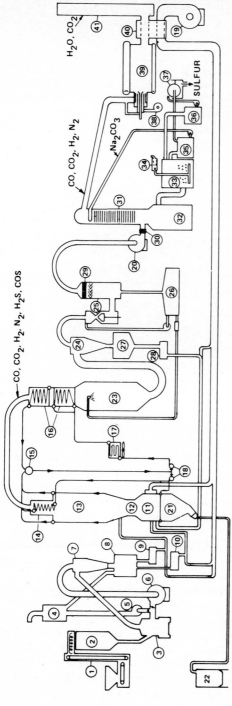

FIG. 3-3 Schematic flow diagram of the C-E PDU. Area 100: 1, coal-unloading equipment; 2, coal-storage silo. Area 200: 3, pulverizer; 4, bag filters; 5, vent fan; 6, recirculation fan; 7, pulverized-coal cyclone; 8, pulverized-coal storage bin; 9, reductor pulverized-coal feeder; 10, combustor pulverized-coal feeder. Area 300: 11, gasifier combustor; 12, gasifier diffuser; 13, gasifier reductor; 14, primary cooler; 15, mixing sphere; 16, secondary coolers; 17, water cooler; 18, circulating pumps; 19, forced-draft fan; 20, induced-draft fan. Area 400: 21, ash hopper; 22, dewatering bin. Area 500: 23, sludge spray dryer; 24, char cyclone separator; 25, wet venturi-scrubber chevron separator; 26, sludge thickener. Area 600: 27, char-receiving bin; 28, char feeder; 29, gas reheater; 30, gas cooler. Area 700: 31, H_2S absorber; 32, reaction tank; 33, oxidizer; 34, air compressor; 35, solution tank; 36, sulfur slurry tank; 37, rotary vacuum filter. Area 800: 38, forced-draft fan, 39, incinerator boiler; 40, tubular air heater; 41, stack.

3-53

FIG. 3-4 View of PDU plant.

tively high surface-to-volume ratio in the PDU gasifier. A layer of plastic refractory 4 to 6 in (100 to 150 mm) thick was initially installed throughout the vessel. Then, in order to model the much thinner refractory lining planned for commercial units, one of the 16 wall-tube panels, called the *test panel*, was lined with a thin layer of plastic refractory approximately 1 in (25 mm) thick held in place by short ¾-in (19-mm) metal studs (Fig. 3-6). The test panel extends from the combustor floor to the reductor outlet and is instrumented to provide heat-transfer data for scale-up. Several different types and compositions of refractory were installed in various parts of the vessel for evaluation purposes.

Convective-cooling surface is located at the reductor outlet to cool the gas prior to char removal and sulfur removal. Primary and secondary coolers are provided (Fig. 3-5). The design of these surfaces followed the same practice used for commercial steam generators. Steam soot blowers are provided before and after each bank of coolers to periodically clean any char accumulation.

The vessel and convection-pass tubing are cooled with water at 700 psia (4.7 MPa) such that no steam is generated. This provides for ease of heat-absorption measurements in various sections of the vessel. Heat absorbed in the vessel and convection pass is rejected to the atmosphere in an air-cooled water cooler. All pressure parts are carbon steel, with maximum predicted metal temperatures of 260 °C. Corrosion test probes containing test sections of many different types of materials and coatings are located in various parts of the gasifier vessel and convection pass.

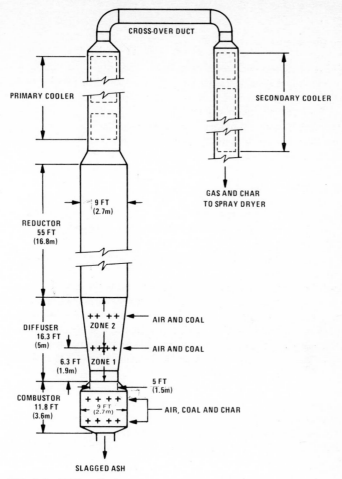

FIG. 3-5 PDU gasifier.

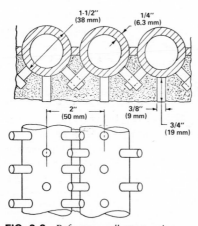

FIG. 3-6 Refractory wall construction.

Auxiliary Systems

The PDU has an indirect coal-feed system. Hot flue gas from a combined incinerator and boiler is used to dry the coal in the pulverizer and transport the pulverized coal to a storage bin. From the storage bin, the pulverized coal is metered with gravimetric feeders and transported pneumatically to the gasifier.

The char-removal system consists of a spray dryer, cyclones, a wet venturi scrubber, and a thickener. Relatively coarse material leaving the gasifier is collected by the cyclones and dropped into a char bin for storage. The fines pass to the scrubber and are collected by the scrubber water and sent to a thickener, where the concentration of solids in the scrubber slurry is raised by settling. The more concentrated slurry is pumped to a spray dryer, where the slurry is dried by the gas leaving the convective-cooling surface of the gasifier and the fine char particles are agglomerated for collection by the cyclones. The char collected in the char storage bin is metered with gravimetric feeders and transported pneumatically to the gasifier.

The ash-handling system at the plant is identical to that used on conventional slagging-bottom boilers.

A Stretford H_2S removal system is used at the plant. In the Stretford system, the char-free product gas is brought into contact with a dilute solution containing sodium carbonate and other chemicals. The H_2S is absorbed and converted to elemental sulfur, which is filtered from the absorbing solution, stored in a hopper, and periodically trucked to disposal. The clean product gas, which leaves the Stretford absorber at low temperatures, is reheated only enough to avoid condensation in the ductwork and is then fired in a standard C-E type A package boiler.

A liquid-oxygen storage, vaporization, and metering system was recently installed to allow the gasifier to operate in an oxygen-enriched mode, yielding a higher gas HHV. Complete testing with oxygen enrichment will permit C-E to offer both low- and medium-Btu gasification systems for a broad range of applications.

The plant is heavily instrumented to provide a wide variety of data. More than 500 data points are logged automatically via computer during operation, permitting extensive evaluation of the process and equipment.

PDU Operation

Coal firing began in December 1977, and gas-making operations began in June 1978. Prior to its shutdown in May 1981, the plant logged more than 5000 hours of coal firing, including 3700 hours of gas-making operation.

All operation to date has been with Pittsburgh seam coal, a highly caking bituminous coal. Table 3-1 shows the proximate analysis, ultimate analysis, and heating value for this coal.

The gas-making history of the plant is shown in Table 3-2, which lists operation by gas-making run. The longest continuous gas-making period (14 days) occurred during run 7, while the longest period of continuous operation (23 days) occurred in run 10.

One important result of PDU testing has been the calibration of a mathematical model of the gasifier. Various empirical parameters in the model involving combustor performance, vessel heat transfer, and chemical reaction rates were evaluated using data from runs 5 through 11. This data spanned a wide range of coal flow, air flow, vessel heat

TABLE 3-1 Pittsburgh seam coal

Proximate analysis, wt %	
Volatile matter	37.5
Fixed carbon	51.5
Moisture (as fired)	0.4
Ash	10.6
Total	100.0
Ultimate analysis, wt %	
Carbon	72.8
Hydrogen	4.9
Oxygen	7.9
Nitrogen	1.3
Sulfur	2.1
Moisture	0.4
Ash	10.6
Total	100.0
HHV,* Btu/lb (kJ/g)	13,160(30.58)

*Higher heating value.

TABLE 3-2 Summary of PDU gas-making operations

Run	Dates*	Total coal firing, h	Total gas-making, h
	1978		
1	June 14–15	34	20
2	June 26–29	81	72
3	Aug. 26–Sept. 1	98	80
4	Sept. 10–16	145	138
5	Oct. 1–6	134	102
	1979		
6	Jan. 22–Mar. 14	846	618
7	May 22–Sept. 5	1021	894
8	Nov. 14–Dec. 20	439	274
	1980		
9	Apr. 22–June 4	484	395
10	Aug. 6–Oct. 6	790	653
11	Nov. 19, 1980– Jan. 10, 1981	453	189
	1981		
12	Mar. 12–May 22	560	311
Total		5085	3746

*Plant operation was not always continuous during these periods.

absorption and gas HHV. Agreement between calculated and measured gas HHVs is generally within ±10 percent. Deviations from agreement can be caused by many factors, including data inaccuracies as well as areas in the model requiring further refinement. In view of the generally good agreement between test data and calculations, the mathematical model is being used in design studies of the C-E gasifier. A more detailed description of the model and model calibration is given in Ref. 1.

Experience with refractories has shown that a thick refractory lining, while showing good durability in the nonslagging areas, will not last in the slagging zones. The design using thin walls that was used on the PDU test panel and that is planned for commercial units has shown good durability in all zones of the PDU gasifier throughout operation.

Experience with the convective-cooling surfaces of the gasifier has been good. The absence of tars and oils in the gas and an entering gas temperature below the ash-softening temperature result in a very dry powdery type of char. In the bare tube sections, gradual char accumulations are easily removed with periodic soot blowing.

During gasifier construction, metallurgical test specimens were installed along the test panel in the combustor and reductor zones and within the convective-cooling surfaces at the reductor outlet. The materials being tested include typical materials that are used for boiler pressure parts, such as carbon steel, low-alloy steels (T-9, T-11, and T-22), and austenitic steels (TP 304 and TP 347). Several coatings for the carbon and low-alloy steels are also being tested. Maximum metal temperatures for the samples are estimated to range from 425 to 540 °C. Corrosion of the samples has been monitored periodically throughout plant operations. Results in all cases showed that wastage was more extensive in the carbon steels (coated and uncoated) and low-alloy steels, while the stainless steels were only slightly affected. Several coatings showed some degree of corrosion resistance as well.

The ash-handling system has performed routinely. Typical analyses of the slag show a carbon content of 1 percent by weight or less, amounting to about a 0.1 percent loss of input carbon. Leachate tests indicate the C-E process produces a sterile, inert ash.

After some modifications, the char-removal system has performed adequately. Data provided by pilot plant operation will allow better selection of equipment for future plants.

The performance of the Stretford system has been satisfactory throughout plant operation. Efficiencies of H_2S removal have routinely measured 98 to 99 percent removal. Some corrosion of the carbon-steel solution piping has been experienced, though mainly in the areas of stagnant flow or in long horizontal runs where undissolved solids may settle. Stainless steel or fiberglass reinforced plastic (FRP) materials used elsewhere in the system without effect are considered possible alternative materials. In order to further improve total sulfur removal, processes for COS removal are being investigated.

Accomplishments

The process demonstrated the following:

- Application of utility-boiler design practice to coal gasification; these included suspension firing, pneumatic coal transport, and atmospheric pressure operation.
- Char collection and complete recycle.
- Feasibility of slagged ash removal for coal gasification.

- High efficiency of H_2S removal.

During operation, the process:

- Completed more than 3700 h of gas-making.
- Used a completely integrated system operation.
- Demonstrated operability of all components.
- The testing of materials and equipment demonstrated the durability of a thin studded refractory lining.

PROCESS ECONOMICS

New Plant Applications

Recent economic studies by DOE, EPRI, C-E, and others have shown that coal gasification is cost-competitive with conventional means for generating electric power from coal and appears to offer significant cost advantages when incorporated into a combined-cycle plant.

Detailed C-E studies performed during Phase II of the development program[2] compared the costs of a conventional coal-fired plant with stack-gas scrubbers to the costs of both conventional and combined-cycle plants using C-E air-blown gasifiers operated at atmospheric pressure. Each plant had a nominal rating of 600 MW_e. In all plants, the steam cycle used was the conventional 2400/1000/1000 cycle.

The conventional steam plant with scrubbers is shown schematically in Fig. 3-7. The net heat rate of the plant is 9456 Btu/kWh. The conventional steam plant integrated with a C-E gasifier is shown in Fig. 3-8. The net plant heat rate is 9666 Btu/kWh, with approximately 27 percent of the total steam to the turbine generated in the gasifier. A combined-cycle plant integrated with a C-E gasifier is shown in Fig. 3-9. The net plant heat rate is 8223 Btu/kWh, using heat-recovery steam generators and gas turbines with an inlet temperature of 1200 °C. These gas turbines are expected to be available in the mid-1980s.

Projected plant-investment cost and electric-generating cost for each of the plant studies

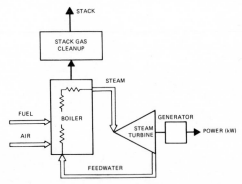

FIG. 3-7 Conventional steam plant.

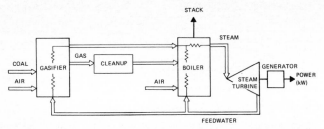

FIG. 3-8 Conventional steam plant integrated with a C-E atmospheric gasifier.

are shown in Table 3-3. All costs are shown in constant 1981 dollars, and were developed on a consistent basis with other DOE and EPRI studies. The table shows that a conventional steam plant integrated with C-E gasifiers results in an electric-generating cost comparable with conventional plants using stack-gas scrubbers, while a combined-cycle plant integrated with gasifiers results in electric-generating costs that are about 9 percent less than those for conventional plants with scrubbers. The high thermal efficiency of the combined cycle more than offsets the slightly higher capital cost. Also, the costs of the plants with gasifiers were shown to be relatively insensitive to gas HHV; a ±25 percent variation in gas HHV from the design value changes capital and operating costs less than 1 percent.

Retrofit Applications

The C-E process offers promising potential for retrofit of existing oil or gas-fired boilers. Conversion of these boilers to direct coal firing would be very costly and require severe unit downrating. C-E has studied the retrofit of a new 510-MW$_e$ oil-fired unit of C-E design with an integrated low-Btu coal-gasifier system and with a nonintegrated medium-Btu coal-gasifier system. These options were compared with replacement of the existing boiler with a new pulverized-coal–fired unit with scrubbers. In the case of the integrated low-Btu system, gasifier steam generation makes up for the reduction in boiler steam-generating capac-

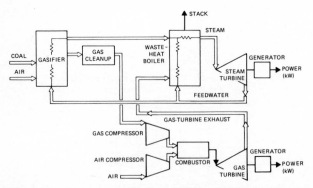

FIG. 3-9 Combined-cycle plant integrated with a C-E atmospheric gasifier.

TABLE 3-3 Cost evaluations of new 600-MW coal-fired electricity generating plants*

	Conventional steam plant with scrubbers	Integrated steam plant producing low-Btu gas	Integrated combined-cycle plant producing low-Btu gas
Net heat rate, Btu/kWh	9456	9666	8223
Capital cost, $/kW	1052	1079	1120
Cost of electricity, mills/kWh	70.6	68.0	64.3

*Based on 1981 dollars and on a levelized coal cost of $3/$10^6$ Btu.

Source: Adapted from Ref. 2 with permission and updated to reflect present costs.

ity that is required for low-Btu gas firing, and thus the steam-turbine load is maintained. All boiler items including heat-transfer surfaces and materials, fans, ducts, and other equipment remain unchanged, with the only boiler modifications required involving the addition of low-Btu gas-firing capability to the fuel-firing system. In the case of the nonintegrated medium-Btu system, an oxygen-blown gasifier is used, with medium-Btu gas supplied to the boiler over the fence. This requires only slight modifications of the firing system.

Projected plant-investment cost and electric-generating cost for the three retrofit options considered are shown in Table 3-4, along with the estimated electric-generating costs for the existing oil-fired unit. These figures show an economic incentive for converting existing oil- and gas-fired units to coal, provided sufficient economic life remains in the plant. Also, the study shows that an integrated low-Btu gasification system is the most attractive option.

Industrial Applications

While the primary emphasis of the C-E gasification development program has been on applications for electric power plants, the existence of a substantial industrial market has always been recognized. The industries most suitable for applications of low- and medium-Btu coal gasification are the petroleum refining, chemical, primary metal and steel, paper and pulp, food processing, and glass.

TABLE 3-4 Cost evaluations of retrofitting a 510-MW$_e$ oil-fired steam generator*

	Oil†	Pulverized coal with scrubbers	Integrated low-Btu gas	Nonintegrated medium-Btu gas
Net heat rate, Btu/kWh	9170	9545	9771	11,620
Capital cost (incremental), $/kW	——	616	590	941
Cost of electricity, mills/kWh	98.9	63.0	57.9	75.3

*Based on 1981 dollars on a levelized oil cost of $10/$10^6$ Btu, and on a levelized coal cost of $3/$10^6$ Btu.

†Base of comparison.

In a recently completed study, C-E investigated industrial applications for coal gasification.[3] From the industries listed, four site-specific case studies were selected. With the cooperation of the companies involved, preliminary plant designs were developed and capital and operating costs were prepared for each. Two of the cases selected, calcining of soda ash and paper drying in a paper mill, were retrofit applications in which the coal-gasification system provided fuel to replace oil or gas. The other two cases, steam flooding of an oil field and a coal-to-methanol plant were new plant applications. The coal-to-methanol plant had an oxygen-blown gasifier, while the other applications had air-blown gasifiers.

In order to determine economic viability, the cost of the fuel from coal was compared to the cost of the competing fuel (oil or gas). For the coal-to-methanol plant, the cost of methanol produced from the gasification plant was compared to methanol produced by conventional methods.

Results showed that for oil-field steam flooding and paper-mill paper drying, the gasification system was not an economic alternative, because the cost of competing fuel was low. For soda ash calcining and the coal-to-methanol application, the gasification system offered an economically attractive alternative at the present time.

1. Koucky, Robert W., and Herbert E. Andrus: "Applications of Test Experience from Combustion Engineering's Process Development Unit to Mathematical Modeling of Entrained Flow Coal Gasification," *16th Intersociety Energy Conversion Engineering Conference, Aug. 9-14, 1981,* Atlanta; Combustion Engineering Publication no. TIS-6891.

2. Hargrove, M. J., S. T. Jack, G. N. Liljedahl, and B. K. Moffat: "Study of Electric Plant Applications for Low-Btu Gasification of Coal for Electric Power Generating," U.S. DOE Report no. FE-1545-59, National Technical Information Service, Springfield, Va., August 1978.

3. Atabay, K., M. Baldassari, and P. R. Thibeault: "Industrial Application of C-E's Coal Gasification Process," U.S. DOE Report no. FE-1545-86, National Technical Information Service, Springfield, Va., January 1981.

BRITISH GAS/LURGI SLAGGING GASIFIER

C. TERRY BROOKS
British Gas Corporation,
Westfield Development Centre,
Cardenden, Fife, Scotland

HENRY J. F. STROUD
KEITH R. TART
British Gas Corporation,
Midlands Research Station,
Solihull, West Midlands, England

INTRODUCTION

Gas-making processes using fixed beds of fuel have been used in the gas industry from its inception.[1] Early processes used horizontal and vertical retorts, coke ovens, and water gas plants, while engineering advances have led to the more recent continuous fixed-bed processes such as those developed by Woodall-Duckham, Wellman, Wilputte, and Riley-Morgan.[2] A radical advance in fixed-bed gasification was the highly successful oxygen-blown Lurgi gasifier which works at high pressure and rejects the ash through a revolving grate.[3]

A recent step forward in fixed-bed technology has been the successful engineering by the British Gas Corporation of systems that liberate the coal ash as a liquid slag[4] and has resulted in the process known as the *British Gas/Lurgi Slagging gasifier*. This process has many advantages which include the following:

- Complete conversion of coal
- High throughputs per gasifier
- High thermal efficiency
- Stability in operation
- Use of comercially proved fixed-bed design concepts
- Low consumption of gasification agents
- Production of only environmentally acceptable effluents

The development of the process over 20 years has involved a 100-ton/day (91-Mg/day) experimental unit and a 350-ton/day (320-Mg/day) unit and has shown that the above advantages are achievable. Moreover, the reliability and acceptability of the process was demonstrated on the latter unit, in which nearly 100,000 tons (88,000 Mg) of coal has been processed in 7000 h of running. In one run lasting over 2000 h, over 30,000 tons (27,600 Mg) of coal was gasified. The process is ready for commercial exploitation.[5]

PROCESS DESCRIPTION

The Slagging gasifier contains a bed of fuel which is maintained at a constant depth. Near the bottom of the bed a mixture of steam and oxygen enters the gasifier at a high velocity through nozzles called *tuyères* so that a low-density region of solids entrained in turbulent gases, called a *raceway*, is formed. The steam-to-oxygen ratio is set to give temperatures sufficiently high to release the mineral matter in the coal as a molten slag, which drains into the slag pool at the base of the gasifier. This slag is readily removed through a slag tap and is quenched in water to form a glassy frit. The hot gases from the raceway are cooled by their passage through the fixed bed, and the products leave the gasifier at the top of the bed.

Individual lumps of coal are progressively heated and gasified as they slowly descend the reactor shaft to the raceway, where carbon gasification is completed.

This countercurrent flow of gases and solids in the fixed bed means that the gaseous products leave the reactor at a modest temperature, so avoiding the need for high-grade heat recovery. This countercurrent flow also results in complex temperature profiles in the bed. The rapid drop in gas temperature near the bed top is caused by the drying and devolatilization of continuously added fresh coal. The products of devolatilization range from gases such as carbon oxides and hydrocarbons, through naphtha, to oil and tar. Minor products include ammonia, phenols, and hydrogen sulfide.

The char remaining after devolatilization passes to the lower part of the bed and is gasified at increasingly higher temperatures; the very high temperatures generated by the oxygen-carbon reaction in the raceway are rapidly moderated by the ensuing strongly endo-thermic steam-carbon reaction.

This apparent complexity of the fixed-bed process, which has been described in detail elsewhere,[6] does in fact lead, paradoxically, to a system which is very stable in operation.

The Slagging gasifier can handle a large variety of fuels ranging from cokes to highly caking and swelling bituminous coals, which can contain a significant amount of fines. [In this chapter fines are defined as coal particles less than ¼ in (6 mm) in size.] A wide range of ashes can be handled, although sometimes flux addition is necessary. By injection at the tuyères, by-product tar and oil can be gasified completely, as can pulverized coal itself. Slagging operation is normally obtained by using a steam-to-oxygen ratio (v/v) in the range of 1:1 to 2:1. The potential for complete conversion of the coal carbon to gas is comple-mented by the rejection of coal ash as an environmentally acceptable glassy frit and by low yields of liquor, which are readily treated by conventional methods to give environ-mentally acceptable effluents. The Slagging gasifier has reached an advanced stage of devel-opment as an excellent first stage for processes that not only convert coal into a number of valuable products including medium- and high-heating-value (MHV and HHV) fuel gases, methanol, ammonia, and Fischer-Tropsch liquids but also generate power in com-bined-cycle schemes.

DEVELOPMENT OF THE PROCESS

The development of the Slagging gasifier goes back to the 1950s. The first gasifier was built by Lurgi, in collaboration with Ruhrgas, at Oberhausen-Holten. Runs at low pressure commenced in 1953, but the project was prematurely terminated before significant prog-ress was achieved. At the same time work at the Midlands Research Station (MRS) by British Gas (or the Gas Council as it was called then) in a small high-pressure gasifier allowed a quantitative measure of operation under slagging conditions to be made.[7] In 1955 British Gas acquired the redundant Lurgi experimental gasifier, which was erected at the MRS and used to carry out initial exploratory research[8] into slagging gasification using coke and operating only at modest pressures [75 psig (520 kPa)].

Encouraging results led to the gasifier and its support services being extensively modified for operation on coal at higher pressure to give a design output of crude gas of 5 MM stdft³/day. Upgrading the gasifier involved fitting a coal stirrer and distributor, refractory lining of the 3-ft (0.9-m) shaft, and, more significantly, changing from a side slag offtake to a slag tap in the center of the hearth. Work on this gasifier between 1962 and 1964 successfully demonstrated[9] the gasification of coal at pressures of 300 psig (2100 kPa) at very high loadings [1168 lb(daf)/ft²h (5.7 Mg(daf)/m²h) or 8 MM stdft³/day (230,000 m³(n)/day)*], limited only by the rate at which coal could be charged. The tapping of slag

*m³(n) symbolizes the normal cubic meter; that is, the volume of one cubic meter of gas measured at zero degrees Celsius and 1.01 × 10⁵ pascals pressure.

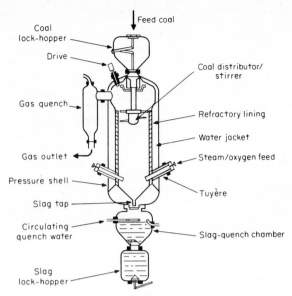

FIG. 4-1 The British Gas/Lurgi Slagging gasifier.

at instantaneous rates of 10,000 to 14,000 lb/h (4.5 to 6.4 Mg/h) under automatic control was also demonstrated.

The next step was to build a still larger gasifier, but circumstances in the United Kingdom prevented this from becoming a reality until the resurgence in coal gasification in the mid-1970s. In 1974 British Gas set up its Westfield Development Centre (Scotland), the site of a full-scale coal-gasification plant based on Lurgi dry-ash gasifiers that had been making town gas which was no longer required owing to the availability of North Sea natural gas. Under the coordination of Conoco Inc., 15 North American gas pipeline and oil companies with EPRI (Electric Power Research Institute) sponsored a program that extended over 3 years to develop further the Slagging gasifier.

The Westfield Slagging gasifier was built by converting one of the existing conventional Lurgi gasifiers to slagging operation. Its main features are shown diagrammatically in Fig. 4-1. The original gasifier was lined to reduce its shaft diameter from 9 ft (2.7 m) to 6 ft (1.8 m), because the gasifier throughput was limited by the total output of the existing oxygen plant, which had previously served several Lurgi gasifiers. A second gas offtake was added, together with an associated downstream cooling system, to match the greater output. Following the installation of the slag-tapping, hearth, and tuyère systems; the addition of a quench chamber, new instrumentation, and control equipment; and the elimination of the grate, the remainder of the Lurgi gasifier continued to be used in its previous role. The gasifier operates at a maximum pressure of 350 psig (2500 kPa) and consumes 350 tons of coal per day (320 Mg/day). The gasifier house and associated equipment are shown in Fig. 4-2.

In this 3-year development period over 20,000 tons (19,800 Mg) of coal was gasified, and the program culminated in a successful 23-day test which demonstrated the commercial viability of the gasifer and confirmed its excellent performance. This program marked the beginning of formal cooperation between British Gas and Lurgi.

FIG. 4-2 Photograph of gasifier house and associated equipment. Key: lock gas holder (1), tar-oil-liquor treatment (2), gasifier house (3), liquor storage tank (4), benzene stripper (5), ammonia recovery (6), benzene absorber (7), air-separation plant (8).

As a result of this successful project, Conoco, with the support of British Gas and Lurgi, responded to an invitation from the U.S. government to submit a proposal to build and operate a coal-based substitute natural gas (SNG) plant utilizing new technology.[10] Their proposal was accepted, and Conoco, with British Gas and Lurgi, used Foster Wheeler as the engineering contractor to carry out the first phase of the engineering of a demonstration plant based on the British Gas/Lurgi Slagging gasifier. This phase, entirely U.S. government–funded, involved a technical support program which included gasifier trials carried out on the Slagging gasifier at Westfield. During this program Pittsburgh No. 8 and Ohio No. 9 coals, both having high-caking and -swelling characteristics and having high sulfur contents, were gasified. The program achieved all the objectives set for it and was successfully completed in August 1978.[11]

Since that time British Gas has been carrying out further development work on the 6-ft (1.8-m) gasifier to meet several objectives:

1. To perfect operating procedures
2. To obtain performance data on a wide range of British coals
3. To develop systems for the injection of pulverized coal at the tuyères
4. To carry out a long demonstration run
5. To demonstrate processing of the crude gas, including its upgrading to SNG using the British Gas HICOM (High Carbon Monoxide) process[12]

The program will involve the continued operation of the Westfield gasifier with the 6-ft (1.8-m) diameter shaft and the construction, commissioning, and operation of a prototype commercial gasifier with an 8-ft (2.4-m) shaft. During this British Gas program, a 3-month interruption was made in 1979 to accommodate a three-run project, sponsored by EPRI, aimed at demonstrating the potential of the Slagging gasifier for electric power generation

TABLE 4-1   Summary of Westfield Slagging gasifier projects (April 1975 to December 1981)

Project	No. of runs	Hours on line	Fuel Gasified	
			Tons	Mg
Sponsors' program	27	1,508	21,780	19,760
DOE program	15	981	12,170	11,040
EPRI trials	3	415	4,370	3,960
British Gas program	25	4,264	58,900	53,430
TOTAL	70	7,168	97,220	88,190

in a combined-cycle plant using Pittsburgh No. 8 coal. The gasifier was shown to be well able to meet all the load-following and other requirements set by EPRI, and some of the results are described quantitatively in the next section.

Up to December 1981, nearly 100,000 tons (88,000 Mg) of coal had been gasified at Westfield in 70 runs covering 7000 hours of operation (see Table 4-1).

PERFORMANCE OF THE SLAGGING GASIFIER

Typical performance data for British and eastern U.S. coals, obtained from the Westfield gasifier, are shown in Table 4-2. The similar performance of the Slagging gasifier for widely different coals is noteworthy. The very low steam demand of the Slagging gasifier with its favorable and fairly constant oxygen demand is clearly evident. Countercurrent operation and a low steam-to-oxygen ratio result in the modest outlet temperatures. The almost complete decomposition of process steam within the gasifier leads to a very small liquor yield and, compared with a dry-ash gasifier of the same diameter, a much greater thermal output. The low yield of CO_2 leads to a crude gas having a high H_2S/CO_2 ratio.

These gasification characteristics result in considerable reductions in capital costs in the process areas of steam raising, oxygen production, gasification, effluent treatment, and desulfurization, while the high thermal efficiency of the gasifier gives lower operating costs because of the lower coal-feed requirements. The Slagging gasifier operates with steam-to-oxygen ratios (v/v) in the range of 1:1 to 2:1 with satisfactory slagging conditions, and over this range the thermal efficiency of the gasifier varies little but there is some effect on product-gas composition as shown by the summary of extensive experimental data in Table 4-3.

The ability of the Slagging gasifier to accept many ash types has been proved at Westfield. Analyses for the range of ashes handled is shown in Table 4-4. Coals with refractory ashes need fluxing, which is achieved at Westfield by mixing either blast furnace slag or limestone with the coal as it enters the coal lock. The amount of flux added depends on the composition of the ash, the amount of ash in the coal, and the temperature of the molten slag in the gasifier. This temperature is a function of the steam-to-oxygen ratio.

The fixed-bed gasifier generates tar and oil, which are carried out from the gasifier in the crude gas. They can be recycled to extinction through the tuyères of the Slagging gasifier with little effect on gasifier efficiency. Performance data for Pittsburgh No. 8 coal with and without tar injection are shown in Table 4-2. Phenols from liquor cleanup and,

TABLE 4-2 Performance data for British Gas/Lurgi Slagging gasifier

Coal (daf)	Frances	Rossington	Manton	Ohio No. 9.		Pittsburgh No. 8
Origin	Scotland	England	England	United States	United States	United States
Size, in	1/4–1	1/4–1	1/4–1 1/4	1/4–1		1/8–1
Proximate analysis, %:						
Fixed carbon	54.0	54.7	57.1	41.4	50.2	48.3
Volatile matter	32.9	31.2	31.5	33.6	34.1	36.1
Moisture	8.7	9.5	4.1	6.1	5.0	4.7
Ash	4.4	4.6	7.3	18.9	10.7	10.9
Ultimate analysis, %:						
C	83.0	83.5	85.1	79.6	83.7	83.7
H	5.5	4.9	5.1	6.1	5.7	5.5
O	9.2	7.7	5.5	7.4	6.9	7.1
N	1.4	1.7	1.6	1.2	1.6	1.7
S	0.5	1.7	2.3	5.6	2.0	1.9
Cl	0.4	0.5	0.4	0.1	0.1	0.1
Coal calorific value, Btu/lb (average)	12664	12589	13281	10868	10616	10598
BS swelling no.	1½	1½	6½	4½	7	7
Caking index (Gray King)	B	E	G6	G	G6	G6
Operating conditions:						
Pressure, psig	350	350	350	350	335	335
Steam-to-oxygen ratio, v/v	1.34	1.29	1.39	1.25	1.22	1.13
Outlet gas temperature, °C	480	480	513	410	516	521
Crude gas composition (main components), vol %:						
H_2	28.6	27.2	28.1	28.7	28.0	29.5
CO	57.5	58.1	56.8	53.2	56.4	55.8
CH_4	6.7	6.8	6.8	6.9	7.1	5.8
C_2H_6	0.4	0.5	0.7	0.3	0.3	0.6

C$_2$H$_4$	0.2	0.2	0.2	0.2	0.1	0.2
N$_2$	4.2	3.9	3.4	4.0	4.2	4.1
CO$_2$	2.3	2.9	3.5	5.5	3.0	3.5
HHV, Btu/scf	355	355	357	342	350	347
By-product yields, lb/ton of coal:						
Tar and oil	192	138	90	149	122	
Ammonia	11	11	6.4	8.9	6.2	
Phenols	9.1	1.6	2.5	NM	1.5	
Fatty acids	14	0.21	0.64	NM	0.2	
Naphtha	40	13	29	28	34	
H$_2$S	7.6	25	34	73	28	
COS	1.8	3.7	4.7	14	2.8	
CS$_2$	0.02	0.04	0.03	0.15	0.02	
Thiophene	0.06	0.09	0.03	0.08	0.04	
HCN	0.06	0.08	0.02	NM†	0.02	
Derived data:						
Coal gasification rate, lb/(ft^3)(h)	852	848	841	664	816	592
Thermal output, MM Btu/(ft)(h)	10.6	10.6	11.0	7.8	10.0	8.0
Steam consumption, lb/lb of coal	0.41	0.40	0.46	0.39	0.39	0.42
Oxygen consumption, lb/lb of coal	0.54	0.56	0.61	0.57	0.57	0.64
Liquor production, lb/lb of coal	0.20	0.21	0.15	0.21	0.17	0.17
Gasifier thermal efficiency, %						
Case 1‡	94.3	93.9	95.7	91.5	97.2	97.1
Case 2§	83.4	82.1	82.7	79.7	85.1	83.7

*For this case alone tar was injected (at the rate of 93 lb per ton of coal feed) through the gasifier tuyeres.   †Not measured.

‡Defined as the total product-gas thermal output (based on HHV, including tar, oil, and naphtha) divided by corresponding thermal input of coal feedstock.

§Defined as total product-gas thermal output (based on HHV, including tar, oil, and naphtha) divided by corresponding thermal input of coal feedstock and the fuel equivalent of the steam and oxygen used.

TABLE 4-3 Performance of the Slagging gasifier at various steam-to-oxygen ratios for Rossington coal*

Operating conditions:					
Steam-to-oxygen ratio, v/v	0.93	1.15	1.39	1.54	1.78
Outlet-gas temperature, °C	529	471	476	468	473
Crude-gas composition, vol %:					
H_2	25.8	26.5	27.3	28.1	28.8
CO	61.2	60.7	57.2	56.3	53.8
CH_4	6.9	6.5	6.6	6.8	7.2
C_2H_4	0.5	0.6	0.6	0.6	0.5
C_2H_4	0.2	0.2	0.2	0.2	0.1
N_2	3.7	3.5	4.0	3.3	3.6
CO_2	1.3	1.6	3.7	4.1	5.5
H_2S	0.4	0.4	0.4	0.6	0.5
Derived data (lb/lb of coal):					
Steam consumption	0.299	0.348	0.426	0.460	0.540
Oxygen consumption	0.593	0.563	0.570	0.553	0.562
Liquor production	0.16	0.18	0.19	0.21	0.27

*Coal expressed as dry, ash-free.

more importantly, pulverized coal entrained in a suitable carrier gas can also be injected directly into the reaction zone via the tuyères. The results of tests in which 15 percent of the total coal was fed in this way are shown in Table 4-5, and higher amounts are possible. Other ways of handling coal fines yet to be experimentally investigated are the pumping of coal slurries through the tuyères and addition of briquettes or extrudates made from coal fines into the top of the bed. The Westfield gasifier has been operated perfectly satisfactorily with Pittsburgh No. 8 coal containing 25 percent fines and Manton coal containing 35 percent fines being fed directly to the top of the fixed bed. Thus in most situations the Slagging gasifier will be able to consume run-of-mine coal.

TABLE 4-4 Range of mineral matter components of coals used in the Westfield Slagging gasifier

Component (as oxide)	Range, %
Silica	32–52
Alumina	21–30
Calcium oxide	1–10
Magnesium oxide	0.8–4
Iron oxide	4–32

TABLE 4-5 Performance of Slagging gasifier with 15% fines injected through tuyères

Coal*	Markham
Origin	England
Size, in	¼–2 or pulverized

Proximate analysis, %:

Moisture	7.18
Ash	4.40
Volatile matter	33.38
Fixed carbon	55.04

Ultimate Analysis, %:

C	83.54
H	4.61
O	7.98
N	1.78
S	1.62
Cl	0.47
BS swelling no.	1
Caking index (Gray King)	D

Operating conditions:

Steam-to-oxygen ratio, v/v	1.18
Outlet gas temperature, °C	546
% coal feed gasified as fines	15

Crude gas composition, vol %:

H_2	27.5
CO	55.6
CH_4	5.7
C_2H_6	0.4
C_2H_4	0.1
N_2	7.2
CO_2	3.1
H_2S	0.4

Derived data (lb/lb of coal):

Steam consumption	0.403
Oxygen consumption	0.633
Liquor production	0.22

*Coal expressed as dry, ash-free.

TABLE 4-6 Gasifier load-following capability

% normal load change required	Response time required for combined-cycle operation, min	Fastest demonstrated response time at Westfield, min
15	3	2
30	10	2
50	37	2
70	100	3

The load-following performance of the gasifier was well quantified in the 1979 EPRI trials, mentioned earlier in this chapter, which were aimed at confirming the suitability of the Slagging gasifier for use with combined-cycle power-generation systems. After an initial run on Rossington coal, Pittsburgh No. 8, a highly caking eastern U.S. coal, was chosen for the tests, which were particularly oriented toward establishing the ability of the gasifier to respond quickly to load changes and to run steadily at a variety of loads. These objectives were successfully achieved, with the gasifier's ability to respond to load changes more than matching the requirements. Table 4-6 highlights the results from over 50 controlled load changes. The gasifier ran stably at all loads between 30 and 110 percent of a standard load, and its load could be changed rapidly within this range. There were no significant transients during load changes, and the gas composition remained substantially constant for all loadings. The Slagging gasifier also has important start-up and shutdown characteristics. It can be started up from an empty state in 4 h. Transition from gas production to hot standby can be achieved in a matter of minutes and held for at least 48 h. When maintenance is required, the gasifier can be shut down and emptied to allow proprietary equipment to be repaired or replaced and then returned to gas making within 7 days. It may be shut down and cooled, kept full of fuel, and yet restarted in 2 h. The level of management supervision is comparable to that normally found, for example, in electricity utility power stations or an ammonia plant.

These characteristics stem from the presence in the gasifier of relatively large quantities of fuel and lead to stable and safe operation.

DESIGN OF A COMMERCIAL PLANT

Commercial-scale slagging gasification plants,[13] whether producing fuel gas, synthesis gas, or SNG, have certain sections in common:

1. Feedstock preparation
2. Gasification and slag handling
3. Gas cooling and tar-oil-liquor separation
4. Gas-liquor treatment
5. Steam raising
6. Oxygen production

Typically the main features of sections 1 to 3 are set up as described in the text that follows. Gas-liquor treatment, steam raising, and oxygen production are not described, as these are of standard design and widely used elsewhere. A representative material balance is given in Table 4-7.

Feedstock Preparation

The purpose of the feedstock preparation area is to provide sized coal, flux, and start-up fuel for the gasifiers. Run-of-mine coal from the primary stockpile (of about 20 days' capacity) is reclaimed and conveyed to a crusher, where it is reduced in size to less than 2 in (5 cm). Screens are then used to separate fines only to the extent necessary, and the resultant feed can be stored in a secondary stockpile. Finally this coal is conveyed to coal bunkers situated above each gasifier.

Gasification and Slag Handling

The main components of the gasifier system are a coal lock-hopper, gasifier, quench vessel, and slag lock-hopper, which are arranged vertically. Coal and a fluxing agent (limestone or blast-furnace slag) are stored in bunkers located above the coal lock-hopper into which they flow by gravity.

Since the gasifier operates under pressure, a lock-hopper system is needed. The system used is identical to that developed over the past 40 years for the Lurgi dry-ash gasifier. It uses special cone-shaped valves which are hydraulically operated. The depressurizing, filling, and repressurizing cycle is automatically controlled and takes about 4 min to complete.

Tar from the tar-oil-liquor separation area is recycled to both the coal distributor and the gasifier tuyères. Tar fed to the top of the bed mixes with the coal feed and serves to reduce dust carry-over. Oil may also be recycled from the tar-oil-liquor separation area to the tuyères, where with the tar it is completely gasified in the raceway, thus eliminating any net tar production from the gasifier.

Molten slag is removed from the hearth through the slag tap and falls into a quench vessel containing water, where it immediately solidifies to form a granular frit. This frit then falls through the quench vessel into the slag lock-hopper, which operates in a similar fashion to the coal lock-hopper, slag and water mixtures being periodically discharged to the slag-handling area. The lock-hopper is then refilled with recirculating water from the slag-handling system before the valve isolating the lock-hopper from the quench vessel is reopened. The water in the quench vessel is circulated and cooled to remove the sensible and latent heat released from the molten slag. The whole system is automated.

Gas Cooling and Tar-Oil-Liquor Separation

The main features of gas cooling and tar-oil-liquor separation are shown in the process flow diagram (Fig. 4-3). The crude gas leaving the gasifier immediately enters a wash cooler. The gas is cooled and saturated by liquor recycled from the sump of the waste-heat boiler and by separated liquor from the tar-oil-liquor separation area. The heavier components, such as tar and associated coal dust, are separated from the gas. The quenched gas is further cooled to provide low-pressure steam and boiler-feedwater preheat.

Tarry gas liquor from the sump of the waste-heat boiler is cooled and then flashed in

TABLE 4-7 Material balance for eastern U.S. bituminous coal*

| Stream no.† | 1 | 2 | 3 | 4 | 5 |
Name	Coal	Flux‡	Gasifier steam	Gasifier oxygen	Wet slag
DAF coal§	51467	——	——	——	——
Moisture	9120	——	——	——	7115
Ash/flux	6080	4681	——	——	9041
Steam	——	——	23701	——	——
Oxygen	——	——	——	32383	——
Nitrogen	——	——	——	142	——
Total	66667	4681	23701	32525	16156
Temperature, °C	25	25	400	135	70
Pressure, psia	15	15	565	545	30

| Stream no. | 6 | 7 | 8 | 9 | 10 |
| | Dusty gas | Oily gas | | | Slag quench |
Name	liquor	liquor	Recycle liquor	Final liquor	water
NH$_3$	80	188	——	268	——
Tar	3832	——	——	——	——
Oil	168	673	——	2	——
Chloride	23	3	——	26	——
Phenols	269	44	——	313	——
Fatty acids	29	13	——	42	——
Water	38000	26139	47765	16374	11672
Dust	733	——	——	——	——
Total	43134	27060	47765	17025	11672
Temperature, °C	175	70	70	40	25
Pressure, psia	460	440	500	50	500

| Stream no. | 11 | 12 | 13 | Stream no. | 14 | 15 |
Name	Injection tar	Recycle tar	Oil	Name	Crude gas	Lock gas
Tar	1626	2206	——	H$_2$	3157	——
Oil	——	——	839	CO	81360	——
Dust	75	658	——	CO$_2$	11077	7500
Total	1701	2864	839	CH$_4$	5030	——
Temperature, °C	70	70	70	C$_2$H$_6$	417	——
Pressure, psia	500	500	500	C$_2$H$_4$	130	——
				N$_2$	741	——
				H$_2$S	1738	——
				COS	200	——
				CS$_2$	58	——
				HCN	8	——
				Naphtha	802	——
				Steam	129	——
				Total	104847	7500
				Temperature, °C	30	25
				Pressure, psia	443	500

*All flows in lb/h.

†Stream numbers refer to streams designated by numbers in diamond-shaped boxes in Fig. 4-3.

‡Includes carbonate lost on ignition.

§Dry, ash-free coal.

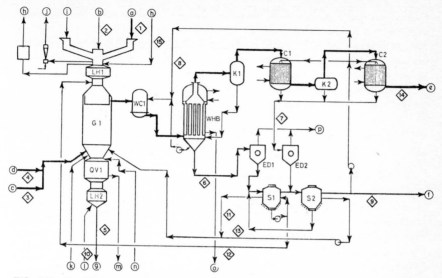

FIG. 4-3 Simplified process flow diagram for gasification, gas cooling, and tar-oil-liquor separation stages. Equipment key: BFW heater (C1), final cooler (C2), tarry-gas-liquor expansion drum (ED1), oily-gas-liquor expansion drum (ED2), gasifier (G1), crude gas KO pot (K1), final KO pot (K2), coal lock-hopper (LH1), slag lock-hopper (LH2), quench vessel (QV1), tar separator (S1), oil separator (S2), wash cooler (WC1), waste-heat boiler (WHB). Key to streams: sized coal (a), flux (b), superheated HP steam (c), oxygen from air separation (d), cooled crude gas (e), gas liquor to phenol extraction (f), slag and water to slag handling (g), lock-hopper gas (h), start-up fuel (i), lock-hopper gas purge to atmosphere (j), HP N_2 from air separation (k), filling water from slag handling (l), vent gas to atmosphere (m), cooling water (n), blowdown (o), flash gas (p). Numbers enclosed in diamond-shaped boxes are explained in Table 4-7.

an expansion drum. The liquor then passes to a tar separator, where tar and dust are separated by gravity from the aqueous liquor. The tar product may be recycled to the gasifier.

The remainder of the condensate, known as "oily liquor," is also cooled and flashed in a second expansion drum, the gases evolved being combined with those from the tarry-liquor expansion drum. Oil and small amounts of tar are then separated in the oil separator and may be recycled to the gasifier. Part of the resulting liquor, essentially free of tar and oil, is pumped to the wash cooler and the final coolers in the gas-cooling area.

The total aqueous liquor from these process areas is sent to the gas-liquor treatment area. *Gas liquor* is a generic term used to describe the aqueous product resulting from gas cooling. It comprises mainly water from the coal moisture and unreacted process steam. It contains varying amounts of inorganic salts, phenols, and fatty acids as well as dissolved ammonia, carbon dioxide, hydrogen sulfide, hydrogen cyanide, and traces of other gases.

APPLICATIONS OF THE SLAGGING GASIFIER

Among the potential uses of the Slagging gasifier is the production of fuel gases (medium- or high-heating-value), chemicals, or electricity by the combined-cycle route. Normally, a clean desulfurized gas is required as the end product (in the case of fuel gas) or as an intermediate stream. As indicated in the previous section, the gas from the Slagging gasifier carries with it impurities such as oil, tar, phenols, and nitrogen and sulfur compounds. The bulk of these can be easily removed by cooling the crude gas so as to separate the tar, oil, and aqueous liquor. The net liquor produced can be processed through stages such as solvent extraction (dephenolation), ammonia stripping, and effluent treatment before discharge or perhaps can be used within the plant; for example, it can be added to the gasifier via the tuyères to replace some of the live high-pressure steam. Separated tar, oil, naphtha, and phenols can be recycled to the gasifier or marketed as valuable by-products. The cooled gas is then treated to remove hydrogen sulfide, carbonyl sulfide, ammonia, hydrogen cyanide, aromatics, naphtha, and organic sulfur compounds. The established Rectisol process, in which the gas is brought into contact with refrigerated methanol, can be used to remove these impurities. Alternatively, several other processes can be considered, depending on the precise constraints. The low CO_2 content of the gas means that even the use of established techniques for acid-gas removal gives an effluent gas with a high concentration of H_2S, which is very suitable for conversion to sulfur by the Claus process. This clean, dry, H_2S-free gas, consisting essentially of CO, H_2, and CH_4 gases, can be used either directly as a fuel of calorific value around 360 Btu/stdft³ [14 MJ/m³(n)], for SNG manufacture, or for synthesis of chemicals.

SNG Production

SNG can be made at high thermal efficiency from Slagging gasifier gas, partly because a substantial proportion of the methane in the SNG is made in the gasifier. The remainder is produced from the CO and H_2 in the gas by reaction with steam over an active-nickel

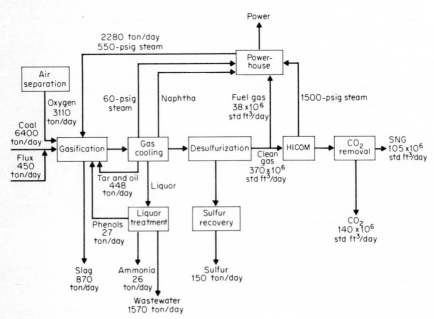

FIG. 4-4 SNG scheme based on eastern U.S. bituminous coal.

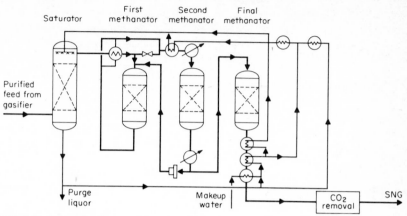

FIG. 4-5 Schematic diagram of the HICOM process.

catalyst. British Gas has developed an upgrading route, called the HICOM (High Carbon Monoxide) process,[12] that has been specifically tailored to take advantage of the particular composition of Slagging gasifier gas (high CO content and low CO_2 and steam contents). In this way the high efficiency of the gasifier is not dissipated in the following stages and an overall coal-to-SNG efficiency of about 70 percent is possible. The combination of Slagging gasifier and HICOM also offers a relatively low capital cost. A block flow diagram of one of the possible routes for the production of SNG from an eastern U.S. bituminous coal is shown in Fig. 4-4. In this route purified Slagging gasifier gas is processed by the HICOM stage, one configuration of which is illustrated in Fig. 4-5. Feed gas first enters a saturator in which it is brought into contact with a countercurrent flow of water that has been heated by indirect heat exchange with gas streams within the HICOM process stage and also with the crude-gas cooling system. This arrangement utilizes otherwise wasted low-grade heat for provision of process steam and makes a substantial contribution to the high efficiency of the HICOM route to SNG. Saturated feed gas then undergoes direct methane synthesis in a series of catalytic reactors, the outlet temperatures of which are controlled by recycle of cooled product gas. The heat released from the highly exothermic synthesis reactions is used to generate high-pressure steam for export to other process stages. Final methanation, gas cooling, carbon dioxide removal, and drying yield an SNG which can contain less than 3 percent hydrogen and 0.1 percent carbon monoxide. This process has been proved on a pilot scale and will soon be demonstrated on a semicommercial scale at Westfield.

Synthesis of Chemicals

As far as the chemical industry is concerned, one of the main problems when natural gas begins to run short will be making synthesis gas (i.e., a mixture of CO and H_2), most of which is presently produced by steam re-forming of natural gas.

At present such synthesis gas is used throughout the world on a huge scale, and possible chemical uses of the gas have been reviewed by Wender[14] and more recently by Denny and Whan.[15] Many products can be manufactured from synthesis gas, and Fig. 4-6 indicates

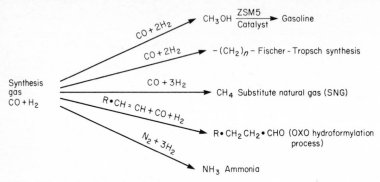

FIG. 4-6   Applications of synthesis gas.

the chemistry of its use. Other possibilities include production of oxygenated compounds such as ethylene glycol, but as yet these are only in the exploratory stages.

All these processes operate at high pressure so that use of a high-pressure gasification process reduces or, in some cases, eliminates the need for further compression of the intermediate gas. This is a major cost- and energy-consuming item when atmospheric-pressure gasifiers are used.

At present the major end use for synthesis gas is ammonia production. Currently about 70 percent of the world ammonia production is natural gas–based. The salient steps in the route are high-temperature methane re-forming followed by extensive CO-shift conversion. As purified Slagging gasifier gas is somewhat like a partly re-formed methane stream, it can directly replace both the natural gas used as process feedstock and that used as fuel (about one-third).[16] Such retrofitting of existing plants has a minimum risk technically and economically and could be implemented rapidly if desired. A study made of five gasifier types for this kind of scheme showed that the Slagging gasifier was the most efficient and had nearly the lowest capital cost.[17]

In a new coal-based ammonia plant built specifically for this purpose, the purified Slagging gasifier gas would be passed directly to a CO-shift unit and then the gas composition would be adjusted by removal of CO_2 to give the required hydrogen content. Studies of the relative capital cost and coal consumption for such schemes using four gasifier types show that the British Gas/Lurgi Slagging gasifier and the Texaco coal gasification process are more attractive than the Winkler and the Koppers-Totzek processes.

	Relative capital cost	Relative coal consumption
British Gas/Lurgi	114	100
Texaco	100	117
Koppers-Totzek	138	154
Winkler	124	115

One of the reasons why the Slagging gasifier route is more expensive than the Texaco route is the need to re-form the methane present in the purge gas. If this methane could be separated from the other components and sold as SNG or even recycled to the gasifier, then the Slagging gasifier option might be even more attractive. A suitable way in which the methane could be removed would be to adapt the cryogenic liquid-nitrogen wash unit which frequently features in ammonia synthesis plants.

Methanol is already an important chemical with many uses. Many have speculated that the methanol market will increase rapidly in future years as new uses develop. For example, it could be used as a clean liquid fuel in its own right, as an intermediate for gasoline production (e.g., the Mobil process), and perhaps as a feedstock for propylene plants.[18] Chemical uses demand a high product purity (greater than 99.5 percent). However, fuel-grade methanol need be only about 98 percent pure, the balance being water and higher alcohols, and so it can be made in a simpler plant.

Methanol-synthesis catalysts are highly sensitive to sulfur, and so the feed gas must be extensively purified and a chemical guard bed, for example of zinc oxide, may be mandatory. Basically the synthesis gas should comprise H_2 and CO at a molar ratio of 2:1, so Slagging gasifier gas with a much higher CO content must be extensively CO-shifted. The purge-gas rate from the synthesis loop has to be sufficient to prevent a buildup of the nitrogen and methane which are present in the purified Slagging gasifier gas. This purge gas can be used as fuel in the plant or passed to an adjoining methane-synthesis plant to produce SNG as a coproduct. An evaluation by EPRI[19] showed that the British Gas/Lurgi process had the lowest capital and operating costs of the four methanol-synthesis routes studied (Table 4-8).

Although the original Fischer-Tropsch (F-T) process for producing higher hydrocarbons from CO and H_2 was first announced over 45 years ago, it has undergone various stages of development in the United States and especially, by Sasol, in South Africa where large quantities of vehicle fuels and oxygenated chemicals are produced by this route. Sasol uses both fixed-bed and circulating-bed F-T reactor systems, each yielding different product distributions. The reactant again must be hydrogen-rich with a H_2/CO ratio of about 2:1, so the highly purified gas from the Slagging gasifier has to be subjected to the CO-shift reaction.

The H_2/CO ratios used in the full-scale F-T reactors at Sasol are 1.7:1 and 2.8:1, respectively. The purge gas from the F-T reactor can be used for SNG production or can be fractionated to give a methane-rich gas and a hydrogen-rich gas suitable for an ammonia

TABLE 4-8 Economics of coal-derived methanol*

Process	Total capital requirement, $ MM	Levelized cost, $/MM Btu†	Overall efficiency, %
Foster Wheeler	1723	5.18	55.5
British Gas/Lurgi	1580	5.00	57.7
Koppers-Totzek	2342	6.75	52.4
Texaco	1925	5.70	58.2

*Based on early 1977 dollars, a 90 percent operating factor, a coal cost of $1.02/MM Btu, and the production of methanol product equivalent to 315,000 MM Btu/d.

†Levelized methanol cost based on HHV.

Source: From *Screening Evaluation: Synthetic Liquid Fuels Manufacture,* Electric Power Research Institute Report, AF523, 1977.

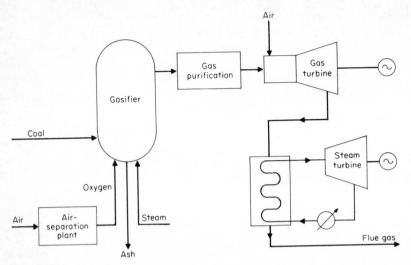

FIG. 4-7 Coal gasification with combined-cycle power generation.

synthesis unit. Additionally some of the purge gas can be re-formed catalytically or with oxygen to give a gas that can be returned to the synthesis reactor.

Power Generation

The atmospheric emission of sulfur oxides from conventional coal-fired power stations is becoming less acceptable. Its avoidance by direct coal desulfurization or stack-gas treatment is technically difficult and expensive. It is more attractive to gasify the coal and then to remove sulfur (mainly in the form of hydrogen sulfide) from the crude product gas, particularly if the gas is at a high pressure. Also, by allowing the use of advanced power-generation cycles, gaseous fuels can be used more efficiently than the initial feedstock, giving improved thermal efficiency for power generation. Thus electricity generation using combined cycles, that is using an optimized combination of gas and steam turbines to drive alternators (Fig. 4-7), can result in an overall efficiency, including gasification, of about 40 percent. This compares with less than 35 percent for conventional steam-cycle power plants fitted with stack-gas cleanup devices.

The use of oxygen-blown gasifiers is not essential in this application but can be economically desirable, since it reduces the cost of the compression, gasification, and gas purification stages. The low CO_2 and steam contents of the Slagging gasifier product are advantages in the combined-cycle route. Little CO_2 is removed with the H_2S, and the capital and operating costs of the sulfur-recovery plant are minimized. Processes which produce gases with higher CO_2 content will also lose pressure energy during removal of acid gas, which would otherwise be recovered in the gas-turbine expander.

An evaluation of combined-cycle schemes has been undertaken by EPRI for five different gasifiers.[20] Their estimates of both plant capital costs and power-generation efficiency are summarized in Table 4-9, which for comparison includes a conventional coal-fired case employing flue-gas desulfurization. It clearly shows the advantage of the British Gas/Lurgi

TABLE 4-9   Capital costs and efficiencies for power-generation systems*

Process	Capital cost requirement, $ MM	Overall efficiency of coal to power, %
British Gas/Lurgi Slagging	711	40.6
Foster Wheeler (air-blown)	705	40.5
Texaco (slurry feed)	816	38.7
Combustion Engineering	931	38.1
Lurgi (air-blown)	906	35.0
Coal-fired plus stack-gas desulfurization	838	34.4

*Based on mid-1976 dollars, a 70 percent operating factor, and coal at $1/MM Btu.

Source: From *Economic Studies of Coal Gasification Combined Cycle Systems for Electric Power Generation,* Electric Power Research Institute Report, AF 642, 1978.

TABLE 4-10   Some key parameters of a large combined-cycle facility using the British Gas/Lurgi gasifier

Total power output, MW	1200
Coal-feed rate (moisture-free), t/h	400
Gasifier pressure, psig	320
HHV of crude gas (dry), Btu/scf	379
Gas-turbine inlet temperature, °C	1315
Gas-turbine pressure ratio	17:1
Gas-turbine exhaust temperature, °C	610
Steam conditions, psig/(°C)(°C)	1450/(480)(540)
Steam-turbine power output, MW	385
Overall system efficiency based on HHV coal, %	40.6

Source: Adapted from *Economic Studies of Coal Gasification Combined Cycle Systems for Electric Power Generation,* Electric Power Research Insititute Report, AF 642, 1978.

Slagging gasifier. Some key parameters for a large combined-cycle facility using the British Gas/Lurgi Slagging gasifier are given in Table 4-10.

When a coal gasifier is used for power generation in a combined power cycle, it must respond quickly to the fluctuating fuel demands of the gas turbine over the full operating range. The EPRI tests described earlier showed the ability of the Slagging gasifier to meet this requirement.

ENVIRONMENTAL ACCEPTABILITY OF SLAGGING GASIFICATION

The effluents produced by slagging gasification are similar to those produced by coke ovens and town gas works already operated successfully. The Slagging gasifier, however, has the capability of recycling several significant by-products to extinction by injection through the gasifier tuyères. This is an advantage if the aim is to maximize gas production and minimize by-product storage and handling facilities.

TABLE 4-11 Major by-products and
effluents from plant producing 105 million
stdft^3/day of SNG

Substance	Quantity, t/d
Slag	870
Sulfur	150
Sludge	2–3
Waste gasifier liquor	1570
General aqueous effluent	6600
Carbon dioxide	8100
Nitrogen	10200
Ammonia	26

The major by-products and effluents for a plant consuming 6400 tons/day (5800 Mg/day) of Illinois No. 6 coal to produce 105 million stdft^3/day (2.97 Mm^3/day) of SNG are listed in Table 4-11.

Methods of treating liquors which contain phenols have been successfully used in Lurgi gasification plants. The Slagging gasifier produces a similar effluent in smaller quantities but of higher concentration. A study of the treatment of this liquor, carried out by British Gas in a project sponsored by the International Energy Agency, showed that acceptable liquor treatment was possible using conventional methods.

There is increasing interest in the pathways of trace metals in the environment, including the trace metals released during coal processing. Work is still continuing on identifying the levels of trace metals which might be significant. The concentrations of most toxic metals in the effluent from the liquor treatment plant are likely to be in the range of hundredths to tenths of one part per million.

There are other potential sources of effluent. The slag quench water is innocuous and contains even lower levels of contaminants than those in ash quench water from dry-ash processes. Depending on the process units used on site there may be chemical solutions which require periodic disposal.

The slag frit is a clean, black, glassy, granular material which separates completely from the quench water and is easily handled. It has several potential uses; for example, it can be used as a road fill or as a component of construction materials. Because of the frit's glassy character, its long-term leaching is negligible.[5] Some or all of the material may be marketed, but any disposal to landfill should present no environmental problems.

Most of the sulfur contained in the feed coal leaves the gasifier as hydrogen sulfide, but some also occurs as carbonyl sulfide as well as traces of carbon disulfide, and organic sulfur compounds. The removal of these compounds does not present any engineering or technical difficulties, as the required technology is already proven and available. Elemental sulfur of high quality can be recovered by the Claus process.

The air-separation plant produces large quantitites of nitrogen, and in some applications, an acid gas removal plant produces carbon dioxide. These must be discharged in a safe manner, usually via a stack.

The high-pressure plant itself is so designed that it should not give rise to any significant releases of odorous or hazardous substances to the atmosphere except for small quantities released during maintenance under careful supervision. Monitoring around the British Gas/

Lurgi Slagging gasifier has shown that the level of polycyclic aromatic hydrocarbons was much lower than in urban atmospheres. This is consistent with British Gas experience with coal gasification in these kinds of plants for over 20 years.

There are likely to be many sources of noise—coal handling, coal charging, compressors, vents, high-pressure gas flow, turbines, and pumps. The coal handling and charging are not any noiser than present methods used at power stations, and new developments in design and operation may lead to reduced noise levels in the future.

FUTURE PERSPECTIVE

The British Gas/Lurgi Slagging gasifier is at an advanced stage of development and can now be regarded as ready for commercial exploitation. It is particularly suitable for high-volatile, unreactive, bituminous coals and should find wide application in the areas of ammonia, methanol, and F-T synthesis; SNG manufacture; combined-cycle power generation; and MHV gas production.

The process offers complete gasification of coal at high thermal efficiency. High reactor throughputs are attainable with extremely good operational controllability and flexibility. Investigations have shown the process to be environmentally acceptable, and indeed it now provides a technology which enables the use of high-sulfur coals without atmospheric pollution.

British Gas is involved in a number of design studies for large-scale demonstration of the Slagging gasifier and is able to provide full commercial and performance guarantees for gasifiers up to 8 ft (2.4 m) in shaft diameter. The operation of an 8-ft (2.4-m) gasifier at Westfield in 1983 will provide further confidence in operating gasifiers of this size and even larger ones.

REFERENCES

1. Stewart, E. G.: *Town Gas: Its Manufacture and Distribution*, Science Museum, H. M. Stationery Office, London, 1958.

2. Hebden, D., and H. J. F. Stroud: "Coal Gasification Processes," *Chemistry of Coal Utilization*, vol. 2 (suppl.), M. A. Elliott (ed.), Wiley, New York, 1981, pp. 1599–1752.

3. Rudolph, P. F. H., C. Hafke, and P. K. Herbert: "Lurgi Coal Gasification," *Synthetic Fuels, Status and Directions*, San Francisco, October 1980, pp. 13–16.

4. Hebden, D., and C. T. Brooks: *Westfield—The Development of Processes for the Production of SNG from Coal*, Institution of Gas Engineers, Communication no. 988, 1976.

5. Sharman, R. B., J. A. Lacey, and J. E. Scott: "The British Gas/Lurgi Slagging Gasifier," *Synfuels International Conference*, Frankfurt, Germany, May 1981.

6. Tart, K. R., and T. W. A. Rampling: "Fixed Bed Slagging Gasification—A Means of Producing Synthetic Fuels and Feedstocks," Institution of Chemical Engineers Symposium Series no. 62, 1980.

7. Hebden, D., R. F. Edge, and K. W. Foley: *Investigations with a Small Pressure Gasifier*, Gas Council Research Communication, GC14, 1954.

8. Hebden, D., and R. F. Edge: "Experiments with a Slagging Pressure Gasifier," *Trans. Inst. Gas Eng.*, vol. 108, 1958–1959, pp. 492–527.

9. Hebden, D., J. A. Lacey, and A. G. Horsler: *Further Experiments with a Slagging Pressure Gasifier*, Gas Council Research Communication, GC112, 1964.

10. Sudbury, J., J. R. Bowden, and W. B. Watson: "A Demonstration of the Slagging Gasifier," *8th Synthetic Pipeline Symposium*, Chicago, October 1976.

11. Sudbury, J.: *Technical Support Programme Report—Phase 1, The Pipeline Gas Demonstration Plant*, FE 2543-13, 1978.

12. Tart, K. R., and T. W. A. Rampling: "Methanation Key to SNG Success," *Hydrocarbon Process.*, April 1981, pp. 114–118.

13. *Phase 1: The Pipeline Gas Demonstration Plant Process Design*, Conoco, FE 2542-28, June 1980.

14. Wender, I.: "Catalytic Synthesis of Chemicals from Coal," *Catal. Rev. Sci. Eng.*, vol. 14, 1976, p. 97.

15. Denny, P., and D. A. Whan: "Heterogeneously Catalysed Hydrogenation of Carbon Monoxide," *Catalysis*, vol. 2, C. Kenball and D. A. Dowden (eds.), Chemical Society, London, 1978, p. 46.

16. Timins, C.: *The Future Role of Gasification Processes*, Institution of Gas Engineers, Communication no. 1112, 1979.

17. Brown, F.: "Make Ammonia from Coal," *Hydrocarbon Process.*, vol. 56, 1977, pp. 361–366.

18. Anthony, R. G., and B. B. Singh: "Olefins from Coal via Methanol," *Hydrocarbon Process.*, vol. 60, 1981, pp. 85–88.

19. *Screening Evaluation: Synthetic Liquid Fuels Manufacture*, Electric Power Research Institute Report, AF 523, 1977.

20. Chandra, K., et al., *Economic Studies of Coal Gasification Combined Cycle Systems for Electric Power Generation*, Electric Power Research Institute Report, AF-642, 1978.

THE KBW COAL-GASIFICATION PROCESS

H. Z. DOKUZOGUZ

KBW Gasification Systems, Inc.
Pittsburgh, Pennsylvania

J. F. KAMODY
H. J. MICHAELS

Koppers Company, Inc.
Pittsburgh, Pennsylvania

D. E. JAMES
P. B. PROBERT

The Babcock & Wilcox Company
Barberton, Ohio

INTRODUCTION

In October 1980, Koppers Company, Inc., and The Babcock & Wilcox Company (an operating unit of McDermott, Inc.) formed a joint venture, KBW Gasification Systems, Inc., to serve the expanding synthetic fuels market (see Fig. 5-1). The objective of KBW is to engineer, design, market, fabricate, construct, and service coal-gasification systems worldwide. KBW is offering commercially an atmospheric-pressure, oxygen- and steam-blown, slagging-type, entrained-flow gasification system.

From the outset, KBW recognized the fact that the existing commercial entrained-flow gasification systems could not economically meet the high synthesis-gas production requirements of the currently proposed synthetic fuel plants. Therefore, the KBW commercial coal-gasification system was designed to offer the synthetic fuels industry an efficient, reliable, and advanced system that utilizes modern and proven technology. More specifically, the KBW gasification system has the following important features:

- It can gasify any rank of coal. Anthracite, bituminous, and subbituminous coals and lignite have been successfully gasified. This includes both eastern and western U.S. coals. The caking properties of the coal do not affect the gasification process. The KBW gasifier can handle wide variations in ash quantity, ash fusion temperature, and sulfur content.

- It can treat 100 percent of the mine output. The maximum size is limited only by the conveying and measuring equipment that feeds to the pulverizer. All of the coal including the fines is pulverized.

- It has major environmental advantages. Tars, phenols, or other hydrocarbons which must be removed from the product gas are not produced in the KBW gasifier because of the high gasification temperature. The ash is discharged as a stable, granular, quenched slag. In some cases, the slag can be sold for concrete aggregate and road underbase.

- It does not produce methane. There is no need for methane re-forming, which may be required in some of the synfuel processes.

- It is based on a wealth of design data, knowledge, and experience possessed by Koppers and Babcock & Wilcox in the areas of coal preparation and handling, entrained-flow coal gasification, slag handling, mass transfer, heat transfer, equipment fabrication, and plant construction.

- The KBW gasifier has a larger internal volume than existing commercial entrained-flow gasifiers. This results in a throughput rate for the KBW gasifier that is more than twice that seen in existing commercial entrained-flow gasifiers for the same residence time.

- Both the gasifier and heat-recovery boiler utilize components that have been proved through years of fabrication and service.

- The KBW gasifier utilizes membrane walls constructed of vertical water-cooled tubes which have been widely used in boilers. This feature enables the gasifier to produce high-pressure steam. The water-cooled tubes can withstand much higher heat fluxes than the jacket-type cooling system while assuring nucleate boiling. With the water-cooled tubes, the fluid circulation is well-defined and the differential stresses are minimized.

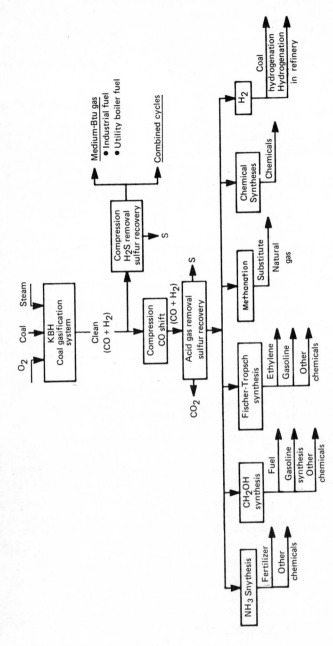

FIG. 5-1 Applications of KBW coal gasification system.

- The KBW gasifier uses relatively thin refractories in the gasification zone instead of thick refractories, which have been proved to be unworkable in slagging gasifiers (i.e., they are easily eroded).

- The KBW gasifier does not require a water-spray quench and hence has a higher thermal efficiency.

GENERAL PROCESS DESCRIPTION

A simplified block flow chart for the KBW gasification process is given in Fig. 5-2. The KBW gasification system takes run-of-mine coal delivered to the plant and produces cool particulate-free synthesis gas. Oxygen and steam are used in the process. Ash from the coal is discharged as a stable quenched slag. This material has various uses such as road underbase.

Coal received at the plant is pulverized and simultaneously dried. The dry pulverized coal is conveyed to the gasifier and injected through the burners in the walls of the lower portion of the gasifier. Oxygen and steam are also injected through the burners, and the pulverized coal is entrained in the gas and gasified as it is carried upward through the gasifier.

The reaction temperature in the lower portion of the gasifier is kept high enough so that the ash from the coal is melted. This material runs down on the gasifier walls and is tapped out as a molten slag through a drain opening in the sloping gasifier floor. The molten slag is quenched in water to produce a material having the consistency of gravel and is removed by a conveyor for disposal.

The walls of the gasifier are made up of vertical water-cooled tubes. The lower portion of each tube is covered with welded steel studs and slag-resistant refractory to limit the heat transfer until the gasification reactions are essentially complete. The wall tubes in the upper portion of the gasifier are left bare to cool the gaseous products to below the ash softening temperature and to prevent slag deposits in the heat-recovery boiler.

In the heat-recovery boiler the gas mixture is cooled further to a temperature suitable for the gas cleanup system. High-pressure steam is generated in the walls of the gasifier and in the heat-recovery boiler. The steam is then superheated in the heat-recovery boiler. The steam pressure and temperature are matched to the requirements of the large drive turbines and various process uses throughout the synfuels plant.

After they leave the heat-recovery boiler, the product gases pass through the gas cleanup

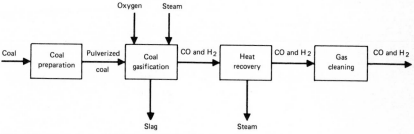

FIG. 5-2   Flow chart of KBW coal-gasification process.

equipment, where particulate matter is removed. The gases then pass to process units for gas compression, CO shift, and H_2S removal as well as other processing stages required for synthesizing the final product(s).

PROCESS CHEMISTRY

In entrained-flow gasification, finely pulverized coal is intimately mixed with oxygen and steam and dispersed into a gas stream of relatively high velocity. The required fineness of the coal is influenced by the content of volatile matter and the ash-fusion temperature of the coal. Experience has shown that for practical furnace sizes and residence time, it is necessary to pulverize the coal until approximately 70 percent can pass through a 200-mesh (74-μm) or finer sieve in order to achieve good mixing and to result in complete reaction. When prepared to this fineness, pulverized coal ignites and reacts readily.

In an entrained flow–type gasifier, first the solid coal rapidly reacts with gaseous oxygen and steam to form mainly CO, H_2, CO_2, and H_2O (gasification reactions). These major product gases then continue to react among themselves as the gas mixture flows upward and its temperature decreases.

In the high-temperature, or gasification, zone of the gasifier, the gasification reactions are essentially complete in less than one-half of one second. The most significant factor influencing the degree of completion of this step (i.e., carbon conversion) is the quantity of oxygen present in the gasification zone (oxygen-to-carbon ratio).

In an entrained flow–type gasifier the carbon conversion is also affected, but to a lesser degree, by the temperature at which the gasification reactions are carried out. Data from pilot plant gasifiers show that the relative heat losses and presence of large quantities of steam also reduce carbon conversion.

Since the oxygen-to-carbon ratio is the most significant factor influencing carbon conversion, it is logical to increase the ratio to maximize carbon conversion. This is true to a point, after which increasing the oxygen-to-carbon ratio results in excessive formation of carbon dioxide and, therefore, a product gas of poorer quality. Thus it is necessary to consider the cost of fuel and oxygen when determining the oxygen-to-carbon ratio.

The concentrations of the major constituents (CO, H_2, CO_2, and H_2O) in the gas mixture can be predicted approximately from the following equilibrium relationship:

$$K = \frac{[H_2]\,[CO_2]}{[CO]\,[H_2O]}$$

In this equation, K is the water-gas equilibrium constant and the quantities in brackets represent the mole or volume fraction of each constituent. Actual gasifier operating experience has shown that the reaction rates among these constituents decrease significantly as the temperature decreases, and essentially freeze at about $1100\ °C$. In addition to the water-gas equilibrium, several other equilibrium relationships representing the reactions among the minor constituents in the gas mixture must also be considered. Therefore, the final composition of the product-gas mixture is determined through iterative computations in which an overall mass balance for the system and all of the equilibrium relationships mentioned above are simultaneously satisfied.

PROCESS PERSPECTIVE

The entrained-flow coal-gasification process was first demonstrated in the United States approximately 30 years ago. In 1948, Koppers Company built a demonstration plant for the U.S. Bureau of Mines in Louisiana, Missouri. This plant utilized an atmospheric-pressure, oxygen- and steam-blown, slagging-type, entrained-flow gasifier which had a coal throughput rate of 2300 lb/h (1040 kg/h). The plant was put into operation in 1949 and was jointly operated by the bureau and Koppers until April 1950.

In 1951, the Babcock & Wilcox Company supplied an atmospheric-pressure, oxygen- and steam-blown, slagging-type, entrained-flow gasifier to the U.S. Bureau of Mines at Morgantown, West Virginia. The gasifier was capable of gasifying 500 lb/h (227 kg/h) of coal (see Fig. 5-3). Babcock & Wilcox participated in the operation of this pilot plant,

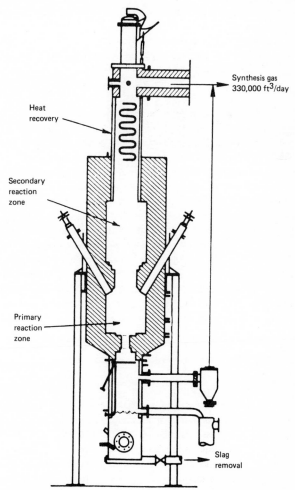

FIG. 5-3 U.S. Bureau of Mines atmospheric pressure gasifier. *(Babcock & Wilcox.)*

which was used to study the gasification of a wide variety of coals from anthracite to lignite. In addition, Babcock & Wilcox supplied to the U.S. Bureau of Mines at Morgantown, West Virginia, a pressurized, oxygen- and steam-blown, slagging-type, entrained-flow gasifier (see Fig. 5-4). This gasifier was capable of gasifying 1000 lb/h (454 kg/h) of coal. It was placed in operation in 1951 and was operated as required for approximately 10 years to study the effect of pressure on gasification reactions. The gasifier operated at pressures up to 300 lb/in² (2.07 MPa).

In the early 1950s, Babcock & Wilcox supplied a semicommercial-size, atmospheric-pressure, oxygen- and steam-blown, slagging-type, entrained-flow gasifier to E. I. Du Pont de Nemours at Belle, West Virginia (see Fig. 5-5). This unit was designed to gasify 3000 lb/h (1360 kg/h) of coal. Successful operation of this semicommercial-size unit led to the installation of a commercial-size gasifier at the DuPont plant at Belle, West Virginia (see

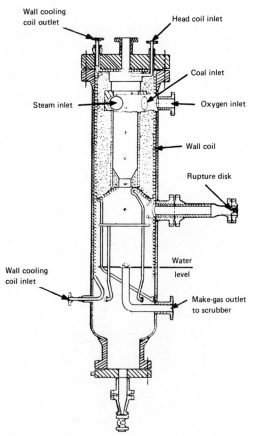

FIG. 5-4 U.S. Bureau of Mines high-pressure gasifier. *(Babcock & Wilcox.)*

Fig. 5-6). The gasifier was designed to gasify 34,000 lb/h (15,400 kg/h) of coal. It went into service in 1955 and had operated for approximately 2 years when inexpensive natural gas became available at the Belle plant and made producing synthesis gas from coal uneconomical.

In the middle 1950s, Babcock & Wilcox performed engineering studies and experimental work on air-blown, slagging-type, entrained-flow gasification for combined gas turbine–steam turbine cycles. A project operated jointly by Babcock & Wilcox and General Electric Company over a 3-year period in the early 1960s resulted in engineering and design studies and construction and operation of an air-blown gasification pilot plant at the Babcock & Wilcox Alliance Research Center (Fig. 5-7). The project demonstrated that low-Btu gas suitable for firing a gas turbine could be produced, but fuel costs at the time were

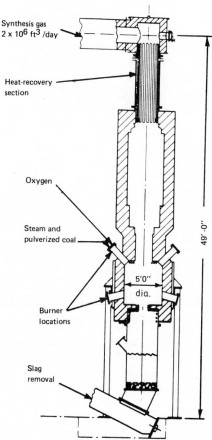

FIG. 5-5 DuPont atmospheric-pressure gasifier—semi-commercial-size. *(Babcock & Wilcox.)*

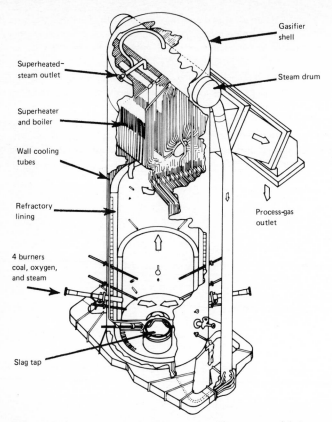

Gasifier
shell

Superheated-
steam outlet

Steam drum

Superheater
and boiler

Wall cooling
tubes

Refractory
lining

Process-gas
outlet

4 burners
coal, oxygen,
and steam

Slag tap

FIG. 5-6 DuPont atmospheric-pressure gasifier—commercial-size. (*Babcock & Wilcox.*)

such that the coal-fired combined cycle could not be economically justified for the production of electric power.

Recently Babcock & Wilcox constructed the gasifier for the Bi-Gas pilot plant at Homer City, Pennsylvania. This pilot plant, sponsored by the U.S. Department of Energy, was designed to operate at pressures from 1100 to 1500 lb/in^2 (7.58 to 10.3 MPa) and to gasify 10,000 lb/h (4540 kg/h) of coal (see Fig. 5-8). The gasifier has been in operation since 1976.

In the mid-1950s, the vast discoveries of oil and gas in this country spelled a temporary demise for the economic practicality of coal gasification in the United States. However, in Europe the need for expansion and modernization of agriculture required the use of coal to produce ammonia. In 1952, the first commercial-size entrained-flow gasification plant was installed in Finland. Since then, 16 commercial-size entrained-flow gasification plants have been installed by others in the eastern hemisphere.

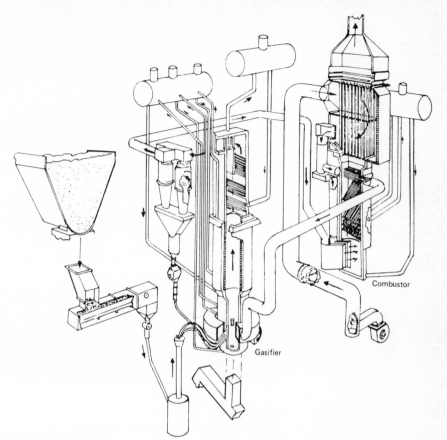

FIG. 5-7 Babcock & Wilcox and General Electric air-blown gasifier. *(Babcock & Wilcox.)*

By today's standards, these initial plants were small in size and inherently simple in design, lacking in the areas of overall thermal efficiency and coal throughput capacity. They employed long residence times and small internal volumes, had limited flexibility to operate efficiently with a wide variety of coals, and experienced refractory erosion problems. During the 1960s and 1970s, several design and hardware modifications were implemented to improve the earlier gasification systems. As a result, several commercial entrained-flow gasifiers utilizing steam-generating cooling jackets, shorter residence times, larger internal volumes, and improved refractories were designed and installed.

As of September 1981, the sponsors of seven commercial synfuels projects, which are aimed at converting coal or peat into methanol and/or gasoline, have chosen KBW to supply the gasification systems for their commercial plants. The sponsors of these projects have applied for loan and/or price guarantees from the U.S. government. Currently, they are awaiting action by the U.S. Synfuels Corporation on these guarantees.

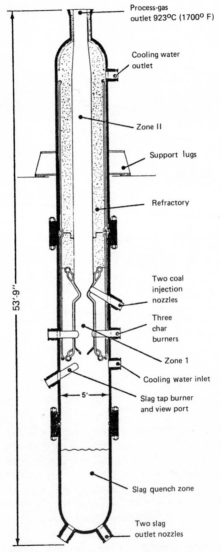

Process-gas
outlet 923°C (1700° F)

Cooling water
outlet

Zone II

Support lugs

Refractory

Two coal
injection
nozzles

Three
char
burners

Zone 1

Cooling water inlet

Slag tap burner
and view port

Slag quench zone

Two slag
outlet nozzles

53'-9"

5'

FIG. 5-8 Bi-Gas pilot plant gasifier.

DETAILED PROCESS DESCRIPTION

The KBW gasification process can be described in terms of three major stages (see Fig. 5-9):

• Coal preparation and feeding
• Coal gasification and heat recovery
• Gas cooling and cleaning

Coal Preparation and Feeding

The KBW system begins at the coal storage yard. Coal, crushed to 1¼ in by 0 in (32 mm by 0 mm), is conveyed to a bunker which feeds the pulverizer. The pulverizer is air-swept to provide drying and transport of the coal. With eastern bituminous coal, the typical product has a fineness of 70 to 85 percent passing through a 200-mesh (74-μm) sieve and a moisture content of approximately 2 percent.

From the pulverizer, the pulverized coal passes to a storage bin with a cyclone separator at the inlet. The cyclone separator removes the pulverized coal from the conveying air and discharges it into the bin. The moisture-laden air is vented to the atmosphere through a bag filter, which removes traces of pulverized coal from the vented air.

From the storage bin, the dried pulverized coal is transported with nitrogen to the service bins located adjacent to each gasifier. Each service bin is equipped with a cyclone separator and a bag filter to remove the pulverized coal from the conveying nitrogen. The nitrogen is vented to the atmosphere, and the pulverized coal is stored under an inert atmosphere.

From each service bin, the pulverized coal passes through two parallel weigh feeders to two smaller feed bins, where it is stored under an inert atmosphere. The feed bins maintain a constant supply of coal at the inlet of variable-speed screw feeders, which regulate the flow of pulverized coal to each burner of the gasifier. There is one feed bin and one screw feeder for each of the eight burners. At the exit of the screw feeder, the pulverized coal is picked up by a stream of oxygen and steam and is transported a short distance to the burner in the gasifier wall. The velocity of the mixture of coal, oxygen, and steam is maintained above the flame propagation velocity at all times, and the screw feeder is designed to maintain a gastight plug of pulverized coal to prevent backflow of oxygen into the coal feed system.

Coal Gasification and Heat Recovery

The KBW gasifier is a square column nominally 15 ft (4.6 m) on a side and 61 ft (18.6 m) high (see Fig. 5-10). The premixed reactants enter the gasifier through eight burners located two each near the bottom of each side wall. The burners are arranged vertically, one over the other, and are offset from the center of the wall so that the gases form a vortex in the gasifier to promote good mixing. The gasification reactions take place as the pulverized coal that is entrained in the hot gas passes upward through the gasifier. The temperature in the "gasification zone" is above the ash fluid temperature so that the ash in the coal is melted, finds its way to the gasifier walls, and runs down as a molten slag to the sloping hearth floor. It then drains continuously through a drain opening for slag in the floor.

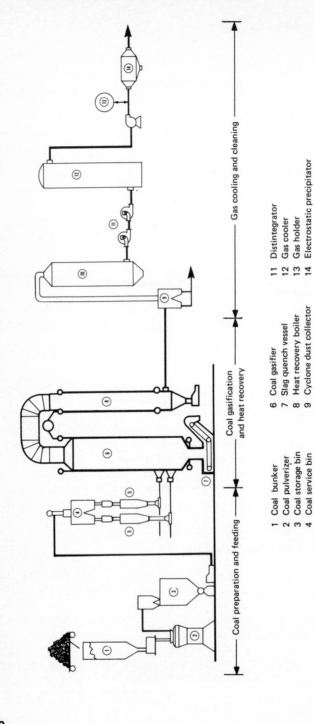

FIG. 5-9 Flow chart of KBW coal gasification process.

1 Coal bunker
2 Coal pulverizer
3 Coal storage bin
4 Coal service bin
5 Coal feed bin

6 Coal gasifier
7 Slag quench vessel
8 Heat recovery boiler
9 Cyclone dust collector
10 Gas cooler

11 Distintegrator
12 Gas cooler
13 Gas holder
14 Electrostatic precipitator

Coal preparation and feeding

Coal gasification and heat recovery

Gas cooling and cleaning

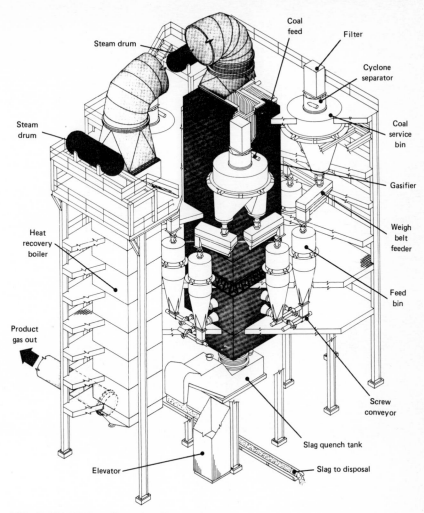

Coal
feed

Filter

Steam drum

Cyclone
separator

Steam
drum

Coal
service
bin

Gasifier

Heat
recovery
boiler

Weigh
belt
feeder

Feed
bin

Product
gas out

Screw
conveyor

Slag quench tank

Elevator

Slag to disposal

FIG. 5-10   KBW gasifier and auxiliary equipment.

Below this opening is a water-filled slag-quench vessel, where cold water shatters the molten slag into a solid granular material resembling black gravel. This material is removed continuously from the slag-quench vessel by a conveyor for disposal.

The gasifier shell consists of water-cooled membrane tube walls continuously welded to form a gastight enclosure (see Fig. 5-11). Experience over approximately 40 years on commercial combustion furnaces, including slag-tap furnaces, has proved that this water-cooled wall construction will provide long periods of continuous on-stream operations with low maintenance. The individual tubes are continuous from bottom to top. All the tubes begin at headers at the bottom of the gasifier and are arranged to form the floor, or hearth, and

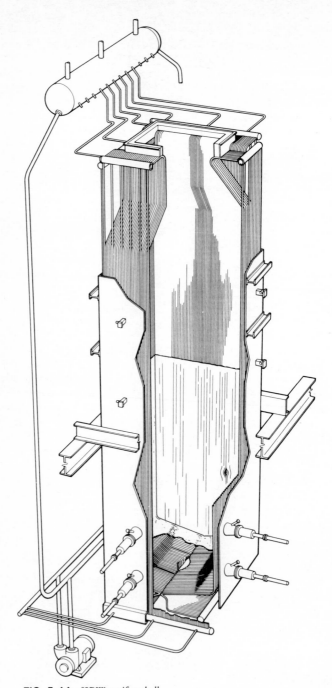

FIG. 5-11 KBW gasifier shell.

the slag outlet. They are then bent to form the vertical gasifier walls. And finally, they are bent in to form the top of the gasifier and the gas outlet before being terminated in headers at the top of the gasifier.

Water enters the individual tubes through the lower headers. As the water flows upward through the tubes, it boils and a portion of the water turns to steam. The mixture of steam and water is collected in the upper headers and flows to the steam drum, where steam is separated from the water and is sent to a superheater located in the heat-recovery boiler. The saturated water is returned through downcomers to the recirculating pumps, where it is pumped to the lower headers to make another pass up through the gasifier tubes. Boiler feedwater makeup is supplied to the steam drum through an economizer section in the heat-recovery boiler.

The tubes are typically 2½-in (64-mm) OD on 3-in (76-mm) centerline to centerline spacing with steel bars continuously welded between adjacent tubes. Groups of tubes and bars are welded together in the fabrication shop to form complete wall panels of "membrane-welded" tubes. These shop-assembled panels are shipped to the job site for field assembly into gastight furnace enclosure walls.

The tubes (Fig. 5-12) in the lower portion or gasification zone of the gasifier have many small steel studs, ⅜ in (10 mm) in diameter by ½ in (13 mm) in length, welded to them. A dense slag-resistant refractory lining is applied around the studs and over the tubes to a thickness approximately equal to the length of the studs. The studs lock the refractory in place and provide cooling for it. The temperature of the furnace face of the refractory is below the ash-fusion temperature, and a layer of frozen slag forms on the refractory. The molten slag flows down over this layer of frozen slag. The frozen slag offers protection to the refractory and tubes while minimizing the heat loss from the gasification zone.

The gasification reactions are essentially completed at an elevation approximately 15 ft (4.6 m) above the upper row of burners, where the gas temperature is approximately 110 °C above the ash-softening temperature. Above this elevation, the studs and refractory are discontinued and the remainder of the gasifier walls are bare membrane-welded tubular walls. The gas is cooled by radiation to the walls in this "cooling zone" as the gas temperature passes from where the ash is clearly molten to approximately 980 °C, where it is dry particulate fly ash. At intermediate temperatures, the ash is plastic and can form deposits. In this temperature range, it is handled best with cold bare steel from which the ash tends to shed off spontaneously. In addition, retractable soot blowers are located in the walls throughout the upper portion of the gasifier to assist in dislodging any deposits that may build up. These deposits fall to the floor of the gasifier, where they melt and flow out through the slag drain opening with the rest of the molten ash from the coal.

The gas with some unreacted char and fly ash at approximately 980 °C leaves the gasifier and passes through the crossover flue to enter the top of the heat-recovery boiler. It flows downward over the banks of horizontal convection surfaces comprising the superheater and economizer. The gas leaves the heat-recovery boiler at the bottom at approximately 230 °C and flows to the gas-cooling and gas-cleaning equipment downstream.

The walls of the heat-recovery boiler are of bare membrane-welded construction identical with the upper portion of the walls of the gasifier. The horizontal convection superheater and economizer surface is supported by castings welded to the wall tubes similar to the way the horizontal convection surface is supported in commercial boilers. Water flows upward through the wall tubes, where a portion of the water turns to steam. The mixture of steam and water is collected in headers at the top of the heat-recovery boiler and flows

Gasification Zone

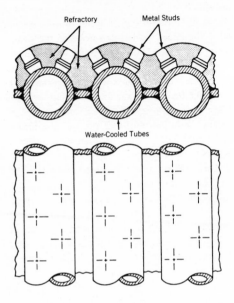

Cooling Zone

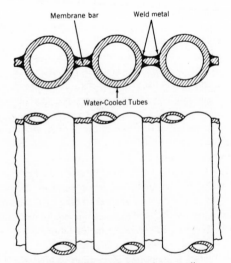

FIG. 5-12 KBW gasifier membrane walls.

to the steam drum. In the steam drum, the steam is separated from the water, the water is recirculated, and the saturated steam flows to the primary superheater, where it flows in a counterflow direction to the gas flowing over the tubes. The steam flows from the outlet header of the primary superheater to an exterior spray attemperator and on to the secondary superheater. Feedwater is sprayed into the steam in the attemperator to control the steam temperature at the exit from the secondary superheater. The steam flows through the tubes of the secondary superheater and is then transported to the various mechanical-drive turbines throughout the plant.

Feedwater from the feedwater pump flows through the tubes of the economizer counterflow to the gas and then flows to the steam drum. This water picks up heat from the gas that is being cooled thereby maximizing heat recovery from the gas.

Gas Cooling and Cleaning

Each gasifier is equipped with a gas-cooling and gas-cleaning train. Gas from the heat-recovery boiler enters a multicyclone dust collector which removes about 90 percent of the entrained particulates.

Gas leaving the cyclone dust collector flows through a saturator cooler, where the gas is adiabatically cooled and saturated with water. Discharge water from the saturator coolers flows to a clarifier and then to a disposal area.

Gas leaving the saturator cooler enters two disintegrators in series, where the entrained particulates in the gas are reduced to about 1×10^{-2} gr/ft^3 (24 mg/m^3). The gas then enters a spray-type final cooler, where it is cooled to about 40 °C by direct contact with water.

From the final cooler, gas flows to a gas blower, which maintains the pressure in the system. From the blower, the gas passes through an electrostatic precipitator. The precipitator reduces the entrained-particulate load to about 1×10^{-4} gr/ft^3 (2.4×10^{-1} mg/m^3). A gas holder is provided to absorb system surges. Gas from the precipitator flows to the plant battery limits.

For reference, typical flow diagrams and mass and heat balances for Illinois No. 6 coal are included as an example (see Figs. 5-13 and 5-14). Different coals and different plant conditions could have quantities that deviate somewhat from the typical values shown in these figures.

SYSTEM PERFORMANCE CHARACTERISTICS

One of the advantages of the KBW gasification process is its versatility in accepting a wide range of feedstocks. Feeds as diverse as peat through all ranks of coal to petroleum coke can be gasified.

The gasifier is standardized with a square internal cross section that is 14 ft 3 in (4.3 m) on a side. Gasifier operating conditions are dictated by the characteristics of the coal being gasified, but typical ranges are given in Table 5-1.

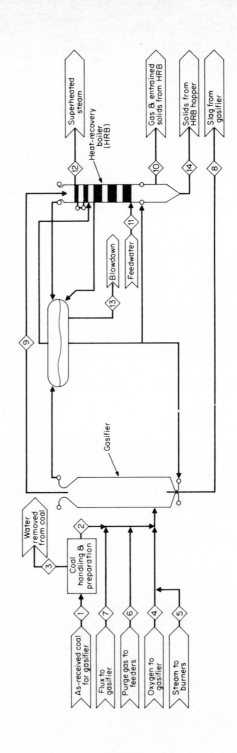

Superheated steam

Heat-recovery boiler (HRB)

Gas & entrained solids from HRB

Solids from HRB hopper

Slag from gasifier

Blowdown

Feedwater

Gasifier

Water removed from coal

Coal handling & preparation

As-received coal for gasifier

Flux to gasifier

Purge gas to feeders

Oxygen to gasifier

Steam to burners

3-106

Figure 5-13 Mass balance (Illinois No. 6 coal)

Line no.	Stream description — Gas	mol.wt	① As-received coal for gasifier lb/h	wt%	② Coal feed to gasifier lb/h	wt%	③ Water removed from coal lb/h	④ Oxygen to gasifier lb/h	mol/h	vol%	⑤ Steam to burners lb/h	⑥ Purge gas to feeders lb/h	⑦ Flux to gasifier wt%	⑧ Slag from gasifier lb/h	⑨ Flow to HRB lb/h	⑩ Gas & entrained solids from HRB lb/h	mol/h	Gas volume % Wet	Dry	⑪ Feed-water lb/h	⑫ Super-heated steam lb/h	⑬ Blowdown lb/h	⑭ Solids from HRB hopper lb/h
1	CO	28.010														94437	3371.56	57.311	61.630				
2	H₂	2.016														2973.	1474.52	25.064	26.953				
3	CO₂	44.010														20576.	467.53	7.947	8.546				
4	H₂O	18.015					9091.				5120.	637.				7427.	412.27	7.008	0.0				
5	H₂S	34.076														3539.	103.86	1.765	1.898				
6	COS	60.070														468.	7.79	0.132	0.142				
7	N₂	28.013						4.	0.15	0.01		0.				770.	27.49	0.467	0.503				
8	Ar	39.948						396.	9.91	0.49						396.	9.91	0.168	0.181				
9	HCl	36.461														82.	2.26	0.038	0.041				
10	CH₄	16.043														47.	2.94	0.050	0.054				
11	HCN	27.026														32.	1.18	0.020	0.022				
12	NH₃	17.030														20.	1.18	0.020	0.005				
13	CS₂	76.131														22.	0.29	0.005	0.005				
14	SO₂	64.059														4.	0.06	0.001	0.001				
15	NO	30.006														2.	0.06	0.001	0.001				
16	O₂	31.999						64000.	2000.00	99.50						2.	0.06	0.001	0.001				
17	Total gas						9091.	64400.	2010.05	100.00	5120.	637.			130797.	130797.	5882.93	100.000	100.000				
18																							
19	C		51200.	57.47	51200.	64.00								256.	4134.								730.
20	H		3280.	3.68	3280.	4.10																	
21	O		5200.	5.84	5200.	6.50																	
22	N		800.	0.90	800.	1.00																	
23	S		3600.	4.04	3600.	4.50																	
24	Cl₂		80.	0.09	80.	0.10																	
25	H₂O		10691.	12.00	1600.	2.00																	
26	Ash		14240.	15.98	14240.	17.80								4272.	8473.								1495.
27	Total coal		89091.	100.00	80000.	100.00																	
28																							
29	Total flux												0.										0.
30	CaCO₃												100.00										
31	H₂O												0.0										
32																							
33	Total solids		89091.		80000.									4528.	14832.	12607.							2225.
34																							
35	Total flow		89091.	100.00	80000.	100.00	9091.	64400.			5120.	637.	0.	4528.	145629.	143404.				190030.	191075.	955.	2225.
36																							
37	Temperature, °F				100.			220.0			260.0		770.	2490.	1800.	450.0				350.0	800.	497.	450.
38	Pressure, psig				0.0			20.0			1000.0		0.0	0.6	0.6	0.5				675.0	625.0	650.0	0.5

Total flow divided by 0.01 equals flow per gasifier

Barometric pressure = 14.40

Stream comments: "8" The slag in this stream is in a liquid phase.

FIG. 5-13 Mass balance (Illinois No. 6 coal).

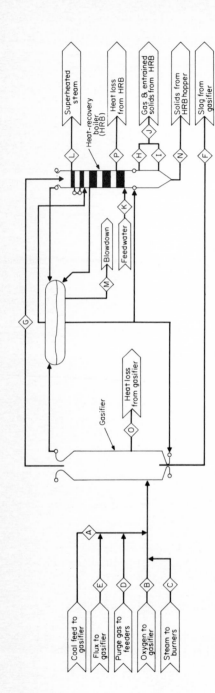

Line no.	Stream number	Stream description (Total heat)	A Coal to gasifier	B Oxygen to gasifier	C Steam to burners	D Purge gas to feeders	E Flux to gasifier	F Slag from gasifier	G Flow to HRB	H Gas from HRB	I Entrained solids from HRB	J Gas & entrained solids from HRB	K Feedwater	L Superheated steam	M Blowdown	N Solids from HRB hopper	O Heat loss from gasifier	P Heat loss from HRB
1		Chemical heat (HHV) 10^6 Btu/h	922.367				0.0	3.608	689.097	620.536	58.276	678.813				10.284		
2		Sensible heat above 77 °F 10^6 Btu/h	0.599	2.033		0.003	0.0	3.419	88.383	16.125	1.117	17.242				0.197		
3		Latent heat of water 10^6 Btu/h							7.808	7.808		7.808						
4		Total 10^6 Btu/h	922.966	2.033	5.754	0.033	0.0	7.027	785.288	644.469	59.393	703.863	53.268	260.140	0.420	10.481	1.384	0.709

Stream comments:

TOTAL HEAT DIVIDED BY 0.01
EQUALS HEAT PER GASIFIER

Reference: 25 °C (77 °F)
Liquid water
1 atmosphere

FIG. 5-14 Heat balance (Illinois No. 6 coal).

TABLE 5-1 Typical performance characteristics

Coal feed rate, t/d (Mg/d)	800–1200 (726–1089)
Gas production, ft³/min (m³/h)	32000–45000 (51440–72340)
Gas composition (dry), vol %	
CO	53–65
CO_2	8–11
H_2	25–35
Trace compounds	1–2
Higher heating value (dry, acid and gas-free), Btu/ft³ (MJ/m³)	290–300 (10.9–11.3)
Oxygen-to-carbon, mass ratio	1.15–1.32
Carbon conversion, %	86–98
Steam pressure, lb/in² (MPa)	700–1200 (4.9–8.3)
Steam temperature, °C	260 to 482

WASTES AND EMISSIONS

The environmental impact of the KBW gasification process will be minimal as a result primarily of the high-temperature slagging operation of the gasifier. No tars, oils, or other condensable hydrocarbons are produced during gasification, and therefore, no liquid effluent streams are produced that contain these compounds.

The KBW process typically produces three different streams of solid-waste products:

1. Slag from the bottom of the gasifier
2. Fly ash from the cyclone separators
3. Filter cake from the gas cleanup system

Slag from the bottom of the gasifier is stable and displays negligible leaching characteristics, since the material has been through a high-temperature molten phase in the gasifier and is resolidified as it is water-quenched in the slag-quench tank. This material can be used as a road underbase.

Fly ash from the cyclone separators is removed in dry form and is quite similar to boiler fly ash. This material can be disposed of in landfills or, alternatively, can be blended with coal for combustion in an auxiliary steam boiler. The latter application is particularly suited for eastern bituminous coals, since the fly ash contains approximately 30 to 40 percent carbon.

Filter cake from the gas cleanup system is also very similar to boiler fly ash and can be disposed of in landfills.

No gaseous emissions are produced during normal steady-state operation of the gasifier. Start-up and shutdown of the gasifier will result, however, in flaring of raw-sulfur-containing synthesis gas for very short periods of time.

PROCESS ECONOMICS

The economics of synthetic fuels production via coal gasification is highly dependent upon factors such as plant site, plant size, plant product(s), feedstock cost, environmental regulations, and financing assumptions. Several economic studies for synthetic fuel projects utilizing the KBW coal-gasification process have been performed based on plant locations in both the eastern and western part of the United States. In addition, similar economic studies based on the KBW process have been carried out for various synthetic fuel plants planned in other parts of the world. It is beyond the scope of this chapter to present a full discussion of these studies.

In an effort to give the reader some representative data, the economics of producing medium-Btu fuel gas and methanol is discussed below for two intermediate-size plants that each contain six KBW entrained-flow gasifiers. The estimates for plant investment are complete turnkey numbers and include items such as site development and general facilities and offsites as well as all coal- and gas-processing equipment. The investments and operating costs are based on eastern U.S. locations and assume eastern bituminous coals as feedstocks.

A six-gasifier plant producing 77.3×10^9 Btu/day of desulfurized medium-Btu gas is estimated to cost \$450 million in mid-1981 dollars. Plant operating costs based on coal feedstock at \$1.30/10^6 Btu (\$30/ton) are \$95 million annually, or \$3.72/10^6 Btu. Since coal costs are a major portion of operating costs, a reduction in delivered coal cost to \$0.87/10^6 Btu (\$20/ton), for example, would decrease the operating costs to \$3.04/10^6 Btu.

A KBW-based methanol plant producing 10^6 gal/day (23,810 bbl/day) is estimated to cost \$770 million in mid-1981 dollars. Operating costs amount to \$130 million annually or \$0.39/gal (\$6.00/10^6 Btu) assuming a coal cost of \$30/ton. Plant operating costs would decrease to \$0.34/gal (\$5.23/10^6 Btu) if coal were available at \$20/ton.

In the above cases, the contribution of the capital cost of the plant to the total cost of the product (i.e., minimum required "selling price") has not been estimated, since it is greatly dependent on factors such as capitalization (debt-to-equity ratios), interest rates, and tax treatments. Since these items vary significantly not only on a worldwide basis but even among potential owners in a given country or state, it is left to the reader to apply his or her own analysis to this portion of the total cost.

SUMMARY

KBW is structured to provide a single-source capability for turnkey entrained-flow coal-gasification plants worldwide from the initial feasibility studies through the start-up of the completed plant. The KBW coal-gasification system is based on the proven capabilities of Koppers and Babcock & Wilcox in entrained-flow coal gasification and in various technologies associated with the processing and utilization of coal and its products. This combination of expertise and experience enables KBW to offer a gasification system of an efficient, reliable, and advanced design that utilizes modern, proven technology.

THE FLUIDIZED-BED COAL GASIFICATION PROCESS (WINKLER TYPE)

FRIEDRICH BÖGNER
KARL WINTRUP

Davy McKee Aktiengesellschaft
Köln, Federal Republic of Germany

INTRODUCTION

Fluidized-bed operation was introduced by Dr. F. Winkler (I. G. Farbenindustrie AG, German Patent 437970, 30 September 1922) as an improvement for gasification of coal fractions of all sizes including coal fines. As a result of its simplicity and high reliability it has provided the largest number of commercial-scale gasification process units of any gasification process. A cumulative 14×10^{12} stdft^3 of coal gas has been produced from 250×10^6 tons of coal in about 70 generators (Table 6-1). Most of the plants were built by Bamag Verfahrenstechnik, which is now owned by Davy McKee.

The coal gas has been used, when blown with air, as low-Btu gas (LBG) for heating and driving gas-powered machines. When the coal gas is enriched with oxygen or operated with pure oxygen (up to 98 percent), medium-Btu gas (MBG) can be produced which is suitable for use as fuel gas[1] to drive gas turbines. The coal gas can also be further processed to synthesis gas that can be used for ammonia,[2] methanol,[3] and oxoalcohol production, to reducing gas for use in steel plants,[4] or to hydrogen for synfuels production.[5]

Capacities up to 3.73MM stdft^3/h of raw gas have been achieved in a single generator with an inside diameter of 18 ft, operating at atmospheric pressure. When the generator is operated at higher pressure (U.S. Patent 4,017,272, April 12, 1977), even higher capacities can be achieved per unit. Suitable pressurized coal-feeding and ash-discharging systems for fluidized-bed generators are commercially available and have been, for example, demonstrated successfully by Bergbauforschung at Essen at up to about 600 lb/in^2 (4000 kPa). Bergbauforschung is also investigating the use of nuclear-waste heat for fluidized-bed coal gasification.[6]

Because of its high reliability and on-stream factor with continuous operating periods of 9 months and more without the use of any spare unit, this process is very economical, especially when inexpensive low-grade carbonaceous materials are available. Applicable inexpensive feeds include high-sulfur coke, peat, brown coal, residual oil, and char.

INFLUENCE OF FEEDSTOCK PROPERTIES

The ultimate analysis of carbonaceous material gives a first indication of the suitability of the feedstock for economical gasification. Figure 6-1 shows the Grout-Apfelbeck diagram, which indicates various types of carbonaceous materials and their principal elements, carbon, hydrogen, and oxygen. Since a material having a high carbon content also has a high heating value, such material is favored for combustion and steam generation. On the other hand, a material having a lower carbon content and a higher oxygen content has a higher reactivity, and thus these feedstocks are favored for gasification. The large arrow in the diagram indicates the wide range of coals that have been economically gasified in commercial Winkler plants. The area within the arrow indicates that subbituminous and brown coals and peat are very reactive materials, and therefore they are usually considered for most gasification projects.[7]

Fluidization can be maintained with any solid material. Therefore even materials with an ash content of more than 50 percent have been economically gasified in Winkler fluid-bed generators,[8] which would include handling of discards.

Preferred particle size is in the range between 0 and ⅜ in. The process has been specifically developed to handle coal fines, which are produced in excess amounts from highly mechanized coal-mining operations.

TABLE 6-1 Plants using Winkler-type fluid-bed gasifiers

Operator and location	Time period	Operating Conditions Temperature, °C	Operating Conditions Pressure, psi (kPa)	Feedstock	Final product	Gasifier capacity, 1000 stdft³/h	Number of gasifiers in use
Test and Demonstration Plants:							
BASF, Ludwigshafen, FRG	1925 to 1958	800 to 1100	15 (105)	Brown coal, bituminous coal	Gas	75	1
HT. Winkler, Rheinbraun, Köln, FRG	1978 to present	950	140 (1000)	Brown coal (Rhineland)	Synthesis gas, reducing gas	50	1
Bergbauforschung, Essen, FRG (Indirect-heated fluid bed)	1976 to present	850	600 (4000)	Caking bituminous coal (Ruhr)	Synthesis gas	100	1
Commercial plants:							
Leuna-Werke, Merseburg, German Democratic Republic	1928 to present	1000	15 (105)	Brown coal, salt coal, coke	Ammonia, methanol, LBG	2240 / 3730	5 / 7
BRABAG, Böhlen, German Democratic Republic	1938 to present	1000	15 (105)	Low-temperature coke	Hydrogen for synfuel	1120	3
BRABAG, Magdeburg, Germany	1938 to 1945	1000	15 (105)	Low-temperature coke	Hydrogen for synfuel	1230	3
Yahagi, Japan	1937 to 1960	1070	15 (105)	Semicoke	Ammonia	330	1
Dai-Nihonginzo-Hiryo, Japan	1937 to 1959	1050	15 (105)	Caking subbituminous coal	Ammonia	520	2
Nippon Tar, Japan	1937 to 1960	1050	15 (105)	Caking subbituminous coal	Ammonia	520	2

Plant/Location	Period			Coal type	Product		
Toyo-Koatsu, Japan	1938 to 1969	1050	15 (105)	Caking subbituminous coal	Ammonia	750	2
Fushun, Manchukuo	1939 to ?	1050	15 (105)	Subbituminous coal	Syngas for Fischer-Tropsch fuel	750	4
Kunming, People's Republic of China	1950 to present	830	15 (105)	Brown coal	Ammonia	700	2
BRABAG, Zeitz, German Democratic Republic	1941 to present	1000	15 (105)	Low-temperature coke	Hydrogen for synfuel	850	3
Treibstoffwerke, Brüx, Czechoslovakia	1943 to 1972	1000	15 (105)	Low-temperature coke	Hydrogen for synfuel	1000	5
Most, Czechoslovakia	1954 to 1973	1000	15 (105)	Low-temperature coke	Hydrogen for synfuel	1200	2
Salawad, U.S.S.R.	1950 to ?	1000	15 (105)	Lignite	Water gas	860	7
Baschkirien, U.S.S.R.	1950 to ?	1000	15 (105)	Lignite	Water gas	860	4
Dimitrovgrad, Bulgaria	1951 to present	1000	15 (105)	Brown coal	Ammonia	670	4
Stara Zagora, Bulgaria	1962 to present	1000	15 (105)	Brown coal	Ammonia	1120	5
AZOT, Gordzde Yugoslavia	1953 to present	1100	16 (115)	Brown coal	Ammonia	260	1
Calvo Sotelo, Spain	1956 to 1970	970	15 (105)	Caking bituminous coal	Ammonia	350	1
Puertollano, Spain	1959 to 1970	970	15 (105)	Caking bituminous coal	Ammonia	350	1
UK-Wesseling, Rheinbraun, FRG	1958 to 1967	950	15 (105)	Brown coal (Rhineland)	Ammonia	630	1
UK-Wesseling, Rheinbraun, FRG	1962 to 1967	950	15 (105)	Brown coal (Rhineland)	Methanol	630	1
AZOT, Kutahya, Turkey	1959 to present	950 to 1075	15 (105)	Low-grade lignite	Ammonia	450	2
Lignite Co., Neyveli, India	1965 to 1979	850 to 1150	15 (105)	Lignite	Ammonia	785	3

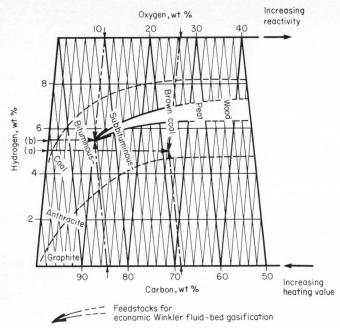

FIG. 6-1 Grout-Apfelbeck classification diagram of carbonaceous materials. On the basis of C + O + H = 100%, brown coal (*a*) has a percentage composition of C = 68.8, O = 26.2, and H = 5.0, and bituminous coal (*b*) has a composition of C = 84.6, O = 10.0, and H = 5.4.

Any type of ash, basic or acidic, can be handled. The Winkler process is the least sensitive of available gasification processes, even when various impurities are fed with the coal. Fluidization is maintained in the solid phase at temperatures just below the ash deformation point. Therefore no reaction can occur with the refractory lining of the generator. And plants which have been in operation for more than 20 years are still working with their original refractory linings. If the ash deformation point is very high, for example above 1300 °C, the correspondingly higher operating temperature allows even less reactive coals to be gasified at economic efficiencies.

For economic reasons the moisture content should not be more than 6 to 10 percent. It is always cheaper to dry the coal separately than to dry the coal by burning it with oxygen in the gasifier. Another alternative, using cheap nuclear waste heat to dry the coal, is being investigated by Bergbauforschung at Essen.[6] For materials with a high moisture content (more than 30 percent), a combination of pressurized dewatering and gasification is under consideration.[9]

KINETICS AND REACTIONS OF FLUID-BED GASIFICATION

When Winkler was investigating the phenomenon of fluidization of fine solid particles, he soon realized that turbulent fluidization optimizes mass and heat transfer. And this is especially important for the gasification of coal, which is a raw material of unique and heterogeneous nature. The main elements of coal—carbon, oxygen, hydrogen, nitrogen, and sulfur—react in several parallel and successive steps during the conversion of coal from the solid phase into the gaseous phase. The main reactions between carbon, oxygen, and hydrogen are shown in Table 6-2. As is also shown, nitrogen and sulfur tend to form NO_x, NH_3, HCN, HCNS, H_2S, COS, CS_2, SCN, mercaptans, and thiophenes. Coal also contains a long list of mineral components in ash.[10] It is an advantage of fluid-bed operation that, as a result of the moderate temperature level, the ash remains unconverted and no reaction occurs. On the other hand, an operating temperature held just below the ash deformation point is generally high enough to achieve an efficient conversion of carbon to usable gaseous products, and then the gas shows only traces of unwanted phenols, pyridines, or other organics.

Because the operating temperature of a fluid bed is limited by the ash fusion profile, the most economical gasification results from the use of more reactive carbonaceous materials with a higher oxygen content and lower carbon content (see Fig. 6-1) as the feedstock. As

TABLE 6-2   Reactions of coal during gasification

	Enthalpy, kJ/mol	
	Exothermic	**Endothermic**
Main reactions:		
For upgrading		
$C + O_2 \rightarrow CO_2$	405	
For gasification		
$C + CO_2 \rightleftharpoons 2CO$ (Boudouard equilibrium)		161
$CO + H_2O \rightleftharpoons CO_2 + H_2$ (Water-gas equilibrium)	42	
$C + H_2O \rightarrow CO + H_2$		119
Side reactions:		
$C + 2H_2O \rightleftharpoons CO_2 + 2H_2$		78
$2C + 2H_2O \rightarrow CO_2 + CH_4$		13
$2C + O_2 \rightarrow 2CO$	245	
$2CO + O_2 \rightarrow 2CO_2$	567	
$2H_2 + O_2 \rightarrow 2H_2O$	482	
$C + 2H_2 \rightarrow CH_4$	84	
$CH_4 + 2O_2 \rightleftharpoons CO_2 + 2H_2O$	801	
$2CH_4 + O_2 \rightleftharpoons 2CO + 4H_2$	74	
$CH_4 + H_2O \rightleftharpoons CO + 3H_2$		207
$CH_4 + 2H_2O \rightleftharpoons CO_2 + 4H_2$		163
$CH_4 + CO_2 \rightleftharpoons 2CO + 2H_2$		248
$CH_4 + 3CO_2 \rightarrow 2H_2O + 4CO$		239
$2C + H_2 + N_2 \rightarrow 2HCN$		
$2C + 2NO \rightleftharpoons N_2 + 2CO$		
$S_2 + 2H_2 \rightarrow 2H_2S$		
$S_2 + 2CO \rightarrow 2COS$		
$2S + C \rightarrow CS_2$		
$3H_2 + N_2 \rightarrow 2NH_3$		
$N_2 + O_2 \rightarrow 2NO$		

indicated in the legend to Fig. 6-1, the carbon, oxygen, and hydrogen contents of brown coal (*a*) and bituminous coal (*b*) are within the boundaries of such an economical fluid-bed operation.

The operating pressure mainly influences the formation of methane, which for example can be increased from approximately 1 percent to more than 10 percent in the range of 1 to 600 lb/in^2 (1 to 4000 kPa); this is an advantage when a gas of high heating value is required. Operation at elevated pressure offers the additional advantage of increased capacity, which reduces the number of gasifiers and capital cost, especially in the case of large "mega-plants."[3]

That effective fluidization can be maintained in the generator by injection of steam and of air and oxygen has been illustrated by the fact that the original test generator of 3-ft diameter was scaled up to the first commercial unit of 18-ft ID in one step without any essential problem.

The maximum specified load per unit cross section is approximately 9.000 stdft^3/h/ft^2 of gas when blown with oxygen and approximately 15.000 stdft^3/h/ft^2 when blown with air at atmospheric pressure, which is the highest figure achieved by any coal gasification system till now.

Under conditions of maximum load the effective gas velocity in the cross section of the generator is approximately 15 ft/s; this value does not apply to the different velocities of the manifold coal and ash particles. It was found that fluidization can easily be maintained if a sufficient quantity of particles larger than 0.05 in is available; then 30 percent of the particles can be an even smaller size. Particles larger than 0.5 in should be avoided, because bigger particles sink quickly to the bottom of the fluidized bed without reacting and tend to rest and form clinker in the shadow of the oxygen nozzles. Since steam reduces the temperature and oxygen increases the temperature, the amounts of the gasifying media are adjusted so that the temperature in the fluidized bed does not exceed the softening point of the ash.

However, the fine particles coming in with the coal feed, and those produced by attrition and gasification in the fluidized bed, are entrained (about 70 to 80 percent of the ash) and are subjected to further gasification while in suspension by the injection of additional amounts of air and oxygen into the space above the fluidized bed. Since the suspension zone of the reactor can be operated at temperatures 200 °C higher than the fluid bed, gasification in this zone improves the carbon-conversion efficiency, especially when less-reactive coals are used.

Winkler gasification is basically an autothermal process, i.e., the heat required for the prevailing endothermic reactions (Table 6-2) is produced *in situ* by burning coal.

In order to avoid the carbon dioxide formed during gasification, Bergbauforschung has introduced a heat exchanger into the fluid bed that uses nuclear waste heat.[6] However this solution creates new, mainly mechanical problems, and it is recommended only if the feed-stock coal is very expensive compared to nuclear power. The velocity of the various reactions depends mainly on the reaction temperature but also on the properties of the carbonaceous material.

In order to avoid the many problems involved in handling molten ash, the operating temperature of the Winkler process is limited to temperatures below the ash melting point. This results in the high reliability and excellent performance which make the Winkler process economically superior in operation to many other coal gasification processes, even if they show some better coal consumption figures.

It was found that recycling the ash with the unconverted carbon particles has a minor effect on carbon-conversion efficiency because residence time in the reactor is sufficient. Although the hot cyclone that is required created mechanical problems, gas recycling was useful when heavy feedstocks had to be handled. The long residence time and the inherent safety of the large content of carbon in the fluid bed are key factors that should be considered, particularly in oxygen-blown gasification.

There has been much research done on the specific kinetics and reactions of the individual particles moving in and out of the fluid bed, and a variety of theories derived mainly from statistics and practical experience have been proposed.[11] But final conclusions await further studies.

DETAILED PROCESS DESCRIPTION

One of the characteristic features of the Winkler fluidized-bed process is that it can handle run-of-mine coal. In addition, because of its simplicity, it has low operating and maintenance costs. The general process scheme is shown in Fig. 6-2. A view of a Winkler plant is shown in Fig. 6-3, and a view into the bottom of the gasifier and the fluidized-bed area is shown in Fig. 6-4.

If the moisture content of the coal is just below 10 percent and if the particle size is in the range of 0 to ⅜ in, the coal can be fed directly into the gasifier. If the moisture content (e.g., of lignites) is greater than 10 percent, the coal should be predried to a moisture level of 10 percent or less for economic reasons and for trouble-free flow. A particle-size spectrum of 0 to ⅜ in can easily be achieved by screening and crushing of oversize particles.

The coal to be gasified is fed to the generator (3) by a screw conveyor (2). (The numbers in parentheses refer to Fig. 6-2.) If the generator is operated above atmospheric pressure (German patent 3234623.9, Sept. 18, 1982 and German patent 3229842.0, Aug. 11, 1982), the coal is pressurized in a lock-hopper system (1). Strongly caking coals are fed pneumatically to the reactor. The difficulty with caking coals has occurred mainly in the feeding screw. Once the coal has entered the fluidized bed, it mixes quickly with the ash-rich contents of the bed and thus does not tend to cake. The lock-hopper operation is cyclic, and nitrogen, carbon dioxide, or synthesis gas is used to pressurize the system. The same gases are also used for pneumatic feeding.

The generator (3) is a carbon-steel vessel with a refractory-type lining. No corrosion or erosion problems have been encountered in the gasifier. A heat-resistant material is used for the water- or gas-cooled nozzles (4) for the several rows of injectors that inject steam and/or air and oxygen into and above the fluid bed. The carbon-steel shell is not cooled, so that visual inspection affords safety control of the operation at any time. The thickness of the refractory lining is selected such that the temperature of the shell will remain above the dew point of the gases in the reactor.

Earlier generators had a grate at the bottom with a water-cooled scraping arm. The fluidized medium was injected through the grate. The newer generators have a conical bottom and no moving parts inside the reactor. This improvement prevents gas channeling, hot spots, and clinker formation and results in a high on-stream factor and low maintenance cost.

A radiant boiler (6), in the form of water tubes along the inner walls of the reactor, is installed near the top of the generator (U.K. Patent GB 2012933 A, 14 January 1978).

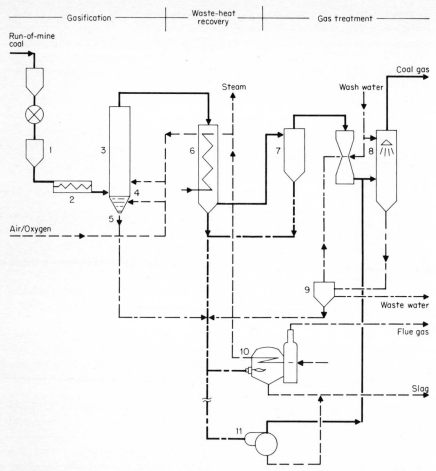

FIG. 6-2   General process scheme. Key: 1, lock-hopper system; 2, screw conveyor; 3, generator; 4, nozzles; 5, discharge screw; 6, radiant or waste-heat boiler; 7, cyclone; 8, scrubber; 9, settling vessel; 10, auxiliary boiler; 11, generator.

This boiler reduces the temperature of the gas before it leaves the gasifier and thus resolidifies any particles of the ash that may have partially melted. Thus it is possible to operate the suspension zone of the reactor at higher temperatures than obtain in the fluid bed, and thereby improve the carbon-conversion efficiency, mainly when less reactive coals have to be gasified. The total height of the gasifier is about 60 ft, and the height of the fluid bed is maintained at a level of approximately 10 ft by a cooled discharge screw (5).

In any case, the gas leaves the top of the gasifier at temperatures below the ash melting point so that no molten ash particles can plug the tubes of the waste-heat boiler (6), which recovers the considerable amount of sensible heat in the gas. Generally the boiler is designed to generate steam up to 700 lb/in² (5000 kPa) and is used as a reaction-steam superheater, boiler-feedwater heater, and air preheater. The steam is also used for internal consumption, and excess steam is exported.

FIG. 6-3   View of Winkler coal gasification plant.

After the heat-recovery stage, the gas passes through a cyclone (7) in which most of the solid particles settle out in the form of a fine, granular dry ash, together with the ash from the bottom of the waste-heat boiler, and, in pressurized operation, are discharged by a lock-hopper system. This ash still contains some carbon which did not react at the temperature of the gasifier; however, this carbon can still be used.

Ammonia or methanol plants, for example, operate with an energy deficit. The dry fly ash can easily be used as combustion fuel in an auxiliary boiler (10) to produce high-pressure steam (e.g., 10 MPa, superheated). In substitute natural gas (SNG) or fuel-gas plants, which have a surplus of steam, the carbon of the ash can be further treated at higher temperatures in a separate generator (11), such as Vortex-type gasifier (German Patent P 3130031.6, 30 July 1981), and the additional gas produced can be added to the main gas stream. When the ash is used in either of these ways, the conversion of carbon to usable products

FIG. 6-4   View into the bottom of the fluid-bed area.

TABLE 6-3 Gas and wastewater composition under various conditions of gasification

	Feedstock			
	Bituminous Coal		Brown Coal	
Gasification medium (pressure)	Oxygen (105 Pa)	Oxygen (1000 Pa)	Oxygen (105 Pa)	Air (105 Pa)
Coal fed to gasifier:				
C, wt %	57.9	56.6	51.9	56.3
O, wt %	6.9	9.7	14.9	15.0
H, wt %	3.7	3.4	3.6	4.3
N, wt %	1.4	1.6	1.6	0.7
S, wt %	0.7	0.7	1.2	3.0
Ash, wt %	25.8	18.0	22.8	12.7
Water, wt %	3.6	10.0	4.0	8.0
LHV, MJ/kg	22.7	20.4	19.4	22.1
Initial ash deformation, °C	990	1250	980	1050
Raw-gas composition (dry):				
H_2, vol %	43.6	34.8	40.0	12.6
CO, vol %	30.8	38.9	36.0	22.5
CH_4, vol %	2.0	4.0	2.5	0.7
CO_2, vol %	20.6	17.8	18.5	6.7
N_2 + Ar, vol %	1.8	3.2	1.7	55.7
H_2S + COS, vol %	0.2	0.3	0.3	0.8
Other CHONS† components, vol %	<1	<1	<1	<1
Oxygen and air consumption, $m^3(n)$/Mg of coal‡	280	343	280	1950
Raw gas produced, $m^3(n)$/Mg of coal*	1220	1300	1330	2780
Wastewater analysis				
pH value			8.2	
H_2S (free), g/m^3			5.8	
HCN (free), g/m^3			0.02	
NH_3 (total), g NH_4/m^3			82	
CO_2 (total), g/m^3			1091	
Phenols (total), g/m^3			0.2	
COD, g O_2/m^3			81.1	
BOD, g O_2/m^3			38.8	

*$m^3(n)$ symbolizes the normal cubic meter, that is, the volume of one cubic meter of gas measured at 0 °C and 1 01k kPa pressure.

†Figure 6-2, outlet settler 9.

becomes nearly 100 percent efficient, and the ash is converted into inert nonleaching slag, which is suitable for street pavements or clinker manufacture. However, the fly ash can also be used directly as feedstock in cement factories.

Removal of the remaining fine particulates of residual ash is accomplished in a scrubber (8) combined with a Theisen washer at atmospheric pressure or a venturi scrubber at elevated pressure. The water from this system is recycled via a settling vessel (9). If the moisture content of the sludge is reduced, say, by a filter, it can be burned together with other combustible residues in the auxiliary boiler to generate steam. Most of the impurities of the gas are adsorbed by the fly ash. The final dust content in the raw gas can be maintained below 10 ppm (dry basis), and the raw gas is suitable for further processing.

Typical gas and wastewater analyses for various feedstocks and applications are shown in Table 6-3.

TECHNOECONOMICS

Economy and production costs are not only the sum of raw material and utilities consumption, labor and maintenance costs, and capital costs but also the sum of the experience and skills of all the people involved and the actual performance of the process, which becomes evident after start-up and is shown by the annual plant availability. Less emphasis should perhaps be placed on values for thermal efficiency, or so-called gasification efficiency, whatever that means. Although the coal gasifier is usually only a small part of a total plant, its integration and joint operation with all parts of the plant, including coal preparation, gasification, gas treatment and processing, the auxiliary units, and the energy system, determine the economy of the plant.

Figure 6-5 shows estimated production costs of some major products from both air- and oxygen-blown coal, with the raw materials ranging from brown to bituminous-type coal. The production costs, of course, vary over a wide range depending on local conditions. Based on the U.S. Gulf coast situation for a completely grassroots plant, the left-hand scale (*a*) indicates production costs in U.S. dollars per metric ton of ammonia or

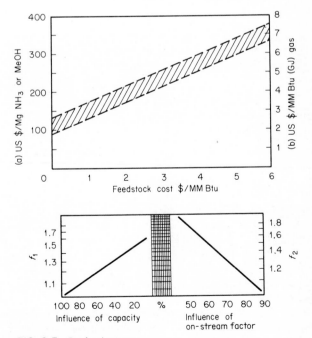

FIG. 6-5 Production costs.

metric ton of methanol, if the plant capacity (100 percent) is approximately 1200 Mg/day of ammonia or 2500 Mg/day of methanol.

If the industrial financing method is used to calculate the costs of producing methanol or ammonia, the cost is US$200/Mg of product and is split in the following way:

$$\text{Cost of coal} = 30\%$$
$$\text{Capital charge} = 50\%$$
$$\text{Remaining operating costs} = 20\%$$

The gasification units represent less than 20 percent of the total plant investment.

The effect on production costs of reducing the capacity is indicated by factor f_1. How the production costs increase if annual plant availability and the on-stream factor are less than 100 percent, is indicated by factor f_2. Both curves have been straightened by using the logarithmic scale, as shown in the middle of the table.

The right-hand scale (b) indicates the total production costs of gas plants, of approximately 50×10^9 Btu/day capacity, that produce synthesis, reducing, or fuel gas.

SUMMARY

Because of the heterogeneity of the carbonaceous materials and the various requirements of different applications no process can claim to be the best for all feedstocks and products.

The characteristics of the Winkler Fluidized-Bed Gasification process (FBG) can be summarized as follows: A wide range of coals and coke can be gasified. Run-of-mine coals including the fines are accepted. Even low-grade coals with ash contents up to 50 percent or more can be economically gasified. Young fossil fuels like peat and wood are possible feedstocks. Because the fluidized bed operates below the ash melting point, there are no slagging problems. Under these conditions, more-reactive coals like brown coals yield a higher conversion efficiency than less-reactive bituminous coals. The total loss of carbon, however, is under any circumstances only a few percent, since the carbon remaining in the dry fly ash can be utilized. The fly ash can be handled in an auxiliary boiler to produce additional steam or it can be gasified in a relatively small, separate high-temperature gasifier such as a Vortex gasifier.

Because there is no serious slagging problem in either the gasifier or the waste-heat boiler, most of the sensible heat is recovered, which ensures a high thermal efficiency and a high on-stream factor—in excess of 90 percent. Operation and maintenance are simple. The gasifier is an empty vessel with no internal mechanical parts and is easily controlled from a central panelboard.

High capacities have been achieved. Single-stream units operating at atmospheric pressure have been installed with capacities up to 3.7MM stdft³/h if air-blown and with capacities up to 2.3MM stdft³/h if oxygen-blown. At elevated pressures, even higher capacities can be reached, and generators with double the capacities that are now in use are being considered.

Operation of the gasifier is flexible. The gasifier is capable of operation with a turndown ratio of 5:1.

Important environmental consequences include the absence of by-products such as phenols and liquid hydrocarbons and the ability of fly ash with its high adsorption properties to support wastewater cleaning. Several commercially proven processes are available for gas cleaning.

Because of its simple operation and the good mixing characteristics of the fluidized bed, this process is able to accommodate various types of coal with different ash properties. The product gas is suitable for many applications. It can be used as low-Btu and medium-Btu gas and as a fuel gas for heating and for driving gas-powered machines. It can easily be enriched so that its higher heating value is of SNG stand. It can also be used as synthesis gas for production of ammonia, methanol, other chemicals, and synfuels.

REFERENCES

1. Bögner, F., T. K. Subramaniam, and R. Vangala: "Fuel Oil Substitution by Enriched Medium Btu Gas Produced from Coal Using Winkler Fluid-bed Gasifiers," *8th COGLAC* (Coal Gasification, Liquefaction and Conversion to Electricity International Conference) *at University of Pittsburgh*, Pittsburgh, Pa, August 1981.

2. Anwer, J., F. Bögner, E. E. Bailey, and T. K. Subramaniam: "Winkler Coal Gasification Process and Its Application in Ammonia Synthesis Plants," *Ammonia from Coal* (Symposium at Muscle Shoals, Alabama, May 1979), *Tenn. Val. Auth. Proc.*, pp. 51–62.

3. Kendron, T., K. Wintrup, and P. Wood: "Techno-Economics of Mega-Methanol Plants from Coal," *Coal Technology Europe 81, Köln, Federal Republic of Germany*, June 1981, *Proc. Coal Technology Europe*, Rotterdam, pp. 439–476.

4. Anwer, J., and F. Bögner: "New Applications of Coal Gasification (Winkler) Process for Direct Reduction of Iron Ores," *The Indian Iron & Steel Annual 1976*, Thanjavur, South India, July 1976, pp. 1–7.

5. Sabel, F.: *Ten Years of Oxygen Gasification*, British Int. Obj. Subcommittee (BIOS) Report no. 199, 1945.

6. *Mehr Energie aus Kohle und Uran* (More Energy from Coal and Uranium), Brochure of Bergbauforschung GmbH, Essen-Kray, Federal Republic of Germany, May 1981.

7. Anwer, J., F. Bögner, and H. Roth: "Industrial Utilization of Young Fossil Fuels by Gasification in a Fluidised-Bed," *IMTG Symposium on New Technologies of Peat Utilization, Bad Zwischenahn, Federal Republic of Germany*, November 1979, Proceedings of DGMT.

8. Bögner, F., T. Kendron, and K. Subramaniam: "Experiences with the Gasification of Low-Grade Coals in Fluidised-Bed Winkler Generators," *Proceedings of the 7th COGLAC at Pittsburgh University*, Pittsburgh, Pa, August 1980, pp. 217–232.

9. Evans, G., and W. Siemon: "Dewatering of Brown Coal before Combustion," *J. Inst. Fuel*, Melbourne, Australia, October 1970, pp. 413–419.

10. Fleming, K.: "Acid-gas Removal Systems in Coal Gasification," *Ammonia from Coal* (Symposium at Muscle Shoals, Alabama), May 1979, *Tenn. Val. Auth. Proc.*, p. 144.

11. Kunii, D., and O. Levenspiel: *Fluidization Engineering*, Wiley, New York, 1969.

LURGI
COAL GASIFICATION
(MOVING-BED GASIFIER)

PAUL F. H. RUDOLPH

Lurgi Kohle und Mineralöltechnik GmbH
Frankfurt am Main,
Federal Republic of Germany

PROCESS PRINCIPLES

The Lurgi gasifier is a reactor that performs countercurrent gasification of coal in a moving bed at high pressure. The resulting coal gasification is an autothermic process which utilizes steam or steam and CO_2 as raw materials in the gasification reactions and oxygen or air as the source of reaction heat in endothermic gasification reactions.

Countercurrent operation using oxygen to supply the reaction heat yields maximum heat recovery (i.e., low oxygen consumption) and a high carbon-conversion efficiency. The application of greater than atmospheric pressure:

- Increases the thermal efficiency as a result of the formation of methane
- Effects a relatively low solids carry-over, which also allows the use of coals having a relatively wide range of sizes
- Allows for operation at a high capacity per unit of reactor volume

The Lurgi gasifier and its operating properties during coal gasification are shown in Fig. 7-1 for a steam and oxygen-blown gasifier with dry-ash removal. As the coal moves downward, it is successively preheated, dried, devolatilized, and gasified, and eventually the residual char is completely burned in a narrow combustion zone with oxygen supplying the necessary heat for the endothermic reactions in the gasification zone. The resulting ash contains virtually no carbon.

The countercurrent flow of gas and solids results in a temperature profile with high temperature in the bottom part and low temperature in the top part whereby the sensible heat of the gas is used to perform the endothermic gasification reactions up to a level that is determined by the final reaction temperature. The balance of sensible heat is still used for coal preheating, drying, and carbonization. The resulting low temperature of the materials leaving the gasifier—gas and ash—contribute to the high efficiency of the reactor system.

As illustrated in Fig. 7-1, the progress of gasification and the type of gas produced are dependent upon the type of coal and the composition of the gasification agent (see discussion under "Thermodynamics and Chemistry"). The formation of methane is mainly a result of the heterogeneous reaction

$$C + 2H_2 \rightarrow CH_4$$

In this reaction "C" is not carbon as such but rather a result of interactions between carbon-containing devolatilization products and gasification products. It is important to note that the bulk of the exothermic methane formation takes place in the upper part of the gasification zone, which includes a part of the carbonization zone, thus raising the temperature level in this zone. The increase in temperature improves the performance of other gasification reactions.

It can also be seen that the effective temperature in the combustion zone is lower than the maximum temperature which could be obtained if only exothermic combustion were taking place. This phenomenon is caused mainly by a commencement of endothermic gasification in the combustion zone. The difference between the effective and maximum temperature is influenced by the reactivity of the coal.

Figure 7-2 represents a typical heat and material balance.

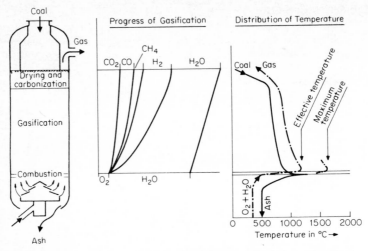

FIG. 7-1 Lurgi pressure gasifier and its operating properties during coal gasification. *(Lurgi.)*

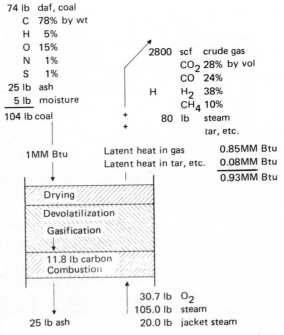

FIG. 7-2 Heat and material balance of coal gasification. *(Lurgi.)*

GASIFIER

In 1934 Lurgi pioneered the design of a moving-bed gasifier that operates under pressure and utilizes oxygen under drastic reaction conditions. The first commercial plant was built in 1936, and ever since the reactor and the gasification process have been constantly improved and refined, benefiting from the rapid progress made since then in chemical and mechanical engineering. The reactor presently used in commercial-scale operation represents the third generation of the Lurgi gasifier. Figure 7-3 shows the gasifier.

Coal Feeding

Above the reactor unit is a coal bunker, which normally contains approximately 3 hours' supply of coal. Coal is fed into the gasifier by a lock-hopper, which is a pressure vessel equipped with top and bottom valves for control of the coal flow and with pressurizing and depressurizing valves. The lock operation is cyclic with an interval of 5 to 15 min or more, depending upon the load of the gasifier. Normally, pressurizing is achieved with downstream product gas, but can also be done with nitrogen, carbon dioxide, or methane. Gas vented from the lock-hopper can be either recompressed or used as fuel. Coal lock operation is fully automated.

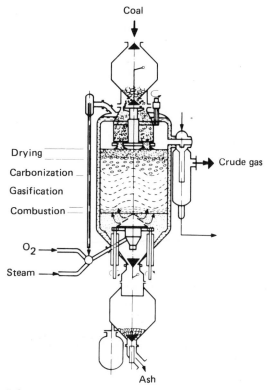

FIG. 7-3 Lurgi gasifier.

Reactor

The reactor is a water-jacketed pressure vessel in which the reactions take place as described earlier under "Process Principles." The outer shell is pressurized and with the inner shell forms the water jacket, whose inner wall is kept at relatively low temperatures through the generation of steam. This obviates the need for any refractory lining, and since the jacket steam is reinjected into the gasifier through the gasifying-agent line, a low pressure differential between the water jacket and reactor is assured.

An intermediate storage bin is located on top of the gasifier to compensate for fluctuations caused by the cyclic feeding and to provide a sure supply of coal which, along with the large amounts of coal in the actual reactor, is a reliable safeguard for the continuity of the operation; it also compensates for unavoidable fluctuations in the coal properties.

Preferably, the coal is fed into the reaction zones by a rotating coal distributor. To accommodate caking coals, blades which rotate within the fuel bed are mounted on the distributor. These blades are multipurpose: They agitate the fuel bed within the caking zone, prevent an undue amount of agglomeration, break up loose agglomerates, and mix fresh coal with previously formed char. This system enables the processing of caking and swelling coals.

Grate

The grate is a revolving device which supports the fuel bed, extracts the ash, and introduces and distributes the gasification agent. The ash zone protects the grate and allows for cooling of the ash.

Ash Removal

At the outer end of the revolving grate are ploughs which extract and transport the ash into the ash-lock chamber, a pressure vessel equipped with two valves; the top valve is normally open, but is closed for the cyclic discharge of ash from the hopper into an ash-transport system. During this time the bottom valve is open.

Crude-Gas Scrubbing

The crude gas leaves the gasifier at temperatures between 250 °C (for example, gas from lignite that has a high moisture content) and 550 °C (for example, gas from bituminous coals that have a low ash and moisture content). It contains some dust and tarry products. The dust is removed in a crude-gas scrubber which is combined with a waste-heat recovery unit (see Fig. 7-9, which appears later in this chapter under "Lurgi Coal-Gasification Process"). The hot gases are intensively washed by circulating gas liquor. This tar-containing condensate is an excellent medium for effective dust removal. The sensible heat that is given up is transformed into latent heat as the gas becomes saturated with water at temperatures between 170 and 205 °C. This gas transfers the heat to the waste-heat exchanger in which condensation of gas-liquor vapors takes place. The heat is recovered and used to generate low- to medium-pressure steam. The crude gas leaving this simple system is virtually free of dust as well as inorganic components such as chlorine and fluorine.

Gas-Liquor Separation and Tar Recycle

The degree of cooling, which also includes cooling of downstream gas, results in a net production of tarry gas liquor, which is passed to a gas-liquor separator (see Fig. 7-9, later in this chapter.) Tar, oil, and phenolic liquor are recovered after settling in these specially designed but simple separators. The dust-containing tar is recycled to the gasifier. This operation can be performed to extinction of all tar. The phenolic liquor is passed to the Phenosolvan plant (see "By-Product Recovery").

THERMODYNAMICS AND CHEMISTRY

Reactions

The principal components of the product gas are formed by a combination of the following reactions:

Oxidation

$$C + O_2 \rightarrow CO_2$$

Gasification

$$C + H_2O \rightarrow CO + H_2$$
$$CO + H_2O \rightarrow CO_2 + H_2 \text{ (CO shift)}$$
$$C + 2H_2 \rightarrow CH_4$$

The oxidation reaction takes place in the combustion zone but can be neglected in an evaluation of the gas composition, since this reaction proceeds to completion with respect to oxygen disappearance. Conversion to carbon monoxide and hydrogen represents the major gasification reaction and occurs in parallel with the CO shift. Methane formation can be considered primarily as a heterogeneous reaction.

The gasification reactions occur on the surface of the char grains. They are a result of interactions between devolatilization products and gasification reagents. The thermodynamics of these reactions cannot be used to predict rates of reactions, but can be used for a thermodynamic analysis within a range of variables for which experience and data from actual operation are available.

Process-Control Variables

The two primary process-control variables are the oxygen rate and the steam-to-oxygen ratio. The relationship between process-control variables and gasification performance, i.e., product yield and gas composition, is as follows:

Oxygen rate

Reaction of oxygen with residual char in the combustion zone produces the heat necessary to support the steam-char reaction. By regulation of the oxygen flow the gasification rate can be varied from 10 to 100 percent of the reactor capacity. Oxygen rates are inversely

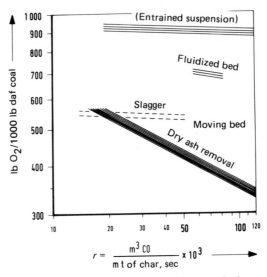

FIG. 7-4 Oxygen consumption as a function of r (reactivity). *(Lurgi.)*

proportional to the coal reactivity, which is a measure of the temperature at which the reactions at the surface of the downward-moving grains of coal and char commence. Coal reactivity is difficult to define and to measure. A simpler way to determine reactivity is to measure it in terms of the volume of CO generated upon reacting char with CO_2 per gram of char per interval of time. The char used is generated under gasification conditions, and the reactivity test is done under pressure. Results from such measurements have been correlated with actual experience from the operation of a commercial plant and are shown in Fig. 7-4, in which oxygen consumption expressed as pounds of O_2 per 1000 lb of daf (dry, ash-free) coal is plotted vs. "reactivity."

The lowest value for oxygen consumption for very reactive coals (such as Wyoming coal) is around 330 lb of O_2 per 1000 lb of daf coal; whereas less reactive coals (such as Illinois coal or Ruhr coals) consume up to 550 lb of O_2 per 1000 lb of daf coal.

For reactive coals, oxygen consumption is only slightly influenced by the steam-to-oxygen ratio; whereas low-reactive coals need a reasonable temperature in the combustion zone. If a low-reactive coal also has a low ash melting point, a gasification process that includes liquid-slag removal (such as that using a British Gas/Lurgi Slagging gasifier) may be the better option.

Steam-to-oxygen ratio

Steam is the primary reactant and moderates the temperature in the combustion zone so that it remains low enough to avoid ash fusion. The minimum steam-to-oxygen ratio is consequently determined by the ash fusion properties. Figure 7-5 shows the maximum combustion temperature, t_{max} vs. the steam-to-oxygen ratio. t_{max} is the maximum attainable combustion temperature which would result if the residual carbon entering the combustion zone were only converted to CO_2. However, t_{max} is, in fact, never reached because gasifi-

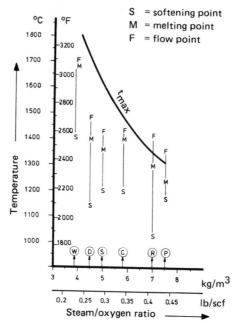

FIG. 7-5 Temperature in combustion zone. *(Lurgi.)*

cation reactions also occur in the combustion zone. The difference between the effective temperature and t_{max} is influenced by the reactivity of the char, grain size, ash content, and residence time, which generally is between 3 and 10 min in a narrow zone with a height which is about 5 to 10 times the diameter of the char grain.

Figure 7-5 also shows ash-melting values of coal vs. the steam-to-oxygen ratio used during commercial operation. The ash produced under these conditions was only lightly sintered, although t_{max} was far above the flow point (F).

Gas Composition

Typical gas analyses are shown in Tables 7-1 and 7-2.

The CO_2 content and the H_2/CO ratio are strongly influenced by the steam-oxygen ratio but are also influenced by the properties of the coal. This is illustrated in Fig. 7-6, in which the various curves represent various coals. All the curves converge at an H_2/CO ratio of about 0.4 and at a steam-to-oxygen ratio of 1.

If CO_2 is added as a component to the gasification agent, the H_2/CO ratio is reduced. One method that has been used is to partially replace steam with recycled CO_2. This method is an efficient way to adjust the H_2/CO ratio to any required value (see Table 7-2) and its application improves the efficiency of a gasification plant.

Figure 7-7 shows the formation of methane as a function of pressure. These data come from the operation of the "Ruhr 100" plant at Dorsten (FRG) where operational pressures of up to 1200 lb/in² confirmed previous predictions. Moreover, the mode of operation

TABLE 7-1 Typical performance data (steam-oxygen operation)

		Subbituminous Coal			
	Lignite[a]	A[b]	B[c]	C[d]	D[e]
HHV,[f] Btu/lb	12,000	12,600	12,700	14,000	14,400
O_2, scf/scf of crude gas	0.12	0.11	0.14	0.15	0.17
Steam (excluding jacket steam), lb/scf of crude gas	0.048	0.038	0.042	0.035	0.038
Crude gas, scf/1000 lb of daf coal	32,339	33,185	38,434	37,588	38,603
Tar,[g] oil, naphtha, lb/1000 lb of daf coal	79	81	27	84	80
Gas analysis (dry and sulfur-free crude gas), %					
CO_2	31.0	29.1	29.4	24.7	30.0
CO	15.8	18.4	21.7	24.5	17.4
H_2	40.6	38.6	38.5	39.0	41.6
CH_4	10.9	12.6	9.8	9.5	9.3
C_nH_m	1.2	1.0	0.3	1.1	0.8
N_2	0.5	0.3	0.3	1.2	0.9

[a] Volatile matter = 48.8%.

[b] Volatile matter = 48.4%.

[c] Volatile matter = 32.0%, Fischer tar = 4%.

[d] Volatile matter = 39.0%, weakly caking.

[e] Volatile matter = 39.0–45.0%, caking.

[f] Higher heating value.

[g] Without tar recycle.

(steam-to-oxygen ratio) and the type of coal influence the methane yield. (Under slagging conditions methane derives mainly from devolatilization.) Subbituminous C coal (for example, Wyoming coal) shows the highest methane yield. Tar recycle contributes only marginally to methane formation.

INFLUENCE OF COAL PROPERTIES

Commercial operations and pilot plant tests have demonstrated that all kinds of coal can be gasified, ranging from peat to anthracite and coke.

Moisture

Coal handling requires that surface moisture be avoided. Otherwise there are no restrictions up to a moisture content of 32 to 38 percent (depending upon the ash content), which are typical values for lignite. Countercurrent operation makes it possible to utilize the waste heat of the production gas to dry the coal within the process without a loss in efficiency.

TABLE 7-2 Typical performance data for other modes of operation

	Mode of Operation		
	CO₂ recycle	Air-blown	
Coal	Subbituminous*	Subbituminous*	Subbituminous†
HHV‡, Btu/lb	12,700	12,700	14,400
O_2, scf/scf of crude gas	0.12		
Air, scf/scf of crude gas	—	0.51	0.55
CO_2, scf/scf of crude gas	0.27		
Steam (excluding jacket steam), lb/scf of crude gas	0.019	0.012	0.011
Crude gas, scf/1000 lb daf coal	44,021	62,223	59,768
Tar, oil, naphtha, lb/1000 lb daf coal	27	72	50
Gas analysis (dry and sulfur-free crude gas), %			
CO_2	37.2	14.0	10.3
CO	32.9	15.8	21.2
H_2	21.3	25.0	20.1
CH_4	7.6	5.0	4.6
C_nH_m	0.2	0.2	0.3
N_2	0.8	40.0	43.5

*Volatile matter = 32.0%, Fischer tar = 4%.

†Caking, volatile matter = 39.0–45.0%.

‡Higher heating value.

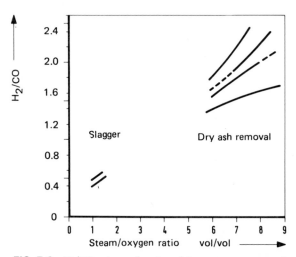

FIG. 7-6 H_2/CO ratio as a function of the steam-to-oxygen ratio.

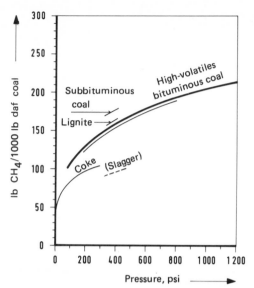

FIG. 7-7 Formation of methane as a function of pressure.

Ash

Coals with an ash content up to 50 percent have been gasified in commercial plants. For economic reasons it is advisable to keep the ratio of ash to fixed carbon below 1:1.

The melting properties of ash determine the steam-to-oxygen ratio of the gasification agent, which controls the temperature in the combustion zone (see Fig. 7-5). For the dry-ash removal system, coals with higher ash melting temperatures yield more economic results. However, highly reactive coals also show a high rate of steam-supported decomposition if the gasifier is operated with a high steam-to-oxygen ratio. On the other hand, for coals with low reactivity and low ash melting temperatures, a gasification process that includes liquid-slag removal (such as that in a British Gas/Lurgi slagger) is more efficient.

Volatile Matter

A general rule is that reactivity increases with the content of volatile matter.

Caking Properties

For processing of caking coals the coal distributor is equipped with blades which rotate in the fuel bed. Even highly swelling coals can be gasified. Carry-over of dust is higher (up to 2 percent of the coal feed) than with noncaking coals. But by recycling of dusty tar this dust is consumed within the reactor.

Sulfur

There are no limitations on the sulfur content of the coal feedstock. Coals with sulfur contents of up to 10 percent have been gasified.

Coal Size

Generally, a moving-bed gasifier has to be fed with graded coal. Minimum size is dictated by the capability of coal screening, i.e., ⅛ in. Maximum size is 3 in, with a limited amount of up to 4-in lumps. The ratio of largest size to smallest size should be 10:1 or less. The following concession can be made: in the case of noncaking coals 5 percent of the coal as undersized particles can be tolerated. The size ratio for most coals with a high content of volatile matter can be greater than 10 because larger coal pieces break up into smaller pieces rather than to dust during drying and carbonization; i.e., the size range becomes more uniform during processing. For caking coals the amount of fines can be much higher; even run-of-mine coal has been gasified. This is possible because in the top part of the fuel bed caking causes the coal fines to agglomerate into larger pieces. Commonly, coal fines are consumed as fuel for the generation of steam and power. In a self-supporting plant, from 20 (in substitute natural gas plants) up to 32 percent (in ammonia plants) of the total coal is used as fuel. Another method is agglomeration of coal fines and utilization of agglomerates as feed for gasification.

Summary

Essentially all kinds of coal can be gasified; however, with all gasification processes the wide variety of coal properties has an influence on the design and the economics of a plant. Fortunately, a wealth of experience on the effect of different coal properties is available for the Lurgi process.

PERFORMANCE DATA

Gasifier Performance

Tables 7-1 and 7-2 show typical performance data for various types of coal and process conditions. These data represent commercial, large-scale plant operation. Table 7-1 presents data from conventional steam-oxygen operation, whereas the data in Table 7-2 show the effect of using an air-blown gasifier, the impact of CO_2 recycle, and the result of operation under slagging conditions (gasification in a British Gas/Lurgi slagger). The pressure is 350 lb/in² (2400 kPa). The impact of a higher pressure on the methane yield can be seen from Fig. 7-7 (data from operation of the Ruhr 100).

The crude gas leaving the gasifier contains sulfur compounds, devolatilization products (tar, oil, naphtha, phenols, ammonia, fatty acids, etc.), and steam (coal moisture and undecomposed steam). Sulfur compounds are found mainly as H_2S. Because of the moderate temperatures in the gasifier the amount of organic sulfur (COS for example) is low and is only 4 to 6 percent of the H_2S content. The amount of HCN is also low.

Stepwise cooling and the resulting fractionated condensation remove all condensable components and components which dissolve in the condensate (gas liquor). Tar is condensed in the first cooler, where the tar-containing liquor provides an excellent medium for dust removal. The mixture of tar and dust which settles out in a separator is recycled to the gasifier. By this method all dust is recovered, and if required, tar can be utilized for gasification up to extinction. Oil and gas naphtha are removed by cooling. Phenols, fatty acids, and ammonia are dissolved in the H_2O condensate and form a gas liquor (see "By-Product Recovery").

Most of the nitrogen in the coal is found as NH_3. For coals with more than 15 percent (daf basis) volatile matter, the ammonia yield is about 90 percent. If the content of volatile matter is less, the ammonia yield decreases. For anthracite with 5 percent volatile matter and for coke, the ammonia yield approaches zero. Ammonia is scrubbed out of the gas in the downstream-gas cooling process and is a valuable component in the gas liquor, since it lifts the pH of the gas liquor to 8.5 or higher, a condition which prevents corrosion.

Most of the chlorine and fluorine originally contained in the coal is found in the gas scrubbed out by the scrubber and neutralized by the ammonia. Higher amounts of chlorine in the coal make it necessary to use a cladded material for the water jacket of the gasifier. The amount and type of other trace components depend upon the composition of the coal.

The special advantage of a moving-bed gasifier is the extremely low content of inorganic compounds in the liquid effluents, which facilitates purification of these streams.

Capacity and Availability

The Lurgi standard gasifier is available in three sizes:

Gasifier, type	Shaft diameter, ft
3-mØ	9
4-mØ	12
5-mØ	15

Capacity data are listed in Table 7-3. The specific capacity is influenced by the type of coal.

Both the gasifier and its attached equipment are extremely reliable even under the severe reaction conditions under which gasification has to be performed. The reactor system can easily cope with load variations, and since the principles of operation are simple, the avail-

TABLE 7-3 Gasifier capacity

Specific capacity:	
Crude gas (dry), scf/(ft²)(h)	20,000
Coal (daf), lb/(ft²)(h)	600
Capacity per unit, GJ in product gas/h:	
3-m-Ø gasifier	300–500
4-m-Ø gasifier	500–800
5-m-Ø gasifier	700–1200

FIG. 7-8   Lurgi gasification plant at Secunda. *(Lurgi.)*

ability factor (the ratio of operative capacity to installed capacity) is high and amounts to 86 percent. Included in this factor are the annual shutdown for general overhaul and inspection and the semiannual short overhaul; i.e., the gasifier is in full production for 314 days per year. More than 150 gasifiers have been delivered (as of 1980); 80 of them represent the most modern design, and more are on order. Figure 7-8 shows the gasification plant at Secunda, which produces 56×10^6 stdft^3/h (1.5×10^6 m^3/h) of raw gas. The raw gas is upgraded to 41×10^6 stdft^3/h (1.1×10^6 m^3/h) of synthesis gas for the Sasol Fischer-Tropsch plant.

LURGI COAL-GASIFICATION PROCESS

While gasification of coal is the key process, it is only one step of an overall scheme that also comprises processes for gas conditioning, gas purification, and by-product and effluent treatment. These other processes have to be tailored to conditions which are governed in a very specific way by the properties of the coal, by the gasification process, and by the product requirements (fuel gas, syngas, etc.).

Figure 7-9 shows a general process scheme for production of gas from coal. The steps include the following:

Coal gasification

Crude-gas shift conversion (conditioning of a crude, impurity-laden gas)

Rectisol gas purification (a convenient method for purification of a crude gas to the very stringent degree of purity required by synthetic processes)

Phenosolvan process (treatment of gas liquor derived from coal gasification)

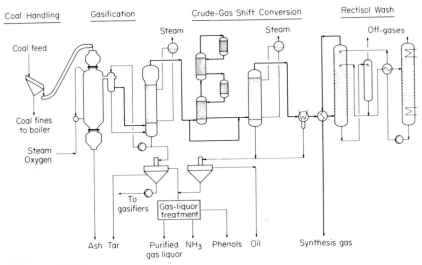

Coal Handling   Gasification   Crude-Gas Shift Conversion   Rectisol Wash

FIG. 7-9   General process scheme for production of synthesis gas.

The gas leaving the gas purification section can be used as pipeline gas of medium-Btu quality or as a feedstock for various synthetic processes.

Coal Gasification

It has been demonstrated in commercial plants that a multiple reactor system is necessary to smooth out fluctuations and to increase the availability of the gasification section close to 100 percent and the availability of the overall plant—if it is a single-train concept—to 92 percent.

Gas Conditioning

The crude gas leaving the gasifier is intensively washed in a scrubber, and its sensible heat is recovered in a waste-heat boiler. The wet scrubbing under pressure with a gas liquor containing hot tar eliminates all problems which particulates may otherwise create.

If the H_2/CO ratio has to be increased, the gas passes through a crude-gas shift conversion which is also a Lurgi process that is used in four commercial plants. The conversion reaction is performed by means of the steam contained in the crude gas, thus eliminating both the expensive cooler-saturator system (which is often applied in conventional shift-conversion processes) and the consumption of additional steam. The crude gas contains sulfur compounds and products originating from coal devolatilization, such as tar and naphtha. The catalyst used is not affected by these impurities; moreover, it possesses hydrogenation properties which improve the quality of by-products.

By-Product Recovery

Steam and tarry products are condensed by cooling the gas partly in waste-heat boilers and partly in air or water coolers. The resulting gas liquor is first treated in a unit that separates

TABLE 7-4   Typical analysis of gas naphtha, oil, and tar

	Gas naphtha	Light oil	Tar	
Specific gravity at 15 °C	0.83	0.97	1.124	
Distillation range, °C			Distillation range, °C	
5%	81	183		
20%	86	190	1.3%	210
40%	91	211	1.7%	210–230
60%	99	235	9.9%	230–270
80%	116	275	7.0%	270–300
95%	160	350	29.1%	300–350
			51.0%	(residue pitch)
			100.0%	
Paraffins, wt %	26			
Olefins, wt %	8			
Aromatics, wt%	66			
Tar acids, wt %		17	7.0	
Sulfur, wt %	0.08	0.3	0.8	

the tar liquor from the gas liquor and is then dephenolized by the Lurgi Phenosolvan process by extraction with an organic solvent (butyl acetate or isopropyl ether). The dephenolized water is biotreated and is then available for re-use in the plant. By-products are tar, oil, gas naptha, and phenols. The Phenosolvan process also provides for the removal of ammonia, which can be made available as anhydrous ammonia by the Chemie Linz-Lurgi process (CLL process). Some of the properties of the resulting tar, oil, and naphtha by-products are listed in Table 7-4. These are valuable by-products, which can either be sold or be consumed as fuel within the plant.

Gas Purification

Gas from coal gasification contains a large amount of CO_2 and H_2S, organic sulfur, and other impurities.

The Rectisol process is based on the ability of cold methanol to absorb all gas impurities, thus achieving the complete purification of the gas in a single process unit. Since the absorption capacity of methanol increases with decreasing temperatures, temperatures below 0 °C are used.

A Rectisol unit for the purification of coal gasification gas consists of three process units. A prewash unit removes gas naphtha, unsaturated hydrocarbons, and other high-boiling impurities. The following two process units remove H_2S, organic sulfur, and CO_2, the extent of CO_2 removal being adjusted to whatever the particular requirements are. The extremely high purity of the gas from Rectisol purification makes the gas suitable for use in any type of synthesis, including those syntheses employing very sensitive catalysts.

Methanol is regenerated by depressurization and distillation. The off-gases from the various stages of flashing and from the regeneration column have to be desulfurized before they can be released to the atmosphere. Various processes are available for desulfurization, e.g., the Claus process for producing sulfur from off-gases rich in H_2S and the Concat process for the production of sulfuric acid.

EXAMPLES OF APPLICATION

General

The combination of front-end process units described in the previous section is the common starting point for the following applications:

1. Direct use as pipeline gas of medium-Btu content (formerly called town gas) or by boosting the heating value via methanation, as substitute natural gas (SNG)
2. Feed gas for the synthesis of methanol, Fischer-Tropsch products, and ammonia
3. Gas for iron ore reduction
4. Fuel gas

For the processes listed under 2 and 3, the methane from coal gasification has to be reformed, thus increasing the amount of primary $(CO + H_2)$ by about 40 percent. This figure takes into account the fact that a portion of the re-former feed gas is used to maintain the endothermic re-forming reaction.

Methanol from Coal

Methanol from coal is one possible synfuel route. Figure 7-10 shows a block flow diagram in which the block labeled "gas production" corresponds to the general process scheme shown in Fig. 7-9. The total plant is self-supporting and is fed by a Wyoming coal. The single-product route (left half of Fig. 7-10) produces only methanol and achieves a thermal efficiency [based on a low heating value (LHV)] of 50 percent. Another possibility is not to re-form the methane but to make it available, for example, as SNG. This simultaneous production of liquids plus SNG is shown in the right half of Fig. 7-10. It yields a higher thermal efficiency and reduces the specific investment cost (see Table 7-5).

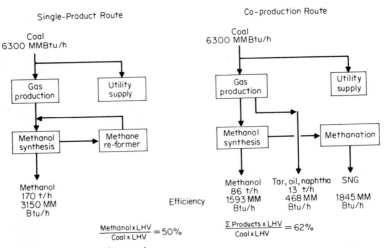

FIG. 7-10 Methanol from coal.

TABLE 7-5 Efficiency and specific investment cost of coal conversion plants (1980 dollars)

Product	Thermal efficiency (η), %	Specific investment cost, US$/(MM Btu)(yr)
SNG*	67	19
SNG†	62	21
Methanol†	50	31
Methanol + SNG†	62	23
Gasoline‡	45	38
Gasoline + SNG‡	61	28
Gasoline + diesel§	40	48
Gasoline + diesel + SNG§	52	31
Electricity¶	38	28

*Based on high heating value.

†Based on low heating value.

‡Produced from methanol using the Mobil MTG (methanol-to-gasoline) process.

§Produced via the Fischer-Tropsch reaction.

¶Produced by a conventional power plant at a capital cost of US$770/kW.

Fuel Gas from Coal

Lurgi gasification is also an excellent process for supplying low-Btu fuel gas from coal, in which the application of the air-blown version reduces the investment cost (see Fig. 7-11). The scheme in Fig. 7-11 shows coal gasification as the starting process for a combined power cycle. The crude gas leaving the gasifier is intensively scrubbed in an attached wash cooler to remove impurities to the degree required by gas turbines. The gas, which has been saturated with steam at 160 °C, is cooled in a heat-recovery system (cooler and sat-

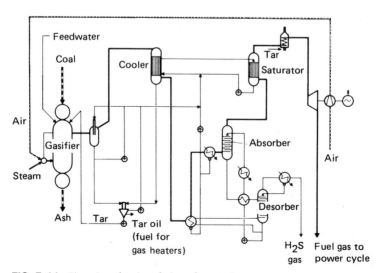

FIG. 7-11   Flow sheet for clean fuel gas from coal.

urator); H₂S is removed by a modified hot potassium carbonate absorption process. The flow sheet shown in Fig. 7-11 illustrates one version of the process described for producing fuel gas from coal and is similar to the STEAG combined-cycle power system operating at Lünen, F.R.G. Of course, the process can be modified, depending upon the intended use of the gas.

THERMAL EFFICIENCY AND INVESTMENT COST

Table 7-5 presents information on thermal efficiency and investment cost. These data allow for an evaluation of production cost. They are based on grassroots self-supporting plants with a coal consumption of about 2.5 million tons/year. These plants receive coal and water and do not import or export electricity. They are equipped with all the necessary processes for dealing with environmental protection regulations. The first column shows the thermal efficiency, η, which is expressed as a percentage and is defined as

$$\eta = \frac{\text{products} \times \text{LHV}}{\text{coal (as feed and fuel)} \times \text{LHV}} \times 100$$

where LHV = lower heating value

The second column specifies the specific investment cost related to the annual production in U.S. dollars per million Btu per year and is based on an on-stream factor of 8000 h/yr (cost basis is 1980). These figures are derived from experience in building and operating commercial plants.

Table 7-5 shows that the thermal efficiency is highest for SNG (and is even higher for town gas or syngas) and decreases sequentially for methanol and gasoline. It can also be seen that when liquid synfuels and SNG are coproduced, the efficiency is considerably better than when only a single liquid synfuel is the product. The specific investment costs follow the same pattern. It is interesting to note that the specific investment cost for a conventional power plant is higher and its thermal efficiency is lower than those of an SNG plant.

The data in Table 7-5 can be used to evaluate the product cost by applying the following equations:

$$\text{Cost of product (US\$/MM Btu)} = \text{fixed cost} + \text{variable cost}$$
$$\text{Fixed cost} = \text{specific investment}$$
$$[\text{US\$/(MM Btu in product)(yr)}] \times f$$

f includes the following items, expressed as a percentage of investment cost: interest + depreciation, taxes, insurance, personnel (ca. 3 percent), and maintenance (ca. 3 percent).

$$\text{Variable cost} = \frac{\text{cost of coal (US\$/MM Btu in coal)}}{\eta}$$
$$+ \text{cost of chemicals, catalyst, and water (US\$/MM Btu)}$$

The cost of chemicals, catalyst, and water is relatively low.

Production Cost of SNG (Example)

If it is assumed that the cost of coal is US\$1/MM Btu and f is 25 percent, the cost of SNG (based on HHV) is US\$6.75/MM Btu, calculated as follows:

$$\text{Fixed cost element} = \frac{\$19}{\text{MM Btu/yr}} \times 0.25 = \$4.75/\text{MM Btu}$$

$$\text{Variable cost element} = \frac{\$1/\text{MM Btu}}{0.67} = \$1.50/\text{MM Btu}$$

$$\$6.25/\text{MM Btu}$$

$$\text{Cost of chemicals, catalysts, water, etc.} = \$0.50/\text{MM Btu}$$

$$\text{Cost of SNG} = \$6.25 + \$0.50 = \$6.75/\text{MM Btu}$$

This example shows the strong influence on production cost of the fixed costs, which are mainly capital costs. This is a feature of coal conversion in general. However, once a plant is built and is operating, the influence of the ongoing price escalation is very slow.

ENVIRONMENTAL ACCEPTABILITY

The Lurgi coal gasification plant has a negligible impact on the environment. This can be seen by analyzing the impact of the various streams.

The product gas is clean, dry, and practically sulfur-free and can be used for several purposes, as has been previously discussed. If the product gas is used as a fuel gas, there will be no problem in meeting the clean air laws of the various government authorities.

Gaseous emissions from vessels that breath and/or from expansion gases are negligible because the whole plant is connected to an incinerator via a closed vent system.

The sulfur that is reclaimed from the sulfur-recovery unit is a salable, high-quality solid. Should the demand for sulfur be low, it can be stored without problems.

Tar, oils, phenols, and naphtha are recovered by processes which do not present any engineering or technical difficulties, as the required technology has been used in gasification plants and related industries for many years and the processes have proved to be environmentally acceptable. The recovered by-products either can be used inside the plant as fuel or can be upgraded to high-level products which are salable.

Water that leaves the plant is treated in a biological wastewater treatment unit having an activated-carbon adsorption step. Since this water is treated to the extent that it could also be reused as cooling water, it can safely be released into the environment.

Gasifier ash is considered to be a nonhazardous waste.

There are likely to be sources of noise—coal handling, coal charging, compressors, vents, high-pressure gas flows, turbines, and pumps—but due to design improvements these emissions are lower than present noise from power stations.

Fugitive emissions from gasification plants are mainly influenced by the maintenance philosophy of the user and are not type-specific for gasification plants.

An energy analysis, "A Comparison of Coal Use by Gasification vs. Electrification and Liquefaction," was published by The American Gas Association on Mar. 26, 1982; this document gave the data in the following table in terms of air-pollution emissions, water and land required, and solid wastes produced. They show that coal gasification would contribute significantly less to environmental degradation than does, for example, conventional coal-fired electricity generation. The data are based on a plant producing 250 MM stdft^3/d SNG from coal.

Air emissions	SNG plant producing 250 MM scf/d	Coal-fired electricity-generating plant providing the same amount of space heat
Particulates, t/yr:	590	4,830
Gases, t/yr:		
SO_x	2870	9,720
NO_x	8440	66,480
HC	180	3,680
CO	640	1,110
Water requirements, mm gal/d	6	22
Solid wastes, t/yr	1390	3,490
Land requirements, acres	810	2,700

FOSTER WHEELER-STOIC PROCESS

ROBERT G. BRAND

Foster Wheeler Synfuels Corporation
Livingston, New Jersey

DELLASON F. BRESS

Foster Wheeler SPEC, Inc.
Livingston, New Jersey

INTRODUCTION

The Foster Wheeler-Stoic (FW-Stoic) gasification process employs a two-stage, air-blown, fixed-bed gasifier to produce gas with a heating value of 150 to 170 Btu/stdft^3 (5.9 to 6.7 MJ/m^3) from noncaking solid fuels, including wood. In excess of 90 percent of the energy in coal is recovered as producer gas and good-quality fuel oil. The process meets today's environmental standards, and there is no foul aqueous effluent.

The primary advantage of two-stage gasification is that the coal-tar hydrocarbons produced are not cracked. Thus, the tar oil recovered can be stored at 70 °C without polymerization. Also, the product gas is superheated with respect to hydrocarbons, but condensate resulting from two-stage gas will not polymerize and foul the gas lines and burners.

FW-Stoic gasifier sizes and capacities are given in the following table.

Gasifier Diameter		Energy Output*	
Ft-in	m	10^6 Btu/h	GJ/h
12-6	3.8	94.8	100
10-0	3.0	59.5	62.8
8-6	2.6	42.7	45.0
6-6	2.0	24.3	25.6

*The energy output includes the tar oil and thermal energy in the gas.

Coal fed to the FW-Stoic process can be a relatively poor grade, with heating values as low as 6000 Btu/lb (14.0 MJ/kg). The coal fed must have a maximum free swelling index of 2.6, a minimum ash-softening temperature of 1200 °C under oxidizing conditions, and a size range of ¾ to 1½ in (19 to 38 mm) with no more than 10 percent finer than ¾ in (19 mm). A larger or smaller size range can be used but the maximum-to-minimum size ratio should not exceed two. Using smaller coal will decrease the gasifier output from its rating for the ¾ in. to 1½ in. size range.

Process Options

The FW-Stoic gasifier discharges the producer gas in two streams. One is top gas containing the moisture from the coal plus both light and heavy hydrocarbons that result from devolatilization of the coal. The heavier hydrocarbons are present in the top gas as a liquid mist. The second stream is bottom gas at 590 to 650 °C and contains only a small amount of CH_4 and virtually no other hydrocarbons. The three options for blending the two gas streams all pass the bottom gas through a cyclone for particulate removal. The flow of bottom gas is then controlled to maintain the top gas at a temperature of 120 °C to ensure devolatilization of the coal. The three methods for mixing the two gas streams follow.

Hot raw gas

With this system, the top gas is passed through a cyclone to remove oil droplets. Since the cyclone is somewhat limited in its ability to remove particulates, a fraction of the oil mist passes through and is mixed with the hot bottom gas. As a result of the presence of the liquid mist in the top gas, this option produces product gas with the highest heating value. This system is suitable for meeting a relatively constant load where the user is connected directly to the gasifier to minimize liquid dropout in the gas piping.

Hot detarred gas

In this option, the top gas is passed through an electrostatic precipitator to completely remove the liquid hydrocarbons present before mixing it with the hot bottom gas. The product gas is thus superheated by about 260 °C above the hydrocarbon dew point, permitting distribution to several users without fear of liquid condensation.

Although the product gas from this scheme has a slightly lower heating value than that from the hot raw-gas scheme, the overall efficiency of the hot detarred gas plant is the same and the system is much more flexible and easier to operate.

Cold clean gas

If the coal used contains sulfur that has to be removed, the hot detarred gas from the second option is further cooled to permit cleanup. Sulfur which is present as H_2S and COS can be removed by the hydrated iron oxide process or by alkaline scrubbing.

The cold clean-gas option should also be considered if the product gas is to be distributed over an extensive network or the burner application requires a gas pressure which calls for a product-gas blower.

Commercial Applications

The gas made has a heating value of only about 160 Btu/stdft³ (6.3 MJ/m³), but since its principal combustible components are CO and H_2, the flame temperatures are similar to those obtained with natural gas or fuel oil. In addition, the volume of the combustion products is only slightly greater than the equivalent amount from the other fuels. The FW-Stoic process gas can therefore be used to supply any heat requirement normally supplied by natural gas or fuel oil in both new and retrofitted applications utilizing burners specifically designed for combusting this type of gas.

The two-stage gasifier has been widely used in Europe and South Africa but has only recently been introduced in the United States. Table 8-1 is a listing of installations of FW-Stoic process gasifiers and illustrates the wide range of applications for which it can be used. Applications include direct firing of a lime kiln, a brick kiln, a heat-treatment furnace, and a bake oven. Indirect heating applications include generation of steam and re-forming of ammonia. Figures 8-1 to 8-5 show both overall views and details of two of these installations.

The boiler retrofit listed in Table 8-1 is an FW-Stoic plant constructed at the University of Minnesota's Duluth campus. It includes a 10-ft (3.8-m) diameter gasifier and produces hot detarred gas which is burned in two 25,000-lb/h (11,340-kg/h) boilers retrofitted to burn FW-Stoic gas. The project was constructed under a cost-sharing cooperative agreement between the university and the U.S. Department of Energy (DOE). In addition to supplying the campus heating loads, the installation is demonstrating the operating relia-

TABLE 8-1 FW-Stoic gasifier installations

Location	Company	Gasifier Units No.	Size, ft-in	Application	Start-up
Driefontein, SA*	Brickor	1	10-0	Brick kiln	1975
Lydenberg, SA	JCI Mining	1	8-6	Metals drying	1977
Livingston, N.J.	Foster Wheeler	1	2-0	Test unit	1979
Duluth, Minn.	University of Minnesota	1	10-0	Boiler retrofit	1979
Columbus, Ohio	Battelle	1	2-0	Test unit	1979
Pretoria, SA	Associated Bakeries	1	4-6	Bake ovens	1979
Nigel, SA	Power Lines	1	8-6	Metals heating, melting	1979
Capetown, SA	Fedmis	3	10-6	In ammonia re-former furnace	1981
Pretoria, SA	Metal Box	1	8-6	Paint and metal drying	1981
Vereeniging, SA	Rand Water Board	2	6-6	Lime kiln	1981
Howick, Natal, SA	South African Rubber	1	4-6	Rubber manufacturing	1981
Johannesburg, SA	South African Railways	2	8-6	Heat treating	1982

*Republic of South Africa.

bility and suitability of such a gasifier in the United States. The plant was started up late in 1978, and after an initial shakedown period has been in continuous operation.

Process Chemistry

Figure 8-6 shows the zones inside the FW-Stoic gasifier. Fresh coal is dried and devolatilized in the distillation zone by controlling the flow of hot gas upward into that zone from the lower, or gasification and combustion, stage. As the coal is heated and dried, the heavier hydrocarbons in the ascending gas condense and leave as a mist in the top gas. The descending coal is further heated, and above 400 °C it softens and pyrolyzes, emitting gases including CO_2, considerable amounts of H_2, and hydrocarbons starting with CH_4 but including such high-molecular-weight aromatic compounds as naphthalene and anthracene. Above 480 °C, the coking reactions are complete and the coal—now essentially coke—resolidifies and enters the lower gasification and combustion zone. Here, the devolatilized coal first reacts at 650 to 1100 °C with oxygen-free gas rising from the combustion zone. The principal reactions which occur in the presence of water vapor, along with their heats of reaction at 18 °C, follow:

$$C + CO_2 \rightleftharpoons 2CO \quad (+172.7 \text{ kJ}) \tag{1}$$

$$C + H_2O \rightleftharpoons CO + H_2 \quad (+131.4 \text{ kJ}) \tag{2}$$

FIG. 8-1 Battery of three FW-Stoic gasifiers producing gas for heating a re-former furnace making ammonia synthesis gas. The coal bunkers are above the coal feed system, with the gasifiers below, bunker for coal fines in the foreground, gas cleanup equipment on the right, and an ash bunker and oil storage tank at the right rear. Installation is at the fertilizer complex of the Fedmis Division of Sentrachem, Capetown, South Africa. *(Fedmis)*

FIG. 8-2 Exterior of FW-Stoic gasifier installation at the Duluth campus of University of Minnesota. Gasifier installation is in tall glass-fronted building, and on left is the coal-unloading and ash-loading building. On right is the original building housing boilers retrofitted to fire producer gas and tar oil. *(FW-Stoic)*

FIG. 8-3 The coal hopper and coal scale above the Duluth gasifier. Note the nuclear gage on the coal hopper for signaling a low coal level. *(FW-Stoic)*

A gaseous reaction at about 650 °C fixes the CO/H_2 distribution in the bottom gas:

$$H_2O + CO \rightleftharpoons H_2 + CO_2 \quad (-41.3 \text{ kJ}) \tag{3}$$

Air and steam pass through the grate and are preheated in an ash layer before reaching the fire zone, where sufficient carbon is reacted to supply the heat required for the process. The steam maintains the fire zone at a temperature low enough to prevent excessive ash

FIG. 8-4 A poking operation in Duluth. Note the steam-ejector poke holes which induce air flow into the gasifier when opened. *(FW-Stoic)*

FIG. 8-5   An ash bucket at the base of the Duluth gasifier. *(FW-Stoic)*

fusion, which can result in the formation of large clinkers. In addition to absorbing sensible heat, the steam moderates the temperature via the endothermic reactions shown in Eq. 2 above and in Eq. 7 that follows. The fire zone reactions which occur at temperatures of 1500 to 1800 °C in the presence of gaseous water together with their heat of reaction at 18 °C follow:

$$C + \tfrac{1}{2}O_2 \rightleftharpoons CO \qquad\qquad (-110.5\ kJ) \qquad\qquad (4)$$

$$C + O_2 \rightleftharpoons CO_2 \qquad\qquad (-393.6\ kJ) \qquad\qquad (5)$$

$$CO + \tfrac{1}{2}O_2 \rightleftharpoons CO_2 \qquad\qquad (-283.1\ kJ) \qquad\qquad (6)$$

$$C + 2H_2O \rightleftharpoons CO_2 + 2H_2 \quad (+90.1\ kJ) \qquad\qquad (7)$$

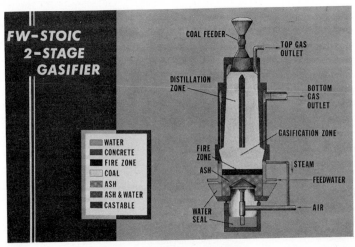

FIG. 8-6   Schematic showing gasifier zones. *(FW-Stoic)*

The exothermic reactions between carbon, or CO, and H_2 to produce CH_4 are favored by high pressure and low temperature. The high temperature and substantially atmospheric pressure in the gasification and combustion zone are not conducive to the formation of CH_4.

Depending upon the form of the sulfur in the coal, part of it will be discharged with the ash. Most of the sulfur found in the gas will be generated in the combustion zone as SO_2, but in the devolatilization zone the SO_2 will react with carbon, CO, and H_2 to produce H_2S, traces of elemental sulfur, and COS. These compounds are found along with sulfur dioxide in the product gas.

In the United States, coals with a free swelling index suitable for the FW-Stoic process are mainly western coals having fairly low sulfur content in the range of 0.3 to 0.6 percent. Thus, for most applications sulfur removal is not required.

Finally, the gasification process will result in trace amounts of N_2O and HCN in the gas.

Process Perspectives

Two-stage gasifiers are described in late nineteenth century literature, and many such plants were installed in Europe between 1920 and 1950. From 1950 onward, cheap Middle East oil caused coal gasification to languish except in South Africa, where the old plants continued to operate. Later, the oil price increases and oil embargos of the seventies led to renewed interest in coal-gasification technology, and since Stoic Combustion Limited Pty of Johannesburg, South Africa, had many innovative ideas concerning two-stage gasification, Foster Wheeler entered into an exclusive license with Stoic to market their technology.

The FW-Stoic design incorporates improvements over earlier gasifiers, the most important being a deep ash seal that results in a product-gas pressure of 1.0 psig (108 kPa), which often eliminates the need for a hot-gas booster blower.

Another improvement is the use of a castable refractory lining and top-gas collection conduits instead of the intricate firebrick assembly previously used.

A combined steel and cast-iron grate holder has been developed to replace the cast-iron grate holder which supports the cast-iron grate elements. This whole assembly—including the ash pan, grate holder, and grate—rotates, and lobes on the holder serve to move the ash downward and grind clinkers. (See the gasifier outline drawing in Fig. 8-7.) The grate holder is subject to severe mechanical and thermal stresses, and the new design using a steel cylinder with bolted-on cast-iron lobes requires lower maintenance and can be repaired without disassembling the grate system.

Although simple gasifiers of the FW-Stoic type were operated in South Africa during the years from 1950 onward, no new plants were built until 1975. In addition, rules concerning the environment and the working conditions of operators had changed, particularly in the United States. These factors required eliminating leakage of the producer gas into the environment, and the venting of noxious gas during start-up or shutdown.

DETAILED PROCESS DESCRIPTION

Table 8-2 shows a mass and energy balance for a 10.0-ft (3.0-m) FW-Stoic gasifier recovering 59.5×10^6 Btu/h (62.8 GJ/h) as hot detarred gas and oil at an overall efficiency of over 92 percent.

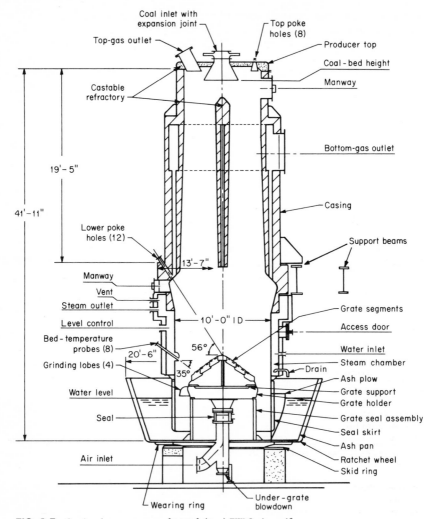

FIG. 8-7 Sectional arrangement of 10.0-ft (3-m) FW-Stoic gasifier.

Figure 8-8 is a detailed flow diagram for this process showing major instrumentation, and Figure 8-7 illustrates construction details of the gasifier itself.

Coal Reception and Storage

Coal storage at the gasifier will depend upon the specific situation. If a reliable coal supplier is assumed, a 3-day storage bunker to care for holiday weekends should be sufficient. The coal storage hopper (TK-111) can be at grade or elevated above the gasifier. The coal reception and feeding system includes unloading hopper TK-110, belt feeder CR-130, coal elevator CR-132, magnet S-130, and polishing screen S-133, used to remove fines gener-

TABLE 8-2 Overall mass and energy balance of FW-Stoic gasification process [10'-0" (3-m) diameter gasifier; total energy output 59.5MM Btu/h (62.8 GJ/h)]

Element or compound	Coal* wt %	Coal* lb/h	Water lb/h	Air, lb/h	Hot Detarred Gas lb/h	Hot Detarred Gas lb mol/h	Hot Detarred Gas vol %	Tar Oil lb/h	Tar Oil wt %	Ash lb/h	Ash wt %
C	64.66	3648.0						157.7	91.74	109.3	16.18
CO_2					1,513.0	34.38	4.07				
CO					6,193.0	221.09	26.14				
H_2	4.41	248.8			288.3	143.01	16.91	14.2	8.26		
N_2	1.08	60.9		9,639.1	9,700.0	346.23	40.95				
O_2	11.80	665.8		2,898.7							
S	0.41	23.1								12.0	1.78
H_2S					11.8	0.35	0.04				
CH_4					285.5	17.80	2.11				
Ash	8.60	485.1								485.1	71.83
H_2O	9.04	510.3	2140.0	106.8	1,479.9	82.14	9.71			69.0	10.21
Oil vapor†					107.8	0.59	0.07				
Total	100.00	5642.0	2140.0	12,644.6	19,579.3	845.59	100.00	171.9	100.0	675.4	100.00
Temperature	15.5 °C		15.5 °C	15.5 °C	346 °C			121 °C		93 °C	
Pressure			5 psig	Atmospheric	1.0 psig			Atmospheric		Atmospheric	
Higher heating value (at 15.5 °C)	11,444 Btu/lb				165.1 Btu/scf			16,000 Btu/lb			
MM Btu/h over 15.5 °C	64.567 (at 15.5 °C)		0	0	56.733 (at 346 °C)			2.767 (at 121 °C)		1.564 (at 93 °C)	

Energy Summary	MM Btu/h	%
Useful output: gas + tar oil	59.500	92.15
Other: Ash	1.564	2.42
To vaporize 2071 lb/h feedwater	2.577	4.00
Heat loss and excess steam vented	0.926	1.43
Total input	64.567	100.00

*Coal is from the Southern Fuel Co. mine in Saline, Utah with the following properties.

Proximate analysis, wt %		Free Swelling Index	1½	
		Fusion temperature	Reducing, °C	Oxidizing, °C
Volatile matter	36.89			
Fixed carbon	45.47	Initial deformation °C	1160	1204
Ash	8.60	Softening (1)	1193	1229
Moisture	9.04	Softening (2)	1204	1243
Total	100.00	Fluid	1221	1277

†Oil vapor composition is 8.34 wt % hydrogen and 91.66 wt % carbon.

ated in the transport and handling of the properly sized coal before it enters storage hopper TK-111. A vent cyclone (S-134) and vent blower (B-131) maintain the equipment under negative pressure in order to eliminate emission of coal dust. This system also evacuates the coal feed lock-hopper so that the suction of the main air blowers (B-130A and B) can be applied to the discharge from the vent blower in order to recycle noxious gases to the process.

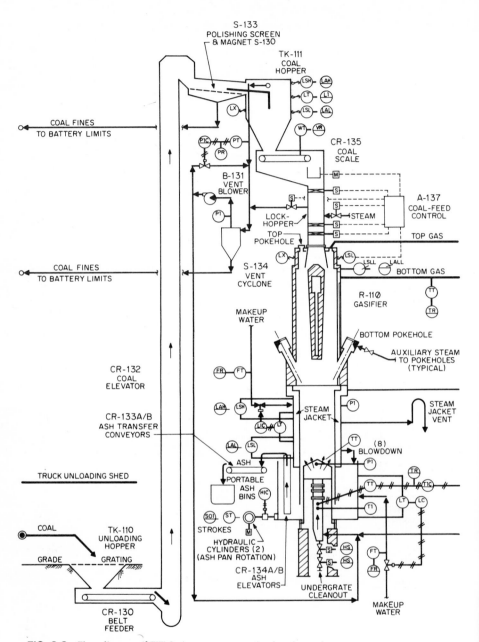

FIG. 8-8 Flow diagram of FW-Stoic process recovering hot detarred gas.

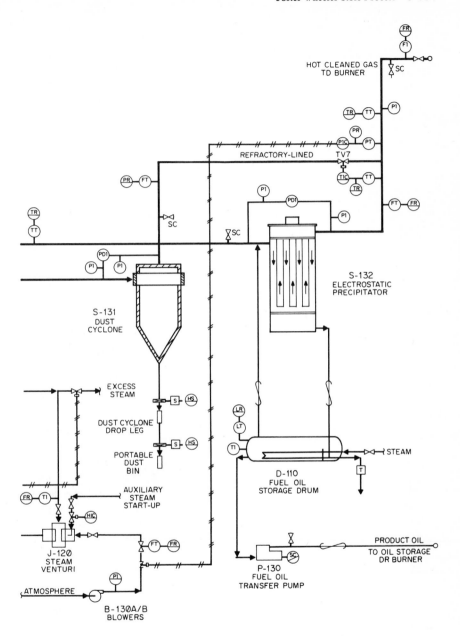

Coal Feed and Gasification

The coal scale (CR-135) removes coal from the storage hopper (TK-111) on a batched weight sequence based on the level of the bed in gasifier R-110 and transfers the coal to the lock-hopper, which then feeds coal into gasifier R-110. The lock-hopper is alternately pressurized and evacuated during the feeding sequence to prevent emission of producer gas. The coal-feed control (A-137) regulates the sequence of the feed operations. The reactions, which take place as the coal passes downward through the upper refractory-lined section of gasifier R-110, have been described under "Process Chemistry."

Coal is fed into the top of the gasifier, and part of the fuel-gas product (top gas) is removed from the top of the gasifier. The other portion of the product (bottom gas) exits from the lower zone. At the base of the gasifier, ash is removed through a water seal.

Air from blower B-130A or B-130B mixes with steam in venturi J-120 and passes through a rotating grate into the base of the coal bed. The mixture of air and steam exchanges heat with the hot ash and rises into a 4- to 10-in- (10- to 25-cm-) deep combustion or fire zone operating under oxidizing conditions. Hot gas from the fire zone rises to a reducing section, where the gasification reactions take place.

A portion of the gas, that is called "bottom gas" and is made in the lower gasification zone, is withdrawn and the remainder rises through the descending bed of coal, where the coal is gradually devolatilized. In order to ensure complete devolatilization and coking of the coal, the top-gas temperature is controlled at 120 °C by resetting the flow rate of the bottom gas. Too high a top-gas temperature may crack the oils, and too low a temperature may force the oils out with the hot bottom gas, causing them to crack. The bottom gas flows upward through flues in the refractory-lined wall. As the pressure drop in the coal bed is only about 1 lb/in^2 (7 kPa), the gasifier includes internal partitions to collect gas across the entire bed and thus assure uniform distribution in the lower zone.

The upper section of the refractory lining is tapered so that the cross section increases as the coal descends in order to allow swelling of the coal as it is heated. The refractory lining ends just above the fire zone. A steam jacket surrounds the fire zone, keeping the walls around the fire zone cool and thus resistant to ash sticking to the surface. The jacket also serves to produce steam for use in the process. As explained, steam prevents clinker formation, which could prevent ash removal and cause maldistribution of the ascending gas, resulting in an uneven, erratic fire zone.

Another vitally important requirement in maintaining a uniform fire zone is that the coal be uniformly sized as specified and free of excessive fines, water, or ice.

Below the grate the ash is collected in a water-sealed pan which rotates with the grate. Ash at the bottom of the pan is ploughed into the ash-removal elevators (CR-134A and B). These elevators lift the ash up over the side of the seal pan and discharge it onto two parallel ash-transfer conveyors, CR-133A and B. Handling of the ash from this point will depend upon local conditions. The ash at this stage is warm, slightly moist, and easy to handle.

Poking

The coal bed in the gasification and combustion zone is periodically agitated, and its condition checked by thrusting a steel poke rod through holes in the side of the gasifier down to the apex of the grate. This manual operation determines not only ash level but also the temperature, width, and location of the fire zone by the color, width, and position of the glowing area on the poke rod. Poking also reveals incipient clinker formation due to insufficient steam in the air. This is indicated when the force required to insert the poke rod increases.

Since the gasifier is under positive pressure in the lower zone, CO-bearing gas could escape when the holes are opened. The FW-Stoic gasifier therefore is equipped with specially designed steam-ejector poke holes. There is a linkage between the poke-hole cover and a steam valve so that opening the poke hole also opens a valve in the steam supply to the jet ejector. The ejector sucks outside air down into the poke hole, preventing outward leakage of producer gas.

Foster Wheeler and Stoic Combustion conducted experimental poking tests in which the rod was mechanically inserted, and both the force required and the bed temperature were recorded on instrumentation attached to the mechanical poke rod. A system patterned after the test unit can be offered on future installations.

The use of bed thermocouples to locate the fire zone and follow trends in its movement has been successful at the Duluth gasifier, and thus poking to locate the fire zone is not a necessity.

Product-Gas Treatment

With the hot detarred gas arrangement, the hot, relatively hydrocarbon-free bottom gas is passed through a refractory-lined dust cyclone (S-131) for particulate removal. The top gas, which includes light and heavy hydrocarbons in both the liquid and vapor state, is passed through electrostatic precipitator S-132. The electrostatic precipitator removes the heavy hydrocarbon liquid, which then drains to the heated fuel-oil storage drum (D-110).

Depending upon the volatile matter in the coal fed, the FW-Stoic process produces a stable, storable fuel oil with physical and combustion properties similar to no. 6 fuel oil at a yield of 3 to 6 percent by weight of the coal feed. By reducing the top-gas outlet temperature from the normal value of 120 °C, the yield increases and the oil made is more similar to a no. 5 fuel oil. Typical properties are given in Table 8-3.

The fuel oil derived from coal has only limited miscibility with no. 6 fuel oil (up to about 10 percent by weight), and a separate storage tank is required for it. A pump (P-130) is provided to transfer the coal-derived fuel oil to storage. The top gas, having a temperature controlled at 120 °C and saturated with heavy hydrocarbons, leaves the electrostatic precipitator and is mixed with the 600-to-650 °C hydrocarbon-free bottom gas to produce a stream of product gas at 340 to 400 °C. This hot, cleaned gas product has an oil dew point of 104 to 115 °C, depending upon the relative flows of the two gas streams. Thus, there is no condensation problem, provided the feed line to the user is insulated. The product gas is generated at 1 psig (7 kPa), in the Duluth plant, permitting the operation of cleanup devices and boiler combustion controls as well as transfer to the boilers without compression of the hot gas.

Controls

The temperature of the top gas is controlled by throttling the amount of bottom gas withdrawn from the gasifier. The gasification rate is regulated by utilizing the pressure in the product-gas line to control the amount of air sent to the process. The temperature of the mixture of air and steam controls the amount of steam injected into the air feed. Other controls as well as indicating and recording instruments for monitoring the process are shown on the flow diagram in Fig. 8-8.

TABLE 8-3 Tar-oil properties

The following analysis was made on a sample of oil collected from the University of Minnesota FW-Stoic gasifier installed at Duluth. The feedstock was a Colorado subbituminous coal with 36% volatile matter.

Specific gravity at 15.5 °C	1.0855
Viscosity at 50.0 °C	555.4 cSt
Viscosity at 98.9 °C	21.35
Viscosity at 121.1 °C	9.40

ASTM Distillation Corrected to Atmospheric Pressure from Test Made at 1.2–1.4 mmHg, abs.

Fraction	Temperature, °C
Initial boiling point	239
2%	263
5	288
10	310
20	342
30	366
40	388
50	410
60	431
70	454
80	488
93	571

Sulfur	0.22 wt %
Pour point	26.6 °C
Flash point, C.O.C.*	157 °C
Carbon residue, Ramsbottom	4.10 wt %
Sediment by extraction:	
With benzene	0.78 wt %
With carbon tetrachloride	4.16 wt %

Ultimate analysis

Component	Wt %
C	85.92
H	7.92
O	4.18
N	1.05
S	0.22
Ash	0.11
Moisture	0.60
Total	100.00

Heating value	16,397 Btu/lb (38.1 MJ/kg)

* Conradson open cup.

Utility Requirements

A 10-ft, 0-in (3-m) FW-Stoic gasifier plant with an output of 59.5×10^6 Btu/h (62.8 GJ/h) (see Table 8-2 for a mass and energy balance) will require the following utilities: electric power, boiler feedwater, and steam.

Electric power

Electric power is used at a rate of 75 kW with the main consumer being the blowers, B-130A and B, supplying undergrate air.

Boiler feedwater

Boiler feedwater is supplied to the gasifier steam jacket at a rate of 2400 lb/h (1088 kg/h).

Steam

The FW-Stoic gasifier can be designed to be independent of outside steam requirements, but if an outside source of steam at about 25 psig (274 kPa) is available, about 1200 lb/h (544 kg/h) is helpful for short intermittent periods during start-up and poking operations.

Products, By-Products, Waste Streams, and Product Gas

Table 8-2 gives a typical analysis for a mixed product gas with a higher heating value of 165.1 Btu/stdft^3 (6.50 MJ/m^3). The gas composition and heating value vary with the coal composition and the amount of water required in the undergrate air to prevent clinker formation. As the water content in the undergrate air increases, the CO content decreases and the H_2 content increases. The heating value of the gas remains relatively constant between a weight ratio (water to steam) of 0.1 to 0.2. Beyond the higher value, the heating value begins to drop precipitously and there is a danger of putting out the fire in the combustion zone.

Fuel oil

Table 8-3 gives the results of an inspection of fuel oil produced at the University of Minnesota, Duluth, gasifier.

Ash

The expected ash composition is as shown in Table 8-2. There are no other wastes or emissions from this gasifier.

PROCESS ECONOMICS

A 59.5×10^6-Btu/h (62.8-GJ/h) FW-Stoic gasifier plant producing hot detarred gas and fuel oil including coal reception and storage facilities will cost about $4 million. Figure 8-9 is a cost-of-production analysis showing a cost of $3.72/$10^6$ Btu of output. The production cost is computed using a coal price of $40/ton (about $1.75/$10^6$ Btu). The coal cost is the largest component, amounting to $1.90/$10^6$ Btu, and is more than half of the total cost of production. Increased usage of coal should reduce its cost as new mines are

Plant: 10'-0" dia FW-Stoic gasifier 59.5 MMBtu/h   Location: Midwest U.S.A.
Capacity: 500 × 10⁹ Btu/yr   Operating days/yr: 350
Capital cost: **$4.0 million including coal storage. Allowing for investment tax credit, investment is reduced to $3.2 million.**
Plant life: 20 yrs.

Cost items	Unit	Unit Price	Unit consumption	Cost per unit product
Raw Materials		$/ton	tons/yr	$/MM Btu
Coal	ton(s)	40.00	23,695	1.896
Total raw materials				1.896
Utilities				
Electric power	kWh	$0.05	6.3×10^5/Yr	0.063
Boiler Feedwater	kgal	$0.40	2.4×10^3/Yr	0.002
Intermittent steam (3 min/hr)	klb	$3.25	5.04×10^2/Yr	0.003
Total utilities				0.068

Direct operating costs		
Labor Operating: 6 operators @ $15,000/yr ---	$90,000/yr	0.180
Maintenance: @ 1% of captial cost	$40,000	0.080
Supervision: 2 supervisors @ $20,000/yr	$40,000/yr	0.80
Supplies: @ 10% of maintenance		0.008
Total direct operating costs		0.348
Indirect operating costs		
Payroll overhead @ 20% of operating + supervisory labor		0.052
General overhead @ 50% of total direct operating cost		0.174
Insurance and local taxes @ 2% of total capital cost	$ 80,000/yr	0.160
Interest, 20 Yrs @ 15% Average Interest on $3.2 million	$351,237/yr	0.702
Depreciation, capital $3.2 million, 20 yrs =	$160,000/yr	0.320
Total indirect operating costs		1.408
Credits — As noted, a 20% investment tax credit has been applied to the capital cost.		
Net production cost		**$3.72**

FIG. 8-9 Production cost analysis

opened and distribution facilities are constructed. It is noteworthy that the second largest factor in the production cost is the interest on the capital investment, which for a 20-year life and 15 percent interest rate averages out at $0.70/10^6 Btu.

SUMMARY

The FW-Stoic two-stage gasifier is an ideal solution to a situation requiring a clean gaseous fuel such as a direct-fired kiln, smelter, or glass furnace. Economic factors will govern each case, but a battery of as many as 12 gasifiers with an output of 27.3×10^9 Btu/day (28.8 TJ/day) is feasible.

Both new and retrofit applications are possible, and the coproduction of a liquid fuel permits continuous production while maintaining the gasifier.

The two-stage design is free from the fouling experienced in the gas-distribution piping and burners handling producer gas from single-stage gasifiers of similar capacities.

While the heat content of the product gas precludes its transport over long distances, each case should be analyzed carefully, since the two-stage gasifier is probably the least costly and most efficient of the known coal-gasification technologies.

The most important area for improvement would be the development of mechanical or other means to permit the gasification of caking coals.

THE WD-IGI TWO-STAGE COAL GASIFIER

DAVID M. JONES

Babcock Woodall-Duckham Ltd.,
Crawley, West Sussex,
England

DESCRIPTION OF THE WD-IGI PROCESS

Essential Features

The essential feature of the WD-IGI two-stage coal gasification process is that the first stage *(distillation retort)* produces a coke free of tar and oil, for completely clean gasification in the second stage *(gasification section)* (see Fig. 9-1).

Coal descends in the distillation retort through gradually increasing temperature zones. All tar and volatile matters are expelled until only carbon and ash remain to enter the gasification section.

The tar produced is of good quality, is fluid, and is easy to handle because it has not been subjected to high temperatures or radiation—which in single-stage gasifiers causes troublesome cracking and formation of pitch and gum components.

In the gasification section a mixture of air and steam or oxygen and steam ("blast") enters through a rotating grate and—apart from a small residue in the ash—the carbon in the distillation coke is completely gasified. The gas produced is known as *clear gas*; it is a mixture of CO, H_2, CO_2, N_2, and CH_4 and is completely free from oil and tar.

Most of the heat for the distillation retort is provided by part of the clear gas, which joins with distillation vapors to leave the retort at 100 to 150 °C as *top gas* having a higher heating value of 180 to 210 Btu/stdft^3 [7 to 8.4 MJ/m^3(n)].* The remaining part of the clear gas having a higher heating value of 136 to 147 Btu/stdft^3 [5.4 to 5.9 MJ/m^3(n)] flows through channels in the retort brickwork and leaves the retort at 650 to 700 °C.

Steam required for addition to the blast to control combustion-zone temperatures is generated in a water jacket around the gasification section. The use of this water jacket avoids the use of refractories in the hot zone of the gasifier and prevents clinker from adhering to the gasifier wall. The features just mentioned are illustrated in the cross section of a WD-IGI gasifier shown in Fig. 9-2.

Process Configurations

Depending upon the application, a plant can be designed in several ways that differ according to blast-mixture composition, mode of gasifier operation, treatment of gasifier product gas, and heating value of product gas. The process options are shown in Table 9-1. Both low-Btu gas (LBG) and medium-Btu gas (MBG) can be made.

The configurations that have practical importance are numbered 1 through 8 in Table 9-1 and will be discussed further. The configurations labeled 0 are not used because the gas produced by these configurations has to be cooled and cleaned before it can be used, as is explained under "Cyclic MBG" and "Oxygen-Blown MBG" for reasons which are stated under configurations 4 and 5 respectively.

The two-stage coal-gasification technology described in this chapter is available from:

Impianti Gas Internazionali, Via Pompeo Litta, 9-20122 Milano, Italy

Integral Engineering Industriebedarf, Grosse Neugasse 8, A1041 Wien, Austria

Le Gaz Integral, Nanterre, 100 Av. W.I. Lenine, Paris, France

*m^3(n) symbolizes the normal cubic meter, that is the quantity of gas occupying one cubic meter measured at 0 °C and 101.3 kPa pressure.

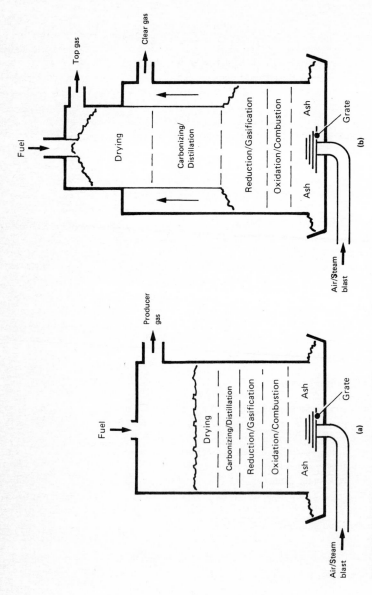

FIG. 9-1 Principles of (*a*) single-stage and (*b*) two-stage producers.

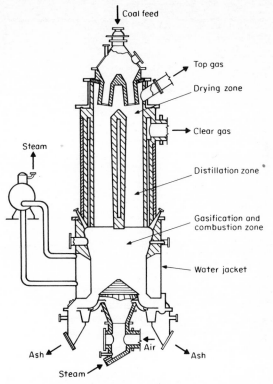

FIG. 9-2   Two-stage gasifier cross section.

Hot raw gas

See Fig. 9-3 and configuration 1 in Table 9-1. This is the simplest, most efficient, and cheapest plant and can be selected for many applications where LBG that is hot and contains tar and oil can be accepted. It cannot be selected when pressure boosting or flow splitting is required.

The air blast, with sufficient steam added for clinker control in the combustion zone, is supplied continuously through the bottom grate of the gasifier.

TABLE 9-1   Two-stage gasification process configurations

Product gas specification	LBG	MBG	MBG
Blast	Air, steam	Air, steam	Oxygen, air, steam
Mode	Continuous	Cyclic	Continuous
Hot raw gas	1	0	0
Hot detarred gas	2	0	0
Cold clean gas	3	4	5
Desulfurized gas	6	7	8

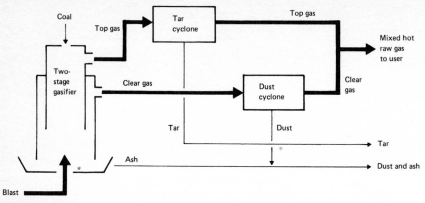

FIG. 9-3 Hot raw gas.

The hot clear gas containing coke dust but no tar passes through a cyclone which removes the dry dust.

The relatively cool top gas containing light tar as vapor and mist but negligible dust passes to a cyclone in which large droplets are removed. The clear gas and the top gas streams are then mixed so that the high temperature of the clear gas vaporizes the tar mist of the top gas, and the mixed gas at an intermediate temperature can be carried considerable distances to users in externally lagged steel pipes without forming troublesome tar deposits. If excessive cooling does occur, the tar that condenses is quite fluid and easily dealt with.

The principle advantages of the air-blown hot raw-gas plant are:

1. It offers the highest thermal efficiency*—about 90 percent.
2. There is freedom from losses by deposition of tar in mains—in contrast to single-stage gasifiers.
3. There is no need to interrupt gas supply to clean and burn out mains—no deposition of tar in mains.
4. There are no by-products or effluents.
5. The plant is simple and cheap.

The disadvantages are:

1. The gas pressure cannot be boosted.
2. The gas flow cannot be split in a controlled manner.
3. The gas cannot be desulfurized.

---

*Thermal efficiency is defined as the sum of the higher heating value + sensible heat + the latent heat of the entire gas, tar, oil, and steam mixture divided by the higher heating value of coal.

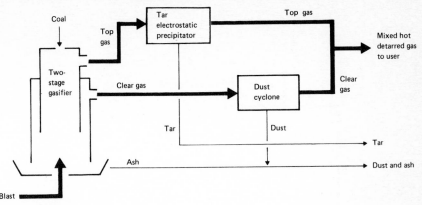

FIG. 9-4   Hot detarred gas.

Hot detarred gas

See Fig. 9-4 and configuration 2 in Table 9-1. The hot detarred-gas plant is very similar to the hot raw-gas plant except that the top gas passes through an electrodetarrer which removes the tar mist before the top gas mixes with the clear gas. The reduced quantity of tar in the mixed gas confers extra advantages on the hot detarred-gas plant compared with a hot raw-gas plant. The hot detarred-gas plant has the following advantages:

1. The plant has a high thermal efficiency—nearly as high as a hot raw-gas plant if the heating value of the separated tar is added to the heat in the mixed gas.
2. There is enhanced freedom from deposition in the gas mains.
3. Except for the separated tar, which is of high quality, there are no effluents.
4. The plant is still quite simple and inexpensive.
5. The gas pressure may be boosted.
6. Flow splitting is feasible.

The disadvantage is that the gas cannot be desulfurized.

Cold clean gas

See Fig. 9-5 and configuration 3 in Table 9-1. This configuration is significantly more complicated than the previous two and is only selected when an LBG is required that has to be conveyed very long distances or where the process being fueled may be sensitive to the heavier, condensible components of hot, raw, or detarred gases. The cold clean-gas configuration is also needed for plants that are followed by a desulfurization unit.

The top gas from an air-blown continuous gasifier passes to an electrostatic tar precipitator followed by a top-gas cooler and then an electrostatic oil precipitator.

The clear gas passes through a dust cyclone followed by a waste-heat boiler and then a clear-gas cooler.

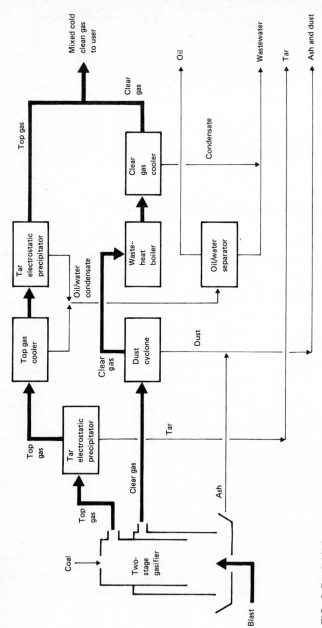

FIG. 9-5 Cold clean gas.

The top gas and clear gas, both clean and cold, are mixed and delivered to the user, possibly via a desulfurization unit.

As shown in Fig. 9-5, the cooling and electrostatic-separation operations give rise to tar, oil, and water condensates.

The cold clean-gas plant has the following advantages:

1. Cold clean gas can be conveyed in uninsulated mains for long distances limited only by pressure drop, and the gas pressure can be boosted if necessary.

2. The gas is amenable to the finest degree of flow control and is suitable for very small burners.

3. Its components are not likely to harm delicate processes.

4. The gas is suitable for desulfurization.

It has the following disadvantages:

1. The plant is relatively complicated and therefore expensive.

2. Removal of tar and oil and cooling to ambient temperature lower the thermal yield and efficiency of the gas plant. The cold-gas efficiency is typically 75 to 78 percent, depending on the quality of the coal.

3. Liquid by-products including contaminated wastewater require disposal. The tar and oil products are, however, useful fuels in their own right and can be turned to advantage if they are stored to cover periods of gas-plant outage.

Cyclic MBG

See Fig. 9-6 and configuration 4 in Table 9-1. Plants designed in this configuration are referred to briefly as "cyclic plants" and are selected when MBG is desired and when oxygen is not available or is uneconomic. The operation occurs as a succession of cycles made up (typically) of three phases, two of which generate useful gas. A gas holder is therefore inserted between the gas plant and the users to act as a buffer and provide a continuous flow to the users.

The presence of this gas holder mandates that the gas must be cooled and cleaned in equipment similar to that used in configuration 3.

The phases of a three-phase cycle are known as the "blow," "up-run" or "integral-gas," and "back-run" phases. The cyclic operation is achieved by hydraulically operated valves controlled automatically in the desired sequence and periods.

Blow phase   The air blast passes from the grate up through the coke layers to make blow gas, which passes out of the clear-gas outlet through the regenerative steam superheater, combustion chamber, and waste-heat boiler to the stack. Blow gas is a weak producer gas, and air is added to it at the steam-superheater and combustion-chamber inlets in order to burn it and maintain the desired temperature conditions.

During this blow phase, gas does not pass up through the distillation retort, but the coke bed in the gasification section is raised in temperature.

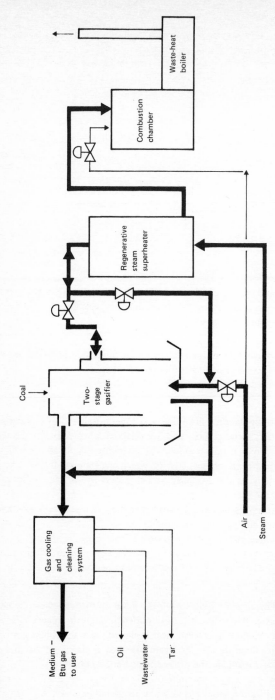

FIG. 9-6 Cyclic operation.

3-178

Up-run phase This phase is a gas-making phase and takes place from bottom to top. Steam is admitted to the bottom of the regenerative steam superheater and leaves from the top at about 600 °C. It then passes through the gasifier grate and up through the coke bed to form water gas. This gas at about 600 °C passes up through the distillation retort and releases the volatile matter of the coal in the usual way of a two-stage gasifier. The mixture of water gas, steam, and coal distillation products is known as *integral gas* and passes from the top of the gasifier to a gas-cooling and -cleaning system.

Back-run phase This phase is also a gas-making phase during which steam from the superheater passes downward through the coke bed in the gasification section and out through the grate. The back-run water gas then passes to the gas-cooling and -cleaning system.

Purges Between the main phases just outlined there are brief steam purges to ensure that potentially explosive situations do not develop in the plant.

The mixture of integral gas from the up-run phase and water gas from the back-run phase, after cooling and cleaning, has a higher heating value of typically 330 Btu/stdft^3 [13.2 MJ/m^3(n)].

The advantage of a cyclic MBG plant is that the process makes MBG without using oxygen.

The disadvantages are:

1. Cold-gas efficiency is rather low—typically 63 percent—compared with continuous air-blown LBG plants—but to evaluate a system properly it is also necessary to consider the utilization efficiencies of MBG and LBG.

2. The plant is more complicated than a continuous air-blown LBG plant.

3. By-products and effluents are similar to those from other cold clean-gas plants. In addition the products of combustion of the blow gas are discharged via the waste-heat boiler stack.

4. The plant requires a small gas holder.

Oxygen-blown MBG

See configuration 5 in Table 9-1. This configuration is considered where MBG is desired and oxygen usage is acceptable.

The block flow diagram for this configuration is nearly the same as Fig. 9-4 for the cold clean-gas plant. Descriptions of the two configurations are also much the same. Slight differences occur because of the extra steam which has to accompany the oxygen-enriched blast. The large excess of steam present in the gas mandates that it be cooled and cleaned completely. The result is a gas with higher heating value (HHV) of 280 Btu/stdft^3 [11.2 MJ/m^3(n)].

The advantages of this type of plant are:

1. The process makes an MBG.

2. The plant is simpler than a cyclic plant.

3. The capacity of gasifier is greater than when a plant is operated cyclically.

The plant's disadvantages are:

1. It requires oxygen.
2. It requires steam to be supplied from outside.
3. The gas heating value attained is lower than that attained by cyclic operation.
4. It produces a large volume of aqueous effluent.

Desulfurized gas

See configurations 6, 7, and 8 in Table 9-1. In the configurations described so far, typically half or more of the total sulfur in the coal will appear in the product, largely as H_2S.

Where a desulfurized fuel gas is desired, one of the cold, clean-gas configurations (3, 4, or 5) is chosen and is followed by a desulfurization unit, typically a Stretford process unit (see Chap. 3-13, Fig. 13-6).

PROCESS PERSPECTIVE

The GI (Gas Integrale) and similar two-stage producers were well established by the 1950s and represented the culmination of about 100 years of efforts to improve producer-gas practice and make it attractive for use with high-volatile, bituminous coals.

History of Gas-Producer Development

This is stated briefly in the following table.

Date	Inventor	Development
1839	Bischoff	The first fixed-grate, updraft shaft producer was introduced.
1852	Kirkham	Cyclic operation to make water gas was invented.
1857	Siemens	A built-in type, cross-draft producer for use with intermittent vertical-chamber carbonizing plants (IVCs) was designed and built.
1867		The first gas engines were put into operation, which caused a big increase in the demand for gas producers.
1875		This was the beginning of the industrial application of the cyclic operation to make water gas.
1883	Mond	The introduction of high-steam-content blast facilitated recovery of by-product ammonia, which was of considerable importance before the production of synthetic ammonia.
1895		Suction producers were introduced to eliminate the gas holder between the producer and engine.

Date	Inventor	Development
1895	Prof. Strach, Vienna	Strache invented a two-vessel gasifier for cyclic water-gas production. One vessel was superimposed on the other, whereas other designs had two vessels side by side. Strache's first gasifier was at Pettau (Austria) and made town gas from lignite (1895). Strache's invention was exploited initially by the VIAG organization and later by the GI group of companies.
1900		By 1900 two-thirds of United States gas production was carburetted water gas.
1900–1910		This was a period of great progress in gas producers, reaching as far as "the mechanical producer" with rotating grates and water jackets.
1920		The mechanical producer in its essentials had already reached its present form and was subsequently improved only in its details.
1920s		In Europe, the production of water gas from coke expanded strongly at a pace determined by the coke market. The synthesis of ammonia and methanol from water gas grew strongly, especially in the United Kingdom—a trend which continued through to 1947. A similar trend was seen in the United States until the discovery of natural gas.
1930–1940		Two-stage gasifiers appeared in 1930, and there was much gasifier development activity in central Europe during this decade.
1950s		By the 1950s there were numerous models of basically the same design of two-stage producer, with the majority of gas-plant builders offering them. A list of names of the period would include Il Gas Integrale (now IGI) Milan, Le Gaz Integral (Paris), IBEG (now Integral) Vienna, Heurtey, Stein-Motala, Koppers, Lurgi, Power-Gas, Otto, Kollergas, Pintsch-Bamag, Tully, Wellman, Stoic, & I.F.E. Many of the names listed above were licensees of one of the few principal independent designs. The GI design for example was licensed to companies in Germany, the United Kingdom, Japan, and South Africa.
1965 onward		The 20 years following World War II saw oil and natural gas displace coal from many of its former uses. Coal ceased to be the basis of manufacturing town gas, synthesis gas, industrial fuel gas, etc., so that in most parts of the world except South Africa new two-stage gasifier plants were no longer installed and existing ones were closed. In South Africa existing plants were retained and new ones built.

Date	Inventor	Development
1970s		The altered price and availability relationships between oil and coal beginning in 1973 increased general interest in all aspects of coal utilization including two-stage gas producers. However, outside South Africa very little plant building has resulted. One of the reasons is the requirement of two-stage gasifiers for a closely specified lump coal. Suitable coals are not universally available and therefore can command a premium price. Even in South Africa where there is an unbroken history of applying two-stage gasification, there is now strong pressure to move toward technologies which could feed on the more available, small-size, lower-quality coals.

Commercial Plants

Over 100 commercial GI-design plants have been installed in Europe, South Africa, and Australia.

The plants have operated in the continuous air-blown, oxygen-enriched, and cyclic configurations for the production of low-Btu and medium-Btu fuel gases.

Plants vary in size from the smallest using one 4-ft- (1.2-m-) diameter gasifier consuming 11 tons/day (10 Mg/day) of coal and generating 200 million Btu/day (210 GJ/day) to large installations comprising multiple 12-ft- (3.6-m-) diameter gasifiers consuming together several hundred tons per day of coal (see Table 9-2).

Availability of Technology and Information

Two-stage gasification of coal is a proven, mature technology available from many firms.

If the coal available to a project is the same as, or fairly similar to, one previously gasified, plants can be designed, their performance estimated, and full commercial guarantees offered.

In all cases laboratory coal tests are performed to determine the coal properties. This

TABLE 9-2 Capacities of standard-size gasifiers producing cold and clean low-Btu gas

Nominal Diameter		Gas Output		Coal Consumption	
ft	m	10^9 Btu/d	GJ/d	t/d	Mg/d
12	3.6	1.9	2000	106	96
10	3.0	1.4	1400	74	67
8½	2.6	1.1	1200	61	55
6½	2.0	0.6	640	33	30
4½	1.4	0.3	320	16.5	15
4	1.2	0.2	210	11	10

information is then used to firm up the plant design and guaranteed performance data. Coal testing is discussed in more detail later under "Specifications for Gasification Coals."

In the case of unfamiliar and/or unusual coals more extensive testing may be undertaken either in a pilot-scale gasifier or in an industrial gasifier.

Operability

The low-pressure, fixed-bed features provide good operability.

Start-up

It is an advantage to have a supply of coke available for start-up because the lower section of a gasifier normally operates on coke. If coal is used in the lower half of a cold plant at start-up, it is possible that tar could find its way to places where it could cause trouble.

If the refractories of a plant that is being started have already been dried, one-half to one day is required to reach a reasonable output. Full output and normal gas quality are reached after about two days.

Standby

A gasifier can be left on hot standby or "banked" and depending on its size will require no attention for ½ to 2 days. After this period the fire temperature can be restored by a short air blast once or twice a day. During a banked period it can be an advantage to keep the retort section supplied with coke rather than coal in order to minimize the nuisance from gases that will be vented during the air blows.

After a period on standby normal operation can be regained within a couple of hours. Thus 5 days of operation per week is quite usual for this type of gas plant.

Flexibility

The response to increased demand for gas is immediate and automatic, and gas quality does not change significantly. The normal operating range is from 25 to over 100 percent of the design capacity.

GASIFIER-PRODUCT PROPERTIES AND UTILIZATION

Typical compositions and properties of product gases, by-product tars, and by-product oils are given in Tables 9-3, 9-4, and 9-5 respectively.

Gas Utilization

LBG has mainly been used for industrial heating, and MBG from cyclic plants has mainly been used in town-gas production works. Table 9-6 lists typical applications and the type of gas used.

Utilization Efficiency

The potential heat in a fuel that is recovered by a consuming system depends on several factors. Dominant among these are the temperature at which the combustion products leave the system and the heating value of the fuel.

TABLE 9-3 Product gases—typical compositions and properties

	Continuous Operation		Cyclic Operation
	Air blast	Oxygen and steam blast	Air and steam blast
Composition, vol %			
CO_2	4.4	16–18	6–10
CO	27.2	40–41	28–32
H_2	15.4	38–40	48–54
CH_4	3.0	1–2	6–8
C_nH_m	0.4	0.2–0.4	0.5–1
C_nH_{2n+2}	——	0.4–0.6	
O_2	——	0.0–0.2	0–0.3
N_2	49.6	1–2	2–8
HHV*			
Btu/scf	170†	283	315–335
$MJ/m^3(n)$	6.8	11.3	12.5–13.4
LHV*			
Btu/scf	162	260	288–304
$MJ/m^3(n)$	6.45	10.4	11.5–12.1

*Values are for cold clean gas.

†For hot raw gas, including sensible heat and tar and oil, the HHV = 200 Btu/stdft^3 [8 MJ/m^3(n)] and for hot detarred gas, including sensible heat and oil, the HHV = 182 Btu/stdft^3 [7.3 MJ/m^3(n)].

TABLE 9-4 By-product-tar properties*

Physical properties	
API	3.25°
Specific gravity (15 °C)	1.05
Viscosity (50 °C)	10 Engler
	94 cP
Distillation fraction	
230 °C	2%†
270 °C	16.5%†
330 °C	38%†
370 °C	53.5%†
Pitch	46.5%
Composition	
Phenols—fraction to 270 °C	70%
Paraffins	18.5%
Pyridine	1.1%
Naphthalene	0.14%
Ash and insolubles	1.1%
Water	2.0%
Higher heating value	16,000 Btu/lb (37 MJ/kg)

*For storage and handling, the tar produced from a two-stage gasifier is similar to bunker C oil.

†Percentage is a cumulative value.

TABLE 9-5 By-product-oil properties

Physical properties	
API specific gravity	17.5°
Specific gravity (15 °C)	0.95
Viscosity (20 °C)	3° Engler
	20 cP
Distillation fraction	
95 °C	Initial boiling point
198 °C	20%*
210 °C	40%*
220 °C	60%*
250 °C	80%*
288 °C	98%*
Phenols	24%
Higher heating value	16,700 Btu/lb (39 MJ/kg)

*Percentage is a cumulative value.

TABLE 9-6 Low-Btu gas applications

Industry	Gas type used*	Remarks
Iron and steel		
Open-hearth regenerative furnaces	A, B	
Soaking pits	A, B	With air and/or gas preheating
Batch and continuous reheating furnaces	A, B	With regeneration. Richer gas with less sulfur preferred
Heat-treatment furnaces	A, C	Richer, cleaner gas preferred
Glass		
Melting in continuous regenerative tank furnaces	A	May only need regeneration on air
Forehearths	C	
Annealing lehrs	C	
Optical glass in pot furnaces	B	
Ceramics (refractories, bricks, etc.)		
Magnesite refractories	C	Used in tunnel kilns
Bricks, tiles, pipes, pottery	A, B, C	
Other industries†		
Lead & copper smelting		
Tin & chrome core reduction,		
Iron ore sintering & pelletizing		

*A = hot raw gas, B = hot detarred gas, C = cold clean gas.

†Low-Btu gas from two-stage gasifiers used.

For ordinary boilers with quite low stack-gas temperatures, efficiencies using the usual fuels exceed 80 percent. For gas fuels the curve of efficiency vs. higher heating value of the fuel shows a maximum of about 87 to 88 percent for gas heating values around 250 to 450 Btu/stdft^3 [10 to 18 MJ/m^3(n)]. When the gas heating value drops to 150 Btu/stdft^3 [6 MJ/m^3(n)], the boiler efficiency will be only 83 to 84 percent.

Besides loss of efficiency the use of LBG generally requires larger equipment than MBG or oil to achieve the same duty.

The efficiency obtained with gases much richer than MBG, such as natural gas with a higher heating value of 1000 Btu/stdft^3 [40 MJ/m^3(n)], is almost 85 percent, better than LBG but not quite as high as MBG.

The differences between different fuels become much more pronounced as the user efficiency decreases as a result of high flue-gas exit temperatures. The result is high-temperature processing without waste-heat recovery. So for a furnace with a stack-gas temperature of 1400 °C the efficiencies on LBG, MBG, and natural gas are 17, 33, and 27 percent respectively.

To compensate for this effect and to provide higher flame temperatures most applications of LBG use air and/or gas regenerators and recuperators which preheat the fuel gas and/or combustion air by extracting heat from the furnace waste gas.

The choice between LBG with its high efficiency of production and low capital cost of production and its low efficiency of utilization and high cost of utilization and MBG with its low cost of utilization is not easy and requires study of each individual case.

Because modern industrial processes have been developed on the basis of rich fuels such as oil, natural gas, and LPG, the usual outcome of studies is to favor MBG. In many cases MBG can be substituted for previous fuels with negligible alteration to the consuming process, whereas a change to LBG probably requires major alterations and interruption of production.

Cleaned LBG can be used in reciprocating gas engines and in gas turbines. It can be used in large diesel engines provided some oil is injected to act as an igniter for the gas. The engine rating will be less than on oil, and the derating increases as the hydrogen content of the gas increases.

Provided fuel cleanliness requirements are met and the appropriate combustion chamber design is used, LBG or MBG from gas producers appears to be a reasonable fuel for gas turbines and combined-cycle power-generation schemes.

Hot raw gas, containing all the tar distilled from the coal, burns with a highly luminous flame and has been found very satisfactory in open-hearth steel furnaces.

Visibility of the flame is important for example in glass-melting furnaces, where the burners have to be aimed to place the flame where good performance is obtained without damage to the furnace roof.

SPECIFICATIONS FOR GASIFICATION COALS

A range of coals is suitable for gasification; however, there are certain restrictions if optimum performance is to be realized. In general, the swelling index of the coal should be in the range of 1 to 2½ (in the case of coke and lignites it can also be 0). Agglomerating properties and particle size (*granulometry*) are also important. Normally, tests are made on representative samples to establish the suitability of the coal and to assess the expected performance of the plant. It is preferable to exclude coals which are badly sized, dusty or too high in fines, too wet, too friable, or decrepitating, or which produce ashes that are too fusible. Quantitative specification of limiting values for these properties is not realizable, since the performance of the plant depends on all of them. Good values for some properties can compensate for unfavorable values of others. Where laboratory tests show that the coal's properties are significantly different from familiar coals, a trial would be arranged for the candidate coal in an industrial-scale or semi-industrial-scale gasifier. From the results of such trials the plant performance could be estimated with good confidence.

Swelling Index

A swelling index determined in a laboratory according to a national standard procedure (e.g., BS 1016, Part 12, 1980) on a crushed sample is not a wholly reliable guide to the behavior of a coal in a gasifier. However, it is a necessary first step in characterizing a gasifier coal. Other small-scale heating tests have been tried but have not become established.

The British standard (BS) swelling test referred to or its equivalent in other national or international standards is one of the tests used for the evaluation of the caking power of coal.

One gram of coal is heated in a special crucible to 1000 °C under standard conditions, and by comparing the profile of the coke obtained with a series of reference profiles, a "swelling index" can be defined. The test is also described as a "free swelling test" because the only resistance which the swelling meets is that offered by the walls of the crucible and, sometimes, as an "agglomerating test" because it indicates the tendency of the particles to fuse together and form a compact mass and allows a clear limit to be drawn between caking and noncaking coals.

Coal-Particle Sizing

Coal for two-stage gasifiers should be double-screened, and preferably the undersize screen aperture should be not less than half the oversize screen aperture. Thus, typical preferred coal sizes are described as follows:

Small nuts	1½–¾ in (20–40 mm)
Large nuts	2½–1¼ in (30–60 mm)
Cobbles	3–1½ in (40–80 mm)

Less closely graded coal can be used at the expense of reduced plant capacity. Fines (adhering to the sized particles) should not exceed 10 percent.

It may be uneconomic to provide such closely sized coal. As an extreme example, if a coal sized ¼ to 1½ in (6 to 40 mm) is considered for gasification, it is necessary to considerably down-rate the gasifier capacities, compared with the typical figures in Table 9-2. Having a low undersize inevitably means an increased proportion of fines—which aggravates the bad effects of the poor sizing.

The adverse effect of poor coal sizing is worse for the cyclic plant than for other configurations so that sizes below a ½ in (15 mm) are not normally considered for cyclic plants.

Ash Fusion Properties

Steam added to the blast to the gasifier has two important effects. It moderates the temperatures in the combustion zone and hence controls the degree of clinkering. It realizes the temperature-moderating effect by participating in the endothermic carbon-steam reaction, which has the beneficial side effect of generating CO and H_2 and thus improving gas quality and cold-gas efficiency.

The hemisphere temperature and the reducing atmosphere in the BS 1016, Part 15, Ash Fusion Test should be above 1200 °C and preferably should be above 1250 °C. If the ash fusion temperature is too low, excessive steam will have to be used to control clinkering, so that efficiency and gas quality will be decreased and carbon-in-ash will be increased and excessive quantities of wastewater will have to be dealt with in cold clean-gas plants.

Coal Heating Value, Moisture Content, and Ash Content

While these properties do affect performance, no limits are defined. Moisture and ash contents that are too high or coal heating values that are low naturally degrade the gasifier performance, and as a rough guide the gasifier capacity figures in Table 9-4 are reduced in proportion to the coal heating value for values below 11,700 Btu/lb (27.2 MJ/kg).

Fischer-Schrader Assay

The properties or tests already discussed mainly concern the general suitability of the coal for gasification and the capacity of a gasifier. The designer requires more detailed information in order to calculate the product-gas composition, precise heat, material balances, etc.

Central to this area of coal evaluation is the Fischer-Schrader assay, which is carried out in accordance with the description given by W. A. Selvig and W. H. Ode (*Low Temperature Carbonization Assays of North American Coals,* U.S. Bureau of Mines Bulletin 571, 1975).

In the standard method a 50-g charge of coal is heated in an aluminum retort to 500 °C over a period of 45 min and held at 500 °C for a further 15 min. Semicoke, tar, water, light oil, and gas are collected from the low-temperature carbonization, and the yield of each is measured. The phenols in the tar are determined. This test thus provides information useful in predicting the products of the upper-distillation retort section of a two-stage gasifier.

ENVIRONMENTAL, HEALTH, AND SAFETY ASPECTS

Emissions

Emissions to the air

Coal-feed lock-hopper operation normally results in the release of minor quantities of gas to the atmosphere. The plant layout is arranged so that operating areas are well ventilated to provide a safe working environment.

In the case of cyclic plants the products of combustion of the blow gas that contain some sulfur dioxide are discharged from the waste-heat boiler stack.

Wastewater

In configurations where the gas is cooled, the steam that it contains is condensed. The resulting wastewater stream contains phenolic compounds typically at a level of 5 to 15 g/L. Before discharge to public sewers the phenols should be reduced to say 5 mg/L. In addition to the phenols the wastewater contains ammonia (typically 7 g/L) and ammonium sulfide.

In a typical plant about 50 gal of wastewater is produced per ton of coal (0.2 m³/Mg). The hot raw-gas and hot detarred-gas configurations do not of course yield a wastewater stream.

The gasifier "blast" has steam added to it as a means of moderating the temperatures and clinkering tendencies in the combustion zone of the gasifier. The possibility exists therefore of using the phenolic wastewater as the source of this steam, and plants have operated with blast saturators fed with wastewater.

Other Aspects

Carbon monoxide

Carbon monoxide is a major component of the gases produced. Carbon monoxide is a colorless, tasteless, odorless, and very poisonous gas.

Provided the necessary precautions are taken—especially to provide good ventilation and to maintain the plant leak-free—operations are safe and healthy.

The same precautions should obviously extend to the gas-distribution system and the gas users' premises.

Gasifier ash

Ash from gas producers is lightly clinkered, and there is usually a demand for it as a filler for roads, playing fields, tracks, etc.

PROCESS ECONOMICS

Gas production costs

Capital and operating costs and gas production cost calculations for a typical hot raw-gas plant and a typical desulfurized cold clean-gas plant are given in Tables 9-7 and 9-8, respectively. For both cases the plant uses a single 12-ft (3.6-m) nominal diameter WD-IGI two-stage gasifier and the design coal is an English high-volatile, very weakly caking bituminous of 2 to 1-in size grading (Dawmill Washed Doubles). A Stretford unit desulfurizes the cold cleaned gas in the second case. Capital costs were valid in mid-1980.

Costs for configurations such as hot detarred gas and cold, clean but not desulfurized gas are intermediate between the cases tabulated. The production of MBG in a cyclic plant is more costly than LBG. The economics of oxygen-blown MBG can be competitive with LBG in cases where oxygen can be supplied at low cost. Brief economic data for cyclic operation are provided under "Summary."

TABLE 9-7 Production cost estimate for hot raw gas from a 12-ft (3.6-m) WD-IGI two-stage gasifier

Plant capacity: 2.33×10^9 Btu/d (2450 GJ/d) of hot raw gas
Battery limits capital cost (BLCC): $2,061,000 (mid-1980)
Load factor: design capacity operation for 7650 h/yr
Annual gas production: 743 billion Btu (784 TJ)

	Annual consumption	**Price**	**Annual cost**
Coal	32,600 t	$73.5/t	$2,397,000
Power	1033 MWh	$51/MWh	52,700
Process water	2.8×10^6 gal	$0.60/1000 gal	1,600
Operating costs			
Labor + supervision			169,000
Maintenance: 4% of BLCC			83,000
Subtotal (operating costs)			252,000
Direct overhead: 15% of labor + supervision			25,400
General plant overhead: 60% of operating costs			151,200
Insurance: 1% of BLCC			20,700
Depreciation: 20% of BLCC			412,200
Interest on capital: 15% of BLCC			309,200
Total cost of production			$3,622,000

Unit cost of gas production: $4.87/$10^6$ Btu ($4.62/GJ)

Labor requirements

Two workers are usually required to run a gasifier. The senior supervises the control panel on the charging-level platform and ensures that the bunkers are supplied with coal.

The junior on the poke-hole platform checks and controls the thickness of ash on the grate and the height of the fire with the aid of a steel sounding rod introduced down into the fuel bed in the lower half of the gasifier. He or she may also poke the fire if necessary.

The extent of manual intervention in the running of the gasifier is affected by the coal properties, the plant configuration, and the rate at which the plant is driven.

Part-time supervision of the operator and his or her assistant would be additional.

For multiple-gasifier plants, more economical use can be made of labor. As an example, the estimated requirements for an oxygen-blown MBG plant comprising an oxygen plant and eight gasifiers consuming about 1100 tons/day (1000 mg/day) of coal are as follows:

	Daywork	**Shift work**
Plant manager	1	
Shift superintendent		1
Assistant		2
Operator		3
General labor	3	

TABLE 9-8 Production cost estimate for cold clean gas from a 12-ft (3.6-m) WD-IGI two-stage gasifier

Plant capacity: 1.72×10^9 Btu/d (1810 GJ/d) of cold clean gas
Battery limits capital cost (BLCC): $3,461,000 (mid-1980)
Load factor: design capaicty operation for 7650 h/yr
Annual gas production: 548 billion Btu (578 TJ)

	Annual consumption	Price	Annual cost
Coal	32,600 t	$73.5/t	$2,397,000
Chemicals			19,400
Power	1481 MWh	$51/MWh	75,800
Cooling water	100×10^6 gal	$45/$10^6$ gal	4,500
Process water	6.7×10^6 gal	$0.60/1000 gal	4,000
Operating costs			
Labor + supervision			219,600
Maintenance: 4% of BLCC			138,400
Subtotal (operating costs)			358,000
Direct overhead: 15% of labor + supervision			32,900
General plant overhead: 60% of operating costs			214,900
Insurance: 1% of BLCC			34,600
Depreciation: 20% of BLCC			692,300
Interest on capital: 15% of BLCC			519,300
Total gross cost of production			4,352,700
Credits for by-product steam, tar, & oil			−976,000
Total net cost of production			$3,376,700

Unit cost of gas production: $6.16/$10^6$ Btu ($5.84/GJ)

SUMMARY

The WD-IGI two-stage gas producer is a mature technology which has been applied commercially worldwide. It was formerly especially popular in Europe, and currently is being used in South Africa.

The gasifier can be used in several process configurations to give LBG or MBG. The processes are used to make fuel gas for industrial-process heating and to make town gas.

Plant capacities

On the basis of their size and other factors, individual gasifiers consume from 10 to 100 tons per day of coal. Plants containing multiple units gasify from tens and hundreds of tons of coal per day up to a thousand tons per day, and in this range of coal utilization will usually produce a gaseous fuel at a lower cost than any other process, assuming there is a ready supply of suitable coal. The coal should be weakly caking or very weakly caking and should be closely size-graded, nut and cobble sizes being preferred. For less than optimum coals, operation may still be feasible but the gas output from a gasifier of any given size is decreased.

Efficiencies

The thermal efficiencies of two-stage producer-gas plants are high, varying from better than 90 percent when delivering hot raw LBG to about 60 percent when delivering cold, clean MBG from a cyclic plant.

Performance and economic data

Summaries for two common configurations, a cyclic plant producing cold clean gas and a continuous air-blown plant delivering hot raw gas, are given below.

Cyclic plants are more demanding of coal quality, and to achieve the results quoted the coal should at least conform to the following specification:

Coal HHV	>11,700 Btu/lb (27.2 MJ/kg)
Swelling index	<2
Ash fusion temperature	>1200 °C
Size	1–2 in (25–50 mm)

From coal of this quality a cyclic plant would produce cold clean gas with HHV of 325 Btu/stdft^3 [13 MJ/m^3(n)]. The inputs required per million Btu or per gigajoule in the higher heating value of product gas are as follows:

	Requirement	
Input	Per 10^6 Btu	Per GJ
Coal	132 lb	57 kg
Boiler feedwater	25 US gal	90 L
Cooling water	180 US gal	640 L
Power*	2.5 kWh	2.4 kWh

*Assuming motor-driven air blowers.

Outputs beside the gas are as follows:

	Output	
Substance	Per 10^6 Btu, lb	Per GJ, kg
Tar and oil	16	7
Export steam	44	19
Ash	16	7

The capital cost of a plant producing 800 million Btu/day (840 GJ/day) would be about $2.7 million. This figure excludes the coal stockyard, ash handling, and a gas holder. Requirements for these all depend strongly on the circumstances of the individual case.

A continuous air-blown plant delivering hot raw gas can perform satisfactorily on somewhat poorer coals than the cyclic plants, but the following data are based on a coal having the following properties:

Coal HHV	12,490 Btu/lb (29 MJ/kg)
Swelling index	½
Ash fusion temperature	1250 °C
Size "Doubles"	1–2 in (25–50 mm)

From coal of this quality a continuous air-blown plant would produce hot raw gas with a higher heating value (including the heating value of the tar and oil vapors and the sensible heats of all components) of 186 Btu/stdft³ [7.4 MJ/m³(n)].

The inputs required per million Btu or per gigajoule in the higher heating value of product gas are as follows:

| Input | Requirement | |
	Per 10^6 Btu	Per GJ
Coal	88 lb	38 kg
Boiler feedwater	4 gal	14 L
Cooling water	Negligible	Negligible
Power	1.4 kWh	1.3 kWh

Outputs beside the gas are as follows:

| Substance | Output | |
	Per 10^6 Btu, lb	Per GJ, kg
Export steam	32	14
Ash	5.5	2.4

The capital cost of a plant producing 2.3 billion Btu/day (2450 GJ/day) of hot raw gas would be about $2.3 million, with the same exclusions as previously mentioned for the capital cost of a cyclic plant.

THE SAARBERG/OTTO
COAL-GASIFICATION
PROCESS

DR. RER. NAT. REINER MUELLER
DIPL. MATH. HARTMUT PITZ

Saarberg + Dr. C. Otto
Gesellschaft für Kohledruckvergasung mbH
Saarbruecken, Federal Republic of Germany

GENERAL PROCESS DESCRIPTION

The Saarberg/Otto process is based on the principle of entrained-bed gasification at high pressure and at high temperatures.

Figure 10-1 shows a block flow diagram of the Saarberg/Otto process. The feed coal is ground and dried and then conveyed into the gasifier by a pneumatic coal-feeding system. The gasifier is a vertical cylindrical pressure vessel consisting of three stages:

Stage I: Primary gasification zone

Stage II: Post-gasification zone

Stage III: Cooling zone

Stages I and II are protected on the inside by water-cooled finned tubes; stage III is refractory-lined.

In the primary gasification zone (stage I), the feedstock reacts with the gasification media—oxygen and steam or oxygen-enriched air mixtures or air—at temperatures between 1650 and 2400 °C. As a result of the high temperature, liquid slag collects in the lower part of the pressure vessel and serves several purposes:

- The molten slag acts as a heat shield, which allows high flame temperatures; thus high energy densities and high conversion rates are obtained.

- The molten slag ensures safe ignition and flame stability.

- The molten slag permits the gasification of relatively large particles; thus the costs for coal preparation are reduced.

- The molten slag permits flexible operation with regard to throughput and feedstock.

Surplus slag flows through a central overflow, is granulated in a water vessel underneath the gasifier, and is then discharged via a lock-hopper system.

Stage II serves as a post-reaction zone.

In stage III the product gas is cooled to approximately 850 °C by means of cold recycle gas and leaves the reactor at this temperature.

A cyclone removes the majority of the entrained solids from the gas. The gas then passes to a waste-heat boiler, where the sensible heat of the raw gas is used to generate superheated high-pressure steam. Part of this superheated steam serves as process steam. After the fines have been removed, the gas is cooled in a spray cooler and then passed to a desulfurization unit and to a unit for further gas treatment.

The cooling water of the spray cooler is circulated via a heat exchanger. Part of this water is blown down and treated in a conventional wastewater system to meet permissible emission standards.

Caking properties, grain size, and ash melting characteristics of the feedstock do not affect this process. All types of coal having an ash content up to 40 percent by weight and residues from coal-liquefaction plants and refineries can be gasified.

The high pressure and high operating temperatures used in this gasification process accelerate the reaction and enable the plant to be sized favorably.

The gas output per gasifier increases in proportion to the absolute operating pressure. Moreover most downstream users of coal-gasification gas operate at high pressure, so

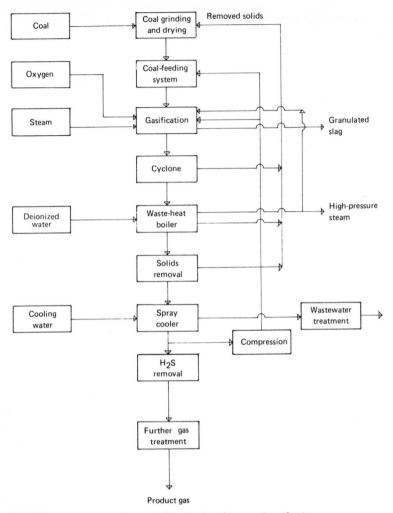

FIG. 10-1 Block flow diagram of the Saarberg/Otto coal-gasification process.

compression costs are eliminated or minimized by operating the gasifier at high pressure. The high gasification temperature assures the nearly complete conversion of the carbon feed and the production of a gas that is free of oils, tars, and phenols.

The sulfur in the gas can be removed by conventional desulfurization processes. The inert granulated slag can be dumped or used in road construction. The amount of wastewater is small relative to the amount of coal converted. Impurities can be removed by conventional methods.

In summary, the Saarberg/Otto process is characterized by a high carbon-conversion rate, a high output of specific gas, and an operation which has only a mild environmental impact.

PROCESS CHEMISTRY AND THERMODYNAMICS

The heat necessary for the gasification process is produced by burning a portion of the coal feedstock.

$$C + O_2 \rightarrow CO_2 \qquad \Delta H = -406.0 \text{ kJ/mol} \qquad (1)$$

$$C + \tfrac{1}{2}O_2 \rightarrow CO \qquad \Delta H = -123.1 \text{ kJ/mol} \qquad (2)$$

The chemical reactions during coal gasification can be described by the following equations:

$$C + H_2O \rightarrow CO + H_2 \qquad \Delta H = +118.5 \text{ kJ/mol} \qquad (3)$$

$$CO + H_2O \rightarrow CO_2 + H_2 \qquad \Delta H = -42.3 \text{ kJ/mol} \qquad (4)$$

$$CO_2 + C \rightarrow 2CO \qquad \Delta H = +160.2 \text{ kJ/mol} \qquad (5)$$

The following reaction also takes place, but because of the high gasification temperature of the Saarberg/Otto process it is of less importance.

$$C + 2H_2 \rightarrow CH_4 \qquad \Delta H = -87.5 \text{ kJ/mol} \qquad (6)$$

These reactions appear to be quite simple but are actually quite complex, because they are mixed-phase reactions between a solid and a gas. Nevertheless, the large number of possible reactions can be represented by Eqs. (1) to (6).

High temperatures favor the production of carbon monoxide over carbon dioxide. Above 820 °C the formation of carbon monoxide is predominant. At the high temperatures in the gasification zone of the Saarberg/Otto gasifier nearly all hydrocarbons are converted to carbon monoxide, hydrogen, and carbon dioxide.

The effects of pressure on the coal gasification process are well known. Reaction (3) is retarded by pressure. Reaction (4) is not affected, and reaction (6) is accelerated by pressure gasification. Thus when a bituminous coal is gasified, it seems impossible to cause an essential increase in the reaction velocity and throughput by raising the pressure to high values. Therefore high pressure has more technical than chemical or thermodynamic significance.

DEMONSTRATION-PLANT DESCRIPTION AND DATA

A pilot plant designed for a coal throughput of 291 tons/day (264 Mg/day), an operation pressure of 362 lb/in² (2.5 MPa), and a gas production of 20MM stdft³/day (528,000 m³ (n)*/day) has been in operation since December 5, 1979. Figure 10-2 shows a view of this pilot plant, which is in Voelklingen/Fuerstenhausen, FRG.

Figure 10-3 shows the process arrangement of the pilot plant. From the storage yard the coal is delivered to a primary screen for separating foreign matter, and then a column-flow conveyer transports it to a grinding and drying facility. A hammer mill grinds the coal to a grain size of less than 0.12 in. By means of hot flue gases the coal is dried to a moisture content of less than 2 percent (lignite to less than 12 percent).

---

*m³(n) symbolizes the normal cubic meter, that is, the volume of one cubic meter of gas measured at 0 °C and 1.01 kPa pressure.

FIG. 10-2 View of the Saarberg/Otto coal-gasification demonstration plant in Voelklingen/Fuerstenhausen. Right: coal-grinding and coal-drying facility followed on the left by the coal-feeding system; the storage bin is above the housing of the coal-feeding system. Middle: in the foreground is the control center and in the background (hidden by pipes) are the gasifier, cyclone, waste-heat boiler, and fibrous filters. Left (of control center): effluent treatment installations.

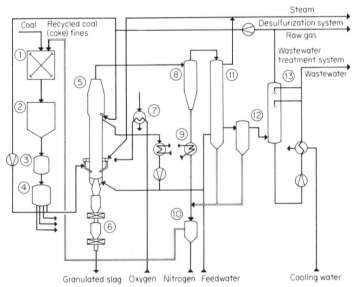

FIG. 10-3 Process flow diagram of the Saarberg/Otto coal-gasification demonstration plant in Voelklingen/Fuerstenhausen. Key: 1, grinding and drying facilities; 2, storage bin; 3, lock-hopper; 4, coal-feed tank; 5, gasifier; 6, slag-discharge lock-hopper; 7, oxygen preheater; 8, cyclone; 9, cooler for char and slag particulates; 10, hopper for recycled fines; 11, waste-heat boiler; 12, hot-gas filter; 13, spray cooler.

From the grinding and drying facility the dry, pulverized coal is pneumatically conveyed (by the flue gas) to a storage bin. This storage bin is under nitrogen at atmospheric pressure and has a capacity of 165.3 tons (150 Mg) of ground coal.

From the storage bin the coal dust is passed by gravity into a lock-hopper, where it is pressurized and then fed to a feed tank. The coal-feeding system continuously supplies controlled quantities of dry coal dust to the gasifier via four feed pipes. Recycled product gas is used as the carrier.

The gasifier that is installed in the demonstration plant is a vertical cylindrical pressure vessel 52.5 ft (16 m) high, with a 4.9-ft (1.5-m) ID. The gasification zone (stage I) and the post-gasification zone (stage II) of the gasifier are protected on the inside by water-cooled finned tubes.

The forced-circulation cooling system of the combined boiling and condensation type is a new development. A heat exchanger is integrated into this cooling system and converts the heat from the reactor into low-pressure steam, which is used for various purposes in the process.

A comprehensive measuring and control system has been installed to protect the reactor and cooling cage.

The cooling zone (stage III) which is on top of the gasifier is refractory-lined. In this zone the gas is cooled to about 850 °C by means of cold, recycled product gas.

Coal, oxygen, and steam are fed into the gasifier through four two-channel, tangential nozzles that are slanted downward. The oxygen is preheated in an oxygen heater by saturated steam from the waste-heat system. The process steam is superheated steam from the high-pressure steam system. Oxygen and steam are mixed externally in a ring main before they flow through the annular portion of the nozzle to the reaction zone. Coal and carrier gas are fed through the center portion of the nozzles. The purpose of using recycled gas as the conveying medium rather than the stream of steam and oxygen is twofold. First the coal can be fed into the reaction zone at the desired velocity independent of the steam-to-carbon and oxygen-to-carbon ratios, resulting in better control. Second, the possibility of a flashback is avoided because there is no flammable mixture outside the gasifier.

The coal stream and the stream of steam and oxygen exiting the tangential nozzles impinge upon the molten slag, imparting a rotational motion to the slag bath. The interactions and reactions between the coal, steam, oxygen, and slag in stage I are typical features seen with the Saarberg/Otto gasifier.

As the slag inventory in the gasifier builds up as a result of the ash content of the solid feed, the excess slag overflows through the central, raised taphole at the base of the gasifier. This excess molten slag is quench-granulized in a water vessel underneath the gasifier and then intermittently discharged via a lock-hopper system.

The gases generated in stage I rise with a rotational motion imparted by the tangential feed system. These gases carry entrained particles of fly coke and slag. The rotational velocity is accelerated by a throat in the gasifier between stage I and stage II. This acceleration of the rotational velocity causes the fly coke and slag particles to be centrifugally thrown against the wall of the gasifier and dropped back into stage I. The layer of molten slag thus formed replaces the refractory mass which initially covered the surface of the cooling tubes and protects the cooling system.

The finer coal particles remain in the gas and flow to stage II, which provides additional retention time at high temperatures for the further gasification of the fine coal particles. This results in further conversion of sensible heat in the gas to chemical energy.

Stage III has an enlarged cross-sectional area which reduces the velocity of the gas to allow disengagement of char and slag particles from the gas stream. Recycled gas is introduced into stage III to reduce the temperature of the outlet-gas stream below the melting point of the slag.

The gas leaving the gasifier has a temperature of approximately 850 °C and a dust load (carry-over) of about 3 percent of the total solid-matter input. More than 90 percent of the carry-over (fly coke and fly ash) are removed by a hot-gas cyclone. Thus the dust load of the gas can be reduced to about 60 mg/stdft^3.

A fly-coke cooler cools the particulate matter removed by the cyclone to approximately 240 °C. The dust is then pneumatically returned to the gasification process.

The sensible heat of the raw product gas is used in a waste-heat boiler to generate superheated high-pressure steam.

The vertical raw-gas cooler consists of a vaporizer with an upstream superheater and a downstream economizer.

The waste-heat boiler decreases the temperature of the raw gas to approximately 280 °C. Part of the steam that is raised is used as process steam. The surplus steam is given to a power plant located in the neighborhood of the demonstration plant. The separated dust is pneumatically returned to the process.

The finer solid particles still present in the gas are removed by high-temperature fibrous filters, the bags of which are cleaned by reverse-gas flow responding to a given cleaning pulse frequency. Thus the dust load of the raw gas is decreased to less than 0.3 mg/stdft^3. In order to gasify any unconverted carbon, the particulate matter separated by the fibrous filters is also pneumatically recycled to the process. The process unit that removes the fines from the product gas is one of those parts of the demonstration plant that is newly developed and which therefore has to be thoroughly checked.

After the gas leaves the filters, it enters a spray-type cooler, where it is cooled to approximately 40 °C by direct contact with cooling water injected in several spraying stages. During the cooling process part of the water vapor present in the gas is condensed and the remaining solids are largely washed out. In addition, the cooling water is saturated with the soluble gas components. The cooled gas then flows to a de-mister, where entrained droplets are separated. Behind the de-mister one portion of the gas production is branched off and recycled to the process. The rest is passed to the desulfurization plant.

The cooling water, heated up in the process, is pumped back to the top of the spray cooler. The surplus water is expanded via a valve into the flash tank, the operating pressure thus being reduced to atmospheric pressure.

The gases released during the expansion of the surplus water leave the flash tank and flow to a vapor condenser. The water coming from the flash tank is acidified by addition of diluted sulfuric acid. As a result, the minerals contained in the water—especially sulfides and cyanides—are converted into gases and the gases are released.

Typical reactions are

$$Na_2S + H_2SO_4 \rightarrow Na_2SO_4 + H_2S$$

$$2NaCN + H_2SO_4 \rightarrow Na_2SO_4 + 2HCN$$

The acidified water is then pumped to a stripper via an effluent heater.

In the stripper almost all the gas still dissolved in the water is driven off by means of steam. The steam required for stripping is generated in a reboiler which is directly connected to the stripper. Low-pressure steam is utilized for heating the reboiler. The vapors leave the top of the stripper and pass into the vapor condenser, where the major portion of the water vapor is condensed. The cooled vapors are transferred to the desulfurization plant.

The purified surplus water leaves the stripping column via an overflow and a downstream liquor-sealed collecting tank. From this tank the water is passed to a cooler, where it is mixed with cool water and then discharged into the sewer system.

In the desulfurization plant the product gas and the stripper vapors are desulfurized. H_2S is removed according to the following reaction:

$$2Fe(OH)_3 + 3H_2S \rightarrow Fe_2S_3 + 6H_2O + 62.4 \text{ kJ}$$

The hydrocyanic acid from the stripper vapors reacts with $Fe(OH)_3$, forming iron cyanides.

PROCESS PERSPECTIVE

In the early sixties, test runs with a single-shaft, slag-bath Rummel-Otto gasifier, operating at atmospheric pressure, were carried out by Otto at Union Kraftstoff Wesseling near Cologne. In 1964 the plant was shut down owing to the availability of low-cost oil, naphtha, and natural gas. In February 1976, Saarbergwerke AG and Otto agreed to a joint-development testing program of a pressurized [362 lb/in² (2.5 MPa)] coal-gasification system, which is known today as the *Saarberg/Otto process*.

Although the demonstration plant in Voelklingen/Fuerstenhausen, FRG, is one of the largest coal-gasification pilot plants in the world [coal throughput of 291 tons/day (264 Mg/day)], it was designed, built, and commissioned in 3½ years. The costs for design and construction amounted to approximately DM54MM (US$28.4MM, in 1980 US$1 = DM1.9), 75 percent of which was sponsored by the German Federal Ministry of Research and Technology. On December 5, 1979 coal gasification was started. The testing period was finished at the end of 1982. The costs of the test runs were also sponsored by the same ministry.

An extensive test program has included running of the plant at various pressures to determine characteristic process data, maximum plant throughput, and conversion efficiencies. These test runs are also used to optimize the conversion efficiency and process economics of the conversion process for different coals. The knowledge gained can serve as the basis for the design, construction, and operation of prototype and commercial-scale gasification plants for use as energy and raw material suppliers of the future.

For commercial plants the diameter of the gasifier would be increased stepwise. With a diameter of 7 ft (2.1 m) the gasifier would have double the capacity [a gas production of about 42MM stdft³/day (1.1 × 10⁶ m³(n)/day)] of the pilot plant. The goal is a gasifier having a diameter of about 10 ft (3 m). A gasifier of this size would produce about 83MM stdft³/day (2.2 × 10⁶ m³(n)/day) of gas.

PRODUCT SPECIFICATIONS

The products of the Saarberg/Otto coal-gasification process are raw gas and steam. The superheated high-pressure steam produced in the waste-heat system of the demonstration plant has a temperature of about 280 °C and a pressure of about 652 lb/in² (4.5 MPa).

3-204 Coal Gasification

Although a pressure of 652 lb/in² is not typical of the Saarberg/Otto process, it is the pressure desired in the steam from the particular plant.) The low-pressure steam produced by the cooling system of the gasifier has (in the demonstration plant) a temperature of about 150 °C and a pressure of 72 lb/in (0.5 MPa).

TABLE 10-1   Analysis of a German bituminous coal (dry basis)*

Component	Wt %
Volatile matter	34.70
Ash	9.15
C	74.49
H	4.95
O (calculated)	8.54
N	1.48
S	1.01
Cl	0.38

*Lower heating value of the coal = 27,690 Btu/kg (29,210 kJ/kg).

TABLE 10-2   Analysis of a raw Saarberg/Otto synthesis gas (dry basis)*

Component	Vol %
CO	62.08
H_2	28.13
CO_2	8.56
N_2	0.81
CH_4	0.11
H_2S	0.31

*Feedstock, bituminous coal; mode of operation, oxygen-blown; lower heating value of the gas, 10,340 Btu/m³ (standard pressure) [10,908 kJ/m³ (standard pressure)].

Up to now, raw gas has been produced by gasifying a bituminous coal and using oxygen-blown operation. A typical analysis of this coal is shown in Table 10-1. A raw gas having the analysis shown in Table 10-2 is produced under the following operation conditions:

Gasification pressure, lb/in² (MPa)	246.2 (1.7)
O_2/coal, scf/kg	32.0
Steam/coal, scf/kg	4.0
O_2/steam, scf/kg	4.0

For the air-blown operation of the demonstration plant we only give calculated values. A gas having the analysis shown in Table 10-3 is produced by gasifying the bituminous

TABLE 10-3 Analysis of a raw low-Btu
Saarberg/Otto gas (dry basis) *

Component	Vol %
CO	29.3
H_2	11.2
N_2	48.7
CO_2	9.5
Ar	0.6
H_2S	0.2
CH_4	0.2
Other compounds	0.3

*Feedstock, bituminous coal; mode of operation,
air- and oxygen-blown (85/15); lower heating value
of the gas, 4810 Btu/m^3 (standard pressure)
[5074 kJ/m^3 (standard pressure)]

coal shown in Table 10-1 and using as a gasification medium a mixture of 15 percent by volume of oxygen and 85 percent by volume of air under the following reaction conditions:

Gasification pressure, psi (MPa)	144.8 (1.0)
(Oxygen + air)/coal, scf/kg	96.0
Steam/coal, scf/kg	6.0
(Oxygen + air)/steam, scf/scf	16.0
Produced raw gas/coal, scf/kg	131.5

UPGRADING TECHNOLOGY

As a result of its high content of hydrogen and carbon monoxide the product gas can be used—following conventional treatment such as dedusting, desulfurization, carbon monoxide–shift conversion, carbon dioxide wash, and carbon monoxide and carbon dioxide methanation—for a series of chemical syntheses, e.g., for ammonia, methanol, and oxo syntheses. It can also be used for the synthetic production of liquid hydrocarbons and as the hydrogen supply in coal-hydrogenation plants.

For the production of town gas, the treatment of the raw gas consists of dedusting, desulfurization, methanation (the gas should have a methane content of about 28 percent), conversion of carbon monoxide, and carbon dioxide washing.

The production of substitute natural gas (SNG) requires methanation of the synthesis gas. The other treatment steps are similar to those for production of town gas.

Since it has a low moisture and carbon dioxide content, Saarberg/Otto gas is well suited—after dedusting and desulfurization—for the direct reduction of iron ore.

Moreover either the gas produced by oxygen-blown operation or the gas which is produced by using air or oxygen-enriched air as a gasification medium can be used—after dedusting and desulfurization—in a combined-cycle gas and steam turbine process for the generation of electric power.

WASTES AND EMISSIONS

The following wastes and emissions are produced when a bituminous coal is gasified using an oxygen-blown mode of operation in a Saarberg/Otto plant:

Solid: Granulated slag

Liquid: Effluent

Gaseous: Flue gases from the drying and grinding facility and from the flare, nitrogen from the dust-recycling system and from the flare

A typical composition of the granulated slag is shown in Table 10-4. The composition of the effluent is shown in Table 10-5.

TABLE 10-4 Analysis of granulated Saarberg/Otto slag

Component	Wt %
SiO_2	48.13
Al_2O_3	29.27
CaO	9.17
FeO	5.52
MgO	5.10
C	0.50
N	0.03
TiO_2	0.94
MnO	0.44
K_2O	1.67
Na_2O	0.36
ZnO	0.03
SO_3	0.42
Cl^-	0.02
F^-	0.006

TABLE 10-5 Analysis of Saarberg/Otto effluent

Component	mg/L
Solid matter	180
pH	6.3
SO_4^{2-}	128.0
NH_4^+	7.2
HCO_3^-	58.0
CN^-	0.55
S^{2-}	Trace

In the demonstration plant, nitrogen (99.5 percent pure) is used for recycling the separated dust and for flushing the flare.

The test runs showed that varying the operating conditions without changing the fuel feed has only a minor impact on the composition of the granulated slag and on wastewater. Varying the fuel feed should surely affect the composition of both the slag and the effluent, but since only one type of bituminous coal has been gasified so far, a detailed investigation of this subject has still to be performed.

PROCESS ECONOMICS

The capital costs for a 307MM-stdft^3/day (8MM-m^3/day) synthesis-gas plant, calculated on a U.S. location and a German cost basis (at mid-1980 rate of exchange: US$1 = DM1.9), are estimated at $102MM. This amount includes coal drying and grinding, the coal-feeding system, the gasifier, waste-heat recovery equipment, the system for removing fly coke and dust, the gas scrubber, effluent-cleaning facilities, piping, steel structures, measuring and control devices, foundations, buildings, erection, offsites, engineering, and interest. The investments for an oxygen plant and for the further processing of the raw synthesis gas are excluded.

It is further assumed that in this plant a bituminous Illinois No. 6 coal of the analysis shown in Table 10-6 is gasified and that a raw synthesis gas of the composition shown in Table 10-7 is produced. Other assumptions include the following: The production rate of gas is 65,350 stdft^3/ton (1900 m^3(n)/Mg) of coal, the production rate of the plant is 90 percent (8000 h/year), and the financing is a 20-year loan at a 9 percent per year interest rate. The costs of the production of raw synthesis gas thus calculated, based on the heating value of the gas are shown as a function of the unit cost of the coal feed in Fig. 10-4.

TABLE 10-6 Analysis of a bituminous Illinois No. 6 coal (dry basis) *

Component	Wt %
Ash	10.56
C	71.69
H	4.92
O	8.24
N	1.42
S	3.12
Cl	0.05

*Lower heating value of the coal = 27,633 Btu/kg (29,149 kJ/kg).

For a cheap strip-mined coal ($1/MM Btu, as available in the United States) the costs for the raw Saarberg/Otto synthesis gas would amount to about $3.9/MM Btu.

For deep-mined coal ($4/MM Btu, as available in the Federal Republic of Germany) the raw Saarberg/Otto synthesis gas costs about $8/MM Btu.

TABLE 10-7 Analysis of a raw Saarberg/Otto synthesis gas (dry basis) *

Component	Vol %
H_2	29.2
CO	58.3
CO_2	10.2
H_2S	1.0
CH_4	0.2
N_2	0.7
Residue	0.4

*Feedstock, bituminous Illinois No. 6 coal; mode of operation, oxygen-blown.

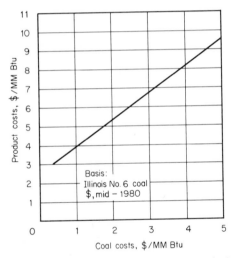

FIG. 10-4 Costs of Saarberg/Otto raw synthesis gas.

SUMMARY

The Saarberg/Otto process is a "second-generation," high-pressure, high-temperature entrained-bed gasification process. Its characteristics and advantages are as follows:

- The process is suitable for the gasification of nearly any type of coal having an ash content up to 40 percent, without having to take into account the coal's caking properties, ash-melting behavior, or grain size, and for the gasification of residues of coal-liquefaction plants and refineries.

- The carbon-conversion rate is about 99 percent; i.e., the carbon feed is nearly completely converted into gas.

- Because of the high gasification temperatures the product gas is free of hydrocarbons such as oils and tars, as well as phenols.
- The operation has only a mild environmental impact. Product gas, steam, a nonleachable inert granulated slag and a small amount (relative to the amount of coal converted) of effleunt are produced. Raw gas and effluent can be treated by conventional methods. The granulated slag can be conveyed to a landfill or used in road construction.

The Saarberg/Otto process is currently being tested and improved in a pilot plant in Voelklingen/Fuerstenhausen, FRG.

The extrapolation of data from the performance of the pilot plant has yielded the following Saarberg/Otto process efficiencies (based thus far on gasification of bituminous German coal only):

$$\text{Cold-gas efficiency} = \frac{\text{LHV gas}}{\text{LHV coal}} = 73\%$$

$$\text{Thermal efficiency} = \frac{\text{LHV gas} + \text{sensible heat of gas}}{\text{LHV coal}} = 88\text{--}94\%$$

$$\text{Total efficiency} = \frac{\text{LHV gas} + \text{sensible heat of gas}}{\text{total heat input to gasifier}} = 75\%$$

where   LHV = lower heating value

THE WESTINGHOUSE COAL-GASIFICATION PROCESS

DAVID P. DOMINICIS
CHARLES K. HOLT

Westinghouse Electric Corporation
Madison, Pennsylvania

INTRODUCTION

Westinghouse Electric Corporation has developed a pressurized fluidized-bed coal-gasification system for the production of synthetic fuels. The gasifier offers the following features:

- It processes a wide range of coals including highly caking coals.
- It processes run-of-mine coal (including fines).
- It produces virtually no tars or oils in the product gas.
- It utilizes carbon efficiently.
- It separates ash effectively.
- It operates safely, reliably, controllably, and economically.

This technology, combined with other available commercial processes for coal handling, gas cleaning, and waste processing, provides efficient and reliable systems for the production of synthesis gas. Applications for the gas range from fuel for power generation, using combined-cycle systems, or for repowering and refueling of existing power plants to synthesis gas as a chemical intermediate in the production of chemicals and fuels.

The Westinghouse gasification process was developed with support from both the U.S. Department of Energy (DOE) and the Gas Research Institute (GRI). The technical feasibility of the gasification system has been successfully verified by the operation of the Westinghouse pilot plant at Waltz Mill near Pittsburgh, Pennsylvania. The gasifier system has been in operation since 1975, and by 1982 had logged over 8000 hours of hot operation. The testing program has evaluated both air and oxygen as the oxidizing medium at pressures of up to 245 psia (1687 kPa). A wide variety of coals and chars has been tested at capacities ranging from 15 tons/day (13.6 Mg/day) with air to 35 tons/day (31.8 Mg/day) with oxygen.

The objective of the technology program has been to develop the process and hardware designs necessary for a coal-gasification plant to achieve commercial use in synthetic fuel applications. The technology development program is being directed toward the enhancement of the data base for the design of commercial-scale hardware. Reactor scale-up is being studied by integrating the results of laboratory modeling with results from process-development units and with the results from a commercial-scale semicylindrical cold-flow gasifier model. This cold-flow scale-up facility, which is 3 m in diameter, permits full front-face viewing of the fluidized bed through a glass window, thereby allowing detailed study of jet behavior, solids circulation, and other phenomena within the bed.

The engineering, economic, and environmental aspects of the Westinghouse coal-gasification technology are being studied for use in many applications throughout the world. Preliminary design efforts are underway to evaluate using medium-Btu gas to fuel an existing Westinghouse W501D combustion turbine, to evaluate building a coal gasification combined-cycle (CGCC) plant that would generate 140 MW of electric power, and to evaluate building a synthetic fuels plant with a total capability of producing 100,000 bbl/day (15,898 m³/day) of methanol from coal.

The most recent development in the technology program involves the detailed engineering of a 1200-ton/day (1089-Mg/day) demonstration coal-gasification system slated for initial operation in early 1984. This gasifier is part of a joint demonstration project to be located at a major synthetic fuels facility.

PROCESS DESCRIPTION

The Westinghouse gasification system, shown schematically in Fig. 11-1, utilizes a single-stage, pressurized, fluidized-bed gasifier followed by particulate recovery, heat recovery, and gas cleaning. The gasifier can utilize either air or oxygen as the oxidant to produce a low- or medium-Btu gas, respectively. By-products from the gasification system include sulfur, ammonia, and ash agglomerates which can be disposed of as landfill or utilized as construction material and are expected to conform to regulations for leachates in the U.S. Environmental Protection Agency's National Pollution Discharge Elimination System guidelines.

Coal is unloaded and either conveyed to the coal preparation area or stored in a pile to ensure a full supply for the plant during interruptions. The coal is then crushed to minus ¼-in (0.635-cm) size and, if necessary, dried to a level to ensure that the coal is free-flowing.

The crushed coal is pneumatically transported from lock-hoppers into the gasifier, where it is reacted with air or oxygen and with steam. The combustion of a portion of the incoming coal provides the heat necessary to devolatilize the remainder of the incoming coal so that it forms a char and to react the char with steam to form hydrogen and carbon monoxide.

The char particles form a fluidized bed in the gasifier. High recirculation in the fluidized bed causes the carbon in the char to be consumed by combustion and gasification, leaving particles that are rich in ash. The unique fluid-dynamic design of the gasifier allows the

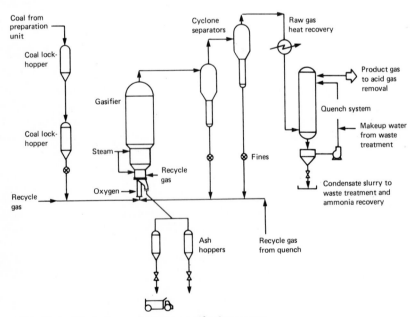

FIG. 11-1 Westinghouse single-stage gasification process.

ash-rich particles to agglomerate with each other and form larger, denser particles which defluidize from the bed and collect in the ash annulus. These agglomerates are cooled and removed through an ash lock-hopper system. The agglomerates typically contain a small percentage of the feed carbon, such that upwards of 95 to 98 percent overall carbon conversion can be achieved in the gasification system depending on the coal feedstock.

Raw product gas exits from the top of the gasifier and passes through refractory-lined cyclones which recover fine entrained particles for recycle to the gasifier. The gas next passes through the raw-gas heat-recovery area, where it is cooled by a series of heat exchangers that generate steam for process use. The final cooling of the gas and removal of particulates from it are accomplished by direct water quench in a venturi scrubber prior to desulfurization in the acid-gas removal system.

Operating experience in the Westinghouse pilot plant has proved that run-of-mine coals (including fines) from lignite to highly caking, highly volatile bituminous coals can be processed. Coals with a wide range of sulfur and ash content and with a total moisture content ranging from 2 to 24 percent have been successfully processed.

The Westinghouse gasifier provides for efficient operation by converting a high percentage of the carbon in the coal, by requiring relatively little water or steam to moderate the reactions, and by consuming less oxygen than gasifiers operating at higher temperatures. An advanced design char-ash separator integrated into the gasifier vessel contributes to high carbon efficiency by providing independent control over carbon content in the by-product ash.

The gasifier utilizes small quantities of water relative to many other gasification systems, since the coal is fed pneumatically and the ash is withdrawn dry. Gasifier steam requirements, which are minimized by the use of oxidant and recycle gas as the fluidizing medium, also contribute to low water usage.

Moderate operating temperatures improve the heating value of the product gas because less combustion of coal is needed to superheat the steam and gas flows to the temperature level required for gasifier operation. Since a smaller portion of the coal is combusted, gasifier oxidant requirements are reduced. An added benefit of the moderate operating temperatures is extended refractory life.

The Westinghouse system produces virtually no tars or oils, even when consuming a highly volatile bituminous coal. Unlike the fixed-bed processes which operate at lower temperatures, the Westinghouse process avoids the production of tars and oils and the toxicity hazards attendant thereon as well as the loss of efficiency that tar and oil cleanup systems cause. Desulfurization of the product gas is accomplished downstream by gas cleanup systems. These systems absorb H_2S formed during gasification and chemically convert it to stable elemental sulfur.

Operational ease results from the large carbon inventory of the bed, which provides operating inertia, while the plug flow nature of the gas flow in the gasifier allows rapid response to load demand changes and good turndown capability. A refractory-lined gasifier vessel with no moving parts simplifies the system's operation and maintenance and simultaneously provides significant reliability benefits.

PILOT PLANT OPERATING EXPERIENCE

The focus of the Westinghouse coal-gasification development program has been directed toward the process and hardware design necessary for a coal-gasification plant. The pilot plant, a major facility in this program, is located on the Westinghouse Waltz Mill site, near Pittsburgh, Pennsylvania. The unit contains the main gasification subsystems for the process and has been an effective tool for process development, scale-up, hardware design, and operating procedure development since its commissioning in January 1975.

The pilot plant had logged over 8000 hours of hot operation by 1982 and has successfully operated with a wide variety of coal feedstocks. These feedstocks are tabulated in Table 11-1 and include 11 different coals and 5 different chars derived from coal. The properties of these feedstocks varied over a wide range with regard to caking properties (as measured by the free swelling index), ash content, ash deformation temperature, moisture, volatile matter, reactivity, and heating value. The range of these parameters for the coals tested is shown in Table 11-2.

All of the feedstocks previously identified were tested in the process-development unit with no pretreatment other than size reduction and the reducing of surface moisture to accommodate pneumatic feeding. Because of existing mining processes, the coals mined underground were generally washed and the strip-mined coals were processed unwashed and as run-of-mine. The Ohio coals were tested both washed and unwashed where the unwashed coal contained a considerable quantity of clay matter that increased the "ash" content from a nominal 16 percent to over 22 percent.

Char feedstocks were tested as received. The FMC Char Oil Energy Development (COED) process chars were extremely fine, having an average particle size of 200 μm, which is about one-tenth the average particle size of 2000 μm for typical gasifier coal feedstocks. (COED is now a portion of the COGAS process.)

All of the feedstocks tested were readily gasified, and the ash from them was successfully agglomerated. There was no evidence of tars or oils in the exit of the gasifier, cyclone separators, or quench scrubber.

The compositions and heating value of product gas are related to the chemical com-

---

TABLE 11-1   Westinghouse gasifier feedstocks

Coals:
   Texas lignite (Monticello mine and Big Brown mine)
   Wyoming sub-C—subbituminous
   Montana Rosebud—subbituminous B
   Indiana No. 7—bituminous
   Illinois No. 6—bituminous
   Western Kentucky No. 9—bituminous (Hamilton mine)
   West Virginia—bituminous (Big Mountain mine, Boone Co.)
   Ohio No. 9—bituminous
   Pittsburgh Upper Freeport—bituminous
   Pittsburgh No. 8—bituminous
   Republic of South Africa coal (Bosjesspruit colliery)

Chars:
   Coke breeze—from coking operations
   Pittsburgh No. 8—from Westinghouse devolatilizer
   Indiana No. 7—from Westinghouse devolatilizer
   Utah—from FMC Char-Oil Energy-Development pilot plant
   Kentucky—from FMC Char-Oil Energy-Development pilot plant

TABLE 11-2   Gasifier feedstock properties

Free swelling index	0–9
Ash content, %	2–22
Ash-deformation temperature, °C	1080 to 1380
Moisture, %	2–24
Volatile matter, %	1–35
Relative reactivity*	1–50
Heating value† MJ/kg	16.3–29.5

*Kinetic ratio based on coke breeze as unity reference.

†As-received basis.

position of the coal processed. The energy requirements of the gasifier, including energy for elevating the temperature of reactants, energy for endothermic reactions, and energy for process heat losses, are provided for by the combustion of some of the coal. Therefore, coals of high quality (lower moisture or ash) allow for higher overall system efficiency, since less energy is wasted by heating nonreactive ash and excess moisture. In general, the CO_2 level in the product gas increases with increasing moisture or ash in the coal as a consequence of the proportional increase in combustion. Conversely, the CO_2 level will decrease with lower moisture or ash content.

Experience at the pilot plant has shown that moderate changes in coal composition such as those which occur in the typical coal supplies from the same mine should have little impact on the composition of the product gas and a negligible effect on the system's operability.

PROCESS PERSPECTIVE

The use of coal gasification is a particularly flexible approach for widespread application to the production of synthetic fuels. Medium-Btu gas, for example, can be utilized to produce synthetic natural gas, gasoline, methanol, and a broad slate of petrochemical products. Additionally, it can be used as an industrial fuel gas in direct-firing processes. Either low- or medium-Btu gas may be used for the production of electric power and process energy.

Those applications of the Westinghouse coal-gasification process which will be discussed in greater depth in this chapter include power generation using a CGCC system and methanol synthesis. Westinghouse, as mentioned earlier, is involved in evaluating the feasibility of using its coal-gasification system for application in both of these areas.

Coal Gasification Combined-Cycle (CGCC) System

As a major supplier to the electric utility industry, Westinghouse developed a coal-gasification technology to generate electricity in an efficient and environmentally acceptable manner. The CGCC system combines the technology of generating a fuel gas from coal (coal gasification) and a modern combustion turbine. The hot exhaust gases from the combustion turbine are used to produce steam, which is used in a bottoming-cycle steam turbine to generate more electricity (combined cycle).

Westinghouse chose the CGCC system for development for two reasons: electricity can be generated more efficiently in a combined cycle than in a standard fossil fuel–fired boiler and steam-turbine cycle and this method of using coal is consistent with the need to reduce the consumption of oil and natural gas while maintaining strict environmental standards.

In the basic combined cycle, the particulate-free desulfurized clean gas exits from the acid-gas removal system and enters the combustion-turbine combustor, where it is burned with air to provide a hot gas that powers the turbine. The hot gas is expanded through the turbine and produces the work required to drive both the combustion turbine's air compressor and its associated electric generator. The exhaust gas from the turbine is then utilized in a final heat-recovery steam generator to generate superheated steam for use in the main steam-turbine system. The most common configuration of the combustion turbine plus the system for recovery of heat from the exhaust gas (basic combined cycle) is shown in Fig. 11-2.

The Westinghouse CGCC design includes a variety of design features and systems to ensure that federal, state, and local environmental requirements are met. Some of the main environmental features include the following:

- Removal of sulfur compounds via the Selexol acid-gas removal system. A Claus, Beavon, Stretford system arrangement provides for recovery of elemental sulfur and for treatment of tail gas prior to venting gases from the plant. Virtually all of the sulfur originally contained in the coal can be removed by these systems.

- Reduction of NO_x emissions as a result of (1) the relatively lower flame temperatures, which produce less NO_x when the clean low-Btu gas is burned and (2) the removal of most of the nitrogen contained in by-product NH_3 as anhydrous NH_3 from the coal during gas cleanup via a PHOSAM system.

- Production of ash agglomerates, which can be disposed of as landfill or utilized as construction material.

- Virtually no toxic compounds in liquid-waste streams such as boiler blowdown and sludge from water treatment. Thus, it is possible to use standard wastewater treatment

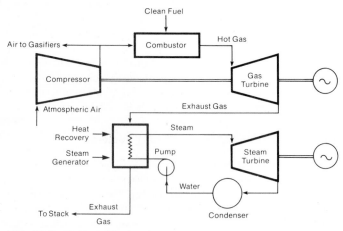

FIG. 11-2 Basic combined cycle.

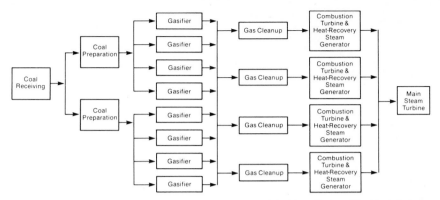

FIG. 11-3 Overall philosophy for CGCC plant system module. Note: Each gasifier block includes a coal-pressurization system, a gasifier, a particulate-removal system, and a raw-gas heat-recovery system.

for the processing of liquid wastes. The Westinghouse gasifier produces virtually no tars or oils, which is a significant advantage, since the absence of these compounds simplifies the liquid-effluent treatment systems.

- Lower quantities of cooling and makeup water, since a majority of the electric power is produced by the combustion turbine and steam requirements for the gasification process itself are low.

A study, completed in June 1981, evaluated the Westinghouse CGCC system for electric power generation. Figure 11-3 is a block flow diagram for the nominal 650 MW–rated plant. Two independent coal preparation trains support the entire gasification process. Each train in this area has a 1500-ton (1359-Mg) coal-storage silo that supplies feedstock to the gasifiers during maintenance of the system.

Eight 30-ton/h (27.2-Mg/h) air-fed gasifiers processing Illinois No. 6 coal are employed in this application to power four 100-MW$_e$ combustion turbines. A train arrangement is used, with each train containing two gasifiers. Each gasifier has its own lock-hopper system for receiving ash, its own particulate-recovery and cyclone system, and its own system for raw-gas heat exchange and steam generation.

Each train has associated gas-quench and acid-gas removal areas for the removal of NH_3 and H_2S, respectively, from the gasifier product gas. A facility for treating process condensate recovers the NH_3, and a sulfur-recovery system recovers elemental sulfur. An area for treating tail gas is also provided. Four combustion-turbine exhaust-heat-recovery steam generators and a single steam-reheat turbine-generator set complete the bottoming cycle.

During normal operation, each gasifier train operates independently of the others. However, in order to provide increased reliability and availability during major equipment outages, the gasifier product streams can be manifolded together downstream of the area for particulate recovery and raw-gas heat exchange. In most cases, multiple trains of major plant systems are provided for increased availability and reliability.

Material and energy balances are provided in Tables 11-3 and 11-4, respectively.

The study evaluated five CGCC configurations, each having an increasing degree of design complexity (Fig. 11-4).

TABLE 11-3   Material balance for Westinghouse combined-cycle plant

	10^3 kg/h
System inputs	
Illinois No. 6 coal	218
Air	5,080
Water	803
Total	6,101
System outputs	
Ash	24
Sulfur	29
Ammonia	4
Solid waste	4
Exhaust gas	5,312
Evaporation	728
Total	6,101

TABLE 11-4   Energy balance for Westinghouse combined-cycle plant*

System input	
Illinois No. 6 coal to gasifiers	6.195×10^9 kJ/h
System outputs	
Electricity (combined cycle)	666,918 kW
Heat rate = 9289 kJ/kWh	
Thermal efficiency = 38.7 percent	

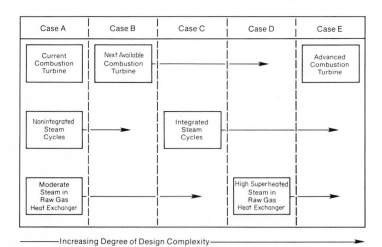

Increasing Degree of Design Complexity

FIG. 11-4   Summary of the configurations of the five CGCC cases.

- In case A, which serves as the base case for the study, the following are incorporated: (1) a commercially available combustion turbine, (2) no integration of the gasification-process and electric-power steam cycles, and (3) a moderate level of superheated steam in the raw-gas heat exchanger. In each of the following cases, one of the key items is varied to asses its effect on efficiency and economics.

- In case B, the mid-1980s combustion-turbine version was utilized.

- In case C, integration of the gasification-process and electric-power steam cycles was assumed.

- In case D, a high degree of superheated steam was assumed to be generated in the raw-gas heat exchangers.

- Finally, in case E, an advanced combustion turbine is assumed.

For comparison with these CGCC cases, an average station energy efficiency of 35.9 percent is used for the conventional coal plant. Similar but slightly lower efficiencies for conventional coal plants have been used in Electric Power Research Institute (EPRI) studies comparing CGCC and conventional plants.

The results of the efficiency calculations are shown in Table 11-5. As expected, efficiency improves from 37.6 percent for case A to 40.9 percent for the case E advanced-turbine configuration. When proceeding from case A to case E configurations, a 5 to 11 percent improvement in efficiency exists over that of the conventional coal plant, depending on which case is chosen. A more detailed look at Table 11-5 indicates that the largest improvement in efficiency occurs between cases C and D, in which the change is the generation of higher levels of superheated steam in the raw-gas heat exchangers. Integration, non-integration, and combustion-turbine inlet temperature appear to be less significant factors with regard to efficiency improvements in CGCC plants.

The results of the cost-of-electricity analyses are shown in Figs. 11-5 and 11-6, which illustrate that all of the CGCC configurations are less expensive to own and operate than conventional coal plants for all capacity factors of interest for baseload applications, that is, greater than 50 percent. In comparison with the conventional coal plant, cases A, B, and C are projected to have lower capital costs as well as better efficiencies, and thus their cost-of-electricity curves are lower for all plant capacities, that is, 0 to 100 percent.

Cases D and E are projected to have higher capital costs than conventional coal plants. This is a result of the higher temperature and pressure in the raw-gas heat exchanger and the advanced combustion turbine. Actually, at very low capacity factors, cases D and E

TABLE 11-5 Comparison of the efficiencies of CGCC and conventional plants

	Efficiency, %	Heat rate, kJ/kWh
Conventional	35.9	10,021
Case A	37.6	9,569
Case B	38.1	9,446
Case C	38.7	9,289
Case D	40.3	8,936
Case E	40.9	8,797

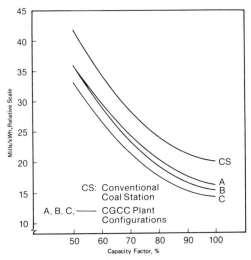

FIG. 11-5 Cost of electricity for conventional and CGCC plants (cases A, B, and C).

also have higher costs of electricity. However, because of the interplay of capital costs and efficiencies in the cost-of-electricity computations, cases D and E have a lower cost of electricity for the plant capacities of interest.

Of all the cases evaluated, case C, which entails a moderate degree of design complexity, has been found to be the most economical. It incorporates the next-available version of

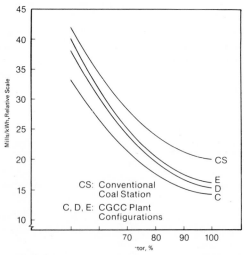

FIG. 11-6 Cost of electricity for conventional and CGCC plants (cases C, D, and E).

the combustion turbine, integration of the gasification and electric-power steam cycles, and moderate levels of superheated steam in the raw-gas heat-exchange area.

Methanol Synthesis

By utilizing a commercial methanol technology in conjunction with the Westinghouse single-stage gasification technology, a preliminary conceptual design study was completed on a Pittsburgh seam coal typical of bituminous coals found in Pennsylvania. The results of the study indicate that this combination of technology will yield about 0.79 Mg of fuel-grade methanol per metric ton of as-received Pittsburgh seam coal in a 10,000-bbl/day (1590-m³/day) commercial unit. This figure includes coal required for steam generation. Subsequent plant expansion to 100,000 bbl/day (15,898 m³/day) could be achieved by the addition of similar process trains. Figure 11-7 is a block flow diagram illustrating this coal-to-methanol conceptual design.

The synthesis gas to produce 10,000 bbl/day (1590 m³/day) of methanol is supplied by two Westinghouse single-stage, oxygen-blown, fluidized-bed gasifier modules, with each module gasifying 30 tons/h (27.2 Mg/h) of Pittsburgh seam coal. The product gas is adiabatically humidified in a venturi quench scrubber to achieve an H_2/CO molar ratio of 0.64, which is needed to accomplish the water-gas shift reaction. The gas is converted by the shift reactor into a hydrogen-rich synthesis gas with an H_2/CO ratio of 2:1.

The shift-conversion reactor, an industrially proven, fixed-bed converter, offers a low-pressure drop across the bed and operates with a commercially proven shift catalyst.

The acid-gas removal system, a Benfield Hi-Pure, removes acid gases by means of chem-

TABLE 11-6 Material balance for Westinghouse methanol plant
[10,000 bbl/day (1590 m³/d) process train]

	10^3 kg/d
System inputs:	
Coal (as received) to gasifiers	1,263
Coal (as received) to steam plant	444
Oxygen to gasifiers	784
Combustion air	5,450
Raw water	7,340
Total	15,281
System outputs:	
Fuel-grade methanol	1,349
Carbon dioxide	1,727
Cooling-water losses	5,291
Flue gas	5,886
Ammonia	3
Sulfur	25
Waste solids	222
Steam losses	504
Miscellaneous losses	274
Total	15,281

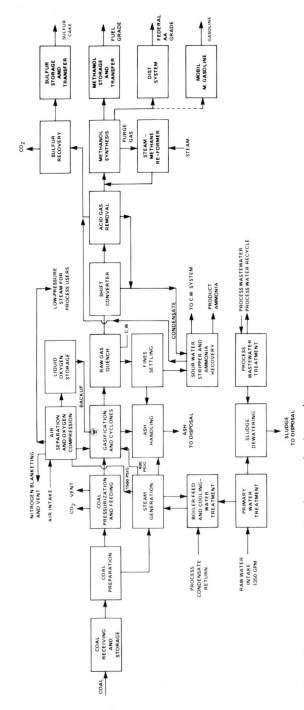

FIG. 11-7 Westinghouse coal-to-methanol process showing product spectrum.

TABLE 11-7 Energy balance for Westinghouse methanol plant
[10,000 bbl/day (1590 m³/d) process train]

	10^9 J/h
System inputs:	
Coal (HHV*) to process	1,685
Coal (HHV) to steam plant	590
Electricity (11.3 MW)	40
Total	2,315
System outputs:	
Fuel-grade methanol (HHV)	1,220
By-products (HHV)	
Ammonia	4
Sulfur	11
Subtotal with by-products	1,235
Consumption and losses	1,080
Total	2,315
Overall thermal efficiency = 53.4 percent	

*HHV = Higher heating value.

ical absorption. The acid gas leaving the top of the regenerator tower is fed to a Stretford unit for the conversion of H_2S to elemental sulfur.

In the process for methanol synthesis a stream of purge gas containing CH_4 is combined with steam and fed to the re-former reactor to produce H_2 and CO. Re-forming the purge gas results in a higher yield of methanol per ton of coal feed.

Sour process water containing NH_3, H_2S, COS, and CO_2 is collected and treated. The acid gas that is captured is sent to the Stretford system for sulfur recovery; NH_3 is recovered as liquid anhydrous NH_3.

A summary of the material and energy balances for this design is given in Tables 11-6 and 11-7, respectively. The plant has a thermal efficiency greater than 53 percent. This efficiency is attributable to the unique characteristics of the fluidized-bed gasification process and the modern low-pressure methanol synthesis process.

SUMMARY

Multiple End Uses

The Westinghouse development program has resulted in the successful demonstration of the pressurized fluidized-bed gasification process. The process has been shown to be capable of filling a number of coal-conversion applications including the following:

- Substitute natural gas
- Chemical feedstocks

- Industrial fuel gas
- Fuel-cell power generation
- Combined-cycle electric power generation
- Steam boiler refueling and repowering

Environmental Impacts

Analysis of the environmental aspects of the system shows that it can be designed to meet federal and state environmental criteria. Atmospheric emissions, which are controlled by existing technology, are low. The ash is nontoxic and can be disposed of by normal methods.

The Westinghouse design encompasses a philosophy of minimum water discharge, and wastewater is treated and reused. In addition, since the process produces no tars or oils, environmental concerns regarding handling and disposal of such materials are not applicable.

FIG. 11-8 Synthetic Fuels Division, Waltz Mill site, Madison, Pennsylvania. A wide variety of coals and chars has been tested on the Westinghouse process-development unit at capacities ranging from 15 tons/day (13.6 Mg/day) with air to 35 tons/day (31.8 Mg/day) with oxygen. (*Westinghouse*)

There are significant environmental advantages, for example, over conventional coal-fired power generation. Atmospheric emissions of SO_2 and NO_x will be only about one-twentieth the standards for power generation. Water use will be only about 65 percent that of a conventional coal-fired generating facility. The former consideration results from application of EPA atmospheric emission requirements, and the latter results from the very low water use of the Westinghouse coal-gasification system.

Integrated Development Program

The Westinghouse facilities at the Waltz Mill site consist of a coal-gasification pilot plant (Fig. 11-8), support laboratories, a commercial-scale fluidized-systems facility, and a test and development center. This is one of the few facilities in the world where the operation of a pilot plant has been integrated with a test and development center in order to permit combustion tests using coal gas as a fuel at typical combustion-turbine delivery pressures.

During the mid-1980s and the early 1990s, it is anticipated that coal conversion will play an important role in producing clean fuels and chemical feedstocks.

THE KILnGAS SYSTEM

ANTONIO V. SORIANO

Allis-Chalmers Coal Gas Corporation
Milwaukee, Wisconsin

GENERAL PROCESS DESCRIPTION AND BLOCK FLOW DIAGRAM

The KILnGAS® process uses a pressurized, ported, rotary kiln to accomplish the combustion and gasification of coal. The process concept is based on Allis-Chalmers commercial background with high-temperature pyro-processing systems in the iron ore and cement industries. The process concept is shown in Fig. 12-1. In this diagram, the process may be thought of as a four-zone process, namely, drying, preheating, devolatilization, and combustion-gasification.

Coal enters the gasifier at the higher end. As the gasifier rotates, a tumbling motion is imparted to the coal bed. As the coal travels down the incline, it passes through the four zones and the ash is discharged at the lower end.

A portion of the gases produced (overbed gases) travels through the gasifier in a direction opposite to the direction of coal travel and passes to the gas cleanup train from the coal-feed end of the gasifier. The remainder of the gases leaves the gasifier at the discharge end and passes to the gas cleanup train. This gas split gives rise to the term, "Bi-Flow operations."

DEMONSTRATION PLANT DESCRIPTION AND DATA

A demonstration plant, known as the KILnGAS commercial module (KCM), was constructed at the Wood River station of the Illinois Power Company in the town of East Alton, Illinois. Ground breaking was held on October 31, 1980. Figure 12-2 shows the schedule of the KILnGAS demonstration program.

A photograph of the KCM is shown in Fig. 12-3. The Wood River plant of Illinois Power is shown adjacent to the KCM.

The Wood River station of the Illinois Power Company is located at the junction of the Wood River and the Mississippi River on a 498-acre tract of land in Wood River Township, Madison County, Illinois. Coal is delivered to this facility by unit trains. Facilities for

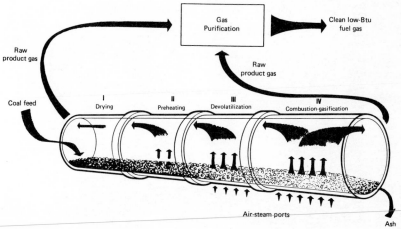

FIG. 12-1 Concept of a ported, rotary kiln for the coal-gasification process.

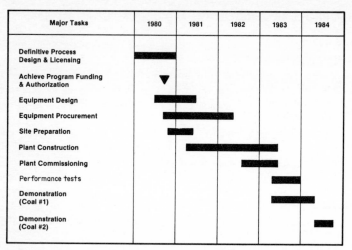

Major Tasks	1980	1981	1982	1983	1984
Definitive Process Design & Licensing	▬				
Achieve Program Funding & Authorization	▼				
Equipment Design	▬▬				
Equipment Procurement	▬▬▬				
Site Preparation	▬				
Plant Construction		▬▬▬▬			
Plant Commissioning			▬▬		
Performance tests				▬	
Demonstration (Coal #1)				▬▬	
Demonstration (Coal #2)					▬

FIG. 12-2 Schedule of KILnGAS commercial module demonstration program.

coal unloading, handling, and storage are presently available on site, with the capacity to service the coal requirements of the power plant plus the additional KCM coal-feed requirements.

The Wood River generating station is a fossil-fueled generating plant consisting of five steam-powered electric generating units with a total capacity of 650 MW.

This plant is capable of burning oil, natural gas, or coal as fuel. The conveyors which supply coal to units 4 and 5 will ultimately service the KCM.

FIG. 12-3 An aerial photo of the KILnGAS commercial module.

Demonstration Plant Description

The KCM will supply approximately 400×10^6 Btu/h (100 kcal/h) of low-Btu gas to the existing steam-powered electric generating unit no. 3. This unit is rated at 50 MW and is presently using oil to fire its boiler. The KCM will employ all of the process steps required to fully demonstrate the production of clean low-Btu gas using the KILnGAS process. These include gasification, recovery of process heat, supplying of air and steam, gas cleanup and purification, liquor processing, ash disposal, and wastewater treatment. The KCM is sized to process 600 tons/day (22,800 kg/h) of coal. The system will operate at a pressure of about 60 psig (415 kPa). This capacity is sufficient to provide data for scaling up the process and equipment to subsequent 400-MW commercial applications with a high degree of confidence.

Specific Data

The KCM is intended to be a "limited-life" (about 15 years) plant that will be used for the purpose of demonstrating technology rather than a plant designed to have continuing economic value. The reference coal for design purposes is Illinois No. 6. However, the plant is expected, ultimately, to be run on a variety of midwestern, eastern, and western coals in order to demonstrate broad feedstock compatibility. The plant is also designed to demonstrate gas cleanup and the applicability of commercially available sulfur-removal technology (i.e., the Holmes Stretford process is to be used for meeting new source performance standards). The more stringent gas purity requirements that must be met by fuel used for gas turbines in combined-cycle applications will also be attained.

The scope of the total plant is shown schematically in the block diagrams of Figs. 12-4 and 12-5. Figure 12-4 is a simplified diagram of the system, and Fig. 12-5 shows the more detailed relationships between the plant systems. It can be seen from the figure that all the systems required for the production of clean gas are included in the plant.

Table 12-1 presents a performance summary of the KCM. Note that the gas-to-coal factor (i.e., ratio of chemical and sensible heat in the gas to heat content of the coal) is approximately 80 percent. Note also that the design goals for emissions are substantially below the "permitted" values in order to ensure that emission limitations are achieved with some margin.

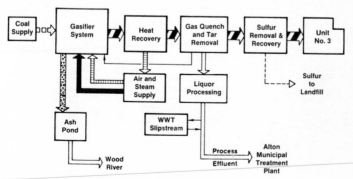

FIG. 12-4 Block flow diagram of KILnGAS commercial module.

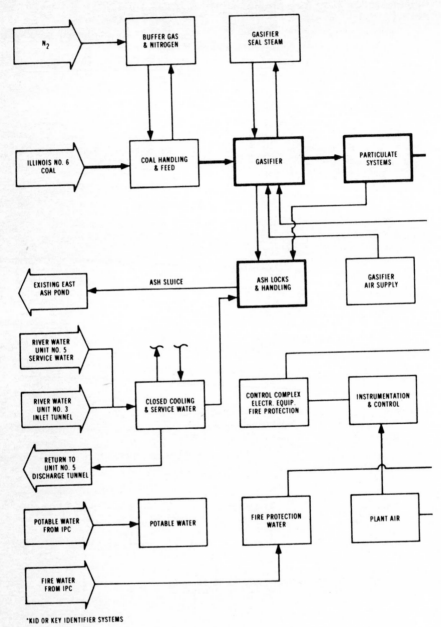

*KID OR KEY IDENTIFIER SYSTEMS

FIG. 12-5 KILnGAS commercial module plant system.

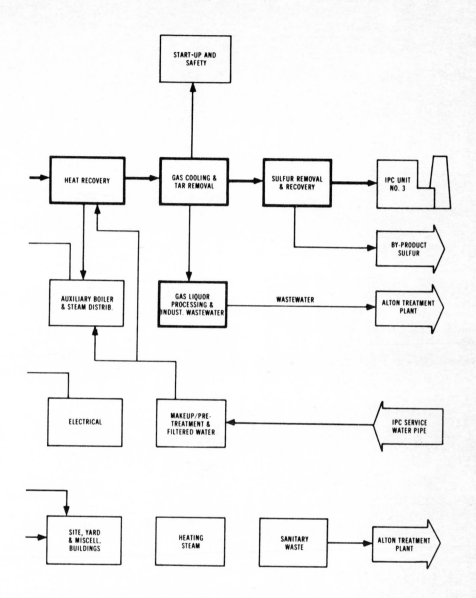

TABLE 12-1   Performance summary of the KILnGAS commercial module

Coal, t/d (10^6 Btu/h)	600
	502
LBG*, 10^6 Btu/h	409
LBG* (HHV†), Btu/scf	158
Effluent	
Ash, t/d	96
Sulfur, t/d	16.7
Process wastewater, gal/d	115,000

	Emissions		
	SO$_2$	NO$_x$	Particulates
Unit no. 3 boiler			
Design, lb/10^6 Btu	0.81	0.1	<0.001
Permitted, lb/10^6 Btu	1.8	0.9	0.1
KILnGAS plant			
Design, t/yr	94	44	3
Permitted, t/yr	100	100	100

*Low-Btu gas.

†Higher heating value.

Objectives of the demonstration program

The demonstration program will

1. Demonstrate the operation of the process in a commercial environment
2. Establish the technical and economic basis on which to offer plants under normal commercial terms in sizes up to 2000 to 5000 tons/day (76,000 to 190,000 kg/h) of coal feed.

Costs and funding participants

The KILnGAS demonstration program has a budget of $155 million, which includes design, construction, operation, and two demonstration runs. The following are the funding sponsors:

Sponsor	Contribution $ in millions
Allis-Chalmers Corporation	90
State of Illinois	18
Twelve electric utilities	41
EPRI*	6
Total	155

*Electric Power Research Institute.

The following are the twelve electric utilities participating in the program:

Baltimore Gas and Electric	Ohio Edison
Central Illinois Light	Potomac Edison
Consumers Power	Public Service Indiana
Illinois Power	Public Service Co. of Oklahoma
Iowa Power and Light	Union Electric
Monongahela Power	West Penn Power

PROCESS PERSPECTIVE

The KILnGAS system has been under development by Allis-Chalmers Corporation since 1971.

The five principal phases of the KILnGAS development program were:

1. Conceptualization and feasibility assessment
2. Basic engineering development
3. Independent assessments and demonstration planning
4. Demonstration
5. Commercialization

This program is shown in Fig. 12-6, with the demonstration portion divided into two major activities.

In the early 1970s, Allis-Chalmers' R&D organization recognized a close similarity between the technology of large-scale processing of minerals and the technology of high-volume, high-temperature coal gasification. In addition, Allis-Chalmers was aware of a

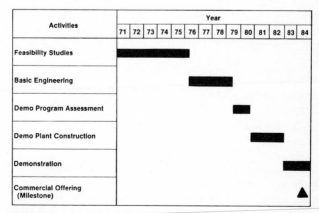

FIG. 12-6   Overview of KILnGAS technology development program.

growing need for clean fuel derived from coal in the electric utility industry, because of growing environmental concern. This prompted Allis-Chalmers to proceed with further research into coal gasification. Furthermore, Allis-Chalmers has for years furnished electric generating equipment to the electric utilities and is familiar with utility needs and with all aspects of marketing and supplying capital equipment to the utilities.

From 1971 to 1975 Allis-Chalmers invested more than $4 million in the R&D program. Results were sufficiently encouraging to warrant further investment of $11.7 million by Allis-Chalmers and a group of 11 electric utilities in a follow-on basic engineering program to complete development.

Basic Engineering Program

The basic engineering program started in 1976, with these objectives:

1. To conduct a more comprehensive technical and economic assessment of the KILnGAS system's commercial potential by means of studies of large-scale prototype plants
2. To test and demonstrate selected aspects of the process and process equipment in order to identify and reduce performance and design risks
3. To establish an engineering data base adequate to support the design of follow-on commercial plants

The studies of commercial prototype plants focused on the KILnGAS system and on combined-cycle applications. All studies pertaining to new generation capacity showed that KILnGAS/combined-cycle plants are competitive with conventional coal-fired plants with scrubbers.

A major element of the basic engineering development program was the specification, design, and implementation of several major test tools that were needed to reduce the risks of the process and the breakdown of critical mechanical components. The major test tools include the following:

1. An entirely new gasifier having a 2-ft (60-cm) ID (inside diameter).
2. A 6-ft (1.8-m) diameter, batch-mode, ported rotary-kiln section known as the deep-bed gasifier.
3. A 120-psig (827-kPa) tube-furnace reactor.
4. A seal-testing facility capable of simulating all gasifier motions and operating temperatures.
5. A 50-ton/h (45,500-kg/h) solids-handling test facility for testing mechanical components.

Testing of the 10-ft (3.0-m) diameter gasifier seal began in February 1979. Measurements found that leakage at design temperatures [550 °F (290 °C)] was acceptable. Leakage was further reduced to levels corresponding to the "optimistic" end of the design calculations.

Solids-handling tests were conducted at full scale [50 t/h (45,500 kg/h)] to introduce coal into the pressurized gasifier.

TABLE 12-2 Electric utility participants in the 3-year basic engineering program

Central Illinois Public Service	Public Service Co. of Indiana, Inc.
Consumers Power	Public Service of Oklahoma
Illinois Power	Potomac Edison Co.
Iowa Power & Light	Union Electric
Monongahela Power	West Penn Power
Ohio Edison	

The electric utilities which participated in the 3-year basic engineering program are shown in Table 12-2.

This program was successfully completed in December 1979, with the following accomplishments:

1. Definition of several prototype plant concepts and numerous application studies that provide additional evidence for the competitive potential of KILnGAS technology.

2. Confirmation of the validity of the basic process concept, based in part on a 10-day continuous pilot plant run on Illinois No. 6 coal at 96 percent availability.

3. Successful demonstration of a 10-ft (3.0-m) diameter gasifier seal.

Demonstration Assessment

Scientific Design, an independent engineering consultant, was charged with the task of objectively assessing the technical feasibility and commercial viability of the KILnGAS process and system design. In general, Scientific Design's conclusions were positive and basically agreed with the findings of a task-force assessment team.

The demonstration assessment cost $3.6 million and was funded by Allis-Chalmers, the 11 electric utilities in the basic engineering program, and 5 additional electric utility partners.

Demonstration program

The demonstration stage is intended to introduce the KILnGAS system into the marketplace and establish an ongoing, profitable growth business. The three stepping stones that will lead to the first normal commercial booking are:

1. Design and construction of a commercial module of the KILnGAS system

2. Operation of the KCM long enough to realistically appraise the system performance, operating, and maintenance characteristics and to collect base-line design data

3. Operation of the KILnGAS system in a normal electric utility load-following environment

Commercialization

KILnGAS plants will be offered in 1984 under normal commercial terms. These plants will be offered as retrofits to boilers presently fueled by oil or natural gas, as retrofits to fuel existing combined-cycle plants, and as fuel sources to fire new-capacity combined-cycle plants.

DETAILED PROCESS DESCRIPTION

A general process block diagram which presents the major process steps in a KILnGAS plant is shown in Fig. 12-7. The KILnGAS gasifier represents the major process block in a commercial KILnGAS plant.

Gasification

In the gasification sequence, coal from the storage bin is fed to the gasifier and elevated to process pressure through two rotary gaslocks. The area between the two gaslocks is kept slightly evacuated, while the gasifier side of the downstream lock is pressurized above gasifier pressure with clean fuel gas. With this arrangement, the downstream gaslock takes the large pressure drop, while the other lock takes very little. Fuel gas is prevented from leaking to the atmosphere, and air cannot flow into the gasifier.

Coal from the gaslocks and recycled hot particulates enter the gasifier separately through parallel rotary-screw conveyors. These solids fall into the gasifier and are mixed by tumbling action within the rotating gasifier. This mixing of coal and hot particulates partially dries and preheats the coal, one of the necessary steps before the gasification reactions can take place. The gasifier itself is mounted at a small angle of inclination, and the solids pass through the gasifier from the uphill (feed) end to the downhill (discharge) end as a result of the combined effects of rotation and slope.

The gasifier operates in the Bi-Flow mode, meaning that the gases produced in the process exit through both ends of the gasifier. A portion of the gases flows counter to the coal movement, and the remaining gases flow cocurrently with the coal movement. As the solids progress down the reactor, they are preheated by the counterflowing hot gases passing over the bed and reach a temperature of 800 °F (430 °C) at a point somewhat past the midpoint of the gasifier. At that point, controlled amounts of air and steam are added to the solids bed through ports in the reactor shell in order to partially combust both carbon and volatiles. This combustion causes a rapid increase in bed temperature and causes more volatiles to be released. During this heat-up combustion period, the volatile components in the coal are pyrolyzed and produce a broad range of components which include H_2, CO, CO_2, H_2, CH_4, benzene, toluene, xylene, naphthalene, H_2S, COS, CS_2, phenols, NH_3, HCN, and tars and pitches. Other components, including complexes of sulfur and nitrogen, are produced in much smaller quantities. All of these components created from coal volatiles join the counterflowing gases and exit from the gasifier through the coal-feed end.

When the bed reaches 1800 °F (1000 °C), volatilization is essentially complete and the remaining char (mixture of carbon and ash) undergoes gasification. Simplistically, the following two reactions take place in the gasification of the char:

$$H_2O + C \rightarrow CO + H_2 \qquad \text{(endothermic)}$$
$$2O_2 + 3C \rightarrow 2CO + CO_2 \qquad \text{(exothermic)}$$

The ratio of air to steam is controlled in order to balance the endothermic and exothermic reactions and maintain the proper bed temperature [2000 °F (1100 °C)] necessary to achieve the best conversion rate and product-gas quality. Air and steam are introduced through ports in the reactor shell. As the reactor rotates, the port valves are opened by a

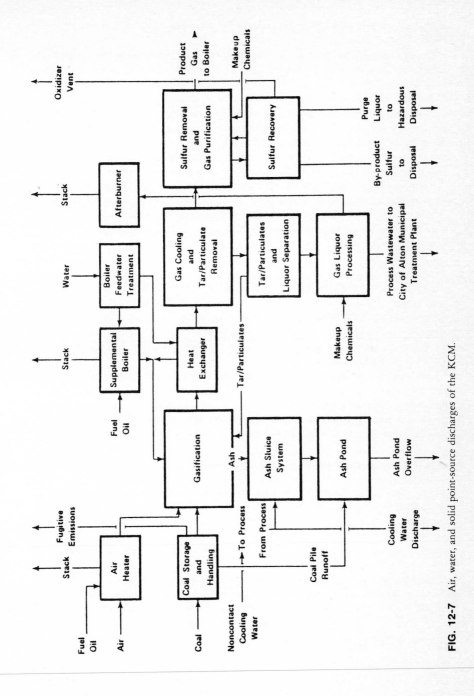

FIG. 12-7 Air, water, and solid point-source discharges of the KCM.

mechanism as they pass below the bed. As the ports pass over the bed, the valves are closed.

Not all of the gaseous products of char gasification are required to preheat the incoming solids. Therefore, some of these gases, primarily CO, CO_2, H_2, H_2O, and N_2, are allowed to flow out of the discharge end of the reactor. This off-gas stream is at approximately 2000 °F (1100 °C) and contains no hydrocarbons or other products of devolatilization.

Both gas streams contain particulate matter entrained from the solids bed. The particulates in the gas stream from the feed end contain a relatively high proportion of carbon. Those in the gas stream from the discharge end contain a high proportion of small ash particles released upon conversion of carbon in the char. In order to recover these particulates, the gas streams leaving both ends of the gasifier pass through independent banks of cyclones. All of the particulates recovered from the gas stream leaving the feed end (carbon-rich) are recycled to the feed end of the gasifier. Only a part of the particulates recovered from the discharge end (ash-rich) is recycled to the gasifier, and the remainder is discarded to the ash-disposal system (through a depressurizing lock-hopper).

Heat Recovery

Both gas streams leaving the gasifier cyclones are at sufficiently high temperature to make heat recovery highly desirable. The off-gas stream from the discharge end is cooled to approximately 300 °F (150 °C) in a heat-recovery steam generator, where high-pressure steam is generated for process uses. This gas stream is free of volatilization products, and the only contaminants are residual particulate matter carried through the cyclones and H_2S and COS generated by the process.

The gas stream leaving the feed end contains tar and other hydrocarbons. The extraction method used here is to pass the feed-end gases through a two-step heat-recovery unit. The first section cools the gas to 600 °F (315 °C) by direct contact with liquor. In the second step heat is recovered from the gas stream between 600 and 240 °F (315 and 115 °C); a heat exchanger is used to reheat clean product gas prior to delivery to the power plant.

Steam Supply and Air-Compressor System

Steam is used in the KILnGAS plant for gasification and for process-heating needs. The majority of this steam is produced at high pressure in the discharge end of the heat-recovery system. At partial load, steam is produced in a supplementary boiler fired on low-Btu gas exhausted from the buffer-seal gas system which supplies the rotary gaslocks. The supplementary boiler can also be fired on oil, which will be necessary in order to supply steam during start-up.

The high-pressure steam from these two sources is first used to run the turbine drive of the air compressor and is then sent to the gasifier and other process areas as needed. Air for the process is supplied from steam and motor-driven air compressors.

Washer and Cooler Subsystems

Upon leaving the heat-recovery units, the gas streams from the feed and discharge ends are sent to separate washer-cooler systems, which together make up the first stage of gas cleanup. Briefly, this stage serves to cool and dehydrate the gas stream, remove tar and particulate matter, and remove much of the water-soluble impurities including ammonia and phenols.

When the gas stream from the feed end leaves the heat-recovery system, it passes through a number of steps in the washer-cooler subsystem. First, the gas is quenched with hot recirculating liquor in a pipeline contactor. Second, entrained quench liquor, tar, and particulates are knocked out of the gas stream in a cyclone. Third, the saturated gas is passed up through a washer-cooler column, where it is brought into contact with counterflowing cold recirculating liquor. The washer-cooler column removes the water-soluble components including NH_3. The gas must be cooled to a temperature of 100 °F (38 °C) in the washer-cooler because of the inlet-temperature requirements of the sulfur-removal step. The liquor from the cyclone and the net liquor from the washer-cooler column are sent to a tar decanter equipped with a bottom scraper. This device allows removal of the tar and particulate matter that settle from the liquor solution, and the tar and particulates are then recycled back to the gasifier for gasification. The decanted liquor is recycled to the pipeline contactor and the feed end of the heat-recovery quench stage.

When the gas stream from the discharge end leaves the heat-recovery system, it passes through a set of steps that are parallel to those in the washer-cooler subsystem for the gas from the feed end. The washer-cooler column serves to cool the gas to 100 °F (38 °C) as required for sulfur removal and also removes residual particulates carried through the pipeline contactor and cyclone. Particulates are separated from the recirculating liquor in a decanter and then sluiced to disposal in an ash pond along with the gasifier-ash discharge. Any excess water condensed from the gas stream is passed directly to the wastewater treatment system.

Gas-Liquor Processing System

The wash liquor that is contacted with the feed-end gas contains several components that cannot be processed directly in the wastewater treatment system or released to the atmosphere. These components are removed by steam stripping the gas liquor and then are disposed of in an incinerator. The stripper is a column in which NH_3 (as well as very small quantities of H_2S and CO_2) is removed from the liquor with steam and is then passed to an incinerator in which N_2 and H_2O are produced from the controlled combustion of NH_3, with a minimum production of NO_x. The liquid bottoms from the stripper contain phenols and other components not vaporized by steam. This stripped liquor (essentially ammonia-free) is partially recirculated to the washer-cooler and the net liquor buildup is discharged to the wastewater treatment system, where the remaining contaminants are removed.

Acid-Gas Removal and Sulfur-Recovery System

Sulfur in low-Btu gas occurs primarily as hydrogen sulfide (H_2S comprises 90 to 93 percent of the total sulfur in the coal), with carbonyl sulfide and carbon disulfide (COS and CS_2) making up the remainder. The Stretford process is used to meet the sulfur requirement and is designed to remove 99 percent of the H_2S from the gas stream and to convert it to yield elemental sulfur as a by-product.

The two functional stages of the Stretford process are H_2S absorption and sulfur recovery. In the first stage, the low-Btu gas is brought into contact with an alkali solution containing vanadium salts and anthraquinonedisulfonic acid. This solution absorbs and reacts with the H_2S from the gas stream, and the cleaned gas can then be passed directly to the heat-recovery unit.

In the second stage of the Stretford process, the solution is oxidized to yield elemental sulfur while the alkali solution is regenerated for use in further H_2S removal. In order to accomplish this, the solution is passed through oxidizing tanks, where air is sparged through the solution and elemental sulfur is recovered through flotation.

Wastewater Treatment System

The wastewater treatment system uses a series of physical and chemical steps to produce treated water that meets effluent regulations. The major load on the treatment system is from wash liquor used in the washer-coolers. In the treatment system, flocculation with dissolved air and filtration are first used to remove particulates and oils. The resulting sludges are stored in decant basins. Activated carbon is then used to absorb soluble organics, and cyanides are oxidized. Finally, ion exchange is used to remove NH_3, after which the fully treated wastewater is used for ash sluicing. Sludge from the water-treatment system will be periodically removed from the decant basins.

Solid-waste disposal system

The solid-waste disposal system operates by sluicing ash and particulates to ash-settling ponds. These solids originate from the gasifier discharge, the cyclones for the discharge-end gas, and the washer-cooler subsystem for the discharge-end gas. Makeup water for the sluicing system is supplied by bleed-off from the cooling tower and by treated wastewater.

Energy Requirements

Energy in the form of coal and electricity will be consumed during the operation of the KILnGAS system. For cold starting, oil or natural gas will be used. Figure 12-8 summarizes the quantity of energy input and the quantity of energy output expected as product gas in the KCM. A conversion of coal heat input to gas of 81.5 percent is expected.

Feedstock requirements

The materials and their rate of consumption based on operation of the KCM at full load are shown in Table 12-3. It must be noted, however, that these rates of consumption will be entirely different in a commercial plant.

FIG. 12-8 Estimated energy balance for KILnGAS commercial module.

TABLE 12-3 Feedstock requirements of the KILnGAS commercial module

Raw material	Quantity
Coal	600 t/d (22,800 kg/h)
Steam	576 t/d (21,900 kg/h)
Air	1500 t/d (57,000 kg/h)
Fuel oil for starting	400 gal/h (1,515 L/h)
Nitrogen	220 scf/min (63 m³/h)
Sulfuric acid	Negligible
Caustic soda (76%)	3.0 t/d (115 kg/h)
Sodium carbonate	1.6 t/d (61 kg/h)
Anthraquinonedisulfonic acid	29 lb/d (31.1 kg/d)
Trisodium citrate	35 lb/d (16.0 kg/d)
Sodium ammonium vanadate	21 lb/d (9.5 kg/d)

Raw coal composition   The initial test coal for the demonstration will be Illinois Coal No. 6, with the following composition:

Proximate analysis, wt %	
Fixed carbon	0.425
Ash	0.360
Moisture	0.113
Volatiles	0.102
	1.000
Ultimate analysis, wt %	
C	0.612
H	0.043
O	0.093
N	0.007
S	0.030
	1.000
HHV*, Btu/lb (kcal/kg)	11,011 (562)

*Higher heating value.

PRODUCT AND BY-PRODUCTS

The major product of the KILnGAS system is a low-Btu fuel gas. There is a potential to recover several by-products including ammonia, sulfur, phenol, and a mixture of tar, oil, and hydrocarbons.

Major constituents of the product gas will be CO, H_2, CO_2, CH_4, and N_2. Typical contaminants in the gas will include trace amounts of various sulfur and nitrogen compounds and heavy metals. Sulfur species will include H_2S, COS, and CS_2, while the primary nitrogen

TABLE 12-4 Composition of clean gas from the KILnGAS system

Constituent	Concentration, vol %
CO	20.0
CO_2	10.0
H_2	18.9
H_2O	1.4
N_2	47.8
CH_4	1.5
Hydrocarbons	0.3
Sulfur compounds	0.1
Gas heating value, Btu/scf	158
(kcal/m³)	(1405)

species will be NH_3. Nickel, mercury, lead, cadmium, and arsenic may be among the heavy-metal constituents. The composition of the clean gas projected for the KCM is shown in Table 12-4.

Elemental Sulfur

Sulfur removed from the raw product gas will be recovered as pure elemental sulfur in the form of a wet filter cake. Trace quantities of the absorption liquor from the sulfur-recovery process (containing sodium salts of anthraquinondisulfonic acid, as well as sodium meta-vanadate, citrate, thiosulfate, and thiocyanate) will be present in the cake but will be removed when the sulfur is converted to a molten state. This residual liquor will be separated by gravity from the molten material and returned to the absorption system. The purified molten sulfur will be resolidified and stored for disposal.

Tar, Oil, and Hydrocarbons

Tar, oil, and hydrocarbons will be recovered from the quench condensate of the raw product gas in a tar decanter. Oil will be recovered from the aqueous condensate in an air-flotation unit. These materials will consist of a wide range of high-boiling, fused-ring aromatics containing some heterocyclics, e.g., nitrogen-, sulfur-, and oxygen-containing ring structures, and lesser amounts of aliphatics and hydroaromatics. The plans are to recycle these materials to the gasifier, where their heating value will be recovered.

WASTES AND EMISSIONS

During operations, air discharges will come from the combustion of low-Btu gas in the electric utility generating plant. Liquid and solid-waste discharges come from the KILnGAS plant. During maintenance, liquid discharges will result from the cleaning of pipes, heat exchangers, or reaction vessels. The commercial KILnGAS system will meet all existing U.S. Environmental Protection Agency (EPA) limits on air and water standards.

Liquid Discharges

These are listed below.

1. Coal-pile runoff
2. Ash-sluice water
3. Cooling-water discharge
4. Sanitary wastewater
5. Process quench condensate

The first four all have a direct counterpart in the utility industry. Only the process quench condensate will be unique in the integrated gasification and power plant complex.

Ash-pond overflow

The ash-pond overflow will consist largely of the coal-pile runoff and the ash-sluice water. The miscellaneous blowdowns (i.e., blowdowns from the boiler and treatment of intake water) will represent an insignificant contribution to this flow. The potential contaminants of interest in the ash-pond overflow will be the trace elements that may have leached from the coal and/or ash.

Process quench condensate

The process quench condensate will be contaminated with a range of organic and inorganic species. A treatment for wastewater consists of on-site pretreatment, including dissolved-air flotation and ammonia stripping.

Solid Discharges

Solid discharges will include the process ash, by-product elemental sulfur, and purge liquor of the sulfur-recovery process. The latter will be considered a solid discharge, since it may contain as much as 30 percent solids in the form of sodium salts.

The process ash will be discharged to an ash pond.

Fugitive Emissions

Potential occupational health hazards associated with the KILnGAS system could include exposure to a range of feedstocks and process materials. These exposures are expected to be minimal during steady-state process operation but have the potential to be significant during plant upsets or shutdowns for maintenance and/or repair.

PROCESS ECONOMICS

The long-term application of the KILnGAS system is to fuel new combined-cycle (CC) electric power plants. Combined-cycle power plants, particularly when used for intermediate-load service, offer favorable economics for power generation. During the basic engineering stage and the following demonstration assessment phase of the KILnGAS development, a series of commercial prototypical KILnGAS/combined-cycle power plants were investigated and compared with the only identified competitor in the marketplace—the conventional coal-fired power plant fitted with a flue-gas desulfurization (FGD) system. These studies have continually shown that the KILnGAS/combined-cycle plant will be economically competitive in terms of total cost of power generation.

The criteria for estimating capital requirements and costs of services were obtained from generally accepted procedures used in the chemical-processing industry and the electric utility industry.

In order to illustrate the economic advantages of the KILnGAS system and combined-cycle plants over conventional coal-fired plants fitted with FGD equipment, we studied a site-specific power plant in a midwestern state. The plant is rated at 500 MW and planned to be in operation in the early 1990s. One alternative is to build a coal-fired plant with FGD equipment. The other alternative is to build a combined-cycle plant fueled by a KILnGAS system.

With the exception of a few parameters which are applicable only to combined-cycle plants (e.g., combustion-turbine temperatures), we used the following assumptions for both alternatives:

Net power output:	500 MW
Capacity factor:	60 percent
Location:	Midwestern U.S.
Combustion-turbine temperature:	2200 °F (1204 °C)
Coal cost (Illinois No. 6)	$1.50/MM Btu
Fixed-charge rate	20 percent
Discount rate	13 percent
Plant life	30 years
Inflation	9 percent/year
Dollar basis	1982

Comparative Economics of a KILnGAS/Combined-Cycle Power Plant and a Coal-Fired Plant With a Flue-Gas Desulfurization (FGD) System

Details of cost breakdown of the KILnGAS system are proprietary to Allis-Chalmers. The breakdown of costs is, therefore, not shown in this economic summary.

KILnGAS/combined cycle plant vs. conventional plant with FGD system

Parameter	KILnGAS/CC	Steam/FGD
Power output, MW	500	500
Heat rate, Btu/kWh (kcal/kWh)	9350 (2356)	9892 (2493)
Fuel (Illinois No. 6), $/MM Btu ($/MM kcal)	1.50 (5.95)	1.50 (5.95)
Inflation rate, %/yr	9.0	9.0
Plant life, yr	30	30
Capacity factor, %	60	60
Total capital requirements, $ in millions	463	512
Fixed-charge rate, %	20	20
Discount rate, %	13	13
Capital cost, $/kW	926	1023
First-year, O & M, $ in millions	13.9	17.2
Levelized annual bus bar cost, $ in millions	204.5	226.3
Levelized annual bus bar cost, mills/kWh	77.8	86.1

It is recognized that the preceding comparisons are too general to be considered accurate for a specific application. They indicate, however, the potential economic advantages of the KILnGAS/combined-cycle power plant over a conventional coal-fired power plant fitted with FGD equipment.

SUMMARY

The KILnGAS process for coal gasification uses a ported, pressurized rotary kiln to accomplish the drying, preheating, devolatilization, and combustion-gasification of coal. The process concept is based on Allis-Chalmers' commercial background with high-temperature pyro-processing systems in the iron ore and cement industries.

The KILnGAS system has been under development by Allis-Chalmers since 1971. In 1976, 11 electric utilities joined Allis-Chalmers in sponsoring a three-year, $11 million basic engineering program to establish an engineering data base adequate to support the design of a follow-on commercial plant. In addition various commercial prototype plants were investigated through definitive design and cost analysis.

At the end of the basic engineering program in mid-1979, a 1-year demonstration assessment was conducted at a cost of $3.6 million. The purpose of this effort was to evaluate the technical and economic viability of the KILnGAS system prior to proceeding with a commercial demonstration of the technology. The highlight of this program was the evaluation by an independent engineering consultant which confirmed both the technical feasibility of the process and its economic competitiveness with conventional technologies and other coal-gasification systems.

On October 31, 1980, ground breaking for the $137 million KCM (KILnGAS commercial module) took place. The sponsors of the demonstration program include Allis-Chalmers, the State of Illinois, EPRI, and 12 electric utilities. The KCM will process 600 tons/day (22,800 kg/h) of Illinois No. 6 coal to produce 400×10^6 Btu/h (10^8 kcal/h) of low-Btu gas [160 Btu/stdft^3 (1410 kcal/m^3)] for direct firing of an existing 50-MW oil-fired boiler of Illinois Power Company. The plant was commissioned in early June 1983, and low-Btu gas was generated before the end of that month. The objectives of the demonstration program are the following:

1. To demonstrate process operation in a commercial environment
2. To establish the technical and economic basis for proceeding with plants under normal commercial terms

In the KILnGAS process (Fig. 12-9), coal is transported through the kiln via a tumbling action resulting from the kiln slope and rotation. The coal is dried and preheated by the counterflowing hot overbed gases. The porting system is used to introduce air and steam reactants into the coal bed. Raw low-Btu gas can be extracted from both ends of the gasifier, giving the Bi-Flow configuration to the KILnGAS process. Gas from the feed end passes through a cyclone, where particulates are collected and returned to the gasifier with the coal feed. The raw gas passes through a heat-recovery unit, where steam is produced for use in the system's auxiliaries. The cooled raw gas then passes through a quench and tar-removal unit, where hydrocarbons are separated and recycled back to the gasifier. The raw gas now free of hydrocarbon goes through the sulfur-removal and -recovery system. At the same time, raw gas extracted from the discharge end of the gasifier passes through

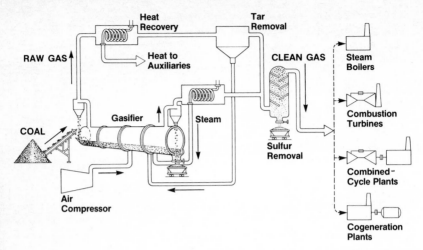

FIG. 12-9 Schematic diagram of the KILnGAS system.

another heat exchanger, where steam is generated for introduction to the gasifier. The raw gas from the discharge is tar-free and is sent directly from the heat exchanger to the sulfur-removal and -recovery system. Clean low-Btu gas coming out of the sulfur-removal and -recovery system is now available for use.

The by-products of the KILnGAS system are primarily elemental sulfur and anhydrous ammonia. The KILnGAS system will meet EPA limits on air and water standards.

The long-term market for the KILnGAS system will be for fueling new capacity power plants for the electric utility industry. The KILnGAS system offers the following unique advantages to the electric utilities:

1. Large-scale materials processing
2. Good process turndown capability
3. High carbon conversion
4. Ability to process difficult coals in an unsized state
5. Low steam requirements
6. Environmental acceptability

The only identified competitor in the marketplace for the gasification-combined-cycle power plant is the conventional coal-fired steam power plant fitted with an FGD system. Comparing the economics of these two options shows that the KILnGAS/combined-cycle plant will be economically competitive in terms of plant investment, operation, and total cost of power generation.

CHAPTER **3-13**

THE PURIFICATION OF GASES DERIVED FROM COAL

COLIN H. TAYLOR

Babcock Woodall-Duckham Ltd.,
Crawley, West Sussex,
RH10 1UX, England

INTRODUCTION

In determining the optimum sequence of processes to purify the gas obtained from a coal gasification plant, many factors must be considered. First, the composition obtained from the coal gasification plant must be taken into account, since the gases produced by the different gasification processes have widely differing compositions; some processes yield a gas containing a small number of components, while the gas produced from others contains a very large number of high-molecular-weight compounds, some of which may be present in very small concentrations but which can be very significant in the choice of downstream processes.

Second, the composition of the required product is obviously of great significance to the purification stream. In this respect it is important that unnecessarily severe specifications are not put on the quality of the product gas. In many cases the cost of removing undesirable components increases rapidly as the permitted level of impurity in the product is reduced.

Third, after the specifications of the feed and product gas have been fixed, the optimum gas-purification system is then selected on the basis of the usual criterion of the lowest cost consistent with factors such as environmental limitations, demonstrated reliability, flexibility with respect to turndown, and variations in feed composition.

In this chapter, the types of processes which can be considered are briefly described. Quantitative data on individual processes is not included; these should be obtained from the respective process licensors. In selecting the optimum process for each step, it is essential that the individual processes be considered not in isolation but rather in the context of the total plant economics.

COMPOSITION OF PRODUCT GASES

The products of coal gasification may be used as a source of the following synfuels:

Low-Btu (LBG) or medium-Btu gas (MBG)

Substitute natural gas (SNG)

Gasoline synthesized by the Fischer-Tropsch reaction

Methanol

Methanol may be used as a substitute fuel, either by itself or in a mixture with gasoline, or may be converted to gasoline by the Mobil MTG (methanol-to-gasoline) process.

Low-Btu and Medium-Btu Gases as Fuels

The simplest way to use coal-derived gas as a fuel is by direct combustion in a furnace adjacent to the gasifier without any purification other than removal of entrained solid and liquid particles. However, removal of the bulk of the hydrogen sulfide content will usually be necessary to satisfy environmental legislation requirements.

If low-Btu gas is derived from gasification with air, it is always generated close to the point of use. However, medium-Btu gas is frequently distributed over long distances, which may necessitate a reduction in the concentration of toxic carbon monoxide. It may also be economically advantageous to increase the calorific value by removal of carbon dioxide.

Substitute Natural Gas

The production of SNG requires removal of carbon from the gas in the form of carbon dioxide, followed by the reaction between carbon monoxide and hydrogen to form methane:

$$CO + 3H_2 \rightleftharpoons CH_4 + H_2O$$

Methanation catalysts are very sulfur-sensitive, and the sulfur content of the gas must be reduced to a very low level prior to methanation.

Fischer-Tropsch and Methanol-Synthesis Gas

Both methanol and liquid hydrocarbons can be obtained by catalytic reaction of carbon monoxide and hydrogen at elevated pressure. The Fischer-Tropsch reaction, as used at Sasol, is carried out at 20 to 25 atm (2000 to 2500 kPa), whereas pressures above 50 atm (5000 kPa) are required for methanol synthesis. Both reactions require a feed gas with a high volumetric hydrogen-to-carbon monoxide ratio; the optimum ratio for methanol synthesis is 2.5 to 2.8, while that for the Fischer-Tropsch reaction is lower, in the range of 1.3 to 2.2.

As shown in Table 13-1, fixed-bed processes yield a gas having a suitable H_2/CO ratio, whereas in the gases produced by fluidized bed and entrained-flow processes the ratio is much lower and some CO must be converted to H_2 and CO_2 by reaction with steam.

Reduction of H_2S content to a low level is necessary to avoid poisoning the synthesis catalyst, and excess carbon is rejected from the system as carbon dioxide.

GAS-TREATMENT PROCESSES

Tar and Dust Removal

In the case of fixed-bed gasifiers, the raw gas contains significant quantities of tar and higher hydrocarbons including benzene, toluene, xylene, and higher paraffins (naphthas). The high-pressure fixed-bed processes incorporate a quench vessel in which the gas is cooled by direct contact with recirculated tarry liquors. This is followed by a waste-heat boiler arranged vertically so that the downward flow of oily condensate prevents accumulation of deposits on the tube walls. A continuous stream is bled from the circulating liquor and separated into dusty tar, water, and oily fractions. A typical flow sheet is shown in Fig. 13-1.

The low-Btu, low-pressure gas generated by fixed-bed producers is most efficiently used in the hot raw state, so that advantage can be taken of the sensible heat of the gas and steam and the potential heat of the tar and oil vapors. Cyclones are used to remove dust and entrained tar droplets. If the gas must be desulfurized, it is cooled and passed through electrostatic precipitators or wet scrubbers to eliminate oil and tar mist, which could cause frothing in the desulfurization unit.

TABLE 13-1 Typical composition of gas produced by different types of coal gasifiers*

Component	Type of Gasifier Pressurized fixed-bed (Lurgi)	Atmospheric fluid-bed	Atmospheric entrained-flow
CO_2	32.0	19.0	12.1
CO	15.3	35.3	57.3
H_2	38.9	41.4	28.2
CH_4	9.8	2.4	0.1
N_2	0.3	0.6	1.2
H_2S	1.0	1.2	1.1
Organic sulfur	0.1	0.1	
C_nH_{2n+2}	0.6	Trace	
Olefins	0.4	Trace	
NH_3	0.7	Trace	
Tar	0.2	Trace	
BTX†	0.6	Trace	
Phenols	0.1		
HCN	Trace		

*Values are in mol % and are based on a dry product gas.

†Benzene-toluene-xylene.

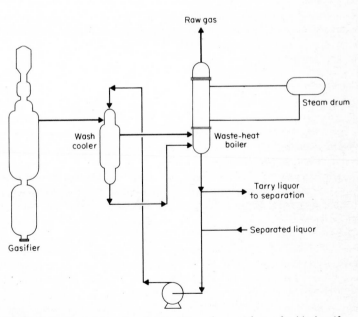

FIG. 13-1 Typical flow sheet for quenching of raw gas from a fixed-bed gasifier.

Most fluid-bed and entrained-flow gasifiers produce a gas which is tar-free. Cyclones are used to recover entrained char particles from the gas produced from fluidized gasifiers; sometimes treatment in a water-scrubbing system follows. The gas produced by entrained-flow gasifiers is at a much higher temperature, and a waste-heat boiler which is specially designed to cope with entrained slag particles is frequently used to recover heat from the gas before it is scrubbed with water.

In all cases, the removal of solid and liquid particles is closely integrated with the system for cooling the raw gas. The developers of most gasification processes regard raw-gas cooling and particulate removal as part of the gasification system. Each has developed a gas-cooling and particulate-removal system which has been shown to meet the requirements of its own gasification process.

Conversion of Carbon Monoxide

Unless the gas has to be transmitted for long distances, it is not essential to reduce the carbon monoxide content of medium-Btu fuel gas. For most other applications—except SNG production using the HICOM (High Carbon Monoxide) process—it is necessary to convert part or all of the carbon monoxide in the crude gas to carbon dioxide so that carbon in excess of the stoichiometric quantities required for synthesis can be rejected from the system.

In the conversion reaction, carbon monoxide reacts with steam over a catalyst to form carbon dioxide and hydrogen:

$$CO + H_2O \rightleftharpoons CO_2 + H_2$$

The reaction takes place at elevated temperatures (200 to 475 °C) and is slightly exothermic. The degree of conversion is limited by chemical equilibrium, with high conversion (i.e., low carbon monoxide content in the product) being favored by low temperatures. Pressure has no effect on the reaction equilibrium, but increasing the pressure reduces the quantity of catalyst required.

Conventional CO shift

Conventional conversion catalysts are based on iron oxide with certain promoters (e.g., chromia) and operate at the higher end of the temperature range (280 to 574 °C). For practical purposes, therefore, the CO content of the gas can be reduced to only about 1 to 2 percent by conventional catalysts.

Iron oxide–based catalysts are tolerant of sulfur compounds such as H_2S, as the catalyst is partially converted by high H_2S concentrations to FeS, which is also active in promoting the CO-shift reaction. Frequent variation between high and low sulfur contents should be avoided, as rapid recycling between Fe_2O_3 and FeS will reduce the strength of the catalyst.

Raw-gas shift

However, the activity of an iron oxide catalyst is significantly impaired by the presence of unsaturated hydrocarbons and nitric oxide, which form gum deposits on the catalyst surface. Conventional iron oxide is therefore not satisfactory for use with the crude gas from fixed-bed coal-gasification processes.

For this application cobalt-molybdenum catalysts have been developed. This catalyst also promotes the hydrogenation of unsaturated hydrocarbons as well as HCN and COS.

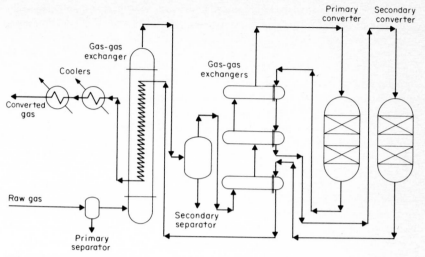

FIG. 13-2   Raw-gas CO shift.

The catalyst is designed to utilize the steam content of the hot crude gas from the gasifier, and operates at high temperatures (380 to 470 °C). A typical flow sheet is shown in Fig. 13-2. A pre-reactor is usually employed to retain carbon-containing residues.

Low-temperature shift

If very low carbon monoxide contents are required in the converted gas, e.g., for ammonia synthesis, it is necessary to employ a catalyst which is active at lower temperatures (200 to 280 °C). Copper-based catalysts are widely used for this purpose. These catalysts are very susceptible to poisoning by sulfur compounds. The inlet gas must therefore be completely desulfurized, and it is normal to install a guard bed of zinc oxide to protect the catalyst from sulfur slip from the upstream sulfur-removal unit.

Removal of Acid Gases

The two acid gases which must be removed from the gas mixture are hydrogen sulfide and carbon dioxide. Hydrogen sulfide must be removed if the gas is to be used for the synthesis of ammonia, methanol, methane, or higher hydrocarbons, since the catalysts used to promote the synthesis reactions are poisoned by sulfur. If the gas is to be used as a fuel, reduction of the sulfur content will in most cases be necessary to prevent atmospheric pollution. It is generally more economical to remove hydrogen sulfide gas than to remove sulfur dioxide from the products of combustion. Carbon dioxide removal is necessary to reject carbon from the system, thus achieving the correct carbon-to-hydrogen ratio for subsequent synthesis, as well as to increase the calorific value of fuel gas in order to reduce transmission costs or to give interchangeability with other gases.

Both gases may be removed by countercurrent absorption, at elevated pressure, in a circulating liquid in a packed or plate column. The solution is then regenerated by stripping, usually with steam generated in a reboiler, at near-atmospheric pressure in another column. The acid gases stripped from the solution leave the stripper at the top. Such processes may be divided into chemical and physical systems, depending on the mechanism of absorption.

Chemical absorption systems

Chemical systems make use of an alkaline solution which reacts chemically with the acid gases. The process flow sheet can be very simple (see Fig. 13-3), consisting of an absorber, a stripper, a circulating pump, a reboiler, and an overhead condenser system to recover evaporated solution as reflux. In some processes the stripper operates at a higher temperature than the absorber, and there is heat exchange between the rich- and lean-liquor streams. Other process variations have been developed to reduce the concentration of acid gases in the purified stream and to reduce the energy requirement for reboiling.

Some of the best-known chemical absorption processes are listed in Table 13-2. The majority are based on potassium carbonate or an alkanolamine as the alkaline constituent of the solution. Monoethanolamine (MEA) and diethanolamine (DEA) are stronger alkalies than potassium carbonate, hence they can reduce the acid gas content of the purified gas to lower levels for a given absorber pressure. The absorptive capacity of the solution for acid gases is also greater. Various additives are used to increase the capacity of potassium carbonate in the proprietary processes.

On the other hand, because of the greater affinity of MEA and DEA for acid gases, more heat is required to regenerate the solution. Also, the alkanolamines (especially MEA) are more likely to undergo side reactions with minor constituents of the gas, forming byproducts which degrade the solution and increase corrosion rates.

While alkaline solutions have a greater affinity for H_2S than for CO_2, the difference is not sufficient for the established chemical absorption processes to be regarded as truly selective. However, the tertiary amines do exhibit a degree of selectivity for H_2S, and this

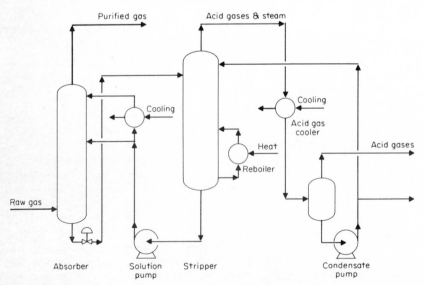

FIG. 13-3 Chemical absorption process for removal of acid gas.

TABLE 13-2 Some widely used chemical absorption processes

Process	Absorbent (in aqueous solution)	Licensor
Amine-based processes		
Adip	Diisopropanolamine	Shell Development Co., Houston, Texas Shell Internationale Research Mij. BV, The Hague, Netherlands
Amine Guard	MEA + corrosion inhibitors	Union Carbide Corp., Tarrytown, N.Y.
DEA	Diethanolamine	
Econamine	Diglycolamine	Fluor Corp., Los Angeles, California
MEA	Monoethanolamine	
Carbonate-based processes		
Benfield	K_2CO_3 + additives	Union Carbide Corp., Tarrytown, N.Y.
Catacarb	K_2CO_3 + additives	Eickmeyer & Associates, Overland Park, Kansas
Giammarco-Vetrocoke	K_2CO_3 + additives	Giammarco-Vetrocoke SpA, Marghera, Italy

feature has been employed in the SCOT process. The Union Carbide Corp. has recently developed the Ucarsol HS process that uses a tertiary amine which is claimed to be truly selective. The process operates in stages with the flow of absorbent to each stage controlled to take advantage of the higher reaction rate of the H_2S absorption. The absorber incorporates proprietary trays specially designed for the process.[1]

Physical absorption systems

In physical systems, there is no reaction between the acid gases and the liquor. Absorption is solely by physical solution, and the quantity of acid gas which dissolves is approximately proportional to its partial pressure. At low partial pressures the absorptive capacity of the solvent is considerably lower than the capacity of chemical absorbents, but at high partial pressures very large quantities of gas can be dissolved (see Fig. 13-4). Therefore physical absorption systems are used chiefly for high-pressure applications. However, it should be noted that the physical solubility of gases is greatly increased by a reduction in temperature. The Rectisol process, using refrigerated methanol as the solvent, has been widely used for the removal of acid gases from Lurgi gas; in this case the cost of refrigeration is justified by the other advantages of the process.

One advantage of physical absorption processes is selectivity. Because it is much more soluble than CO_2, H_2S can be absorbed selectively from a gas containing both. On the other hand, all gases are soluble to some extent in the physical solvents, including those gases which are desired products. To avoid losses of product, the rich liquor is frequently flashed in one or more stages before it enters the stripper, and flashed gases are recycled to the absorber.

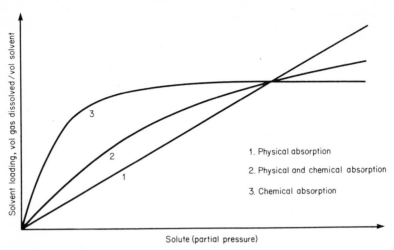

FIG. 13-4 Equilibrium curves illustrating the difference between physical and chemical absorption processes.

Another advantage of physical systems is their ability to remove trace quantities of other undesirable components—such as higher hydrocarbons, HCN, and organic sulfur compounds—without the formation of unwanted by-products. They are therefore particularly suited to the treatment of "dirty" gas.

The inclusion of flash stages, recycle, refrigeration in some circumstances, and selective H_2S removal stages can make the flow sheet for a physical absorption system very complex (Fig. 13-5). However, the low operating temperatures and lack of chemical activity result in low corrosivity, and hence most equipment can be made of carbon steel. Moreover, the much weaker attachment between the solvent and the solute is reflected in the much lower energy requirement for regenerating physical systems compared with chemical processes. In some applications it is possible to eliminate the reboiler by using air, nitrogen, or an inert gas as a stripping agent for regenerating the liquor.

The solvents employed in a number of physical absorption processes are listed in Table 13-3. Table 13-4 compares some of the pertinent features of chemical and physical absorption processes.

Physical-chemical systems

The Shell Sulfinol process employs as the absorbing medium a mixture of a chemical solvent (diisopropanolamine) and a physical solvent (tetrahydrothiophene dioxide). The process is claimed to combine the advantages of the physical and chemical systems.

Wet oxidation systems

The physical and chemical absorption systems described in the preceding sections can be used for removal of H_2S and CO_2, which are rejected as an off-gas from the stripper. Wet oxidation systems are used only for the removal of H_2S, which is converted to elemental sulfur. These systems have been used commercially over a wide range of pressures from normal atmospheric to 65 atm (6500 kPa).

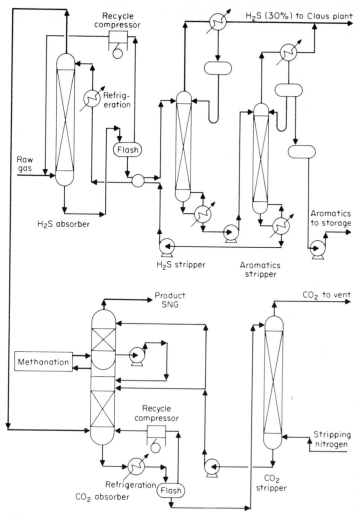

FIG. 13-5   Physical absorption process for selective H_2S and CO_2 removal.

The processes employ an alkaline solution (usually sodium or potassium carbonate) containing an oxygen carrier. A typical flow sheet is shown in Fig. 13-6. The H_2S is absorbed by countercurrent scrubbing in a tower which usually contains an open type of packing (e.g., timber grids). The HS^- ions in solution react with oxygen donated by the oxygen carrier, yielding a dilute suspension of sulfur:

$$2HS^- + O_2 \rightarrow 2OH^- + 2S$$

The solution flows to an oxidation tank in which air is bubbled through the liquid. The air carries the sulfur particles to the top as a froth and reoxidizes the oxygen carrier. The

TABLE 13-3 Major physical absorption processes

Process	Absorbent	Licensor
Estasolvan	Tri-*n*-butyl phosphate	Friedrich Uhde GmbH, Dortmund, Federal Republic of Germany Institut Francais du Petrole, Paris
Fluor Solvent	Propylene carbonate	Fluor Corp., Los Angeles, California
Purisol	*N*-methyl-2-pyrrolidone	Lurgi Mineralöltechnik, Frankfurt, Federal Republic of Germany
Rectisol	Methanol	Linde AG, Munich, Federal Republic of Germany Lotepro Corp., New York Lurgi Mineralöltechnik, Frankfurt, Federal Republic of Germany
Selexol	Dimethyl ether of polyethylene glycol	Norton Chemical Process Products, Morristown, N.J.
Sepasolv	Methylisopropyl ethers of ethylene glycol	BASF AG, Ludwigshafen, Federal Republic of Germany

TABLE 13-4 Comparison of chemical and physical absorption processes*

Item compared	Amine processes	Carbonate processes	Physical processes
Flow sheet	Simple	Very simple	Complex
Utility consumption	High	Medium	Low
Selectivity for H_2S	Low	Low	High
Effect of minor gas constituents	May react irreversibly	Some may contaminate	Most may be recovered
Potential for corrosion of carbon steel	High	Medium	Low
Relative unit cost of chemicals	Medium	Low	High
Loss of required product	Low	Low	Potentially high
Typical applications	Low to medium pressure	Medium to high pressure	High acid-gas partial pressure
	Low residual acid-gas content	Bulk acid-gas removal	Selective H_2S removal
	Clean feed gas		Dirty feed gas

*This comparison is indicative only. Individual processes may incorporate features which result in departures from these guidelines.

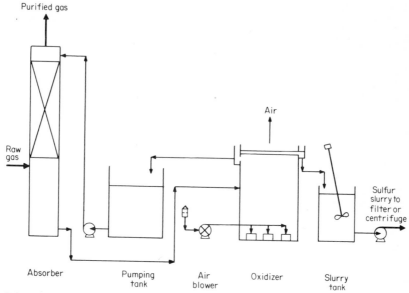

FIG. 13-6 Wet-oxidative process for H_2S removal.

sulfur froth is run off over a weir, and the regenerated solution is pumped back to the absorber.

The flow of air in the oxidizer is not sufficient to strip CO_2 from the liquor. The CO_2 content of the liquor therefore reaches equilibrium with the CO_2 in the gas, and no significant removal of CO_2 occurs. The H_2S content of the gas can be reduced to very low levels, but organic sulfur compounds are not removed. Three commercially available processes are listed in Table 13-5.

Methanation

The removal of small quantities of carbon oxides by reaction with hydrogen over a catalyst to form methane has been standard practice in ammonia plants for many years. The catalyst

TABLE 13-5 Wet-oxidative processes for H_2S removal

Process	Absorbent and reactant (in aqueous solution)	Licensor
Giammarco-Vetrocoke	Potassium carbonate + arsenate	Giammarco-Vetrocoke SpA, Marghera, Italy
Stretford	Sodium carbonate + anthraquinone disulfonic acid (ADA) + vanadium	British Gas Corporation, London
Takahax	Sodium carbonate + naphthaquinone sulfonic acid	Tokyo Gas Co., Tokyo

used contains 20 to 30 percent nickel, and the reaction is highly exothermic. However, the carbon oxides concentration in the feed is so low that a simple, once-through, fixed-bed reactor is satisfactory.

Methanation to SNG

In an SNG plant, the methanation stage is very different from the trace methanation step in an ammonia plant. The carbon oxide content of the feed gases is high, and to meet the product-gas specifications it is desirable that the carbon monoxide and hydrogen contents of the outlet gas be as low as possible.

A once-through reaction system is not suitable in these circumstances. A considerable amount of heat must be removed from the system to maintain the temperature within acceptable limits. In addition, conventional methanation catalysts deactivate rapidly under these conditions and special high-activity catalysts must be developed.

The problem of heat removal may be solved by recycling gas from the outlet of the methanator to the inlet through a waste-heat boiler. The recycle-to-fresh feed ratio required may be 5:1 or more. The feed-gas composition is usually adjusted by partial CO shift and CO_2 removal so that the H_2/CO ratio is 3:1, which is the stoichiometric ratio for the methanation reaction:

$$CO + 3H_2 \rightleftharpoons CH_4 + H_2O$$

In order to achieve equilibrium at the maximum temperature and thus maximize the calorific value of the product, one or more adiabatic stages may follow the main methanation reactor. A simplified flow sheet is shown in Fig. 13-7.

All methanation catalysts are sensitive to poisoning by sulfur, so a bed of zinc oxide is usually provided upstream of the primary methanator to eliminate sulfur compounds which slip through the main H_2S-removal system.

A number of demonstration plants have been operated to test this process. The first, at

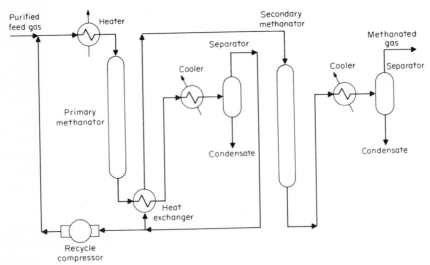

FIG. 13-7 Process for methanation to SNG.

Westfield, Scotland, methanated gas from a conventional Lurgi dry-ash gasifier, and the SNG produced was successfully distributed in the area through the natural gas grid.[2]

Other approaches have been proposed to deal with the problem of heat removal from methanators. These include reactors in which the catalyst is in particle form in a fluidized bed, is in the form of a slurry, or is in the form of a coating on the surface of tubes. These concepts have been demonstrated on a small scale and may have advantages which could result in their future utilization in commercial installations.

The HICOM process

It is a feature of the gas generated by the British Gas Corporation/Lurgi Slagging gasifier that the CO content is high and the steam content low compared with the gas from other processes. To obtain a gas with an H_2/CO ratio of $3:1$, it would be necessary to add a significant quantity of steam in order to convert CO to CO_2 and H_2.

The British Gas Corporation has therefore developed the HICOM (High Carbon Monoxide) process for the methanation of gases with a high CO content. Equilibrium dictates that the gas leaving the process contains little CO but a high percentage of CO_2, which is removed by conventional means.

The process route incorporates the recycling of methanated gas to remove heat from the system, as in the conventional processes. The relatively small quantity of steam required for the reaction may be added to the gas by direct contact with hot water in a saturator tower. Since the water can be heated by exchange with a low-temperature gas stream, low-grade heat can be recovered which would not be usable in the conventional route and the overall efficiency of the plant is increased.[3]

OPTIMIZATION OF THE GAS-PURIFICATION TRAIN

Alternative Process Routes

In the preceding sections, the various types of processes which may be included in a gas-treatment train have been briefly described. Within each category, there may be a number of different proprietary processes which differ in capital cost, energy requirements, chemical consumption, etc. Moreover, the individual processes may be assembled in the gas-treatment train in a number of different ways. Some of the available alternatives for SNG production are shown in Figs. 13-8 to 13-11.

In Fig. 13-8, the gas (scrubbed and cooled if necessary to remove particulates and tar) is partially shifted, and CO_2 and H_2S are removed by a nonselective process before methanation. If the gas is produced in a fixed-bed process, a physical solvent system is probably preferred because of its ability to remove organic sulfur compounds, HCN, and hydrocarbons in a prewash section. The mixed acid gas contains too little H_2S for satisfactory operation of a Claus unit, so a wet-oxidative scrubbing system is used to recover sulfur.

Figure 13-9 shows an alternative process route which incorporates a selective acid gas–removal system. Two streams of acid gas are produced; one of these contains all the H_2S and is sufficiently concentrated for satisfactory operation of a Claus unit. A tail-gas cleanup system is required in most cases to reduce the sulfur content of the tail gas to an environmentally acceptable level.

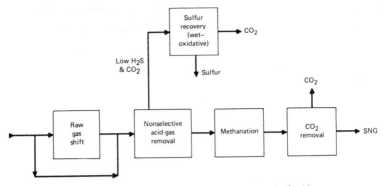

FIG. 13-8 Route to SNG incorporating nonselective removal of acid gas.

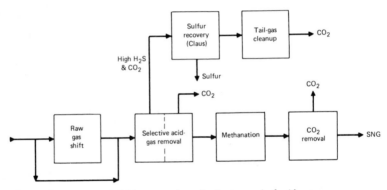

FIG. 13-9 Route to SNG incorporating selective removal of acid gas.

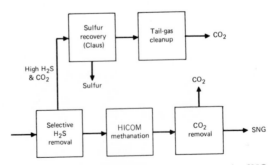

FIG. 13-10 The HICOM process as a step in SNG production.

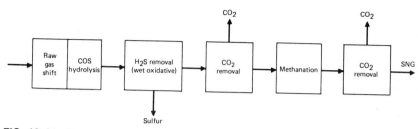

FIG. 13-11   Route to SNG incorporating a wet-oxidative process for sulfur removal.

A typical flow sheet incorporating the HICOM methanation process is shown in Fig. 13-10. This system is being developed for cases where the raw gas has a high CO content and insufficient water vapor for its complete conversion to CO_2, as in the case of the British Gas/Lurgi Slagging gasifier.

It should be noted that satisfactory operation of the Claus unit is susceptible to variation in the sulfur content of the coal. If the coal fed to the gasifier contains substantially less sulfur than design specifies, the H_2S content of the gas fed to the Claus unit will fall, possibly to a level too low for satisfactory performance.

In certain circumstances, especially at relatively low operating pressures or with gas having a relatively low H_2S content, it may be economically advantageous to use a wet-oxidative process for primary H_2S removal, with direct recovery of sulfur, as shown in Fig. 13-11. Since the absorption capacity of the solution is lower than that of conventional chemical and physical absorbents, the solution circulation rates will be high, but hydraulic turbines can be used to reduce power requirements. It should be noted that wet-oxidative processes do not remove organic sulfur compounds, so a hydrolysis reactor will usually have to be included.

Selection of Process Steps

Clearly, a number of process routes are feasible for the production of synfuels from coal-derived gases. In many cases, particularly for the removal of acid gases, there are also a large number of proprietary processes which may be suitable for each process step. Selection of the optimum combination of process steps for a particular set of circumstances forms a major part of the conceptual design effort for a synfuels project.

The particular requirements of an individual project greatly affect the selection of the processes to be employed. A comparison by Christensen[4] of a number of acid gas–removal processes demonstrates that the relative economics of the processes depends on such factors as the operating pressure, acid gas concentration, and presence of minor constituents in the raw gas. The relative costs and availability of utilities and energy will also vary considerably from one project to another, depending on the gasification process employed, location, accounting procedures, etc. It is essential that gas-treatment processes be considered in the context of the complex as a whole.

Reliability and Operability

It is also important that the gas-treatment system not be selected in terms of capital- and operating-cost considerations alone. The gas-treatment system represents a considerable proportion of the total capital cost of the plant, so there is an incentive to design for minimum capital cost. However, the number of gas-treatment trains is frequently much smaller than the number of gasifiers, and standby trains are not normally provided. Hence the loss of production that might result from trouble in the gas-treatment system is potentially several times greater than the loss that might result from gasifier problems.

In the conceptual design of a synfuels project, the gas-treatment system should therefore be considered in as much detail as the gasification stage, and selection of the gas-treatment processes must be made in terms of the economics and operability of the total complex.

REFERENCES

1. Sigmund, P. W., K. G. Butwell, and A. J. Wussler: "HS Process removes H_2S selectively," *Hydrocarbon Process.*, vol. 60, no. 5, May 1981, p. 118.

2. Landers, James E.: "Review of Methanation Demonstration at Westfield, Scotland," *Sixth Synthetic Pipeline Gas Symposium,* Chicago, Oct. 28–30, 1974, p. 297.

3. Tart, K. R., and T. W. A. Rampling: "Methanation key to SNG success," *Hydrocarbon Process.*, vol. 60, no. 4, April 1981, p. 114.

4. Christensen, K. G. and W. J. Stupin: "Comparison of Acid Gas Removal Processes," U.S. Department of Energy Report no. FE-2240-49, April 1978.

PART 4

SHALE OIL

Commercial production of shale oil from oil shale by retorting began essentially simultaneously in Australia, France, and Scotland in the mid-nineteenth century and continued until the middle of this century. By 1947, annual Australian production of gasoline from oil shale reached 100,000 bbl/year (16 $\times$ 10^6 L/year). The maximum French oil-shale throughput was 550,000 tons/year (500,000 Mg/year), which was reached in 1950. Scotland, at its peak in 1913, processed about 3,000,000 tons (2.7 $\times$ 10^6 Mg) of oil shale, with an annual shale oil production of approximately 6000 bbl (10^6 L).

Shale oil production was initiated in a number of countries in the 1920s and 1930s including China, Estonia, Spain, Sweden, Germany, South Africa, and the United States. Estonia first developed its deposits in the 1920s, and production continues to the present time. In 1978, reported oil-shale production in Estonia was approximately 14 $\times$ 10^6 bbl/year (2.2 $\times$ 10^9 L/year). The only other presently existing shale-oil production works is in China, where the production rate was about 6000 bbl/day (10^6 L/day) in 1979.

All of this production of shale oil involved pyrolysis or distillation of oil shale in a vessel called a "retort" to which heat can be supplied either directly through combustion within the retort or indirectly by performing the combustion outside of the retort and contacting hot gases or solids with oil-shale feed.

Contemporary directly heated processes are fired by burning fuel gas in the retort. This is shown schematically in Fig. I-1. This type of retorting has a high thermal efficiency, but the shale-oil recovery is only about 80 to 90 percent of the Fischer assay.

Most modern oil-shale technology involves variations of indirectly heated retorting. Many processes combust the gaseous products of the retort in an external furnace, with the combustion gases used to heat the retort as shown in Fig. I-2. These systems, which produce a carbonaceous spent shale and a medium-heat-content gas product, provide

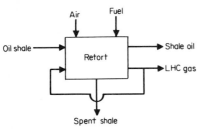

Air   Fuel

Oil shale →

Retort

→ Shale oil

→ LHC gas

Spent shale

FIG. I-1   Directly heated retort.

relatively low thermal efficiency because energy is not recovered from the residual carbon. However, oil recovery is high, 90 to > 100 percent of the Fischer assay.

Another major class of indirectly heated retorts involves the heating of a solid carrier, either spent shale or ceramic materials, by combustion of retort gas or residual coke. If the retorted shale is used as the heat carrier, the energy content of the residual coke is also combusted in the furnace as shown in Fig. I-3. Processes based on this system yield a medium-heat-content product gas and achieve about 100 percent of the Fischer assay.

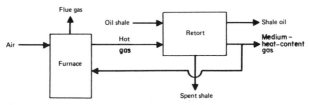

FIG. I-2   Indirectly heated retort, gas to oil shale.

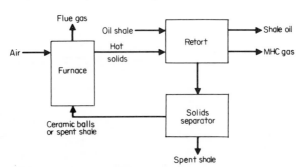

FIG. I-3   Indirectly heated retort, solid to oil shale. (Solids separator not required for Lurgi-Ruhrgas and Galoter processes.)

Retort structures may also be formed underground by a combination of explosive fracturing and mining. This is termed *in situ processing*. The necessary heat is provided by injecting air or air and steam to sustain movement of the fire front in the oil-shale formation in either a horizontal or vertically downward mode, causing the shale oil to collect at the bottom of the *in situ* retort, whence it is pumped to the surface. Thus, this would be a form of directly heated retort.

The major modern oil-shale technologies are listed in the following table together with some key process parameters.

Shale-oil production technologies

Process	Shale size, in (mm)	Retort	Heating mode	Exchange method
Lurgi-Ruhrgas	<0.25(<6)	Aboveground	Indirect	Solid (spent shale) to shale
Galoter	<1(<25)	Aboveground	Indirect	Solid (spent shale) to shale
TOSCO II	<0.5(<13)	Aboveground	Indirect	Solid (ceramic balls) to shale
Petrosix	Similar to Paraho	Aboveground	Indirect	Combustion gas to shale
Kiviter	1–4(25–100)	Aboveground	Indirect	Combustion gas to shale
			Direct	Combustion in retort
Superior	<3(<80)	Aboveground	Indirect	Combustion gas to shale
Union Oil Retort B		Aboveground	Indirect	Externally heated product gas to shale
Paraho	0.2–3(5–75)	Aboveground	Direct or indirect	Combustion gas to shale
Fushun Retort	0.3–3(8–80)	Aboveground	Direct	Air and water injection to retort
Geokinetics		True *in situ,* retort formed by explosive fracturing	Direct	Horizontal-moving fire front
Occidental MIS		Modified *in situ,* retort formed by expansion into mined voids	Direct	Vertical downward-moving fire front

The Lurgi-Ruhrgas, TOSCO II, Paraho, Occidental Modified *In Situ,* and Geokinetics *In Situ* processes are described in separate chapters. The remaining technologies are discussed in a summary chapter at the end of this section. This section also includes a chapter on oil-shale pre-beneficiation technology.

LURGI-RUHRGAS PROCESS

DR. ING. ROLAND W. RAMMLER
DR. ING. HANS-JUERGEN WEISS

Lurgi Kohle und Mineralöltechnik GmbH
Frankfurt am Main, Federal Republic of Germany

INTRODUCTION

The production of synfuels from coal, oil shale, and tar sands will certainly play an important role in the future. There are ample deposits of these three raw materials in many countries, and they are more evenly distributed throughout the world than crude oil.

Pyrolysis is an interesting alternative for the processing of these raw materials for synthetic oil production, and it may be a more attractive solution compared with other, sometimes more complex processing schemes. Hence, under favorable local conditions pyrolysis can be expected to be superior to other processes.

The Lurgi-Ruhrgas (LR) Pyrolysis Process was originally developed for the carbonization of coal. The process's great flexibility with regard to the properties of the feed material permits it to be applied to all three raw materials. The LR system is a typical example of flash pyrolysis. It features the use of circulating, fine granular heat carriers to supply the heat for the endothermic pyrolysis reactions and uses the pyrolysis residue itself as this heat carrier.

CHEMICAL AND PHYSICAL REACTIONS DURING PYROLYSIS

Coal Carbonization

Carbonization of bituminous coal with slow temperature rise

The structure of coal has been described as consisting of at least three basic complexes— the wax and resin complex, the oxyhumin complex, and the dehydrohumin complex. Of these, the wax and resin complex is the sole tar former. Not all the wax and resin material is recovered as tar and gas; some remains in the primary coke. This residual quantity increases with the degree of coalification and has been assumed to govern the caking capacity of the coal as well as the mechanical properties of the coke.

In contrast to previous views, primary decomposition of the coal is considered to be not only a depolymerization of the molecule but also a distillation of the wax and resin complex.

According to the thermobitumen theory, the high-molecular-weight high-carbon fractions of the coal change only a little during heating, while the low-molecular-weight hydrogen-rich fractions form a plasticizer for the nonfusing coal residue. This plasticizer is a thermal decomposition product of the coal (thermobitumen).

When bituminous coal is heated in the absence of air, drying proceeds and is then followed by devolatilization in three phases that involve the continuous fragmentation and dehydrogenation of the coal. Primary devolatilization extends to the softening of the coal. As the temperature increases further, major devolatilization takes place, which chiefly covers the plastic range, where primary tar and primary gas are formed. Resolidification toward the end of the plastic range is the result of evaporation and partial decomposition of the liquid phase. During the third phase, further devolatilization above 500 °C, gaseous products, primarily methane and hydrogen, evolve from the semicoke and the solid residue simultaneously shrinks. Heating beyond 900 °C produces little more gas but the carbonaceous residue continues to change in structure.

Major devolatilization is initially accompanied by contraction of the coal, which finally leads to much swelling when the coal has good coking properties. The devolatilization rate, plasticity, and swelling reach their maximums at almost the same temperature.

Oxidation of the coal reduces its plastic properties substantially. Even slight oxidation, for instance, during storage in the open, may have a distinct effect even before changes in the chemical composition can be noted. The pyrolysis of oxidized coals yields less tar but more CO_2, CO, and gas liquor.

Different inert components have different effects on the plasticity of coals under comparable conditions.

The pressure and composition of the gas environment are of major importance for the carbonization of bituminous coal. They influence the caking properties as well as the quality and quantity of the devolatilization products, which are also dependent on the coal rank as well as on other process parameters, such as temperature, heating rate, and gas velocity.

The behavior of sulfur during carbonization has been discussed in numerous publications. It depends, among other things, on the type of sulfur compounds originally present.

Flash carbonization of bituminous coal

Flash carbonization implies heating and destructive distillation of coal within a few seconds, substantially without secondary decomposition.

Flash carbonization can be described as an evaporation process. When the coal particle is heated, primary decomposition occurs and the liquid primary bitumen is separated from the solid residue. The primary bitumen has a boiling range around 350 °C. When the particle's surface temperature has reached this level, the primary bitumen begins to evaporate, while lower temperatures prevail inside the grain and primary decomposition proceeds toward the center. Evolved gases and thermal expansion force the liquid primary bitumen to the outside. As a result of the intensive heat supply it is largely evaporated before it can decompose. Only when no further primary bitumen passes to the surface is evaporation displaced to the particle's interior. In the final phase the outer zones of the particle become further heated and the semicoke continues to carbonize with the evolution of gas.

Observations made during flash carbonization of semibituminous coal [13.8 percent volatiles, maf (moisture- and ash-free)] seem to indicate that low carbonization temperatures cause a growth in grain size, while high temperatures effect a size reduction.

Questions regarding the behavior of the grain during flash carbonization require further investigations.

Carbonization of brown coal (lignite)

The organic components of brown coal are usually divided into three complexes. The first complex is humic matter, which is the bulk of the fraction that is insoluble in a benzene-alcohol mixture and is the chemically nonuniform degradation product of plants. The humic matter largely determines the specific physical properties of brown coal. The second complex includes cellulose and lignin, which are undecomposed or slightly altered plant substances and constitute substantial proportions of the xylite. The third complex, which is particularly important for thermal decomposition, includes the bitumen extractable with organic solvents and consists mainly of wax, resin, hydrocarbons, and oxyacids. These compounds stem chiefly from the outer protective layers of pollens, spores, and leaves; the cork fabric; and the resin of the original plants.

Raw brown coal contains up to 65 percent water. Therefore, drying is an important primary step in carbonization and takes place in two phases. In the first phase the drying rate is almost constant and the bulk-water flow through the capillaries prevails over diffusion to the surface. In the second phase inner diffusion governs the drying rate and the evaporation rate drops considerably.

The carbonization reactions are much more complicated than drying. Destructive distillation gradually commences when heating raises the temperature beyond the drying range. At first, deoxidation takes place followed by desulfurization. Fixed water, CO_2, and H_2S are set free ("Bertinierung"). Moreover, evolution of methane and homologues ("Carburitierung") occurs at temperatures above 280 °C. The course of CO_2 evolution depends on the degree of coalification and the petrographic composition of the coal. At the same time, exothermic reactions take place which are more pronounced the higher the oxygen content of the coal. In the same temperature range the bitumen formers are converted to bitumen. At 300 to 500 °C the original bitumen and the newly formed bitumen are converted to primary tar and high-Btu gas is produced. The primary tar vapors are partly decomposed. Evolution of tar terminates at about 520 °C. The evolution of gas, which contains an increasing concentration of hydrogen, continues with a further rise in temperature.

Decomposition of the coal in the temperature range of low-temperature carbonization is accompanied by a heat release of about 120–140 kcal/kg (500–600 kJ/kg) of maf coal.

During carbonization, the humic substances yield chiefly coke and gas and very little or no tar. Cellulose furnishes much more tar and decomposition water as compared to lignin, which yields more than twice as much coke. A high content of bituminous matter favors the tar yield. In particular, waxes furnish much more tar than resins, from which the tar yield is comparatively low.

When bituminous coal is heated, it passes through a temperature range where it becomes plastic; high-volatile subbituminous coal and brown coal, however, do not display plasticizing and caking during coking under atmospheric pressure. The devolatilization of brown coal and the ensuing structural changes are accompanied by a shrinkage of the solid residue. New cells are not formed during coking as is the case with bituminous coal, but instead a structure with numerous micropores remains. While bituminous-coal coke has a fused structure, brown-coal coke has a shrunken structure.

There is a relationship between coke quality and shrinkage caused by cohesion forces. The higher the shrinkage during carbonization the better the quality of coke produced under comparable conditions. Up to 600 °C linear shrinkage is usually between 11 and 17 percent.

The mechanical properties of the coke are affected by both the extent of shrinkage and the shrinkage rate.

Brown coals with a high tar content furnish coke of lower strength. Inert components, such as sand or clay, reduce shrinkage and coke strength, sand to a greater extent than clay. A sand content as low as 3 percent can cause the collapse of binderless brown-coal briquettes during pyrolysis.

Dried brown coal has a specific surface area of 0.5 to 5.0 m^2/g. It does not change during drying. It increases 10-fold during primary devolatilization up to about 400 °C, mainly as a result of the formation of macropores. During the major devolatilization phase micropores are preferentially formed and the BET surface increases to about 250 m^2/g. In the high temperature range (above 900 °C), there is a decrease in the specific surface area. Increasing the heating rate seems to increase the specific surface area.

Flash carbonization of brown and high-volatile subbituminous coals, under conditions for maximum tar yield, furnishes only 100 to 150 percent of the tar indicated by the Fischer assay versus 170 to 190 percent for the flash carbonization of bituminous coal.

Retorting of Oil Shale and Coking of Tar Sand

Oil shales are generally sedimentary rocks containing various amounts of organic matter. They are mostly of marine origin. The organic matter in oil shales is primarily an organic polymer called *kerogen*, which was formed principally from algal deposits. The organic material is almost insoluble in typical organic solvents and can normally only be extracted by pyrolysis.

Unlike oil shales, tar sands are sand or sandstone deposits which are impregnated with a heavy and viscous petroleum, or bitumen, which can be almost completely extracted by organic solvents. Coking of this bitumen involves reactions that are similar to kerogen pyrolysis.

A generally accepted model for kerogen pyrolysis is that the conversion of kerogen to oil is a multistep process in which bitumen is an intermediate. However, it should be emphasized that the bitumen formed from kerogen decomposition is not chemically identical to the naturally occurring bitumen. It has been shown that oil production can be fitted accurately into two consecutive first-order rate equations if a finite heating time is taken into account. It is worthwhile noting that for these rate expressions, the kerogen-to-bitumen reaction is the rate-determining step above 500 °C.

Many studies have focused on the amount of gaseous and solid products produced by each of the two reactions. One interpretation is that most of the carbonaceous residue is formed during the kerogen-to-bitumen conversion. In contrast, most of the gas is evolved during the bitumen-to-oil conversion, although small amounts of H_2S and CO_2 precede hydrocarbon evolution, e.g., with Australian kerogen.

Since liquid hydrocarbons historically have had a higher economic value than gaseous or solid hydrocarbons, special attention has been directed toward maximizing liquid yields. There are three basic chemical mechanisms which reduce oil yield from shale. They are coking, cracking, and combustion.

Coking can be defined as a liquid-phase condensation or polymerization reaction that results in the fusion of two or more molecular species, with the ultimate formation of a high-polymer carbonaceous product plus minor amounts of lower-molecular-weight gases and liquids.

Cracking involves vapor-phase bond-fission reactions that eventually lead to a distribution of smaller molecular units plus, possibly, some carbonaceous residue. An important difference between coking and cracking is the distribution of final products. As defined, cracking reactions lead to a greater evolution of lower-molecular-weight hydrocarbon fragments (CH_4, C_2H_6, etc.), whereas coking reactions produce mainly a high-polymer carbonaceous residue.

Combustion is for the most part a mass-transfer-limited reaction resulting from the coexistence of oxygen and organic material at high temperatures. Under fuel-rich conditions, combustion and cracking become closely related. Oil which escapes combustion at these high temperatures will undergo thermal cracking. In addition, oxygen is a good initiator of free-radical cracking, since it is a free radical.

As discussed in the previous section, oil production from bitumen can be described quite well by a single first-order rate expression.

The physical description of this process is as follows. The bitumen decomposes to oil, gas, and a carbonaceous residue. The oil exists in two physical states: as a liquid oil and as an oil vapor.

Oil in the liquid phase is subjected to two competing processes: vaporization, which leads to the production of oil, and coking, which leads mainly to a carbonaceous residue. For high liquid yields, one must hold down the residence time of the oil in the liquid phase in order to minimize oil coking. This can be accomplished either by high retorting temperatures and high gas sweep rates or by high heating rates. The cause for the increase in oil yield with an increase in heating rate is twofold. As a result of the kinetics of the reaction the average temperature of oil evolution depends on the heating rate. Higher heating rates result in the oil being evolved at a higher average temperature, hence a smaller fraction of the oil is in the liquid phase. Furthermore, higher heating rates increase the rate of oil vaporization and thus reduce liquid-phase coking. These two effects increase oil yield by reducing coke formation.

It can also be stated that high retorting temperatures increase liquid yield by decreasing oil coking. However, too high a temperature will decrease oil yield because of secondary vapor-phase oil cracking.

The conditions under which oil shale is retorted have a significant effect on the properties of the oil. It has been determined for Colorado oil shale that loss in the yield of oil by coking causes an increase in the H/C ratio and a decrease in the nitrogen content of the remaining oil. Associated with the increase in hydrogen content is a decrease in the density of the oil.

In contrast to oil coking, oil cracking causes a decrease in the H/C ratio and an increase in the nitrogen content of the remaining oil. Also the pour point of the oil (which is dominated by the long-chain alkane components) drops sharply with cracking.

These differing trends with coking and cracking can be easily understood by considering the following statements. Oil coking is most likely caused by condensation and polymerization reactions involving nitrogen-containing compounds. Since the nitrogen is contained mostly in aromatic rings, the removal of these hydrogen-deficient compounds causes an increase in the H/C ratio. In addition, the lower temperatures for oil evolution associated with coking cause a decrease in the alkene-to-alkane ratios in the oil.

In contrast, oil cracking causes a conversion of alkanes to alkenes and aromatics. This causes a decrease in the H/C ratio in the remaining oil, a result which is most likely a function of both temperature and conversion. Of further interest is the increase in nitrogen content with cracking. The reason for this increase is that the aromatic nitrogen compounds in shale oil are resistant to cracking. As the alkanes and alkenes are partially converted to gases, the nitrogen compounds are selectively concentrated. For example, the result of a 50 percent conversion is roughly a doubling of the nitrogen content.

LURGI'S EXPERIENCE

Coal Carbonization

When Lurgi started its activities in upgrading coal in 1925, the primary target for the devolatilization of coal was the production of liquid hydrocarbons, such as tars, middle oil, light oil, and naphtha. Later, in the thirties, the target shifted, and the production of coke for industrial use as well as for household purposes from low-quality coals attracted increasing interest.

The Lurgi Spülgas Kiln

The first contribution by Lurgi was the so-called Spülgas (recycled product gas) Kiln, which became available commercially some time between 1925 and 1926. The kiln consists of two shafts, one set upon the other. Several zones are incorporated into the kiln; they are the drying, retorting, and coke-cooling zones (see Fig. 1-1). The coal is fed in at the top and travels through the various zones, and the coke is taken out at the bottom. The coal is dried and carbonized by recycled hot product gas, the so-called Spülgas, which moves in counterflow.

Initially, brown coal was the only feed. The coke from the brown coal was mixed with a binder and then formed into briquettes which were marketed as smokeless coke. Plants

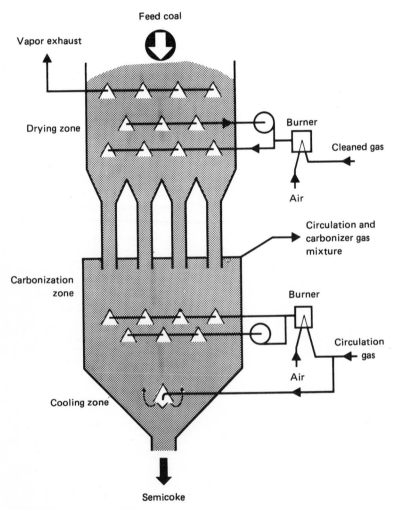

FIG. 1-1 Lurgi Spülgas kiln.

for this type of application have been designed and constructed for clients in the United States, Canada, New Zealand, Korea, and Japan.

In 1940 a plant with 80 Spülgas kilns was commissioned at Most, Czechoslovakia. This plant processed 25,000 tons/day (22,700 Mg/day) of hard brown coal and produced approximately 1,100,000 tons/year (1,000,000 Mg/year) of tar, which was hydrogenated to gasoline in an adjacent plant. The by-product coke was gasified for the production of hydrogen, which in turn was needed by the hydrogenation plant.

Around 1934 the Spülgas kiln was adapted for use with bituminous coal. At the same time the throughput capacity of the kiln was gradually increased and reached 500 tons/day.

During the years from 1935 to 1945 Lurgi designed and constructed in central Germany 98 Spülgas kilns for the production of liquid hydrocarbons from brown coal. In addition, 27 units were designed and erected for the production of tar and coke from bituminous coal. The tar produced was hydrogenated to diesel fuel and gasoline.

Despite the completely changed energy situation that occurred in the years after 1950, a few Spülgas kilns were still designed and erected, e.g., two units at Modderfontein in South Africa and 9 units at Neyveli, India (see Fig. 1-2).

FIG. 1-2   Lurgi Spülgas carbonization plant at Neyveli, India.

The Krupp-Lurgi carbonization process

The principle of this process is to heat high-volatile caking coal up to 600 °C in steel chambers. The steel chambers are heated from the outside by fuel gas mixed with recycled product gas. The chambers are sealed from the outside air. They are filled and emptied in a batchwise operation. Volatile products leave the steel chambers as gases and vapors and are condensed in a subsequent separator. The main advantage of this process over the conventional coke-oven process is a higher yield of liquid hydrocarbons.

Retorting of Oil Shale

As was mentioned earlier in this chapter, Lurgi started its coal carbonization activities in 1925 and branched out to retorting oil shale in 1931.

Lurgi has developed several processes for the carbonization of coal and the retorting of oil shale over the years. These developments, which finally led to the Lurgi-Ruhrgas process, are briefly described below.

The Spülgas kiln

The Lurgi Spülgas Kiln, of which more than 250 units were designed, built, and successfully operated for the carbonization and devolatilization of brown and bituminous coal, has been slightly modified for the retorting of oil shale (see under "The Lurgi Spülgas Kiln" in a previous section of this chapter). The shale travels in the Spülgas kiln from top to bottom. The shale is dried and retorted by recycling product gas (Spülgas) moving in counterflow to the shale. The Spülgas is internally preheated by cooling the spent shale in the lower shaft section, and it is finally superheated by being mixed with hot combustion gases from an external product-gas combustion chamber. This kiln yields a highly diluted gas.

Attempts were made to avoid dilution of the product gases by reheating the Spülgas in an external tube heater, but this concept was abandoned because of (1) the high cost of this heater, (2) its low availability, (3) sulfur corrosion, and (4) fouling by cracking products.

In 1944 a 3000-ton/day plant for retorting oil shale by using the Spülgas kiln was engineered for Spanish Puertollano oil shale, but because of the situation after World War II Pumpherston retorts were installed at Puertollano, Spain, instead of Spülgas kilns.

In 1958 a 300-ton/day Spülgas kiln including a residual-carbon gasification zone was installed near Kisangani for Zaire oil shale. The spent oil shale, a white product, was used in the cement industry and was therefore regarded as the main product, whereas the by-product shale oil and gas were used for underfiring the Spülgas kiln.

The maximum throughput of a single Spülgas kiln for oil shale is approximately 500 tons/day. The throughput per kiln is limited by the requirements for a uniform distribution of gas and solids over the shaft area of the kiln. Oil yields of up to 93 percent in relation to the Fischer assay have been obtained by this retort.

The tunnel kiln

The feed material in a tunnel kiln is retorted in wagonettes moved forward periodically through a tunnel (see Fig. 1-3). The heat carrier is retort gas, which is passed through tubular heaters and then through the shale bed in the wagonettes. The kiln is particularly suitable for very high-grade shales, which temporarily become plastic during retorting.

Two tunnel kilns, each with a throughput of 450 tons/day, were installed for the retorting of oil chalk at Heide, Germany before World War II. Four kilns of the same size were later erected for processing Estonian shale. The average oil yield from processing oil shale in a tunnel kiln reaches 95 percent of the Fischer assay.

Lurgi-Schweitzer process

The principle of this process is the retorting of lumpy low-grade oil shale in a descending gas stream. The Lurgi-Schweitzer retort consists of a cylindrical shaft with no refractory lining and is operated batchwise. After the retort is charged, the upper shale layer is ignited like a tobacco pipe and air is sucked through the shale bed from top to bottom, whereby

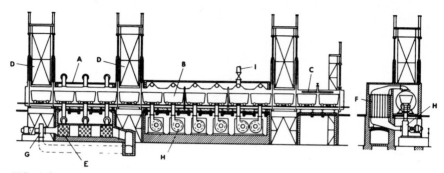

FIG. 1-3 Lurgi tunnel kiln. Key: A, drying zone; B, distillation zone; C, cooling zone; D, lock gates; E, tube heater of drying zone; F, tube heater of distillation zone; G, circulation blower of drying zone; H, circulation of blower of distillation zone; I, distillation-gas outlet.

the combustion zone moves slowly downward. The hot gases retort the shale below the combustion zone so that the heat required for the process is produced only by combustion of the residual carbon. On completion of the retorting process, the retort is tilted, discharged, and subsequently recharged. A general scheme of such a plant is depicted in Fig. 1-4.

A commercial plant for retorting oil shale, comprising 28 units, was designed and installed by Lurgi at Dotternhausen, Germany in 1942 and was in operation until 1949.

This process gives a maximum oil yield of 65 percent in relation to the Fischer assay.

The Hubofen

The original idea was to develop retorts with a circulating grate, but this operating method proved to be complicated and costly at the time this idea was being considered. The concept, however, led to the development of the "Hubofen," which consists of an inclined oscillating grate on which the shale is moved constantly and slowly downward. The air

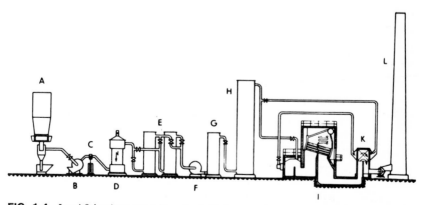

FIG. 1-4 Lurgi-Schweitzer retort. Key: A, distillation oven in operating condition; B, dust separator; C, gas-pressure controller; D, electrostatic precipitator; E, gas cooler 1st stage; F, blower; G, gas cooler 2d stage; H, gas scrubber; I, steam-boiler unit; K, preheater; L, stack.

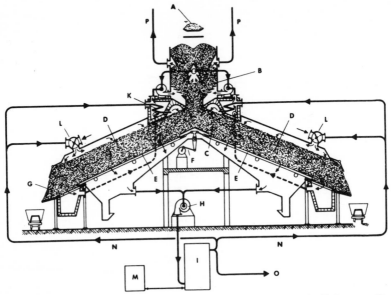

FIG. 1-5 Lurgi Hubofen. Key: A, conveyor belt; B, predryer; C, predistillation zone; D, distillation and gasification chamber; E, moving grates; F, grate drive; G, final combustion grate; H, gas blower; I, condensation; K, heat exchanger; L, control air and recycle gas; M, oil tank; N, recycle gas; O, to power plant; P, condensate.

sucked through the shale bed inclines the combustion and retorting zones in such a way that retorting is just completed at the grate end (Fig. 1-5).

This retort is remarkable for its simple setup, its substantially open construction, and its favorable heat economy. In 1944 two 60-ton/day pilot plants based on this process were built, one at Dotternhausen, Germany, and the other at Messel, near Darmstadt. In 1954 a third plant of the same size was constructed at Kohlscheid, near Aachen, FRG, for the gasification of washery refuse.

Retorting in situ and in piles

During World War II Lurgi also participated in the efforts toward developing *in situ* retorting of oil shale in Württemberg, Germany. Chambers were formed in the shale seam by making parallel adits. The lower layer of the seam was mined, and the upper part of the seam was blasted by explosive charges. The retorting process was initiated by suction in one adit, and after ignition of the full chamber length, retorting proceeded to the parallel adit. Because much infiltrated air was present, the oil recovery was low and the hazard of explosions great. Furthermore, it was not possible to enter the adits later on, as the hazards of poisoning and explosion could not be sufficiently suppressed.

Because of problems with uneven grain-size distribution in the *in situ* retorts that resulted in poor control and gas distribution, retorting in piles was tried. This form of retorting consists of piling more uniformly crushed, lumpy oil shale aboveground over long, perforated exhaust pipes, that are resting on the ground, and igniting the surface of the pile. Because of a side wind, retorting was not sufficiently uniform and oil recovery was moderate. The poor results obtained with retorting *in situ* and in piles during World War II showed that retorting of oil shale was economical and safe only in appropriately designed aboveground retorts.

LURGI-RUHRGAS PROCESS

Development of the LR Process

LR pilot plants

The LR process evolved over many years to its present capability for handling a wide variety of feeds, ranging from coal to liquid hydrocarbons, tar sands, oil shale, and diatomaceous earth. At the end of the forties and the beginning of the fifties there was a shortage of town gas (medium-Btu gas) in Germany. The problem of how to produce additional quantities of town gas without bringing coke into the market at the same time was considered. The Lurgi and Ruhrgas companies joined forces in 1949 and established a scheme for devolatilizing and carbonizing coal prior to its being fired in a power plant and thus prepared the ground for the development of the process that today is known as the *Lurgi-Ruhrgas process.*

The development work was done jointly by the two companies and was originally aimed at the carbonization of coal at high temperatures. The initial version of the retort used ceramic balls as heat carriers, a pyrolysis drum, and a ball heater. Because of technical problems the heat carrier was changed from ceramic balls to produced fine char. Smaller heat carriers had the following advantages:

1. They improved the heat transfer of the circulating material.
2. They permitted the pneumatic lifting and conveying of the circulating material.
3. They eliminated bothersome mechanical elevators.
4. They simplified reheating of the circulating material by combustion of some of the char with the rising air.

The modified process was tried out in a 12-ton/day (11-Mg/day) pilot plant at Herten, FRG, over several years (see Table 1-1). The devolatilization of coals ranging from low-volatile bituminous coal to lignite was studied at temperatures between 450 and 900 °C.

Parallel with the experimental work at Herten a larger demonstration plant was operated from 1958 to 1960 at Dorsten, FRG. (See Table 1-1.) This pilot plant processed 240 tons per day of high-volatile bituminous coal with moderate caking properties at 900 °C, in close combination with the power station of the Fürst Leopold colliery. The main purpose of this demonstration plant was the production of town gas and coal tar. This unit is illustrated in Fig. 1-6. The tests in Dorsten were stopped in 1961 because by then the increasing availability of natural gas from the Netherlands and of Middle East crude oil made the production of gas by devolatilization of coal economically unattractive.

During this development work the high flexibility of the Lurgi-Ruhrgas process system and the wide range of possible feedstocks were recognized. It was found that by using sand as the circulating heat carrier, the process could be used to crack naphtha or crude oil to olefins. This variation was investigated at length in a pilot plant operated by Bayer

TABLE 1-1 Plants using the Lurgi-Ruhrgas process

Client	Location	Product(s)	Circulating material(s)	Feed capacity (feedstock)	Operating period
Ruhrgas AG	Herten, FRG	Various (pilot plant)	Ceramic balls (initially); sand or char (later)	0.5 Mg/h (coal, oil shale, tar sand)	1949–1968
Ruhrgas AG	Dorsten, FRG	Town gas, coal tar	Char	10.0 Mg/h (bituminous coal)	1957–1961
Bayer AG	Leverkusen, FRG	Olefins (pilot plant)	Sand	1 Mg/h (crude oil)	1955–1957
Erdölchemie GmbH	Dormagen, FRG	C_2–C_4 olefins	Sand	10 Mg/h (naphtha)	1958–1969
VEB Leuna Werke	Leuna, German Democratic Republic	C_2–C_4 olefins	Sand	2×10 Mg/h (naphtha)	1965–present
Maruzen Petrochemical Co. Ltd.	Chiba, Japan	C_2–C_4 olefins	Sand	18 Mg/h (naphtha)	1964–1971
Duperial S.A.I.C.	Rosario, Argentina	C_2–C_4 olefins	Sand	8 Mg/h (naphtha)	1964–present
Lanzhou Petrochemical Complex	Lanzhou, People's Republic of China	C_2–C_4 olefins	Sand	26.5 Mg/h (crude oil)	1968–present
Coke-oven Works Lukavac	Lukavac, Yugoslavia	Coke-oven feed	Char	66.7 Mg/h (lignite)	1963–1968
Ruhrkohle AG	Bottrop, FRG	Char (for Form coke)	Char	14.6 Mg/h (bituminous coal)	1974–1978
Bergbauforschung GmbH	Essen, FRG	Various (pilot plant)	Various	0.3 Mg/h (oil shale, tar sand, coal)	1970–1980
British Steel Corp.	Scunthorpe, England	Char (for Form coke)	Char	33.3 Mg/h (bituminous coal)	1978–late 1979
Lurgi Kohle u. Mineralöltechnik GmbH	Frankfurt am Main, FRG	Various (test plant)	Various	1 Mg/h (coal, oil shale, tar sand)	1980–present
Getty Oil Co.	Bakersfield, Calif., U.S.A.	Raw oil from diatomaceous earth	Spent diatomaceous earth	9 Mg/h (diatomaceous earth)	1981–present

FIG. 1-6   LR demonstration plant at Dorsten, Federal Republic of Germany.

at the works at Leverkusen, FRG, from 1955 to 1957 (see Table 1-1). The commercial units built on the basis of this development became known as *sandcrackers*. Around the same time the 13-ton/day pilot plant at Herten was being operated on a large variety of feedstocks including tar sands and oil shales and thus demonstrated that the Lurgi-Ruhrgas process is very suitable for the production of oil from these materials.

Coal carbonization

In 1961, after the developmental work on coal carbonization had been completed, two commercial LR units with a throughput capacity of 1765 tons/day (1600 Mg/day) of predried lignite were designed by Lurgi, erected at Lukavac, Yugoslavia, and commissioned in 1963 (see Table 1-1). Figure 1-7 shows the plant which was built to produce fine-grained, low-temperature char—having a volatile-matter content of 15 to 22 percent—for use as a blending char in a conventional coke-oven works.

In conjunction with the research and development work on a form coke process, three LR plants were designed and built (see Table 1-1). A pilot plant for this process, including the LR unit, exists at the research facilities of Bergbauforschung at Essen, FRG. A demonstration plant with a feed of bituminous coal to the LR unit of 387 tons/day was started up at the Prosper colliery at Bottrop (Ruhr area) in 1974. A second demonstration plant

FIG. 1-7 LR coal-carbonization plant at Lukavac, Yugoslavia.

with a feed of bituminous coal to the LR unit of 900 tons/day was operated at Scunthorpe, England until 1979. This unit is depicted in Fig. 1-8.

Production of olefins

The LR process variation by which naphtha and heavier hydrocarbons are cracked to olefins became commercial in 1957, when the first sandcracker was commissioned at Dormagen, FRG (see Table 1-1).

Further sandcrackers with various improvements were built in the German Democratic Republic (2 units), Argentina, Japan, and China (see Table 1-1). The sandcracker in the People's Republic of China is shown in Fig. 1-9. This unit produces 72,000 tons/year (65,000 Mg/year) of C_2–C_4 olefins from crude oil and represents the largest LR system built so far. A unit of this size applied to the retorting of high-volatile bituminous coal would permit a feed rate of almost 3000 tons/day.

Processing oil shales, tar sands, and diatomaceous earth

As was already mentioned, the 13-ton/day test unit at Herten processed not only coal of various types but also oil shales and tar sands (see Table 1-1). Oil shales and tar sands were

FIG. 1-8 LR demonstration plant at Scunthorpe, United Kingdom.

also processed at the 9-ton/day test unit at Essen. During these tests shale samples of up to 200 Mg were processed.

Since early 1980 Lurgi has been operating a 26-ton/day LR pilot plant at its research and development center in Frankfurt am Main (see Fig. 1-10 and Table 1-1). The purpose of this pilot plant is to study the behavior of various feedstocks during retorting, to improve the process as well as certain components, and to gain data for the design of large-scale units.

Since late 1981 a 220-ton/day LR demonstration plant has been operating at Bakersfield, California (see Table 1-1). This plant processes diatomaceous earth and is shown in Fig. 1-11.

Carbonization of Coal

Process description

Figure 1-12 shows the flow diagram of the LR coal-carbonization plant at Lukavac, Yugoslavia, which was designed to operate independently of a boiler unit. The feed coal from the bunker is blown to the reactor by cooled carbonization gas; there it meets the hot char, which flows continuously from the separating and collecting bin through a controlling and distributing mechanism to the reactor. The injection nozzles and the char inlet are arranged

FIG. 1-9 LR sandcracker at Lanzhou, People's Republic of China.

so as to mix the coal and char as thoroughly as possible. Because of the intensive heat exchange the coal is heated very quickly. The gases and condensable carbonization products set the upper layer of the fuel bed in the reactor into a gentle motion which promotes the intermixture. The mixture of circulated and freshly produced char flows to the lift pipe, where it is lifted pneumatically to the collecting bin by air or combustion gases and is heated at the same time. The lift pipe is heated with carbonization gas or, preferably, by partial combustion of the char in the presence of conveying air preheated to about 370 °C. The surplus char that accumulates in the recycle system of the heat carrier is discharged continuously from the collecting bin and passed to a char-cooling section.

The waste gas that has been freed from most of the char in the separating and collecting bin flows through the air preheater and then through the multicyclone at approximately 390 °C. It is then supplied to an adjacent raw-coal drying plant, where its low heating value and the remainder of its sensible heat are utilized.

The carbonization gas set free in the reactor is dedusted in a cyclone and cooled in a spray cooler to about 105 °C. It is cooled chiefly by evaporation of injected gas liquor in the presence of flushing tar, which is recycled continuously through the spray cooler and the scrubbing cooler beneath. In the scrubbing cooler the residual dust is removed from the carbonization gas by the flushing tar. The electrostatic precipitator downstream separates the tar mist from the gas; its bottom section serves at the same time as a collector for the tar. In the tubular cooler the gas is cooled indirectly to about 30 °C. The conden-

FIG. 1-10 LR pilot plant at Frankfurt am Main, Federal Republic of Germany.

FIG. 1-11 LR demonstration plant at Bakersfield, California.

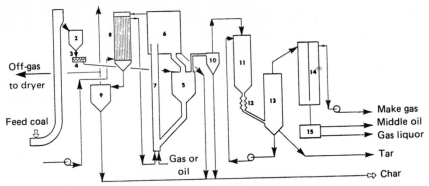

FIG. 1-12 Flow sheet of an LR carbonization plant. Key: 1, vertical conveyor; 2, feed bin; 3, downcomer; 4, dosing screws; 5, carbonization vessel; 6, separating bin; 7, lift pipe; 8, air preheater; 9, multicyclone; 10, product-gas cyclone; 11, injection cooler; 12, uniflow scrubber; 13, electrostatic precipitator; 14, tubular cooler; 15, oil and water separating tank.

sation products—middle oil and gas liquor—are separated in the separating tank and piped to storage tanks.

Most of the char produced is withdrawn from the collecting bin. Smaller amounts accumulate in the mulyicyclone and cyclone. All of the char is passed through a cooling unit. The LR coal carbonization process can be operated at temperatures between 430 and 900 °C.

In units with high capacities and in plants processing caking coal, the coal is intermixed with the hot heat carrier in a double-shaft mixer of special design. Such a mixer is also employed when maximum tar yields are desired, as it enables the residence time of the tar vapor in the hot zone of the retort to be relatively short. The mixer is located between the collecting bin and the carbonizer shaft. Pyrolysis is accomplished chiefly in the mixer, and the carbonization gas and vapors are drawn directly from the mixer into the condensation equipment.

Material balances

Material balances for various pilot plant runs are shown in Table 1-2. These balances refer to test runs performed in the pilot plant at Herten with various high-volatile coals. The operating conditions were set to achieve maximum tar yield; i.e., the carbonization temperature was around 600 °C. The high yields of liquid hydrocarbons as compared with the Fischer assay numbers show the flash carbonization effect which failed to show up in the processing of the coal sample from Pakistan, which may be described as a hard brown coal.

Products

Table 1-3 shows typical operating data from the LR carbonization plant at Lukavac, Yugoslavia. The composition and properties of low-temperature tars from pilot plant runs with various feed coals are shown in Table 1-4.

As a result of the flash heating, the *tars* contain more high-boiling fractions than tars produced at low heating rates. Consequently, they have higher densities and solidify at

TABLE 1-2 Material balances of pilot plant runs

	Run No.			
	1	**2**	**3**	**4**
Coal				
Type	High-volatile bituminous coal A	Subbituminous coal A	Subbituminous coal A	Subbituminous coal B
Place of origin	Germany	Arizona	Utah	Pakistan
Moisture, %	3.3	8.9	6.0	14.0
Ash, %	5.8	4.9	13.3	5.9
Volatile matter, %	34.6	38.5	36.7	37.5
Volatile matter (maf), %	38.1	44.6	45.5	46.8
Tar (Fischer assay), %	14.7	11.3	11.9	10.5
Caking index (Damm)	~20	6	<3	0
Grain-size distribution, wt %				
>2 mm	9.9	10.5	8.8	27.2
2–1 mm	20.0	23.7	23.0	22.8
1–0.5 mm	19.1	27.3	26.4	19.2
0.5–0.2 mm	22.5	20.5	18.8	14.5
0.2–0.1 mm	12.1	8.8	8.0	6.5
<0.1 mm	16.4	9.2	15.0	9.8
Carbonization temperature, °C	583	590	599	649
Products, kg/Mg of feed coal				
Circulated material	511	186	310	193
Dust in carbonization gas	31	122	133	162
Dust in flue gas	104	178	124	100
Carbon burned in lift pipe	10	11	4	48
Coke	656	497	571	503
Tar (dust-free)	178	151	132	42
Oil (dry, dust-free)	19	42	32	31
Oil in gas liquor	2	5	4	1
Gas naphtha	7	15	13	16
Liquid hydrocarbons	206	213	181	90
% of Fischer assay	140	188	152	97
Gas liquor	81	144	115	178
Water in oil	—	6	4	—
Water in carbonization gas	1	1	1	11
Water	82	151	120	189
Carbonization gas	51	108	89	211
Post-devolatilization gas	5	31	39	7
Gas	56	139	128	218
Total	1000	1000	1000	1000

TABLE 1-3 Typical operating data from LR plant at Lukavac, Yugoslavia

	Run No.	
	1	2
Volatile components in the coke (water-free), %	20.2	16.9
Feed coal		
Water content, %	7.0	6.9
Ash content (water-free), %	8.1	7.9
Feed rate, Mg/h	26.9	33.4
Consumption		
Heat (for firing), kcal/kg of feed coal	250	310
Electricity, kWh/h	355	380
Steam (450 kPa · g), kg/h	850	850
Cooling water (Δt = 15 °C), m³/h	435	540
Temperature, °C		
Hot air	360	371
Coke at outlet of carbonizer shaft	434	447
Off-gas at outlet of collecting bin	501	534
Off-gas downstream of air preheater	376	392
Carbonization gas at outlet of carbonizer shaft	439	465
Carbonization gas downstream of injection cooler	101	101
Carbonization gas downstream of cross-tube cooler	25	31
Product, kg/Mg of feed coal		
LTC coke* + tar (water-free)	698	679
Middle oil	13	13
Carbonization gas including gasoline (for firing)	73	83
Carbonization liquor (calculated)	176	178
Miscellaneous	40	47
Heat output†, %		
Coke and tar	86.8	84.4
Middle oil	2.2	2.2
Carbonization liquor	0.3	0.3
Off-gas	5.5	7.9
Cooling water	4.2	4.2
Heat losses	1.0	1.0

*Low-temperature carbonization coke.

†Calculated on the basis of the heat content of the feed coal and H_0.

higher temperatures. Data for the middle oil and gas naphtha produced are also given in Table 1-4.

The heavy tar, which represents the major portion of the total liquid hydrocarbon product, is contaminated with dust as a consequence of the inevitable carry-over of dust from the retort into the condensation section. The conversion of the tar to syncrude or marketable products by hydrotreating requires either an upstream dedusting step or an alternative hydrogenation process which tolerates dust in the feed. Instead of being dedusted in a separate plant, the heavy tar may be recycled to the mixer of the LR carbonizer for redis-

TABLE 1-4 Composition and properties of low-temperature tars from pilot plant runs with various feed coals

	Tar Fraction from Subbituminous Coal A*			Tar Fraction from Subbituminous Coal A†			Tar Fraction from Lignite‡		
	Heavy tar	Middle oil	Gas naphtha	Heavy tar	Middle oil	Gas naphtha	Heavy tar	Middle oil	Gas naphtha
Percentage of total tar, wt %	70.7§	22.4	6.9	72.7§	20	7.3	64.9§	33.7	1.4
Ultimate analysis, wt %									
Ash	1.57	0.16		3.28	0.007		1.95		
C	81.81	83.36		79.9	83.09		81.38	84.65	84.65
H	5.98	9.76		5.8	8.92		8.88	10.28	10.28
S	0.37	0.43		0.48	0.51		0.46	0.93	0.79
N	1.01	0.33		1.20	0.66		0.78	0.50	0.55
O	9.26	5.96		9.34	6.75		6.55	3.64	3.73
Density at 20 °C, kg/L	1.259	0.95	0.738	1.234	1.00	0.728	1.0535	0.892	0.780
Dust content, wt %	9.3	0.02		6.7	3.83		3.01	0	0
Conradson residue (in dust-free tar), wt %	45.3	1.06		45.7	8.91		15.5		
Flash point, °C	305	61		251	70				
Softening point, °C (Krämer Sarnow)	129			116				18	
Pour point, °C		−19			+2				
Viscosity, cSt/°C	51.7/200	7.7/20		45.8/220	40.7/20		41.7/120	2.8/20	
Initial boiling point, °C		85	33		118			85	45
Final boiling point, °C		318	139		329		471	296	182
Distillate, vol %		89.5	97.0		81		71.28	92.7	95.8
Last runnings, vol %		1.5	0.5		1.5		27.98	6.7	3.8
Residue and loss, vol %		9.0	2.5		17.5		0.74	0.6	0.4

*Volatile matter (maf), 44.6 wt %; tar content (maf), 13.1 wt %.
†Volatile matter (maf), 45.5 wt %; tar content (maf), 14.8 wt %.
‡Volatile matter (maf), 55.1 wt %; tar content (maf), 21.9 wt %.
§Based on dust-free tar.

tillation and coking. The coke retains most of the dust, and the distillate represents a higher-quality tar.

Because circulating granulated heat carriers are used rather than combustion gas, the *carbonization gas* does not become diluted with combustion gas and consequently has a low nitrogen content. Table 1-5 shows the composition of the low-temperature carbonization gas from the LR coal-carbonization plant at Lukavac, Yugoslavia, at various devolatilization temperatures.

TABLE 1-5 Composition of carbonization gas at Lukavac, Yugoslavia

Component, vol %	Approximate Carbonization Temperature, °C		
	435	450	470
$CO_2 + H_2S$	51.4	48.0	44.9
C_nH_m	2.9	3.7	4.0
CO	11.8	11.5	11.4
H_2	10.8	13.1	15.2
CH_4	22.1	23.0	23.7
N_2	1.0	0.7	0.8

Char is obtained at three locations: (1) from the heat-carrier circulation system from which it is withdrawn as circulated material, (2) from the flue-gas cyclones, as dust, and (3) from the carbonization-gas cyclones, as dust. These solid flows can be kept separate if this is desirable for further processing, or they can be combined into one total char flow. The analytical data in Table 1-6 refer to such mixtures. The content of volatile substances in the char is determined by the temperature in the collecting bin, which is normally 100 to 150 °C higher than the carbonization temperature. The grain size of the feed coal affects the particle distribution of the char, but it depends primarily upon the tendency of the coal to disintegrate during pyrolysis and upon the mechanical stability of the individual grains.

The volatiles content of the char is normally 5 to 8 percent, which makes the char highly reactive. It is therefore well suited for combustion and gasification. Generally the char should be used in the same condition in which it leaves the retort. This applies to both the temperature and to the grain size.

The *gas liquor* contains up to 30 g/L of monovalent and polyvalent phenols, as well as fatty acids, free and fixed ammonia, H_2S, and other contaminants. Accordingly, high BOD (biological oxygen demand) and COD (chemical oxygen demand) values are found. The pH range of the gas liquor is around 8 to 9. The liquor needs thorough treatment, e.g., phenol extraction by the Lurgi Phenosolvan process and biological treatment, before it can be discharged.

Retorting of Oil Shale

The LR process can be applied to the carbonization of coal, to the direct coking of tar sands, to the retorting of oil shale, and to the production of syncrude from diatomaceous earth. The following section deals in detail with the application of the LR process to the retorting of oil shale. The differences in the application of the process to coal have already been described, and the application of the process to tar sands will be discussed in a later section.

TABLE 1-6 Composition of total char from pilot plant runs

	Run No.*			
	1	2	3	4
Carbonization temperature, °C	583	590	599	649
Char				
Proximate analysis, wt %				
Ash	9.8	10.8	28.5	13.0
Volatile matter	7.4	6.4	4.3	4.9
Fixed carbon	82.8·	82.8	67.2	82.1
Ultimate analysis, wt %				
Ash	9.8	10.8	28.5	13.0
C	81.2	82.7	66.6	
H	1.7	1.6	1.2	
S	1.1	0.4	0.5	5.0
N	1.6	1.1	0.8	
O	4.6	3.4	2.4	
Grain-size distribution, %				
>2 mm	2.2	1.8	5.1	2.8
2.0–1.0 mm	6.3	12.4	18.3	15.3
1.0–0.5 mm	49.3	30.0	24.8	18.5
0.5–0.2 mm	26.4	31.2	21.3	22.2
0.2–0.1 mm	6.1	12.2	10.1	17.4
0.1–0.063 mm	9.7	5.1	6.2	7.5
<0.063 mm		7.3	14.2	16.3

*Run no. in this table corresponds to the run no. and use of the same coal feedstock as in Table 1-2.

Process description

The LR process for the retorting of oil shale resembles the scheme for coal carbonization and is described here with Figs. 1-13 and 1-14.

The LR process consists principally of the circulation system, which includes the lift pipe (1), collecting bin (2), mixer (3), and surge bin (4), the product-gas section, and the flue-gas section.

Circulation system (LR-loop)

The circulation system, or "loop," consists of the following elements (please refer to Fig. 1-13):

• A lift pipe (1) to convey and heat the circulating fine-grained heat carrier

• A collecting bin (2) to separate the combustion gas from the hot heat carrier

• A screw mixer (3), which induces retorting by mixing the hot heat carrier and the raw shale feed

• A surge bin (4), which provides surge capacity and time to complete retorting

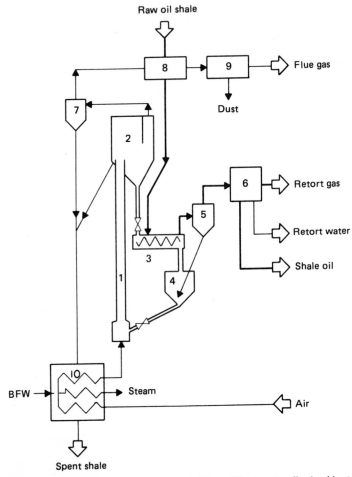

FIG. 1-13 Simplified LR process scheme. Key: 1, lift pipe; 2, collecting bin; 3, screw mixer; 4, surge bin; 5, cyclone; 6, cooling and condensation system; 7, cyclone; 8, shale preheaters; 9, electrostatic precipitator; 10, fluidized bed cooler.

On the product-gas side of the loop are the cyclone (5) and the cooling and condensation facilities (6) for the oil vapors.

On the flue-gas side of the loop are the cyclone (7), the shale flash preheater (8), and the electrostatic deduster (precipitator) (9). Hot spent shale is withdrawn from the collecting bin (2) and cyclone (7) and cooled in a fluidized-bed cooler (10) where process air is preheated and steam is generated.

In the loop, raw oil shale, crushed to approximately ¼ in and preheated to approximately 150 to 200 °C, is fed into the screw mixer, where it is mixed with two to four times as much hot (650 to 750 °C) spent shale from the collecting bin. The raw shale is thereby flash-heated to about 500 to 540 °C and is retorted within a few seconds. Retorting is

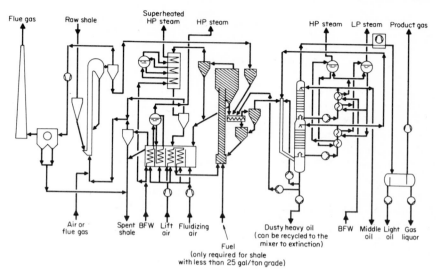

FIG. 1-14 LR process scheme for the retorting of oil shale.

enhanced by self-fluidization of the solids in the mixer because of the gases and vapors evolved from the raw oil shale.

The mixture of fresh spent shale and heat carrier leaves the mixer and passes to the surge bin, where retorting is completed, and is then transferred to the lower section of the lift pipe. Combustion air, preheated to about 450 °C, is introduced at the bottom of the lift pipe. The air simultaneously conveys the shale to the top of the lift pipe and burns the residual carbon from the spent shale.

The combustion gas and reheated spent shale are separated at about 650 to 750 °C in the collecting bin. The spent shale is returned to the screw mixer, thereby closing the loop. The overall process is outlined in more detail in Fig. 1-14.

Condensation system

The *condensation system* (6) shown on the right side of the LR loop consists mainly of a cat-cracker fractionator-type condensing tower and a final air cooler-condenser. The condensing tower may serve several LR trains and consists of two stages. In the lower stage a heavy-oil fraction is condensed at approximately 200 to 260 °C and residual dust is removed. Heavy-oil dust concentrations can easily be kept low by recycling heavy oil to the mixer. If heavy oil is recycled to extinction, a completely dust-free oil product is obtained at the expense of a 3 to 5 percent lower liquid yield and a somewhat higher coke and gas production. In the upper stage of the condensing tower the product vapors are cooled to 100 to 150 °C and a water- and dust-free middle oil is condensed. In the final air cooler-condenser unit, light oil and gas liquor (retort water) are condensed, leaving a high-Btu retort gas at approximately 50 °C, which can be used as feedstock for a downstream hydrogen production or as fuel.

Shale preheater

The *shale preheater* (8) consists of a vertical riser in which the feed shale is pneumatically elevated and preheated by the hot LR flue gases. Coarser particles separated from the gas upstream of the main cyclone that collects the feed shale may be recycled to the riser. Pressure at the bottom of the riser is controlled by a flue-gas blower downstream of the cyclone that collects the feed shale. In order to adjust the heat demand of the shale preheater to the heat supply from the LR flue gas, a small steam generator may be incorporated into the LR-flue-gas duct.

The flue gas is then finally dedusted in an *electrostatic precipitator* (9) before being discharged to the atmosphere.

The 650-to-750 °C hot spent shale withdrawn from the LR system is cooled to approximately 200 °C in a multistage *fluidized-bed cooler* (10) in which the LR process air is preheated to about 450 °C and additional steam is produced. If the spent shale still contains some combustible carbon left unburned from the LR system, this carbon can be combusted in the first stage of the fluidized bed so that no potential fuel is left unused in the process.

Material and heat balances

Typical material and heat balances for the LR process, based on 1 Mg of raw oil shale, are given in Tables 1-7 and 1-8, respectively, for a 23-gal/ton Colorado oil shale. The balances refer to recovery of heat from the flue gas and spent shale down to a temperature of approximately 200 °C and to air cooling in the condensation section.

As can be seen from these balances, the oil yield is 100 percent of the Fischer assay, a high-Btu retort gas is obtained, and, in addition, superheated steam is produced.

The total thermal efficiency η_{LR} of the LR process is defined as the amount of oil and gas produced per metric ton of oil shale multiplied by their respective lower heating values (LHV) plus the recoverable useful heat for steam production, all divided by the LHV of 1 Mg of feed shale. In the example shown in Table 1-8, the total thermal efficiency as a percentage value is 92.7 percent.

TABLE 1-7 Overall material balance for Colorado oil shale*

Input, kg	
Oil shale (1.7% moisture)	1000.0
Total air	533.4
C_3 − gas	5.6
Total input	1539.0
Output, kg	
Total oil + naphtha (C_4+)	87.1
C_3 − gas	28.9
Gas liquor (sour water)	10.0
Spent shale (dry)	754.3
Flue gas	658.7
Total output	1539.0

*All figures based on 23-gal/ton Colorado oil shale and 100% Fischer oil recovery.

TABLE 1-8   Overall heat balance for Colorado oil shale*

Input, kJ/Mg	
Oil shale	4 702 000
Total air	5 000
C_3 − gas	124 000
Total input	**4 831 000**
Output, kJ/Mg	
Total oil + naphtha (C_4+)	3 492 000
C_3 − gas	642 000
Heat for steam prod	347 000
Spent shale (at 200 °C)	149 000
Flue gas (at 210 °C)	130 000
Heat losses	71 000
Total output	**4 831 000**
Thermal efficiency η_{LR}†	0.927

*All figures based on 23-gal/ton Colorado shale, 100% Fischer oil recovery, and lower heating values.

$$\eta_{LR} = \frac{3{,}492{,}000 + 642{,}000 - 124{,}000 + 347{,}000}{4{,}702{,}000} = 0.927.$$

Product properties

The typical composition and properties of LR shale oils and retort gases derived from a 30-gal/ton Colorado (United States) and a 20-gal/ton Timahdit (Morocco) oil shale are given in Tables 1-9 and 1-10, respectively. Note that the Colorado oil shale tested had a total sulfur content of approximately 1.4 percent and a carbonate CO_2 content of approximately 18 percent, while the Timahdit oil shale had a much higher total sulfur content of approximately 2.7 percent and a somewhat lower carbonate CO_2 content of approximately 14 percent.

LR-process reactions

In the LR system there are several typical reactions which have a positive effect on the quality of the LR flue gas, the retort gas, the shale oil, and the spent shale. These reactions can be subdivided, by their location, into LR lift-pipe and LR-mixer reactions.

Lift-pipe reactions

Most oil shales, such as Colorado and Timahdit oil shale, contain calcium and magnesium carbonates which are partly decomposed at the lift-pipe temperatures of 650 to 750 °C, according to the following reaction:

$$CaCO_3 \rightarrow CaO + CO_2$$

The oxides thus formed have a strong sulfur-binding characteristic and react as shown in the following reaction:

$$CaO + SO_2 + \tfrac{1}{2}O_2 \rightarrow CaSO_4$$

TABLE 1-9 Properties of LR shale-oil (excluding naphtha fraction)

Property	Colorado oil shale	Timahdit oil shale
Density (at 20 °C), g/cm^3	0.905	0.971
Grade API	24	14
Viscosity, cSt		
At 20 °C	17	11
At 40 °C	8	6
Pour point, °C	+17	−23
Ultimate analysis, wt %		
Carbon	85.30	78.00
Hydrogen	11.10	9.70
Oxygen	1.17	4.36
Nitrogen	1.71	1.47
Sulfur	0.72	6.47
Boiling analysis, °C		
IBP*	72	78
5 vol %	134	148
10 vol %	160	167
20 vol %	200	195
30 vol %	239	227
40 vol %	282	268
50 vol %	324	308
60 vol %	370	341
70 vol %	410	377
80 vol %	450	423

*Initial boiling point.

TABLE 1-10 Composition of LR retort gas*

Component, vol %	Colorado oil shale	Timahdit oil shale
N_2	3.3	3.6
CO_2	28.8	26.4
H_2	23.1	26.8
CO	3.6	5.8
CH_4	15.2	18.0
C_2H_4	6.6	2.7
C_2H_6	8.2	6.3
C_3H_6	6.8	5.5
C_3H_8	4.1	3.5
H_2S	0.3	11.4
	100.0	100.0

*Carbon-containing compounds contain carbon chains no longer than C_3.

Gypsum is formed and the flue gas is desulfurized as a result of this reaction. Therefore, if enough carbonates are available in the raw shale, low concentrations of SO_2 are obtained in the flue gas as a result of these two reactions. At normal operating conditions a 30 percent carbonate decomposition rate has been obtained for both Colorado and Timahdit oil shales, which leads to very low SO_2 concentrations in the LR flue gas.

Another important point is that, as a result of a reaction which is similar to the formation of cement, fine granular spent shales with a high CaO or MgO content and a low content of organic carbon will exhibit good cohesive strength when the spent shales are moistened.

Another reaction that is environmentally significant is the formation of NO_x from air and shale nitrogen in the LR lift pipe. The thermodynamic equilibria for the formation of NO_x in air show very low NO_x concentrations at normal lift-pipe temperatures.

Lift-pipe temperature, °C	NO_x equilibrium concentration in air, vol %
600	1.7×10^{-6}
800	1.8×10^{-4}
1000	3.1×10^{-3}
1200	2.0×10^{-2}
1400	7.5×10^{-2}

Indeed the NO concentration in the flue gas—NO_2 could not be detected—is normally below 100 ppm (vol/vol).

Mixer reactions

Two typical reactions can normally be observed in the LR mixer. The first reaction, which decreases the concentration of H_2S in the retort gas, depends on the content of carbonates in the raw shale and on the degree of carbonate decomposition in the lift pipe. This reaction can be written as

$$CaO + H_2S \rightarrow CaS + H_2O$$

The unstable CaS is later oxidized to $CaSO_4$ in the lift pipe. The second reaction, which decreases the viscosity and the pour point of the shale oil, is based on the catalytic effect of hot spent shale in the LR mixer.

Most spent shales contain a considerable amount of silica (SiO_2) (Colorado and Timahdit spent shales both contain approximately 40 percent), which, in contrast to clay Al_2O_3, has a pronounced visbreaking effect. A simplified explanation is that SiO_2 catalytically tends to break C — C bonds while Al_2O_3 tends to break C — H bonds in the shale oil.

Advantages of the LR process

As was discussed previously under "Oil-Shale Retorting," Lurgi has gained experience in retorting oil shales with five different processes over a period of 50 years. The desirable features of a modern oil shale retort are listed below and then, using Lurgi's previous experience as a basis, they are compared with the features of an LR oil-shale retort.

1. *High oil yield* Flash retorting in the LR process, results in oil yields of up to 105 percent as compared to the Fischer assay.

2. *High thermal efficiency* Since residual carbon on the spent shale is burned in the LR lift pipe and is therefore used as retort fuel, thermal efficiencies of approximately 90 percent are achieved.

3. *High-Btu retort gas* The LR process yields an undiluted high-Btu retort gas which is suitable for the production of hydrogen.

4. *Utilization of 100 percent of mined shale* Fines do not have to be rejected in the LR process.

5. *Quality requirements of the oil shale* Shale quality requirements are low for the LR process. This was demonstrated in pilot plant tests with more than ten different oil shales mined in America, Africa, Europe, and Australia.

6. *Low water requirements* The production of an undiluted vapor product, results in cooling requirements that are low and can essentially be met by air cooling. Water is required only for cooling and moisturizing spent shale.

7. *Minimum SO_2 and NO_x emissions* The low process temperatures and the combustion of residual carbon in the lift pipe in the presence of the calcium oxide in the spent shale minimize the SO_2 and NO_x emissions in the LR process.

8. *Organic carbon in spent shale* The combustion of residual carbon on the spent shale causes the content of organic carbon in the spent shale to be very low.

9. *Disposal characteristics of the spent shale* If the carbonate content of the shale is sufficient, cementation reactions cause the spent shale from the LR process to exhibit good cohesive strength when moistened.

10. *Pour point of the shale oil* The catalytic effect of the silica in the spent shale during retorting causes LR shale oil to normally exhibit a lower pour point than shale oils produced via other retorting processes.

11. *Sulfur content in the retort gas* The presence in the mixer of calcium oxide from the spent shale is instrumental in causing a low H_2S concentration in the retort gas.

12. *Utilization of simple and demonstrated process equipment* The LR system uses simple and commercially demonstrated process equipment, including only one moving part in the hot section, i.e. the mixer.

Direct Coking of Tar Sands

Process description

Figure 1-15 shows an overall process flow diagram for direct coking of tar sands. This process is much like that for oil shale.

In the loop, raw tar sand of approximately less than ⅜ in is fed into the mixer, where it is mixed with approximately two to four times as much hot spent sand from the collecting bin at a temperature of 650 to 750 °C. The raw tar sand is then flash-heated to about 500 °C and coked within a few seconds.

The spent tar sand leaves the mixer and passes to the surge bin, where coking is completed, and is then transferred to the lower section of the lift pipe.

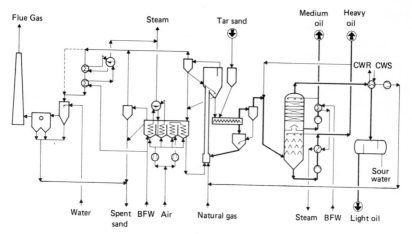

FIG. 1-15 LR process scheme for direct coking of tar sands.

Combustion air, preheated to about 450 °C, is introduced at the bottom of the lift pipe. The air conveys the material to the top of the lift pipe and simultaneously burns the residual coke from the spent tar sand.

The combustion gas and the reheated spent tar sand are separated from each other in the collecting bin at about 650 to 750 °C. The spent tar sand is returned to the mixer, thereby closing the loop.

The condensation system shown on the right side of the LR loop consists mainly of a cat-cracker fractionator-type condensing tower and a final air cooler-condenser. The condensing tower may serve several LR trains and consists of two stages. In the lower stage a heavy-oil fraction is condensed at approximately 200 to 300 °C and residual dust is removed. The required cooling is achieved by production of steam.

The heavy-oil dust concentration can be kept low by recycling heavy oil to the mixer. If heavy oil is recycled to extinction, producing lighter fractions of a higher quality at the expense of an approximately 5 percent lower liquid yield, no dedusting of the heavy oil is required.

In the upper stage of the condensing tower the product vapors are cooled to 100 to 150 °C by preheating boiler feedwater; water- and dust-free medium oil are condensed.

In the final cooler-condenser unit, light oil, naphtha, and sour water are condensed, leaving a high-Btu gas at approximately 50 °C, which is used, for example, as fuel in the lift pipe.

The hot spent tar sand that is withdrawn from the LR system at 650 to 750 °C is cooled in a multistage *fluidized-bed cooler* to approximately 200 °C. In the fluidized-bed cooler, LR process air is preheated to about 450 °C and additional steam is produced. If the spent sand still contains some unburned combustible coke from the LR system, this coke can be combusted in the first stage of the fluidized bed so that no potential fuel is left unused.

The flue gas from the LR loop and the off-gas from the fluidized bed are cooled in a *flue-gas waste-heat recovery system* to a temperature of approximately 200 °C, whereby high-pressure steam is generated.

The flue gas is then finally dedusted in an electrostatic precipitator before being discharged to the atmosphere.

Material and heat balances

Some typical material and heat balances for the application of the LR process to an Athabasca tar sand are given in Tables 1-11 and 1-12, respectively.

These data allow a first rough assessment of the LR technology but should not be extrapolated, since large changes may occur as a result of different feed properties or modifications of the overall process setup.

TABLE 1-11   Overall material balance for Athabasca tar sand

Input, kg	
Tar sand (11.5% bitumen, 4.5% water)	1000
Air	704
Natural gas	13
Limestone	10
Total input	1727
Output, kg	
Total oil (C_4^+)	98
Gas liquor	47
Spent sand (dry)	851
Flue gas	731
Total output	1727

TABLE 1-12   Overall heat balance for Athabasca tar sand

Input, kJ/Mg	
Tar sand	4,494,000
Air	7,000
Natural gas	610,000
Total input	5,111,000
Output	
Total oil (C_4^+)	3,925,000
Heat for steam and hot water production	611,000
Spent sand (at ~200 °C)	176,000
Flue gas (at ~200 °C)	152,000
Heat losses	247,000
Total output	5,111,000
Thermal efficiency (η) *	0.89

$$^*\eta = \frac{3,925,000 + 611,000}{4,494,000 + 610,000} = 0.89$$

Product properties

Tables 1-13, 1-14, and 1-15 show the properties and composition of the oil, the retort gas, and the waste stream respectively for an Athabasca tar sand.

UPGRADING OF OILS PRODUCED VIA THE LR PROCESS

Nonconventional refinery feedstocks like heavy crude oils and oils obtained from retorting oil shale or from the direct coking of tar sand require upstream processing prior to conventional refinery operations. Since the final products are to be transportation fuels, the primary upgrading delivers either a syncrude (distillate) or prerefined distillate fractions ready for finishing or blending. A syncrude should be capable of being processed in blends with conventional crude oils or with the corresponding conventional crude-oil fractions.

TABLE 1-13 Composition and properties of oil from Athabasca tar sand

Density (at 60 °F), g/cm³	0.963
Grade API	15.3
Viscosity, 10^{-6} m²/s	
At 40 °C	56.5
At 60 °C	22.4
Conradson carbon, wt %	4.54
Pour point, °C	-30
Bromine number, g Br/100 g oil	25.9
Ultimate analysis, wt %	
Carbon	84.07
Hydrogen	10.79
Oxygen	0.87
Nitrogen	0.65
Sulfur	3.62
LHV,* kJ/kg	40.030
Boiling analysis, °C	
Initial boiling point	123
5 vol %	224
10 vol %	254
20 vol %	309
30 vol %	346
40 vol %	390
50 vol %	428
60 vol %	463
70 vol %	498
80 vol %	534
Yield, wt %	75.5
Distillation residue, wt %	23.9
Loss, wt %	0.6

*Lower heating value.

TABLE 1-14 Composition and properties of retort gas from Athabasca tar sand

Component, vol %	
N_2	3.0
CO_2	22.6
H_2	24.1
CO	4.1
CH_4	16.0
C_2H_4	8.2
C_2H_6	8.6
C_3H_6	7.6
C_3H_8	3.1
H_2S	2.7
	100.0
Density, $kg/m^3(n)$ *	1.13
LHV†, $kJ/m^3(n)$ *	29.170
NH_3, ppm (vol/vol)	200

*$m^3(n)$ symbolizes the normal cubic meter, that is, the volume of one cubic meter of gas measured at 0 °C and 101.3kPa pressure.

†Lower heating value.

TABLE 1-15 Composition and properties of waste streams from Athabasca tar sand

Spent sand	
Grain size, wt %	
>200 μm	8.5
100–200 μm	79.2
63–100 μm	8.6
30–63 μm	2.2
<30 μm	1.5
	100.0
Organic carbon, wt %	0.2
Gas liquor	
pH value	5.6
Phenols (extractable), kg/m^3	0.85
Free ammonia, kg/m^3	0.0
Fixed ammonia, kg/m^3	0.79
Fatty acids, kg/m^3	2.0
Flue gas	
CO_2, vol %	8.0
N_2, vol %	75.9
O_2, vol %	8.5
H_2O, vol %	7.6
	100.0
SO_2, ppm (vol/vol)	~250
NO_x, ppm (vol/vol)	~50

Characteristic properties of shale oils and tar-sand oils obtained by LR retorting or direct coking are shown in Table 1-16. The range of properties within each group clearly show the comparative differences. In comparison with conventional crude oils, shale oils are characterized by lower hydrogen content and relatively high and widely varied heteroatom content (particularly of oxygen and nitrogen) and high amounts of unsaturates referring to the origin by pyrolysis. Another characteristic is the high content of trace elements, which are represented by arsenic for the group of elements that includes As, Sb, Sn, P, and Bi. Arsenic is present in amounts which are 10^3 times higher than those in conventional crude oils. The oils from tar sands show less but still substantial amounts of unsaturated compounds and higher average molecular weights than those of shale oils. That tar-sand oils more closely resemble petroleum crude oils than do shale oils is shown by the absence of impurities from the arsenic group and somewhat better H/C ratios.

The primary upgrading of shale oils comprises the removal of the trace elements and yields a series of distillates that are ready for catalytic hydrotreating. The upgrading of raw shale oils can be carried out by a noncatalytic liquid-phase hydrogenation step, particularly in the presence of inert solids which are carried over from the LR retort. The conversion of the residue content can be enhanced by using a solvent that is a hydrogen donor.

A simplified flow diagram for upgrading oils produced via LR retorting of oil shale is shown in Fig. 1-16. The liquid streams from the LR retorting section are separated by distillation into two fractions <200 °C and >200 °C. The >200 °C fraction is subjected to partial removal of solids so that 1 to 5 percent by weight of the finest solids remain dispersed in the feed for the hydrovisbreaker. Atmospheric and vacuum distillation of the hydrovisbreaker effluent yield three fractions <200 °C, 200 to 500 °C and >500 °C. The >500 °C fraction contains the remaining solids and—in the case of shale oils—the trace elements. This vacuum residue fraction may be recycled to the LR mixer, coked elsewhere, or used, for example, for the production of road construction materials. Where commercial

TABLE 1-16 Characteristic properties of LR shale oils and tar-sand oils (200 °C+ fraction)

	Shale Oil			Tar Sand Oil		
	Maroc	Puertollano	Rundle	Madagascar	Athabasca	Californian diatomite
Specific gravity (20 °C), g/mL	1.0109	0.9840	0.9062	0.9483	0.9729	1.0080
Ultimate analysis, wt %						
H	9.11	9.50	11.24	11.53	10.55	9.50
C	80.7	87.2	85.7	86.7	84.4	85.8
S	7.33	0.49	0.54	0.45	3.79	1.10
N	1.54	1.19	0.91	0.46	0.24	0.70
O	1.32	1.45	1.33	0.73	0.69	2.90
Mol wt (average)	250	256	244	399	359	330
Bromine no., g Br/100 g	73	48	47	23	24	35
Arsenic, ppm	31	20	9	<1	<1	<1
Residue > 500 °C, wt %	15	17	7.5	24.5	19	15

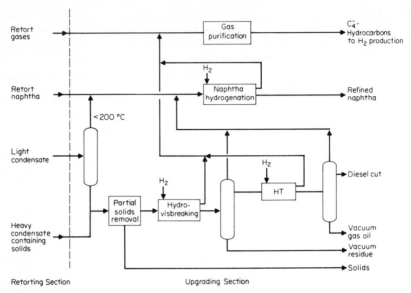

FIG. 1-16 Process scheme for upgrading LR shale oils.

refinery techniques modified by the use of appropriate catalysts can be applied, hydrotreating units handle the naphtha and gas-oil fractions. The removal of heteroatoms (particularly nitrogen) requires specific care, since they are for the most part incorporated into the heterocyclic bond structures of the retort oils.

Characteristic data from the processing of LR retort oils via the Lurgi hydrovisbreaking process are shown on Table 1-17. These data have been collected from continuous bench-scale tests run without the use of a hydrogen-donor solvent.

Typical results obtained by standard hydrotreating of the respective hydrovisbreaker distillates are shown in Table 1-18. The refining of shale oils to commercial products requires adjustments for nitrogen removal and for increasing the cetane number of the diesel fraction.

SUMMARY

The well-known flash pyrolysis technology which has been applied for many years to coal can now also be used to produce synfuels from oil shale, tar sand, and diatomaceous earth via the Lurgi-Ruhrgas process.

The LR process has been used in large-scale commercial installations to carbonize coal and to crack crude oil and naphtha into olefins; it is now ready to be applied to the production of synfuels from oil shale, tar sands, and diatomaceous earth.

The LR process is based on the simple and versatile principle of flash heating the feedstock in a mixer by means of circulated solid hot heat carriers. The heat carrier is (usually) the feed residue and is produced by the process itself.

TABLE 1-17 Hydrovisbreaking of LR retort oil (200 °C + fractions)

	Shale oil (Puertollano)	Tar Sand Oil (Madagascar) Mild	Severe
Yield, wt % liquid feed			
Inorganic gases	0.5	0.5	0.6
C_5 — hydrocarbons*	0.7	1.4	2.2
C_6 to 200 °C fraction	4.5	10.8	19.8
200 to 530 °C fraction	78.8	73.9	71.8
Residue (530 °C + fraction)	15.7	13.9	6.3
Properties of 200 to 530 °C distillate			
Specific gravity (at 20 °C), g/mL	0.9528	0.9159	0.9121
Ultimate analysis, wt %			
H	9.93	11.91	11.72
C	87.5	87.0	87.0
S	0.49	0.32	0.32
O	1.3	0.4	0.4
N	1.0	0.3	0.3
Mol wt (average)	219	261	243
Bromine no., g Br/100 g	29	18	17
Arsenic, ppm	<1	—	—

*Carbon chain length ≤ C_5.

TABLE 1-18 Hydrotreating of hydrovisbreaker distillate (200-to-500 °C fractions)

	Shale oil (Puertollano)	Tar sand oil (Madagascar)
Yield, wt % of liquid feed		
Inorganic gases	1.0	1.1
C_5 — hydrocarbons*	0.5	0.7
C_6 to 170 °C fraction	6.0	7.1
170-to-380 °C fraction (diesel oil)	80.5	74.9
380-to-500 °C fraction (vacuum gas oil)	12.0	17.8
Properties of total effluent		
Specific gravity (at 20 °C), g/mL	0.8736	0.8616
Ultimate analysis		
H, wt %	12.16	13.05
C, wt %	87.4	86.7
S, ppm	120	141
O, wt %	<0.2	<0.1
N, ppm	90	8
CCL of the diesel oil fraction	40.5	47

*Carbon chain length (CCL) ≤ C_5.

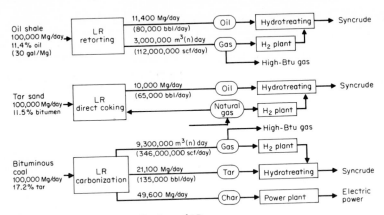

FIG. 1-17 Three major applications of LR process.

Reheating and pneumatic transport of the heat carrier are achieved simultaneously in a vertical lift pipe by combustion of coke in the residue.

The three major applications of the LR process are illustrated in Fig. 1-17.

BIBLIOGRAPHY

Marnell, P.: "Lurgi/Ruhrgas Shale Oil Process. Process Proved by Plant Performance," *Hydrocarbon Process.*, September 1976, pp. 269–271.

Rammler, Roland W.: "Betriebserfahrungen mit dem Lurgi Verfahren zur Entgasung feinkörniger Brennstoffe," *Erdöl und Kohle, Erdgas, Petrochem.*, February 1966, pp. 117–121.

————: "One Process Retorts Three Feeds for Synthetic Crude," *Min. Eng.*, September 1970, pp. 63–66.

————: "The Production of Synthetic Crude from Oil Sand by Application of the Lurgi-Ruhrgas Process," *Can. J. Chem. Eng.*, vol. 48, 1970, pp. 552–560.

————: "The Lurgi-Ruhrgas Process for Carbonization of Coal," *Oil Gas J.*, Nov. 9, 1981, pp. 291–306.

————, H. J. Weiss, A. Bussmann, and T. Simo: "Oil Recovery by Distillation of Oil Shale and Tar Sand," *Ger. Chem. Eng.*, vol. 4, 1981, pp. 241–250.

————: "Lurgi-Ruhrgas Process—New Application on Tar Sands and Heavy Oil," *Second UNITAR Conference*, Caracas, February 1982.

————: "Lurgi Coal Pyrolysis for the Production of Synthetic Fuels," *Energy Prog.*, vol. 2, no. 2, 1982, pp. 121–129.

Weiss, Hans. J.: "The Retorting of Oil Shale by the Lurgi-Ruhrgas (LR) Process," *6th IIASA Resource Conference, Golden, Colo.*, June 1981.

————: "Production of Synfuels from Oil Shale, Tar Sands, and Coal by the LR-Process," *Synfuels' 1st Worldwide Symposium, Brussels*, October 1981.

THE TOSCO II PROCESS

C. S. WAITMAN
R. L. BRADDOCK
T. E. SIEBERT

Tosco Corporation
Los Angeles, California

GENERAL DESCRIPTION OF THE TOSCO II PROCESS

The TOSCO II oil-shale retorting process is based upon the principle of pyrolysis, using an inert heat carrier (ceramic balls) to pyrolyze the kerogen in the shale matrix. The heat carrier is intimately mixed with the shale in a rotary-drum retort. Heat transfer is relatively rapid, and complete pyrolysis of the shale is assured. The process recovers all of the hydrocarbons that can be volatilized from the shale through pyrolysis. In addition to the rotating-drum retort, steps are provided for (1) separation of the spent shale from the heat carrier, (2) recirculation and heating of the heat carrier, (3) recovery and utilization of heat from volatile hydrocarbons, spent shale, and flue gases, (4) condensation of the oil product, and (5) treatment of spent shale for disposal. A flow diagram is shown in Fig. 2-1. The process has been successfully demonstrated in a 1000-ton/day (900-Mg/day) semiworks unit, which has processed more than 220,000 tons (200,000 Mg) of Colorado oil shale. On the basis of the success of this demonstration program, Tosco Corporation has completed preliminary engineering for the eventual construction of a shale oil project including six parallel 11,000-ton/day (10,000-Mg/day) TOSCO II retorts at its Sand Wash Project site in northeastern Utah. These retorts are scheduled to be in operation by the late 1980s. Tosco and several licensees also are engaged in studies and design activities which call for additional TOSCO II retorts to be contructed at other project sites in the United States and abroad. Additionally, the technical applicability of the TOSCO II process to various domestic and foreign shales, as well as other hydrocarbonaceous materials such as coal, scrap rubber tires, and tar sands, has been demonstrated at Tosco's research facilities, which include a 25-ton/day (23-Mg/day) TOSCO II pilot unit.

In its research programs, Tosco has made considerable gains in improving process efficiency by adding steps to utilize the residual carbon on the spent shale as a fuel to generate steam, to dry and preheat raw shale, and to heat the recirculating heat carrier. These steps combine the demonstrated retorting technology of the TOSCO II process with fluidized-bed technology for combustion or gasification of spent-shale residues. Test units have been

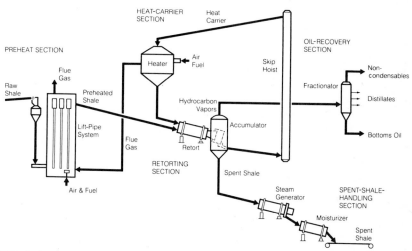

FIG. 2-1   Basic configuration of the TOSCO II retorting process.

operated utilizing these steps, and a larger, more advanced pilot plant is being planned. Utilization of the residual carbon is of particular interest in processing moist, low-grade, high-carbon-content shales.

PROCESS CHEMISTRY AND THERMODYNAMICS

In the retorting section, oil shale which has been crushed, dried, and preheated is mixed with the inert heat carrier. Sufficient heat is transferred from the heat carrier to the oil shale to bring it to approximately 500 °C. At 500 °C, complete pyrolysis of the oil shale is assured, and the kerogen contained in the shale is vaporized. The heat carrier, which reaches thermal equilibrium with the shale, is separated from the spent shale by screening and elutriation. The heat carrier is lifted and fed to a heater. Once it is heated to a temperature above 600 °C, the heat carrier is recirculated to the retort. Flue gases from the heater are utilized to preheat and dry the crushed raw oil shale before it is fed to the retort. Spent shale, which leaves the retorting unit at approximately 500 °C, is cooled in two stages and moisturized for disposal.

Hydrocarbon vapors, which leave the retorting section at approximately 500 °C, are condensed in an oil-recovery section which includes a conventional fractionator. The oil-recovery unit can be designed to produce either oil products of the optimum cut points (boiling ranges) or a whole oil for upgrading or refining.

Emissions from the process consist principally of flue gases containing particulates, hydrocarbons, and combustion products. Tosco has demonstrated the applicability of conventional emission-control processes such as incineration and wet scrubbing to reduce the level of pollutants in the flue-gas streams to meet the requirements of stringent emission and air-quality regulations. Emission of oxides of sulfur and nitrogen is controlled by the selection of fuels used in the process and, in the case of nitrogen oxides, the firing temperatures and configurations of process heaters.

Tosco is continuing to develop additional processes to provide alternatives to the basic configuration. These are designed to utilize the residual carbon on the spent shale. In one alternative, the spent-shale-handling unit and preheat unit can be modified as shown in Fig. 2-2. Here, a conventional fluid-bed boiler is used to burn the carbon from the spent shale to generate steam. Hot flue gases from the fluid-bed boiler are utilized to provide the additional heat that is required to evaporate moisture contained in the raw shale feed.

Progress is also being made toward development of the new process called the *hydrocarbon solids process* (HSP), which is shown in Fig. 2-3. In it, the heat carrier and spent shale are fed to a fluid-bed combustor and heater, in which the residual carbon on the spent shale is burned. The heat generated from combustion of the spent shale is used to heat the recirculating heat carrier. Waste heat from the combustor-heater can be utilized to generate steam and/or to preheat shale.

PILOT PLANT DATA

Tosco has done extensive work with its basic process configuration in both the 25-ton/day (23-Mg/day) pilot plant and the 1000-ton/day (900-Mg/day) semiworks unit. It also has worked to develop the analytical tools necessary to characterize shales and the correlating factors that will forecast their performance in the process with minimum testing. The

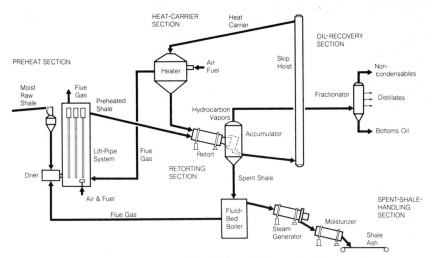

FIG. 2-2   TOSCO II retorting process designed for moist low-grade shales.

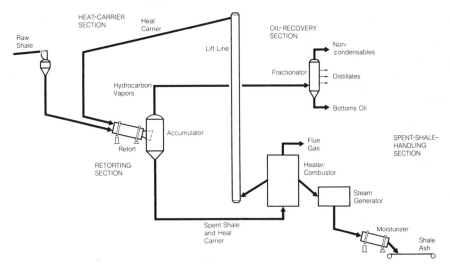

FIG. 2-3   TOSCO II retorting process with integral spent-shale combustion.

Tosco material balance assay (TMBA), which was derived from the basic Fischer assay procedure, can be used by Tosco to make accurate projections of the commercial yields from its process by the analysis of very small samples of any oil shale. Subsequently, pilot plant tests are run, normally utilizing about 100 tons (90 Mg) of shale, in order to verify yields, to identify important retorting characteristics, and to determine the properties of the retort gases, shale oil, and spent shale. Additionally, the pilot unit can be used to determine the behavior of a shale in each important section of the process. TMBA results, pilot plant data, historical semiworks data, and commercial-design parameters for a typical rich Colorado shale are compared in Table 2-1. Note particularly the close correlation of the percentage recovery of TMBA hydrocarbons; the properties of retort gases, shale oil,

TABLE 2-1 Comparison of TMBA*, pilot plant, and semiworks data with commercial-design parameters

	TMBA	Pilot plant†	Semiworks‡	Commercial§
Yields, kg/kg TMBA				
C_5+ oil fraction	1.0	0.987	0.932	0.929
Gas	1.0	1.031	1.462	1.458
Total hydrocarbon	1.0	0.991	1.017	1.014
Water of retorting	1.0	1.000	1.000	1.000
Spent shale	1.0	1.001	0.995	0.998
Gas composition, mole %				
Methane	17.7	18.8	19.6	19.4
Ethene	1.9	4.4	8.1	8.1
Ethane	6.5	6.9	7.6	7.5
Propene	2.8	4.5	7.1	7.1
Propane	2.9	3.1	3.3	3.3
Total C_4 gases	3.1	4.7	5.1	5.2
Total C_5 gases	1.9	1.4	3.0	3.3
Nonhydrocarbons	63.2	56.2	46.2	46.1
Oil properties				
Elemental analysis, wt %				
Carbon	85.50	85.02	84.25	84.25
Hydrogen	11.83	11.16	11.10	11.10
Nitrogen	2.09	2.13	2.14	2.14
Sulfur	0.78	0.79	0.81	0.81
Distillation fractions, °C				
IBP,¶	38	49	42	42
10%	138	177	174	174
30%	288	321	306	306
50%	399	432	438	438
Density, g/cm³	0.907	0.935	0.928	0.928
Spent-shale properties				
Organic carbon, wt %	4.93	5.60	4.65	4.65
Mineral carbon, wt %	5.70	4.03	5.72	5.72
Size distribution				
% passing through				
4.699 mm	—	94.5	95.3	95.3
0.833 mm	—	70.2	82.6	82.6
0.147 mm	—	48.0	55.2	55.2
0.044 mm	—	28.9	31.0	31.0
Preheat emissions				
Dust to scrubber, % of feed		0.7	0.7	1.3
Hydrocarbon, ppm (by weight)		92	—	97

*Tosco material balance assay.
†Feed rate of 25 t/d (23 Mg/d).
‡Feed rate of 1000 t/d (900 Mg/d).
§Feed rate of 11,000 t/d (10,000 Mg/d).
¶Initial boiling point.

and spent shale; and the emission factors for the process in the pilot plant and semiworks unit.

Tosco has operated the pilot plant unit to both test and optimize the process for new shales as well as to provide samples of retort products or spent shale. The pilot plant has been utilized to retort not only Colorado oil shales with assays ranging from 10 to 50 gal/ton (42 to 210 L/Mg) but also oil shale from Morocco, coal, and scrap tires.

Tosco's development work has also included bench-scale upgrading tests on crude shale oil. The technique tested most extensively is severe hydrotreating, which converts crude shale oil into a high-quality synthetic crude-oil product.

PROCESS PERSPECTIVE

The TOSCO II process was selected for development in the mid-1950s. In 1957, after bench-scale studies of the TOSCO II process were completed, the 25-ton/day (23-Mg/day) pilot plant was constructed in Denver, Colorado. This plant later was relocated to its present site at Tosco's research center in Golden, Colorado.

Successful operation of the pilot plant and a favorable initial design and cost estimate for a commercial-scale facility led to the construction of the 1000-ton/day (900-Mg/day) semiworks plant shown in Fig. 2-4. This plant is located at the site of the Colony Project (formerly jointly owned by Tosco and Exxon Corporation but wholly owned by Exxon since mid-1982) on Parachute Creek, 15 miles north of the town of Parachute in northwestern Colorado. It was completed in 1965. Testing of the semiworks plant from 1965

FIG. 2-4   Colony semiworks plant. Key: A, mine mouth; B, crusher; C, crushed-shale storage; D, retort.

to 1967 confirmed operability of the process and showed that the process recovers 100 percent of the Fischer assay hydrocarbons. At the conclusion of the initial operating period, Tosco retained an engineering contractor to review the results of operations and to prepare designs for the major equipment items to be included in a commercial-scale oil-shale complex. This design study, completed in 1968, verified the viability of the process and resulted in recommendations for modifications of the detailed design of some of the equipment in the semiworks plant in order to reduce scale-up uncertainties and improve pollution-control performance.

A second operating phase began in 1969 with modifications of the plant implementing the recommendations of the 1968 design study. This second operating period, successfully completed in April 1972, included acquisition of further data for the design of the commercial-scale retorting units. As noted earlier, during the two operating periods, the plant processed some 220,000 tons (200,000 Mg) of shale and produced some 180,000 bbl of crude shale oil. In addition, another 1,000,000 tons (900,000 Mg) of oil-shale ore were extracted from the pilot mine at the Colony site during full-scale tests of mine design, equipment, and techniques for the commercial project.

Definitive design studies for the 66,000-ton/day (60,000-Mg/day) Colony Project were initiated in 1972. These led to development of a mine plan for the property and to designs for ore-handling facilities, crushing systems, the TOSCO II retorting facility, and oil-recovery and pre-refining facilities. Additionally, all data necessary to obtain critical environmental permits were developed. In 1974, as a result of unprecedented inflation in construction costs and the enactment of ambiguous federal regulations, design and field construction activities on the Colony Project were suspended. By 1979, however, the final Environmental Impact Statement had been approved and the U.S. Environmental Protection Agency (EPA) had determined that the Colony Project would meet the strict air-quality guidelines dictated in its Prevention of Significant Deterioration (PSD) regulations. In 1980, Exxon, after an extensive review of technical, environmental, and economic data for the project, purchased the Atlantic Richfield Company's 60 percent share in the project and joined Tosco in reactivating design activities. Civil engineering work for the project commenced contemporaneously. The work included construction of a plant access road which climbs 2000 ft (600 m) from the base of the Parachute Creek valley to the plant site atop the adjacent plateau, and the excavation of nearly 6×10^6 yd³ (4.5×10^6 m³) to develop a level plant site. By the end of 1981, mechanical specifications for engineered equipment in the TOSCO II retorting unit were completed. Additionally, procurement of key crushing and material-handling equipment had commenced, process design of all facilities within the complex was at an advanced stage, and a detailed construction plan was established. In May 1982, however, Exxon suspended the Colony Project, and Tosco subsequently sold its 40 percent interest in the project to Exxon.

Studies and engineering designs for utilization of the TOSCO II process also have commenced for Tosco Corporation's solely owned leases in Utah. Tosco's 17,000 acres (6.9 $\times 10^7$ m²) at the Sand Wash Project site are capable of supporting a 66,000-ton/day (60,000-Mg/day) retorting facility for 30 years.

In addition to these projects, numerous companies with large domestic oil-shale deposits are considering the application of the basic TOSCO II process to their properties. Also, as a result of work done in conjunction with agencies of the Moroccan government, sufficient testing and design work have been completed to confirm the applicability of the TOSCO II process to oil shale of the Timahdit deposit. Process additions for the combustion of spent shale to generate steam and provide heat for drying are applicable to the moist Moroccan shale.

In proceeding with the development of process steps to utilize the carbon on the spent shale, Tosco has installed a fluid-bed boiler at the 25-ton/day (23-Mg/day) pilot plant to test the process scheme shown in Fig. 2-2. In addition, more than 3000 h of testing on the HSP configuration shown in Fig. 2-3 have been conducted in a 6-ton/day (5-Mg/day) pilot unit. To date, data from these tests indicate that favorable yields, efficiency, and environmental acceptability can be attained. Plans are being made to construct a larger pilot plant and conduct more definitive tests.

DETAILED PROCESS DESCRIPTION

Basic TOSCO II Process Sections

The basic configuration of the TOSCO II process, as described earlier, consists of a retorting section, a heat-carrier section, a preheat section, a spent-shale handling section, and an oil-recovery section.

Retorting

The retorting section, shown in Fig. 2-5, consists of a rotary-drum retort, a pyrolysis-accumulator vessel, a trommel, and an elutriator. Within the retort, the heat carrier [which consists of nominal ½-in (0.013-m) high-density ceramic balls] is mixed with ½-in (0.013-m) and finer preheated oil shale. The retort operates in an inert atmosphere at a pressure sufficient to move the pyrolysis vapors through the pyrolysis accumulator and the oil-recovery unit. The retort pressure and the inert atmosphere of the retort are maintained through the use of circumferential carbon ring seals which are precisely aligned with the retort drum through a proprietary mechanism. To account for the changing properties of oil shale during retorting, Tosco has developed mathematical formulas for the transport of oil shale and heat carrier through the retort.

The pyrolysis accumulator serves three basic functions. First, the level of spent shale which it contains acts to provide a seal between the pressurized atmosphere in the retort

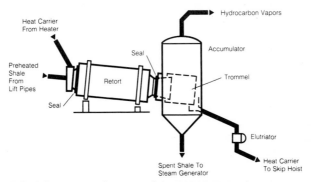

FIG. 2-5 Basic configuration of the TOSCO II retorting process—retorting section.

accumulator and the normal atmospheric pressure in the spent-shale handling circuit. Second, the system assures high efficiency in the separation of entrained shale dust from the pyrolysis vapors. Third, a rotating trommel contained in the pyrolysis-accumulator system screens the spent shale from the heat carrier. The elutriator, which is located outside of the accumulator, is utilized to remove any remaining shale fines.

Heat carrier

The heat-carrier section, shown in Fig. 2-6, consists of a mechanism for lifting the heat carrier and feeding it to a heater. Serious consideration was given to various equipment for elevating the heat carrier: bucket elevators, pneumatic lifts, and skip hoists. Power requirements for a skip hoist are moderate compared to those for a pneumatic lift, and reliability and capacity of a skip-hoist system exceed those of bucket elevators for some specific applications. The electrical system, in-service inspection system, and lubrication system to be utilized for the skip hoist would be designed to maximize its reliability.

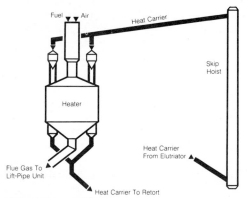

FIG. 2-6   Basic configuration of the TOSCO II retorting process—heat-carrier section.

The heater is basically a moving-bed pebble heater. Fuel and air are fed to a large combustion chamber, and the hot combustion gases flow with the heat carrier through a moving bed. The heat carrier is withdrawn from the bottom of the bed through a network of equally spaced withdrawal pipes and is fed to the retort. Hot flue gas is withdrawn from the heater by means of a series of disengagers which are arranged to ensure uniform gas flow through the bed.

Preheat

The preheat section, shown in Fig. 2-7, is designed to serve two functions. First, it recovers waste heat contained in the flue gas exiting from the ball heater; second, it preheats the raw shale, thus minimizing the heat-transfer requirements in the retorting unit. The mechanism for exchanging heat between the hot flue gas and the shale consists of a series of lift pipes. In each lift pipe, shale is entrained by the flue gas at the bottom and heat is transferred from the flue gas to the raw shale as they flow cocurrently to the top of the lift pipe. The flow of flue gas and raw shale, although cocurrent in each lift pipe, are countercurrent in the system. The entire preheat section can raise the temperature of the

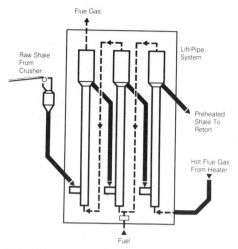

FIG. 2-7 Basic configuration of the TOSCO II retorting process—preheat section.

raw shale from ambient temperature to above 200 °C while cooling the flue gas from above 600 °C to approximately 120 °C.

Spent-shale handling

In the basic configuration of the TOSCO II process, the spent-shale handling section, shown in Fig. 2-8, contains a rotary-tube steam generator designed to recover sensible heat from the spent shale. The mechanical design of the steam generator is based upon that developed for steam-tube dryers utilized widely in the soda-ash industry. Spent shale discharges from the rotary-tube steam generator and is fed to the rotary-drum moisturizer. The moisturizer is designed to carry out the quench cooling of the shale and to effect a uniform moisture content of the spent-shale product. The moisturizer utilizes aqueous sludges and wastewaters, thus minimizing the requirements for raw water makeup to the unit.

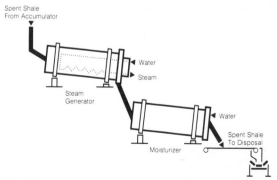

FIG. 2-8 Basic configuration of the TOSCO II retorting process—spent-shale handling section.

Oil recovery

Oil vapors from the retorting section are condensed, producing a bottoms stream and distillates. The bottoms stream contains shale fines not removed in the pyrolysis accumulator and is comparable in boiling range to a conventional vacuum residuum (500 °C+). Cut points for the distillate fractions can be optimized for the downstream upgrading or refining processes. Noncondensable vapors from the oil-recovery circuit can be compressed and treated to refinery-gas standards for use as either plant fuel or hydrogen-plant feed; these vapors can also be fed to a hydrocarbon condensate recovery system for recovery of valuable LPG.

Process Options

In the basic configuration of the TOSCO II process, all of the hydrocarbons that can be volatilized through pyrolysis are recoverable as high-quality liquids or gases either in the oil recovery unit or through subsequent processing of the hydrocarbon-rich gases. The process produces high-pressure steam in excess of its demands, but it imports electricity and requires liquid or gaseous fuel. The thermal efficiency and net oil production for the basic TOSCO II retorting process when it is processing rich Colorado shale are tabulated in Table 2-2. Note that when rich Colorado oil shale is processed, the net liquid yield (including LPG, but after deducting hydrocarbons required to fuel the process) exceeds the standard Fischer assay yield.

TABLE 2-2   Yields and thermal efficiency of TOSCO II process with rich (34.8-gal/ton; 145-L/Mg) Colorado shale

	% of dry raw shale	Heat content, % of HHV* of dry shale
Raw materials		
Raw shale (1.5% moisture) (A)	101.5	100.0
Pyrolysis yield		
Gas	4.06	17.4
Oil	12.56	70.6
Utilities		
Fuel required	3.44	13.3
Power required (B)	—	0.8
Steam produced (C)	7.24	2.6
Net yields		
Recoverable LPG (D)	0.62	4.0
Naphtha (E)	2.74	15.9
Gas Oil (F)	6.74	38.1
Bottoms (G)	3.07	16.7

$$\text{Thermal efficiency} = \frac{(C) + (D) + (E) + (F) + (G)}{(A) + (B)} = 76.2\%$$

$$\text{Net liquid recovery} = \frac{(\text{LPG} + \text{naphtha} + \text{gas oil} + \text{bottoms})\dagger}{34.8 \text{ gal/t}} = 103\%$$

*Higher heating value

†Volumetric yield of each.

Several important characteristics of low-grade shales, which are typical of shales found for example in Australia, the People's Republic of China, Israel, Jordan, and Morocco and of the Devonian shales found in the eastern United States, differ from those of the rich Colorado shales in the following important ways:

1. The low-grade shales often are moist, a condition which results in a higher heat demand for the retorting process.

2. Direct yields of oil and gases from low-grade shales are generally lower than those from rich Colorado shales.

3. Following pyrolysis, the spent shales from low-grade shales contain a higher amount of residual organic carbon.

The properties have led to Tosco's designing options to optimize the application of its processes to such shales. The process shown in Fig. 2-2, in which a conventional fluid-bed boiler is added to burn the residual carbon on the spent shale and flue gases from the boiler are utilized to dry the raw shale, offers considerable improvements in efficiency. The net yields of the TOSCO II process with and without an add-on fluid-bed boiler are compared in Table 2-3 for a typical low-grade shale. For this case the add-on fluid-bed boiler improves thermal efficiency by 22 percent.

For the HSP process shown in Fig. 2-3, which utilizes an integral spent-shale combustor and heater, preliminary data indicate that higher yields and improved product properties, as well as improved thermal efficiency, can be achieved. In order to capitalize upon the benefits of this process, Tosco is developing the process for application to both low- and high-grade shales.

PRODUCT AND BY-PRODUCT SPECIFICATIONS

The properties of raw shale oil from Colorado, Morocco, and the People's Republic of China are listed in Table 2-4. As noted in Table 2-4, the nitrogen content of the shale oil

TABLE 2-3   TOSCO II process yields with moist, lean (20-gal/ton; 83.5-L/Mg) foreign shale*

	TOSCO II basic configuration	TOSCO II with fluid-bed boiler
Raw materials		
Raw shale (10% moisture)	110.00	110.0
Pyrolysis yield		
Gas	2.33	2.33
Oil	7.21	7.21
Utilities		
Fuel required	3.21	2.50
Net yield		
Oil	6.34	7.04
Net liquid recovery†	82	91

*Values are expressed as % of dry raw shale

†Net liquid recovery $= \dfrac{\text{volumetric oil yield}}{20 \text{ gal/t}} \times 100.$

TABLE 2-4 Properties of TOSCO II raw shale oil

	Shale Origin		
	Colorado	Morocco	China[*]
Elemental analysis, wt %			
Carbon	84.25	80.62	
Hydrogen	11.10	9.54	
Nitrogen	2.14	1.68	1.17
Sulfur	0.81	5.99	.41
Oxygen	1.24	1.86	
Density, g/cm^3	0.928	0.999	0.887
Fractions, vol %			
Naphtha (C_5 to 190 °C)	14.0	9	10
Furnace oil (190 to 360 °C)	24.0	36	32
Gas oil (360 to 515 °C)	38.0	—	40
Residuum (515 °C+)	24.0	—	18
Pour point, °C	24	−7	27

[*]People's Republic of China.

typically 1 to 2 percent by weight. Most conventional crude oils have a nitrogen content of 0.3 percent by weight or less. The arsenic content of crude shale oil is typically about 50 ppm, far greater than the arsenic content of conventional crude oils.

The presence of these contaminants requires that shale oil be treated before being fed to conventional refinery process units such as hydrocrackers, catalytic crackers, or re-formers. Such treatment likely would include hydrotreating at high pressure. The hydrotreated shale oil is a desirable product for use as a refinery feedstock, as a raw material for production of petrochemicals, or as an environmentally clean fuel. Upgrading raw shale oil also improves its piplining properties, thus reducing the cost of transportation to marketing centers.

By-products from a shale-oil project most often consist of sulfur and ammonia. Sulfur is recovered by treating both the noncondensable retort gases and the hydrotreater off-gas. As a result of the relatively high nitrogen content of most shale oils, ammonia is also a by-product of the shale-oil hydrotreating process. Typical production rates for sulfur and ammonia for a rich Colorado shale and for a low-grade high-sulfur foreign shale are tabulated in Table 2-5.

UPGRADING TECHNOLOGY

The Sand Wash Shale Oil Project is expected to utilize an upgrading scheme which includes delayed coking of the bottoms oil and hydrotreating of the middle distillates and light distillates in separate hydrotreating stages. Numerous upgrading options are available which may be applied to TOSCO II shale oil, including fluid coking or FLEXICOKING and whole-oil hydrotreating. Extensive upgrading tests have been performed on Colorado

TABLE 2-5 Production of sulfur and ammonia by the TOSCO II process with severe shale-oil hydrotreating*

	Rich Colorado shale	Lean foreign shale
Sulfur		
t/d	195	873
Mg/d	177	792
Ammonia		
t/d	139	60
Mg/d	126	54

*Feed rate is 66,000 t/d (60,000 Mg/d) of raw shale.

shale oil derived from the TOSCO II process. Table 2-6 summarizes the results of whole-oil hydrotreating tests conducted by Gulf Oil Corporation.

WASTES AND EMISSIONS

The Colony Project, which is designed to use six parallel 11,000-ton/day (10,000-Mg/day) TOSCO II retorts, is the subject of a detailed final Environmental Impact Statement completed in 1977. The project received all the critical environmental permits. These include the Environmental Protection Agency's PSD permit (the key federal air-quality permit) and the Colorado Mined Land Reclamation permit (the key permit required for spent-shale disposal). The principal environmental concerns of this project, which are likely to be representative of other commercial shale-oil projects in the United States, are discussed below.

Particulate Emissions

The control of particulates from the preheat system used in the TOSCO II process has received considerable attention. In 1968, after a review of the characteristics of the dust from the preheat system, a study team recommended that a wet scrubber be installed on the stack of the semiworks preheat system. During the 1969–1972 operation of the semiworks plant, detailed performance data were gathered on this scrubber [20,00 stdft^3/min (570 m^3/min)] and on a small slipstream scrubber [60 stdft^3/min (1.7 m^3/min)], and isokinetic sampling was performed to establish the quantities and characteristics of the dust from the preheat system.

These studies defined with certainty the design parameters for commercial-scale scrubbers and provided an opportunity to measure corrosion and erosion rates.

Spent-Shale Disposal

Some 82 percent by weight of the raw crushed shale ore charged into the Colony Project retorts was to be reduced to spent shale. The spent shale is a fine-grained, dark-colored material containing about 4.5 percent by weight organic carbon in an inorganic matrix.

TABLE 2-6 Upgrading of TOSCO II whole shale oil by the Gulf Oil Company

Operating conditions:

Reactor pressure, $N \cdot m/m^2$ (psig)	14(2100)	
Gas circulation, m^3/m^3 (scf/bbl)	890(5000)	
Average catalyst temperature, °C (°F)	399(750)	
Space velocity	0.5	

Yields, % of HDS charge:

H_2S, wt %	0.72	
NH_3, wt %	2.05	
H_2O, wt %	1.45	
C_1 to C_4 fraction, wt %	2.09	
C_5+ Syncrude oil, vol %	102.6	
Chemical H_2 consumption, m^3/m^3 (scf/bbl)	224(1260)	

Determinations:	Feed	Syncrude
Density, g/cm^3	0.930	0.868
Nitrogen, wt %	1.99	0.32
Sulfur, wt %	0.70	0.05
Oxygen, wt %	1.32	0.03
Hydrogen, wt %	11.14	12.84
Pour point, °C	24	21

Fractions:

Naphtha (C_5 to 190 °C)

Yield, vol % of syncrude oil	16.7	
Density g/cm^3	0.764	
Nitrogen, ppm	0.5	
Sulfur, ppm	0.5	

Furnace oil (190 to 360 °C)

Yield, vol % of syncrude oil	43.7	
Density, g/cm^3	36.0	
Nitrogen, wt %	0.23	
Sulfur, wt %	0.05	
Aniline point, °C	65	
Pour point, °C	-1	

Gas oil (360 to 515 °C)

Yield, vol % of syncrude oil	24.3	
Density g/cm^3	0.892	
Nitrogen, wt %	0.43	
Sulfur, wt %	0.05	
Aniline point, °C	87	
Pour point, °C	35	

Residuum (515 °C+)

Yield, vol % of syncrude oil	15.3	
Density g/cm^3	0.919	
Nitrogen, wt %	0.68	
Sulfur, wt %	0.05	
Nitrogen equivalents, ppm	1.0	

Source: H. C. Stauffer and S. J. Yanik, "Shale Oil: An Acceptable Refinery Syncrude," *ACS Division of Fuel Chemistry Reprints,* vol. 23, no. 4, 1978.

The principal constituents of spent shale are dolomite (calcium magnesium carbonate), calcite (calcium carbonate), silica, and silicates. About 50 percent by weight of the spent shale is finer than 100 mesh (0.147 mm). Size reduction occurs in the retorting process because retorting removes organic material which holds together the finely divided sedimentary particles that form rich oil shales.

Spent shale from the Colony Project would be disposed of through the formation and revegetation of a compacted embankment in a network of gulches. These gulches are of sufficient volume to contain all of the spent shale resulting from operations of the Colony Project. The disposal techniques to be used for the Colony Project were developed in experimental and field-demonstration programs carried out over a period of more than 10 years. These programs provided the information needed for assurance that the final embankment will be stable and that it can be revegetated.

Some important findings of the semiworks program relating to spent shale embankment design are discussed below:

1. The moisture content of processed shale as it leaves the oil-shale production facility should be in the range of 10 to 15 percent by weight, with a desired optimum value of 12 percent by weight. At this moisture content, there was no generation of dust during the transport of spent shale by the belt conveyor, and hauling, dumping, and spreading of the processed shale by heavy equipment could be done satisfactorily.

2. Embankment stability is achieved by spreading moisturized shale in thin layers and then compacting it. For much of the embankment, a compacted density of 85 lb/ft^3 is satisfactory. This density is achieved during travel over the spent shale by the haul rigs used to spread it. Where a higher density is required for critical portions of the embankment (e.g., frontal slopes), self-propelled, segmented-wheel compaction machines were found to be the most cost-effective equipment for achieving the required compaction.

3. Newly deposited spent shale forms an erosion-resistant surface crust as it dries out.

4. In order to ensure embankment stability, the slope of the front of the embankment should not exceed about 15 degrees. A low slope also facilitates equipment travel and is an aid to revegetation.

Field investigation of spent-shale revegetation began in 1966 with construction of test plots to evaluate plant-growth factors and plant species. The first field-demonstration plot was constructed and seeded in 1967. Field revegetation work was expanded in 1970 and thereafter with construction of additional demonstration plots to evaluate the effects of exposure and elevation on the growth of various plant species. These plots produced a good growth of grass and have remained in good condition. Monitoring of all the demonstration plots is part of an ongoing revegetation program.

Water Pollution

The Colony Project was designed to avoid pollution of the existing watershed and underground aquifers. This objective would be achieved by (1) compacting the spent-shale embankment to cause it to be, for practical purposes, impermeable to the flow of water through it, (2) using water produced in the process units for moisturizing the spent shale, thus avoiding discharge of water into local streams, and (3) constructing a dam downstream of the disposal embankment and plant complex to collect water runoff for use in the process units or for moisturizing spent shale.

SUMMARY

Commercialization of the TOSCO II process is proceeding in conjunction with the Sand Wash Shale Oil Project. The process has been fully tested, recovers all of the volatile hydrocarbons contained in oil shale, and meets or exceeds the requirements of all applicable environmental regulations. Improvements in efficiency can be achieved by burning the residual carbon in the spent shale. This can be done by adding a fluid-bed boiler to the TOSCO II process or by using the HSP process which Tosco is developing. Such improvements are particularly important if moist, low-grade, high-carbon-content shales are to be processed.

PARAHO OIL SHALE RETORTING PROCESSES

JOHN B. JONES, JR.
JOSEPH M. GLASSETT

Paraho Development Corporation
Englewood, Colorado

INTRODUCTION

Paraho is a contraction of Portuguese words meaning "for the good of mankind." It symbolizes the application of Paraho's patented technology to the recovery of oil from shale.

Paraho Development Corporation is an oil-shale technology company organized in 1971 and based in Grand Junction, Colorado. Paraho, with its fully patented oil-shale-processing technology, has been able to produce a large amount of crude shale oil in long, continuous operations. For that reason, Paraho is often referred to as a leader in oil-shale development.

Paraho has conducted a number of major oil-shale research and production programs. They include the Paraho Oil Shale Demonstration; the Department of Defense (DOD) and Department of Energy (DOE) 100,000-bbl (15,900-m³) production program; and research and development, including the processing of oil shale from several foreign countries. These successfully completed programs have demonstrated the environmental and economic acceptability of the Paraho technology.

HISTORICAL BACKGROUND

Development Engineering, Inc. (DEI) was founded by John B. Jones, Jr., who, as one of the co-inventors and the project engineer, assisted with the development of the U.S. Bureau of Mines (USBM) gas-combustion retort and subsequently served as a consultant to U.S. and foreign oil-shale projects. DEI invented hardware and process techniques that were applied to the construction of three commercial vertical-shaft lime kilns in the late 1960s. The kilns are noted for their low fuel requirement and high-quality lime product, and they are still in production.

Following DEI's success with limestone kilns and some experience with U.S. and Brazilian oil shale, company officials organized Paraho Development Corporation, with DEI as a subsidiary, to apply the process and its mechanical innovations to oil-shale retorting. Through DEI, a lease was obtained from the U.S. Department of the Interior, in May 1972, for the use of the USBM oil-shale facility at Anvil Points near Rifle, Colorado, to demonstrate the Paraho retorting technology. The first program was a 3-year, $10 million demonstration called the Paraho Oil Shale Demonstration Program. It was underwritten by 17 participating companies.

Sohio Petroleum Company

Gulf Oil Corporation

Shell Development Corp.

Standard Oil Co. (Indiana)

The Carter Oil Co. (Exxon)

Sun Oil Company

Chevron Research Company

Marathon Oil Company

Mobil Research and Development Co.

Southern California Edison Co.

The Cleveland-Cliffs Iron Co.

Atlantic Richfield Co.

Texaco, Inc.

Arthur G. McKee and Co.

Phillips Petroleum Co.

Kerr-McGee Corporation

Webb-Gary-Chambers-McLorraine Group

These companies received the right to enter into nonexclusive license agreements for commercial use of the Paraho oil-shale technology. Two retorts were built and operated. One unit—the pilot plant—had a 30-in (0.76-m) inside diameter (10) and a capacity of about 20 bbl/day (3.18 m³/day) of oil. The semiworks unit had a 102-in (2.59-m) ID and a capacity of about 200 bbl/day (31.8 m³/day). From 1976 to 1978, Paraho continued to work with the U.S. Navy's Office of Naval Research (ONR) and DOE in producing shale oil, increasing storage capacity at Anvil Points, installing an automated data-acquisition system, monitoring the environment, and generally refurbishing the Anvil Points area. During these two research efforts the old USBM room-and-pillar mine was reopened and, in all, some 200,000 tons (181,000 Mg) of oil shale were mined and over 110,000 bbl (17,500 m³) of crude shale oil were produced.

In June 1980, Paraho entered into an agreement with the DOE to design and estimate the cost of an oil-shale facility based on a single commercial-sized Paraho retort module and its support facilities. This design project required 18 months to complete at a cost of about $7,600,000. Project funding came from the DOE, Paraho, and 14 sponsoring companies involved in energy, mining, and engineering. The module, which has a design capacity of 10,520 bbl/day (1670 m³/day) of oil, will be built at Paraho's state-owned oil-shale lease near Bonanza, Utah.

Also in 1980, Paraho began planning the construction of a commercial oil-shale plant of 31,560 bbl/day (5,018 m³/day) capacity. Planning for the module was completed in December 1981, and that for the commercial plant in April 1982.

In the pilot-plant operation Paraho has also been active in testing oil shale from other deposits, both domestic and foreign. The company plans to continue in its consulting, testing, advising, research, and development roles.

THE PARAHO DIRECT HEATED PROCESS

Process Description

The Paraho Direct Heated Process is the result of Paraho Development Corporation's many years of innovative and successful experience in the production of shale oil through the use of patented, proven, aboveground retorting technology. The basic Paraho retort is a vertical carbon-steel vessel lined with refractory material. It is equipped with shale- and gas-handling devices which include a number of patented features for the precise control of the process materials streams.

Among the direct-heated retort's important characteristics are its low-cost, low-attrition, moving-bed, gravity-feed system; combustion of residual carbon in the moving shale bed; unique thermal exchange between descending shale and ascending recycle gas; and minimal use of moving parts. Since the Paraho retort utilizes its own heat energy, no external heat source is required other than that needed for initial cold start-up.

Crushed shale is fed into the top of the retort, where, as a moving bed, it begins its journey downward through four barrier-free but strictly functional zones. Cool off-gas carrying crude shale oil as a suspended mist is collected at the top of the retort and piped to the oil-and-gas–recovery system. Retorted shale exits at the bottom through the proprietary Paraho grate, which controls the shale flow. A more detailed description of the direct-heated process follows (Figs. 3-1 and 3-2).

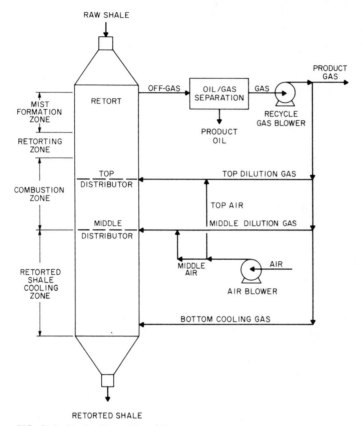

FIG. 3-1   Paraho Direct Heated Process.

Raw shale, crushed and screened to a feed size ranging from 0.25 to 3 in (0.6 to 8 cm), moves by conveyer to the feed system atop the retort. In a continuous moving bed, the shale flows downward consecutively through the *mist-formation, retorting, combustion,* and *cooling zones.* As the shale descends, it engages in an efficient heat exchange with a countercurrent flow of recycle gas, which is introduced into the retort at different levels by three specific-purpose gas and air and gas distributors. Near the top of the retort (in the mist-formation zone), the ambient temperature shale is warmed by rising hot oil vapors and gas which, in turn, are cooled to form an oil mist entrained in the gas.

As the shale continues downward into the retorting zone, it is heated by rising hot gases until its temperature reaches about 482 °C, the point at which kerogen is pyrolized into oil vapor, gas, and carbonaceous residue. This residue, which remains on the shale, becomes a fuel when the shale bed descends into the combustion zone, where combustion is achieved by the injection of air (diluted with recycle gas) through the middle and top distributors. This combustion provides the heat for the retort process.

Below the combustion zone the descending shale is cooled to about 149 °C by passing through a stream of cool recycle gas which is injected into the shale bed by the bottom gas distributor. The cooled retorted shale is discharged at the bottom through the hydraulically operated Paraho grate mechanism, which regulates the rate and uniformity of descent of shale through the retort. Displaced shale drops into a collection hopper and passes through a rotary seal and out of the retort to be conveyed to the retorted shale-disposal site.

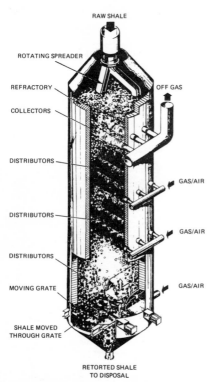

FIG. 3-2   Paraho Direct Heated Retort.

One of the products of the process is a low-Btu gas, part of which is recycled to the retort to serve primarily as a heat carrier and as a cooling agent. The means by which this is accomplished is described in the following summary of the *oil- and gas-recovery system*.

At the top of the retort, the off-gas from the mist-formation zone is collected from the shale bed and routed through an equipment train which includes a knockout drum, a coalescer, and an electrostatic precipitator. The separated liquid oil, which contains some water, is piped to the rundown and settling tanks, where the oil and water separate.

Part of the product gas is recycled to the retort to serve as both cooling agent and heat carrier. The bottom distributor delivers cool recycle gas to the base of the retort shale bed, where it cools the descending hot retorted shale and then becomes a part of the hot-gas train as it ascends. Additional recycle gas, along with a carefully proportioned amount of air, is injected through the middle distributor to the lower part of the combustion zone, where the carbon residue ignites with the oxygen and raises the temperature of the stream of ascending gas.

The top distributor also delivers an air-gas mixture to the upper region of the combustion zone, thereby supporting further combustion and producing additional heat to be utilized in the retorting zone above. Through the use of two distributors in the combustion zone, temperatures are kept below 760 °C, the point at which significant heat-absorbing carbonate decomposition takes place. The portion of the gas that is not recycled—the product gas—can be used to generate electric power and steam used in the shale plant. Excess electric power is available for sale.

Process Function and Advantage

Gravity feed is an important feature of the Paraho retort. This design concept provides the distinct advantage of very few moving parts within the retort, which results in low capital cost and a reliable, low-maintenance system. In addition, the retorting process utilizes a double countercurrent heat-exchange system. At the bottom of the retort, descending hot retorted shale warms injected cool gas. This gas rises and extracts more heat from the shale. Thus, the retorted shale is cooled internally without the addition of water. At the top of the retort, descending cool shale is heated to retorting temperatures by rising hot gases and the retort off-gas is cooled to about 66 °C, thereby conserving heat and eliminating external condensers. These two gas-shale heat exchanges give the Paraho retort a high thermal efficiency. And since the process requires no external heat source, it is energy-self-sufficient. There is also an advantage in processing shale in solid form, because the lumps-in, lumps-out concept makes retorted-shale disposal a simpler and a cleaner operation. Figure 3-2 is a cutaway illustration of the direct-heated retort.

Products and By-products

The liquid oil recovered in the direct-heated retort's day tanks has averaged 89 percent of Fischer assay through a wide range of operations. An additional 6 percent (of Fischer assay) of naphtha can be recovered from the gas. For the operating conditions that are expected to be used in a commercial plant, the liquid oil recovery is expected to be between 94 and 100 percent. Gas also is recovered and utilized, so that the *net* oil-plus-gas recovery is 103 percent of the oil-plus-gas yield of the Fischer assay. Since the retort's heat requirements are met by combustion of carbon residue in the retorted shale, as opposed to an external energy source, the *gross* energy yield of the retort is 114 percent of the Fischer assay oil plus gas.

Table 3-1 shows the yield and properties of Colorado-sources shale oil from a Paraho Direct Heated Retort after gravity separation of water. The oil is similar to shale oil from other Paraho processes. The pour point is high, but pour point depressants have been identified and with their use the oil can be pipelined to transportation points or refineries.

TABLE 3-1 Paraho direct heated retort oil yield and properties, Colorado oil shale

Yield, gal/ton	
Shale assay	27.4 (114 L/Mg)
Retort production	24.4 (102 L/Mg)
Percent of assay	89 to 94
Physical properties	
Specific gravity, °API (density)	21.4 (925 kg/m³)
Viscosity, SUS at 130 °F (54 °C)	83
Viscosity, SUS at 210 °F (99 °C)	48
Pour point, °F	85 (29 °C)
Water, wt %	0.3
Sediment, mL/100 g	0.1 (0.001 L/kg)
Analysis, wt %	
Carbon	84.8
Hydrogen	11.4
Nitrogen	2.0
Sulfur	0.6
Oxygen	1.2

TABLE 3-2 Direct-heated Paraho retort gas properties, Colorado oil shale

Yield, EO,* gal/ton	
Shale assay	2.1 (8.8 L/Mg)
Retort production	6.0 (25 L/Mg)
Percent of assay	285.0
Analysis, dry gas, vol %	
H_2	5.5
N_2	61.0
O_2	-0-
CO	2.9
CH_4	2.4
CO_2	24.2
C_2H_4	0.7
C_2H_6	0.6
C_3	0.6
C_4	0.6
C_5+	0.6
H_2S	0.3
NH_3	0.6
Total	100.0
Water vapor, vol %	17.5
Heating value, Btu/stdft^3, dry	140 (5.2 MJ/m^3)

*EO = equivalent oil

Table 3-2 shows the yield and properties of the product gas after the oil and water have been separated. The gas includes a significant amount of carbon dioxide, not only from the oxidation of carbon but also from the decomposition of a portion of mineral carbonates in the shale. The major component of the gas is nitrogen, which, along with the CO_2, dramatically lowers the heating value to only 140 Btu/stdft^3 (5.2 MJ/m^3). After removal of hydrogen sulfide, ammonia, and water vapor, the gas can be burned on site for electric power generation.

THE PARAHO INDIRECT HEATED PROCESS

Process Description

The Paraho Indirect Heated Retort is similar in many physical characteristics to the direct-heated system, but it differs significantly in the method of heat generation, the heating value of its product gas, and its net energy-recovery efficiency.

Among the numerous similarities of the two systems are the following: Both have few moving parts and make use of a low-cost low-maintenance gravity descent of oil shale through barrier-free zones that are strictly functional. Both depend upon a countercurrent flow of shale and gases during a unique thermal exchange that results in the efficient retorting of shale for the recovery of desirable products and the simultaneous cooling of retorted shale to facilitate disposal. Both use recycle gas instead of water for cooling purposes—

an important consideration in some of the arid and semiarid regions in which many oil shale deposits are found.

The most obvious difference between the two systems is the method by which the desirable retorting temperatures are obtained. Since there is no combustion within the indirect-heated retort, the heat is supplied by recycling part of the gas through external heaters and then injecting the heated gas into the retort through two levels of hot-gas distributors. At the same time, cool recycle gas is injected into the bottom of the retort in much the same manner and for the same purpose that cool gas is utilized in the direct-heated process.

Because fuel must be supplied to the heaters, the efficiency of the indirect-heated process is somewhat lower than that of the direct-heated method (see Products and By-products). It is still quite high when compared to the efficiencies of many other industrial processes, including oil refineries and coal-fired power plants. In comparison with the direct-heated process, the indirect-heated system produces a product gas that has a much higher Btu content and is both free of nitrogen and lower in carbon dioxide; it is also of special value for industrial purposes.

Figure 3-3 illustrates functions of the Paraho Indirect Heated Retort, which, like the direct-heated system, is an aboveground vessel of carbon steel lined with insulation and refractory brick.

Crushed raw shale, ranging in size from 0.25 to 3 in (0.6 to 8 cm), is fed into a hopper at the top of the retort. From there it passes through a rotary seal that permits shale entry

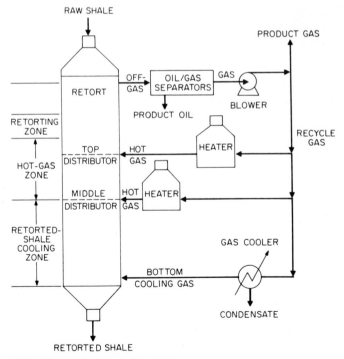

FIG. 3-3   Paraho Indirect Heated Process.

while retaining retort gases. A solids distributor uniformly spreads the shale, which, as a moving bed, continues to descend by gravity through the unrestricted zones of the retort. The shale, fed into the retort at ambient temperatures, is heated by a rising stream of hot gas and oil vapor which is condensed to a mist as it meets the relatively cool shale in the *mist-formation zone*. As the shale descends, it encounters preheated gas at different levels and eventually is heated to about 482 °C, the temperature at which kerogen decomposes into gas and oil vapor and a carbonaceous residue in the *retorting zone*. (A comparison of Figs. 3-1 and 3-3 shows that the indirect-heated retort has a hot-gas zone instead of a combustion zone.) The carbon residue (which is burned in the combustion zone of the direct-heated retort) remains on the shale during the indirect-heated process.

Below the retorting zone the shale is cooled to about 204 °C before being withdrawn at the bottom. The vertical descent of shale is blocked by a horizontal retarder plate. A hydraulically operated pusher bar reciprocates horizontally and displaces a controlled amount of shale from the edge of the plate to pass through a rotary seal or star valve, which prevents loss of gaseous products. Thus, by controlling the volumetric rate of discharge, the proprietary mechanical grate also controls the rate of processing the shale through the retort.

Product gas is recycled and injected into the retort at three levels. This gas is not for combustion purposes; it serves instead as a heat carrier. Cool recycle gas is injected into the bottom of the retort, where it is preheated by hot retorted shale. As the gas rises, it cools the retorted shale and accumulates more heat. Recycle gas is heated in external heaters and is injected at the middle and top distributors. No air is added, so combustion does not take place in the retort. The gas is externally heated to about 649 °C in heaters which burn product gas, retorted shale, coal, or other fuel. The hot gases rise and entrain the oil-gas mist retorted from the shale and exit via the off-gas collector at the top of the retort. The products are separated by using standard mist agglomeration, extracting, and precipitation from the gas stream.

Process Function and Advantage

There are distinct advantages in using gravity flow of shale in the Paraho retort. Because the retort has few moving parts, the capital costs of the system are low, as are maintenance requirements. The indirect-heated system also takes advantage of a countercurrent flow of shale and gas both above and below the center of the retort. At the bottom, descending hot retorted shale meets ascending cool gas; at the top, descending cool shale meets ascending hot gas. This countercurrent heat exchange has the advantages of great thermal efficiency and internal cooling of retorted shale so that water is not required for cooling either the retort off-gas or the retorted shale. That the retorted shale is in the form of lumps is an advantage in waste disposal.

Products and By-products

The indirect-heated retort generally yields 91 to 96 percent of Fischer assay oil. Gas also is recovered and utilized, so that the rate of oil-plus-gas recovery is 95 to 98 percent of Fischer assay. Since some fuel is required for the furnaces that heat the recycle gas, that energy must be deducted as an expense against the system. As a result, the net yield from the process is 82 to 85 percent of Fischer assay.

TABLE 3-3   Indirect-heated Paraho retort oil yield and properties, Colorado oil shale

Yield, gal/ton	
Shale assay	27.4 (114 L/Mg)
Retort production	24.9 (104 L/Mg)
Percent of Fischer assay	91 to 96
Physical properties	
Specific gravity, °API (density)	20.2 (932 kg/m³)
Viscosity, SUS at 130 °F (54 °C)	80
Viscosity, SUS at 210 °F (99 °C)	44
Pour point, °F	75 (24 °C)
Water, wt %	1.0
Sediment, mL/100 g	0.1 (0.001 L/kg)
Analysis, wt %	
Carbon	84.9
Hydrogen	11.4
Nitrogen	1.9
Sulfur	0.6

The yield and properties of shale oil extracted from Colorado oil shale by the Paraho Indirect Heated Retort are shown in Table 3-3. While the shale oil is similar to that produced by other Paraho processes, there are some areas of difference. The pour point is slightly lower than that of oil from the direct-heated retort. Available pour point depressants make it possible to pipe the product to refining or transportation sites. The oil contains sulfur and nitrogen, which must be removed by hydrotreating.

Gas produced by the indirect-heated process (Table 3-4) differs significantly from the gas from the direct-heated process. Since there is no inner-retort combustion, the gas is nitrogen-free and is lower in carbon dioxide. This results in a relatively high gas heating value of about 1036 Btu/stdft³ (38.6 MJ/m³), which is a great advantage of the indirect-heated retort. The gas contains hydrogen sulfide and ammonia, which must be removed before the gas can be utilized.

PILOT-PLANT RETORT

During a 36-month period beginning in 1973, Paraho conducted the Paraho Oil Shale Research Project, which involved the construction and operation of the pilot-plant retort and the semiworks retort at Anvil Points (Fig. 3-4.) The pilot plant was an important part of the successful $10 million privately funded program which was organized to demonstrate the Paraho technology in the retorting of oil shale. It was constructed between December 1973 and April 1974, and the first test run was made in early June 1974. The semiworks retort was placed in operation in August 1974.

The pilot-plant retort and the semiworks retort were operated independently and sometimes concurrently, with each complementing the other at various stages of the research, development, and demonstration program. The cross-sectional area of the pilot-plant retort

TABLE 3-4 Paraho indirect-heated retort gas yield and properties, Colorado oil shale

Yield, EO,* gal/ton	
Shale assay	2.1(8.8 L/Mg)
Retort production	3.1(13 L/Mg)
Percent of Fischer assay	148
Analysis, dry gas, vol %	
H_2	23.0
N_2	0.4
O_2	-0-
CO	2.7
CH_4	30.2
CO_2	14.8
C_2H_4	11.3
C_2H_6	5.6
C_3	3.0
C_4	0.8
C_5+	4.4
H_2S	2.7
NH_3	1.1
Total	100.0
Water vapor, vol %	28.8
Heating value, Btu/stdft³, dry	1036 (38.6 MJ/m³)

*EO = equivalent oil

is 4.91 ft² (0.0456 m²); the inside diameter is 30 in (0.76 m); and the capacity is about 20 bbl/day (3.2 m³/day) of shale oil. Because of its smaller size, the pilot retort was less expensive to modify. It also provided a means of experimenting with a variety of oil shales, both foreign and domestic, without consuming large quantities of shale. This was especially important in the case of foreign shales, which are costly to transport to Colorado. Although the pilot plant was normally scheduled to run for a brief period of time, one operation continued for 77 days before being terminated at the conclusion of scheduled tests. Later, development of the indirect-heated process was conducted in the pilot retort. When the semiworks retort was operated by using the indirect-heated process, the pilot retort was operated in the direct-heated mode to provide a supply of low-Btu gas to purge the retort equipment.

In late 1977 and early 1978, selected process-variable tests were performed in the pilot retort. These tests included operation with rich shale having a grade of 31 to 36 gal/ton (129 to 150 L/Mg) by Fischer assay. In September 1979 and again in September 1980, Israeli shale was tested in the pilot retort. These tests were compared with results of a standard run of Colorado shale during the same period to confirm material and energy balance data. In January 1981, Moroccan shale was tested in the pilot retort. All of these studies showed that the pilot retort was functionally equivalent in its operations to the semiworks retort and provided confidence in plans to scale up retort size.

FIG. 3-4   Paraho pilot plant and semiworks retorts at Anvil Points, Colorado.

SEMIWORKS RETORT

The semiworks retort has an inside diameter of 8.5 ft (2.6 m) and a normal feed rate of about 300 tons/day (272 Mg/day) of oil shale. During the research program at Anvil Points, 31 runs involving 100 test periods were performed in this vessel; they resulted in a yield of approximately 10,000 bbl (1590 m^3) of raw shale oil. One continuous operation during this program lasted 56 days. The raw shale oil produced during the run was refined at the Gary Western Refinery (near Fruita, Colorado) into gasoline, heavy fuel oil, and other products.

Further research contracts with the DOE and the ONR resulted in the production of approximately 74,000 bbl (11,800 m^3) of crude shale oil for conversion to military fuel at the Toledo, Ohio, refinery of Standard Oil Company of Ohio. In 1977 the semiworks retort was operated 24 h/day, 7 days a week to produce 12,500 bbl (2000 m^3) of shale oil. No shutdown was required, so a new on-stream record of 105 continuous days was made.

A summary of the semiworks operation for 1977–1978 is given in Table 3-5. The retort operated for the full year of 1977 and had an oil-plus-gas yield of 103 percent of Fischer assay and an operating factor exceeding 90.8 percent. The shale grade average was 27.4 gal of oil per ton of raw shale (114 L/Mg) as measured by the Fischer assay. The average

TABLE 3-5   Semiworks operation summary

	1977			1978	
	Average	Low	High	Low	High
Shale mass rate					
lb/hr·ft²	461	416	486	375	537
Mg/m²·h	2.25	2.03	2.37	1.83	2.62
Shale assay					
Oil, gal/ton	27.4	25.2	29.7	21.1	36.6
Oil, L/Mg	114	105	124	88	153
Oil production, bbl/day	160 (25.4 m³/d)				
Yield (oil + gas, FOE),					
vol % assay*	103				
Retort operating					
factor, %	90.8				90.7 (avg)

*Net yield after supplying heat for the retort

oil production rate was 160 bbl/day (25.4 m³/day), and the average mass rate was 461 lb/h·ft² (2.25 Mg/h·m²).

Since the semiworks retort was designed for research and not for production-type operations, spares were not provided for any of the auxiliary equipment. Maximum use was made of the auxiliary equipment available from previous retorting work at the government's Anvil Points facilities. Some of this equipment was first used in the 1940s. A list of equipment items affecting the retort operating factor is given in Table 3-6.

TABLE 3-6   Semiworks equipment contributing to downtime, 1977

Equipment	Downtime, h
Retort vessel and distributors	0.0
Raw shale feed conveyor	2.8
Top seal	3.8
Grate*	143.3
Bottom seal†	0.0
Retorted shale conveyors	0.0
Oil-gas separators‡	543.2
Oil-handling system	0.0
Gas blower‡	203.3
Air blower	126.5
Instrumentation	0.0
Electrical supply	64.0
Miscellaneous	11.3

*Hydraulic control problems

†Repairs made while working on other equipment

‡Oil-gas separation problems

Most of the retort downtime in 1977 was caused by problems with the oil-gas separation equipment. Special studies showed that the separator's operating efficiency decreased rapidly when the oil production rate exceeded 180 bbl/day (28.6 m^3/day). Although the retort operates well at rates of 200 bbl/day (31.8 m^3/day), 1978 production rates were limited to those giving good separator efficiency, and the retort operating factor was thereby increased. In all, the semiworks and pilot plant retorts have processed more than 200,000 tons (180,000 Mg) of oil shale and produced more than 110,000 bbl (17,500 m^3) of crude shale oil.

Refining of Paraho Raw Shale Oil

Under terms of a contract with the U.S. Navy Paraho arranged for The Standard Oil Company of Ohio (Sohio), to conduct investigations at its research facilities in Warrensville, Ohio, to determine coking and hydrotreating characteristics of Paraho shale oil. The objective of the study was to develop process data on a small scale to be used as guidance for the 10,000-bbl (1590-m^3) run at Gary Western's refinery. Pilot work was carried out within the capabilities and limitations of the Gary Western facilities. The studies were conducted toward two goals: (1) evaluation of the processibility of the shale-oil fractions within the constraints of the available refinery operating limits and (2) estimation of the probable product yield and quality.

Gary Western test run

During the test run at Gary Western, 10,000 bbl (1590 m^3) of Paraho raw shale oil were processed to produce naphtha, light gas oil, heavy gas oil, heavy fuel oil, coke, and gas. The shale oil was charged into a delayed coker fractionator at a rate of 2500 bbl/day (400 m^3/day). Residuum produced in the bottom of the fractionation column was sent to a delayed coker. Representative yields for the operation are tabulated below.

	bbl	m^3	%
Shale oil feed	9956	1583	100
Products:			
Gasoline	1180	188	12
Jet fuel	1660	264	17
Diesel fuel	2080	331	21
Heavy fuel oil	2360	375	24
Subtotal liquids	7280	1158	74
Coke, 1000 lb	424	67	——
Wt %	——	——	13

Some of the observed results included (1) the severe coking operation and hydrotreating of the jet fuel feedstock produced a larger quantity of gas than expected; (2) storage limitations and allocation of much naphtha to jet fuel manufacture resulted in a smaller quantity of gasoline than expected; (3) the use of acid treatment rather than hydrotreating decreased the quality of diesel fuels; and (4) the production of heavy fuel oil was greater than expected. On the basis of these results, two specific refinery improvements were recommended: high-pressure [2000 to 3000 lb/in² (13.8 to 20.7 MPa)] hydrotreating and clay treating.

Sohio refinery tests

In a later shale-oil refining program, Sohio conducted extensive refinery tests on Paraho shale oil in a joint program with DOE and DOD, managed by the Navy. The program was designed to refine up to 100,000 bbl (15,900 m³) of Paraho shale oil into military fuels by using existing refinery facilities (Fig. 3-5). The shale oil was transported by railroad tank cars from Anvil Points to Toledo.

FIG. 3-5 Paraho shale oil enroute to Sohio.

When compared with conventional petroleum, shale oil has several characteristics that require special considerations in handling and processing. These include high nitrogen and oxygen content, low hydrogen-carbon ratio, low yields of 343 °C− material, moderate arsenic and iron, and suspended ash and water content. The high nitrogen content is probably the area of greatest concern to oil refiners.

Sohio's Toledo refinery had a twofold objective in the first phase of its preparations for processing Paraho shale oil: (1) to develop and demonstrate a method for producing, in existing refinery facilities, specification and stable military fuels from crude Paraho shale oil and (2) to maximize yields of jet fuel (JP-5) and diesel fuel marine (DFM) while minimizing yields of 343 °C+ bottoms (residual fuel). A schematic diagram of the special process developed for the refining program is shown in Fig. 3-6.

The Toledo hydrotreating run lasted from November 8 to December 4, 1978. The hydrotreated shale oil was fractionated into naphtha, jet fuel, diesel fuel marine, and bottoms (residuum, or No. 6 fuel oil).

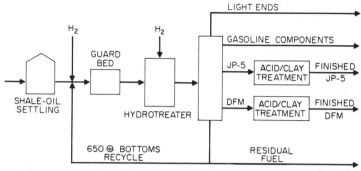

FIG. 3-6 Shale-oil-refining process.

The gasoline or naphtha fraction was utilized within the refinery as feed to another hydrotreater and gasoline re-former. The jet fuel and diesel fuel were stored separately in tankage provided by the refinery. Most of the 343 °C residuum not consumed by recycle was utilized internally. A portion was loaded into railroad cars for future combustion testing by the Electric Power Research Institute, and 15,000 gal (57 m³) was shipped to a facility designated by the EPA for industrial gas-turbine testing.

The stored raw jet fuel (JP-5) was tested with acid and clay during the month of January 1979. Approximately 6200 bbl (990 m³) of finished JP-5 was shipped to Rickenbacker AFB via rail for testing in aircraft engines by the Navy. Approximately 18,200 bbl (2890 m³) of raw DFM was treated and distributed for testing. In addition to the JP-5 and DFM, a 500-bbl (80-m³) railroad car of JP-8 was produced for the Air Force on rundown during the hydrotreating run. This material is similar to JP-5 except that it has a lower freeze requirement. The treated JP-8 passed all military specifications. Numerous samples of crude intermediate and product streams were taken for health-effect studies by DOD and DOE. Conclusions growing out of the tests included:

1. Fuels meeting military specifications and possessing good storage-stability characteristics can be produced from crude shale oil.

2. The processing scheme requires: (*a*) settling and a guard bed to protect the hydrotreating catalyst, (*b*) hydrotreating to remove hetero-atoms, increase the hydrogen-carbon ratio, and improve the 343 °C— liquid yield of shale oil, and (*c*) treating with acid and clay or additional treatment with hydrogen to meet thermal and storage stability requirements of the jet and diesel fuels.

3. It was demonstrated that crude shale oil can be processed in commercial refinery equipment. The preliminary refinery yields were reported as follows:

Product	Liquid yield, vol %
Gasoline, C_6–163 °C	11
JP-5, 163–249 °C	26
Diesel fuel, 249–343 °C	31
Fuel oil, 343 °C+	32

A brief summary of the Sohio refinery run follows.

Gasoline component   Hydrotreated gasoline (C_6 to 157 °C) was paraffinic and therefore had a low octane rating. The octane rating was improved by re-forming over a commercial catalyst. The unleaded product experienced a significant gain in octane rating to 93 research octane number (RON). The final leaded product was somewhat unstable, probably from a nitrogen-lead reaction and the severe re-forming conditions.

Jet fuel   The raw JP-5 fuel was treated with sulfuric acid and clay contacted in a final finishing step. Acid treatment reduced the nitrogen to less than 20 ppm, a reduction which was more than sufficient to meet gum and thermal-stability requirements (JFTOT*). Sohio subjected the fuel to an additional storage-stability requirement not contained in the military specifications. The low-nitrogen JP-5 fuel (12 ppm) passed the JFTOT after aging at both 260 °C and 288 °C, indicating a very stable fuel. The high-nitrogen JP-5 fuel (189 ppm) also passed the JFTOT before and after aging at 260 °C but failed after aging at 288 °C. This fuel is still acceptable, but is less stable than the low-nitrogen fuel.

Diesel fuel marine   Acid treatment was required to reduce the nitrogen content so that this fuel would meet the stability requirement. The fuel is thermally stable, but it experiences gum problems when exposed to light. The light instability is not viewed as a major problem, because these fuels are produced, stored, and utilized without exposure to light.

Residuum, or No. 6 fuel oil   The residuum, or 343 °C+ bottoms, was not treated, but the material was found to be very clean—without ash, metals, or sulfur and with a relatively high hydrogen content. It does have a high pour point, however, and it might cause some handling difficulties if it were utilized in an unmodified No. 6 fuel oil system.

Results   In general, the refinery run produced fuels which met military specifications and had good storage stability.

10,520-BARREL-PER-DAY MODULE

Paraho's extensive shale-retorting experience in the pilot plant and semiworks operations at Anvil Points provided a basis for technical criteria to be used in the design of a commercial-scale oil shale facility. In a cooperative agreement effective June 15, 1980, Paraho was designated as the participant, on a cooperative basis with DOE, to design a 10,520-bbl/day (1670-m^3/day) oil-shale module using Paraho Direct Heated Process technology. Also included in the agreement was the design of the supporting mine and other support equipment.

Costs of the module design program were shared by DOE and 14 large energy, mining, and engineering companies which joined the designated participant (Paraho) as sponsors: Arco Coal Company, Chevron Research Company, Cleveland-Cliffs Iron Company, Conoco Inc., Davy McKee Corporation, Husky Oil Company, Mobil Research and Development Corporation, Mono Power Company (Southern California Edison), Phillips Petroleum

*Jet Fuel Thermal Oxidation Test

Company, Placid Oil Company, Sohio Shale Oil Company, Sunoco Energy Development Company, Texaco Inc., and Texas Eastern Synfuels, Inc. Three of these sponsors—Cleveland-Cliffs Iron Company, Davy McKee Corporation, and Sohio Shale Oil Company—also served as subcontractors to Paraho in the project. Other subcontractors were Aero-Vironment and VTN, Consolidated. Woodward-Clyde Consultants, Holme Roberts and Owen, Moon Lake Electric Association, and Hercules Incorporated were consultants.

The module project involved 18 months of planning and designing by Paraho and its subcontractors at a budgeted cost of $8,077,555. During the module project, Paraho's goals were:

1. To produce detailed engineering and fabrication drawings for a single module retort and oil shale mine

2. To prepare a definitive construction cost estimate ($\pm$10% accuracy) of the retort facilities

3. To develop preliminary designs and capital costs for support facilities

4. To develop an operating plan for the module that would demonstrate the technical suitability of Paraho technology

5. To estimate the operating costs for the module and support facilities

6. To prepare plans for evaluating the environmental impact

7. To prepare a conceptual design and capital- and operating-cost estimate for the production of 100,000 bbl/day (15,900 m³/day) of crude shale oil from Green River Formation oil shale

In achieving the objectives of the Paraho Oil Shale Module project, engineers developed a comprehensive design for a full-scale Paraho retort module, along with associated technical, environmental, health, safety, operating, and economic criteria for a 2-year demonstration program. The module is designed to produce 10,520 bbl per stream day (1670 m³/day) of crude shale oil and to demonstrate the engineering performance, operating characteristics, environmental-control requirements, and economics of a commercial-size Paraho shale-oil production facility. The processing units and supporting systems are shown in the block flow diagram of Fig. 3-7.

Location and Lease

The Paraho Oil Shale Module is designed for a 582-acre (2.36-Mm²) site which was leased from the State of Utah for a 20-year period. Located in Sec. 32, Twp 9S, R25E, the site is about 50 mi (80 km) southeast of Vernal, Utah; 35 mi (56 km) southwest of Rangely, Colorado; 3.5 mi (5.6 km) southwest of Bonanza, Utah (which is served by paved Utah State Highway 45); and just a few miles west of the Utah-Colorado boundary. Core holes at the site have provided information for reserve calculations, rock strength, formation water, and fractures. Estimated resources range from 108,424 bbl/acre (4.26 m³/m²) of 28.1-gal/ton (117-L/Mg) oil shale for a 55-ft (16.8-m) mine-room height to 147,405 bbl/acre (5.79 m³/m²) of 27.1-gal/ton (113-L/Mg) oil shale for a 77-ft (23.5-m) mine-room height.

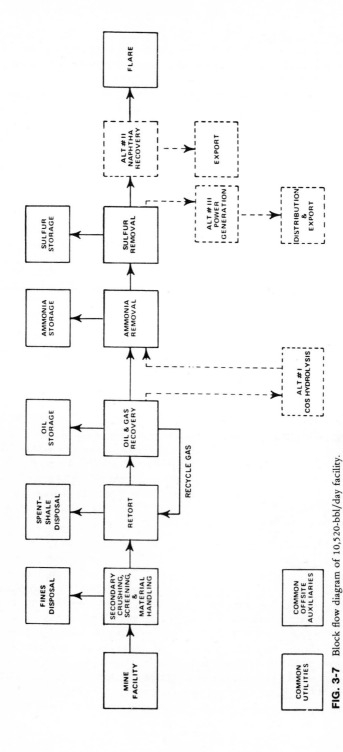

FIG. 3-7 Block flow diagram of 10,520-bbl/day facility.

Mining and Shale Handling

The module mine is designed as a room-and-pillar, heading-and-bench operation to remove a 60-ft (18.3-m) layer of oil shale with an average grade of 28 gal/ton (117 L/Mg). The operation will be about 500 ft (150 m) underground, with access through a vertical shaft, horizontal adit, and inclined ramp. The mine is designed to produce 23,000 tons/day (20,900 Mg/day) of shale.

Front-end loaders and trucks move the loose shale to the in-mine primary crusher, which breaks the rock to -12 in (-0.30 m). This shale exits the mine via an inclined conveyor to secondary and tertiary crushers which reduce the shale to -3 in (-0.08 m). The prepared shale is distributed to the retort hoppers by tripper and shuttle conveyors. Weigh belts maintain volume control to the retort.

Retort Section

The commercial-scale retort (Fig. 3-8); constructed of carbon steel and lined with refractory brick, has an inside cross section 24 ft (7.3 m) wide and 138.5 ft (42.2 m) long. The raw shale charging floor is about 100 ft (30 m) above the retort's concrete base. It is designed to process 17,730 tons/day (16,080 Mg/day) of oil shale and to deliver (from 28 gal/ton oil shale) a total of 10,520 bbl/day (1670 m³/day) of raw shale oil. An additional product will be 128 million stdft³/day (3.6 million m³/day) of low-Btu gas.

The feed to the retort consists of oil shale ranging in size from 0.38 in (0.01 m) to 3 in (0.076 m), with more than 94 percent by weight of the particles ranging between 0.525 in (0.013 m) and 3.0 in (0.076 m). A description of the retorting operation, beginning with the rock feed at the weigh scale and conveyor belt, follows.

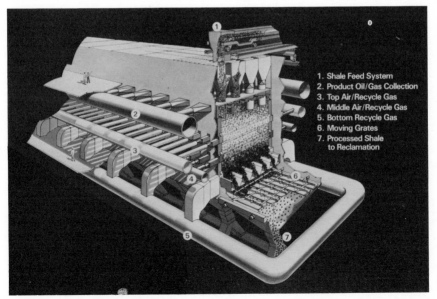

1. Shale Feed System
2. Product Oil/Gas Collection
3. Top Air/Recycle Gas
4. Middle Air/Recycle Gas
5. Bottom Recycle Gas
6. Moving Grates
7. Processed Shale to Reclamation

FIG. 3-8 Paraho commercial module retort.

The weigh scale measures and controls the feed to the retort at the rate of 738.8 tons/ h (670 Mg/h) of oil shale. The weigh scale and conveyor feed the rock to the tripper conveyor. The shuttle conveyor moves perpendicular to the tripper conveyor, and the conveyors' programmed motions allow rock to be fed to the individual hoppers. A computer program controls the feed system by adjusting the rock feed rate, preselecting the conveyor's position, and monitoring the feed rates. Redundant systems allow the operation to continue feeding oil shale even after the loss of primary and secondary measurement loops. In the event the control system senses too much feed to the shuttle conveyor, the bypass gate activates and sends the rock to the dribble and bypass bin and to the recycle conveyor, which returns the shale to the retort feed system.

The shale rock feeds by gravity into the top of the retort through feed tubes. Each tube supplies a designated portion of the retort cross-sectional area. There are several Paraho grate units, each independently controlled, which discharge the retorted shale from the retort and at the same time facilitate precise movement of the shale bed through the retort. The grate-mechanism pusher bars are the principal moving parts in the Paraho retort.

Recycle gas—a low-Btu, oil-free gas—enters the retort through the bottom gas distributor above the Paraho grate-retarder plates. It cools the descending retorted shale, and in extracting heat from the shale, the gas becomes a part of the countercurrent heat exchange within the retort as it ascends. An air-gas stream enters the lower part of the combustion zone through the middle distributor, and another air-gas stream is delivered to the upper region of the combustion zone by the top distributor. Maximum gas temperatures occur in a zone just above the upper air-gas distributor. This heat is sufficient to retort the shale above it, driving off the volatiles. The retorted shale continues downward to the combustion zone, where the combustion of residual carbon in the processed shale provides additional heat to the rising gas stream. The shale oil in the retorting zone is driven off by the heat exchange between the hot gas and the shale at about 482 °C. Temperature control is important: extremely high temperature decomposes the carbonates in the rock and produces an excess of carbon dioxide; low temperature results in lost hydrocarbon yield.

The recycle air-gas mixture, depleted of its oxygen content, leaves the retort through a plenum at the top of the retort at a temperature of about 60 °C. This retort off-gas carries the crude shale oil as a suspended mist to the oil-gas separation units.

Zone control of the retort is regulated by the speed of the Paraho grate-pusher bars at the bottom, by the overall feed of recycle gas to the bottom distributors and air-gas to the middle and upper distributors, and by the action of individual control valves on the upper air-gas distributors. The off-gas lines at the top of the retort have individual pressure-control valves on each of the headers. Manually adjustable valves—on the middle air-gas distributors and on the bottom gas distributors—provide further flow adjustment and distribution.

The retorted shale leaves the retort by gravity at a bulk volume rate set by the Paraho discharge grate. The hot retorted rock, cooled by the exchange of heat with the incoming bottom-distributor recycle gas, leaves the retort at approximately 140 °C through a purged, double-rotary gas-lock system. Retorted shale descends to one of two parallel belt conveyors, which are equipped with an emergency water-spray system to protect the belts in the event the shale has not been adequately cooled before leaving the retort. The duplicate conveyor is a backup to assure continued retort operation in the unlikely event one belt is inoperable.

Oil and Gas Recovery

Off-gas from the top of the retort, laden with fine oil mist, is directed via four separate streams to systems consisting of knockout drum, packed-bed coalescer, and electrostatic precipitator, where 99 percent of the oil and some of the water are removed from the off-gas. The recovered oil and water are pumped to day tanks for further separation by gravity. The oil is pumped to storage, and the water is pumped to wastewater treatment facilities. Much of the gas, which is circulated by centrifugal compressors, is recycled to the retort. The remainder is piped to the gas-cleanup area as net product gas. Ammonia and some carbon dioxide and hydrogen sulfide are removed in packed-bed absorbers. Further processing separates ammonia and water. Solid sulfur is then recovered from the ammonia-free gas and pumped in a molten state to a loading facility.

Products and By-products

The primary objective of the Paraho Oil Shale Module project is the production of quality shale oil in facilities representative of a commercial-scale operation. The properties of Paraho raw shale oil include:

Specific gravity (density)	21.4° API (925 kg/m³)
Viscosity	89.9 SSU at 54.4 °C
	46.5 SSU at 98.9 °C
Pour point	29.4 °C
Nitrogen content	2.0 wt %
Sulfur content	0.6 wt %
Arsenic content	20 ppm
Basic sediment	0.5%

The major by-product of the Paraho process is the mist-free off-gas from which 99 percent of the crude shale oil and much of the water have been removed. Most of this gas is recycled to the retort as the heat-transport medium. The remaining, or net product, gas is processed to yield ammonia, sulfur, and a low-Btu product gas. The low-Btu gas can be burned to generate electricity. Water is recovered at several points in the oil- and gas–handling system. After being sent to the wastewater treatment system, it is available for a variety of purposes including dust control and fire protection.

Retorted-Shale Disposal and Raw-Shale Fines Storage

The designs for retorted-shale disposal and raw-shale fines storage are based on studies made by Development Engineering, Inc., a subsidiary of Paraho Development Corporation, and the USBM. Designs for disposal and storage areas incorporated special provisions to prevent leaching, auto-ignition, erosion, dust, and other problems. Approximately 10 million yd³ (7.6 Mm³) of processed shale will be disposed of during the module operation. Hilly terrain makes the module site suitable for the construction of a cross-valley dam which, along with the retaining area, will be compacted to prevent leachates from reaching ground water or local streams. Reclamation measures, including covering with topsoil followed by revegetation, are among the design features.

Paraho completed the engineering design and cost estimates of its 10,520-bbl/day (1670-m³/day) module in December 1981 at a cost of about $7,600,000. Goals for the next step included construction and operation of the module retort and support equipment for a period adequate to confirm technical, economic, and environmental-control capabilities and to demonstrate the Paraho process technology at a scale representative of a commercial operation. A commercial-size plant, visualized as a future development of the project, would consist of three or more Paraho 10,520-bbl/day (1670-m³/day) modules. Figure 3-8 is an artist's rendering of the Paraho commercial module retort.

Through the construction and operation activities of the module program Paraho plans to demonstrate its ability to scale up proven Paraho oil-shale retorting technology, confirm the application of this technology to commercial-scale operations, develop the confidence of the industrial, governmental, financial, and environmental communities to endorse and to proceed with oil-shale commercialization, and provide facilities for developing future process improvements and expansions.

31,560-BARREL-PER-DAY COMMERCIAL PLANT

In the latter part of 1980, the DOE awarded a $3,111,809 grant to Paraho for a commercial feasibility study covering the expansion of the single-retort module to a commercial shale-oil plant capable of producing 31,560 bbl/day (5018 m³/day) of crude shale oil, or 33,860 bbl/day (5383 m³/day) of upgraded (hydrotreated) oil per day. This study was completed in April 1982.

Location and Resources

The commercial plant design (see artist's conception, Fig. 3-9) is site-specific for a tract of land in Uintah County, Utah, at an elevation of approximately 5500 ft (1680 m) above sea

1. Administrative Building
2. Mined Shale Storage
3. Conveyor from Mine
4. Crushing
5. Crushed Shale Storage
6. Retorts
7. Oil/Gas Product Separation
8. Electrical Generation
9. Processed Shale Management
10. Mine Shaft Headframe
11. Mine Ventilation
12. Oil Upgrading
13. Oil Storage/Transportation
14. Water Intake

FIG. 3-9 Plan of 31,560 bbl/day Paraho-Ute commercial shale-oil facility.

level. The mining elevation is approximately 5100 ft (1550 m) above sea level. The minable zone is about 300 to 600 ft (91 to 183 m) below the overburden of rock and leaner shale. The site is about 27 mi (43 km) by road from Rangely, Colorado, and 50 mi (80 km) by road from Vernal, Utah.

The shale tracts provide five functional sites: the mine access, the plant site, the administration areas, the disposal site, and temporary construction areas.

Design Features

The design for the commercial-size facility takes advantage of the modular concept used in designing the 10,520 bbl/day (1670 m³/day) facility: it consists of three full-scale module retorts. The process area will comprise three retort trains, gas-cleanup units, a flare, an oil-upgrading unit, and a power generation unit. Each train will be made up of a retort and its dedicated oil- and gas-handling system. The design includes appropriate capacity in the mining, shale-handling, shale-disposal, fines-storage, and other areas.

A timetable for the commercial plant includes permit approvals and completion of site-specific engineering in one year, first retort start-up two years later, and second and third retort start-ups in the fourth and fifth years, respectively. This sequential start-up for the three retort modules matches the oil-shale tonnages that can be made available from the mining operations.

Mining and Shale Handling

A room-and-pillar, heading-and-bench mining operation will be capable of delivering 64,290 tons/stream day (58,300 Mg/day) of 28-gal/ton (117-L/Mg) shale to the retorts, with adequate added capacity to compensate for off-shifts and scheduled maintenance. The mine design and shale resource determination are based on a 77-ft (23.5-m) mining height. Initial crushing of the oil shale will be done underground in the mine. All crushed material, including raw-shale fines, will be conveyed to the surface. Final crushing and screening will be done above ground at the plant site in the shale-preparation area. Particles larger than 3 in (0.076 m) will be returned to the tertiary crushers. Standby crusher(s) and stockpiling will ensure uninterrupted raw-shale feed to the retorts. The raw-shale supply system is designed with the capability of feeding raw shale to the retort directly from the crushers or from a 24-h covered surge inventory of screened shale.

Distribution

Feed-belt conveyors will supply raw shale to hoppers located at the top of each retort. Stationary distributor legs from the hoppers will maintain a constant bed level with minimal shale segregation by size. Each leg will distribute raw shale over designated retort areas. The distributor legs may be purged to provide a seal against hydrocarbon emissions.

Paraho Retorting Process

Each of the three retorts in the commercial facility will be a vertical carbon-steel structure lined with refractory material. Employing the basic principles of Paraho's proven above-ground retorting technology, the retorts feature a moving-bed, gravity-feed system; barrier-free zones with specific functions, energy self-sufficiency through use of recycle gas as a heat-transport medium and carbonaceous residue in the shale bed as fuel for retorting combustion; efficient thermal exchange between descending shale and ascending air and recycle gas; and the proprietary Paraho bottom-grate system which precisely controls shale movement through the retort. The commercial plant has the capability of running with one, two, or three retorts operating.

Air–recycle gas distributors

The distributors provide the required heating and cooling zones in each part of the Paraho retort. As the shale travels downward in the retort, it passes through the mist-formation, retorting, combustion, and cooling zones while engaging in a thermal exchange with a countercurrent flow of air and recycle gas supplied by the top and middle distributors and recycle gas supplied by the bottom distributor. (For a more complete description, see the earlier section titled "Paraho Direct Heated Process.") Air and recycle gas are mixed externally to the top- and middle-level distributors to ensure precision in mixing and distribution.

Paraho bottom grate

Each retort will have 16 independently acting Paraho bottom grates, which maintain a uniform temperature profile in the bed by regulating the shale-mass flow through the retort. The bottom grates consist of a series of pusher bars, designed on the basis of Anvil Points experience, with special provision to prevent wall effects in the displacement of shale.

Bottom seals

Retorted shale will exit the bottom of each retort through star-valve bottom-seal devices designed to prevent hydrocarbons from escaping to the atmosphere.

Off-gas collectors

The retorts will have an above-bed off-gas collection system to remove gas and oil mist from the top of each retort. The design will minimize gas channeling in the shale bed. The off-gas will be ducted to the oil-and-gas–recovery system.

Retorted shale

Retorted shale will leave the bottom of the retorts at a normal operating temperature of 149 °C, with possible short-time excursions to 204 °C. Retorted shale will be discharged from the retorts and transferred to the disposal site at a base-case rate of 43,215 tons/stream day (39,200 Mg/day). The design-case rate will be 52,285 tons/stream day (47,400 Mg/day). Retorted shale and raw-shale fines will be disposed of in physically separated sites according to environmentally acceptable designs based on Anvil Points research and experience. Retorted shale will be moistened and compacted to form enbankments and floors impervious to leaching. Exposed surfaces of the retorted-shale piles will be adequately covered with topsoil removed and stockpiled from construction and storage areas. This will be followed by revegetation with plants that grow best in the region.

Base-Case Operations

The base-case shale-mining rate will be 59,100 tons/stream day (53,600 Mg/day), based on mining 7 days/week, 24 h/day. The base-case raw-shale feed rate to each retort will be

17,730 tons/stream day (16,100 Mg/day), for a combined rate of 53,190 tons/stream day (48,200 Mg/day) for the three retorts. Base-case operating conditions are defined as a raw shale quality of 28 gal/ton (117 L/Mg), a retort-mass flux of 455 lb/hr·ft² (2.22 Mg/h· m²), and an 89 percent product oil yield. At the base-case feed rate and operating conditions, each retort will produce 10,520 bbl/stream day (1670 m³/day) of crude shale oil and a net 2270 bbl/stream day (361 m³/day) fuel-oil equivalent (FOE) of low-Btu gas on a dry basis. Possible improved oil recovery, in the form of naphtha, will reduce the number of barrels of gas FOE per stream day. The total commercial plant base-case production will be 31,560 bbl/stream day (5018 m³/day) of crude shale oil and 6810 bbl/stream day (1083 m³/day) (FOE) of low-Btu gas (on a dry basis). Each retort train will have an on-stream factor of 90 percent.

Facility Design Rates

Gas- and oil-handling facilities have design rates in excess of base-case operations. Each retort will be sized for a mass-flux rate of 550 lb/h·ft² (2.68 Mg/h·m²). All shale-crushing and -handling facilities will be sized to deliver 71,440 tons/stream day (64,800 Mg/day) of raw shale to crushing and screening equipment.

Oil- and Gas-Recovery System

Each retort will have a dedicated oil- and gas-handling system. The design for removal of oil mist from the retort off-gas will utilize coalescers and electrostatic precipitators capable of removing oil mist droplets with at least 99 percent efficiency.

Recycle gas   The portion of the mist-free off-gas which is returned to the retorts. The recycle gas at each retort will be distributed to the top, middle, and bottom distributors in a total volume between 12,000 and 16,000 stdft³/ton (0.38 and 0.50 m³/kg) of raw shale.

Crude shale oil   Oil-mist droplets collected from the coalescer and electrostatic precipitator contain water. Day tanks will be provided as separators to settle out the water from the oil to less than 1 percent by weight. Total base-case production rate of crude shale oil from which water has been removed will be 31,560 bbl/stream day (5018 m³/ day).

Upgrading Paraho Shale Oil

A two-train *hydrotreater unit* will upgrade the oil and remove sulfur and nitrogen. The unit provides the necessary turndown capacity and allows continuous operation of one train during catalyst change-out on the other train. It will handle 31,560 bbl/day (5018 m³/day) of oil for a yield of 33,860 bbl/day (5383 m³/day), since hydrotreating results in a slight increase in volume. The design hydrogen-oil ratio is 1800 stdft³ of hydrogen per barrel of oil (320 m³/m³).

A *hydrogen plant* will provide the hydrogen necessary for the hydrotreater. The unit will be single-train, and it will be capable of being turned down sufficiently to operate at 40 percent of full capacity.

Pour-point depressant   After the oil stream leaves the hydrotreater, pour-point depressant will be added. The chemical will be stored in steam-heated vessels and will be injected into the oil stream by a gear pump. After the chemical injection, the oil will be pumped to the hydrotreated-oil storage tanks.

Electric Power Generation

Electricity will be produced by burning the low-Btu product gas in gas-combustion turbines and in waste-heat boilers driving steam turbines. Power generated will be used to operate the plant, and the excess power will be sold to public utilities.

100,000-BARREL-PER-DAY CONCEPTUAL DESIGN

As part of the module engineering study, Paraho developed a conceptual design for a 100,000-bbl/day (15,900-m³/day) oil-shale facility using Paraho surface-retorting technology. Included in the design basis are coarse-shale storage; secondary crushing and screening; retorting; raw (dewatered) shale-oil production; oil storage and loading; off-gas cleanup including by-product recovery; retorted-shale disposal; electric power substation; on-site power generation; and laboratory, maintenance, and administrative facilities.

The design is based on a number of assumptions provided by DOE for planning and comparative purposes. These include a site in western Colorado; shale to be supplied by a room-and-pillar mine located on property leased from the United States and operated by a private mining company; shale rated at 30 gal/ton (125 L/Mg) and crushed to −9 in (−0.23 m) delivered to the plant boundary; water and electric-power transmission lines provided at the battery limits; and design criteria and economic bases provided by SRI International.

The conceptual design for the 100,000-bbl/sd (15,900-m³/day) oil-shale facility evolves, in general, from the 10,520-bbl/sd (1670-m³/day) module design. This is accomplished largely through process augmentation, with processes of the module being increased in capacity to accommodate the larger design. Two processes, naphtha recovery and power generation, which were offered only as alternatives in the module, are fully integrated into the 100,000-bbl/sd (15,900-m³/day) design. The conceptual design also includes appropriate considerations for increased capacities and dimensions in technical, environmental, health, safety, operating, and economic criteria.

Design and Operating Factors

The plant design calls for an on-stream factor of 90 percent or 7885 operating hours per year. Shale-crushing and -handling facilities are sized to deliver 110 percent of retort requirements. Gas- and oil-handling facilities have design rates equivalent to 100 percent of base-case operations. Each of five Paraho direct-heated retorts is a rectangular unit that is 26 ft (7.9 m) wide and has a nominal cross-sectional area of 5512 ft² (512 m²). Design height of the shale bed within the retort is 25 ft (7.6 m). In addition to the battery of Paraho direct-heated retorts, the retorting area includes associated mechanisms for distribution of oil shale, air, and recycle gas for the production and collection of off-gas, from which raw shale oil and other products are recovered.

The Paraho Process

The shale in each Paraho retort flows downward by gravity through mist-formation, retorting, combustion, and cooling zones at a rate of feed and discharge controlled by the proprietary Paraho bottom grate. Temperature control and thermal exchange for the combustion, retorting, and cooling processes are achieved through bottom, middle, and top gas distributors. Retorted shale is discharged at the bottom of each retort through the Paraho grate mechanisms and is conveyed to a disposal site specified in DOE's criteria.

Off-gas collected from the retorts contains oil mist. Raw shale oil, gas, and water are separated from the off-gas in a system of knockout drums, coalescers, and electrostatic precipitators. Wet oil is directed to intermediate storage tanks for separation of water and sediment from the product oil. The product oil is then sent to final tankage facilities capable of storing 1,500,000 bbl (240,000 m³). A portion of the mist-free off-gas is returned to the retort as recycle gas. Product gas not returned to the retort is further processed for the recovery of ammonia, sulfur, and naphtha and is also used for on-site generation of electricity. By-products are sent to final storage facilities capable of storing 15 days of base-case production.

The block flow diagram of the 100,000-bbl/sd (15,900-m³/day) facility is shown in Fig. 3-10.

System Scale-up and Modifications

Essentially all plant systems common to the module and the 100,000-bbl/sd (15,900-m³/day) plant must be modified, and generally the modification is a scale-up. Major systems requiring modification include the following:

Raw-shale crushing and screening

Only a single stage of crushing and screening is included to accommodate the −9-in (−0.23-m) shale from the mine. (Two stages of shale preparation are required for the module to reduce −12-in (−0.30-m) shale from the mine.)

Components added or deleted

Several systems not included in the module are added to the 100,000-bbl/sd (15,900 m³/day) commercial facility. Some systems included in the module are not in the commercial design.

Naphtha recovery This process is more economically attractive for the larger plant than for the module. Five trains, based on the naphtha-recovery system designed as an alternative for the module, recover more than 4000 bbl/sd (636 m³/day) of naphtha, thereby increasing the oil yield.

Briquetting The module has no fines-utilization facilities. The commercial facility's briquette plant is designed to produce a stable, compacted retort feed of acceptable size.

Power generation

An eight-train combustion-turbine system, using commercially available equipment, will produce power for the commercial facility. Low-Btu gas from the naphtha-recovery process will be burned as fuel. Excess electric power will be transferred to the battery-limits utility.

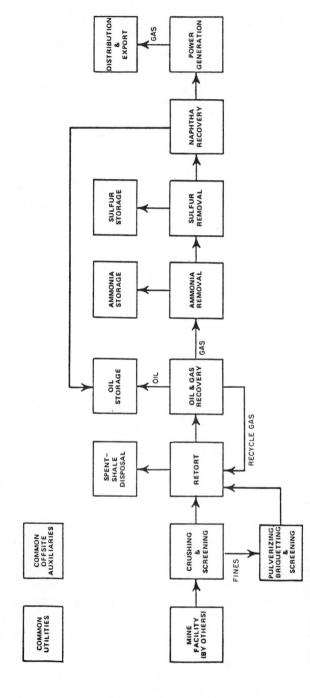

FIG. 3-10 Block flow diagram of 100,000-bbl/day facility.

Miscellaneous systems

Equipment not part of the module design but included in the 100,000-bbl/sd (15,900-m^3/day) design include vibrating activator feeders, vibrating grizzly feeders, gage-mill crushers, mixers (pug mill), and screw feeders.

Deleted equipment

The deleted equipment includes all mine stationary equipment, all mine mobile equipment, bucketwheel stacker reclaimers, belt stackers, blender wheel reclaimer. (According to DOE criteria, the mine is a separate operation.)

Other system and equipment changes

Projections for the commercial plant include, in some cases, separate trains or systems for each of the five retorts. One compressor and an ammonia-removal unit is provided at each retort. Five sulfur-removal units and five naphtha-recovery units are also included.

Retorted-Shale Disposal

The location and technical assumptions for retorted-shale disposal are included in a study completed for DOE by SRI International. Assumptions are related to site capacity and production rates, and they include detailed considerations of the environment.

BIBLIOGRAPHY

Allred, V. Dean: *Oil Shale Processing Technology*, The Center for Professional Advancement, East Brunswick, N.J., 1982, pp. 107–130.

Bartick, H., et al.: "The Production and Refining of Crude Shale Oil Into Military Fuels," Final Report to Office of Naval Research; Contract N00014-75-C-0055 (No. AD-A024 652/OGI), U.S. Department of Commerce, Springfield, Va., NTIS (1975).

Heistand, R. N.: "Product Yields," *Proceedings of the 12th Oil Shale Symposium*, Colorado School of Mines Press, Golden, Colo., 1979, pp. 237–240.

Jones, J. B., Jr.: "The Paraho Oil Shale Retort," *Colo. Sch. Mines Quart.*, vol. 71, no. 4, 1976, pp. 39–48.

Jones, J. B., Jr.: "Technical Evaluation of the Paraho Process," *The 11th Israel Conference on Mechanical Engineering, Technion University, Haifa, Israel, July 1977*.

Jones, J. B., Jr.: "Oil Shale Retorting—The Paraho Process," *International Symposium on Oil Shale Chemistry and Technology, Jerusalem, Israel, October 1978*.

Jones, J. B., Jr., and R. N. Heistand: "Recent Paraho Operations," *Proceedings of the 12th Oil Shale Symposium*, Colorado School of Mines Press, Golden, Colo., 1979, pp. 184–194.

Pforzheimer, H., and S. K. Kunchal: "Commercial Evaluation of an Oil Shale Industry Based on Paraho Process," *172nd National Meeting, American Chemical Society, Petroleum Division, New Orleans, March 1977*.

Pforzheimer, H.: *Processing of Oil Shale, from Rock in the Ground to Finished Petroleum Products*, National Petroleum Refineries Association, Lima, Ohio, March 1979.

OCCIDENTAL MODIFIED *IN SITU* (MIS) PROCESS

ALDRED L. STEVENS
MARNEY D. TALBERT

Occidental Oil Shale, Inc.
Grand Junction, Colorado

INTRODUCTION

Oil can be produced from oil shale by either of two general methods: (1) surface retorting with the accompanying surface or underground mining of the shale or (2) *in situ* retorting. *In situ* is Latin for "in (its original) place."

In all retorting, heat is required to pyrolyze the hydrocarbon in the shale (kerogen) to ultimately form oil, gas, and a carbon residue (char), which remains on the spent shale. Heat may be introduced by direct combustion of this residual carbon, by electrical techniques, by injection of hot gases or liquids, or even possibly by nuclear means.

Oil shale is not a porous rock, and most of its formations are impermeable; therefore, flow paths for process gases must be created by fracturing. In a true *in situ* process, fracturing techniques such as hydrofracturing or using chemical or nuclear explosives create formation permeability without extracting any of the oil shale. In the modified *in situ* (MIS) process, pioneered by Occidental Oil Shale, Inc. (OOSI), a limited amount of oil shale is extracted by mining to create voids, the remaining volume to be processed is fractured with explosives, and the now permeable fractured shale is retorted underground.

History

Until the first oil well was drilled in 1859, the U.S. oil-shale industry flourished with at least 50 commercial plants extracting fuel oil from eastern oil shales.

In the 1920s, an oil-shale boom was triggered when predictions of forthcoming fuel shortages were combined with an announcement by the U.S. Geological Survey (USGS) that large fuel resources were contained in the oil shales of the Green River formation. The boom abruptly ended with the discovery of large oil fields in eastern Texas.

Little oil-shale research and development was conducted in the United States until World War II. In 1944, out of concern for the hazards of reliance on imported energy, Congress passed the Synthetic Liquid Fuels Act, which authorized the U.S. Bureau of Mines (USBM) to establish a liquid fuel supply from domestic oil shale. USBM began a comprehensive program that has continued to the present day, with authority transferred several times and most recently to the Department of Energy (DOE) in 1978.

The Federal Prototype Oil Shale Leasing program was created to encourage private development of oil-shale lands. The first leases were awarded in 1974 with the sale of leases to four tracts in Colorado and Utah. One of those tracts, Colorado-b (C-b), is now leased by OOSI and Tenneco Shale Oil Company (TSOC).

In addition to activities on federal lands, private companies have engaged in exploration and research and development programs on their own lands. In 1953, Sinclair Oil and Gas Company conducted one of the earliest *in situ* experiments. It showed that induced or natural fractures could be used to provide flow paths between holes drilled from the surface and that ignition and combusion of the fractured formation could be accomplished.

Worldwide, Sweden and Germany also used *in situ* retorting. The Swedish oil-shale industry began in the 1920s and ceased in the 1960s. In addition to two types of aboveground retorts, it used an *in situ* process in which the oil-shale deposits were pyrolyzed with electric heaters.

During World War II, to augment its oil supplies in response to wartime fuel shortages, Germany used a horizontal MIS process in the low-grade shale of Würtenberg along with two types of aboveground retorts. These MIS retorts were each 162 ft (49 m) long, 25 ft (7.6 m) high, and 15 ft (4.6 m) wide.

MIS Application

"In 1874, workers on the transcontinental rail line found that rocks picked up from the excavations along the Green River in Wyoming ignited when used to protect campfires from the night winds.... The rocks of interest were pieces of oil shale from the Green River formation."*

Although large amounts of low-grade oil shales can be found in a variety of locations within the United States, the most significant deposits found so far are contained in the Green River formation. They underlie about 16,500 mi² (43×10^9 m²) in Colorado, Utah, and Wyoming. This formation is divided into several geologic basins, the most important of which is the Piceance basin in Colorado (Fig. 4-1).

*An Assessment of Oil Shale Technologies, Office of Technology Assessment, Congress of the United States, GPO, Washington, D.C., stock no. 052-003-00759-2, June 1980, p. 108.

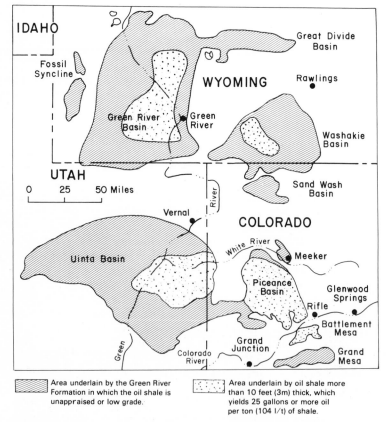

| | Area underlain by the Green River Formation in which the oil shale is unappraised or low grade. | | Area underlain by oil shale more than 10 feet (3m) thick, which yields 25 gallons or more oil per ton (104 l/t) of shale. |

FIG. 4-1 Oil-shale deposits of the Green River formation. This information is divided into four geologic basins, of which the Piceance basin in Colorado contains the richest source. (D. C. Duncan and V. E. Swanson, *Organic-Rich Shales of the United States and World Land Areas*, U.S. Geological Survey Circular 523, 1965.)

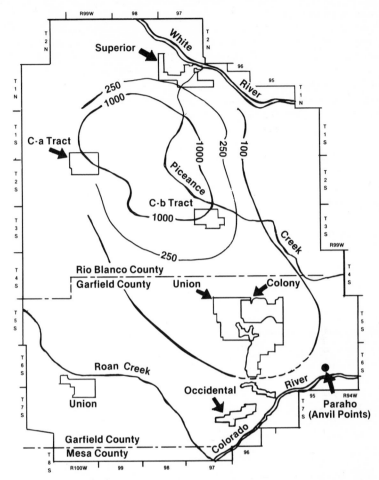

FIG. 4-2 Piceance basin of Colorado showing several oil-shale site locations. The basin occupies most of Rio Blanco and Garfield counties. The oil shale averages more than 25 gal/ton (10.4 $\times$ 10^{-5} m^3/kg) for the thicknesses shown in areas indicated by the contour lines.

The Piceance basin covers approximately 630 mi^2 (1.63 $\times$ 10^9 m^2). It has a saucerlike shape with outer edges that are higher, thinner, and leaner. The center of the basin contains oil shale to a thickness of as much as 2000 ft, and over 1000 ft (305 m) of it contains shale with an average of 25 gal/ton (10.4 $\times$ 10^{-3} m^3/kg) (Fig. 4-2). The oil in place in this one formation is almost 1.8 trillion barrels (286 $\times$ 10^9 m^3). Assuming a recovery of 40 percent, the Piceance basin shale-oil reserves are about 600 billion barrels. Proven oil reserves for the United States and for the world are 28 and 650 billion barrels (4.45 and 103.34 $\times$ 10^9 m^3), respectively.

The MIS process is particularly attractive for application to deep deposits, such as some in the Piceance basin, for a number of reasons. Most significant is that the mining is minimized because approximately 75 percent of the resource remains underground. For processing thick, rich oil-shale, MIS is comparable in cost with aboveground retorting and its associated mining. Also, oil can be economically recovered from the lower-grade shale deposits ranging down to 10 gal/ton (4.17×10^{-5} m^3/kg), which comprise more than half of the U.S. oil-shale reserves. Further, the shale mined to create the voids can be retorted on the surface. Spent-shale cooling and disposal is necessary only for the surface-retorted shale. The *in situ* spent shale remains in place underground.

In 1972, Occidental began work on a modified *in situ* process at its Logan Wash research and demonstration site at the southern fringe of the Piceance basin in Colorado. Eight MIS retorts have been burned at Logan Wash; they ranged in size from 32 ft (9.75 m) square and 70 ft (21 m) high to commercial-sized retorts up to 165 ft (50 m) square and 245 ft (75 m) high. The results of the demonstration work accomplished at Logan Wash will be directly applied to the commercial retort design for Tract C-b.

OOSI, in partnership with TSOC, is currently planning a commercial plant on the Colorado Lease Tract C-b to process oil shale in the Piceance basin with MIS technology. The commercial plant, in addition to processing MIS retorts, will process the mined shale in aboveground retorts.

GENERAL PROCESS DESCRIPTION

OOSI has developed a "modified *in situ*" oil-shale process that has undergone extensive testing since 1972 in field retorts at its Logan Wash research mine north of De Beque in Garfield County, Colorado. OOSI's philosophy in developing this process for producing oil from oil shale is to maximize the recovery of in-place oil while minimizing cost and environmental impacts. Underground retorts are formed by first mining to remove 20 to 25 percent of the shale rock to create voids or "rooms" and then expanding the remaining volume of oil-shale rock into the voids through fragmentation with conventional explosives. Blast holes are drilled into the support pillars in the three void level rooms and into the two shale layers between the three void rooms, as shown in Fig. 4-3. Explosives are placed in these holes and detonated in proper sequence to rubblize the rock and distribute the pillars in the void rooms immediately prior to the main rubblizing blast and to rubblize and expand the shale layer rock into the void rooms. In this way, the void in the three rooms is redistributed throughout the rubble, providing the necessary permeability for flow of process gases.

After rubblization, extensive construction and preignition testing of instrumentation occurs before combustion is initiated. Steel bulkheads are constructed in the retort entry drifts to isolate the retorts from the mine access drifts. After installation of necessary instrumentation and piping, various pressure tests are run to establish safe and efficient operating conditions.

Ignition is initiated at the top of the retort by using an external source of heat to ignite the top of the shale rubble. The objective of the ignition phase is to establish a combustion zone at the top surface of the rubble that will be propagated downward through the shale rubble column during the subsequent processing phase. After a predetermined amount of oil shale has been heated to combustion temperatures, the external heat source is eliminated, and the retorting process is sustained by introducing air to the top of the retort. The reaction carbon in the front moves downward through the retort at the rate of a few

**Mine Out Voids By Removing 15 To
25% Of The Solid Shale.**

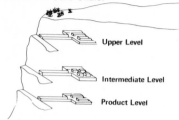

Upper Level

Intermediate Level

Product Level

**Drill Blast Holes In Solid Rock Between
Voids And Load With Explosives.**

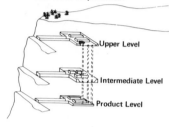

Upper Level

Intermediate Level

Product Level

**Rubblized Retort After Detonation.
Rock Appears Intact But Is Highly Fractured.**

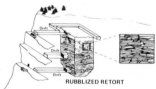

RUBBLIZED RETORT

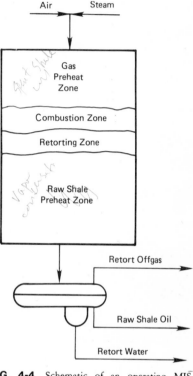

Air   Steam

Gas
Preheat
Zone

Combustion Zone

Retorting Zone

Raw Shale
Preheat Zone

Retort Offgas

Raw Shale Oil

Retort Water

FIG. 4-3 Steps in forming a modified *in situ* (MIS) retort. The retorts are formed by mining three voids and expanding the volume of oil-shale rock into the voids by rubblization.

FIG. 4-4 Schematic of an operating MIS retort showing the four process zones.

feet (meters) per week, and the kerogen contained in the shale is decomposed into oil vapor, low-Btu off-gases, and residual carbon. As shown in Fig. 4-4, four zones occur in an operating retort: (1) the gas preheat and spent-shale (burned-shale) cooling zone, (2) the combustion zone, (3) the retorting zone where kerogen decomposes at approximately 480 °C, and (4) the raw-shale preheat and vapor-condensation zone. The shale oil condenses, flows downward by gravity, and is collected at the bottom of the retort in preconstructed sumps from which it is pumped to storage and/or off-site delivery.

The rates and relative amounts of air introduced at the top of the retort control the rate of flame front advance and the retorting temperature.

In addition to the bulkhead product shale oil, a second liquid product oil in the form of a condensed light shale oil (CLSO) can be recovered from the retort off-gas stream by passing it through contact condensers located in the surface processing area. CLSO has

TABLE 4-1 Properties of CLSO and No. 2 diesel fuel

Property	CLSO*	No. 2 diesel fuel†
API gravity, 60 °F	37.1	32.8
BS&W, %	0.1	0.1
Flash point, P. M., °F	145	168
Gross heating value, Btu/lb	18,756	18,974
Pour point, °F	−30	−10/+20
Kinematic viscosity at 100 °F	1.90	5.54
Total sulfur, wt %	0.50	0.44
Total nitrogen, wt %	0.8	0.223
Ash, wt %	0.005	0.000

*CLSO properties determined by OXY analysis.

†No. 2 diesel fuel properties determined by military specs.

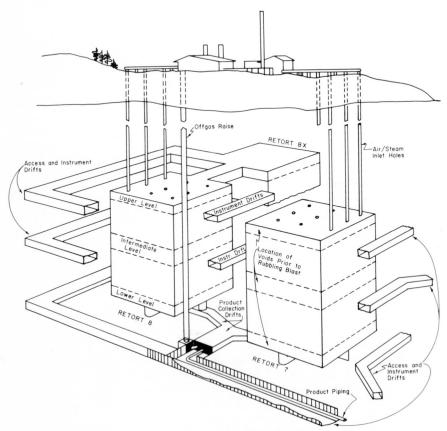

FIG. 4-5 Schematic of retorts 7 and 8 at Logan Wash. Access drifts have been cut away for illustration purposes. Retorts 7 and 8 operated simultaneously during 1982 and demonstrated the commercial viability of the MIS retorting technology.

properties similar to No. 2 diesel fuel, and in a commercial operation a minimum of upgrading may be required (Table 4-1).

Another by-product of the MIS retorting process is in the form of the retort off-gas, which provides a low-Btu energy source.

Shown schematically in Fig. 4-5 are retorts 7 and 8, two commercial-sized retorts operated in 1982 at Logan Wash. Any commercial venture of the future will require the ability to operate simultaneously more than one retort. Thus, retorts 7 and 8 incorporated all the knowledge previously acquired and demonstrated Occidental's MIS process on a commercial scale by simultaneously processing more than one retort (Stevens[1]).

PROCESS CHEMISTRY AND THERMODYNAMICS

The MIS retorting process comprises five process systems: ignition, combustion air injection, the retort itself, oil and water handling, and off-gas handling.

Ignition

Various combustion systems, based on shale oil, diesel fuels, and/or liquid petroleum gas (LPG), have been used to ignite MIS retorts through a series of laboratory tests, pilot plant experiments, and a series of miniretorts. The resulting data promulgated the conclusion that a hot-gas-generator ignition system is the most effective.

Retorts 7 and 8 were ignited using a hot-inert-gas generator (HIGG) as the external heat source (Table 4-2). The retorts were designed to produce a plenum above the rubble surface that allowed for side-entry ignition. The HIGG was designed to deliver inert gas (exhaust gas with less than 0.5 percent oxygen) to the retort plenum through an exhaust- or flue-gas distributor pipe. The plenum acts as a completely stirred tank for distributing the hot gas uniformly over the rubble surface. When ignition temperature was achieved, oxygen (air) was gradually introduced to make the transition from the external heat source to direct combustion within the rubble, and the external heat source was then eliminated. Retorting continued with combustion of residual carbon, which provided sufficient heat to maintain effective retorting conditions downstream of the advancing combustion front. Uniform retort ignition is a vital step in good retort operation and satisfactory oil yields as was demonstrated with retorts 7 and 8 yielding more than 200,000 bbl of shale oil over 10 months of operation.

TABLE 4-2 HIGG operating characteristics

Fuel	Shale oil or diesel
Fuel flow rate	13 gal/min, maximum
Combustion air	18,500 sfcm, maximum
Firing rate	20–106 MM Btu/hr
Flue-gas O_2 content	0.5–15%
Quench water	85 gal/min, maximum
Flue-gas temperature	300–1600 °F

Combustion Air Injection

Oxygen in the inlet air consumes the residual carbon remaining after the oil shale is pyrolyzed and the liquids and gases exit the oil-shale rock. Thus, no external fuel addition is needed to provide the heat for driving the pyrolysis process.

Retort

The retort is divided into four consecutive zones that propagate downward through the rubbled column of oil-shale rock as processing operations continue: (1) air-inlet preheat and spent-shale zone, (2) the combustion zone (650 to 760 °C), (3) the retorting zone (425 to 480 °C), and (4) the off-gas cooling and raw-shale preheat zone. Water and product oil condense in the raw-shale preheat zone and drain to the bottom of the retort for collection in the sump.

Considerable work has been done on the kinetics and mechanisms of kerogen pyrolysis (Hubbard and Robinson,[2] Braun and Rothman,[3] and Cha and Garrett[4]). When oil shale is heated to 425 to 480 °C, the kerogen decomposes to form oil, gas, bitumen, and carbonaceous residue. The bitumen further decomposes to oil, gas, and carbon.

$$\text{Kerogen} \rightarrow \text{gas} + \text{oil} + \text{bitumen} + \text{carbon residue} \rightarrow \text{oil} + \text{gas} + \text{carbon} \quad (1)$$

Robinson[2] reported that when Green River kerogen is pyrolyzed at 500 °C, it yields approximately 60 percent oil, 9 percent gas, 5 percent water, and 20 percent carbon residue. Approximately two-thirds of the organic carbon and hydrogen originally present in the kerogen is contained in the oil product. The principal gaseous constituents are nitrogen (N_2), carbon dioxide (CO_2), hydrogen (H_2), carbon monoxide (CO), hydrogen sulfide (H_2S), ammonia (NH_3), methane (CH_4), ethane (C_2H_6), and alkenes.

Although the actual reaction mechanism is complex (the diffusion rate changes during pyrolysis as porosity develops from products leaving the shale matrix), the reaction has been treated as first-order with respect to weight concentration of kerogen in the formation of bitumen. Also, the pyrolysis of the bitumen is treated as first-order in the subsequent formation of oil and gas. Some carbonate decomposition occurs in oil-shale retorting. Since the reactions are endothermic, they cause reduced energy efficiency in the process. Typical 28 gal/ton (11.7×10^{-5} m³/kg) shale contains 23 percent dolomite [$CaMg(CO_3)_2$], which requires 500 Btu/lb (1.16 MJ/kg) for decomposition, and 16 percent calcite ($CaCO_3$), which requires 700 Btu/lb (1.63 MJ/kg) for decomposition. Dolomite begins to dissociate somewhat below 565 °C and calcite at 620 °C, yielding CO_2 and oxides of calcium and magnesium.

Oil and Water Handling

Retort oil and water are collected in the sump behind the product-level bulkhead where some separation occurs. From there the oil and water are pumped to a separator tank where additional separation occurs. Water is pumped from the separator to a process sump where it can be reused. Oil is piped from the separator to a Heater Treater facility where it is stored for shipment. The design basis for retorts 7 and 8 required the use of a Heater Treater system to treat the full product stream. However, by testing different types of demulsifiers at varying rates early in the retorts' run, it was found that treatment of the product stream could be handled without using the Heater Treater. By introducing a small quantity of demulsifier at the product-level bulkheads, the emulsion was reduced to levels well within specifications by the time it went through the first oil-water separation tank.

Off-gas Handling

The off-gas handling system consists of positive-displacement off-gas blowers that pull off-gas under negative pressure from the bottom of the retort through 36-in-diameter steel pipes into a common steel-cased raise to the surface processing facilities. The off-gas travels through a parallel arrangement of contact condensers into the blower header and is then discharged into the stack through another header.

COMMERCIAL PERSPECTIVE

Occidental Oil Shale, Inc. is a 50 percent partner with Tenneco Shale Oil Company, a subsidiary of Tenneco, Inc., in the Cathedral Bluffs Shale Oil Company (CB) partnership for the development of a commercial project.

The project is an underground oil-shale mine with MIS and aboveground retorts constructed in phases to initially produce 14,100 bbl/day of shale oil.

The project site is a 5094-acre (2.06×10^7 m²) tract known as Tract C-b (Federal Prototype Oil Shale Lease Tract Colorado-b, C-20341) located in Rio Blanco County, Colorado (Fig. 4-6). The tract is in the Piceance Creek structural basin between the Colorado River on the south and the White River on the north. The basin is dominated by a large central plateau which covers more than 75 percent of the basin's land surface. The Cathe-

FIG. 4-6 Cathedral Bluffs commercial operations site showing surface construction. The oil-shale resource is located on federal Tract C-b in the Piceance Creek basin in Rio Blanco County, Colorado.

C-b History

- April 1974, Colorado-b Tract (C-b) was leased under the Federal Prototype Oil Shale Leasing Program for a bonus payment of $118 million.
- Original participating companies were Atlantic Richfield, Ashland Oil, Shell Oil, and TOSCO.
- Occidental became operator of Tract C-b in 1976, with Tenneco Shale Oil Company as an equal partner in 1979.

Facts

- An estimated 2.8 billion bbl of oil in place.
- Tract C-b contains 5094 acres.
- Environmental monitoring was initiated in 1974 to provide baseline information of conditions prior to oil-shale development.
- Ongoing environmental program includes monitoring air quality, water quality, vegetation, and wildlife in great detail.
- Since 1974, over $18 million have been spent for environmental data collection, analyses, and reporting.
- Since 1974, over $9 million have been spent for mitigation of socioeconomic issues (housing, busing, roads, impact coordination, and growth monitoring).

Shaft Information

- Construction of the large production and service shafts started February 1978. The 313-ft production headframe was continuously formed and poured in 26 days.
- The round, concrete-lined production shaft is 29 ft in diameter and 1857 ft deep. The service shaft is 34 ft in diameter and 1758 ft deep.

FIG. 4-7   Tract C-b facts.

dral Bluffs line the western edge of the basin. Elevations on the site vary from 6400 ft (1951 m) in the lowest valley to 7100 ft (2164 m) on the ridges at the southern edge. The climate is semiarid.

The completed work at Tract C-b includes extensive site preparation, water-handling- and surface-facility construction, and three shafts, 34, 29, and 15 ft (10.4, 8.8, and 4.6 m) in diameter (Fig. 4-7). The shafts have attendant concrete headframes and hoisting equipment in the service and production shafts and a conventional steel headframe resting over the ventilation and escape shaft with a ground-mounted hoist (Fig. 4-8). To gain access to the oil-shale resources, the shafts have been sunk to over 1700 ft (518 m).

Because of the great success in 1982 at Logan Wash, OOSI and TSOC jointly filed an application with the United States Synthetic Fuels Corporation (SFC) in January 1983 seeking financial support for a commercial oil-shale project on Tract C-b.

The proposal application is for a production facility utilizing aboveground retorting

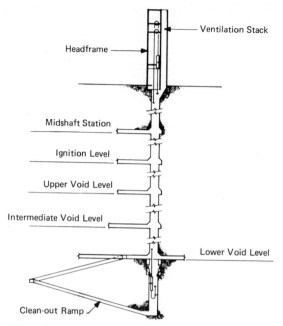

FIG. 4-8   Vertical cross section of the 29-ft-diameter (8.8-m-dia) production shaft for hoisting the mined oil-shale rock to the surface for aboveground retorting.

technology in combination with modified *in situ* technology. The raw oil shale produced will be upgraded on site using hyrotreating technologies to produce a pipelineable syncrude. The project anticipates initial production of approximately 11,000 bbl/day of raw shale oil from the aboveground retort and approximately 2300 bbl/day from four MIS retorts. The oil-upgrading facility would yield approximately 14,100 bbl/day of commercial crude oil substitute. The expected production life of the project is not less than 30 years, during which time the project is expected to produce approximately 150 million barrels of upgraded synthetic fuel.

As of this writing, the project has been judged to have passed the SFC's strength and maturity requirements, and a nonbinding Letter of Intent and Term Sheet was issued by the SFC to the project sponsors on July 28, 1983. Conclusive negotiations are continuing toward the signing of a definitive agreement late in 1983. An award from the SFC would include maximum financial assistance of $2190 million consisting of $1812 million in loan guarantees and $378 million of initial price guarantees at $60 per barrel of syncrude produced (1983 dollars). Rollover provisions provide for a total of $2190 million of price guarantees as guaranteed debt is repaid. The project anticipates mechnical completion of the aboveground retort and upgrading facilities in the second quarter of 1987 and initial production and sale of product in the fourth quarter of 1987. Mechanical completion of the MIS facilities is expected in the second quarter of 1988 (Table 4-3).

TABLE 4-3 Cathedral Bluffs project review

Project location	Piceance basin of Rio Blanco County, Colorado, Federal Lease Tract C-b
Project scope	One aboveground retort and four continuously operating MIS underground retorts; upgrading facilities; pipeline
Product	14,100 bbl/day syncrude
Project schedule	Oil production from aboveground retort in first quarter 1987; oil production from MIS underground retorts in fourth quarter of 1988
Project life	Not less than 30 years
Current status	Nonbinding Letter of Intent with Synthetic Fuels Corporation, July 1983; negotiations continuing for definitive agreement by end of 1983

RETORTS 7 AND 8

Chronology and General Configuration

Retorts 7, 8, and 8X were developed as a subset of a retort cluster to be operated simultaneously to demonstrate the commercial readiness of the MIS technology. Figure 4-5 shows the general configuration of these retorts. Retort 8X, a partial retort, was developed to test pillar stability and rubblization and was not processed.

Construction of retorts 7, 8, and 8X was begun in June 1979. In late 1981 and early 1982, retorts 7 and 8 were ignited (Table 4-4). After ignition, the retorts were operated without interruption until gas injection was terminated and water quenching began in November 1982. In order to extend the life of the retorts, a plan was developed to cool the off-gas and reduce the thermal load on the bulkheads and downstream processing equipment. The off-gas was continuously cooled, and the retorts were never shut down because of the maximum off-gas temperature being reached. As a result, the life of retorts 7 and 8 was extended by about 3 months, and oil yield increased. In early 1983, after the

TABLE 4-4 Chronology of processing for retorts 7 and 8

Event	Retort 8	Retort 7
Rubblizing blast	Apr. 1981	Feb. 1981
Ignition	Dec. 1981	Jan. 1982
First oil recovered	Jan. 1982	Mar. 1982
Steam turned on (design: 20%)	Jan. 1982	Feb. 1982
Inlet gas increased to design flow rate [0.54 stdft^3/(min)(ft^2)]	Mar. 1982	Mar. 1982
Off-gas temperature moved above steam plateau	June 1982	July 1982
Off-gas cooling spray on	July 1982	Aug. 1982
Injection steam removed	Oct. 1982	Oct. 1982
Airflow reduced	Oct. 1982	Oct. 1982
Recycle gas injection	Oct. 1982	Oct. 1982
Recycle gas injection stopped; water quench started	Nov. 1982	Nov. 1982
Blowers turned off	Jan. 1983	Jan. 1983
Off-gas cooling spray off	Jan. 1983	Jan. 1983
Contact condensers turned off	Feb. 1983	Feb. 1983

retorts had been in the quench cycle for several months, the blowers, followed by the contact condensers, were turned off when the retorts reached an oil production rate at which it was no longer practical to operate and the natural draft was sufficient to handle the dwindling flow of gas from the spent retorts.

Surface Processing Facilities

Principal elements of the surface processing facilities (SPF) are shown in Figs. 4-9 and 4-10. Fresh-air blowers and a steam generation plant supplied process air and steam to the retorts through pipes from the surface. Other blowers, operating in the suction mode, pulled the process gases through the retorts and up the off-gas raise. Control valves at the product-level bulkheads permitted maintaining subatmospheric pressure within each retort neutral at the upper-level bulkhead, to eliminate leakage of retorting gases into the mine ventilation system. As the off-gas reached the surface, it was pulled through a system of three contact condensers for condensing water and oil and absorbing ammonia from the off-gas. From there it passed through the blowers and into the stack where it was vented to the atmosphere in accordance with several permits.

A water-treatment pilot plant was constructed adjacent to the boiler building to study techniques for treating process condensate water. The plant consisted of three sections: a steam stripper section, a biological oxidation section with clarifier, and a carbon column section. It was sized to handle up to 6 gal/min. The main objectives of this treatment plant were to develop a process to allow for disposal of excess water (in accordance with permits) and to develop data for use in future full-scale water-treatment plant design.

FIG. 4-9   The MIS retorting off-gas stream is routed through contact condensers for cooling and removal of moisture and oil aerosol from the gas. Fin fans cool the water for recycling through the contact condensers.

FIG. 4-10 Blowers are utilized to supply fresh air to the process and to pull the off-gas from the bottom of the retorts, thus giving a downward process flow through the retorts.

A permit was issued by the state of Colorado in early 1982 for the discharge of treated process water. The water-treatment plant was operated from May through November 1982. Dischargeable water was produced for only a short while; however, these waters were discharged under the first National Pollution Discharge Elimination System (NPDES) permit ever issued for such waters. Retorts 7 and 8 were approximately halfway through the retorting process when operation of the water-treatment plant began.

Process control

The processing operation was monitored in and/or controlled from the process control building located on the surface. Signals from all critical points within the process train were routed to the control panel and displayed in analog form (Fig. 4-11). These signals were also run to the digital data acquisition system for recording and analysis to further evaluate current operating performance. Control of the process operating conditions was accomplished from the control room using current analog data and the analyses from the digital data displayed on two TV monitors in the control room and throughout the site.

PRODUCT OIL PROPERTIES AND PRODUCTION HISTORIES

Oil Properties

Table 4-5 lists the average properties of the dry bulkhead shale oil and the condensed light shale oil over the duration of the processing run. Also shown are the properties of the

FIG. 4-11 The surface-located control panel is similar to that typically found in a refinery. Flow rates, temperatures, pressures, and gas analyses are recorded by computers.

TABLE 4-5 Properties of shale oil from retorts 7 and 8

Parameter	Units	Dry bulkhead oil	Condensed light oil	Whole oil
API gravity, 60 °F	°API	23.5	37.1	24.8
BS&W	%	0.1	0.1	0.1
Fire point				
Cleveland open cup	°F	275	185	260
Flash point				
Cleveland open cup	°F	260	180	240
Pensky-Martin	°F	211	143	190
Gross heating value	Btu/lb	18,400	18,900	18,600
Pour point	°F	25	−30	25
Viscosity, kinematic				
100 °F	cSt	28.17	1.90	18.20
140 °F	cSt	11.22	1.34	8.09
Arsenic	µg/g	12	2.8	12
Asphaltenes	%	0.4	0.03	0.4
Carbon, total	%	83.2	85.3	84.8
Hydrogen, total	%	11.8	12.7	12.0
Nitrogen, total	%	1.5	1.0	1.5
Sulfur, total	%	0.85	0.51	0.83
Oxygen (by difference)	%	1.1	0.5	0.9
Mercury	µg/g	0.08	0.14	0.12
Nitrogen, Kjeldahl	%	1.4	0.8	1.3
Ash	%	0.03	0.005	0.03

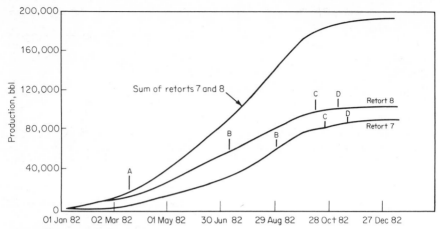

FIG. 4-12   Retorts 7 and 8 cumulative shale-oil production indicating more than 200,000 bbl produced during the 10-month period. Key: A, Inlet gas to design flow rate (0.54 stdft³/(min)(ft²); B, off-gas cooling spray turned on; C, recycle gas injection started; D, recycle gas injection terminated and water quenching started.

whole oil, a proportional composite of the dry bulkhead oil, and the condensed light shale oil.

Because oil-shale grade changes with depth in the retorts and the average temperature within the retort and of the off-gas changes over the life of the retorts, several of the properties of the oil also changed with time.

Production Histories

The combined oil recovery from both retorts totaled over 200,000 bbl, representing 70 percent yield of the Fischer assay oil in place in the rubblized shale rock (Fig. 4-12). This oil production includes oil pumped from the retort bulkheads, oil mists and vapors condensed in contact condensers, condensable C^+, vapors found in the stack gas, and a small amount of oil in the bulkhead water (Stevens and Zahradnik[5]). The partitioning of the total product between these four streams is shown in the following table.

Product	Percent
Bulkhead oil	60.7
Condensed light oil	9.5
Uncondensed light oil	27.5
Oil in water	2.3
Total	100.0

UPGRADING TECHNOLOGY

Shale oil can be used for some purposes in its as-produced state. It can also be readily pretreated and upgraded for direct use as a refinery feedstock to produce gasoline, diesel fuel, and jet fuels. The technology for this upgrading or pretreating is available and has been demonstrated in refinery runs with shale oil.

As-produced shale oil has a low sulfur content ($\approx$ 1 percent) that allows it to be burned directly in a full-scale boiler test without the addition of scrubbers. The results were presented by Skogen[6] at an Engineering Foundation Conference, "Combustion of Tomorrow's Fuels," November 8, 1982. Consumers Power Company of Michigan had burned 5000 bbl of shale oil in 1976 and had been able to meet all emission limits (including NO_x) in effect at that time.

An impediment to refinery processing of shale oil is the high level of organic nitrogen ($<$ 2 percent). Over half of the organic nitrogen is basic, and basic nitrogen is a refinery catalyst poison. The nitrogen content can be reduced by hydrotreating or by passing the shale-oil feedstock over a catalyst in the presence of hydrogen. This reduces not only the nitrogen but also the sulfur and oxygen, while the hydrogen content of the oil is increased by saturating the olefinic compounds. The whole shale-oil feed or selective cuts can be hydrotreated after distillation, depending on the required use of the products. Other methods of nitrogen-level reduction involve liquid SO_2 extraction and organic acid solvent extraction. Neither of these methods has been commercially developed for shale-oil upgrading.

The trace metals, with a typical content on the order of 17 ppm of arsenic and 50 ppm of iron can be removed from the shale oil by utilizing standard refining methods. Removal of these metals is necessary, since they deactivate hydrotreating and reforming catalysts.

The higher pour point ($>$ 15 °C) of shale oils, compared to conventional crudes, presents transportation problems. Without pour-point reduction as part of the upgrading, heated pipelines or alternative transportation methods must be developed since the average ground temperature may be less than 5 °C. Pour-point reduction can be accomplished either by thermal treatment (visbreaking, cracking, etc.) or by addition of pour-point depressants. Modification of the pour point could be accomplished as part of a complete upgrading scheme or by any number of methods used prior to final upgrading.

The oil-upgrading facilities planned for Tract C-b consists of four major units: hydrotreating, hydrogen-producing, sour-water treating, and oil-upgrading sulfur recovery. The hydrotreating unit provides sufficient hydrotreatment of the raw shale oil to produce a high-quality syncrude that will meet specifications for pipelining and conventional refinery feed. Wash water used for control of salt formation in the heat exchangers of the hydrotreating unit is stripped of H_2S and ammonia before being recirculated back into the hydrotreating unit. The H_2S will then be converted to elemental sulfur by conventional conversion technology.

The hydrogen producing unit is a conventional design that utilizes natural gas for feed.

PERMITS AND ENVIRONMENTAL PROTECTION

Occidental and Tenneco are committed to high standards of environmental protection in the development of the commercial facility at Tract C-b. As part of the Department of the Interior's Prototype Oil Shale Leasing Program, commercial development is subject to an Oil Shale Lease which contains extensive provisions for protection of the environment. To achieve this objective, project management has instituted an extraordinary comprehensive monitoring and assessment program in cooperation and consultation with numerous federal, state, and local officials. The program is designed to characterize environmental conditions in the undisturbed state and to develop mitigating measures which will ensure that commercial operations will not result in unreasonable adverse effects on the environment.

Cathedral Bluffs has obtained all the permits required for current operations; the remaining permits needed have been or will be applied for and obtained in accordance with normal agency processing schedules and the project development timetable (Table 4-6).

The principal waste material from the process will be approximately 10,000 dry tons per stream day of retorted or processed shale. The processed shale is principally marlstone (a form of limestone) that will have relatively low residual carbon content not removed during the retorting process. Numerous tests on processed shale from Trace C-b have determined that it passes all the criteria tests established by the Resource Conversion and Recovery Act amendments to the Solid Waste Disposal Act to qualify as a noncorrosive, nontoxic,

TABLE 4-6   Cathedral Bluffs site conditions

Ambient conditions	
Maximum temperature	95 °F
Minimum temperature	−25 °F
Precipitation, average	15 in
Precipitation, intensity	1 in/(h)
	(10 years)
Wind (on Tract C-b)	
Design speed	100 mi/h
Prevailing direction	S-SSW
Snowfall, average	70 in
Relative humidity	28–62%
Elevation	6800 ft
Pressure, atmospheric	11.4 psia
Design conditions:	
Temperature	
Maximum	90 °FDB, 65 °FWB
Minimum	−35 °F (100% RH)
Precipitation	
Average	15 in/year
Intensity	1.5 in/h
Wind	
Speed, maximum	100 mi/h
Direction	S-SSW
Snowfall	
Annual	70 in
Snow pack	36 in
Frost line	48 in
Snow loading	40 lb/ft²
Earthquake zone	4
Lightning storms (number per year)	36
Dust storm provisions (yes/no)	Yes

nonflammable, and nonradioactive material. The project intends to moisturize, place, and compact this material in such a manner as to prevent deep percolation and/or any effluent drainage, to utilize practices that will control fugitive dust, and finally to revegetate the material in accordance with approved State of Colorado Mined Land Reclamation practices.

Other waste products resulting from the commercial operations at Tract C-b will be various sludges and spent catalysts commonly found in any significantly sized industrial petroleum refinery operation. Specific characteristics of these wastes have been tentatively identified at this time, and they will be disposed of in accordance with industry standards and in strict compliance with all applicable laws.

Conventional unit operations are used throughout Tract C-b enabling the use of conventional environmental control techniques as well. Dust emissions are controlled with water sprays on storage piles with bag house systems at solids transfer points. Sulfur emissions are controlled through the use of the proprietary Flue-Gas Desulfurization (FGD), Unisulf, and conventional Claus technologies on the respective fuel-gas and sour-water-treater off-gas streams. These technologies will maintain sulfur emissions within the state emission limitation of 0.3 lb SO_2 per barrel of syncrude. At design capacity the process facility will be essentially in water balance and thus comprises a zero-discharge facility. Any excess mine water will be disposed of in reinjection wells or on spent shale. All runoff water from the spent-shale pile is collected and returned to the water-treatment plant for cleanup and reuse. Fugitive hydrocarbon emissions will be controlled by covering the API separator, tanks, and other emission points. In addition, all emergency vents and pressure-relief valves containing hydrocarbons bearing gases from pressure systems will be routed to a flare stack for incineration of any released hydrocarbons.

Underground operations are regulated by the federal Mine Safety and Health Administration (MSHA). Cathedral Bluffs has worked and will continue to work closely with representatives of MSHA to ensure that the requirements are met or exceeded. This includes the special considerations for meeting gassy mine standards and operation of underground retorts.

Presently the project produces excess water through mine dewatering. However, when full operation is achieved in 1989, the water developed on tract will approximately balance with project requirements. The water rights held by the project are adequate to supply planned needs as well as substantial expansions. In addition to on-site water, the project has rights to water from the proposed Powell Park/Piceance Creek pipeline (Figs. 4-13 and 4-14).

The project has studied the potential effects on groundwater as a result of the processed MIS shale remaining underground. Based on an accumulation of hydrologic, geologic, and analytical chemistry data from the Logan Wash research site and laboratory tests, the project firmly believes that abandoned MIS retorts will not have a significant effect on groundwater or surface-water quality.

SOCIOECONOMICS

The project has been active in socioeconomic impact mitigation since commencing work at Tract C-b in 1977. The socioeconomic efforts of the project were examined from the standpoint of employment and population growth, labor force availability, public facilities

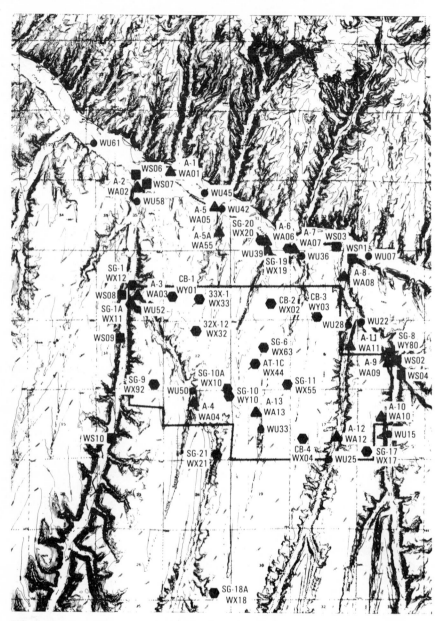

FIG. 4-13 Tract C-b water-monitoring sites for project development. Some monitoring wells are on-tract, while others are monitored near the tract.

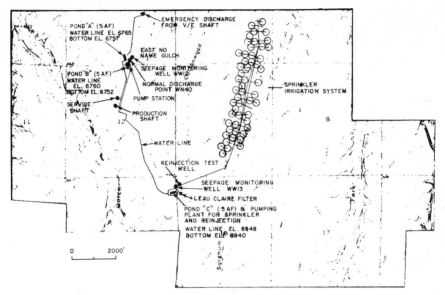

FIG. 4-14 Water management system layout at Tract C-b. Surplus water discharges during mine development are minimized with this system. During production, it is anticipated that the project will be at zero-discharge operation.

and services, housing supply, fiscal impacts, and impacts upon lifestyle and cultural resources. The analysis was done using a PAS model and related assessment techniques developed under the auspices of the Cumulative Impacts Task Force.

The socioeconomic impacts of the project will be primarily upon communities in western Garfield County and eastern Rio Blanco County. The communities in the project area have upgraded their respective capacities to provide public services and facilities. The facility improvements and expansions give the counties and communities of Rifle, Meeker, and Silt a capacity to accommodate substantial growth. The vast majority of this public facility upgrading was accomplished with $103 million of assistance from the state Oil Shale Trust Fund.

Indirect development associated with the project will consist primarily of new housing and related access roads. Labor requirements for the project will result in a construction workforce with a peak of 1800 employees and an operational workforce of approximately 1000. The maximum combined workforce will reach approximately 2500. To date, none of the oil-shale projects operating in western Colorado has experienced difficulty in attracting workers. Sufficient workers are expected to be available in future years because of less competition for labor locally and the greater availability of housing and other amenities within the project environs.

Construction and operation of the project will result in substantially increased personal income for residents in the area. For addition to local supply purchases and tax payments, the project should generate an annual local payroll of about $90 million at the peak of construction activity, and over the long term, annual wages will approximate $35 million.

Since 1977, the project has provided both funding and specialized expertise to local government agencies to help solve specific problems. This type of assistance coupled with the financial support of the Oil Shale Trust funds has proved valuable in rural areas where local government staffs are small and unfamiliar with the problems of rapid growth. The project realizes that this type of skilled technical assistance can prevent costly mistakes and delays in public-sector projects. The project has had a continuous close working relationship with state and local government officials since the inception of the project and recognizes the importance of keeping lines of communication open concerning plans for development in order to avoid conflicts and to foster expansion of public intrastructure.

PROCESS ECONOMICS

Cost of Producing Shale Oil

The combination of Occidental MIS retorting and aboveground retorting is generally believed, within the shale-oil industry, as promising the best economics for shale-oil production from deposits similar to those found at Tract C-b. Nontechnical uncertainties, including government fuel-pricing practices, environmental regulations, and land-leasing policies, as well as the availability and cost of large amounts of investment capital, make it difficult to arrive at firm predictions of shale-oil production costs. In the past, those costs have increased more rapidly than world oil prices. When oil was available at $2 per barrel, shale-oil production costs were estimated to be between $8 and $11 per barrel. With today's world oil prices as they are, shale-oil production is achieving favorable economics.

The reasons for the predicted increased costs of shale-oil production are clear. Over the years, worldwide inflation and the cost of doing business have increased markedly (labor wage rates and salaries, equipment), rates of return demanded by investors have increased from 5 to 6 percent to 15 to 18 percent, and increased environmental and other government regulations and delays in obtaining permits have added substantially to project overhead. Also, research and development costs in the shale-oil industry are significant. Nevertheless, productivity and efficiency improvements have been noted and are expected to continue to reduce the cost of shale oil as the industry grows.

Shale-oil production is very capital-intensive and thus especially sensitive to the time cost of money. The progress of any project will be determined by the rate at which finances are available.

The magnitude of investment necessary far exceeds the net worth and debt capacity of all but a very few of this country's industrial corporations. Consequently, financing is essential. Even to a credible project, the banking community is reluctant to advance a loan unless there is the security of a history of substantial production.

U.S. government loan guarantees are one incentive that will provide prospective lenders with sufficient assurance that obligations will be retired on schedule. The capital investment of oil shale is amortized over a large reserve and for a long time. The reserve on Tract C-b represents one of the ten largest North American oil reserves. The production life of the project is expected to be in excess of 30 years.

Cost of Upgrading and Refining

Upgrading of raw shale oil can be accomplished on a commercial shale-oil production site. To upgrade crude shale oil, hydrogen must be added to remove impurities such as nitrogen and sulfur. Once the impurities are removed, the shale oil is similar to an Arabian petroleum crude oil and can be fed directly to refineries in the United States. Costs associated with upgrading shale oil vary with the process used and the amount of hydrogen required by the particular shale oil.

Once the upgraded shale oil has been transported to a refinery, costs associated with refining are dependent upon the type and size of the refining facilities. Because of the differences among refineries and among shale oils, it is difficult to determine a single per-barrel refining cost. Most opinions presented at industry seminars and conferences on oil shale place the per-barrel costs at $9 to $11 for upgrading raw shale oil on a commercial production site and at $2 to $3 for refining upgraded shale oil.

SUMMARY

Occidental Oil Shale, Inc. has developed a process to recover oil from oil shale in deep, thick deposits such as those found in the Piceance Creek basin in western Colorado. This process, the vertical modified *in situ* process, involves the extraction of a portion of the shale, expansion of the remaining rock into the resulting void by explosive fracturing, and retorting the fractured shale in place underground. Commercial implementation of the MIS technology with companion aboveground retorting is underway at Tract C-b by Cathedral Bluffs Shale Oil Company. Environmental impacts are being minimized through off-gas and water-treatment systems, and Cathedral Bluffs is working with communities surrounding its plant to mitigate socioeconomic problems.

REFERENCES

1. Stevens, A. L.: "Occidental Vertical Modified *In situ* Process for the Recovery of Oil from Oil Shale, Phase II," U.S. Department of Energy Report, June 1983.

2. Hubbard, A. B., and W. E. Robinson: "A Thermal Decomposition Study of Colorado Oil Shale," USBM Report 4744, 1950.

3. Braun, R. L., and A. J. Rothman: "Oil Shale Pyrolysis: Kinetics and Mechanism of Oil Production," *Fuel*, vol. 54, 1975, p. 129.

4. Cha, C. Y., and D. E. Garrett: "Energy Efficiency of the Garrett *In situ* Oil Shale Process," *CSM 8th Oil Shale Symposium*, Golden, Colo., 1975.

5. Stevens, A. L., and R. L. Zahradnik: "Results from the Simultaneous Processing of Modified *In situ* Retorts 7 and 8," *CSM 16th Oil Shale Symposium*, Golden, Colo., 1983.

6. Skogen, Haven S.: "Combustion of Oil Shale," *Engineering Foundation International Conference on Combustion of Tomorrow's Fuels*, Santa Barbara, Calif., November 7–12, 1982.

THE GEOKINETICS
IN SITU RETORTING PROCESS

MITCHELL A. LEKAS

Geokinetics Inc.
Salt Lake City, Utah

GENERAL PROCESS DESCRIPTION

The Geokinetics process is a true *in situ* process for extracting oil from oil shale (Fig. 5-1). The process utilizes direct combustion and a horizontally moving fire front, and it is designed specifically for areas in which the oil-shale beds are relatively close to the surface. There is no requirement for constructing a mine or surface retorts. In applying the process, a pattern of blastholes is drilled from the surface through the overburden and into the oil-shale bed (Fig. 5-2). The holes are loaded with explosives (Fig. 5-3) and fired by using a carefully designed delay system (Fig. 5-4). The blast results in a fragmented mass of oil shale with a high permeability. The void space in the fragmented zone comes from lifting the overburden and producing a small uplift of the surface. The fragmented zone constitutes an *in situ* retort.

The bottom of the retort is sloped to provide drainage for the oil to a sump, from which it is lifted to the surface by a number of oil production wells (Fig. 5-5). Air-injection wells are drilled at one end of the retort (Fig. 5-6), and off-gas wells are drilled at the other end. The oil shale is ignited at the air-injection wells, and air is injected to establish and maintain a burning front that occupies the full thickness of the fragmented zone. The front is moved

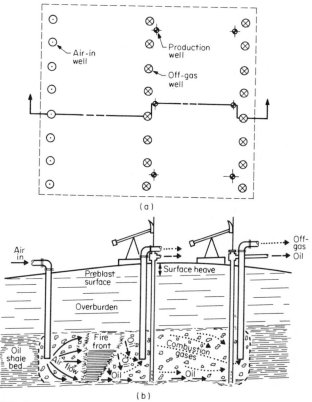

FIG. 5-1 Plan and cross section of a typical horizontal *in situ* retort.

FIG. 5-2 Blastholes are drilled with a large rotary drill.

FIG. 5-3 The blastholes are loaded with Ireco liquid explosive slurry pumped directly from special trucks.

in a horizontal direction through the fractured shale toward the off-gas wells at the far end of the retort. The hot combustion gases from the burning front heat the shale ahead of the front, thereby driving out the oil. The oil drains to the bottom of the retort, where it flows along the sloping bottom to the oil production wells (Fig. 5-7).

As the burn front moves from the air-injection to the off-gas wells, it burns the residual coke in the retorted shale as fuel. A negative pressure is maintained at the off-gas wells to recover the combustion gases. These gases are combustible, and they can be used for power generation. Low-pressure blowers are used to provide air to the retort, and small

FIG. 5-4 Blast of retort.

(a)

(b)

FIG. 5-5 (a) Equipment emplaced on retort. (b) Pressure blowers for injecting air into the retort, and vacuum blower for removing the off-gases.

(a)

(b)

FIG. 5-6 (a) Retort air-injection blowers and wells (b) Off-gas wells, collector pipes, vacuum blower and gas-and-oil–handling equipment.

FIG. 5-7 Oil production wells.

oil-field-type pumps are used to lift the oil to the surface. The off-gases from the retort are moved at low pressure through thin-walled, large-diameter pipe to scrubbing units to remove the sulfur compounds and are then burned to generate electric power.

The horizontal *in situ* process can be utilized in a secondary recovery operation to follow after a room-and-pillar mining operation. In this modification of the process, after normal room-and-pillar mining has been completed in a portion of the mine, the oil shale in the pillars and in low-grade shale remaining above and/or below the mine is fragmented with explosives, thereby creating a rubble-filled retort. This is then retorted by using the basic horizontal *in situ* process. All oil recovered by this process is in addition to oil recovered by the primary mining and surface retorting operation.

PILOT PLANT DESCRIPTION AND DATA

Field testing of the Geokinetics process began in 1975 at a test site near Vernal, Utah. A total of 25 test retorts have been blasted to develop blasting and rock fragmentation techniques, and 16 retorts have been burned to develop retorting and operating data (Table 5-1). This work has established the technical viability of the basic process as follows:

1. It is possible to drill a pattern of blastholes from the surface into the oil shale and fracture the shale with explosives to establish a zone of high permeability with a relatively impermeable zone between the fragmented shale and the surface.
2. It is possible to drill through the rubblized material and construct the various wells necessary for the operation, including air-injection, off-gas, oil-recovery, and instrument wells.
3. A point ignition can be made in the rubblized shale and expanded into a burn front that covers the cross section of the retort.
4. The burn front can be moved down the length of the retort as a cohesive temperature front with satisfactory sweep efficiency.
5. Produced oil can be recovered from wells drilled to the bottom of the rubblized zone.
6. Recovery of up to 50 percent of in-place oil can be achieved.

In addition, basic retorting parameters such as air-injection pressures, air-injection rates, rates of fire front advance, and recovery factors have been established. A specific blasting pattern has been developed and scaled up to full size.

Pilot operations have established the following parameters:

1. Air-injection pressures $\frac{1}{2}$ to 2 lb/in^2 (3.5 to 14 kPa)
2. Fire-front advance rate   1 ft/day (0.3 m/day)
3. Percent recovery of oil in place   40 to 60
4. Heating value of off-gases   75 Btu/ft^2 (667 kcal/m)
5. Oil composition   (See Table 5-2)
6. Air-injection rate   1 ft^3/min (0.003m^3/min) per square foot of cross section
7. Off-gas composition   (See Table 5-3)

TABLE 5-1 Summary of data on retorts 1 through 26 (to 9/1/81)

Retort number	Date blasted	Date ignited	Thickness of shale, ft	Average overburden thickness, ft	Dimensions		Shale oil recovered, bbl
					Width, ft	Length, ft	
1	7/75	9/76	10	0	10	50	56
2	7/75	3/76	3	10	10	30	28
3	1/76	7/76	10	17	20	40	82
4	2/76	2/77	10	16	20	40	146
5	2/76	5/77	11	19	20	81	354
6	Abandoned						
7	11/76	—	10	15	20	50	—
8	11/76	—	23	22	20	83	—
9	12/76	9/77	22	22	40	83	1,007
10	12/76	1/79	11	14	20	50	445
11	3/77	4/77	12	14	20	45	272
12	3/77	—	11	31	30	50	—
13	6/77	—	11	31	30	50	—
14	6/77	2/78	12	29	40	70	384
15	7/77	5/78	20	31	50	75	1,003
16	8/77	8/78	20	41	62	87	2,067
17	5/78	6/79	17	26	72	156	3,700
18	7/78	11/79	17	27	108	156	5,500
19	12/78	—	30	50	126	182	—
20	4/79	—	24	36	40	100	—
21	6/79	—	23	35	40	100	—
22	6/79	—	23	34	50	100	—
23	9/79	3/81	24	36	50	100	—
24	11/79	12/80	28	45	217	230	—
25	7/80	—	28	55	217	230	13,500
26	8/81	—	28	50	217	230	—

PROCESS PROSPECTIVE

The R&D program was completed by the end of 1982. At that time four full-size pilot retorts had been tested.

DETAILED PROCESS DESCRIPTIONS

A production unit for an oil-shale bed 30 ft (9.1 m) thick will consist of 15 to 20 individual retorts operating simultaneously. Each retort will produce from 100 to 150 bbl/day (15,900 to 23,900 L/day) for a total production of about 2000 bbl/day. Air-injection pressures will be about 2 lb/in² (14 kPa), and total air-injection volume will be about 100,000 ft³/min (2800 m³/min). Off-gases will be withdrawn at a vacuum of 4 to 6 lb/in² (0.28 to 0.42 kg/cm²). The off-gases will be moved at low pressure through thin-walled, large-diameter pipe to a gas-treatment plant. After cleanup, the gases, which have a heating value of about 75 Btu (19 kcal), will be utilized for power generation. A plant of this size would be operated by a crew of 50 employees.

PRODUCT AND PRODUCT SPECIFICATION

The process produces crude shale oil and a combustible off-gas. Approximately 60 percent of the total energy produced is in the form of oil and 40 percent is gas. The composition of the oil is shown in Table 5-2, and gas composition in Table 5-3.

UPGRADING TECHNOLOGY

The current process development program is aimed at developing an extraction process to produce crude shale oil. The product will be sold under contract to a refinery that has facilities for treating crude shale oil, or a small upgrader will be constructed to serve a number of production units.

It is expected that a shale-oil refinery will be constructed at Roosevelt, Utah, to process crude shale oil from the area. Our plans are to transport the oil to Roosevelt for processing.

WASTE AND EMISSIONS

Two types of waste products are produced:

1. *Waste gases* The retort off-gases will be treated and burned for power generation. The combusted gases from the power plant will be emitted. An analysis of the raw off-gas as produced at the retort, and prior to any treatment, is given in Table 5-3. The sulfur content of the raw off-gas was 1848 ppm H_2S in retort 18 and 926 ppm H_2S in retort 24 (by volume).

TABLE 5-2 Analysis of oil produced from retort 14, showing change in characteristics of the oil during retorting; 5-day composite samples

Days	API gravity	Pour point, °F	Nitrogen, %	Sulfur, %	Distillate Temperatures, °C											Residue, %
					IBP	5%	10%	20%	30%	40%	50%	60%	70%	80%	90%	
1-5	24.4	55	1.88	0.91	210	398	436	501	548	594	630	670	732	736	742	6
6-10	25.0	60	1.83	0.92	302	384	431	496	552	605	644	689	734	753	—	14
11-15	25.6	66	1.67	0.84	212	382	438	490	550	599	640	672	710	743	—	13
16-20	25.8	65	1.55	0.81	230	416	452	501	548	589	630	680	705	729	750	9
21-25	26.2	64	1.66	0.78	232	407	444	498	546	589	634	685	712	738	753	3
26-30	25.8	69	1.46	0.77	223	416	449	502	547	592	641	672	709	741	755	6
31-35	26.4	60	1.45	0.83	205	408	444	503	555	598	643	692	728	739	755	5
36-40	26.7	63	1.42	0.80	214	406	441	493	543	588	627	670	709	734	755	6
41-45	26.7	65	1.19	0.77	238	383	436	486	529	571	616	657	700	737	754	6
46-51	26.5	64	1.28	0.91	223	439	466	510	548	589	627	670	709	734	748	6
Avg	25.9	63	1.54	0.83	229	404	444	498	547	591	633	676	715	738	754	7

TABLE 5-3 Analysis of retort 16 off-gas, showing change in content during retorting; 10-day composite samples*

Days	N_2	O_2	CO_2	CO	CH_4	H_2	Propane	Propylene	Isobutane	n-Butane	1-Butene	trans-2-Butene	cis-2-Butene	1,3-Butadiene	Isopentane	n-Pentane	1-Pentene	Ethylene	Ethane
1–10	71.1	6.3	15.5	2.2	1.1	2.5	0.109	0.119	0.009	0.029	0.046	0.007	0.005	0.009	0.017	0.104	0.077	0.24	0.21
11–20	67.4	4.8	18.2	3.5	1.5	3.6	0.078	0.069	0.019	0.030	0.034	0.009	0.003	0.003	0.005	0.013	0.006	0.27	0.31
21–30	67.9	4.3	17.7	3.6	1.3	4.2	0.082	0.061	0.012	0.030	0.030	0.006	0.003	0.003	0.006	0.016	0.006	0.30	0.35
31–40	67.2	3.7	18.1	4.6	1.2	4.3	0.068	0.048	0.012	0.032	0.028	0.006	0.004	0.002	0.013	0.019	0.008	0.30	0.38
41–50	67.7	3.7	17.6	4.3	1.1	4.3	0.092	0.062	0.013	0.037	0.030	0.007	0.003	0.001	0.008	0.024	0.008	0.28	0.46
51–60	65.2	2.0	18.6	4.9	1.8	5.9	0.113	0.074	0.017	0.051	0.034	0.010	0.006	0.003	0.013	0.207	0.206	0.24	0.44
61–70	—	—	—	—	—	—	—	—	—	—	—	—	—	—	—	—	—	—	—
71–80	60.3	4.1	13.3	9.9	1.3	10.5	0.084	0.071	0.017	0.054	0.042	0.010	0.005	—	0.006	0.031	0.008	0.03	0.19
81–90	59.1	2.3	17.6	8.5	0.8	10.7	0.140	0.139	0.030	0.072	0.052	0.012	0.006	0.005	0.015	0.052	0.015	0.12	0.31
91–100	62.5	1.9	19.3	5.0	1.1	9.2	0.250	0.189	0.033	0.084	0.065	0.015	0.010	0.003	0.017	0.067	0.020	0.05	0.24
101–110	62.0	1.5	22.0	3.9	1.3	8.1	0.238	0.184	0.046	0.070	0.058	0.027	0.011	0.017	0.015	0.058	0.085	0.13	0.28
111–120	54.5	2.8	24.9	3.8	2.3	10.2	0.283	0.167	0.041	0.096	0.073	0.017	0.017	0.025	0.020	0.059	0.146	0.13	0.52
Avg	64.5	3.4	18.8	4.6	1.3	6.4	0.144	0.112	0.024	0.053	0.045	0.012	0.007	0.009	0.013	0.054	0.050	0.21	0.33

*Results are all percent by volume.

2. *Retort water*   Water is produced with the oil as an oil-water emulsion. The ratio of oil to water is approximately 1:1. The water will be treated and injected into a deep disposal well. An analysis of the raw process water, as produced at the retort and prior to any treatment, is given in Table 5-4.

SUMMARY

The Geokinetics process offers an economically and environmentally attractive method for producing shale oil from oil-shale beds lying relatively close to the surface. Front-end costs for the operation are low, and a commercially viable operation can be established on individual tracts as small as one square mile. The process provides a means of establishing economic production on a small scale with the option of expanding the scope of the

TABLE 5-4   Statistical characterization of retort water

Parameter	Content, mg/L*	Parameter	Content, mg/L*
Alkalinity as CaCO₃	17,836.36	Bromide as Br	0.18
Hardness as CaCO₃	153.72	Boron as B	60.55
pH units	8.56	Cadmium as Cd	0.084
Conductivity, μmho/cm	34,035.91	Calcium as Ca	32.58
COD (chemical oxygen demand)	3,682.36	Chromium as Cr	0.078
Ammonia†	1,270.68	Cobalt as Co	0.56
Bicarbonate as HCO₃	17,174.91	Copper as Cu	0.209
Carbonate as CO₃	2,825.45	Fluoride as F	35.15
Chloride as Cl	3,016.36	Germanium as GE	0.044
Cyanide as CN	13.31	Iron as Fe	13.99
Nitrate†	34.16	Lead as Pb	0.642
Nitrite†	1.33	Lithium as Li	0.179
Phenol	11.56	Magnesium as Mg	17.49
Phosphate total	2.1	Manganese as Mn	0.937
Phosphate, ortho	1.07	Mercury as Hg, μg/L	3.78
Sulfate as SO₄	609.09	Molybdenum as Mo	11.91
Sulfide as S	447.36	Nickel as Ni	1.62
Oil and grease	103	Potassium as K	121.43
Surfactants as MBAS	23.20	Selenium as Se	0.215
Total dissolved solids	22,144.64	Silica as SiO₂	17.95
Gross alpha ±3.5 pCi/L	8.29	Silver as Ag	0.135
Gross beta ±5.0 pCi/L	26.45	Sodium as Na	9,392
Antimony as Sb	0.011	Strontium as Sr	0.002
Arsenic as As	2.55	Tin as Sn	0.168
Barium as Ba	0.54	Vanadium as V	0.43
Berryllium as Be	0.059	Zinc as Zn	0.095
Bismuth as Bi	0.059		

*Unless otherwise specified.
†Determined as nitrogen

operation in a modular fashion to any size. The overall operation can be dispersed over a number of noncontiguous tracts.

The initial costs involved in starting a production operation using the technique are relatively low, and start-up time is short. The process eliminates the need for a mine and related mining equipment, surface retorts, and all rock-moving machinery. The equipment required includes a number of drill rigs to drill the blastholes. The same drills will drill the air-injection wells (Fig. 5-2), the off-gas well, and the oil production wells (Fig. 5-7). Low-pressure blowers (Fig. 5-5) are used to provide air to the retort, and small oil-field-type pumps are used to lift the oil to the surface. The off-gases from the retort are moved at low pressure through thin-walled, large-diameter pipe to scrubbing units to remove the sulfur compounds and are then burned. Supplies used are explosives, diesel fuel for drills, and diesel fuel or electricity to power the blowers.

OIL SHALE PRE-BENEFICIATION

PHILIP C. REEVES
JOSEPH H. ABSIL

Roberts & Schaefer Resource Service Inc.
Rolling Meadows, Illinois

LOGIC OF PRE-BENEFICIATION

For the last 400 years, oil shale has been known to the scientific community, and much time and many resources have been devoted to perfecting a retort which will function economically on shale with a wide range of qualities. These designs have become more complex and sophisticated with each generation. The recent shortage of liquid oil has given birth to a plenitude of new designs and techniques to improve liquid oil availability from oil shale (actually, marlstone with a high kerogen content).

Pre-beneficiation, which is quite common and, indeed, is standard practice in other mineral industries, has been largely overlooked by the oil-shale industry. This has led to many false starts in the practical aspect of actually producing oil from oil shale because of the relatively large capital expenditure required for an operating retort versus the value of the oil actually obtained.

Most retorts function more economically with an oil shale having a content of 25 gal/ton (104 L/Mg) as a minimum. Some retorts are amenable to much higher oil content, e.g., 40 gal/ton (167 L/Mg) or more. The retort, heretofore, has been selected for a given shale, depending to a great extent on oil content expected from a given deposit. Conversely, some shale deposits have been bypassed for lack of an economical process by which they can be retorted.

This section will deal with oil shale which occurs in the western United States and similar worldwide deposits as opposed to that which is called "eastern oil shale." The western shale is not true shale and is generally banded rich and lean in its kerogen content. Eastern shale, on the other hand, is much more homogeneous in nature and, as such, looks less attractive for pre-beneficiation. Preliminary work has indicated that eastern shale can benefit from pre-beneficiation, but it is necessary to study eastern oil shale on a case-by-case basis to determine the economic viability.

Pre-beneficiation depends on a relationship between particles of differing mass. In the case of oil shale, the incidence of oil (kerogen) in the shale is related to the presence of light material. It does not have an exact correlation, such as x gal/ton $= y$ specific gravity, but the lighter fraction of any oil-shale sample has the higher oil content.

Therefore, any oil-shale can be increased in oil content by removal of the heavier constituents. It may not always be of economic advantage to beneficiate a particular deposit, but the oil content can be increased. Each project has its own economic balance which will determine the feasibility of pre-beneficiation.

If a proposed mining plan anticipates removal of only a rich fraction from the total seam, it is likely that pre-beneficiation is of lesser importance. However, even some cases of high-grade mining may result in a feedstock from which can be removed a significant portion of low-grade, e.g., 15-gal/ton (62.5-L/Mg), shale.

The most obvious cost savings in pre-beneficiation is in capital expenditure. Since a retort is volume-sensitive, it behooves the operator to obtain feedstock that has the highest practical oil content. If a retort has the ability to handle 8000 tons/day of feedstock, the same retort can produce nearly twice as much oil if the feedstock is enriched to twice the oil content of the natural shale.

This, then, is the basis for capital expenditure efficiency. A retort to handle the shale feed of 8000 tons/day may, at the 1981 value basis, cost $100 million or more. This is an expenditure of $12,500 per ton per day capacity. A beneficiation plant for the same tonnage rate would cost approximately $5 million. This would amount to about $625 per ton per day capacity. Therefore, if the oil content of a shale sample can be increased from 20 gal/ton (83.5 L/Mg) to 30 gal/ton (125 L/Mg), two retorts can produce the oil otherwise requiring three retorts. This represents a capital savings of about 30 percent. There are some natural shales with which even greater savings can be realized.

Against this capital savings, we must analyze and deduct, where applicable, increased mining costs, beneficiation plant operating costs, and other incidental expenditures. However, to the savings in capital cost may be added the relative ease of disposal of the reject natural shale as compared to the disposal problems of retort residue.

PROCESS DESCRIPTION

Many different specific gravity separating systems have been developed in the coal and industrial minerals fields in the past years. They have ranged from simple classifying systems to somewhat exotic electronically controlled systems. Each has strong points and weaknesses. The simpler systems are generally inexpensive and are functional when the required separation is easy or of limited economic benefit. The more sophisticated systems are best confined to laboratories where technically trained personnel are available for constant adjustment and troubleshooting, or where cost of operating is of minor importance.

In the oil-shale beneficiation industry, it appears the major criteria will be efficiency of separation, with cost of operation, water consumption, and capital expenditure of respectively diminishing significance.

One other important aspect of oil-shale beneficiation, which must be considered, is liberation. Since the western oil shale was deposited in nature in banded form, there are, within the deposits, bands of rich and lean shale. To make an effective separation, these bands must be separated into oil-shale fractions in which some are of higher kerogen content than others. Crushing the oil shale achieves the liberation of rich shale from lean shale, since the shale has a natural cleavage line at the junction of rich and lean portions. Actually, the rich shale tends to have, in nonprofessional terms, a "rubbery" quality, which leaves the rich shale in larger particles than those of the lean, more brittle shale. Some beneficiation does occur simply by crushing and screening the fines out of certain oil-shale deposits. The finer fractions of a crushed sample generally have a lower oil content than the coarse fraction.

This crushing does not have to be ultrafine to effect good liberation. Most shales can achieve effective liberation at a 38-mm (1½ in) top size. Some may require reduction to 10 mm (⅜ in). However, each sample should be examined for the economic point at which maximum oil returns can be achieved while the reject stream is kept within a reasonable oil content level. Generally, it is not economical to feed a retort material with an oil content lower than 20 gal/ton (83.5 L/Mg). Therefore, if we establish a criterion for acceptable reject from a 25-gal/ton (104-L/Mg) feed, which is in the range of 10 to 15 gal/ton (42 to 62.5 L/Mg), the enriched fraction should produce a realistic feedstock for a retort. In the lower-grade natural oil shale [10 to 15 gal/ton (42 to 62.5 L/Mg)] it is generally possible to get some fraction which will yield 25 gal/ton (104 L/Mg) or more. In this case, recovery of retort feed product may be only 10 to 30 percent, but it bears investigation to examine the economic feasibility of each grade in a deposit.

Based on the aforementioned criteria, the heavy-medium cyclone process seems uniquely suitable for application to oil-shale beneficiation. It will accept feed products up to 50 mm (2 in) in top size; it is the most efficient gravity-separation process yet developed; it can be built at relatively nominal cost in industrial-size modules; and it has proved itself in other industries to be economical to operate.

Other gravity-separation techniques may be applicable to certain deposits, but only if a site-specific phenomenon dictates or rearranges normal feasibility criteria.

The general form of beneficiation plants is presented in Fig. 6-1. Since the logic of all beneficiation techniques is similar and since the heavy-medium cyclone process can be more universally applied than the other processes, the general form will anticipate a heavy-medium cyclone process.

The raw oil shale from the mine should be deposited in a storage or surge pile to provide independent operation of the plant and mine. The storage should be developed to ensure that this independence is available. The contingency is not only an occasional interruption at either site but also the possible difference in the number of shifts each will be working. Storage could assure three-shift operation in the beneficiation plant while permitting the mine to work two shifts. Obviously, other schemes can also be accommodated.

After storage, the raw oil shale should be screened. This step may be omitted if the fines, ½-mm (30-mesh) to 1½-mm (14-mesh) top size, have sufficient kerogen content to make an acceptable retort feedstock. In most cases, other than the very rich deposits, fines will be of marginal value and should be eliminated prior to the wet process. Fines carry a much higher proportion of surface moisture because of their high surface area per ton. Therefore, if they can be removed dry prior to the actual beneficiation process, they will not be a factor in overall water consumption.

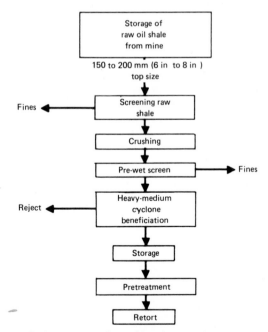

FIG. 6-1 General form of beneficiation plants.

The next step is size reduction. In order to provide proper liberation, the oil shale should be reduced to the size dictated by economies of recovery in the beneficiation facility. The style of crusher can be an important factor in achieving maximum liberation while restricting fines generation. Different types of crushers may be dictated by local conditions, but it is important to overall water consumption that the fines generated be held to a minimum.

From crushing to the required top size, depending on the fines content, the material should go to pre-wetting and screening. If the fines content is low, this entire step can be eliminated. Since the heavy-medium cyclone (and other processes) are wet washing systems, this phase accomplishes two things. First, it gets the material wet so as to lessen the tendency to place a large load on the medium-recovery circuit. Dry feed into heavy medium tends to increase the quantity of adhering medium which must be rinsed off the final product. Second, the pre-wet screen removes the particles smaller than ½ mm (30 mesh) from the beneficiation plant feed. This material tends to pollute the heavy medium and thus requires more extensive medium-recovery equipment. This product is not screened off with the natural fines because in reducing the feed products to 2 in (50 mm) or less, the fines generated will be of somewhat higher kerogen content than the natural fines. This product should be examined to determine if it is suitable for use as retort feedstock. In some cases, it may even be economic logic to consider upgrading this material by froth flotation. Its classification as feedstock, reject, or a candidate for further treatment should be determined on a case-by-case basis.

The next step is the beneficiation plant. The oil shale is introduced into the heavy-medium cyclones along with the heavy medium at a preselected specific gravity. The specific gravity of the heavy medium is controlled to make the gravity separation which has been established by testing to produce the desired end products. The light, oil-rich fraction will be discharged from the vortex (overflow) of the heavy-medium cyclone, while the heavier reject material will be discharged from the apex (underflow). After draining, rinsing, and/or dewatering, the products leave the plant separately: the reject for disposal and the benficiated oil shale to storage.

Storage is recommended here as a further means of assuring flexibility of operations between the beneficiation facility and the retorts. If there are fluctuations in plant feed or feed source, the facility can also be adapted for use in blending. Additional water drainage may also be accompliahed.

Pretreatment here is used as a generic term for any physical alteration between the beneficiation plant and the retort. Some retorrts may require the material to be crushed to a 1/4-in (6-mm) top size; some may want to eliminate fines; some may require bone-dry feedstock; and some may require no pretreatment. Pretreatment is included as a reminder that if a physical requirement is demanded at the retort, this is where it can be met.

The beneficiated oil shale is then distributed as feed to the retorts.

PROCESS PERSPECTIVE

The heavy-medium cyclone cleaning process had its beginning in the Netherlands in the late 1940s. While testing classifying cyclones in the laboratory, the research staff of Dutch States Mines noticed that when shale pollution of the water was very high, the cyclones no longer made a size classification but started to emit coarse, very clean coal from the vortex (overflow) of the cyclone. They began testing to optimize this phenomenon by using ground coal shale, barite, and other high-density products. They finally adopted ground magnetite as the medium. Magnetite satisfied several objectives. It was heavy (± 5.0 sp gr), so the amount required was relatively small; it was magnetic, so it was easy to recover; and it was available in commercial quantities. After some early successes, the coal people began in the late 1950s to use this process in quantity, and cyclone cleaning quickly became the standard of the industry.

Following the successful introduction of the heavy-medium cyclone process into the coal industry, the Dutch States Mines Corporation and others began to adapt the process for use in other fields. Cyclone cleaning grew into the field of mineral-ore pre-beneficiation in the 1960s and into automotive scrap separation in the late 1970s. The heavy-medium cyclone cleaning system has been used successfully for pre-beneficiation of lead-zinc ore, barite, fluorspar, wolfram, diamonds, iron ore, potash, magnesite, and several other different natural deposits.

Roberts & Schaefer Resource Service Inc. (R&S) saw the application to oil shale as a logical step in the growth of the heavy-medium cyclone process. Since R&S has been the licensee of Dutch States Mines in the United States since the mid-1950s, it was fully aware of the potential benefit if the process were applicable to the oil-shale industry.

DEMONSTRATION PLANT DESCRIPTION

The flow sheet of Fig. 6-2 presents the actual flow occurring while oil shale is being processed in the plant. It is important to mention, at this point, that this is not a laboratory or a scaled-down version of a plant. Figure 6-3 is a photograph of the demonstration plant

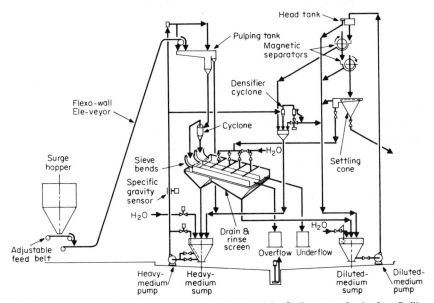

FIG. 6-2 Flow sheet of demonstration plant, Roberts & Schaefer Resources Service Inc., Rolling Meadows, Ill.

FIG. 6-3   Demonstration Plant, Roberts & Schaefer Resource Service Inc., Rolling Meadows, Ill.

built and operated by R&S. Actual full-sized equipment, which might be used in a production plant, is utilized. This enables an interested party to observe first-hand the process as it will be used. Further, no scale-up factor is required in addressing the results to a commercial plant. The results obtained are repeatable not only in additional runs in the demonstration plant but in any eventual commercial plant constructed.

The heavy medium recommended for this process is a mechanical suspension of finely ground magnetite and atomized ferrosilicon. This combination has been found to be very stable, and it gives predictable results. Another advantage is that this material is magnetic and can be readily rinsed off the products and recovered on magnetic separators. It is important that medium recovery be as efficient as practical in a high-capacity plant such as that anticipated for oil shale. Medium losses can become one of the major operating costs of a plant if maintenance or design is deficient.

The flow sheet presented in Fig. 6-2 is, as mentioned, indicative of the flow used while oil shale is being treated. Test material is loaded into the plant surge hopper. The shale is fed by a variable-rate belt feeder into a Flexo-wall Ele-veyor which delivers the material into a pulping tank at the upper floor of the plant. This design is used in the demonstration plant, but it is entirely feasible to pump heavy medium and oil shale directly to the cyclones. Either gravity or pumps can be utilized to feed the cyclones as long as the feed pressure to the cyclones is adequate to achieve the desired separation. This particular demonstration plant is used for many products other than oil shale, and it is preferable to use gravity as the pressure-inducing agent if the material to be treated is friable.

In the pulping tank, raw oil shale is mixed with heavy medium at a predetermined specific gravity and is discharged into the heavy-medium cyclone for separation. In the cyclone, the oil shale is separated into a light concentrate and a heavy reject. The heavy medium used in the plant is adjusted prior to each run so as to demonstrate the variations in product which can be achieved. Every shale sample has slight variations in specific gravity at which a given oil content is available. Each prospective client has its own criteria for the end products. Our process is able to adapt to variations in shales and clients' needs.

The products are discharged onto their own separate sieve bends for partial drainage of heavy medium. The sieve bends discharge onto the drain-and-rinse vibrating screen. (In a large commercial plant, there will be a separate screen for each concentrate and reject product.) On the first section of the drain-and-rinse screen, the product is drained of free heavy medium, and the remainder of the screen is utilized to rinse off any medium still on the product. This is referred to as "adhering media." It is imperative that these screens be sized to achieve proper removal of adhering media and to keep operating costs within economic balance.

The drained and rinsed products are discharged into separate containers for further analysis. The diluted medium, which is a combination of rinse water and adhering medium, is directed into the dilute-medium sump, from which it is pumped to the magnetic separators. The magnetic separators recover the medium solids for reuse in the heavy-medium circuit. The nonmagnetic tailings are discharged into a settling cone for removal of fine, nonmagnetic particles from the spray-water circuit. This permits removal from the system of the natural degraded particles which are the result of handling. This also permits a maximum reuse within the plant of both medium solids and rinse water.

The final circuit indicated on the flow sheet is the densifier circuit, which receives a portion of the heavy medium and removes a controlled portion of medium and diluting water which may come into the plant as surface moisture on the feed. This is done automatically on signal from the nuclear specific gravity sensor. This device controls the specific gravity of separation to within ± 0.005 specific gravity units. This, then, makes available to the process an automatic, consistent, and repeatable specific gravity of separation.

This demonstration facility is a prototype of the plants which, economics indicate, will be in common use in the oil-shale industry in the next decade.

TEST RESULTS

R&S, with the cooperation of Cathedral Bluffs Shale Oil Company, an equal-interest partnership between Occidental Oil Shale, Inc. and Tenneco Shale Oil Company, completed its first set of demonstration runs in the Dutch States Mines heavy-medium cyclone process in December 1980. A 20-ton sample from Cathedral Bluffs Tract C-b was sent to the R&S demonstration facility. The raw sample was a relatively low-grade shale containing an average of 13.2 gal of oil per ton in its natural state. This material was subjected to specific gravity separation ranging from 1.95 to approximately 2.4 sp gr. The recovery of oil-shale particles was extremely efficient: as high as 98.5 percent of the particles which contained more than 7 gal/ton. The overall recovery of shale ranged from 18 percent at the lower specific gravity, of shale containing 42 percent of the available oil in the shale sample, up to 60 percent recovery of shale containing 82 percent of the oil available.

The actual test data developed from five of the first eight trials are presented in Table 6-1. The first three tests were on insignificant weights of samples and were used as an indication of a logical starting point for the other, more significant sample runs. The oil contents of the feed, overflow, and underflow are indicated. What may be of more interest is the last pair of columns. These indicate the quantity of shale and oil in the overflow fraction. They also indicate, at the low specific gravity of test 4, that the recovery was only 6½ percent of the shale but over 19 percent of the oil available in the total shale sample. This is also reflected in the oil content of the overflow, which is over 37 gal/ton (154 L/Mg).

TABLE 6-1 Data from tests on oil-shale samples

		Oil Content*						Recovery, %	
	Sp gr	Feed		Overflow		Underflow			
No.	sep.	Gal/ton	L/Mg	Gal/ton	L/Mg	Gal/ton	L/Mg	Shale	Oil
4	1.975	12.7	53.0	37.6	156.9	11.0	45.9	6.49	19.17
5	2.105	13.4	55.9	31.9	133.1	9.3	38.8	17.94	42.85
6	2.22	13.1	54.7	27.1	113.1	7.7	32.1	27.95	57.72
7	2.25	12.1	50.5	24.9	103.9	6.2	25.9	31.47	64.84
8	2.385	13.1	54.7	18.5	77.2	6.2	25.9	60.35	81.95

*Oil content as determined by Fischer assay method (ASTM-D-3904-80). Results at ±6 gal/ton. One gal/ton = 4.1727 L/Mg.

By interpolation, it can be reliably estimated that, for this particular sample, the logical cut point would be between test 7 and test 8. That would result in the recovery of approximately 40 percent of the shale, containing 70 percent of the oil available.

These data are very significant to those in the oil-shale industry. If we assume that 70 percent of the oil can be extracted from 40 percent of the shale, the addition of heavy-medium cyclone precleaning, in this case, would create more than 40 percent savings in the cost of surface retorts. Since a retort is tonnage- or volume-sensitive, the same retort can produce almost twice as much oil from preconcentrated shale as from raw shale, assuming that the shale is not too rich for the particular retort being used. This, of course, would have considerable impact on the economics of shale-oil extraction. The material rejected from the heavy-medium cyclone would still be in its natural state as raw shale and would not be subjected to the heat of a retort, which may degrade the rock composition. Only the lighter portion of the shale is sent to the retort. Disposal of the natural raw shale may ultimately be more easily accomplished in light of ecological considerations. Each particular shale needs to be tested to determine the optimum specific gravity of separation depending on oil content and size consist. The heavy-medium cyclone can be used to treat any material which is less than 2 in (50 mm) in top size.

That can be put more succinctly: Figure 6-4 represents a rough plot of the oil content vs. specific gravity of several different oil-shale samples. It is obvious from these data that no single specific gravity of separation can be employed for all shale deposits. If separation

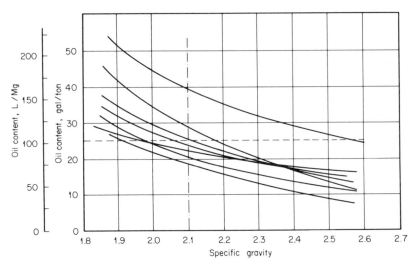

FIG. 6-4   Oil content vs. specific gravity of associated oil-shale samples.

is performed at 2.1 specific gravity, the overflow could be as high as 45 gal/ton (188 L/Mg) or as low as 20 gal/ton (83.5 L/Mg), just from these few samples. It also can be noted that, to get any particular oil content, the range of specific gravity may be enormous from one sample to another. The rich deposits (which are usually identified as the mahogany zone) generally contain 30 gal or more per ton of shale (at least 125L/Mg) on an as-received basis. In this chart, we indicate that to achieve a 25-gal/ton (104-L/Mg) product, the specific gravity of separation may range from 1.9 on up to almost 2.6.

One other interesting aspect of the heavy-medium treatment plant is that since oil-rich shale tends to have a rubbery quality, the fine material, 1 mm x 0 portion, may be, in general, a very low-grade shale—7 gal/ton (29 L/Mg) more or less—and often, as a result, can be economically disposed of in its natural state without expensive flotation and filtration.

Figure 6-5 is a graphic indication of the phenomenon stated above in other terms. In this figure, each specific gravity component has been shown with an indication of its relative oil content. Here it is easy to see why some shale samples can be beneficiated to a greater extent than the normal, rich-shale deposit. In this representation, the number of tons of each specific gravity component in a 100-ton sample is depicted separately. Also shown is the theoretical number of gallons of oil contained in each specific gravity fraction. If all of the fractions which have a specific gravity of 2.3 or less are accumulated, over 80 percent of the oil available in this particular sample is contained in approximately 45 percent of the total sample. The average oil content of the shale that would be discarded is considerably less than 4 gal/ton (16.7 L/Mg). Thus, a poor grade of shale can actually be made into a high-quality raw material.

If the sample to be considered is a rich grade of shale and the retort intended for use can tolerate a higher grade, it might well be worthwhile to investigate the pre-beneficiation process for its economic advantage anyway. We have discussed the application of the process with one prospective client who finds it makes economic sense to upgrade his natural

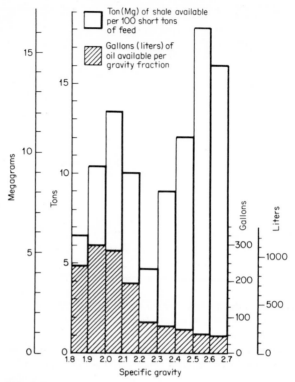

FIG. 6-5   Specific gravity components with relative oil content.

30-gal/ton (125-L/Mg) shale to somewhat over 40 gal/ton (167 L/Mg) by eliminating the heaviest 30 percent of his natural shale.

Subsequent tests have been run, but results cannot be released at this time because of confidentiality agreements. In each of the tests run, however, we have observed that the same type of efficient separation is made at many different gravities and particle sizes.

PRODUCT QUALITY AND OIL RECOVERY

All processes have particular instances in which they are most effective; the heavy-medium cyclone process is no exception. There are also cases in which it is not applicable. However, if beneficiation is required to meet product quality, there are few, if any, processes which can be applied as universally as the heavy-medium cyclone process.

There are exceptions, e.g., beneficiation is not advisable when the particular oil shale is very rich in overall oil content and no portion of it contains less oil than can be economically retorted. That would include the fines fraction, which is usually low in oil content. For instance, a shale sample which contains only 2 or 3 percent fines with an oil content of 10 gal/ton (42 L/Mg) whereas the remainder of the deposit contains 35 gal/ton (146 L/Mg) could hardly justify the investment required.

Each case should be analyzed on its own particular merits. The analysis should take into account variations such as mining costs, beneficiation costs, added material handling, retorting costs of various grades, and upgrading the final product. One other factor to be considered with beneficiation is that the overall reserves of a deposit may be decreased owing to the residual oil which is left in the reject material. However, realistically speaking, reserves which are uneconomical to retort should have never been considered as reserves. Beneficiation may also show the reverse to be true. In some cases, material which was considered to be too low in oil content and was previously categorized as overburden may have an oil content of 15 to 20 percent, which is economical to retort if separated. This would increase reserves if open-pit mining is contemplated.

Actual product quality available from beneficiation can be fairly well established by a float-sink analysis on a representative sample of the oil shale in a deposit. The test should be made on various top sizes to arrive at the point where separation of high-grade from low-grade shale is optimum. Theoretically, a reduction in top size always increases the liberation of rich shale, but practical limits will dictate a median point at which liberation is adequate and operating cost is acceptable.

The float-sink analysis from a given sample is an indication of product quality available. Since conventional heavy organic liquids react slightly with oil shale and the standard modified Fischer assay is considered accurate to $\pm$ 1 gal/ton (4 L/Mg), the float-sink data cannot be considered exact, but they are close enough to establish that a resource property falls within the range of the marketplace. Preliminary results of heavy-medium cyclone testing have indicated an organic efficiency of 92 to 97 percent. Within the constraints of laboratory analysis currently employed, product quality and quantity can be projected with reasonable accuracy.

Actual oil-recovery efficiency is a function of the reject material. If the reject material is of very low grade, the percentage of oil recovery may be very high. Conversely, if upgrading entails only the rejection of marginal material, actual oil-recovery efficiency may be somewhat lower. In any case, the specific gravity of separation can be easily and automatically adjusted to give optimum results.

REJECT PRODUCTS

Reject products have not, to date, been investigated as to long-term effects. The basic reject material from a beneficiation facility will be a mixture of untreated fines and low-grade natural oil shale. It is conceivable that this material can be mixed with the retort reject for improved stability of the disposal site if the oil content of the reject shale permits codisposal without any detrimental effect resulting from chemical or physical deterioration.

It may also be feasible to dispose of the reject material separately. The material which is uneconomical to retort in today's market may at some future time be an economic source of fuel. It would be a convenience for future generations to be able to rerun the reject for additional recovery without the expense of remining the material from a semiconcrete deposit formed by a mixture of retort residue and lean oil shale. It appears there is less ecological impact from natural oil shale in well-compacted disposal sites than may be encountered from retort rejects.

PROCESS ECONOMICS

The economics of beneficiation must include all the extra cost items required as well as the benefits generated. The benefits will be such items as lower capital costs for retorts, less retort reject for treatment and disposal, lower operating cost per barrel of oil from retorts, and possibly other savings which, though real, are difficult to categorize. For instance, if a rich shale is retorted, it is reasonable to assume there is less "dirt" in the crude oil produced. Most retorts produce an oil-shale ash which is entrained in the vapor stream to the oil condenser. A higher-grade shale produces less "dirt" per gallon (liter) of crude oil.

The drawbacks to beneficiation are in the area of extra equipment and labor to prepare material for the retorts. Disposal of rejects from the beneficiation facility will always be an extra cost. Crushing of material may or may not involve extra cost, depending on the tolerance of the retort for larger particles. If the shale feed to the retort has to be reduced to less than ¼ in (6 mm) for economical oil recovery, crushing can no longer be considered a beneficiation extra cost item. In the United States, the beneficiation plant itself can be expected to cost $12,000 to $15,000 per short ton per hour of input capacity. This is not dissimilar to the cost of a coal preparation plant of equal capacity. The cost variation is expected to be due to economic variation in size economies and difference in fines treatment from one deposit to another.

Based on 1981 dollars, operating cost for this type of plant should vary from $0.65 to $1.25 per short ton of feed. This includes operating labor, electric energy (at 4 cents/kW), medium losses of 2½ to 3 lb/ton (1.0 to 1.23 kg/Mg), and maintenance parts and labor. Some contingency is in reserve for expendables such as oil and grease. Water treatment or supply is not included, since it can fluctuate widely and may not be required, depending on total project concept.

From analyses which have been made of the overall economics, it would appear that any project except one involving ultrarich and/or selectively mined oil shale would significantly benefit economically from beneficiation prior to retorting.

SUMMARY

Pre-beneficiation has been used in many industries, but very seldom has its economic advantage, particularly that of heavy-medium cyclone pre-beneficiation, been as great as that indicated for the oil-shale industry. On overall balance, however, every project should be examined closely to determine the possible benefits of pre-beneficiation.

It appears that the heavy-medium cyclone process is one whose time has come. It is uniquely suited to the oil-shale industry. It conserves energy, money, and water, and it reduces environmental shock. It helps produce more oil more readily from less shale.

ADDITIONAL OIL-SHALE TECHNOLOGIES

JOHN WARD SMITH

Consultant
Laramie, Wyoming

INTRODUCTION

The preceding chapters on oil shale present current technologies aimed at producing oil for the energy market existing in the United States today. Most are intended for oil production from oil shales of the Green River Formation of the United States. Several additional current technologies exist, including some in at least semicommercial operation. Those to be outlined include the Union Oil retorts, Superior Oil's rotary grate retorts, Brazil's Petrosix retort, the Chinese operations based on the Fushun retort, the Kiviter and Galoter retorts currently producing oil from Estonian shale, and direct combustion of shale for production of electric power.

Two factors dictate the nature of oil-shale technology. The first is that the processes must be suitable to the characteristics of the resource available. Brazil's Petrobras spent years trying to apply retort technology designed for naturally dry Colorado oil shale to water-loaded (up to 30 percent by weight moisture as mined) Tertiary shale in the Paraiba River valley before adapting their procedures to the drier Irati Formation shales. The second factor is that the nature and scale of production must fit the available product market and the available labor supply. A complex and technically sophisticated oil-shale technology capable of producing 50,000 bbl/day (7250 Mg/day) of crude oil is hardly suitable to a developing country really needing jobs the population can handle and fuel to cook and to pump and heat water and to provide light. Oil shale occurs worldwide and should be exploited to supply energy products needed by the local population. In their current states, all of the technologies listed above have tended to adapt to market requirements and to the characteristics of the oil-shale deposits around which they were developed.

Oil shale has a long history of development, and all of the technologies described in the shale-oil chapters have roots in history. These roots will be examined, as will be the diversity of invention surrounding oil shale. Commercial development has already been successful and the technologies of two of such developments will be discussed. Additional approaches to oil-shale development technologies appear continually. Some of the more promising or spectacular "emerging technologies" will be outlined. These center around minimizing or modifying the materials handling problems inherent in any production, maximizing developable resource, or enhancing product yield or its value.

HISTORICAL OIL-SHALE TECHNOLOGIES

Oil shale has posed a continuing challenge enthusiastically met by potential developers. Since issuance of the initial patent specifically mentioning oil shale in about A.D. 1600, oil-shale technology proposals have accumulated constantly. Early oil-shale products were predominately used for lamp or heating oil, but the use of "ichthyol" from oil shale as a treatment for skin disorders created an oil-shale market persisting into the 1920s. In about 1920 the demand for gasoline created a new burst of enthusiasm for shale oil. Before this burst was curtailed by petroleum discoveries such as the east Texas one, a flood of oil-shale patents was issued across the world. All of the present-day technologies have forerunners in those patents. A few of them will be outlined to illustrate the diversity of approaches taken toward shale-oil production. The White retort, which operated a test successfully in Colorado, is described. Commercially successful historical developments in Scotland and in Union of South Africa are described. Under the Synthetic Fuels Act of

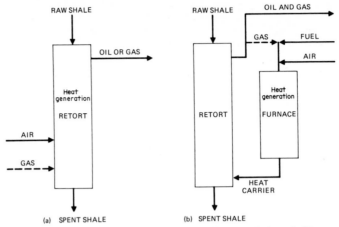

FIG. 7-1 General oil-shale-retorting systems; (a) internally heated; (b) externally heated.

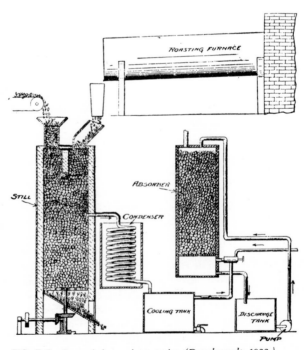

FIG. 7-2 Spent shale as a heat carrier. *(Danckwardt. 1922.)*

1946, Klosky[1] surveyed and published abstracts of patents on oil shale and shale oil collected worldwide. These U.S. Bureau of Mines (USBM) Bulletins are an invaluable source on the history of oil-shale invention.

Antecedents of Current Technology

Techniques for heating oil shale to produce oil have taken many forms, but just two basic patterns have evolved. These are outlined in Fig. 7-1. In the internally heated system the oil shale supplies its own heat by burning part of the organic matter or its products in the reaction chamber. Heat may be supplied internally by burning the residual carbon coke, the gas evolved, part of the oil, or some combination of these. In the externally heated retort heat generation is carried out separately and the heat is then transported to the retort chamber. In one externally heated form the retort shell itself is frequently used as the heat carrier. In other forms hot heat-carrier material joins the raw shale. Proposed heat-carrier material has included ceramic balls, recycle gas, sand, spent shale, and molten metal. Fuel planned to heat the carrier is usually retort off-gas but has included the carbon on spent shale, part of the product oil, locally available coal or natural gas, and even heat from nuclear power plants. In meeting the oil-shale challenge, human ingenuity has produced a remarkably diverse set of interpretations of the generalizations presented in Fig. 7-1.

Paul Danckwardt of Denver, Colorado, received a U.S. patent in 1922 for an externally heated device to roast waste shale (spent shale) and mix the hot waste shale with raw shale to drive off volatile matter (Danckwardt). This patent is illustrated in Fig. 7-2. Although in this patent gravity in a vertical retort was used to achieve mixing instead of the auger drive used in a horizontal retort, the external heating principle is very similar to the Lurgi process and to the USSR Galoter retort to be described later.

A semicontinuous internally heated retort using combustion inside the retort was designed by Charles C. Bussey, whose 1922 patents were assigned to Samuel Darby [Bussey, 1922 (a) and (b)]. Figure 7-3 presents the patent illustration. Translating the patent description from "patentese" indicates that shale was continually introduced from the top and withdrawn from the bottom. The bed was ignited above the grate, and air fed into the bottom drove the gas and oil vapors to the top, where they were removed. Bussey understood well the problems created by trying to get shale lumps to flow and be heated uniformly, because he made the retort larger toward the bottom. This flow problem plagued the USBM gas-

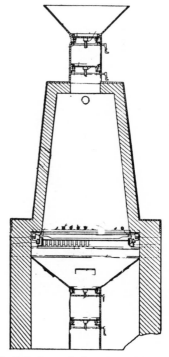

FIG. 7-3 Internally heated continuous retort. *(Bussey, 1922, a and* b.*)*

combustion retort which gave rise to the Paraho and Petrosix retorts. The Pintsch retort, designed by Hans Rosenthal and patented in the United States in 1925 and 1926, added gas combustion to this process both externally (Fig. 7-4) and internally (Rosenthal, 1925, 1926). This resembles the heating principles used by the Paraho and Petrosix processes.

Franz Puening, of Pittsburgh, approximated the TOSCO process in his U.S. patent [Puening 1929(a)]. In Fig. 7-5 solid heat-storing bodies flow into an oil-shale stream to heat the oil shale to carbonizing (a popular patent word) temperatures. Separation and recycling of the heat carriers are shown in Fig. 7-5. Gas from the carbonization was one of the heat sources proposed in this patent.

A technique for shale-oil production similar to the Occidental Oil Shale Corporation's vertical modified *in situ* technique received an Australian patent in 1922 (Fig. 7-6). In this

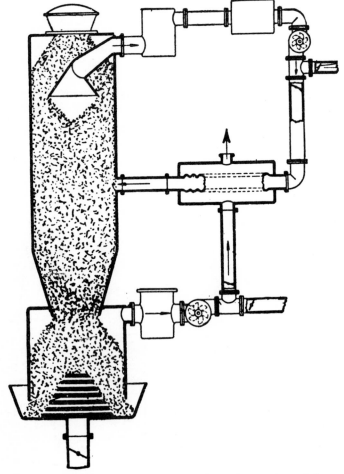

FIG. 7-4 Externally heated gas-combustion (Pintsch) retort. *(Rosenthal, 1925.)*

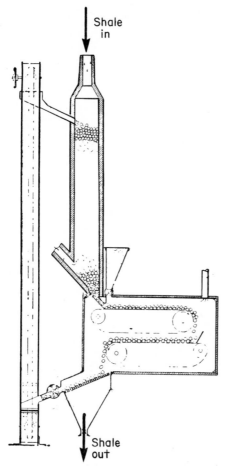

FIG. 7-5   Solid-body heat carrier externally heated retort. *(Puening, 1929a)*

patent, granted to C. F. de Ganahl of New York, an area of a shale bed is shattered "into relatively finely divided pieces" by explosives planted in shale above a mined-out area (de Ganahl). Fuel and air are then admitted until temperatures reach a point at which a portion of the released hydrocarbons will burn. A rather glorious products collection system was incorporated in the patent, which includes variations for operating beneath a level surface as well as the hillside indicated in Fig. 7-6.

The Superior Circular Grate is of a similar nature to the technology shown in Fig. 7-7, copied from a patent issued in 1929 to Franz Puening of Pittsburgh [Puening, 1929(b)]. Puening was one of several individuals who created many retort designs, apparently all on paper. This rotating grate consisted of a heating chamber (J, Fig. 7-7) and a coking chamber (K, Fig. 7-7). Heat from the coking chamber heated the vessels of the heating chamber,

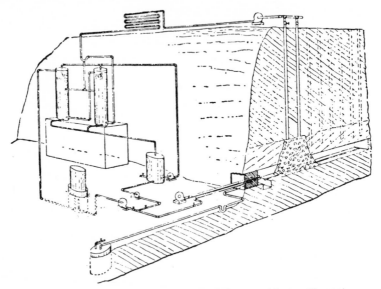

FIG. 7-6 Internally heated retort created from shale in place. *(de Ganahl, 1922)*

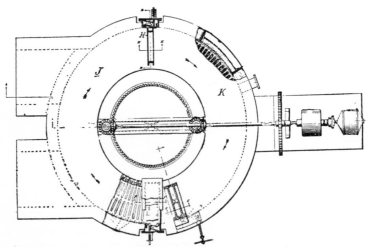

FIG. 7-7 Rotating grate retort apparently internally heated. *(Puening, 1929b)* Key: J = wall-heating chamber; K = coking chambers.

but the patent is a bit vague about how this was accomplished. The design operated continuously.

The Geokinetics *in situ* retorting process is an improvement on a patent issued in 1922 to Wilson W. Hoover and Thomas E. Brown of New York (Hoover and Brown). Figure 7-8 illustrates the installation required. To quote the patent's only claim, this "method of

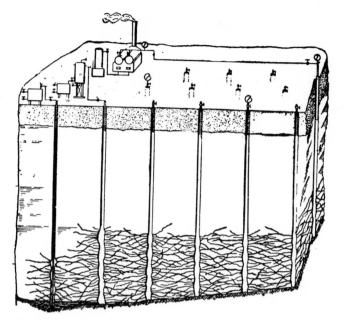

FIG. 7-8 Creating permeability in oil shale in place by explosives installed from the surface. *(Hoover and Brown, 1922)*

working oil shales consists in gaining access to a determinate zone of any desired area in a subsurface deposit of oil shale by drilling a group of contiguous wells, suitable in number, proximity, and depth for the purposes hereinafter set forth; in the introduction through the wells into such zone of charges of explosives; in exploding the same, in thereby so fracturing the formation in such a zone as to produce a series of intercommunicating fractures throughout the zone which shall render the same permeable." The recovery technique didn't match that of Geokinetics, but the rubblization technique is identical.

Figure 7-9 illustrates a retorting system directly related to the Union Oil Company retort tested in Colorado in the early 1950s. David J. L. Davis and George W. Wallace, of New York, patented this system in 1926 [Davis and Wallace, 1926(a)]. Their first claim for this internally heated system reads "The process of distilling oil shale, which consists of feeding a column of shale upwardly through a closed chamber, creating a zone of combustion in the shale at the top of said chamber, passing air downwardly into said zone to maintain it and to produce a zone of distillation immediately below said zone of combustion, and continuously removing the spent shale from the top of the chamber." Another patent apparently issued consecutively to these gentlemen describes some of the mechanics of the retort and its oil-recovery system [Davis and Wallace (1926b)]. Figure 7-9 very specifically defines the existence of separate zones in the combustion retort column.

The inventiveness of inventors accepting the oil-shale challenge was not always restricted to practicalities of function or economics. Some exotic designs were produced. Several had huge collections of moving parts which required exposure to corrosive atmospheres at elevated temperatures. Figure 7-10 illustrates one such externally heated retort which sent

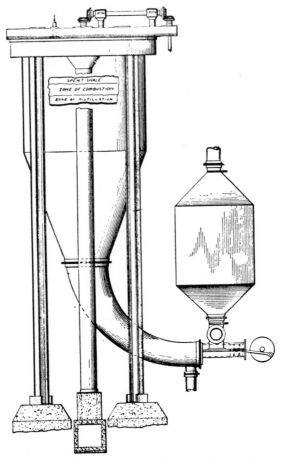

SPENT SHALE
ZONE OF COMBUSTION
ZONE OF DISTILLATION

FIG. 7-9   Internally heated retort with shale driven up. *(Davis and Wallace, 1926.)*

oil shale inserted at the upper right on a circus ride across two solid inclined rotating disks with pockets for moving and carrying the shale. Edward B. Roth of St. Louis, Missouri, received a U.S. patent for his retort in 1926 (Roth). No record of its being tested exists, but it would have given the oil shale an exciting trip. Another early retort incorporating an exotic technique is illustrated in Fig. 7-11. In this remarkable system oil shale was carried down from the top of the retort chamber by a rotating screw inside a tube. The oil shale was released into some unspecified molten metal at the bottom of the central column in Fig. 7-11. The shale's rise through the molten metal was retarded by the helical vane surrounding the conveyor. The whole chamber in Fig. 7-11 was surrounded by a heating oven to keep the metal melted. This system obviously never was tested, or the spectacularly unfortunate results would still be on record. However, in 1921 this system received two U.S. patents, both issued to L. B. Ard [1921(a) and (b)].

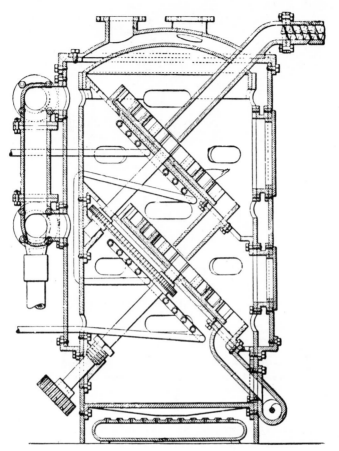

FIG. 7-10   Externally heated retort providing oil shale a circus ride. *(Roth, 1926.)*

The White Retort

Not all the development designs went untested. Gavin[2] listed about 75 proposed processes and stated that several experimental or demonstration plants had been erected in various parts of the United States. Russell[3] points out that the historical oil shale experimentation reached boom proportions during the period 1915–1930. Russell recounts records of oil-shale testing for several projects in some detail. Much of the retorting experience is lost in the maze of stock promotion publicity surrounding the oil shale during this period.

However, an account of the tested performance of the White retort has come to light through the courtesy and reporting of Thomas R. Coffey, now a consulting engineer in Grand Junction, Colorado. As a 10-year-old, Mr. Coffey watched his father construct and operate a novel experimental retort at a site 14 miles up Conn Creek from De Beque, Colo. The company conducting the experiments, called the Washington Shale Oil and Products Company, began operations in about 1920. It initially built a Ginet retort, an externally

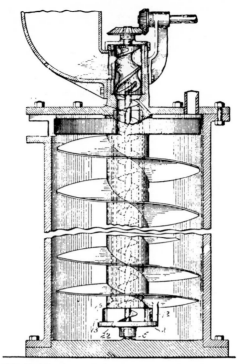

FIG. 7-11 Externally heated retort theoretically injecting oil shale into molten metal. *(Ard, 1921)*

heated system whose complex operation, somewhat analogous to the Salermo retort to be
discussed in the next section, involved shoveling shale through a hot retort with mechanical
shovels. This retort produced only a little oil in testing, and its operation proved difficult.
The company's experience with the Ginet retort convinced C. O. White, company president, that moving parts in the hot zone represented an insuperable hazard to oil-shale
retorting.

A new continuous-flow externally heated retort called the White retort was designed to
have no moving parts in the retort's hot zone. Patents applied for in 1927 and 1928 were
granted to C. O. White for the White retort in 1929 and 1931 (White, 1929 and 1931).
These patents are the source for Fig. 7-12, which presents the essentials of the White retort.
It consisted of a metal slide mounted on a hillside at an angle sufficient to ensure that
finely crushed shale would slide freely. The slide track was flat metal and had short vertical
sides which supported a metal cover, creating a metal box perhaps 24 inches wide and 3
inches thick.

The retort operated by delivering mined oil shale into a pulverizing mill represented by
the box at the right in the upper drawing of Fig. 7-12. Pulverized shale was stored in the
following hopper until it was admitted to the shale slide at a rate controlled by the indicated valve. Valve operation and the shale stored above the valve prevented entry of air.
Shale admitted to the slide spread out in a thin layer across the heated bottom of the slide

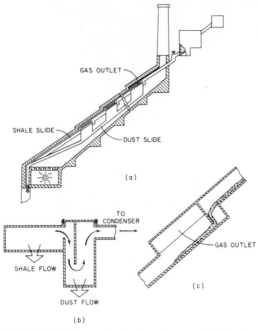

GAS OUTLET

SHALE SLIDE

DUST SLIDE

(a)

TO CONDENSER

GAS OUTLET

SHALE FLOW

(c)

DUST FLOW

(b)

FIG. 7-12 White externally heated retort. *(White, 1929, 1931)*. *(a)* Complete White retort and heating oven; *(b)* trap for entrained dust; *(c)* shale flow pattern around stop plate.

and was heated as it slid. To keep it from continually accelerating so that a very long retort would be necessary, stop plates inserted periodically practically stopped the shale and then dropped it into the next section of the slide to continue its progress. Flow of the shale stream through one of these baffles is illustrated in the lower-right drawing of Fig. 7-12. The shale flow through the retort proceeded to a vertical outlet at the lower end of the retort where a valve arrangement controlled the release of spent shale and prevented entry of air into the retort. Flow rates through the upper and lower valves were coordinated to prevent shale buildup in the retort.

A vapor exit was installed behind each stop plate to remove volatile products from the retorting chamber. These led to an air-cooled condenser pipe, outside the heated area, where oil was separated from the high-Btu product gases which were then available to heat the retort oven. To control shale dust and particles entrained with the exiting vapors, a dust trap was installed in each vapor takeoff. These traps were located inside the hot zone, thereby preventing plugging by oil buildup on dust. The lower-left drawing in Fig. 7-12 is a section of one of these traps. The gases flow up to and around a baffle plate, but dust particles impact the plate and are dropped out of the stream. The collected particles fall into a hot conduit which removes their volatile constituents and then permits the particles to slide down to rejoin the main stream at the spent-shale exit. Figure 7-13 pictures a working laboratory-scale model used to test the practicality of the technique. It illustrates how the retort was constructed.

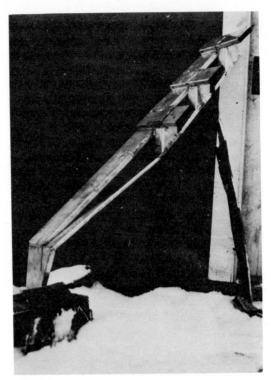

FIG. 7-13 Laboratory model of White retort. Height about 3 ft against the wall. (*Courtesy Thomas R. Coffey, P.E.*).

The entire slide system was installed in a furnace. Heat generated from a gas burner indicated at the lower end of the upper drawing in Fig. 7-12 surrounded the shale slide as the combustion products moved toward the chimney at the top. Heat input was controlled at this burner. In this design only the oil shale moved in the hot zone, and it moved by gravity. No rock pumps or augers were required. No cooling water was required because the condenser was air-cooled. The system seemed so practical that it was built and operated. Figure 7-14 shows a White retort under construction in 1928. It required extensive acetylene torch welding, a rather new technique in 1928.

The equipment was completed in 1929, and a continuous production run was conducted during that summer. According to Mr. Coffey, the run began on a Sunday morning and stopped on Sunday morning 2 weeks later. It was stopped because the 500-bbl oil tank was full! Details of this performance have long since been forgotten. Certainly, the 14-day test run must have developed some problems like shale plugging and dust control. (Mr. Coffey says the operation was not dusty!) However, to produce 500 bbl of oil in 14 days, the retort processed about 50 tons/day (45 Mg/day) [assuming 30-gal/ton (125-L/Mg) oil yield and 100 percent oil recovery]. The 14-day continuous run must have been considered successful, but Mr. Coffey remembers no additional runs. He does remember his father saying they produced oil at $1.50 per barrel. At that time, crude oil sold for less than $1

FIG. 7-14 White retort under construction, 1928. C. O. White at right center. (*Courtesy Thomas R. Coffey, P. E.*)

per barrel. The White retort, though successfully operated, was unable to compete with natural oil in the United States.

South Africa

Oil-shale development projects have occurred in several places over the world—Australia, Canada, Spain, France, Germany, Sweden, etc. In all of them technologies were developed for or adapted to the local deposits. Most of them received governmental encouragement and financial support in the form of tax breaks, production bonuses, and import duties on cheap foreign oil. Although all produced oil and generated technologies, almost all failed to compete successfully.

Two oil-shale industrial developments were able to compete successfully with natural oil, one in South Africa and one in Scotland. The South African one grew up around an oil-shale deposit near Ermelo, Transvaal, where the South African Torbanite Mining and Refining Co., Ltd. (SATMAR), mined an oil shale it called torbanite; SATMAR then retorted it and refined and marketed the oil produced. SATMAR, a privately held company, obtained financing and began development at Ermelo in 1935. Initial retort operations began in November 1935. The company persisted and even paid dividends until it mined itself out of existence in about 1962, when it had exhausted its oil-shale resources. Thorne and Kraemer of the USBM visited this operation in October 1947 as part of the alternate fuels studies generated by the Synthetic Fuels Act of 1944. Their report,[4] which outlined the technology in use in 1947, is summarized here.

The oil shale developed in SATMAR occurred in a layer 10 to 36 in thick and yielded 25 to 145 gal/ton (100 to 600 L/Mg). The oil yields were not uniform either vertically or horizontally, but the shale being processed averaged 56 gal/ton (235 L/Mg). A seam of coal averaging 30 in thick (0.75 m) overlay the oil shale. The room-and-pillar underground mine, which managed 80 percent recovery of shale from the mined area, operated by first removing the coal and then blasting the oil shale. Great care was taken to prevent including any coal in the oil shale because coal severely decreased the quality of the oil produced. Part of the coal produced was used to generate the producer gas burned to help heat retorts. The balance of the coal had little market value, so the oil-shale products paid the mining costs. Mining in 1947 produced about 800 tons of oil shale per day.

In 1947 the SATMAR retorting plant consisted of four Davidson retorts, ten Salermo retorts, an oil-shale crushing and screening plant, and seven gas producers gasifying coal to supply fuel gas. The Davidson retorts received +1-in (+2.54 cm) material, while all smaller material was fed to the Salermo retorts. No fines were discarded. Both the Davidson retorts and the Salermo retorts were externally heated by using noncondensable gases from the oil shale supplemented by producer gas generated from coal.

Davidson retorts had been operated in Estonia as early as 1932, and Davidson had several patents on variations. The SATMAR Davidson retort (simplified in Fig. 7-15) consisted of a cylinder 4 ft in diameter and 75 ft long, sloped slightly from shale inlet to shale outlet. This cylinder rotated slowly in a brick-lined heating chamber. Heat was generated in a combustion chamber at the outlet end of the retort. The gaseous combustion products were passed along a flue to a row of ports into the heating chambers around the rotating retort. Heat admitted to the heating chambers was controlled at the ports. The Davidson retorts at Ermelo operated with two temperature zones—a drying and preheating section kept at about 400 °C in the first third of the retort, and a retorting zone kept at about 630 °C. Capacity of each Davidson retort was about 30 tons/day (27 Mg/day). This feed rate was not very flexible. If too little were admitted, the charge would slide instead of tumbling, and the shale would tend to stick together in cakes. If too much were fed, evolution of oil was not complete. Spent shale was discharged through a double-door arrangement to prevent air entry into the retort. Each Davidson retort produced about 38 bbl/day of

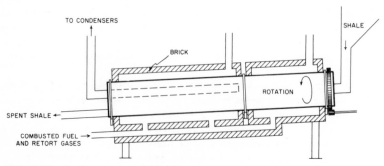

FIG. 7-15   Davidson retort—SATMAR.

oil at a recovery of about 90 percent of assay. Electric energy required to turn the huge retort tube and meet other requirements was 7.3 kWh/ton (8 kWh/Mg). Gas-energy input to the Davidson retort was 2.2 × 10^6 Btu/ton (2.6 × 10^6 MJ/Mg). In 1947 the carbon on the Davidson spent shale was discarded because of the cheap coal available.

In the SATMAR Salermo retort small oil-shale particles were progressively shoveled through a series of semicircular troughs heated from the underside. This retort, called "Salermo" in South Africa, strongly resembles and probably originated from the retort patented by P. M. Salerni in 1928 (Salerni). Figure 7-16 is a partial cutaway view of the Salermo retort as published by Thorne and Kraemer,[4] showing the position of the semicircular troughs on the right and the paddle drive train on the left. Figure 7-17 is a stylized view of the troughs to illustrate shale travel and operation of the paddles. The SATMAR Salermo retort consisted of 36 semicircular mild-steel troughs 16 in (0.4 m) in diameter and 7.5 ft (2.3 m) long. These troughs formed the bottom of the retort. Side, end, and top cover plates isolated the troughs from the atmosphere. A rod with 16 T-shaped paddles attached (Fig. 7-17) revolved in each trough. Each rod was rotated from the outside at 5 r/min. Shale entered the retort and was agitated and moved from trough to trough by the paddle rotation. Below the first nine troughs was a combustion chamber leading to flues which distributed heat along the length of the retort. Noncondensable shale gas was the primary fuel, supplemented by producer gas. As the oil shale was shoveled from trough to trough, it was gradually heated. The first troughs were dryers and preheaters. By trough 14 the shale temperature reached 300 °C, and at the discharge it reached 450 to 500 °C. Small amounts of gas were added and burned along the flues to help raise and maintain the higher temperature toward the discharge end. Evolving gases were led off to condensers for separation.

The Salermo retorts processed an average of 77.5 tons/day (70 Mg/day). Each one operated continuously for 140 days and was then cooled, overhauled, and reheated, a process that took about 20 days. Recovery was about 90 percent of assay, or about 50 gal/ton (210 L/Mg) of oil. During an operating day, each retort produced about 90 bbl (13 Mg) of oil. The electric power required was about 3.2 kWh per short ton (3.5 kWh/Mg). The total heat input to the retort, primarily from 1000-Btu/ft³ (37-J/cm³) retort gas but supplemented by producer gas, was about 1.8 × 10^6 Btu/ton (2.1 × 10^6 MJ/Mg) of shale. Like that from the Davidson retort, the Salermo spent shale contained large quantities (up to 40 percent) of carbon coke. Because waste coal was mined with the shale, no effort was made to recover the energy represented by this carbon.

Crude shale oil produced at Ermelo was shipped 120 miles by rail to a refinery near Johannesberg. In 1947 the retorting plant produced about 800 bbl (116 Mg) of oil per day, whereas the refinery capacity was about twice that. Consequently, the refinery operated intermittently. Thorne and Kraemer[4] give details of the refining techniques. The primary products were gasoline (44 percent) and asphalts (48 percent), both of which were in high local demand.

SATMAR operated an integrated oil-shale operation from mining through refining and marketing shale-oil products. It was a commercial operation which operated at a profit and paid dividends to stockholders. The company received no government subsidy and even tolerated a rather heavy tax (11 U.S. cents/gal) on the motor fuel produced. Two factors helped SATMAR: an inexpensive and inexhaustible labor supply and the high cost of transporting motor fuel from the nearest seaport. This cost was about 10 to 11 U.S. cents per U.S. gallon of gasoline. The cost of crude shale oil laid down by SATMAR at its

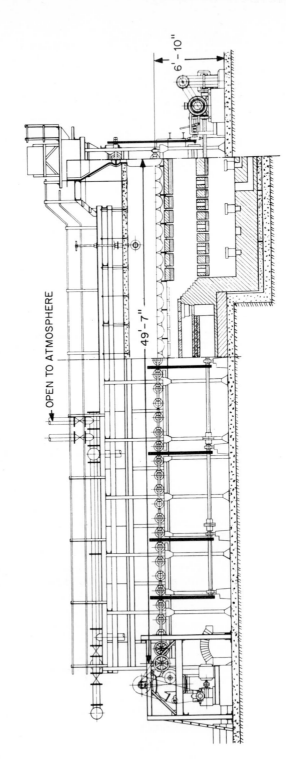

FIG. 7-16 Salermo retort—SATMAR

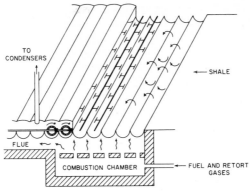

FIG. 7-17 Trough and shovel operation in Salermo retort.

refinery was estimated as about $3 per bbl (0.145 Mg). Retail price of motor fuel in Johannesberg in 1947 was $0.39 per U.S. gallon, including $0.11 tax. SATMAR operated profitably and continually improved its operation until the local resource was exhausted.

Scotland

Scotland developed a commercial oil-shale industry in about 1850, and production continued until 1963. Only in the last stages of its existence did this industry receive government support to help counter competition from cheap natural petroleum. The Scottish industry apparently was launched in about 1850 by James Young, who received British patent number 13,292 (Young, 1850). Its title, "Treatment of Certain Bituminous Mineral Substances," required reading the claims to discover that Young intended to make "Paraffin oil," a fact emphasized in his U.S. patent number 8,833 issued in 1852 (Young, 1852). The British patent's primary claim is revealing: "<u>Improvements</u> in the treatment of certain bituminous mineral substances and in obtaining products therefrom." The underscore was added to emphasize that processing of oil shale was under way in Scotland in 1850. Although Young began work on coals, he discovered and settled on treating a raw material he called variously "parrot-coal," "cannel-coal," "gas-coal," "boghead coal," "Torbanehill material," and finally "torbanite." This was a remarkably rich bituminous rock with an oil yield of more than 135 gal/ton (560 L/Mg). Apparently this material was processed, because by 1862 it was all gone. However, from this development the word torbanite became a rock name widely applied to oil-shale rocks, and a town name, "Torbane," in Australia echoed the name of the Scottish hill where this material was originally discovered.

Although the extrememly rich torbanite deposits were exhausted early, a Scottish oil-shale industry exploiting the Lothians oil shales developed rapidly after Young's initiation in 1850. Over the years more than 140 different companies and individuals undertook oil-shale processing. In 1871 there were 51 active companies. Many of them failed financially or combined with others to continue. By 1910 only six companies survived, and in 1919 five remaining companies consolidated into one company, Scottish Oils, Ltd., which became a subsidiary of the Anglo-Persian Oil Co., subsequently the Anglo-Iranian Oil Co. and now British Petroleum.

Technology of the Scottish industry changed continually over the more than 100 years of its existence, producing many retorts and processing schemes. History of this evolution was outlined by Gavin in 1922.[2] Ultimately, the Scottish oil-shale processing plants evolved to center around the Pumpherston retort. From its operation in Scotland, the Pumpherston retort gained a worldwide reputation as a successful oil-shale retort. It was used on oil shales in several other countries, notably France, where Gavin reports over 200 were in operation in 1905. This retort and its associated technology in Scotland will be outlined.

The Scottish industry introduced two new factors to oil-shale retorting. One was use of water in retorting, and the other was production of marketable ammonia as ammonium sulfate during retorting. Steam was first used in Scotland in about 1861, and its use became common in the industry. It was added to the retorts to speed removing the oil vapors, thus limiting cracking. Steam apparently increased oil yield of the vertical retorts by 25 percent and also increased the wax content of the oils produced. In addition, steam helped heat transfer, made heat distribution more uniform, and eased temperature control problems. It also assisted in the production of ammonia. In 1865, Robert Bell, an engineer at the Broxburn oil-shale plant, noticed that luxuriant plant growth was occurring in a field where wastewater from retorting was being discarded. Ammonia in the wastewater proved to be responsible, and development of ammonia production concurrent with oil production was initiated. The ammonia was marketed as ammonium sulfate and was used for fertilizer. Gavin[2] credits income from coproduction of ammonium sulfate with maintaining the Scottish oil-shale industry, stating flatly that oil income was insufficient to meet costs.

The oil-shale resource supporting the Scottish industry came from the lower Carboniferous calcareous sandstone sequence which came to be called Lothians oil shales. They underlie about 75 mi^2 (200 km^2) south of the Firth of Forth and west of Edinburgh in Scotland. The area is heavily faulted, creating major discontinuities in the oil-shale deposits. This forced development to spread out. The shale beds suitable for mining ranged from 2.5 to 10 ft (0.7 to 3 m) in thickness. In USBM Report of Investigations 4776, Guthrie and Klosky[5] show locations of oil-shale mines and plants in the area in 1947 and also present a generalized stratigraphic column for the Lothians oil-shale group. At different times and places perhaps 20 separate oil-shale layers had been mined. Most mines were underground, and mining was conducted primarily by room-and-pillar methods followed by pillar removal. The mines were labor-intensive, producing only 3 to 4 tons (Mg) per man shift (8 h). In 1919 mining represented about 53 percent of the total cost of producing marketable products from Scottish oil shale. Only low levels of methane were usually encountered in the mines; but because of British regulations arising from gassy coal mines, methane monitoring was continuous. The mines were essentially dry because of the impermeability of the oil shale. Gavin[2] reports that the shale dust in the Scottish mines tended to retard explosions rather than propagate them. The oil-shale ore produced by the mines was crushed to an average size of a thin brick before feeding to the retorts. Fines were normally neither screened out nor discarded.

The Pumpherston retorts were mounted in huge banks because of their limited throughput. The Westwood plant visited by Guthrie and Klosky[5] in 1947 consisted of two benches of 52 retorts each backed up to each other in pairs. Total capacity of the plant was 1040 tons/day (945 Mg/day), or about 10 tons/day (9.1 Mg/day) per retort. Shale processed in this plant in 1947 had an average oil yield assay of 21.4 gal/ton (89 L/Mg). Daily oil production was about 530 bbl (77 Mg).

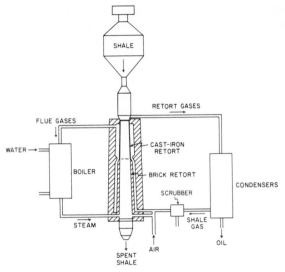

FIG. 7-18 Single Pumpherston retort, 1947.

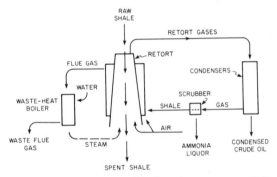

FIG. 7-19 Flow diagram of the Pumpherston process, 1947.

Figure 7-18 is a generalized diagram of a single Pumpherston retort, and Fig. 7-19 presents the flow diagram of the Pumpherston process in 1947. Above each pair of retorts was a large hopper carrying a 30-h oil-shale supply for the two retorts. This fed into the small shale hopper indicated at the top of the retort in Fig. 7-18. Each retort consisted of a cast-iron section 14 ft (4.3 m) long and an oval-shaped and tapered brick section 20 ft (6.1 m) long. The cast-iron part of the retort was surrounded by masonry enclosing a free space for heating the retorts externally by burning noncondensable shale gas. The crushed shale entered the retort at the top, flowed downward by gravity, and emerged at the bottom through a discharge hopper with water-sealed doors. As indicated in Fig. 7-19, shale gas and air were supplied and the gas was burned at the bottom of the retort. The combustion products passed upward around the cast-iron section to preheat the entering shale. Residual heat in the flue gas was recovered in the waste-heat boiler. Steam from this waste-heat boiler together with a regulated amount of air was injected into the interior of the retort at the bottom. The steam generated ammonia and swept the oil vapors upward. In addition, at the high temperatures maintained in the lower part of the retort (1000 °C), the air and the steam reacted with the residual carbon to supply much of the heat required in the retorting process. Virtually all the residual carbon was consumed. Injecting air at the bottom of the retort, a relatively new procedure in 1947, had increased each retort's throughput capacity from 4 to about 12 tons/day (3.6 to 10.9 Mg/day), permitting the use of lower-grade oil shale. Flow of shale through the retort was controlled by the rate of removal of spent shale at the bottom of the retort. Heat was recovered from the exiting spent shale. The spent shale was discarded in large piles called "bings," still in evidence throughout the area of oil-shale processing in Scotland. Although some spent shale was made into brick by adding lime and steam-curing the resulting bricks, no major use for the spent oil shale was ever found.

The gases and oil vapors leaving the top of the Pumpherston retort passed through preliminary water sprays and then through water-cooled, multiple condensers. Most of the ammonia stayed with the condensed water, but the gases were further scrubbed first with water to capture the rest of the ammonia and then with gas-oil to remove remaining naphtha. The condensed oil and ammonia-containing water separated easily. The water was processed in an ammonia-recovery plant on site. The uncondensed gas used in retorting, amounting to about 9300 std ft^3/ton (290 stdm^3/Mg) of shale, had a heating value of only about 150 Btu/ft^3 (5600 kJ/m^3) primarily because of dilution with inert gases and CO_2. However, the Westwood Pumpherston plant was energy self-sufficient with the burning of the residual carbon in the retorts. Each ton of shale processed produced about 160 lb (about 0.5 bbl, or 0.07 Mg) of oil and 27 lb (12 kg) of ammonium sulfate. For processing, the oil was shipped to oil refineries serving several retort plants. Guthrie and Klosky[5] describe the oil-refining scheme which produced the following distribution of products:

Product	Percent
Motor spirit (57 octane)	19.00
Diesel fuel	52.00
Paraffin wax	9.00
Tar (including acids and bases)	11.60
Coke (1 percent ash)	2.75
Resin	1.00
Loss (by difference)	4.65

All of these products found a ready marked in Scotland. Acid was recovered from the acid tars and used in ammonium sulfate production. The tars produced were used as refinery fuel.

The Scottish oil-shale industry had been built for permanence, and the operating personnel had grown up with the industry. It had withstood the impact of cheap oil from the United States, but it could not compete with the low-priced Near East oil. One other factor contributed to the shutdown of the Scottish oil-shale industry: the gradual consumption of the better-grade oil shales. Gavin[2] and Guthrie and Klosky[5] combine to present the following picture of grade decrease of the processed shale.

Year	Average Oil Yield Assay of Processed Shale	
	gal/ton	L/Mg
1886	33.3	139
1890	37.0	154
1900	30.3	126
1910	26.7	111
1918	24.5	102
1947	21.4	89

Although the Scottish industry was discontinued in 1963, it demonstrated emphatically that oil shale was a viable energy industry which couldn't quite compete with natural oil.

CURRENT PRODUCTION TECHNOLOGIES

Production of energy from oil shale is going on now in China and USSR; both countries have oil-shale technologies in place. In addition, one well-worked-out group of retort systems developed by Union Oil and aimed at Green River Formation oil shale is moving toward commercial oil production. Superior Oil Company developed a retort system around a traveling grate. Initially designed to solve problems associated with coproduction of alumina and sodium products from mineralized Green River Formation oil shales of Colorado, this process also seems applicable to nonmineralized oil shale. Brazil has operated the Petrosix retort on a very large pilot scale (2200 Mg/day) for several years. These current production systems were not covered in the previous technology descriptions. In addition, direct combustion of oil shale for electric power generation is being conducted now. These technologies will be briefly described.

People's Republic of China

Production of oil from oil shale in China was initiated about 1926 near Fushun in the northeastern section of China formerly called Manchuria. The province in which this development occurred is now named Liaoning. The deposit at Fushun is Oligocene oil shale 90 to 190 m thick yielding 4.5 to perhaps 8 percent by weight oil. The oil shale overlies a bituminous coal bed that ranges up to 450 ft (137 m) thick. Oil shale is produced as a by-product during the strip-mining of this coal. The Japanese began exploiting the Fushun oil shale during their occupation (about 1926) and liquid fuels from the Fushun plant contributed significantly to their efforts during World War II.

In 1954 the Chinese constructed a second oil-shale processing plant at Fushun. A similar plant was subsequently constructed at Maoming, Guangdong Province, to exploit strip-minable Eocene oil shale. More than 100 oil-shale deposits described as "commercial" have been reported in People's Republic of China, but no resource data have become available. The information that is available on the oil-shale technology of the People's Republic of China is largely extracted from a paper submitted in 1980 to the United Nations Conference on New and Renewable Sources of Energy (UNERG) by Qian of the Petroleum Institute of East China.[6] Additional information was provided by Baker and Hook in the 1979 *12th Oil Shale Symposium Proceedings.*[7]

The Chinese oil-shale industry uses the Fushun retort. Because it has been in operation for 50 years, it has been gradually modified, but the Fushun retort was initially modeled after an internally heated version of the Pintsch retort (Fig. 7-4) tested in Estonia, where it had become an internally heated process. In operation, the Fushun retort strongly resembles the Gas-Combustion and Paraho retorts. The retort used presently in both plants at Fushun as well as at Maoming is a vertical cylinder about 3 m (10 ft) in diameter and 15 m (50 ft) high. Average oil yield of the shale processed in 1979 was about 6.7 percent by weight (~17 gal/ton). Crushed oil shale screened to discard particles smaller than 8 mm (0.3 in), which incurs a 15 percent loss of fines, is fed into the top of the retort. Operation of the retort is continuous since the shale is progressively dried and heated as it moves downward, reaching retorting temperatures in about the middle of the retort. The spent shale moves into the "gasification section," where reaction with hot air and steam gasify and burn the residual carbon. The temperature in this section is 700 to 800 °C. The hot gases travel upward to heat the incoming shale while the remaining ash travels downward to heat the incoming air and water. It then drops into the "ash dish," is cooled to 70 °C, and is removed from the system.

The oil vapors and noncondensable gases leave the top of the retort at 90 to 110 °C en route to the condenser system. The cooling-and-recovery system strongly resembles that described for the Scottish plant. Ammonium sulfate also is recovered here in addition to the oil. A block of 20 to 25 retorts shares a single condenser system. The evolved gases supply heat to the retort and are also used as a source of plant energy. The plant requires no external energy supply. Oil yield of the entire system is about 60 percent of Fischer assay. Each retort processes about 20 Mg/day of oil shale.

To obtain one metric ton (Mg) of shale oil at Fushun, the following inputs to the system are required:

Oil shale	30 Mg
Water	5 to 6 Mg
Electricity	150 to 170 kWh
Steam	2 to 3 Mg

The oil produced is refined in separate plants by thermal processing and hydrotreating to produce gasoline and diesel fuel. Paraffin wax is produced as a by-product. This product slate is nearly identical to the Scottish product slate described earlier.

Estimating the meaning of costs in the People's Republic of China is difficult, but the operators maintain that the cost of production of shale oil is "not higher" than the domestic market price for natural crude oil and is lower than the world oil market price. They must believe this, because although they are now net exporters of oil because of natural oil discoveries in the country, they continue to operate their three oil-shale retorting plants. Production is about 6000 bbl (870 Mg) per day.

USSR

In Estonia, extending toward Leningrad along the south shore of the Gulf of Finland, are Ordovician deposits of an oil shale called kukersite. This resource supports the Estonian oil-shale industry. Although USSR has many oil-shale deposits, the more than 1500-mi^2 (390-km^2) Baltic kukersite shale deposit is the only one under significant development. The Estonian oil shales are rather flat-lying and shallow; they dip toward the south away from the Baltic coast. Initially they were particularly suitable for strip mining. The deposits consist of several rather thin (0.5 to 3 ft; 0.15 to 1 m) but very rich oil-shale beds separated by marlstone and limestone beds. Strip mining at shallow depths has permitted selective mining of the oil-shale beds, producing a run-of-mine ore yielding about 50 gal/ton (208 L/kg). The occurrence of ground water in all of the mining operations required all processing techniques to handle shale containing about 14 percent water. However, since the Estonia-Leningrad area has neither petroleum nor coal deposits, these oil shales were an extremely attractive energy resource in a harsh climate where energy was needed most just when it was most difficult to ship in.

Development of the Estonian oil-shale deposits began about 1920. Gavin[2] reported that two small, privately funded plants were in operation in 1921, with a promise of oncoming development on a larger scale. The Estonian government financed construction of a plant with a 200-ton/day (180-Mg/day) capacity in 1925, using a Pintsch-type retort to produce low-energy gas for local consumption. Swedish, German, and British capital had invested in the enterprise and had plants in production by 1937, when Estonia passed Scotland as the largest producer of oil-shale energy. In 1937, Estonia mined 1,123,860 Mg of oil shale. This weight was slightly less than Scotland produced, but it represented substantially more energy than the Scottish production because Estonia's shale was significantly richer.

In 1937 six plants in Estonia were processing oil shale, and only one plant was government-supported. One, the Port Kunda Company, was mining and burning raw shale for the production of cement, a reasonable maneuver in a limestone-rich but fuel-short area. One, the Kutta Joud Company, mined oil shale and sold it directly for heating fuel. The State Oil-Shale Industry plant operated the largest mine and used Pintsch retorts to produce gas and oil. The company's 1924 model retort produced gas at the rate of 20,650 ft^3/ton (644 m^3/Mg) of feed shale. The gas had a heating value of 139 Btu/ft^3 (1242 kcal/m^3). The extremely rich feed shale (29 percent by weight of oil by Fischer assay) also produced a substantial amount of condensable oils which were processed primarily to motor fuels. The New Consolidated Gold Fields Company mined shale and processed it in Davidson retorts which operated much like the retorts (Fig. 7-15) used in South Africa. The primary difference between the Estonian and the South African Davidson retorts was the Estonian development of an apparatus to burn carbon on the spent shale for process energy. This was not a particularly easy accomplishment, because ash from kukersite tends to fuse at a rather low temperature.

Two compaines, the Kivioli Company and the Eestimaa Olkonsortium, mined shale and operated tunnel kilns like those originating in Sweden for the production of shale oil. Although tunnel kilns are no longer used, they were apparently operated successfully for perhaps 30 years. The Kivioli Company (Estonian Mineral Oil A.G.) erected two tunnel kilns in 1930 and 1931 after a period of testing and research. These had a capacity of 250 tons/day (227 Mg/day). The company was so pleased with the results that two additional tunnel kilns, with a daily capacity of 400 tons (364 Mg), were erected in 1936 and 1937. These plants were essentially oil-shale-bearing retort cars running on tracks through a huge tube furnace. According to Guthrie and Klosky (1951), the 400-ton (364-Mg) retort consisted of a steel shell 174 ft (53 m) long and 8.2 ft (2.5 m) in diameter. Its tracks accommodated 18 cars—3 in the drying section of the tunnel, 13 in the distillation section where oil production occurred, and 2 in the quenching section where the cars and their shale were cooled. Each car held about 2.2 tons (2 Mg) of raw shale spread out in a layer less than 1 ft (0.3 m) thick. Fines were removed before charging, and the shale didn't move after the cars were loaded, so no dust problem occurred either outside or inside the retort.

The operation of the retort was controlled by one man, who loaded cars, operated the locks which isolated the tunnel furnace from the air, and controlled the entry of each loaded car, which simultaneously pushed a car out of the last lock. A car entered the first lock every 8 min; thus 7.5 cars were processed each hour. Since each car had a capacity of 2.2 tons (2 Mg), the plant processed about 400 tons (365 Mg) of raw shale per day. Temperatures in the distillation zone probably never exceeded 600 °C, and the cars survived their periodic heat treatments well, requiring only limited maintenance.

Heat for the process was supplied by directly burning the fines of organic-rich shale screened from the shale feed. The fines were burned in a furnace beside the tunnel oven. Heat was transferred to the drying section with superheated steam and to the retorting area with noncondensable shale gas. A slight positive pressure was maintained in the tunnel to prevent entry of air. The product gases were removed and condensed for oil recovery. The noncondensable gases not diluted by air provided a substantial supplementary heat source.

The tunnel oven operated by the Eestimaa Oklonsortium (called "Oil-Shale Syndicate" by Guthrie and Klosky[5]) was substantially larger than that of the Kivioli Company but operated in a similar fashion. The cars were moved by gears meshing with racks on the car bottoms. Additional moving parts inside a retort operating at higher temperatures must have made this tunnel kiln more difficult to maintain and manage. Fines were screened out and formed into lumps which were returned to the raw-shale feed. The Oil Shale Syndicate burned all of the evolved gases and part of the product oil to heat their furnace. Processing oil shale in Estonian tunnel furnaces apparently was a poorly controlled and energy-wasteful procedure made feasible by the richness of the shale resource.

The Estonian oil-shale industry continued to grow. In 1939 Estonia was taken over by USSR, but this didn't interrupt the shale industry. However, when the Germans took Estonia in 1941, the oil-shale plants were disabled by the retreating Russians. Because of their need for oil and their interest in oil production from shale, the Germans imported technicians to restore the plants. Restoration of the damaged plants with local materials and construction of a large Schweitzer-Lurgi plant of German design was carried on by the Germans until the Russians returned in 1944. In 1940, Estonia produced 1.7×10^6 bbl (250,000 Mg) of crude oil, but, in 1941, production of only 10,000 bbl (1500 Mg) was reported. By 1944 the Germans had restored the plants enough to schedule production of about 550,000 bbl (80,000 Mg) of crude oil. When the Russians retook Estonia, the plants were not destroyed. The Russians apparently also took over the German plans for expansion of the Estonian shale industry. This included research and development of improved production methods.

Shale production in Estonia has steadily increased. In 1978 the reported oil-shale production was 35×10^6 tons (32×10^6 Mg). This is equivalent to about 42×10^6 bbl (6×10^6 Mg) of potential oil per year or more than 100,000 bbl (16,000 Mg) per day, but only about one-third was processed for oil. The balance was burned directly to produce electric power. Shale processing provides 90 percent of Estonian electric power requirements. In a paper submitted to UNERG in 1981, Lyashenko and Nekhorosky[8] reported 14 mines in operation in the Baltic oil-shale deposit. Total production for 1979 was 36,000,000 Mg (40×10^6 tons). This came from seven underground mines and four open-pit mines in Estonia and from three underground mines in the Leningrad province. The open-pit mines produced 14.7×10^6 Mg (16×10^6 tons). The Russian oil-shale mining has been largely mechanized. Lyashenko and Nekhorosky[8] present details of current mining methods, production costs, and environmental and reclamation costs incurred with mining.

The Estonian oil-shale technology gradually evolved. One new plant was built in the 1950s, but processing to produce oil was still conducted in the retorts described above until the late 1960s. According to Baker and Hook,[7] shale processing was redirected to production of chemicals as conventional oil and gas became more available. The electric power plants were the primary oil-shale consumers. However, direct combustion of even the rich Baltic shale had some drawbacks, particularly massive emissions of fine ash particles and sulfurous gases to the air. In addition, the low softening temperatures of some of the ash constituents caused fouling and raised maintenance costs. To help solve this problem, the Russians began upgrading oil shale for direct firing. Lyashenko and Nekhorosky[8] outlined the process which, they report, removed about 38 percent of low-organic refuse and raised the feed-shale heating value 40 percent.

Apparently the enrichment did not solve all of their combustion problems. In addition, energy recovery by direct combustion was low. Consequently, the Russians have conducted vigorous research and development of techniques for producing shale oil, a more adaptable product. Two modern processes have been developed and tested. These center on the Galoter and Kiviter retorts, which will be described. Much of the technology discussion and the two illustrations are based on material presented in 1975 by Resources Sciences Corporation to a group of United States companies interested in the new Russian oil-shale technology. This is supplemented by information from Lyashenko and Nekhorosky[8] and from Tiagunov and Stelmakh.[9]

The Kiviter retort (Fig. 7-20) was designed to process shale chunks ranging in size from 25 to 100 mm. It apparently descended from the Pintsch retort, which the Russians call a gas-generation retort. The Kiviter retort is a vertical system internally heated by combustion of coke residue and noncondensable shale gas. In many respects it resembles the U.S. Bureau of Mines gas-combustion retort and consequently the Paraho retort. Kiviter units with capacities of 200 to 350 tons/day (180 to 320 Mg/day) have been put into commercial operation. The unit has also been successfully tested on low-yield oil shale, which permits faster throughput.

Operation of the Kiviter retort is continuous. Raw oil shale enters from the top (Fig. 7-20) and is heated to temperatures required to decompose the organic matter. Heat is provided by the rising gases supplemented by recycle gas burned in the "heat carrier preparation chamber." The combustion products pass out through the shale and carry the oil vapors and evolving gas with them into the collection chamber. They are delivered to the condensing system. Additional recycle gas and air admitted to the chambers near the

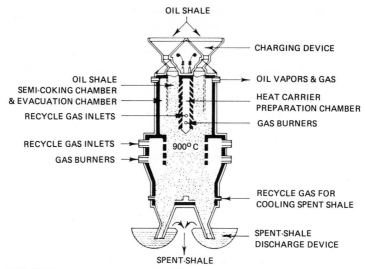

OIL SHALE

CHARGING DEVICE

OIL SHALE
SEMI-COKING CHAMBER
& EVACUATION CHAMBER
RECYCLE GAS INLETS

OIL VAPORS & GAS

HEAT CARRIER
PREPARATION CHAMBER

GAS BURNERS

RECYCLE GAS INLETS

GAS BURNERS

900° C

RECYCLE GAS FOR
COOLING SPENT SHALE

SPENT-SHALE
DISCHARGE DEVICE

SPENT-SHALE

FIG. 7-20 Kiviter retort, USSR, modified from Resource Sciences Corporation report.

900 °C label heat the shale residue to burn off the carbon coke. Recycle gas admitted at the bottom cools the spent shale, which then leaves the retort through a water-sealed discharge system.

The Kiviter retort operating on Baltic shales enriched at the mine to an oil yield of 21.3 percent by weight (56 gal/ton; 233 L/Mg) produces the following materials balance from 1 ton and 1 metric ton (Mg) of wet feed.

Materials	lb/ton	kg/Mg
Input:		
Dry shale	1820	910
Associated water	180	90
Steam	100	50
Air	860	430
Total input	2960	1480
Output:		
Shale oil	360	180
Gas	1120	560
Process water	320	160
Spent shale	1160	580
Total output	2960	1480

The Kiviter process recovers about 75 percent of Fischer assay oil. Electric energy required by the process is 18 kWh per ton of wet feed shale. Gross heating value of the nitrogen-diluted product gas is 1520 Btu/lb (3534 kJ/kg). The Russians designed special burners to use this low-energy gas.

The Kiviter process incorporates several improvements over the older gas-generator retorts. The primary one is continuous operation, which depended on development of a spent shale discharge mechanism. Another improvement depends on the fact that the Kiviter retort (Fig. 7-20) injects gas heated by combustion in the area labeled Heat Carrier Preparation Chamber. This restricts organic decomposition to a single zone, which helps the coking and bridging problems common to the rich Baltic shales. Baker and Hook[7] reported that construction of a 1000-Mg/day (1100-ton/day) Kiviter retort was begun in 1976 in Estonia. Reports from USSR in 1981 say that it is still under construction.

The Galoter retorting process complements the Kiviter retort by using the fine shale particles (<25 mm) discarded from the Kiviter process. Figure 7-21 presents a flow diagram for the Galoter system complete with material flows in USCS units for a system processing 3670 tons/day (3330 Mg/day). Tiagunov and Stelmakh[9] report that two units of this size are nearing completion. One Galoter complex with a capacity of about 500 tons (Mg) per day has been in operation in Estonia since 1964. According to Tiagunov and Stelmakh,[9] by November 1979 this complex had operated for 4100 days (11.2 years). Over this period 1,932,000 tons (1,760,000 Mg) of shale had been treated, yielding 259,000 tons (235,000 Mg) of commercial oil products and 102×10^6 m^3 of high-energy gas having a combustion heat of 10,000 to 11,000 kcal/m^3 (1100 to 1200 Btu/ft^3). Obviously, the Galoter system has been well tested.

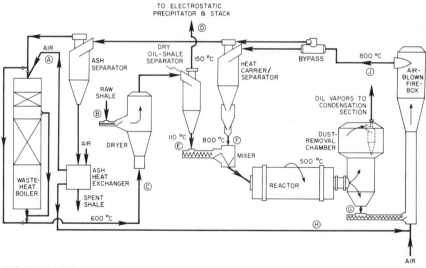

FIG. 7-21   Galoter retort system, USSR, modified from Resource Sciences Corporation report.

The Galoter retort is externally heated. It shares characteristics with the TOSCO II and the Lurgi retort systems. Thermal decomposition of the raw shale to oil and gas at controlled temperatures occurs in a rotating drum as in TOSCO II, and as in the Lurgi process the heat carrier is part of the spent shale heated by combustion of the residual carbon after retorting. In the flow sheet (Fig. 7-21) wet raw shale enters the dryer, where it is dried by a controlled flow of furnace gases. The dried shale is separated from the drying gas and is then fed to a mixer and reactor, where it joins spent shale heated to about 800 °C. The combined flow of about two parts hot spent shale and one part raw shale are tumbled and thoroughly mixed in the reactor, where the raw shale reaches 500 °C and its organic matter is decomposed and volatilized. The oil vapors and gas are led through a dust trap to the condensing section. The carbon-bearing shale is burned in the air-blown firebox. Part of the resulting hot ash is routed to the mixer. The balance of the hot ash and all of the hot gases from the combustion chamber travel through heat-recovery systems before being discarded. The boiler at the left of Fig. 7-21 recovers excess heat as steam.

Material	lb/ton	kg/Mg
Input		
Dry shale	1752	876
Contained water	248	124
Flue gas	4	2
Steam	6	3
Water	348	174
Air	1353	676
Total	3711	1855
Output		
Shale oil	259	130
Gasoline	16	8
Gas	93	46
Process water	46	23
Flue gas	1799	900
Steam	348	174
Spent shale	1150	574
Total	3711	1855

The Russian report through Resource Sciences Corporation gives the following process materials balance on 1 ton and 1 Mg of run-of-the-mine Baltic shale yielding 16.6 percent by weight oil (43 gal/ton; 180 L/Mg) by Fischer assay. Oil recovery by the Galoter system is 85 to 90 percent of Fischer assay. Electric energy used is 24 kWh per ton of wet feed shale.

The Galoter retort system offers a number of advantages in oil-shale treatment. Lyashenko and Nekhorosky[8] listed several:

1. It substantially reduces consumption of labor, electric power, steam, metal, and refractory components compared to older methods.

2. It yields high-quality, low-sulfur shale oil which is available as a substitute for petroleum in fuel or petrochemicals manufacturing.

3. It yields high-energy gas for use as fuel or as chemical feedstock.

4. It locks 92 percent of the raw-shale sulfur in the ash, thereby cutting emissions substantially.

5. The thermal efficiency of the process is high, with heat recovery of about 86 percent.

6. The Galoter retort teams very well with the Kiviter retort to consume all sizes of shale.

7. The system seems adaptable to a wide variety of oil shales.

Tiagunov and Stelmakh[9] report that two 3300 Mg/day (3670 ton/day) Galoter units are "nearing completion" in Estonia. Lyashenko and Nekhorosky[8] report that the first "pilot-plant" Galoter system with a 3300 Mg/day (3760 ton/day) capacity was "commissioned" in 1979. Translation difficulties prevent determining whether this means that the plant was authorized or that the plant was put into service.

Union Retorts

Union Oil Company of California has developed a series of retort designs. The company began acquiring oil-shale property in the Green River Formation in Colorado in about 1920. Among its long-term objectives was development of a "practical method for extracting hydrocarbons from oil shale rock on a commercially significant scale." This development has produced three retort processes, referred to as the Union A, Union B, and Union SGR retorts. The last two are variations on the Union A process incorporating very similar retort designs. Figure 7-22 presents the Union A retort patented by Union in 1950 (U.S. Patent 2,501,153). In 1978 Snyder and Pownall[10] summarized Union Oil Company's experience, apparatus, and plans, which provided the basis for Fig. 7-22.

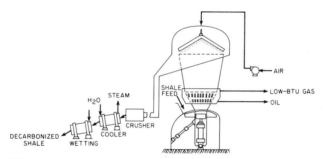

FIG. 7-22   Union A retort. (*From Ref. 10 by permission.*).

The novel part of the apparatus shown in Fig. 7-22 is the rock pump designed to force raw oil shale up into and up through the body of the retort. Relatively few oil-shale retorts have operated this way, although Fig. 7-9 shows one such design. How this rock pump worked is well demonstrated in Fig. 7-23, from T. A. Hendrickson's *Synthetic Fuels Data Handbook.*[11] In step 1 (Fig. 7-23) crushed oil shale is drawn down into the feed cylinder from the feed hopper; in step 2 the feed cylinder is moved and positioned below the retort body; in step 3 the hydraulic ram forces shale up into the retort body; and in step 4 the feed cylinder is moved back in position under the feed hopper ready to take another bite of raw shale.

To reach the capacity required for commercial production, a second form of this system, described by Snyder and Pownall,[10] mounts the solids pump on a movable carriage positioned between two feed hoppers, one on each side of the retort. The feed system is completely enclosed in the feeder housing and immersed in product shale oil. The pump consists of two piston-and-cylinder assemblies 10 ft (3 m) in diameter which alternately feed shale to the retort. While one piston is moving upward charging shale to the retort, the other cylinder is filling with crushed and screened feed shale as the piston moves downward. When this cycle is complete, the carriage moves horizontally on tracks so that the full cylinder is now under the retort, and the empty one is under the other of the two feed

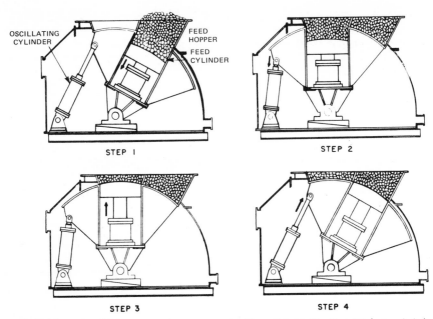

OSCILLATING CYLINDER

FEED HOPPER

FEED CYLINDER

STEP 1

STEP 2

STEP 3

STEP 4

FIG. 7-23   Union Oil Company rock pump operation (*From Ref. 11. Fig. 43, p. 72, by permission*).

bins ready for reloading. Union's experience with the rock pump led the company to conclude that the process advantages of downward flow of gas and oil and the positive-drive upward flow of shale solids justified the equipment required.

In 1958 Fred Hartley, Union Oil Company president, summarized the operation of the Union A retort as "continuous, underfeed, countercurrent retorting." In operating retort A (Fig. 7-22), shale is pumped into the bottom and is forced upward through the expanding cone-shaped chamber. Shale-input cycles are conducted continuously. Air is pumped into the top of the retort and down through the rising shale. By burning residual carbon, this air flow creates and maintains a very hot zone near the top of the retort. This combustion supplies the heat required for retorting. Gases leaving this hot zone transport heat down into the underlying raw shale, heating it to retorting temperatures. The evolved oil vapors travel down with the gas stream into the cooler shale, where much of the oil condenses. The walls of the retort below this condensation section are slotted to allow escape of the oil and gas. They are caught in a chamber surrounding the retort and led off into the product-recovery system. Any shale fines escaping through the slots are fed back to the piston and reintroduced into the retort. The feeding mechanism is filled with oil up to the slot edges to form a seal that prevents air entry but permits oil escape. The oil level is controlled by continuously withdrawing product oil. At the top of the retort hot shale ash at 1100 to 1200 °C continues to rise above the combustion zone, preheating the entering air. As the shale ash overflows the retort, it is collected, conducted to a crusher to break up clinkers, and then to a cooler (Fig. 7-22) which generates steam. The ash is then wetted to control dust during disposal.

Retort A (Fig. 7-22) was used at Union's demonstration plant in Parachute Creek canyon, Colorado, during the 1950s. The company had previously conducted pilot-scale studies at throughput rates of 2 tons/day (1.8 Mg/day) and 50 tons/day (45 Mg/day). The demonstration retort had a 5.5-ft (1.7-m) feed piston and a maximum cone diameter of 17 ft (5.2 m). Its design rate was 360 tons/day (327 Mg/day). Design changes arising from experience and experimentation more than tripled throughput rates, which reached 1200 tons/day (1088 Mg/day).

Union found the retort easy to operate and maintain because the shale under positive drive prevented the solid flow stoppage created by extensive coke formation, a bugbear of the USBM gas-combustion retort, among others. The conical retort chamber also helped ensure complete carbon combustion. Gas produced during the process had a low heating value (120 Btu/ft^3; 2.7 kJ/m^3) because of dilution by nitrogen and carbon dioxide. Although oil yields reached 800 bbl/day (116 Mg/day), retort A recovered only about 75 percent of Fischer assay.

To optimize oil recovery, Union developed two other techniques using the Union A upflow retort design. The SGR variation (for steam-gas recirculation) eliminates the combustion of carbon inside the retort. The carbon-bearing spent shale is reacted with air in separate chambers called "combustors" to produce hot flue gas at about 1600 °F (900 °C). Heat is recovered from this flue gas to run the process and to generate steam. Part of the heat is transferred to recycling product gases. This hot gas replaces the air entering the top of the retort in Fig. 7-22 and provides the heat necessary for retorting without overheating the shale and oil. Union Retort B also is indirectly heated by using recycle gas heated to 950 to 1000 °F (510 to 538 °C) in an external furnace as the heat-transfer agent. Combustion of the high-energy product gas provides most of the process heat. Snyder and Pownall[10] state that the Union B retort obtains high yields ($\sim$100 percent Fischer assay) of better quality oil by avoiding coking and refluxing. The retorting taking place under positive drive at the top of the retort prevents coke buildup and agglomeration of coke and the resulting increase in gas-pressure drop, thereby permitting the retort to operate unimpeded on even the richest shales. They feel that the Union B procedure permits operation at exceptionally high mass velocities. The Union B retort is the form chosen for Union Oil Company's ongoing (1982) oil-shale development.

Superior Circular Grate Retort

Superior Oil Company faced a unique oil-shale retorting problem when it began to examine possibilities for developing Colorado oil shales containing saline minerals. In addition to shale oil, these minerals, nahcolite ($NaHCO_3$) and dawsonite [$NaAl(OH)_2CO_3$], offer more marketable products from the same mined oil shale. The coproducts include nahcolite itself, soda ash, and alumina. The processing methods needed to produce alumina also require production of nahcolite, soda ash, and shale oil, automatically generating an integrated production process. Smith[12] outlined the required process and pointed out that first the nahcolite must be removed and then the shale oil driven out before alumina and soda ash are produced. Thermal reactions of the aluminum-bearing mineral dawsonite in the oil-shale matrix require retorting the oil shale at carefully controlled temperatures in order to produce alumina.

After discovering that onerous royalty and process ownership demands made it impractical to use the existing oil-shale retorts capable of meeting the process requirements, Superior launched a search for a suitable technique among the existing heating and cooling systems in other fields. Moving grates, both straight and circular, offered several attractive advantages. For example, the moving grates had a long and successful operating history of heating and cooling iron ore pellets at large throughput rates. The grate systems permitted independent control of the temperature and gas flow rates, thereby allowing careful management of energy input into separate processing zones. In addition, existing grates were designed to handle soft ore pellets without breaking them up. The pellets don't move against each other or against the grate or grate walls during horizontal travel through the processing zones. This is a significant advantage in processing oil shale, which becomes weaker as its organic matter is driven off. Dust carry-over is also minimized.

Circular grates with water seals had been developed. These isolate the traveling grate, as illustrated in Fig. 7-24, sealing in hydrocarbon process gases and sealing out air. In addition, the water seals permit the retort members to undergo thermal expansion and contraction without developing leaks. Examples of water-sealed traveling grates are operating successfully. Many function at temperatures substantially higher than those required to produce both shale oil and alumina. Superior selected the water-sealed circular grate for development.

Circular kilns had to be adapted to oil-shale–processing requirements. The ore could be heated by blowing hot gas through the traveling grate, but several problems remained. The biggest development necessary was a system to remove and recover the oil and gas evolving from oil shale. In addition, thermal balance and thermal recovery required optimization for oil shale. For processing dawsonite-bearing oil shale, the traveling grate must carry the

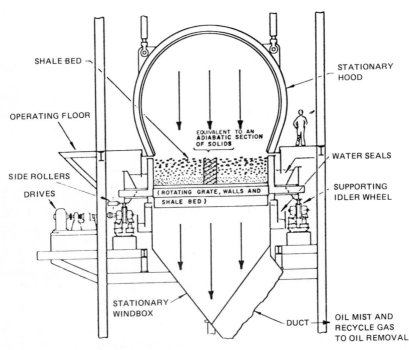

FIG. 7-24   Water seals around a traveling grate.

oil shale successively through a retort zone to produce shale oil, through a carbon-recovery zone to burn the residual carbon at controlled temperatures, and through a cooling zone to recover heat. All these functions must be balanced in length, rate of travel, and energy input and output to achieve thermal efficienty and product-recovery efficiency in processing oil shale.

Superior undertook these studies by using an adiabatic fixed-bed retort designed to duplicate a section of the solids traveling through the processing zones of the circular grate. Testing oil shale with this retort produced an oil-recovery system for the traveling grate. It was based on direct-contact spray-scrubbing with water followed by oil accumulation with an electrostatic precipitator. John H. Knight in a presentation to the Morocco–United States Oil Shale Colloquium in 1980 outlined the process cost factors and improvements developed during this testing. One key process improvement developed was increasing throughput by overlapping the process zones. This improvement, which was patented (Knight, 1977), exploits the natural temperature lag of traveling beds heated and cooled from the top to eliminate interzone leakage and improve carbon energy recovery under controlled temperature conditions. Developed in the fixed-bed retort, this concept was successfully applied in Superior's pilot circular-grate retort. Process improvements centering around the cross-flow device of Knight's 1977 patent markedly improved pilot plant performance. The initial tests on the pilot circular-grate retort in 1976 produced maximum throughput of 60 tons/day (54.5 Mg/day) with an oil recovery of only about 40 percent of Fischer assay. However, in 1977 the improved plant using Knight's patent achieved a throughput rate of 240 tons/day (218 Mg/day) with more than 98 percent oil recovery.

On the basis of these tests, commercial-scale circular grate retorts intended to produce 10,000 to 15,000 bbl/day (1450 to 2175 Mg/day) of oil were designed. Two operational modes were developed by Superior. One, an internally heated system, is fueled by burning product gases inside the retort. The other is an externally heated system fueled primarily by burning product gases. In both designs energy is recovered from the residual carbon by burning it inside the retort. Figure 7-25 is a flow diagram for the internally (direct) heated system, and Fig. 7-26 shows the indirect system. These figures, used with his permission, were extracted from Knight's 1980 presentation in Morocco. They also appear in Knight and Fishback.[13] In these diagrams the circular grate is pictured as a straight string. In each of these modes oil shale is heated by hot gases flowing through the oil-shale beds moving around the retort on the traveling grate. The major difference between the designs is how the hot gas is produced.

In the direct mode (Fig. 7-25) hot recycle gas is burned with air inside the retort. Both the recycle gas and the combustion air are preheated. Passing them through the traveling shale beds preheats these gases and cools the spent shale. In the direct mode the recycle gas is low-energy because of major dilution by nitrogen from the combustion air. This low-energy gas is burned in special burners located above the entering shale beds. The combustion products at 725 to 775 °C carry the heat into the shale bed to generate and drive out oil and gas. This oil and gas ride the gas flow down out of the shale beds to the oil-recovery system.

The traveling grate then moves the shale bed into the carbon-recovery area. Preheated air ($\sim$200 °C) directly entering the top of the spent shale bed burns the carbon. This combustion heats the shale residue. The combustion products join the recycle gas stream. After oil removal, the recycle gases at about 100 °C pass through the heated mineral residue on the moving track. They emerge at 275 to 375 °C. Final cooling of the spent shale

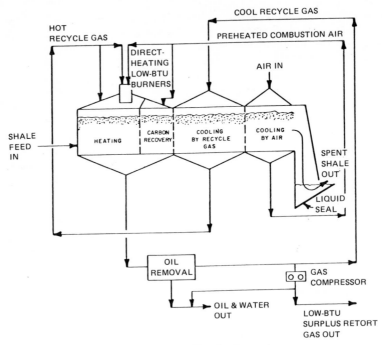

FIG. 7-25 Superior circular-grate retort, direct-heating mode.

is done by passing ambient air through the traveling bed. The cooled spent shale emerges from the retort through a liquid seal.

In the indirect mode (Fig. 7-26) part of the recycle gas is burned in an exterior furnace. Heat exchangers in this furnace heat another fraction of the recycle gas to retorting temperatures. This hot fraction passes through the raw shale bed and delivers enough heat to retort the shale. Both streams of recycle gas are preheated by passing them through the residual shale heated during carbon-energy recovery. Recovery of the energy represented by the residual carbon is managed as in the direct mode. One retort can be adapted to either mode by changing the connections of the series of inlet and outlet pipes around the circular grate.

The inherent lag of heat transmission through an oil-shale bed presents a flexibility problem to circular-grate shale processing. Each grade of shale produces a different temperature profile in the bed. The series of inlet and outlet pipes around the retort adapts to this lag, which increases with increased throughput rate. As the traveling grate moves faster to increase throughput, the lag zones stretch out. Appropriate zone boundaries can be detected by the nature of the exiting gases. Pipe connections can then be modified to maximize isolation, optimize the retorting conditions, and reach steady-state operation. There is some limitation on the adjustment of the retort to shale grade, but run-of-mine shale tends to be quite uniform. Both the direct and the indirect mode avoid the coking and agglomeration problem by controlling the heat input rate. Superior reports successful testing of shale grades ranging from 19 gal/ton (80 L/Mg) to 40 gal/ton (167 L/Mg).

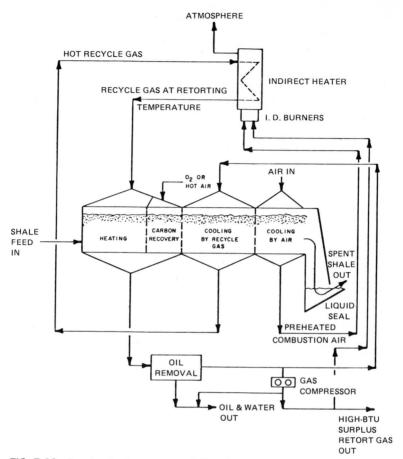

FIG. 7-26 Superior circular-grate retort, indirect-heating mode.

Superior Oil Company incorporated these retorts into its comprehensive technology to produce nahcolite, shale oil, alumina, and soda ash from the same mined rock. Apparently temperature control in the combustion phase of both modes of operation was adequate to preserve alumina solubility. The company also adapted the retorting process to other oil shales. Modifications included addition of a drying zone at the shale entry to permit handling wet shales. Another modification evaluated was lengthening the carbon-combustion zone and adding water or steam to it in order to adapt the process to handle oil shales with high residual carbon after retorting.

Petrosix Process

Brazil has developed the Petrosix retort and built the largest pilot plant to operate in the world outside USSR and China. This is the product of years of development driven by the fact that Brazil has little petroleum and what it has is extremely remote from population centers. Currently in a petroleum-based world economy Brazil produces only about 25 percent of its petroleum consumption. Purchase of the balance creates a heavy economic burden on the country and continually feeds the forces of inflation. This has led to an intense quest for additional sources of domestic energy. Brazil's effort at encouraging alcohol production for fuel is well known. However, this can't completely substitute for petroleum, and Brazil's petroleum demand continues unabated.

One possible petroleum source is Brazil's several oil-shale deposits. In the 1880s the Cretaceous Marahu deposit in the State of Bahia was developed. This shale was extremely rich, yielding up to 600 L/Mg of light oil, and would burn readily. The small resource was completely consumed by about 1920, although records exist of the shale being burned directly in steam locomotives into the 1920s. Also about 1880, a company was organized to process Tertiary oil shale to produce illuminating gas for the city of Taubaté in the State of São Paulo. Twenty indirectly heated Henderson-type retorts were built with the help of Scottish engineers. This same company attempted to develop other Brazilian oil-shale deposits. Whatever economic success the company had disappeared with the development of electric power and the arrival of cheap U.S. petroleum in the 1920s. In 1941 the abandoned plant in the Tremembe-Taubaté area was acquired by a private company which built several other retorts and produced and marketed liquids and waxes. In 1951, the Brazilian government bought the Tremembe plant and established an organization to investigate and develop production of energy from Brazilian oil shale. In 1954 Petrobras, the national oil company 50 percent owned by the Brazilain government, took over this oil shale organization, naming it "Superintendecia da Industrializacao da Xisto" or SIX for short. SIX investigated existing processes for their applicability to Brazilian oil shale and eventually engaged Foster-Wheeler Corporation to design and construct a retort in Tremembe. Foster-Wheeler was heavily influenced by the USBM gas-combustion retort. To provide oil-shale technology background, Cameron and Jones Inc., of Denver were hired as consultants. Both Jones and Cameron had participated in the design and operation of the USBM gas-combustion retort at Anvil Points, Colorado.

Tests at the Tremembe installation demonstrated that the high water content of the Tertiary shales could not be managed economically. In 1957, laboratory and bench-scale tests were initiated at Tremembe on oil shales of the Permian Irati Formation. These met with immediate success, and a short time later design of the Petrosix retort was initiated with the assistance of Cameron and Jones, Inc. Perhaps the startling resemblance of the Petrosix retort to the gas-combustion and the Paraho (Jones-designed) retorts is not so startling.

SIX had simultaneously undertaken detailed investigation of Brazil's oil-shale deposits. The Irati Formation, Permian marine black shale, offered the most promise. It occurred in the well-populated southern end of the country; it was dry; it contained a significant amount of pyrite which could yield substantial sulfur production; and it could be strip-mined. An area of the Irati Formation outcrop around São Mateus do Sul in the State of Parana was selected as a site for testing. The Irati Formation outcrops for 20 km in this area. The normal dip of the beds is about 1.5 percent, and the topography of the area consists of rolling hills with gentle slopes. An area about 4 km wide might be strip-mined before the shale became more deeply buried than 30 m, the estimated economic cutoff depth. The oil shale consists of two layers, the upper, about 6.5 m thick, averages 6.4 percent by weight of oil ($\sim$17 gal/ton), and the lower, about 3.2 m thick, averages 9.1 percent by weight of oil ($\sim$24 gal/ton). These oil-shale beds are separated by a clay-limestone bed averaging 8.6 m thick. Reserves estimated for this area when developed by the Petrosix process were 600×10^6 bbl of oil (87×10^6 Mg), 10×10^6 Mg of sulfur, 4.5×10^6 Mg of LPG, and 22×10^9 m³ of light combustible gas. The world's largest pilot-scale retort system was installed in 1970 on this resource. Initially designed with the flexibility required in an experimental development, this plant with a capacity of 2200 Mg/ day has been operating more or less in a production mode since about 1975.

The Petrosix process, adapted to the needs of Brazil, is shown in flow diagram form in Fig. 7-27. The retort is externally heated to maximize sulfur recovery and eliminate agglomeration problems. Shale sized to 15 cm or smaller and blended to uniform grade enters the top of the retort. Hot recycle gas enters the middle of the retort to heat and drive off oil and gas. Hydrogen sulfide also is generated. The heated shale travels downward and is cooled by recycle gas entering the bottom of the retort, and the shale flow rate through the retort is controlled by the grate discharge mechanism. The spent shale leaves the retort at about 150 °C. It passes through a sampling system which measures discharge flow rate and grabs a sample for Fischer assay. The assay results are used as input to the process-control system. The processed shale enters one of two depressurization vessels operated in parallel. While one is filling, the other is discharging its load into a mechanism which wets and cools the shale. The shale leaves the retort and is discarded in a water slurry. The residual carbon is not used at present.

The recycle gases carrying the evolved products travel through a recovery system which starts with two cyclone separations to remove dust and heavy-oil particles. This operation is followed by electrostatic precipitation. The gases are then compressed, and the recycle

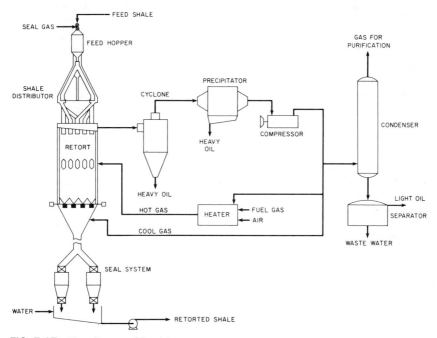

FIG. 7-27   Flow diagram of Petrosix process.

gas streams required by the retort are separated. The balance of the gases passes into a refrigerated condenser which separates light oils. The gases are then scrubbed with sodium carbonate in an absorption tower to remove mercaptans and SO_2. They then enter a second tower, where a stream of diethanolamine removes H_2S. This sulfur is recovered in a Claus unit. The cleaned gases are high-energy, suitable for pipelining. The plant was designed to process 2200 Mg of mined shale per day and produce 1000 bbl/day (145 Mg/day) of oil plus 17 Mg of sulfur and 36,500 m^3/day of high-energy fuel gas. It reached those production levels in 1975 during the demonstration phase, but it has been operating at somewhat lower throughput in commercial production.

In 1977, Brazil designed a production plant which would have a net production of 44,500 bbl/day (6450 Mg/day) of oil. The finished plant would include 20 Petrosix retorts 11 m in diameter and 65 m tall. Cost estimates for construction of this plant in two stages had reached $2 billion in mid-1980.

Direct Combustion

The USSR has the only existing industrial application of direct combustion of oil shale. At present the Russians are operating two 1600-MW power plants by using Baltic oil shale as a heat source. These plants are products of evolution, and their processes have been optimized for the local power market and for the Baltic shale. Lyashenko and Nekhorosky[8] state flatly that "Baltic oil-shale-fired power plants are characterized by reliable and cost-saving operation." Apparatus details available are sketchy, but Lyashenko and Nekhorosky[8] report that the shale combustion furnaces have a "moderate combustion intensity of 1.5 to 2 Gcal/m^3/hr" and operate in two stages. The combustion furnaces heat twin boilers specifically for this application. The two boilers work together to drive 200-MW generator sets. Twin-boiler performance specifications are 2 × 320 Mg (2 × 350 tons) of steam per hour at a pressure of 2000 lb/in^2 (14Mpa) at 570 °C. Burner design has been particularly adapted to minimize ash deposit fouling. Provisions are made for forced steam- and water-cleaning and for vibracleaning of heat surfaces to remove ash deposits.

The Baltic shale fed to the combustion furnaces is upgraded by an enrichment technique. Run-of-the-mine shale has a heating value of 1960 kcal/kg; the feed shale produced by enrichment has a heating value of 2812 kcal/kg. While handling a more energetic fuel undoubtedly helps the combustion process, the rich Estonian shales burn quite adequately without enrichment. The enrichment is carried out for two reasons. The first is that some of the components of the ash have softening points as low as 1000 to 1100 °C. This tends to facilitate ash deposition which enrichment reduces. The second factor is that ash particulates (fly ash) and sulfurous gases resulting from direct combustion have objectionable environmental effects. Enrichment decreases those effects.

Most oil shales of the world offer significantly less heat energy than the Baltic shales. They also usually contain more mineral matter. In a paper presented to the Morocco–United States Oil Shale Colloquium in Rabat, Morocco, 1980, Babcock Corporation presented a study of direct combustion of oil shales. It pointed out that a 900-MW power plant to be fueled by oil shale was under construction in Rumania. This, with the Russian plants, made only three oil-shale-fueled power plants in the world. However, the study likened burning oil shale to operating an electric power plant fueled with low-energy lignite, many examples of which now operate in Europe. The lignites have heating values down to about 900 kcal/kg. Babcock conducted pilot testing and economic studies on burning oil shales for power generation. The conclusion was that each deposit presents individual problems requiring careful design. The fouling problems about which the Russians consistently complain were actually the least troublesome found among oil shales from several countries. Babcock's economic studies indicated that direct combustion of oil shales with net heating values about 800 kcal/kg is cheaper than burning natural petroleum for electric power production.

OTHER DEVELOPMENT APPROACHES

Oil shale has furnished a continuous challenge to imagination and enterprise. Although responses to the challenge have centered on retorts, the inventor's paradise, other approaches have been tried. Several *in situ* production techniques have been proposed, and some have even been tested. Other plans are to undertake major mining to minimize the primary cost of shale-oil production and to maximize resource recovery. Still other plans are to improve resource recovery by getting more of the organic matter out of the rock. These technologies will be described briefly to indicate their nature, leaving the reader the fun of evaluating their significance.

In situ Technologies

Use of oil shale as an energy source faces an immediate basic problem. Even the best of it is a lean ore! In producing oil by surface retorting from a shale assaying 25 gal of oil per ton of rock (104 L/Mg)—a reasonably rich shale—a single ton of rock must be handled five times. It must be mined, gathered up and hauled to the retort, crushed, and retorted; and then the spent rock must be disposed of. The yield from handling the ton of rock five times is 0.6 bbl (0.09 Mg) of oil. The desire to avoid or minimize these handling costs leads to the concept of production directly from the shale *in situ*. But here a different problem appears: the rock tends to be impermeable.

If the rock can be blown through, true *in situ* production might be possible. The Tarfaya deposit of Morocco does have permeability, but most deposits do not. The U.S. Department of Energy attempted to encourage "true *in situ*" oil shale development with a singular lack of understanding or success.

Using existing permeability for heat circulation, as tested by Equity Oil Company, came close to "true *in situ*" processing. The technique depended on the porosity and permeability generated naturally in the oil shales of a part of Colorado's Green River Formation by ground water solution and removal of a soluble mineral, nahcolite ($NaHCO_3$). The remaining crystal casts were waterfilled and interconnected laterally but not vertically. By pumping steam into the holes, Equity experimenters hoped to raise the temperature of the oil shale surrounding the crystal casts high enough to produce oil. They did, but the amount of oil shale accessed could never recover the production costs.

Another technique for recovering oil shale from the shale *in situ* involves removing part of the formation by solution. Unfortunately, oil shales are notoriously resistant to solvents. However, in process patents issued to T. N. Beard in 1973, Shell Oil Company proposed a process to use the water solubility of the nahcolite in Colorado oil shale to develop access to oil shale [Beard, 1973(a) and (b)]. This process was tested with success, but land questions arising from the U.S. government's dog-in-the-manger attitude toward oil shale prevented further experimentation.

Generation of permeability in oil shale by rock fragmentation is the basis of the Occidental and the Geokinetics processes described earlier in Part 4. In the Geokinetics process enough of the shallow rock is moved upward to acquire void space from the air above the ground surface. Project Bronco proposed to do this on a gigantic scale in Colorado oil shale by using a nuclear device implanted deep in the thickest part of the 2000+-ft (600+-m) formation. This was supposed to produce a huge column of broken rock suitable for burning. Happily, cooler heads prevailed when it was pointed out that the rock column would be full of water and that carbonate mineral decomposition around the nuclear explosion would overpressure the chamber and cause it to vent to the atmosphere.

In the Occidental process, a vertical retort is built by removing part of the formation to provide void space sufficient to permit rubblization of the remaining shale. A variation of this idea is to drive drifts along the oil-shale beds, install appropriate piping and equipment, and then rubblize the shale into the drifts. This rubble would then be burned horizontally to produce and drive out the oil. The Germans tested the method in the Lias shale near Württemberg during World War II. They did produce oil, but they did not continue production after the war's end.

Electrical treatment of oil shale has been proposed repeatedly as a way to produce oil from oil shales *in situ*. The best-tested of the electrical methods was tried in Sweden. The Ljungstrom process, named after its inventor, was tested at Kvarntorp, where the shale beds are capped by impermeable limestone, a process asset. Holes were drilled through the limestone cap and into the oil shale. Resistance heaters carrying 88 A at 152 V were inserted in each hole. These heaters were operated on power from hydroelectric plants during times of low demand. The heaters were placed in a hexagonal pattern with a production hole in the middle. According to Guthrie and Klosky[5] each hole produced 40.8 bbl (6 Mg) of oil and each barrel (0.145 Mg) required 950 kWh of electric energy applied over 125 days. Although proponents of the Ljungstrom method claimed that the heating value of the products was about three times the energy used to produce them and production continued for some time, the process did not survive the impact of Near East oil.

Two experiments in the United States attempted to apply radio-frequency energy to heating oil shale *in situ*. One, tested by Raytheon and Texaco in Green River Formation oil shale in Utah, involved radiating microwave energy into oil shale from a borehole. The other, conducted by the Illinois Institute of Technology Research Institute, involved dielectric heating from electrodes planted to enclose a block of oil shale. Both methods generated oil from shale.

Major Mining

Materials handling and resource recovery are the two biggest mining problems facing oil-shale development. Oil-shale mining is a major fraction of the total cost of producing energy from oil shale in surface plants. The room-and-pillar mining method demonstrated at Anvil Points, Colorado, by the USBM produced the required high volumes of oil shale at relatively low cost. Improvements will continue as this technique is used. However, this method of resource recovery can be applied to only about 80 percent of the richest oil-shale sections, making subsequent development of leaner materials even less attractive. Only startling breakthroughs in mining methods can significantly alter this limitation on resource recovery. Two methods have been suggested, indicating that more may appear.

One method, perhaps applicable worldwide, was suggested by Dr. Tell Ertl[14,15] in reference to Colorado's Piceance Creek basin oil shales. He proposed that open-pit-mining the entire 1200-mi², 3000+-ft-deep (3100-km², 910+-m) Green River Formation deposit in Colorado would achieve nearly 100 percent recovery of the oil shales and saline minerals in the entire deposit. Although first reaction to this mind-boggling proposal was negative, it makes sense not only for Colorado but for many of the world's oil-shale deposits. Because this proposal almost requires nationalization of an oil-shale industry and because there are land ownership problems and the government is incapable of thinking on this scale, the development is unlikely. However, the concept illustrates the scale of thinking necessary to maximize resource recovery.

Another proposal to maximize oil-shale resource recovery at minimum cost was made by R. W. Johns. He developed and applied for a patent on a mining concept called the "fixed arch shield," claiming that it would recover nearly 100 percent of an oil-shale deposit (Johns, 1974). The technique is a combination of long-wall and block-caving techniques. Johns proposes construction of parallel tunnels about 1000 ft (308 m) apart, framed and protected with steel arches. These are constructed underground in the bed to be mined. Mining machinery attacking the 1000-ft (308-m) face between the parallel tunnels operates continuously, controlled by one or two operators. The machinery moves on tracks between the two corridors under protection of a fixed arch shield. Oil shale is a bedded material, and mining by this concept can follow and exploit the bedding. Johns extends his concept by permitting the roof to collapse, as in block-caving, and then remining the rubble which has dropped into the same horizon. Although some difficulties are evident in this proposal, it is a strong concept offering the possibility of maximizing resource recovery and perhaps lowering production costs. It also demonstrates the innovative thinking which can improve both resource recovery and the economics of oil-shale production.

Gasification and Hydrogenation

When oil shale is heated, gas is one of the products. If the heating is done in an inert atmosphere, the gas produced has a high energy content. The Russians exploited this gas production extensively with the Estonian shales, and many of the previously described techniques made use of the evolved gas in oil-shale processing. However, some processes have been designed to accent production of high-quality gas. The Institute of Gas Technology of Chicago (IGT) began a program aimed in this direction in 1953. This resulted in a U.S. patent, titled "Process for Pipeline Quality Gas from Oil Shale" which claimed to produce gas having a heating value of 900 to 1100 Btu/ft³ in a process self-sufficient in hydrogen (Inst. Gas Technology, 1972). In this process oil shale was heated to produce crude shale oil and spent shale solids. The crude shale oil was hydrogasified in a fluidized-bed hydrogenator at temperatures well above the temperature required to produce shale oil. This process also produced synthesis gas by reacting the spent shale, the oil shale fines, and the coke from oil gasification with steam and air. The synthesis gas was reacted with water to produce additional hydrogen—all that was required in the process.

The more hydrogen available in the organic matter of an oil shale, the larger the fraction of organic matter which can be volatilized as oil vapors and gas during the retorting. If hydrogen can be added to the organic matter in the retorting process, a larger fraction of the organic carbon in shale can be recovered as oil and gas. Texaco has obtained several patents on such a process, called hydrotorting (U.S. Patent 3,224,954, for example), and laboratory tests of its hydrogenation process did produce increases in oil yield and some reduction in sulfur and nitrogen in the product oil. As an extension of the IGT gas-production process described above, IGT undertook direct hydrogenation of oil shale and achieved organic carbon recoveries in oil and gas as high as 97 percent. This process, now called Hytort, has been tested extensively on the Devonian black shale of central United States. Detailed process descriptions are available from IGT. Efforts at commercialization are under way (1982).

EPILOGUE

This panorama of technologies for oil-shale development illustrates the range of imagination required to develop a more difficult resource. The world depends on oil, is running out of oil, and has not made adequate progress in replacing oil with other energy sources. Since oil is the primary product of oil shale, pressure for its development as the next most economic source of oil will continue to rise. The range of innovation available in oil-shale technology is spectacular. When development occurs, however, innovation becomes a high-cost item and becomes very slow. The experiences of Scotland and Russia demonstrate this quite graphically. Initial selection of oil-shale technology matching the characteristics of the deposit, the nature of the energy market to be served, and the character of the labor force available is an essential requirement of successful oil-shale development.

REFERENCES

1. Klosky, S.: *An Index of Oil Shale Patents*, U.S. Bureau of Mines Bulletin 468, 1949, 650 pp; *Index of Oil and Shale Oil Patents, 1946–56, Part 1, U.S. Patents*, U.S. Bureau of Mines Bulletin 574, 1958, 134 pp; *Index of Oil Shale and Shale Oil Patents, 1946–56, Part 2, United Kingdom Patents*, U.S. Bureau of Mines Bulletin 574, 1958, 75 pp; *Index of Oil Shale and Shale Oil Patents, 1946–56, Part 3, European Patents and Classification*, U.S. Bureau of Mines Bulletin 574, 1959, 62 pp.

2. Gavin, M. J.: *Oil Shale: An Historical, Technical, and Economic Study*, U.S. Bureau of Mines Bulletin 210, 1922, 201 pp.

3. Russell, Paul L.: *History of Western Oil Shale*, The Center for Professional Advancement, East Brunswick, N.J., 1980, 152 pp.

4. Thorne, H. M., and A. J. Kraemer: *Oil Shale Operations in the Union of South Africa, Oct. 1947*, U.S. Bureau of Mines Rept. Invest. 5019, 1954, 31 pp.

5. Guthrie, B., and S. Klosky: *The Oil Shale Industries of Europe*, U.S. Bureau of Mines Rept. Invest. 4776, 1951, 73 pp.

6. Qian, J. L.: *Oil Shale Industry in China*, UNERG, New York, 1981, 12 pp.

7. Baker, J. D., and C. O. Hook: "Chinese and Estonian Oil Shale," in J. H. Gary (ed.), *Twelfth Oil Shale Symposium Proceedings*, Colorado School of Mines Press, Golden, Colorado, 1979, pp. 26–31.

8. Lyashenko, I. V., and I. Kh. Nekhorosky: "Use of Baltic Oil Shales at Thermal Power Plants," *United Nations Conference on New and Renewable Energy Sources (UNERG)*, U.N., New York, 1981, 15 pp.

9. Tiagunov, B. E., and G. P. Stelmakh: *Utilization of Oil Shales and Bituminous Rocks as Promising Energy Sources*, UNERG, New York, 1981, 57 pp.

10. Snyder, G. B., and J. R. Pownall: "Union Oil's Long Ridge Experimental Shale Oil Project," in J. H. Gary (ed.), *Eleventh Oil Shale Symposium Proceedings*, Colorado School of Mines Press, Golden, Colo., 1978, pp. 158–168.

11. Hendrickson, T. A.: *Synthetic Fuels Data Handbook*, Cameron Engineers (now Pace Co.), Denver, 1975.

12. Smith, John Ward: "Alumina from Oil Shale," *Mining Eng.*, vol. 33, no. 6, 1981, pp. 693–697.

13. Knight, J. H., and J. W. Fishback: "Superior's Circular Grate Oil Shale Retorting Process and Australian Rundle Oil Shale Process Design," in J. H. Gary (ed.), *Twelfth Oil Shale Symposium Proceedings*, Colorado School of Mines Press, Golden, Colorado, 1979, pp. 1–16.

14. Ertl, Tell: "Mining of Colorado Oil Shale," *Oil and Gas J.*, vol. 47, no. 24, 1948, p. 116.

15. Ertl, Tell: "Mining Colorado Oil Shale," *Proc. Second Oil Shale Symposium, Color. Sch. Mines Quart.*, vol. 60, no. 3, 1965, pp. 83–92.

PATENT BIBLIOGRAPHY

1. Ard, L. B., 1921(a), U.S. Patent 1,373,689.

2. Ard, L. B., 1921(b), U.S. Patent 1,378,643.

3. Beard, T. N., 1973(a), U.S. Patent 3,779,602.

4. Beard, T. N., 1973(b), U.S. Patent 3,759,573.

5. Bussey, C. C., 1922(a), U.S. Patent 1,432,275.

6. Bussey, C. C., 1922(b), U.S. Patent 1,432,276.

7. Danckwardt, P., 1922, U.S. Patent 1,432,101.

8. Davis, D. J. L., and G. W. Wallace, 1926(a), U.S. Patent 1,607,240.

9. Davis, D. J. L., and G. W. Wallace, 1926(b), U.S. Patent 1,607,241.

10. de Ganahl, C. F., 1922, Australian Patent 7,689.

11. Hoover, W. W., and T. E. Brown, 1922, U.S. Patent 1,422,204.

12. Inst. Gas Technology, 1972, U.S. Patent 3,703,052.

13. Johns, R. W., 1974, U.S. Patent Application 502,296.

14. Knight, J. H., 1977, U.S. Patent 4,058,905.

15. Puening, F., 1929(a), U.S. Patent 1,698,345.

16. Puening, F., 1929(b), U.S. Patent 1,698,347.

17. Rosenthal, H., 1925, U.S. Patent 1,538,796.

18. Rosenthal, H., 1926, U.S. Patent 1,592,467.

19. Roth, E. B., 1926, U.S. Patent 1,598,882.

20. Salerni, P. M., 1928, Australian Patent 12,543.

21. Texaco, 1965, U.S. Patent 3,224,954.

22. Union Oil Co., 1950, U.S. Patent 2,501,153.

23. White, C. O., 1929, U.S. Patent 1,703,413.

24. White, C. O., 1931, U.S. Patent 1,799,268.

25. Young, J., 1850, British Patent 13,292.

26. Young, J., 1852, U.S. Patent, 8,833.

PART 5

OIL FROM OIL SANDS

The terms *tar sands* and *oil sands* are used interchangeably in this handbook as in the majority of synfuels literature. As is pointed out in the second chapter of this section, both terms refer to asphaltic bitumen-containing sand deposits in which the bitumen is too viscous to flow to production wells at commercially acceptable rates. Bitumen may also be recovered from porous carbonate rocks and diatomaceous earth by technologies covered in this section and by the Lurgi-Ruhrgas process described in Part 4.

The bitumen in oil sand has a specific gravity of less than 12° API (0.986 g/mL). Thus, oil sands may be viewed as a source of extremely heavy crude oil.

The approaches for producing syncrude from oil sands are summarized in Fig. I-1. Only about 10 percent of Canadian and U.S. oil-sands reserves can be recovered economically by mining. The recovery of bitumen from the remaining deposits requires the use of *in situ* methods. There are presently only two full-scale commercial oil-sands plants in operation. Both are surface-mining operations in which the bitumen is extracted from the oil sand through the use of caustic hot-water extraction and flotation techniques. This process technology is covered in detail in the first chapter of this section.

A number of alternatives to the hot-water extraction technique have been investigated, including combined water and solvent extraction, solvent extraction, oil agglomeration, and high-temperature retorting. These processes, along with the various proposed *in situ* recovery methods such as steam injection and underground combustion, are discussed in the second chapter of this section.

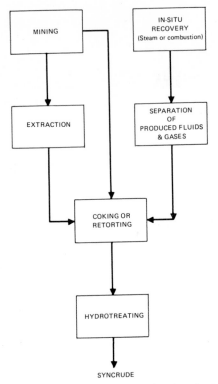

FIG. I-1 Surface and *in situ* oil-sands technologies.

SUNCOR HOT WATER PROCESS

HAROLD L. ERSKINE

Erskine & Associates,
Berwyn, Pennsylvania

INTRODUCTION

The heavy-oil deposits in Alberta comprise one of the world's great hydrocarbon reserves. They cover an area of almost 60,000 km^2 and are contained in four giant deposits:

1. Athabasca 870×10^9 bbl $(138 \times 10^9 \ m^3)$
2. Cold Lake 270×10^9 bbl $(43 \times 10^9 \ m^3)$
3. Wabasca 119×10^9 bbl $(19 \times 10^9 \ m^3)$
4. Peace River 92×10^9 bbl $(15 \times 10^9 \ m^3)$

These reserves of about 1350×10^9 bbl $(214.6 \times 10^9 \ m^3)$ compare with the rest of the world's major heavy-oil reserves in this way

1. Orinoco (Venezuela) uncertain, some estimates go as high as 4000×10^9 bbl $(636 \times 10^9 \ m^3)$
2. Alberta 1350×10^9 bbl $(214.6 \times 10^9 \ m^3)$
3. Olenek USSR 600×10^9 bbl $(95.4 \times 10^9 \ m^3)$

    Other deposits, although large, are more scattered. The United States is estimated to have oil-sands deposits of about 28×10^9 bbl $(4.5 \times 10^9 \ m^3)$ in deposits located from Kentucky to California.[1] These heavy oil formations contain enough hydrocarbon in to satisfy the world's appetite for some years to come. It remains, therefore, to get at it and get it out. And these are no easy tasks.

    Among the largest deposits are the Athabasca oil sands, which stretch along the Athabasca River for about 100 miles in north central Alberta. There outcrops of oil sands appear and have been known since prehistoric times. Some of the early trappers and fur traders observed them in their travels. They were remote and inaccessible: the nearest settlements to these outcrops were Ft. McMurray and Ft. McKay, both small trading posts.

    In the 1930s, the Canadian Government did some preliminary core-hole drilling in the area. Some demonstration plants were set up, and it was determined that about 10 percent of the Athabasca reserves could be recovered by mining the oil sands and processing the resulting ore. At about the same time, the Alberta Research Council and Karl Clark developed the hot-water process for extracting the oil from the matrix.[2] It appeared then that enough of the components were in place to commercialize the oil sands, but nothing happened. There was plenty of oil that was light and sweet, whereas the Athabasca bitumen was heavy and sour. Since there was a plentiful supply of cheap oil, no one wanted to tackle the very real problems that had to be faced in mining and processing oil sands.

THE ATHABASCA TAR SANDS DEPOSIT

Today, a walk along the face of a bench in an oil-sands mine reveals, even to an amateur, that the deposit is not uniform. In some spots, the sand is rich and black; in others, it is

ACKNOWLEDGMENTS

    The author wishes to thank SUNCOR for permission to publish this information, the late Mr. E. W. Dobson, Suntech, Inc., for the material balances, and Mr. D. Johnson, SUNCOR, for the economic calculations.

TABLE 1-1 Typical assays of Athabasca tar sands

Component	Assay, wt %		
	1	2	3
Bitumen	12	18	10
Water	4	2	6
Sand	74	76	64
Fines	10	4	20

gray and shows almost no bitumen. Clay lenses are interbedded between well-saturated areas of sand; there is also some shale.

The geologists find[3] that there are several different geologic strata in these mine pits. As an average, the feed to the SUNCOR plant is about 12 percent by weight bitumen and 4 percent by weight water. The highest grades, normally at the bottom of the lease (in old river channels) comes to about 18 percent by weight bitumen and 2 percent by weight water. Weight saturation below 8 percent bitumen is considered too lean to be processed at SUNCOR and is rejected to overburden. At Syncrude the rejection is 6 percent by weight. The cutoff is determined by economics and the amount of surplus capacity built into the plant. Typical analyses are given in Table 1-1.

One component of the matrix that gives trouble is clay, and it is present in various amounts in what a miner would consider to be random locations. It is one of the principal variables in the control of the extraction process, and its presence determines what quantity of mineral slimes are produced in the process. The principal ingredient of the deposit is sand. When washed clean, it is a fine beach sand with an average particle size of about 130–150 μm. Some rutile, ilmenite, leucoxene, siderite, and other minerals have been identified.

Properties of Bitumen

The properties of the oil sands and bitumen have been explored by Speight and his coworkers at the Alberta Research Council,[4] and Table 1-2 is compiled from their data. One striking property of the bitumen is the shape of the viscosity curve. At −45.6 °C it is greater than 1 million cP (1000 Pa·s). This high viscosity acts like a cement that binds together the bitumen, water, and sand matrix and makes mining very difficult.

Not all bitumen is the same, however. Schulte[5] has measured an increase in the asphaltene content in the bottom half of the deposit and has found that the weight percent of 524 °C distillate decreases with depth. He concludes from other studies that the amount of bitumen recovered by the hot-water process is biased, for better recovery, toward the poorer grades (those deficient in lighter components). It is known that some bitumens separate more easily than others.

Bitumens do have different densities, and the densities vary with location. Different workers have measured various properties, but additional data are needed to clarify the picture. Densities in the north appear to be lower than those in the south, and there appears to be a similar relation between viscosities. That is apparently true of the asphaltene content as well, but the data are less complete. When a lease is cored for exploration properties, the properties of the bitumen are not always measured. If they are, the information is usually confidential.

TABLE 1-2 Properties of oil sands and bitumen (Mildred–Ruth Lakes area)

Gravity, °API	5.9 (1.0298 kg/m³)
Distillation, °F (°C)	
IBP	505 (263)
5	544 (284)
10	610 (321)
30	795 (424)
50	981 (527)
95	—
EP	1030 (554)
% rec.	50
Viscosity	
SUS at 100 °F	35,000
SUS at 210 °F	513
Pour point, °F	+50 (10)
Elemental analysis, wt %	
Carbon	83.1
Hydrogen	10.6
Sulfur	4.9
Oxygen	1.0
Nitrogen	0.4
Hydrocarbon type, wt %	
Asphaltenes	16.0
Resins	34.1
Aromatics	30.0
Saturates	19.0
Trace metals, with ppm	
V	250
Ni	100
Ash, wt % (ignition-free basis)	0.75
Ramsbottom carbon, wt %	10
Conradson carbon, wt %	13.5

It can be seen from Table 1-2 that the bitumen is slightly heavier than water, is high in sulfur, and is low in hydrogen. These characteristics govern subsequent upgrading. As recovered, the bitumen is not considered to be suitable for pipelining for any great distance. Recovery and upgrading facilities have always been located at the site.

RECOVERY OF OIL

Any facility for recovering oil from the Athabasca tar sands by surface mining must go through these steps (Fig. 1-1):

1. Identification of the resource
2. Development of a mining plan and an overburden and stripping plan

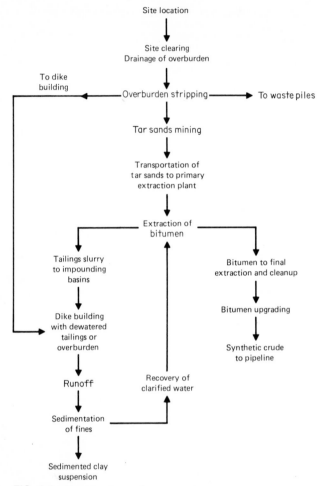

FIG. 1-1   Operational steps for a tar-sands venture.

3. Extraction of the oil from the sands

4. Purification of the bitumen

5. Upgrading of the bitumen, at least to a material that can be transported to a remote refinery

6. Successful disposal of the waste from the extraction operations

7. Provision of sufficient off-site facilities, including steam; water; pipelines; roads, bridges, etc.; living and recreation facilities for employees; and electric power

In the following pages we will be dealing closely with items 1, 2, 3, 4, 6, and briefly with item 5. We will note only that item 7 will be governed by the available facilities and the location of the site. The Athabasca tar sands are located in a remote area with extremes of weather. Temperatures in the winter can fall to -51 °C; in the summer they can rise to $+38$ °C. The land is poor; much of it is covered with muskeg, and small lakes abound. Roads are scarce. Sometimes, until suitable roads are built, one can cover the areas only in the winter when the ground is frozen. Although it is not as wild as it was in 1962, there is still a pioneering aspect about any venture in this area.

Identification of the Resource

In 1962, developing the tar sands was an act of faith. It was known that there was a lot of bitumen along the Athabasca River, but how much of it was there? Was it uniformly distributed? Could it be dug out? Could it be separated? How?

When Great Canadian Oil Sands (GCOS, later SUNCOR) started to develop a plant to extract the bitumen from lease 86, it was necessary to identify the reserves within the 4000-acre property and to characterize the deposit as closely as possible. This was done by coring, examining the cores, making a geological record of what was seen from the cores, and subjecting samples of each section of the core to chemical analysis. Later, about 20 percent of the cores were subjected to mineralogical analysis. In this case the original cores were taken on 600-ft centers and would later be supplemented by cores taken on much closer spacing to define boundaries of clay incursions, pockets of high- or low-grade material, etc. Figure 1-2 shows a core diagram typical of the Athabasca deposit.

The ore body was then contoured at constant bitumen content. On the *average* these charts have held up very well and are still in use today, but the averaging of an ore body such as this does not apply to short-term variations (less than one month). The averages are best applied on an annual basis and supplemented by cores taken ahead of the mining equipment. Normally a core is sampled for its full length and composited at a maximum of 3-ft (1-m) intervals. The compositing is closer if the geologist indicates that the interval chosen will not yield correct results. Usually, however, a 150-ft (45.7-m) core will give 50 samples.

Each sample takes about a day to analyze, and many samples are run concurrently to minimize the time required. This is an appreciable cost item. A lease needs about

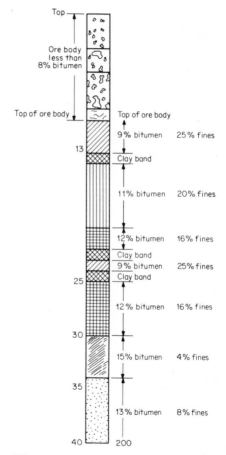

FIG. 1-2 Cross section of a core hole typical of the Mildred Lakes area in north central Alberta.

1000 cores, or 50,000 samples. The cost of analysis and reporting is about $50 per sample,* so about $2,500,000* is required to assay bitumen.

Mineralogical assay is another matter. It is at least as important as bitumen content, but the data are much more difficult to obtain. Reliable clay laboratories are scarce, and 50,000 samples would soak up much of North America's capacity for several years. Laboratories must be regularly standardized, and the cost is high, about $300* per sample. Special preparation and analytical procedures are required. Normally, therefore, only the most important samples are analyzed for clay mineralogical analysis; even then, the time delay is of the order of 6 months. Mineralogical assay is very important, however; the quantity of clays in the feed to extraction determines the recovery and also the amount of clay converted to mineral slime(s).

Mining Tar Sands

The tar sands are among the few oil sources that are mined, and each tar-sands mine is one of the largest such operations in the world. The basic problem is removing the ore and transporting it to the extraction plant as quickly and cheaply as possible. Just one look at a tar-sands mining operation would convince anyone that the digging is easy and cheap, but moving the dirt is difficult. It is, in effect, a vast materials handling job. In North America, digging usually gets first priority because land is plentiful, the overburden is deep, and the ore body is relatively thin. There is also a reluctance, based on cost, to transport overburden away from the digging face. With plenty of land available and good slope stability, the standard practice is to dump the overburden alongside the ore body in the worked-out mine (this technique is called overcasting.) To accomplish this, some of the largest draglines in the world have been built in North America.

The two existing tar-sands operations represent two different design philosophies. SUNCOR uses bucket-wheel excavators (BWEs), which are just devices to remove material from a mining face and immediately transport it away (Fig. 1-3). Syncrude opted for digging with draglines and originally planned to dump the overburden into the worked-out pit alongside the operating face. Unfortunately, the slopes were not as stable as tests indicated, so this scheme was impractical. Now both concerns remove the overburden separately (Fig. 1-4).

Overburden is useful to SUNCOR because in-pit dikes to contain tailings can be built with it; therefore SUNCOR's system was designed to that end. However, it is less efficient to transport the material from a long overburden face to a number of different locations than to route it to a single terminal point. Overburden removal is therefore a more costly operation than mining and transporting the tar sands.

Transportation of the tar sands to the processing plant is achieved by steel-cord, deep-trough conveyor belts operating at speeds of 1000 ft/min (365 m/s). Troughing angles can be 35, 45, or 60 degrees, but not smaller than 35 degrees. Catenary idlers were just being developed when the SUNCOR plant was being built, and they will probably see more application in the future. Idlers are simpler, easier to install, and permit deeper troughing angles than the belts.

A typical, nondragline operation has several benches separated by about 275 m (Fig. 1-5). Each bench contains its own bucket wheel, belt wagon, cable car, and face conveyor. The face conveyors are shiftable and join the trunk or plant-feed conveyors at a transfer

---

*All dollar amounts are Canadian.

FIG. 1-3 Bucket-wheel excavator, belt wagon, transfer station, and cable-reel car working on tar sands—open-cast face. Bench conveyor in foreground.

station. Either one or several conveyors can feed a plant conveyor, which should be doubled for reliability and flexibility. The number of benches is set by several considerations:

1. Cost and availability of large BWEs.
2. How far the operation can be stretched out. Three benches mean at least 800 m between the face of the first bench and the conveyor of the last bench. This may present formidable problems if one wishes to build an in-pit dike behind the mining operation.
3. Depth of the ore body.

FIG. 1-4 Overburden stripping using hydraulic shovel and truck.

FIG. 1-5 Overall view of both mining benches, showing excavators and conveyors. Box cut for lower face in foreground.

Mining in the winter, at temperatures as low as -46 °C, presents some difficult problems which have been mostly overcome. Equipment has to be made of material capable of withstanding these temperatures, and the oil sands must be preblasted to reduce digging forces and tooth wear. Overburden at SUNCOR is not usually mined in the winter if the material is to be used in building dikes. Frozen soil is not a stable building material. Muskeg is often stripped in the winter when it is most workable. Refer back to Fig. 1-1.

Here are some problem areas that still give trouble, and great care must be taken in their treatment:

1. Trucks have a difficult time. The soil is soft, and gravel is not plentiful in the area. Frames tend to wear out faster than is normal. Conveyors and stackers, if they are feasible, provide a less troublesome alternative but at higher cost and long delivery. It is still difficult to compact dike material by using a stacker.

2. Surge piles have annoying characteristics. They work fine in the summer and become a frozen mass in the winter. Reclaimers for surge piles need high specific digging forces which are used only about half the time. An alternative is blasting. When GCOS was designed, surge piles were contemplated. Although their benefits were obvious, their design presented formidable problems. They were therefore avoided, but, surprisingly, the system worked.

3. The design of the conveyor system requires a great deal of attention. Use only steelcord, oil-resistant belt. Detailed design specifications are available at a price.

4. If the lease is small or is covered with mineable ore, then all of the lease area (or most of it) will be dug up. A long-range mining plan is essential. One must plan for the initial

impoundment of the tailings and runoff and, if necessary, go off-lease. If in-pit ponds are required, lay them out well in advance. The plan always can be modified in the light of experience.

5. Provide an adequate coring program which will give not only the bitumen content but also the clay mineral content, particle size, and geological data. Build up a three-dimensional picture of the property.

6. Flexibility in a large earth-moving project like this is desired by everyone but the accountants; there is never enough. The other side of the coin is that the cheapest system is one that gives the fewest options, and this may be better than offering too many choices. It also discourages "high grading." If a bucket-wheel and conveyor system is chosen, then the designers will be forced toward a relatively inflexible layout. Basically the ends of the mine face are fixed and one digs what is in front. Long parallel faces with box cuts are common here; slewing is possible but difficult.

7. Remember the weather. In north central Alberta it is cold in the winter and the season is long. In other areas weather may not be such a problem, but it should never be ignored.

The Hot-Water Extraction Process

In 1981 the accepted method of separating the bitumen from the sand matrix was by the hot-water process. Two plants representing an investment of over \$3 billion (Canadian) are in operation. Several more are planned.

The chemistry and thermodynamics of the hot-water process

The chemistry of the hot-water process is basically the chemistry of clay. First it is necessary to deflocculate the clay, which makes the bitumen easier to float. Water-soluble constituents that have been found in the bitumen[6] may lower surface tensions at the bitumen-water interface, but no positive benefits from these compounds has been found.

Bowman[7] discusses the chemistry and thermodynamics of the bitumen in the hot-water process by looking at interfacial tensions as they affect the free-energy changes. Since data in this area of thermodynamics are sparse, he had to make some simplifying assumptions. He concluded that the desired transformations require a high temperature and high pH except for the initial attachment of gas bubbles to bitumen particles. This agrees with everyone's observations, but the scientific data to flesh out these observations is yet to come.

In his review paper, Bowman[8] notes that measurable quantities of organic material remain in the inorganic matrix even after toluene extraction. This is true even after several kinds of extraction, and the organics can be completely removed only by severe chemical treatment which destroys them. Although organic content is variable, the results from several hundred treatments give the average quantity of organic compounds attached to the inorganic matrix as 3.5 percent by weight of the clay mineral. These organics are not recoverable by any known process; nevertheless, they comprise a substantial portion of the total weight of the ore body.

The hot-water extraction process was first described by Clark in about 1930[9] and amplified in 1944.[2] Clark discovered that the sand is water-wet and that the oil is essentially isolated from it by a thin film of water. This discovery indicated the possibility of mixing

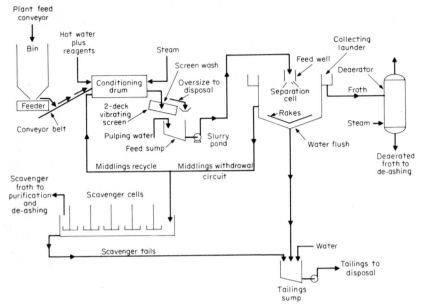

FIG. 1-6 Process flow chart for primary extraction of tar sands.

the sand with water and floating off the oil—and this possibility was realized. Results improved when the mass was heated, which lowered the density and viscosity of the bitumen. Various alternative combinations of water, solvents, and surface-active agents have been tried as recovery processes, but they do not have the simplicity of the hot-water process and have never gained much attention.

Several hot-water-extraction demonstration plants were built by ABASAND in 1935 (destroyed by fire and rebuilt in 1942), by the Alberta Research Council at Bitumont in the years after the World War II, and by Cities Service in the 1950s. These plants demonstrated that the hot-water process was feasible and that bitumen could be recovered in satisfactory yields. Unfortunately, no design data came out of these plants nor was the equipment used in them suitable for large-scale operations (Cities Service data were confidential and were never released.) When GCOS decided to build a commercial plant in 1962, a pilot plant had to be built and run for about two years to obtain the data necessary for scale-up. The process that developed from this work is shown in Fig. 1-6; it represents current commercial practice. First, we will review the numbers to show the task facing the designers of the first plant.

Input

The specified capacity of the plant was 42,500 bbl/day (6757 m³/day) of syncrude. Since the conversion of bitumen to syncrude is 0.75, about 57,000 bbl/day (9062 m³/day) of bitumen was required. Normal operating factors raised this to 65,000 bbl/day (10,334 m³/day), which was the design output. The average grade of the tar sand (lease average) was 12 percent by weight, so that 1 ton of tar sand fed to the plant contained 0.68 bbl (0.108

m^3) of tar. The expected recovery was 88 percent, and so 1.67 tons (1.515 Mg) of tar-sand feed was required to produce 1 bbl (0.159 m^3) of bitumen. To produce 65,000 bbl (10,334 m^3) required 108,550 tons/day (98,466 Mg/day) of tar sand.

One seldom sees an average feed, so a range based on the mine limits is specified. At GCOS this was 8 percent by weight minimum. Table 1-3 shows the leverage working with these limits to produce 65,000 bbl/day (10,334 m^3/day) of bitumen. Note that between the top expected grade and the bottom expected grade, the tonnage throughput increases by a factor of 2. It is not feasible to design a production plant to accommodate such a wide range of feed rates. Both the Syncrude and the GCOS plants were designed to process the feed with three operating lines; a fourth complete line was available for maintenance and to supply excess capacity for short periods as needed. The design capacity of each GCOS line was 1500 tons/h (1361 Mg/h), which accommodates the 12 percent feed. If the fourth line is added, then a feed of 9 percent can be handled for a while, but not on a permanent basis. A feed of 16 percent bitumen requires only two lines.

The flow of material shown in Fig. 1-6 may be traced as follows. Only one line is shown; others would be identical.

1. Run-of-mine feed is introduced into the plant feed bin. One large bin serves all four lines. Withdrawal is achieved by four "pant legs," each equipped with heavy-duty solids feeders. The material is dumped onto a conveyor belt which feeds the conditioning drum.

   The purpose of the feed bin is to convert one or two mine feed lines into four separate flows. Other solutions are possible and may be more desirable, since the GCOS bin is about as large as one would wish to design. It is a mass-flow bin, and the withdrawal of the material is controlled by the speed of the feeder underneath.

   Some hang-ups in the bin have been encountered, mainly when the material consolidates within the bin. Air permeation and air blasts are used to keep the material moving and to break up arches.

2. The bin material, now divided into four equal parts, is fed to a processing line. The first piece of equipment that it encounters is the conditioning drum; here the material is heated to the proper temperature in the presence of sodium hydroxide. In essence,

TABLE 1-3 Bitumen content vs. raw feed for 65,000 bbl/ day (10,335 m^3/day) production

Bitumen Content		Production,	Raw Feed	
wt %	bbl/ton	tons/bbl	tons/day	tons/h
8	0.454	2.50	162,000	6770
9	0.511	2.22	144,300	6013
10	0.567	2.00	130,000	5417
12	0.681	1.67	108,550	4523
14	0.794	1.43	92,950	3873
16	0.908	1.25	81,250	3385

tons = short tons (2000 lb) of tar sand = 0.9071 Mg; bbl/ton $\times$ 0.1752 = m^3/Mg

we condition the ore so that we can separate the bitumen from the mineral. The conditions are:

Temperature, 160 to 190 °F (71 to 88 °C)
Residence time, variable, less than 8 min
Pulp consistency, 70 percent solids
Reagent dosage, variable with amount of clay, from 0 to 200 ppm
pH, greater than 8 and less than 10

The purpose of the reagent is to deflocculate the *in situ* clay.

Conditioning drums are sized to achieve a minimum residence time while taking into account the temperature of the ore, the degree of mixing or the time required to minimize the amount of tar sand rejected to oversize, and the time needed to raise the mass to the required temperature. A set of curves plotting the bitumen recovery vs. residence time then determines the drum size.

The conditioned ore now is discharged over the drum weir onto a two-deck vibrating screen with the holes set at about 2 in (51 mm). Wash water is added here as a spray. Oversize is rejected to a conveyor for disposal; undersize drops into a sump where additional water is added, if necessary, to produce a pulp having a solids content of 50 to 55 percent by weight. This pulp is then sent to the second major piece of equipment, the separation cell, for the recovery of bitumen and discarding of the mineral matrix.

A separation cell is a large tank which, as noted, separates the bitumen from the minerals. For this process, the feed is introduced below the surface via a feed well. If all is well, the bitumen rises to the top and is discharged over peripheral weirs into a collecting launder whence it flows to a deaerator. The mineral material sinks to the bottom, where it is discharged to a tailings sump. Water can be injected into the bottom to give a hydroseparator effect and displace middlings in the tails.

There are three areas or volume sections in the separation cell.

1. *Top*   This contains the floated bitumen.
2. *Middle*   This is the middlings section. It is a mixture of clay, water, and bitumen with a clay to water (C/W) ratio of about 0.1.
3. *Bottom*   The heavy tails are discharged here.

A separation cell resembles a classic thickener but is designed for considerably heavier duty; it is required to handle up to 1800 tons/h (1633 Mg/h) of sand. It is a Stokes law device sized on an area basis to accommodate feed measured in tons per hour per square foot (tonnes per hour per square meter).

Because the separation cell is a Stokes law device, the mineral must settle within the time frame allotted by the design. Accordingly, the viscosity of the middlings must not exceed a certain value. Economic considerations in the design of the vessel dictate a maximum middlings viscosity of 5 cP (0.005 Pa·s). This control is achieved by withdrawing the middlings and/or diluting the feed with water.

If the middlings are not removed, the only place for the material to go is out with the tails, thus creating undesirable bitumen losses. Recycling the middlings without clay removal will build up the viscosity of the middlings to such levels that the sand cannot

TABLE 1-4 Product compositions, percent by weight (air-free)

Source of product	Bitumen	Minerals	Water
Separation cell	55.1	15.0	29.9
Scavenger cell	8.9	14.2	76.9
Combined	34.1	14.6	51.3

sink; then the cell has to be dumped and restarted. This can occur in a surprisingly short time.

Treatment of the middlings has received considerable attention. SUNCOR sends them to a flotation circuit where the oil is recovered and the tails are used to dilute the sand pumped to disposal. Middlings contain 1 to 2 percent by weight of bitumen, but the volumes are considerable (3000 to 5000 gal/min; 11.4 to 18.9 m³/min). A fairly long residence time (about 15 min) in the flotation cells is required to get about 85 percent recovery. If oil is sold at world prices, the volume is justified, and flotation cells having volumes greater than 1200 ft³ (34 m³) are now readily available.

The product of this operation, which pays for everything, is bitumen. Table 1-4 shows a typical product analysis. These data show that there is still considerable mineral and water mixed with the bitumen. Most of it must be removed by additional purification before being converted to synthetic crude oil. As made, the froth contains considerable air and is a compressible fluid. It cannot be pumped by normal means. Both plants using the hot-water process remove the air by steam deaeration and make the froth suitable for pumping.

Hot-Water Extraction Material Balances

Table 1-5 gives a product balance for two different feeds and two sets of operating conditions. Total material balances are given in Tables 1-6 and 1-7. It can be seen that the recovery of bitumen is sensitive to the feed composition and the drag rate. Without a scavenger circuit the recoveries of bitumen would be in the 75 to 78 percent range, and even lower with more fines. The balances in Tables 1-5, 1-6, and 1-7 are based on the assumption that the system is under control and that sufficient caustic is added to deflocculate the clay and maintain stokesian conditions in the cell. If these are not achieved, then dilution water must be added to maintain the cell in an operating condition. Various recovery equations have been proposed and are used. They are, however, specific to particular plant configurations and are generally determined by the statistical analysis of pilot plant data. They are useful for planning. Tables 1-6 and 1-7 give complete material balances for two types of feed. Figure 1-7 shows the stream numbers and the designations. Numbers given in the material balances are typical of these feed grades. Note that in the high-fines case the recovery is much worse and that the middlings drag rate is three times what it is in the low-fines case.

Final Extraction

The product from primary extraction contains 7 to 12 percent mineral and 40 to 60 percent water. Both values are unacceptable for a feed to other processing units. Further purification is necessary, and it is achieved by a two-step centrifuging process. Many processes

TABLE 1-5 Material balance; feed rate = 2000 t/h (1814.2 Mg/h)

Fines	12% bit., 15% fines	12% bit., 15% fines	9% bit. 25% fines	9% bit. 25% fines
Flush water I, gpm	0	1666	0	1666
Middlings drag I, gpm	0	2776.6	2362.9	3812.4
Feed	2000			
Bitumen	240	240	180.0	180.0
Mineral	1700	1700	1720.0	1720.0
Water	60	60	100	100
Total	2000	2000	2000	2000
Froth (primary)				
Bitumen	220.1	209.0	138.3	138.3
Mineral	22.4	21.2	27.7	27.7
Water	130.6	123.9	110.0	110.0
Total	373.1	354.1	276.0	276.0
Tails				
Bitumen	15.1	6.7	14.0	0.7
Mineral	1586.4	1369.1	1354.9	1240.8
Water	753.6	644.6	644.2	584.3
Total	2355.1	2014.4	2013.1	1825.8
Scavenger froth				
Bitumen	0	16.4	14.6	18.5
Mineral	0	19.5	18.7	20.6
Water	0	62.9	60.3	68.6
Total	0	98.8	93.6	107.7
Scavenger tails				
Bitumen	0	9.1	1.7	9.5
Mineral	0	199.1	144.7	219.6
Water	0	673.4	490.8	748.7
Total	0	881.6	637.2	977.8
Drag rate I, gal/t	0	117.5	100	163
Vol. of middlings drag vs. gpm	0	3917	3335	5422
Vol. of water required				
Igal/tTS	201.7	212.7	196.7	196.2
t/tTS	—	—	—	0.818
recovery, %	91.7	93.9	84.9	87.1

Scavenger volume set at 4 ft³/tTS·h

tTS = short tons of tar sand

t = short tons (2000 lb)

bit. = bitumen

Igal = imperial gallons

TABLE 1-6 Primary extraction material balance; feed = 2000 t/h (1814.2 Mg/h), 14% bitumen, 9% fines (very good feed)

Stream number	Bitumen, t/h	Mineral, t/h	Water, t/h	Total, t/h	Solids, %	Notes
1	280.0	1700.0	20.0	2000	—	
2	0.6	6.7	665.0	672.3	—	
3	0	0	72.0	72.0	—	Saturated steam
4		Caustic as required	—	—		
5	280.6	1706.7	757.0	2744.3	62.2	
6	0.2	1.9	190.0	192.1	—	
7	0.3	14.4	0.7	15.4	—	
8	280.5	1694.2	946.3	2921.0	58.0	
9	0	0	0	0	—	
10	1.4	160.3	908.2	1069.9	—	
11	281.9	1854.5	1854.5	3990.9	46.5	
12	1.7	129.9	845.4	977.0	13.3	
13	268.8	20.6	124.1	413.5	5.0	
14	1.4	160.3	908.2	1069.9	15.0	
15	4.8	1658.4	845.4	2508.6	66.1	
16	8.6	145.1	822.2	975.9	15.0	
17	7.8	43.9	248.6	300.3	14.6	Middlings drag rate is 107 gal/tTS
18	0.8	101.2	573.6	675.6	15.0	
19	0	0	0	0	—	
20	0.8	27.7	156.6	185.1	15.0	
21	7.0	16.2	92.0	115.2	14.1	
22a	0.3	2.6	261.5	264.4	—	
22b	0.6	5.6	559.8	566.0	—	
23	5.7	1666.6	1666.7	3338.9	49.9	
24	275.8	36.8	216.1	528.7	6.0	Recovery is 98.5% air-free basis

All values are in short tons per hour; to convert to tonnes, multiply values by 0.9071.

were proposed for this step, but the only one that worked reliably was the process used. It is relatively simple, although it may be mechanically complex. Basically, the froth from the primary extraction is deaerated, diluted so the ratio of diluent to bitumen is 0.4:1, and then passed through a two-stage centrifugal cleaning process. A process schematic is given in Fig. 1-8, and a material balance is given in Table 1-8 for a unit feed of 1000 ton/h (907.1 Mg/h).

Anyone confronted with such a simple process would think there must be some tricks making it work properly, and so there are. Without revealing all, these points are important:

1. It is very necessary to choose the right rotating machinery. Not just any set of centrifuges will do.

2. Having arrived at a choice of machinery, it is necessary to operate it under the right conditions.

TABLE 1-7   Primary extraction material balance; feed = 2000 t/h (1814.2 Mg/h), 6% bitumen, 40% fines (very poor feed)

Stream number	Bitumen, t/h	Mineral, t/h	Water, t/h	Total, t/h	% solids	Notes
1	120.0	1700.0	180.0	2000	—	
2	0.5	5.0	505.0	510.5	—	
3	0	0	72.0	72.0	—	
4	As required					
5	120.5	1705.0	757.0	66.0	66.0	
6	0.2	1.9	190.0	192.1	—	
7	2.4	120.0	17.3	137.7	—	
8	118.3	1586.9	929.7	2664.9	59.6	
9	1.3	12.8	1283.0	2317.3	—	
10	0	0	0	0	—	
11	119.6	1599.7	2212.7	3932.0	40.7	
12	4.5	105.3	548.4	658.2	—	
13	38.6	7.7	30.9	77.2	—	
14	0	0	0	0	—	
15	10.8	1266.5	548.4	1825.7	69.4	Separation cell tails
16	74.7	430.8	2181.8	3013.3	14.3	Mids drag is 329 gal/tTS
17	56.0	62.1	316.3	434.4	14.3	
18	18.7	368.7	1865.5	2252.90	16.4	
19	18.7	368.7	1865.5	2252.9	16.4	
20	2.8	33.5	170.8	207.1	16.2	Discard to tails
21	53.2	28.6	145.5	227.3	12.6	
22a	0.1	1.4	138.5	140.0	—	
22b	0	0	0	0	—	
23	32.4	1670.1	2723.2	4425.7	37.7	
24	91.8	36.3	176.4	304.5	11.9	Recovery = 76.5%

All values are in short tons per hour; to convert to tonnes, multiply values by 0.9071; tTS means tons of tar sand.

3. It is absolutely necessary to have the scroll (or solid bowl) centrifuge first in line, followed by the disk machines.
4. It is standard practice to protect the disk machines with input filters.
5. Tailings are collected and pumped to disposal.

This process yields, on the average, about 99 percent recovery, and it was designed to produce a product having 0.5 percent solids by weight. These requirements are met routinely. Aside from normal maintenance requirements, the process gives little trouble. Sometimes the mine sends considerable quantities of fibrous materials to the plant, and then cleaning the filters becomes a troublesome chore. This is not a frequent occurrence, however. Wear problems have been identified and reduced to manageable proportions.

TAILINGS DISPOSAL

The disposal of the tailings and other waste streams from the hot-water process is an enduring problem. To produce a feed for a hot-water tar-sands-processing plant involves

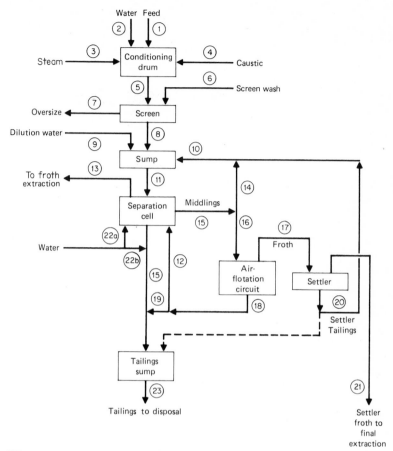

FIG. 1-7 Material balance. Stream numbers correspond to those in Tables 1-6 and 1-7. Line 14 is middlings recycle and may not be required. Streams 22a and 12 may be combined. Streams 13 and 21 are combined and sent to deaeration.

setting up one of the largest earth-moving projects anywhere. When the sand has been processed, there is a large gain in volume of waste products, viz:

Overburden volume gain 5 to 10%   Swell factor 1.05 to 1.10

Tar-sands volume gain 40%   Swell factor 1.4

Thus, for every cubic yard of tar sands mined, 1.4 yd³ of waste must be disposed of. Further, this is not cheap. The sums are of the size that management wants to see spent through a complicated investment decision process, but here there is no "do-nothing case." The problem can be controlled, but it can never be ignored. Its severity is set by the space and volume available to store the tailings, runoff, and sludge and by how quickly the sludge

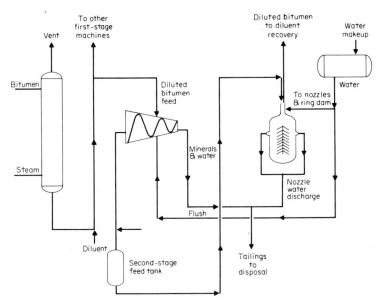

FIG. 1-8 Final extraction or bitumen demineralization.

TABLE 1-8 Material balance final extraction; 1000 t/h total feed

Stream	Bit.	Min.	Water	Diluent	Total
Primary froth	612	78	310		1000
Steam			40		40
Diluent				395	400
Scroll feed	612	78	355	395	1440
Scroll product	607	32	346	393	1378
Scroll tails	5	46	9	2	62
Disk feed	607	32	346	393	1378
Recycle water	3	10	294	3	310
Ring dam water	3	10	302	3	318
Nozzle water	8	27	286	5	326
Final product	599	5	52	388	1044
Combined tails	13	73	295	7	388
Ring dam overflow	TR	TR	8	TR	8

Values are in short tons (2000 lb); to convert to tonnes, multiply by 0.9071.

TR = trace

is made as well as how quickly it consolidates. It is desirable to predict sludge-make as well as consolidation rates, but no accurate method for doing that exists.

We may classify the tailings disposal as follows:

1. Transportation
2. Separation of the sand from water
3. Storage of the wet sand
4. Handling the runoff from the dewatering process
5. Lease abandonment and reclamation

Transportation of Tailings

Tailings from the hot-water process average 40 to 50 percent solids by weight. Most of this is sand, but a small portion that varies with the feed is silt and clay. Roughly, the silt-to-water ratio is about 0.1 and the clay-to-water ratio is 0.05, so that the combined ratio of silt plus clay to water is 0.15. These very fine materials make the pumping of the tails easier, and the higher the ratio of clay plus silt to water, the easier it is to pump the tails.

Since the tailings from the process are already in the form of a sand, clay, silt, and water slurry, they are handled as they come and are pumped directly to disposal. Large neoprene-lined slurry pumps are used with gear reducers and variable-speed drives. Up to six stages of pumping are needed, at present, to overcome friction losses and static head. Line size is 20 in nominal, and each line handles about 10,000 U.S. gal/min (37.9 m³/min). The longest line currently in use is about 15,000 ft (4572 m or 4.572 km), but runs of 10,000 ft (3048 m) with static heads of 190 to 200 ft (58 to 61 m) are common. Four lines are operated and a fifth is used as a swing line (since expansion). Switching from one line to another is achieved by a distributor located in the final tailings pump house. Currently about 80,000 ft (24.4 km) of tailings lines are installed.

As with any slurry-pumping process, the design is so set that the mixture flows at velocities just above the deposit velocity. This minimizes pressure drop but requires adequate control. Published correlations for deposit velocity and pressure drop should not be used; instead, test runs should be conducted in a suitably sized test loop (at least 20 in IPS or 508 mm). This work has been done, and the information is used to design all of SUNCOR's and Syncrude's tailings systems.

Disposal of Sand in Dikes

The slurry from extraction is pumped to a disposal area which can be off-site or in a mined-out area. The sand is used to build dikes to contain the runoff. Using sand to build containment dikes is a relatively new technique first introduced by SUNCOR and its Geotechnical Consultants. Previously, sand was felt to be an unsuitable construction material, but the large quantities available from a hot-water tar-sands extraction process made it necessary to develop techniques to use sand as a dike-building material.

Dikes are built by the cell method. First, sand is pushed up at the edge of the cell to form a temporary dike; a drain box or weir is installed at one corner. Then slurry is run into the cell, where the sand settles and the water drains off through the drain box. As the sand is settled, it is compressed with earth-moving equipment to expel more water. Some of the finer sand escapes this process and is deposited on a beach below the drain box.

Dikes built in this way have void ratios of 0.67 to 0.70 in the compacted section and about 0.80 to 0.85 in the beach. Each cell is raised about 8 to 10 ft (2.44 to 3.05 m) in one operation (called a lift). About 75 percent of the sand goes into the compacted dike and 25 percent escapes to the beach. Each cell is built on a 1:1 outside slope with a maximum height set at 30 ft (9.15 m), or three lifts for this slope.

When a cell is complete, another is formed adjacent to the preceding cell and the process is repeated. In this way one moves down the dike to complete a lift and then goes back to the beginning and starts over. About 5 to 6 weeks must elapse to allow the water to drain before a lift is put on top of a preceding lift.

In the Athabasca region, dikes are built only in the months with good weather, May to the end of October. In this time about three lifts, or 30 ft (9.14 m) of dike, with a 1:1 slope on the upstream side are built. This is not a stable arrangement, so during the winter months sand is overboarded off the edge of the newly built dike sections to stabilize the arrangement. This sand lies quite flat (about 7 to 10 percent slope), and the angle the beaches take depends upon the average particle size of the sand being discharged at that particular time.

The dikes and beaches are quite stable, and SUNCOR has built dikes to 300 ft (91.4 m) by using the technique described here (Fig.1-9). Safety factors exceed 1.5. Dikes are also built of overburden, but this material is more variable. Slopes for overburden run from 3:1 to 4:1 on both upstream and downstream sides (Fig. 1-10). It is sometimes necessary to stabilize an overburden dike on the upstream side by overboarding tailings sand against the face.

The Tailings Ponds

Water runoff from the dike-building operation is collected inside an area confined by the tailings dikes. Sand settles rapidly, the silt more slowly, and the clay more slowly still. The rate of sedimentation is determined by these factors:

1. Impurities dissolved in the water
2. The amount of organic materials adsorbed on the clay
3. The mineralogy of the clay and silt fraction
4. The particle size of the minerals

It has been already noted that, in the tailings, the ratio of clay and silt to water is about 0.15. As discharged to the tailings ponds, this ratio is about 0.10 and the clay fraction is about 0.05. The clay fraction undergoes a slow sedimentation in this way:

1. Small clay particles approach each other until they can act as a unit and then sediment. Elapsed time is 10 to 40 days. Clay-to-water ratio is 0.1.
2. The particles sink to an intermediate layer, called a transition zone, where water is slowly expelled (or, conversely, the particles get closer together). After 4 to 5 years, the clay-to-water ratio is 0.10 at the top and 0.25 at the bottom.

FIG. 1-9 View of Tar Island dike looking north.

FIG. 1-10 Plant view, looking east. From left to right: refinery, cokers, utility building, extraction unit, and tailings pump house. Foreground shows the east-west dike, being build from overburden, in the process of construction; its slope is 3.5 to 1.

3. The particles then enter an equilibrium, or consolidation, zone (or else the processes going on above them force them into this zone), where the average clay-to-water ratio is 0.3. At the bottom of this zone the solids contents increase because of sand inclusions from the overboarding process, but the clay-to-water ratios remain remarkably constant. About 80 percent of the volume of the SUNCOR tailings ponds is filled with this material, and the volume is approaching 2 billion ft^3.

From numerical solutions to the consolidation equations it has been estimated that the time required to consolidate this mass to about half its present volume will take longer than 1000 years. It would seem that it should be possible to consolidate the material much faster, but 15 years of effort have convinced SUNCOR that the problem is very intractable. Further, SUNCOR is not alone, similar problems exist in the phosphate, bauxite, and other industries. We are, then, dealing with a natural phenomenon that has universal application but is compounded in the tar-sands industry by the tremendous quantities involved.

Listed below are some approaches which don't work or are too expensive in application:

1. *Flocculate the runoff.* You get your water back more quickly; but if the pond area is large enough, this step is not needed. It does not reduce the volume of the sludge at equilibrium. (In fact, that volume may be larger.)

2. *Mix the sand with the sludge and contain the sludge within the pore volume.* This could work if there were enough sand. You need particle-to-particle contact to provide some stability, and enough sand is not available. Various techniques have been suggested and some have been tried, but they have all been difficult to implement and seem to be ruinously expensive.

3. *What about freeze and thaw?* Freezing and then thawing a clay gel breaks up the gel and separates the water from the solids. The solids consolidate rapidly. This technique is well known and it works, but there is not enough real estate to implement it. It would take several cycles. If there is under treatment at any one time about 2 billion ft^3 (56.6 MM m^3), then a 5-ft (1.52-m) depth would be adequate and about 9200 acres (37.2 × 10^6 m^2, or 3723 ha), not including dikes, would be required. This area is larger than SUNCOR's entire lease, and no more land is available.

4. *Try electrophoresis or electroosmosis.* This process works in the laboratory, but no one has succeeded in making it work on a large scale. Projected costs have been very high.

5. *Other methods.* People have tried various chemicals, physical treatments, etc., with varying degrees of success. All have failed because of cost or because they did not work.

SUNCOR's main research effort is directed toward solving this problem. Some progress has been made but much still needs to be done.

UPGRADING

Table 1-2 lists the principal properties of Athabasca bitumen typical of the Mildred Lakes area. It is easy to see that the bitumen is not a desirable feed to a standard refinery. The technology for upgrading heavy, sour, hydrogen-deficient crude oils is still in its infancy. However, what technology does exist may be divided into two types:

1. Coking the material
2. Hydrogen processing

In 1962, GCOS (SUNCOR) elected to use delayed coking, a well-known and proven process. Yields of light material average about 75 percent of the bitumen fed to the coker. The coker products are divided into three fractions which are hydrotreated separately to remove sulfur, and these are then combined to make synthetic crude oil. The properties of this material are given in Tables 1-9 and 1-10.

TABLE 1-9 SUNCOR synthetic crude

Cut pt, °F	Cumulative vol crude, %	Cut vol, %	Cut pt, °F	Cumulative vol crude, %	Cut vol, %
50	0.7	0.7	400	31.8	0.0
100	3.4	0.0	420	34.2	0.0
140	5.6	0.0	450	38.2	9.2
145	5.8	5.1	465	40.7	0.0
150	6.3	0.0	480	43.0	0.0
155	6.7	0.0	500	45.9	0.0
160	7.0	0.0	525	49.4	0.0
170	7.8	0.0	550	53.1	0.0
180	8.8	0.0	575	57.0	0.0
190	9.7	0.0	600	60.8	0.0
200	10.7	0.0	625	65.2	0.0
205	11.1	0.0	650	69.5	31.3
210	11.7	0.0	675	73.9	0.0
215	12.1	0.0	700	77.9	0.0
220	12.6	0.0	725	81.6	0.0
230	13.7	0.0	750	85.4	0.0
235	14.1	0.0	775	88.8	0.0
240	14.6	0.0	800	91.7	0.0
250	15.7	0.0	825	94.0	0.0
260	16.7	0.0	850	95.9	0.0
265	17.2	0.0	875	97.3	0.0
270	17.7	0.0	900	98.4	0.0
280	18.9	13.1	925	99.0	0.0
290	19.9	0.0	950	99.4	0.0
300	21.0	0.0	975	99.6	0.0
310	22.1	0.0	1000	99.8	30.3
320	23.0	0.0	1025	99.9	0.0
340	25.2	0.0	1050	99.9	0.0
360	27.2	0.0	1070	99.9	0.0
375	29.0	10.1	1100	99.9	0.0
380	29.5	0.0			

Table 1-10 SUNCOR synthetic crude oil

Crude oil properties			
Gravity, °API	34.7	33.4	35.1
Sulfur, wt %	0.19	0.20	0.20
Salt, lb/MM bbl	2.	2.	2.
Vis., SUS at 77 °F	37.9	40.4	37.0
Vis., SUS at 100 °F	35.0	36.7	35.0
K factor	11.70	11.77	11.71
Flash pt., °F	70.0	40.0	40.0
Pour pt., °F	−32.0	−55.0	−35.0
BS&W, VP	0	Trace	Trace
OD color	74.	53.0	16.0
Reid VP at 100 °F	2.4	1.6	3.1
TAN, mg KOH/g	0.03	0.05	0.06
Acid gases			
H_2S, wt % dissolved	0.0010	0	0
RSH, wt % dissolved	0.0002	0	0
Fresh breakdown of crude oil (MH), vol %			
C_4 and lighter	0.7	0.4	1.8
Debut str run gas	5.1	4.0	3.8
(F-1 clear BON)	73.4	72.6	72.7
Re-former charge stock			
BTX	13.1	10.8	9.4
Motor	10.1	10.2	12.0
Fresh heavy naphtha	9.2	10.2	11.0
Heavy gas oil (TSO)	60.2	62.6	60.9
Vacumn tower distillate	1.4	1.4	0.9
Vacumn tower bottoms	0.2	0.4	0.2
Long-range go, °F	450–1000	450–900	450–1015
Vol % crude oil	61.5	62.6	61.9
BPGC	186.4	188.4	188.7
K factor	11.58	11.45	11.43
Sulfur, wt %	0.31	0.35	0.28
Nitrogen, wt %	0.0736	0.0649	0.0520
Aromatics, wt %	48.4	46.9	44.7
Residuum, °F	1000+	900+	1015+
Vol % crude oil	0.2	1.8	0.1
Vis, SUS at 210 °F		101.3	
Sulfur, wt %		0.57	
Nickel, ppm		1.	
Max. pot. aromatics, vol %	145–280	145–280	145–265
Benzene	0.25	0.20	0.19
Toluene	0.88	1.05	0.84
Xylene	0.99	0.87	0.76
Total	2.12	2.12	1.79

Syncrude followed a similar pattern in choosing fluid coking (which gives better yields of light products) and somewhat different hydrotreating processes. It does have more yield of synthetic crude oil per barrel of bitumen charged than SUNCOR obtains.

Other processes are available. H-Oil® and hydrovisbreaking appear to work, but they have not as yet been chosen for upgrading the bitumen. H-Oil probably gives the most satisfactory range of products suitable (at present) for North American refineries. Hydrovisbreaking upgrades the bitumen to a pipeline grade of material which can be further processed elsewhere.

All processes depend upon Alberta's supplies of abundant (still!) natural gas. When these supplies vanish, natural gas will still be available from the Arctic finds if a pipeline is built to bring them south. Alberta also has abundant supplies of coal which could be used to make hydrogen, although at considerably higher cost. At any rate, one can look for increased costs of treating Athabasca bitumen in the future. Improved processes, tailored to its special properties, are needed to upgrade the bitumen.

ECONOMICS

In times when oil is politics and inflation a way of life, economic calculations have little meaning. Furthermore, the first cost of a plant is not the only capital required. Mining equipment must be replaced periodically; conveyors must be lengthened at $1500 to 2000 per linear foot ($4900 to 6600 per meter), and there are large expenditures for earth-moving equipment to build roads and maintain drainage. A considerable fleet of vehicles must be maintained, and the sand causes considerable wear on all moving parts.

To build a tar-sands plant, considerable up-front work is required to obtain the necessary permits and settle the royalty costs, the tax rate, and the depreciation rates. Canada's laws differ from those of the United States. Depreciation is 30 percent declining balance on plant; there is a reserve depreciation, and projects of this size and scope may be entitled to special incentives. Research into oil sands is entitled to special write-offs.

When all of the underbrush is cleared away, one can start on the actual design. It takes time to spend this kind of money, and normally 5 years is required from start to completion. The capital can be broken down as shown in Table 1-11. The cost of building a new town is not included in Table 1-11. Both SUNCOR and Syncrude enlarged existing facilities at an existing town site (Ft. McMurray).

TABLE 1-11 Capital cost distribution

	% share of total
Mining, overburden, conveying	51
Extraction	12
Upgrading	14
Utilities	5
Service dept.*	18

*Includes buildings, warehouses, roads, bridges, pipelines, etc.

TABLE 1-12 Economic analysis ($\times 10^6$ except as noted)

Year	Capital	Production 10^6 bbl	Price, $/bbl	Revenue	Operating cost	Crown royalty	Incentive	Tax PGRT	Income	Cash flow
1981	500						(50)			(450)
1982	550						(55)			(495)
1983	605						(60)			(544)
1984	1331						(133)			(1198)
1985	1464						(146)			(1318)
1986	805	18.2	70.49	1,286	805	129	(80)	38	6	(417)
1987		40.2	78.95	3,170	1,063	317		169	0	1,621
1988		40.2	88.43	3,550	1,169	355		190	0	1,836
1989		54.8	99.04	5,422	1,715	542		297	623	2,245
1990	1179		110.92	6,073	1,886	607		335	905	1,161
1991			124.23	6,802	2,075	680		378	1,486	2,183
1992			139.14	7,618	2,282	762		427	1,826	2,321
1993			155.84	8,532	2,511	853		482	2,084	2,602
1994			174.54	9,556	2,762	956		544	2,368	2,926
1995	1899		195.48	10,703	3,038	1070		613	2,044	2,039
1996			218.94	11,987	3,342	1199		692	2,908	3,846
1997			245.22	13,426	3,676	1343		780	3,339	4,288
1998			274.64	15,037	4,044	1504		879	3,807	4,803
1999			307.60	16,841	4,448	1684		991	4,321	5,397
2000	3058		344.51	18,862	4,893	1886		1118	3,862	4,045
2001			385.85	21,125	5,382	2112		1259	5,322	7,050
2002			432.15	23,660	5,920	2366		1419	6,094	7,861
2003			484.01	26,500	6,512	2650		1599	6,934	8,805
2004			542.09	29,680	7,164	2968		1801	7,859	9,888
2005	4925		607.15	33,241	7,880	3324		2029	7,230	7,853
2006			680.00	37,230	8,668	3723		2285	9,702	12,852
2007			761.60	41,698	9,535	4170		2573	11,081	14,339
2008			853.00	46,701	10,488	4670		2897	12,586	16,060
2009			955.35	52,306	11,537	5231		3262	14,245	18,031
2010			1070.00	58,582	12,691	5858		3671	16,088	20,274

IRR 25.0%

NPV	15%	C$5624 $\times 10^6$
	20%	1580
	25%	7
	30%	(633)

Are the tar sands a good deal? That depends on the crown royalty, tax rate, depreciation rates, scarcity of other oils, and the price customers will pay for the oil. Table 1-12 gives an economic calculation based on these assumptions:

1. Capital C$4.0 billion spent over 5 years
2. Royalty at 10 percent
3. Syncrude price (C$40) at 2 percent more than inflation
4. Canadian ownership
5. Inflation at 10 percent
6. Start-up takes 3 years

Table 1-12 shows, then, an IRR of 25 percent. Note that there is a tax incentive during the early years which reduces the negative cash flow and improves the IRR. Not all of the incentives are available if the company is not Canadian-owned. Typically, the following breakdown in operating costs holds:

Service	% of budget
Maintenance	37
Operations	50
Overhead	13

The maintenance costs of the mine and overburden operations form 23 percent of the total operating budget. Another way to look at the allocation is by departments:

Department	% of operating budget
Overburden	31
Mine	24
Extraction	8
Process	14
Utilities	7
Overheads	16

These percentages include maintenance. Note that the "front end," which classifies as overburden, mining, and extraction, accounts for 63 percent of the operating budget. These departments also include tailings disposal.

Naturally, there is a question raised about the desirability of such an investment, and it is a difficult one to answer. In the final analysis, the answer depends upon the cost of the plant and the price allowed for the products. To some degree, governmental policies will set the first and certainly the last. Alberta, with its large potential reserves, looks like the place where most of the action will be toward the end of the century.

BIBLIOGRAPHY

Author's note

There is considerable literature on the Athabasca tar sands scattered throughout the technical journals, patent gazettes, and the popular press. One work of general scope which should always be consulted is:

Camp, F. W.: *The Tar Sands of Alberta,* 2d ed., Cameron Engineers, Denver, 1974.

Further, Dr. Camp made an excellent summary of the sludge problem which was presented at the Canadian Chemical Engineering Conference at Toronto on October 6, 1976 and published in the proceedings of that conference.

In the popular press, I can recommend:

Comfort, D. J.: *The Abasand Fiasco,* Friesen, Edmonton, 1980.

Mrs. Comfort provides much interesting historical data and numerous anecdotes.

I have also limited the technical citations to the most interesting and relevant. Much unpublished data resides in the files of SUNCOR, Syncrude, and Sun Co. and is unavailable for publication.

REFERENCES

1. Phizackerley, P. H., and L. O. Scott: "Major Tar Sand Deposits of the World," *Proceedings of the Seventh World Petroleum Congress,* vol. III, 1967. Also C. P. Outtrim and R. G. Evans: *Alberta ERCB Heavy Oil Symposium,* The Petroleum Society of CIM, May 1977.
2. Clark, K. A.: *Trans. Can. Inst. Mining Met.,* vol. 47, 1944, pp. 257–274.
3. Carrigy, M. A., "Mesozoic Geology of the Ft. McMurray Area," in M. A. Carrigy (ed)., *Guide to the Athabasca Oil Sands Area,* Alberta Research Council, Information Series 65, pp. 77–101.
4. Speight, J. G., and S. E. Moschopedis: "The Influence of Crude Oil Composition on the Nature of the Upgrading Process," *Proceedings: the Future of Heavy Oils and Tar Sands,* First International Conference, Edmonton, Alberta, June 1979. (Sponsored by Unitar, AOSTRA, and DOE.)
5. Schulte, R.: "Research for a Mining Venture in the Athabasca Tar Sands," Petroleum Soc. of CIM paper No. 374039, May 7–10, 1974.
6. Moschopedis, S. E., J. F. Fryer, and J. G. Speight: "Water Soluble Constituents of Athabasca Bitumen," *Fuel,* vol. 56, January 1977, pp. 109–110.
7. Bowman, C. W.: "Molecular and Interfacial Properties of Athabasca Tar Sands," *Proceedings of the Seventh World Petroleum Congress,* vol. 111, 1967, pp. 583–604.
8. Bowman, C. W., and J. Leja: "Application of Thermodynamics to the Athabasca Tar Sands," *Can. J. Chem. Eng.,* vol. 46, December 1968, pp. 479–481.
9. Clark, K. A., and D. S. Pasternak: "Hot Water Separation of Bitumen from Alberta Bituminous Sand," *Ind. Eng. Chem.,* vol. 24, 1932, p. 1410.

EMERGING TECHNOLOGIES FOR OIL FROM OIL SANDS

CLEMENT W. BOWMAN
RAYMOND S. PHILLIPS
L. ROBERT TURNER

Alberta Oil Sands Technology and Research Authority
Edmonton, Alberta, Canada

THE RESOURCE BASE

The terms *tar sand* and *oil sand* have been used interchangeably in the literature and normally refer to bituminous sand deposits in which the asphaltic bitumen is too viscous to flow to production wells at commercially acceptable rates. However, bitumen also occurs in other mineral matrices, such as porous carbonate rocks and diatomaceous earth. Since many of the recovery processes are applicable to bitumen in various host rocks, the term oil sand will be used in this chapter to refer to any type of deposit containing bitumen with a specific gravity of less than 12° API (density = 0.986 g/mL).

The location of many of the world's heavy-oil and bituminous sand deposits is shown in Fig. 2-1. No precise figure exists on the magnitude of the total in-place resource, but estimates range from 5×10^{12} to 10×10^{12} bbl. Knowledge of the geological setting and physical chemical properties of these resources is crucial in assessing the technology required for their development. Some of the more important factors which must be considered in evaluating a deposit are illustrated in Fig. 2-2, and are discussed briefly below:

Depth

Oil sands occur in both surface outcrops and deeply buried deposits. The depth determines whether the appropriate recovery method is open-pit mining or an *in situ* technique. A rule of thumb is that open-pit mining is economical at a ratio of overburden to pay zone of 1:1, or less. This sets the depth limit for surface mining for the Canadian deposits at roughly 150 ft ($\sim$50 m). Depth also establishes the limits for various *in situ* recovery technologies, with high-pressure steam injection being applicable from 500 to 3000 ft (50 to 1000 m) and *in situ* combustion being applicable down to 5000 ft (1500 m) or more. Depth determines in large measure the temperature of the reservoir, the average geothermal gradient being about 1 °C per 100 ft.

The minimum depth required for safe injection of fluids into oil sands is not well established, but consideration must be given to the competency of the cap rock and the injection pressure required. If the overburden is too thin, the possibility exists of fluid breakthrough to the surface. Reservoirs with less than 500 ft of overburden should be approached with caution, although a recently issued U.S. patent claims success at depths less than this.[1]

Reservoir factors

In many deposits, the rock matrix is water-wet, and this permits the use of aqueous methods for separating bitumen from mined oil sand. When the host rock is oil-wet, or bi-wetted, hydrocarbon solvents may be required. Thermal cracking techniques, such as retorting, can also be employed.

For *in situ* recovery processes, the porosity and permeability of the host rock are critical. With high bitumen saturations, porosities of 20 to 30 percent are preferred for thermal approaches to ensure that the percentage heat loss to the rock is kept to reasonable limits. Permeabilities of 500 millidarcies or higher are desired to provide adequate production rates.

Many deposits have very low reservoir pressure, and external fluids must be injected to provide the drive energy.

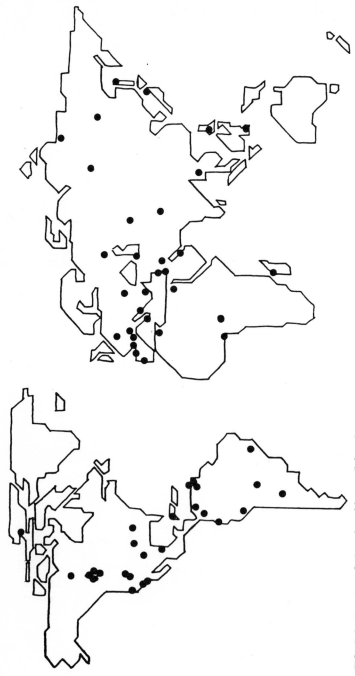

FIG. 2-1 Location of oil sands and heavy-oil deposits.

5-36

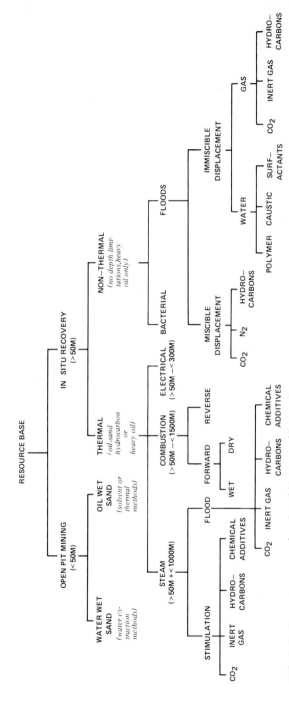

FIG. 2-2 Recovery processes for oil-sand bitumen and heavy oil.

Factors which influence the success of *in situ* processes include the presence of certain types of clays; the fineness of the sand; the location, continuity, and thickness of impermeable bands (e.g., shale stringers) within the sand; and the competence of the overburden and underburden.

Certain types of clay may swell, or otherwise alter, in the presence of steam or hot water and can restrict the flow of fluids. Very fine sand or silt can reduce the porosity and cause production problems. The presence of movable fine material needs to be considered in designing casing and completion procedures.

Shale stringers may restrict the flow of steam within the reservoir, leaving large portions of the reservoir unswept. They may also divert or change the orientation of fractures generated during the injection process.

A relatively thick, competent shale barrier in the immediate overburden will assist in containing the injected fluids and improve the chances of success of the process.

Thick water-bearing zones may make steam stimulation or steam flooding impractical, since the steam preferentially follows such zones. However, a thin high-water-saturated bottom zone (bottom water) may act as a "hot plate" and rapidly heat the overlying oil by conduction and convection. In such cases, steam would be injected at relatively high volumes at pressures below the fracture pressure. Artificially created horizontal fractures can serve the same function.

Bitumen composition and viscosity

Most bituminous deposits known to the authors are asphaltic in nature, rather than waxy. The complex naphthenic aromatic clusters result in very high viscosities, approaching 6 million cP at reservoir temperatures for some of the Canadian oil-sand deposits. The temperature-viscosity curve of the oil is one of the key variables which must be known to enable a prediction of recovery rate using mathematical models.

Once the bitumen is recovered, the properties of the bitumen will determine the appropriate upgrading approach. Most bitumens are hydrogen-deficient compared to conventional petroleums. The hydrogen-to-carbon atomic ratio determines the amount of hydrogen which must be added, or the amount of carbon which must be rejected, to produce an acceptable refinery feedstock. Bitumens frequently contain high levels of heteroatoms, such as sulfur, oxygen, nitrogen, vanadium, and nickel, and these must be removed in the upgrading process.

Most bitumen upgrading processes produce a coke or pitch fuel which may be used to supply energy for the bitumen recovery operation, whether it be surface mining or *in situ* recovery. There are significant opportunities to integrate the energy systems for the entire oil-sand plant, and this requires careful matching of the mining, bitumen recovery, and upgrading operations. Surface-mining plants require the equivalent of between 15 and 20 percent of the feed bitumen as a source of energy; *in situ* recovery plants require the equivalent of 20 to 30 percent of the bitumen, depending on the process selected.

Extensive research and development work is underway throughout the world to develop recovery and upgrading technology applicable to a wide range of oil-sand deposits. The technical targets of these programs and the specific processes being developed to meet these targets are discussed in the following sections.

MINING

Scope of Current Mining Research

Underground mining of deeply buried oil sands has been proposed by World Oil Mining Ltd. of Calgary, but to date no fieldwork on this concept has been attempted.[2] Concern has been expressed about the high anticipated cost of underground mining and the lack of information on the geotechnical properties of unconsolidated oil sands at depth. However, underground mining may prove to be applicable for specific oil-sand formations where the hydrocarbon cannot be displaced by *in situ* methods.

Two mining-related programs are currently underway which will likely influence the course of future mining activities:

1. Overburden removal with hydraulic dredges
2. Emplacement of horizontal wells from shafts and tunnels

A horizontal-well pilot has been proposed by Gulf Canada Resources, and fieldwork may be initiated in 1983.[3]

One investigation on hydraulic dredging is described in the next section.

Overburden Removal with Dredges

In Canadian surface-mining plants, the recovery of 1 bbl of raw bitumen requires the mining of approximately 1 yd^3 of oil sand plus two-thirds yd^3 of overburden. The use of large-scale equipment for mining the oil sand, such as draglines or bucket wheels, can be optimized to reduce the mining cost to the order of US$0.50 to $0.75/yd^3 (US$0.65 to $1.00/m^3). However, handling the overburden is a complex and expensive procedure, mainly because of the need to segregate, stockpile, and handle overburden material for retaining dikes, roads, and land reclamation.

The actual costs for mining and handling the overburden materials in the Athabasca deposit may be greater than US$2.00/yd^3 (US$2.60/m^3). Improved systems for mining, conveying, storing, and reclaiming these materials would have a favorable impact on the cost of synthetic crude oil from surface-mining operations.

One method which has recently received attention is the use of hydraulic-cutter suction dredges.

Cutter suction dredges float on water, with a cutter head submerged in front of the hull (Fig. 2-3). Material that is excavated is removed through a suction line and discharged to a disposal location through a pipeline.

A plan for dredging of overburden from the Athabasca deposit has been developed by H. G. Stephenson (Mining Consultants) Ltd. of Calgary, and is illustrated in Fig. 2-4.[4] In this plan, the mining lease is divided into 2000- by 5000-ft blocks and the blocks are mined in the following sequence:

1. The overburden is mined from the first dredging pond.
2. The dredges are advanced to form the next dredging pond.

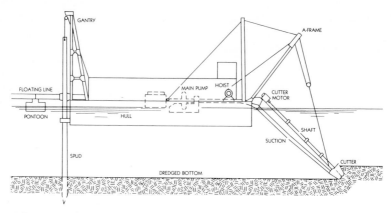

FIG. 2-3 Hydraulic-cutter suction dredges. (Hydraulic Dredging, *John Huston, Cornell Maritime Press, Centreville, Md., 1970.*)

3. The area vacated by the dredge becomes a retention pond and is used as a reservoir to supply water as the dredges advance.

4. Oil-sand mining, using draglines and/or bucket wheels, takes place after the retention pond is drained.

5 Tailings disposal into the mined-out area and revegetation complete the development sequence.

The mining plan shown ensures that no active pond is created adjacent to an area which has not been backfilled.

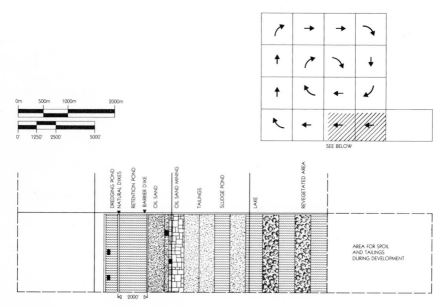

FIG. 2-4 Illustration of a plan for mining of overburden.

Dredging has the following advantages over dry-mining methods:

1. The muskeg within the overburden does not have to be predrained. Currently this must be done two years in advance of mining and when the ground is frozen.
2. A dredge is waterborne, and thus it is highly maneuverable and would have an advantage over heavy land-based equipment which operates on materials that could be unstable or exhibit poor trafficability under adverse climatic conditions.
3. Dredging is a proven, high-volume, low-cost method which can handle all types of soil conditions existing in the overburden.

Dredging has the following disadvantages:

1. When overburden containing swelling clays is being dredged, there may be environmental concerns associated with the ponds and sludge that may occur.
2. When operating with dikes that impound water above the mining level, safety considerations become a factor.
3. Dredging may have to be suspended in winter because of ice or fog formation on the ponds.
4. The method to be used for integrating dredging into mine reclamation is uncertain at this time.

A conceptual plan for a commercial-size dredging scheme has been prepared and is currently being evaluated by a number of companies.

EXTRACTION OF MINED OIL SANDS

Objectives of New Extraction Processes

One of the major problems of the hot-water process is the handling of tailings. The process uses approximately equal quantities of oil sand and water as feed materials. Large ponds are required to allow the finely divided clay in the tailings to settle before the water can be recycled to the plant. A highly dispersing chemical environment is required for effective operation of the bitumen flotation process. Unfortunately, this is precisely the condition which retards or prevents the settling of clay in the tailings ponds.

The sediment discharged into the tailings ponds stratifies into three layers: coarse sand which settles rapidly, an upper, partly clarified water layer which contains less than 5 percent mineral material and which can be recycled to the bitumen extraction plant after a few months, and an intermediate sludge layer which has a solids content ranging from 10 to 30 percent.

This sludge is very difficult to dewater by usual methods and tends to accumulate, leading to a continual increase in the volume of impounded tailings. The search for an effective, low-cost technology for sludge dewatering is being actively pursued by companies such as Suncor Resources Inc. An example of a new research activity in this area is an electrophoretic process for clay removal under development by Xana Engineering Ltd. of Calgary.[5] Detailed information on this process is expected to be released during the next few years.

The search for alternates to the hot-water process has been underway for several decades, and four classes of processes have evolved:

- Sequential water and solvent treatment
- Solvent extraction
- Oil agglomeration
- High-temperature retorting

In processes that use water and solvent sequentially, the solvent reduces the oil viscosity, increases the density difference between the oil and water, and consequently improves the efficiency of gravity separation. These processes can operate at a lower temperature and pH than the hot-water process, conditions which improve the settling characteristics of the tailings solids. This improvement must be balanced against solvent losses and concern for safety.

Solvent extraction, without the addition of water, is an excellent laboratory method of separating raw bitumen from sand. An anhydrous commercial-size solvent system which employed a perforated conveyor belt to separate the diluted bitumen from the sand was described by Cities Service Ltd. during the 1960s.[6] Economic appraisals are currently underway on the use of commercially available countercurrent solvent extractors.

Agglomeration of the oil particles, with wet screening of the agglomerates, is a concept which has been actively studied for over 50 years, starting with a small pilot unit by Fyleman in England in 1921 and continuing with the Phase Exchange process by Weingaertner in 1958, the Sand Reduction process by Imperial Oil, and the Spherical Agglomeration process by the National Research Council of Canada in the 1960s.[7] Further testing of this processing concept, using a wide range of oil-sand types, is now required.

Retorting of the total oil sands and subsequent combustion of the residual coke on the sand release essentially all the hydrocarbon material from the sand. The approach may be relatively insensitive to feed quality and is applicable to oil-wet and cemented sands. To be economically viable, however, heat must be recovered from the spent sand and used to reduce the requirements for external energy.

Specific emerging extraction processes are described in the following sections.

Dry Retorting—UMATAC Process

General process description

UMATAC Industrial Processes Ltd. has developed a direct thermal process which simultaneously cracks hydrocarbons present in oil-sand feed, recovers the liquid-oil fractions as hot vapor, and burns coke deposited on the host sand to provide the major heat requirements of the process.[8] Initial development started in 1975 with batch-scale testing. A continuous pilot plant was constructed in Calgary in early 1978, which operated as a coker during 1978 using a feedstock of reduced crude oil and then operated during 1979 and 1980 on various grades of oil sands.

The processing unit consists of a large, horizontal, rotating vessel which is divided into a series of compartments or zones so that the required functions can be conducted separately (Fig. 2-5).

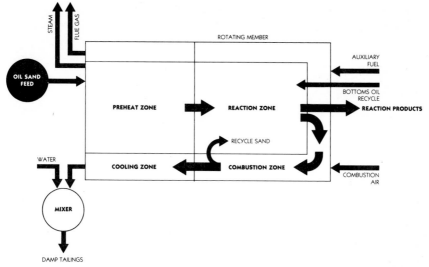

FIG. 2-5 Dry retorting of oil sands by the UMATAC process.

Oil-sand feed is introduced into the preheating section of the processor, where connate water is evaporated as steam, frozen material is ablated, oversize is removed, and oil sands are heated by heat exchange with the hot outgoing tailings sand. The heated oil sands are then transported by the rotating action of the drum into the reaction zone, where they are mized with hot sand recycled from the combustion zone. The temperature in the reaction zone must be sufficiently high to thermally crack the hydrocarbons, leaving a coke residue on the sand. The stream of cracked vapor passes through cyclones to remove fine solids and is processed by conventional refining techniques. A heavy bottoms oil from the fractionating tower may be recycled back to the reaction zone or used to supplement process fuel.

The coked sand moves from the reaction zone to the combustion zone, where preheated air is injected to burn most of the coke off the sand grains to provide heat for the process. Auxiliary burners, which are externally located, provide heat for start-up and trim control. The hot sand from the combustion zone passes through a recycling arrangement that diverts some sand to the reaction zone while allowing the excess sand to move into the outer heat-exchange zone. In the heat-exchange zone, the sand and combustion gases are cooled by the incoming oil-sand feed. The partially cooled sand is then removed from the processor, further cooled and dampened by water addition, and finally transported by conveyors to the tailings disposal area.

Combustion gases are removed from the processor, passed through cyclones to reduce the content of fine solids, and then passed through a wet scrubber that removes most of the remaining fine solids. Sulfur dioxide produced by the combustion of the coke is also removed in the wet scrubber. The liquid from the wet scrubber is used to cool the tailings sand.

Yield pattern of product

The following table shows typical yield patterns from processing oil sands containing 11 percent bitumen.

	Yield, wt %[*]	
	Without bottoms recycle	**With bottoms recycle to 524 °C cut point (estimated)**
Gross coke	15.2	18.6
C_3- off-gas	7.3	8.5
C_4+ liquids	76.7	72.0
Measured losses	0.8	0.9
Total	100.0	100.0

[*]Values are expressed as weight percent of the bitumen in the feed.

When the bitumen content of the feed is reduced to 5 to 8 percent, the coke yield is increased somewhat and the yield of C_4+ hydrocarbons is reduced to 68 to 70 percent (without recycle).

Properties of product oil

The C_4+ product from a "once-through" operation has the following properties and composition:

C_4 product	Vol %	Specific gravity, °API	S, wt %	Density, g/mL
C_4 to 190 °C	19.3	42.3	1.89	0.814
190 to 343 °C	23.2	22.2	2.80	0.920
343 to 524 °C	41.8	12.1	3.87	0.985
524 °C to C_4+	15.7	——	4.48	1.089
Total oil	100.0	16.4	3.40	0.956

Energy balance

Coke is used as the primary source of fuel, but either heavy bottoms oil or the off-gas from the thermal cracking reaction can be used as supplemental fuel. The projected heat requirements are in the range of 500,000 to 600,000 Btu per ton of feed. Calculations indicate that 80 percent of this heat may be obtained from coke combustion and the remainder supplied by supplemental fuel.

Advantages and disadvantages

The UMATAC process has the advantage of combining in one unit the several operations that are required when the hot water process (extraction, froth treatment, and primary upgrading) is used in an oil-sands plant, and it has the potential for reducing the complexity of surface-mining plants. For example, large settling ponds to handle recycle water would not be required. The ability to achieve reasonable oil recoveries from low-grade oil sands is also an important advantage.

A disadvantage is the contamination of cracked gases with combustion gases, which necessitates a more complex plant for gas treating and gas separation. Desulfurization of flue gas will also be needed, but some form of desulfurization will be a future requirement for all processes which utilize residuum for fuel. Methods of handling and storing hot tailings of dry sand must be developed.

The developer has proposed the construction of a demonstration plant, using a processor with a diameter of 18 to 22 ft (5.5 to 7 m).

Modified Hot-Water—RTR Process

The RTR process is being developed by RTR-Rio Tinto TIL Holdings S.A. at a pilot plant operating on the Suncor lease at Fort McMurray, Alberta. A simplified flow sheet is shown in Fig. 2-6. The process has the following main characteristics:

1. Gentle treatment of the oil sand with hot water to avoid excessive disintegration of clay lumps. No sparging steam is used in the digestor, as is employed in the hot-water process.
2. Recovery in a desanding vessel of a diluted froth from which water and solids are removed in a proprietary solids-liquid contactor.
3. Flocculation of the middlings in a thickener and centrifugation of the resulting sludge. Clarified water is recycled in a closed circuit.
4. Rejection of solids in a relatively dry form that is suitable for handling by conveyors.

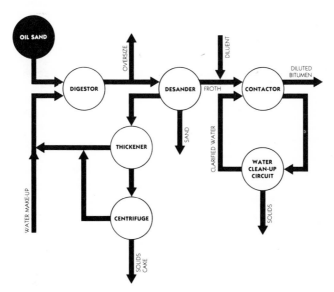

FIG. 2-6 RTR oil-sand extraction process.

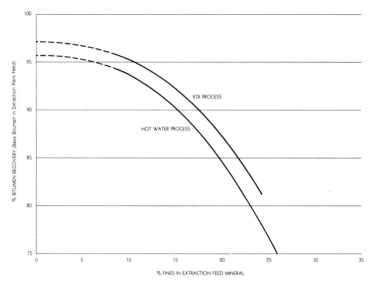

FIG. 2-7   Comparison of oil recoveries from the RTR process and the hot-water process.

The circuits for middlings treatment and contactor water treatment involve a series of chemical additions not shown in Fig. 2-6. These treatments maintain the appropriate flocculation and dispersing environments at each stage of the process.

The developers report a higher oil recovery from the RTR process than from the hotwater process (Fig. 2-7). Additional laboratory and engineering work is required to establish the appropriate method of handling the streams of centrifuge cake and reject sand. Operation of the pilot plant is continuing, with the objective of obtaining further operational experience and data for scale-up.

Sequential Treatment with Water and Solvent—Magna Process

The Magna process involves the slurrying of oil sand sequentially with water and solvent and the recovery of the hydrocarbon phase from the mineral matter in a proprietary separation vessel known as the *Rotary Turbulence Gravity Separator (RTGS) cell*. The diluent acts as a solvent for the bitumen, reducing both its viscosity and density. The major steps in the Magna process are illustrated in Fig. 2-8.

The process is based, in part, on the cold-water process investigated by the Canadian Federal Department of Energy, Mines and Resources during the 1950s. Extensive pilot plant research was subsequently carried out during the 1970s in Toronto by the present developers of the process, Magna International Inc.

The potential advantages of the process include the following:

1. Faster settling of clay and silt, thus reducing the requirement for large settling ponds and increasing the availability of recycle water

2. Higher quality of primary product oil, with lower fines and water content

3. Removal of bulk clay before the feed enters the separation zone, permitting use of lower-grade feedstocks

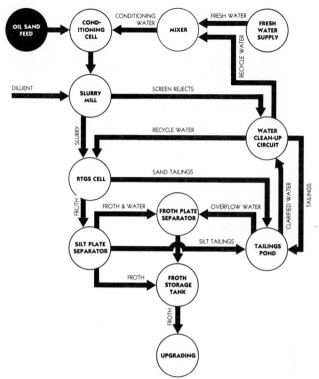

FIG. 2-8   Magna oil-sand extraction process.

The following table lists typical performance figures for the Magna and the hot-water process.

	Magna process	Hot-water process
Product oil		
Fines, wt %	3–5	8–10
Water, wt %	20–30	30–40
Fines content of tailings water, wt %	2–7	10–15

The developer has approached oil-sand leaseholders with a view to constructing and operating a large-scale demonstration facility.

Solvent Extraction and Fines Agglomeration—SESA Process

The spherical agglomeration concept arose through the work of Dr. I. E. Puddington and colleagues at the National Research Council of Canada and initially involved the agglomeration of bitumen in a slurry of oil sand and cold water. The concept was later applied successfully to the agglomeration of fine clay and mineral matter from a suspension of oil sands in a liquid-hydrocarbon solvent. This version is called the solvent extraction–spherical agglomeration (SESA) process.[9]

Test work involved a series of bench-scale tests during the early 1970s in Ottawa and subsequent tests on a large bench-scale facility in Calgary. The following operations are involved in the SESA process, as shown in Fig. 2-9.

1. The dissolution or extraction of the bitumen is carried out with a bitumen-rich naphtha solvent, with an appropriate water content, in a rotating cyclindrical drum, and results in 98 to 99 percent of the bitumen going into solution.

2. The suspension of undissolved solids, bitumen, and solvent is separated by successive screening and thickening, liquor clarification, and washing into the following fractions:

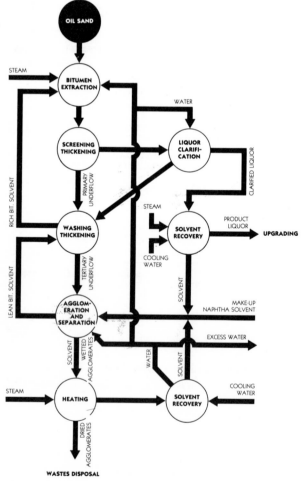

FIG. 2-9 SESA oil-sand extraction process.

(a) a clarified liquor containing some 45 percent of the dissolved bitumen with less than 1 percent solids and (b) a suspension of solids, bitumen, and solvent containing about 55 percent of the dissolved bitumen with about 64 percent solids and 6 percent water.

3. Fresh solvent and controlled small amounts of water are added to the suspension of solids, bitumen, and solvent, and the mixture is mixed in a horizontally mounted, rotating drum. This results in the segregation of the bulk of the fine clay and other solids in the form of spherical agglomerates, which are washed and separated from the remaining bitumen solution.

4. The spherical agglomerates, containing some 85 percent solids, are dried in such a way that over 95 percent of the adhering solvent is recovered.

5. The residual solids, containing about 1 percent bitumen and about 0.5 percent solvent, are disposed of in mined-out areas, permitting full reclamation of these areas.

During the initial pilot tests, bitumen recovery from oil sands ranged from 92 to 96 percent. The clay and mineral components largely agglomerated into dense spheres of 10-mm average diameter. The spheres contained 10 percent water, 4 percent solvent, and less than 1 percent bitumen. Most of the residual solvent (95 percent) was recovered by drying, and the overall solvent loss sustained was 0.5 percent.

One of the major potential advantages of the SESA process is its ability to handle lower-grade feedstocks. Figure 2-10 shows oil recoveries from a range of feedstocks for the SESA and the hot water processes.

At the present stage of development, a larger-scale demonstration of the SESA process is required to optimize the design of a commercial plant, to obtain more accurate estimates of capital and operating costs, and to determine the economic viability of a commercial SESA plant. A scale-up program of this nature is presently being considered.

Extraction Processes for Utah Oil Sands

Utah oil-sand deposits are conservatively estimated to contain 30×10^9 bbl of in-place bitumen, and the Utah deposits differ significantly from the Canadian deposits with regard to bitumen viscosity, bitumen and sand association, sand composition, and size of sand particles. Four specific differences are listed:

1. The bitumen viscosity is at least one order of magnitude greater than the viscosity of the Canadian bitumens.

2. The sulfur content is low (0.5 percent for the Uinta Basin bitumens).

3. The water content is very low, and the bitumen is directly bonded to the sand particles.

4. The sand is mainly alpha quartz with little evidence of the presence of clay minerals. Although the $< 10\text{-}\mu\text{m}$-diameter sand can amount to 5 percent by weight, the low clay content facilitates separation of solids and liquids in the tailings stream, and should permit effective water recycle.

Three different water-based physical separation processes have been developed for Utah oil sands at the University of Utah in a research program under the direction of Dr. J. D. Miller.[10] Characteristic features of these processes are identified in Table 2-1.

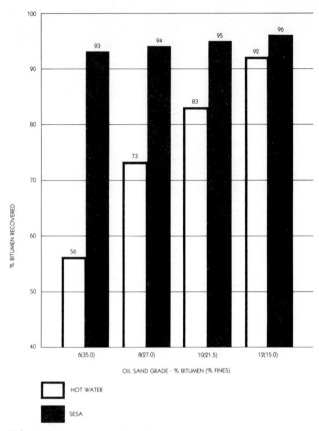

FIG. 2-10 Comparison of the SESA process and the hot-water process.

TABLE 2-1 Extraction processes for Utah oil sands

Process strategy	Phase disengagement	Phase separation	Recovery, %	Bitumen content of concentrate, %
Hot-water process	Digestion takes place under conditions of high shear, high alkalinity, and high temperature.	Modified froth flotation	96	65
Modified hot-water process	Digestion takes place under conditions of high shear, low alkalinity (additives required), and moderate temperature.	Dispersed oil flotation	95	65
Ambient temperature process	Liberation occurs by size reduction (crushing and grinding).	Conventional froth flotation	92	20

The high-intensity force field required in the phase-disengagement step for each process is a reflection of the high bitumen viscosity and the direct bonding of the bitumen to the sand particles. On the other hand, in each case the lack of clay minerals results in good sedimentation of the tailings.

The hot-water process for the Utah oil sands was the first process developed in the late 1970s and has been found to be a successful process strategy for many deposits. Results from these bench-scale studies at the University of Utah have led to the construction in Salt Lake City of a 50-bbl/day pilot plant by Enercor with participation by the state of Utah. This pilot plant began operations in October 1980, and, on the basis of results from this program, a 5000-bbl/day plant is being designed. If this testing program is successful, a 50,000-bbl/day commercial plant will be constructed toward the end of this decade.

Other New Extraction Methods

In addition to the previously described emerging extraction processes, test work is underway on other techniques in various development laboratories throughout the world.

1. Kruyer Research and Development Ltd. of Edmonton, Alberta, is investigating a bitumen agglomeration process which uses an oleophilic sieve to collect the agglomerates.[11]

2. Getty Oil Company is testing an extraction process that uses a pure solvent in a large field pilot project for possible use in the recovery of bitumen from the McKittrick diatomaceous earth deposits in California. Recovery of bitumen from these deposits presents a new challenge to the development of extraction technology.[12]

3. Rompetrol and associated research institutes are developing hot-water and solvent water extraction methods for the Derna-Budoi and Liatita-Pacuresti oil-sand deposits in Romania.

New information from such programs is likely to be released in the patent and process literature during the coming years.

IN SITU RECOVERY

Objectives of *In situ* Recovery Processes

This section deals with oil-sand *in situ* recovery processes which appear to be sufficiently well developed from a technological point of view to have potential commercial application. Also included is a brief discussion of new processes which are still at the laboratory or small pilot study stage.

Little primary production from typical oil-sand deposits can be expected. The reasons for this are the extremely high viscosity of the bitumen at reservoir temperatures and the lack of sufficient reservoir pressure to drive the bitumen to production wells. However, the situation is not entirely discouraging. If the Canadian oil sands are used as an example, several positive features on which to build successful processes can be identified:

1. The deposits have high bitumen saturations (up to 2000 bbl/acre-foot).

2. A moderate amount of heating reduces bitumen viscosity to a relatively mobile range ($<$500 cP at 100 °C).

3. The dissolving of modest amounts of gas in the bitumen also results in a considerable reduction in viscosity.

4. The sand has a very high permeability (1 to 5 darcies, oil-free basis).

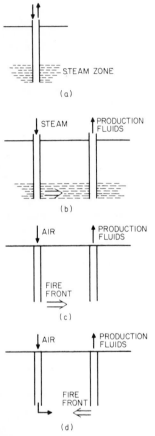

FIG. 2-11 Thermal *in situ* recovery processes. (*a*) Steam stimulation: Injection of steam into well followed by production of water and oil from the same well. Steam injection for ≈ 1 month yields water and oil productions for ≈ 3 to 6 months. (*b*) Steam displacement: Injection of steam into wells and recovery of displaced fluids from nearby production wells. (*c*) Forward combustion: Injection of air into deposit to promote combustion, with displacement of fluids toward nearby production wells. (*d*) Reverse combustion: A variation of the above with the combustion zone moving from the production wells to the injection well.

The following *in situ* recovery techniques may be applicable to the reservoir situations just described:

1. Steam injection by either a stimulation and production cycle or a steam drive between injection and production wells.

2. Injection of air to sustain *in situ* combustion of part of the bitumen to heat the reservoir.

3. Injection of gases or hydrocarbon solvents to reduce bitumen viscosity, possibly with simultaneous injection of steam.

4. Use of an array of horizontal injection and production wells, either directionally drilled from the surface or emplaced from underground tunnels, to increase the effectiveness of the preceding techniques.

5. Electric preheating to reduce bitumen viscosity and to help initiate communication between injection and production wells.

6. Establishment of horizontal communication by hydraulic fracturing methods. [Below some critical depth (approximately 1000 ft) the fractures tend to orient themselves in a vertical plane.]

Thermal approaches (1 and 2 in the preceding list) have been the most actively studied, and are illustrated in Fig. 2-11.

Stages in Process Development

Successful commercial techniques for *in situ* recovery are preceded by extensive deposit evaluations and pilot studies in the field. The usual sequence is illustrated in Fig. 2-12.

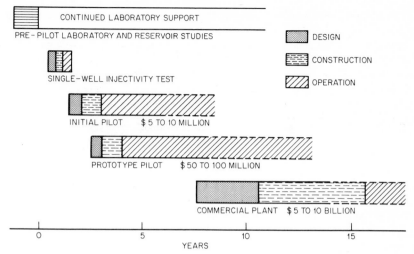

FIG. 2-12   Stages in commercializing an oil-sand reservoir.

Pre-pilot, laboratory, and reservoir studies

Coring and logging programs are carried out in at least one well prior to proceeding to a pilot test. Cores provide data on oil and water saturations, particle-size distribution, porosity, horizontal and vertical permeability, and the mineralogical nature of the clay. Coring above and below the zone of interest indicates the competency of the shale barriers. Logs determine the presence and thickness of water zones and shale stringers. Studies of particle size are required to design liner slots and other potential sand-control methods.

Single-well test

The first pilot study in a new reservoir is commonly a single-well test, along with two or three observation wells. This first well provides data on fluid injectivity, orientation and direction of fractures or flow channels, steam-to-oil ratios, potential production rates, and the physical and chemical properties of the reservoir fluids. It also identifies potential sand-production problems.

Observation wells are located to maximize information with regard to heat movement and fracture orientation. They provide temperature measurements across the zone of interest and data on reservoir pressure. Temperature profiles usually are taken some distance above and below the zone of interest. Production and injection wells also have devices to read bottom-hole temperature and pressure.

Multiwell pattern test

On the basis of the results of the first pilot study a multiwell pattern is tested with a minimum of three and a maximum of nine production and injection wells, accompanied by an appropriate number of observation wells.

This pilot study is designed to test well spacing and configuration on the basis of data on heat movement and fracture orientation that were obtained from the initial test.

Two or three small multiwell pilot tests may be carried out concurrently in this second stage in order to test a variety of completion techniques and operating strategies. It is generally difficult to evaluate some of these factors sequentially in a single pilot test because of contamination from prior operations.

Prototype

This semicommercial stage is normally a multipattern operation and is designed to confirm and refine results from the previous pilot studies. Figure 2-13 illustrates some of the patterns used for prototype operations.

Operating strategies are refined, and cost-saving techniques in construction and operations are investigated. The operation is large enough to eliminate edge effects and reservoir and/or geological variations.

Production from a pilot study of this scale offsets operating costs and, to some extent, capital costs. However, operations are still oriented toward an evaluation of the process, and data gathering is an important goal.

In a small reservoir this stage could be the core or nucleus of the final commercial project and would be designed as such.

The length of time from the initial injectivity test to the production of meaningful results from the prototype is in the order of 5 to 8 years.

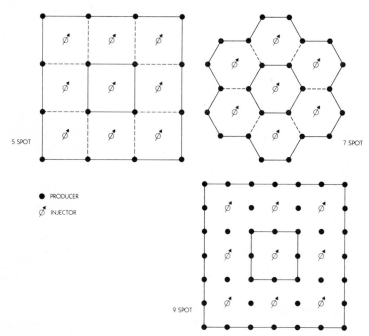

FIG. 2-13 Typical multiwell patterns for prototype *in situ* recovery operations. Producer ●, injector ∅.

Recovery Processes using Steam

The most common fluid used to carry heat to the reservoir is steam, which may vary in quality from 0 to 100 percent steam depending on the requirements of the process being investigated.

The major criterion used to evaluate thermal processes is the overall energy balance. Common practice is to calculate the steam-to-oil ratio, which is the volume of steam injected (expressed as water equivalent) divided by the volume of oil or bitumen produced. Because of the variations in heat content and steam quality of the injected fluids, a more precise figure would be the ratio of energy input to energy output. However, steam-to-oil ratios give a quick, easily understood evaluation of the success of a steam process.

Steam processes (Fig. 2-11) are classed as either cyclic steam stimulation (a) or steam drive (b). Steam stimulation tends to be the most energy-efficient because heat is scavenged by the production of fluid through the heated zone. For steam-drive processes, a large percentage of the injected heat is left behind in the reservoir. However, some heavy-oil operators have been successful in improving recoveries of oil and heat by converting to a water drive once the oil in the reservoir was thoroughly mobilized and heated by steam.

In heavy-oil pools, steam drive has generally given higher recoveries than steam stimulation. In one field, steam drive improved oil recovery from 18 percent of the oil in place to 45 percent, and recoveries of up to 75 percent have been reported. The highest recovery reported in a steam-stimulation project to date is 35 percent.[13]

In addition to the pre-pilot plant work referred to previously, steam processes can be evaluated by the use of physical-stimulation equipment, in which samples of reservoir rock, oil, and water are placed in a vessel which may be up to several feet in diameter. A series of tests are conducted using varying quantities and qualities of steam, with and without additives. The data from these tests are used in mathematical modeling studies and also in the design of pilot plants. Physical simulator tests are more applicable to steam drive than steam stimulation, but with proper interpretation they are useful for either process.

Cyclic Steam Stimulation

In this process, steam is injected into the producing well to heat the bitumen in the immediate area of the well bore. [See Fig. 2-11 (a).] After a predetermined volume of steam has been injected, the well is placed in production and allowed to produce until the effects of the steam injection have been dissipated. The cycle is then repeated as often as results indicate is appropriate. A soak period lasting from a few hours to several days may intervene between the termination of steam injection and the commencement of production. This is to allow time for the steam to condense and to maximize heat utilization.

There are currently 16 active pilot projects in Canada and the United States that are investigating *in situ* recovery of oil-sand bitumen using the cyclic steam process as the initial phase.

A description of a major operating pilot project and a proposed Canadian commercial application follows.

AOSTRA/BP Marguerite Lake pilot project (Fig. 2-14)

Location: Cold Lake field, Alberta
Status: Operational since 1978

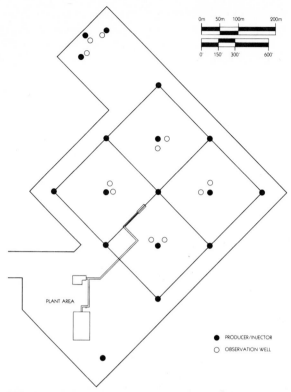

FIG. 2-14 AOSTRA/BP Marguerite Lake *in situ* pilot project. Producer and Injector well ●, observation well ○.

Operator: BP Exploration Canada, Ltd.

Participants: AOSTRA, Hudson's Bay Oil & Gas Canada Ltd., PanCanadian Petroleum Ltd.

Process: Cyclic steam

Depth to Top of Pay: 1500 ft (457 m)

Thickness of Pay Section: 115 ft (35 m)

Pattern Configuration: Four 5 spots

Spacing: 2.5 acres (1 hectare) per well

Operation Period: June 1978 to December 1985

Estimated Gross Cost: C$55 million

Bitumen Specific Gravity (Density): 10° to 12° API (1.000 to 0.986 g/mL)

Reservoir: Clearwater sandstone

Estimated Recovery: 20 to 25 percent of original in place (O.O.I.P.) (cyclic steam phase)

This pilot project is a follow-up to small-scale activity which BP has carried on at its Cold Lake lease since 1965.[14] The well configuration and orientation are based on a northeast and southwest regional fracture trend. No commercial primary recovery is possible from this reservoir, and fluid injection must take place above parting pressure.

Special test wells have been added to this project to test a combustion follow-up to the basic cyclic steam process. Another special feature of this pilot project is the use of seismic techniques to attempt to map the heated zone.

The operator is currently designing a prototype commercial project which will use a cyclic steam process and will produce 7000 bbl/day (1100 m³/day). Well spacing will be 2.5 acres (1.0 hectare) per well.

Esso Resources Cold Lake project (Fig. 2-15)

Location: Cold Lake field, Alberta

Status: Applicant is currently considering phased-development approach.

Operator: Esso Resources Canada Ltd.

Production: 140,000 bbl/day (22,000 m³/day) of synthetic crude oil

Process: Cyclic steam

Depth of Top of Pay: 1500 ft (457 m)

Pay Thickness: 140 ft (43 m)

Pattern Configuration: *   Rectangular

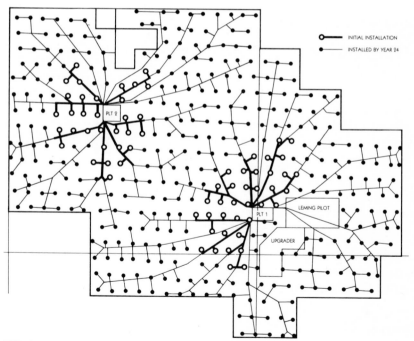

FIG. 2-15   Proposed Esso Resources Cold Lake commercial production system. Initial installation ○—, installed by year 2024 ●—.

*Estimated or forecast.

*Spacing:**   4 acres (1.6 hectares) per well

*Operational Period:**   25 years

*Estimated Capital Cost:**   C$12 billion (including upgrading facility)

Bitumen Specific Gravity (Density):   10° to 12° API (1.000 to 0.986 g/mL)

Reservoir: Clearwater sandstone (Lower Cretaceous)

*Steam-to-Oil Ratio:**   2.5

*Estimated Recovery:**   20 percent O.O.I.P.

Three pilot projects, with the first commencing in 1964, have provided the information upon which the design and proposed operating strategy of the commercial proposal are based.[15]

The original pilot project, designated as Ethel, consisted of three separate well groups containing 3, 14, and 9 production and injection wells and 1, 23, and 14 observation wells, respectively. A number of concepts related to cyclic steam and steam drive were investigated. Other tests made an attempt at reservoir heating by injecting steam into bottom water and mixing gas into the injected steam.

The second pilot project, designated as May, began in 1972 and is still operating. Three 5-spot patterns were constructed on 5-acre (2.0-hectare) spacing. Eight additional infill wells were later drilled to give a 2.5-acre (1.0-hectare) spacing over part of the area. Performance of wells with underlying water and with no underlying water was compared. Bottom water was shown to reduce oil production. Some steam-drive tests were also undertaken.

The third pilot project, designated as Leming, began operation in 1975 in an area with no bottom water. Eight clusters of 7 wells (total of 56 wells) were drilled on 7-acre (2.9-hectare) spacing, with 1 central well in each cluster drilled vertically and the other 6 drilled directionally from the central well. In 1976 one of the patterns was infill-drilled, reducing the spacing to 1.8 acres (0.7 hectare) per well.

A different well-spacing configuration was conceived 2 years later and tested with the drilling of 12 new wells on 4-acre (1.6-hectare) spacing on the northeast edge of the Leming pilot area. Again, the drilling was directional from a central pad, but the subsurface distance between wells was extended in the on-trend direction of preferred communication and compressed in the off-trend direction. This elongated well-spacing geometry is designed to match the shape of the well pattern more closely to the shape of the heated zone around each well and thereby provide better volumetric heating efficiency.

Early performance was encouraging and has led to the development of several additional well clusters with some further experimental modifications in inter-well distances and operating strategies. Other thermal recovery processes, applied as a follow-up to steam stimulation, are also under investigation. All of this work is aimed at further refining design parameters and operational guidelines of commercial projects.

Steam Drive

In this process, steam or hot water is injected continuously into an injection well which may be a central well of a pattern, one of a line of wells, or one of the peripheral wells of a pattern. (See Fig. 2-13.) The injected steam moves an oil bank ahead of it toward the producing wells, which may be the outside wells in a pattern (inverted), the central well in a pattern (normal), or a line of wells paralleling a line of injection wells (line drive).

Steam flooding is generally applied to a reservoir where there is sufficient mobility to allow for economically viable injection rates at pressures below the fracture pressure. In an oil-sands reservoir this normally means that heat communication between producers and injectors must be established prior to implementing a steam flood.

Observations of steam-flood behavior in heavy-oil reservoirs, where the technique is relatively mature and well established, indicate that regardless of where steam enters the reservoir, it will migrate upward. When it reaches a permeability barrier, it will finger out toward the producing well. These fingers spread out to give a broad areal distribution, ideally in a radial form. When the producing well is reached, the steam zone grows downward until the entire zone is heated.

There are currently no successful steam-flood commercial operations in oil-sands reservoirs. Details of two major pilot projects in which steam flooding is being tested are given below.

AOSTRA/Shell Peace River pilot project (Fig. 2-16)

Location: Peace River field, Alberta

Status: Operational since November 1979

Operator: Shell Resources Canada Ltd.

Participants: AOSTRA, Shell Explorer Ltd., Amoco Canada Petroleum Ltd.

Process: Steam drive with cyclic pressuring

Depth to Top of Pay: 1800 ft (550 m)

Pay Thickness: 90 ft (27 m)

Pattern Configuration: Seven 7 spots

Spacing: 3.5 acres (1.4 hectares) per producing well

Operational Period: 1979 to 1985

Estimated Gross Cost: C$170 million

Bitumen Specific Gravity (Density):   8° to 9° API (1.014 to 1.007 g/mL)

Reservoir: Bullhead sandstone

Estimated Recovery: 50 percent O.O.I.P.

This prototype is based in large measure on mathematical and physical simulation.[16] Two small pilot studies and a substantial amount of exploratory drilling were carried out in the period 1962–1967.

The present pilot project is large enough that heterogeneities in the reservoir should be evened out, and it may be possible to proceed directly to a commercial design.

The reservoir contains a thin [6- to 9-ft (2- to 3-m) thick] zone of high water saturation at the bottom, and this is heated by injection of steam from the seven central injection wells. Once the water zone is heated, the reservoir is pressured up by restricting production and continuing steam injection. This mobilizes a substantial portion of the oil which is produced by a normal steam drive. Several cycles are required to fully deplete the reservoir.

The pilot project is currently partway into the initial steam-injection phase.

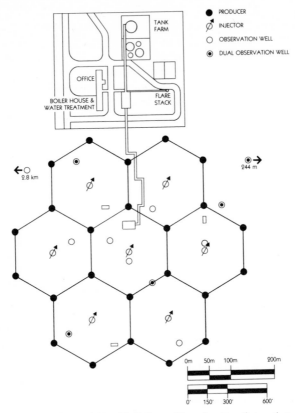

FIG. 2-16 AOSTRA/Shell Peace River *in situ* pilot project. Injector ⌀, producer ●, observation wells ○ or ⊙.

Lagoven Jobo steam-drive pilot project (Fig. 2-17)

Location: Jobo field, Venezuela

Status: Under construction

Operator: Lagoven

Process: Steam drive

Steam Injection Volume per Well:* 1000 bbl of water per day

Depth to Top of Pay; >3400 to 3850 ft (1040 to 1170 m)

Pay Thickness: >45 to 170 ft (14 to 50 m)

Pattern Configuration: Six 7 spots

Spacing: 5 acres (2 hectares) per producing well

Bitumen Specific Gravity (Density): 8° to 9° API

*Estimated or forecast.

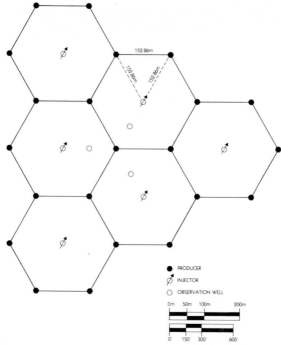

FIG. 2-17 Lagoven Jobo *in situ* pilot project. Injector ⌀, pro-ducer ●, observation well O.

Steam-to-Oil Ratio: 1.5 to 3.0

Estimated Recovery: 30 to 35 percent

This pilot project is designed to test the steam-drive process as a tertiary recovery mech-anism in the southeastern portion of the Orinoco oil belt of Venezuela.[17]

The field is capable of some primary recovery (1 to 2 percent), and steam stimulation in other parts of the Orinoco belt has achieved recoveries of 12 percent or more.

The 23 producing wells are expected to produce about 10,000 bbl/day.

Combustion as a Recovery Mechanism

The second major thermal process for *in situ* recovery of oil-sand hydrocarbon is one of three forms of underground combustion. These are illustrated in Fig. 2-11 (*c*) and (*d*) and include the following:

1. Dry forward combustion

2. Reverse combustion

3. Wet forward combustion

Preferred applications are thin or deep heavy-oil reservoirs, conventional oil reservoirs following waterflooding, and oil-sand reservoirs following steam stimulation and/or steam drive.[18]

Dry forward combustion

In this process high-pressure air is injected into the reservoir. After ignition, the hydrocarbon in the well bore area burns and the hot combustion products and light ends (from distillation and cracking) move through the reservoir toward the producing wells, heating and mobilizing the hydrocarbon. The cracking process leaves behind a residual coke, which becomes fuel for the combustion process. Eventually the mobilized oil reaches the producing well, where it can flow or be pumped to the surface along with combustion gases. A major disadvantage of the forward combustion process is that mobilized and vaporized oil moves into the cold region ahead of the combustion front. Here it cools and may no longer flow. This difficulty led to the development of the reverse-combustion process.

Reverse combustion

It has been found that most reservoirs have some permeability to air. If air is injected continuously until it reaches the producer and is then ignited, the combustion front moves backward from the producer. The combustion gases along with some of the mobilized fluids continue to move toward the producer. This is the reverse-combustion process.

There appears to be no upper viscosity limit to this process, but it has several disadvantages; the fuel used is a lighter fraction, the coke remains unburned on the sand, and the production wells may coke up.

Wet forward combustion

A further disadvantage of dry forward combustion is the large amount of heat remaining in the reservoir behind the combustion front and the low heat-carrying capacity of the combustion gases. By the injection of water simultaneously or alternately with the air, the heat remaining in the reservoir is scavenged and steam is formed which moves through the combustion zone to form a steam bank. This mobilizes some of the oil ahead of the fire front.

Depending upon the ratio of water to air the process can be referred to as normal wet, partially quenched, or superwet.

Combustion mechanisms

Comments which follow apply only to the forward-combustion processes.

Varying zones exist in the reservoir as shown in Fig. 2-18.[18] The rate of air and water injection, the composition of the original hydrocarbon, and the properties of the reservoir affect the behavior of the process.

As can be seen from Fig. 2-18, a variety of drive mechanisms are occurring. Vaporized light ends and steam improve the mobility of the cold oil. The combustion gases provide a gas drive, and after the steam condenses, a hot-water flood exists.

The air-to-oil ratio and air-to-water ratio are important economic and operating parameters in any combustion process. If the air flux (rate of air movement through the reservoir) is too low, the fire front may stall and burn all the hydrocarbon available without mobilizing any of it, or the fire front may not be sustained. If the air flux is too high, some

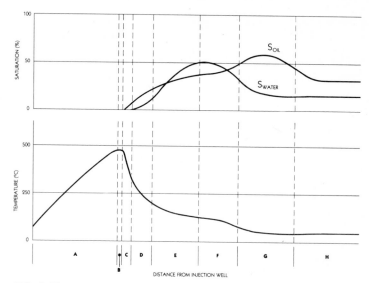

FIG. 2-18   Various zones in a combustion tube. A, burned zone; B, combustion zone; C, cracking region; D, evaporation and visbreaking region; E, steam plateau; F, water bank; G, oil bank; H, initial zone. *(From Ref. 18 with permission.)*

oxygen may bypass the fire front, which could result in low-temperature oxidation downstream.

The combustion reactions produce a wide range of organic acids, some of which are water-soluble. Produced waters associated with this process may be extremely acidic, with a pH as low as 1.5. In addition, the formation of surfactants can lead to stable emulsions, which may be difficult to break with available equipment and chemicals.

Laboratory fire-tube tests, using sand, bitumen, and water from the reservoir, provide data on the effect of water-to-air ratios, fuel lay-down rates, and necessary air-injection rates. The fire tube is assembled to duplicate reservoir conditions as closely as possible with respect to oil and water saturations and reservoir rock and fluids. Other methods of determining fuel content, in addition to fire-tube tests, are available.[19]

The fuel lay-down rate, or fuel content, is the amount of coke deposited on the sand. Too high a fuel content will stall the combustion front unless excessive volumes of air are used. If the fuel content is too low, either the sand will not support combustion and the fire will go out or low-temperature oxidation will be the only reaction occurring. Acceptable limits for the fuel content are in the range of 1 to 2 lb coke per cubic foot of reservoir rock. There is some indication that fuel content decreases with increased water-to-air ratios.[20]

Once fuel content and required air-flux rates have been determined, a field test using air or natural gas is carried out to determine whether sufficient injectivity is available in the reservoir. A design rate is established to ensure an acceptable rate of advance of the heat front. Suggested rates are in the order of 1 to 5 in/day.[20]

Mathematical models have been found to be useful in applying fire-tube test results and reservoir data to pilot design.

One relatively large pilot project using forward combustion is described in the following section.

AOSTRA/Amoco Gregoire Lake Block I pilot project (Fig. 2-19).

Location: Athabasca field, Alberta

Status: Field test completed 1981

Operator: Amoco Canada Petroleum Company Ltd.

Participants: AOSTRA, Petro-Canada Exploration Inc., Shell Canada Resources Ltd., Shell Explorer Ltd., Suncor Incorporated Resources Group

Process: Wet Forward Combustion (COFCAW, or combination of forward combustion and water flood)

Depth to Top of Pay: 1080 ft (329 m)

Pay Thickness: 90 ft (27 m)

Pattern Configuration: Nine 5 spots

Spacing: 2.5 acres (1 hectares) per producing well

Operational Period: 1977 to 1981

Estimated Gross Cost: C$50 million

Bitumen Specific Gravity (Density): 8° to 10° API (1.014 to 1.000 g/mL)

Reservoir: Lower Fort McMurray sandstone

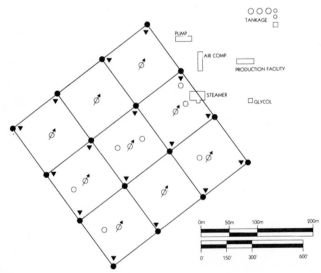

FIG. 2-19 AOSTRA/Amoco Gregoire Lake *in situ* pilot project. Producer ●, stimulation well ▼, observation well O, injector Ø.

This large pilot project was the culmination of some 20 years of activity by Amoco in the Athabasca deposit, during which period a series of pilot studies aimed primarily at testing various types of combustion processes (including reverse combustion) and well-completion procedures were carried out.[21]

The Block I pilot project was designed to test the patented Amoco COFCAW process, which had been developed and demonstrated on heavy-oil reservoirs elsewhere.

The process involves preheating the reservoir using forward combustion, blowing down to reduce reservoir pressure, and then employing COFCAW displacement, in which a specified ratio of air and water are injected concurrently to supply heat and to drive fluids to the producing wells. Heat communication between injection and production wells was not established, and the pilot test was suspended before the COFCAW stage was reached.

Other Emerging *In Situ* Processes

A number of other *in situ* processes, which either are used prior to the use of steam or combustion or are refinements of one or both of these processes, have been proposed and in some cases are in the field pilot stage.

Electric preheat processes

The theory of using electrical resistance heating as a preheat mechanism prior to fluid displacement by more conventional means has been reviewed in a recent publication.[22]

In this process an electric circuit (ac) is established between two or more pairs of electrodes installed in the pay zone, using the oil sand as a part of the circuit.

Although electric power is expensive, the process has the advantage of not requiring any movement of fluid within the formation, and heating occurs throughout the entire conducting zone simultaneously. This is illustrated in Fig. 2-20.

A small pilot project to test electric heating was operated in the field in the Athabasca deposit by Petro-Canada Exploration Inc. on behalf of itself and Esso Resources Canada Ltd., Canada Cities Service Ltd., and Japan Canada Oil Sands Ltd. This pilot study was based on laboratory tests started in 1972.

The pilot project consisted of four electrode wells 1400 ft deep and spaced in a rectangular pattern approximately 60×100 ft in size. The electric preheat and subsequent steam-injection stages took place during the period 1980–1983.

Few details of this test are available, and the operator anticipates no commercial application for several years.

Other electrical methods such as radio-frequency heating and electromagnetic methods have been proposed.[23] Some laboratory work has been done, but field pilot tests will be required to establish the commercial feasibility of such methods for oil-sands applications.

Steam plus additives

Patents have been issued covering the injection of mixtures of light hydrocarbon and steam and mixtures of CO_2 and steam.[24] Mixtures of light hydrocarbon and steam are being investigated by Texaco Canada Resources Ltd. in a field test in the Athabasca deposit. Informal reports are encouraging, but no firm details are available.

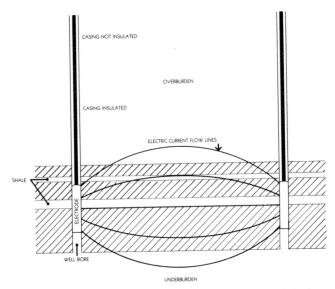

FIG. 2-20 Electric preheat process. *(From Ref. 22 with permission.)*

A patent covering the addition of CO_2 and light hydrocarbons to the steam in a steam drive has been issued on the basis of laboratory work with a physical simulator.[25] No field tests have been carried out as of 1983.

Laboratory physical-simulation tests on mixtures of CO_2 and N_2 (flue gas) with steam have been encouraging. This modification to the steam process could enhance the value of high-pressure down-hole steam generators.

Inter-well heat communication developed by steam or possibly by electrical methods is essential to the success of any process that attempts to recover hydrocarbons from oil sands *in situ*. Additives may improve recoveries, reduce steam-to-oil ratios, or supply other refinements to the basic steam process.

Down-hole steam generation

Down-hole steam generators are intended to be used to generate steam at the bottom of the well and are referred to as (1) low-pressure generators when only the steam is injected and the combustion products return to the surface or (2) high-pressure generators when both steam and combustion gases are combined and injected.

Extensive work by the U.S. Department of Energy in conjunction with Sandia Laboratories and Rockwell Corporation has resulted in the development of prototype models of these devices. Two high-pressure models are currently being field-tested in the Long Beach heavy-oil pool in California.[26]

These devices appear to have application in a steam drive process and may improve the economics of the process by reducing heat loss from the well bore and lessening the environmental impact of burning fuel for steam generation at the surface.

The cost of withdrawing the equipment from the well between cycles seems to preclude its use in processes that use cyclic steam stimulation.

Oxygen-enriched air for combustion processes

The use of oxygen or oxygen-enriched air is being investigated by several operators of combustion projects. This concept could possibly reduce capital expenditures for compression equipment, improve combustion, and reduce the volume of combustion gas to be handled. It does, however, increase corrosion and the danger of an explosion.

Recovery from Bituminous Carbonate Rocks

Devonian carbonate rocks having an average of 10 to 15 percent porosity and impregnated with bitumen over a vertical distance of up to 300 ft have been reported under a wide area of northern Alberta known as the "carbonate triangle." The full geographical extent of these occurrences is at present unknown.

In general, the bitumen-to-rock ratio is only half that of the overlying oil-impregnated sandstones. However, high-porosity reservoirs exist, and small-scale steam and combustion pilot projects operated by Union Oil Company of Canada yielded high rates of oil production.[27] Questions that need to be answered in future work include the following:

1. Is there any evidence of well "sanding" or plugging aroung the well bore?
2. Is there any evidence of generation of CO_2 by rock decomposition?
3. What is the extent of barren or impermeable zones?

OIL UPGRADING

Properties of Bitumen and Heavy Oil

Bitumen and conventional heavy crude oils have many properties in common which render them unsuitable for use as the dominant feed to conventional refineries for the manufacture of transportation fuels. Some of the more significant properties are exemplified by the typical analyses outlined in Table 2-2.[28]

As compared with conventional light- and medium-gravity crude oils, the dominant characteristics of oil-sands bitumen and heavy oils are listed below:

- Extremely high viscosity at ambient temperatures which renders pipeline transportation virtually impossible without the addition of substantial quantities of diluent such as natural-gas condensate or naphtha. (The maximum acceptable viscosity imposed by a major Canadian company with a crude oil pipeline, for example, is 40 to 48 cSt at 38 °C.)
- Deficiency of hydrogen relative to the hydrogen content of conventional light- and medium-gravity crude oils.
- Large percentage of high-boiling-point material which severely limits the volume of virgin transportation fuels that may be recovered by simple separation processes.
- Substantial quantities of resins and asphaltenes which act as coke precursors in high-temperature refining operations.
- High content of sulfur and/or nitrogen, which necessitates very severe hydroprocessing of the distillate fractions in order to produce specification fuels or intermediate products for refineries.

TABLE 2-2 Properties of bitumens and heavy oils

	Canadian			Venezuelan		
	Athabasca	**Cold Lake (clearwater)**	**Lloyd-minster heavy oil**	**Boscan**	**Laguna**	**Jobo II**
Specific gravity, °API	8–10	10–12	10–18	10–12	11.1	8.4
Elemental analysis, wt %						
Carbon	83.1	83.7	82.8	—	—	—
Hydrogen	10.6	10.5	11.8	—	—	—
Sulfur	4.8	4.7	3.4	5.2	3.0	3.7
Nitrogen	0.4	0.2	0.3	0.65	—	—
Oxygen	1.1	0.9	1.7	—	—	—
H/C (elemental)	1.52	1.49	1.70	—	—	—
Molecular weight	570–620	490	410	—	—	—
Hydrocarbon type, wt %						
Saturates	22	33	47	—	—	—
Aromatics	21	29	24	—	—	—
Resins	39	23	17	—	—	—
Asphaltenes	18	15	12	9–17	7.3	8.6
Metals, ppm (wt)						
Vanadium	250	240	100	1,200	390	390
Nickel	100	70	40	150	54	106
Ash, wt %	0.7	—	—	—	—	—
Conradson carbon, wt %	13.5	12.6	8	15	12.8	14.1
Distillation temperature, °C						
Initial boiling point	260	170	80	—	—	—
5 vol %	285	255	175	—	—	—
10 vol %	320	300	230	—	—	—
30 vol %	425	405	310	—	—	—
50 vol %	530	555	—	—	—	—
Viscosity, cP						
Reservoir temperature	5,000,000	100,000	3,000	—	—	—
30 °C	—	—	—	26,200	6,600	62,265
60 °C	3,700	—	—	2,950	1,000	5,535
100 °C	1,000	100	20	—	—	—

Source: Adapted from Ref. 28 with permission.

- High content of metals, particularly of vanadium and nickel, which causes deactivation of downstream cracking catalysts.
- Entrained mineral matter, in bitumens derived from mined oil sands, can plug catalyst beds in upgrading and downstream processes.

Generalized Upgrading System

All of the upgrading plants in existing or pending oil-sands projects are essentially the same in terms of the sequence of the processing steps used to produce a stable synthetic crude oil. This sequence of operations is depicted in Fig. 2-21. In each case, bitumen is fed to a primary upgrading process in which conversion of the high-boiling-range components in the virgin bitumen is achieved. The products of conversion are the following:

- Hydrocarbon off-gas
- Cracked liquid distillates—naphtha and light and heavy gas oils
- A residual fraction in the form of petroleum coke or pitch

The liquid distillate fractions contain significant concentrations of sulfur and nitrogen. These cracked distillates are also highly aromatic, and in the case of the naphtha fraction, high bromine numbers give evidence of substantial olefin formation. These distillates are therefore hydrotreated before being recombined to produce a marketable synthetic crude oil. The consequences of hydrotreatment are the following:

- Saturation of olefins and diolefins to provide a stable synthetic crude oil
- Reduction of sulfur and nitrogen concentrations to levels suitable for downstream refining to finished products
- Limited saturation of aromatic compounds to improve the cetane number and smoke point of diesel and jet fuels
- A shift to more naphtha and middle distillates through hydrocracking of the gas-oil fraction.

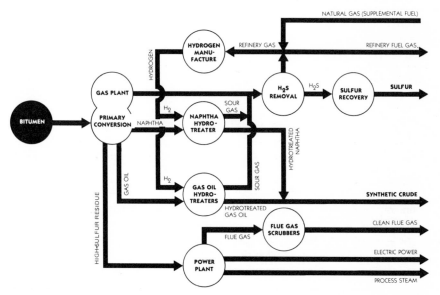

FIG. 2-21 Generalized steps in upgrading bitumen and heavy oil.

The hydrogen sulfide–rich off-gas from hydroprocessing is combined with the off-gas from the primary upgrader; the hydrogen sulfide is removed and converted to elemental sulfur. The resulting sweet gas is then available for use as fuel or as feed to hydrogen-manufacturing facilities.

The residue contains a large part of the sulfur and nitrogen and essentially all of the metals and ash. The obvious end use of this by-product is as fuel to provide steam and electric power for the entire oil-sand facility. Unfortunately, the high sulfur content in the residue makes its use generally unacceptable without some form of sulfur removal, which greatly increases the cost of its utilization as a fuel.

The existing oil-sands plants and those proposed for the immediate future are based on commercially available thermal cracking processes, i.e., delayed coking and fluid coking. This type of process relies on the principle of rejection of carbon in the form of coke, thereby increasing the hydrogen-to-carbon ratio in the remaining liquid-hydrocarbon products. These processes are characterized by relatively low liquid yields. The amount of coke produced by the cracking reaction is proportional to the carbon residue measured in the feed bitumen. Yields of distillate oils therefore have a finite maximum for any given feedstock. Conversion levels, expressed as the percentage by weight of ≥ 525 °C material in the feed which is converted in the process to lower-boiling-range products, are in the range of 60 to 70 percent.

In Alberta, some of the by-product coke is stockpiled and natural gas is used as a preferred refinery and boiler fuel. Because of the availability of low-cost natural gas, this is considered to be the most economical way of satisfying the limitations on sulfur dioxide emissions.

The FLEXICOKING process, licensed by Exxon Research and Engineering, offers a partial solution to the problem of coke utilization by gasifying the net coke produced in the fluid-coking process with air and steam to yield a low-heating-value fuel gas. Hydrogen sulfide contained in the gasifier product may be readily removed, and elemental sulfur recovered. This process has been proposed for the Esso Resources Cold Lake project.[15]

Objectives of New Upgrading Technology

Relative to the currently applied thermal cracking (coking) processes, new upgrading technology in Alberta will be directed toward achieving the following:

- Significant increases in the overall yields of liquid products of a quality acceptable to refiners.
- Improved economics, perhaps through increased liquid yields or through reduced capital and operating costs.
- Reduction in the release of harmful emissions.
- Improved energy efficiency by eliminating residue (coke) stockpiles and/or by achieving a significant reduction in the amount of unconverted residue that is produced. Such a reduction may put the plant out of fuel balance and require the use of external sources of energy such as natural gas or coal.
- Development of processes which may be applied economically to small-scale upgraders.
- Development of processes for the production of petrochemicals, such as olefins and aromatics, directly from bitumen.

Examples of Emerging Upgrading Processes

A number of processes in various stages of development have been proposed to replace the commercially applied coking processes for upgrading oil-sands bitumen. It is doubtful that any of these will satisfy all of the objectives outlined in the previous section for new upgrading technology. However, they offer alternatives to future oil-sands developers and could in fact eliminate many of the problems associated with coking. The more prominent alternative processes are described briefly in the following summary.

Gulf HDS process[29]

Developer
The developer of this process is the Gulf Research and Development Company.

Process type
This process can be characterized as a fixed-bed catalytic hydrogen addition process.

Description
Charge oil is mixed with hydrogen, heated to reaction temperature, and introduced into a multistage system of reactors containing a fixed bed of Gulf catalyst. The liquid product contains unconverted ≥ 525 °C material. However, this unconverted material has a higher API specific gravity than the ≥ 525 °C fraction in the feed, and the sulfur content has been substantially reduced. It is proposed that this high-boiling material be retained in the synthetic crude oil.

Comments
The process has been demonstrated commercially on residual oils. The fixed-bed catalyst is deactivated by metals contained in the charge oil, and the developers predict a relatively high rate of catalyst replacement with the concentrations of metals quoted in Table 2-2.

The fixed-bed configuration is susceptible to plugging as a result of suspended mineral material in the feed. Thus the feed must be filtered to reduce the mineral content to trace amounts. This would be of particular concern when processing bitumen derived from mined oil sands.

Development status
Gulf has more than 10 years' commercial experience in six plants that handle high-sulfur atmospheric and vacuum residues.

CANMET process[30]

Developer
The developers of this process are Petro-Canada Exploration Inc. and the Department of Energy, Mines and Resources of the Government of Canada.

Process type
The CANMET process can be characterized as a thermal hydrogen addition process.

Description

This is a noncatalytic, high-conversion, hydrodemetallization, hydrogen addition process. A special coal-based additive mixed with the charge oil is claimed to promote hydroconversion and inhibit coke formation in the reactor at relatively low hydrogen partial pressures. A small residue of pitch remains. This residue contains the removed metals and any mineral matter which may exist in the feed. Moderate desulfurization of the distillate fraction occurs.

Comments

Conversion of ≥ 525 °C material originally in the feed is reported to be in the range of 85 to 90 percent in pilot plant tests. Pilot tests also show a good distribution of distillate yields.

Development status

Extensive experience has been gained on a variety of feeds in a 1-bbl/day pilot plant over the past 10 years. Plans for a demonstration plant are underway.

Shell Fixed-Bed processes[31]

Developer

The developer of these processes is the Shell group.

Process type

The Shell fixed-bed processes can be characterized as a fixed-bed catalytic hydrogen addition process.

Description

These processes are proposed for hydroconversion and desulfurization of high-boiling-point materials in a variety of trickle-flow, multiple fixed-bed reactor arrangements and specific-catalyst systems. The specific catalyst and reactor arrangements would be chosen to accommodate the characteristics of particular feedstocks. For feeds containing high concentrations of metals and suspended mineral material, the hydroconversion and desulfurization stage is preceded by a bunker-flow guard reactor containing a proprietary hydroemetallization catalyst which selectively removes metals and traps suspended mineral matter. In this bunker-flow reactor, deactivated demetallization catalyst may be withdrawn and fresh catalyst added to maintain catalyst activity. Regeneration of spent catalyst may be economically feasible depending on rates of catalyst consumption.

Comments

The process may be designed to achieve various levels of conversion. In order to operate in a high conversion mode, however, it is claimed that the hydroconversion stage must be preceded by a hydrodemetallization system.

Development status

The process is under development for oil-sands bitumen and heavy oils containing high concentrations of metals. Shell's experience with fixed-bed hydroconversion and desulfurization systems is based on commercial applications to residual oils.

ABC process[32]

Developer
The developer of this process is the Chiyoda Chemical Engineering and Construction Co. Ltd.

Process type
The ABC process can be characterized as a fixed-bed catalytic hydrogen addition process.

Description
This process, which is under development, incorporates a fixed-bed hydroconversion and mild hydrodesulfurization reactor followed by solvent deasphalting of the unconverted residual pitch. This hydroconversion process features a proprietary metals-tolerant catalyst for removal of metals and conversion of asphaltenes. Unconverted ≥ 525 °C material from a once-through operation is fed to a conventional solvent deasphalter. The deasphalted oil is then hydrocracked to distillate fractions, which are combined with the distillates from the hydroconversion stage to yield a synthetic crude oil. In an extension of this process, Chiyoda proposes to recycle precipitated asphaltenes to extinction in the demetallization reactor.

Development status
Pilot plant experience includes tests on a variety of feeds including conventional crude oil residues, bitumen, and heavy oils.

Comments
Suspended mineral material in the charge oil would be capable of plugging the catalyst bed. Chiyoda therefore recommends a solids-removal system and a desalter as a feed preparation step.

Aurabon process[33]

Developer
The developer of this process is the UOP Process Division.

Process type
The Aurabon process can be characterized as a catalytic hydrogen addition process.

Description
The process is designed principally for hydrodemetallization and hydroconversion of asphaltenes in residual oils and heavy crude oils. Some hydrodesulfurization is also achieved. The process consists of a reactor circuit and a separation section. Unconverted asphaltenes are recycled to the reactor section. Published information on the Aurabon process does not describe the catalyst system, but it is believed to be a flow-through material suspended in the charge oil. A drag stream purges accumulated metals and ash from the system. The total liquid yield is high, but contains a high percentage of unconverted nonasphaltenic material that boils above 525 °C and has a high carbon residue.

Comments

Conversion of high-boiling fractions is relatively low, but the high degree of demetallization and of removal of suspended material provides the opportunity for further conversion in conventional fixed-bed systems downstream.

Development status

Extensive pilot plant experience has been acquired over the past 10 years. A semicommercial demonstration plant has been designed.

Veba-Combi-Cracking process[34]

Developer

The developer of this process is Veba Oel Entwicklungsgesellschaft, mbH.

Process type

The Veba-Combi-Cracking process can be characterized as a hydrogen addition process that employs an optional catalyst.

Description

Residue from prefractionated charge oil is mixed with hydrogen and fed to a liquid-phase upflow reactor. A slurried throwaway catalyst may be employed, depending on the characteristics of the feed and the required yields. The distillate fractions from the liquid-phase hydroconversion are separated from the unconverted residue and hydrogenated in the gas phase in a fixed-bed catalytic reactor to complete hydrodesulfurization and hydro-denitrogenation of the distillate product.

Adjustment to the desired yield distribution may also be achieved in the gas-phase reactor. The unconverted pitch residue containing essentially all of the metals is discharged for disposal off site.

Comments

High conversions of 500 °C+ feed material, in the order of 90 percent, are claimed, together with almost total removal of sulfur, nitrogen, and metals in the distillate fractions.

Development status

The process was operated commercially in Germany to upgrade coal liquids to gasoline and middle distillates during the Second World War. Thereafter, it was used until 1964 to convert residual oils. It has recently been reintroduced at the scale of a pilot project for application to bitumen and heavy-oil upgrading.

H-OIL® process[35]

Developer

The developer of this process is Hydrocarbon Research Inc.

Process type

The H-Oil process may be characterized as a catalytic hydrogen addition process using an ebullating catalyst bed.

Description

This is an ebullating bed hydroconversion process which provides for continuous removal of spent catalyst and catalyst renewal. The removal of metals by their inclusion in the pitch residue is essentially complete. Sulfur and nitrogen removal is moderate. The preferred operating modes are a once-through low-conversion approach or a high-conversion mode carried out at low-severity reactor conditions but with recycle of vacuum-bottoms product to the reactor system. With this high-conversion mode, conversions in excess of 90 percent are claimed.

Development status

The process has a long history of development, with commercial experience on conventional oil residues dating back to the mid 1960s. Bench- and demonstration-scale experience is available on a wide variety of feeds.

LC-Fining process[36]

Developer

The developers of this process are Cities Service Research and Development Company and C-E Lummus.

Process type

LC-Fining can be characterized as a catalytic hydrogen addition process that uses an ebullating catalyst bed.

Description

This is a hydroconversion process which uses an ebullating catalyst bed and which also provides for catalyst addition and removal. Metals and ash removal with pitch residue is complete. Sulfur and nitrogen removal is moderate. Conversion levels offered depend on requirements for total liquid yield and on feed properties and may range from 60 to 90 percent.

Development status

LC-Fining is a parallel development of the H-Oil technology from which it emerged. There is shared commercial and pilot plant experience. More recently the licensors have processed oil-sands bitumen in pilot plant operations dedicated to the development of the LC-Fining process.

Eureka process[37]

Developer

The developer of this process is the Kureha Chemical Industry Co. Ltd.

Process type

The Eureka process can be characterized as a carbon rejection process involving thermal cracking.

Description

The feed is heated in charge heaters and is discharged continuously into one of two steam-stripped batch reactors in which cracking occurs. The residence time is controlled by the frequency of reactor switching and quenching. Cracked residue exits the reactor system as a liquid pitch. Yields of distillate products are slightly lower than fluid-coker yields, and the data indicate a rather low yield of cracked naphtha.

Comments

The process utilizes conventional low-pressure equipment. The design is modular and may have an application in small upgraders designed to process production from small-scale (25,000-bbl/day) *in situ* oil-sands projects. However, no significant improvement in distillate yields or in yield distribution over current commercial experience is forecast.

Development status

One commercial plant has been operating in Japan since 1976 to upgrade vacuum residue from Middle East crude oils. Pilot plant experience has been obtained on bitumen and conventional heavy oils.

Dynacracking process[38]

Developer

The developer of this process is Hydrocarbon Research Inc.

Process type:

The Dynacracking process can be characterized as a carbon rejection process-involving thermal cracking, coupled with mild noncatalytic hydrogenation.

Description

This process features a single three-zoned reactor in which coke, formed by thermal cracking of the charge oil, is deposited on an inert fluidized solid carrier in the upper zone of the reactor and is gasified in the lower zone of the reactor in the presence of oxygen and steam. The upper and lower zones are separated by a stripping zone. Gasification yields a hydrogen-rich gas which contacts the cracked oil vapor and causes partial saturation of the olefins and conversion of a portion of the sulfur content to hydrogen sulfide. Metals and ash are deposited on the inert carrier and purged from the system by removal and renewal of the carrier. Pilot plant data indicate overall liquid yields are less than fluid-coker yields but a high percentage yield of naphtha and middle distillate is possible.

Comments

There is no residual coke or pitch. It is claimed that the flexibility of the process permits control over the liquid yield distribution. In spite of low liquid yields, the process may have application in small-scale upgraders because of the elimination of residue. There is a potential hazard due to the introduction of oxygen into the reactor. The process is carried out at intermediate pressures in the range of 400 to 600 psig.

Development status

There is no commercial or recent pilot plant experience. A semicommercial demonstration plant is planned.

ROSE process (Residuum Oil Supercritical Extraction)[39]

Developer

The developer of this process is the Kerr-McGee Refining Corporation

Process type

The ROSE process can be characterized as a carbon rejection process involving precipitation of asphaltenes.

Description

This process has a multistage solvent deasphalting system that is capable of separating asphaltenes, resins, and low-carbon residue oils. In the final stage of separation, the mixture of oil and solvent is heated to supercritical conditions to facilitate economical recovery of solvent. A large percentage of the metals are removed in the precipitated asphaltenes. The yields of distillate oils, comparable in boiling range to fluid-coker distillates, are low, since no conversion of high-boiling-range material occurs.

Comments

The process uses conventional equipment and is based on well-developed solvent-extraction technology. The recovery of solvent is claimed to be more energy-efficient than conventional solvent-stripping operations.

Development status

Three commercial units are in operation. Two additional units are under design.

REFERENCES

1. Britton, M. W., W. L. Martin, J. D. McDaniel, and H. A. Wahl, Jr.: Continental Oil Company, Fracture Preheat Oil Recovery Process, U.S. Patent 4,265,310, May 5, 1981.

2. Johns, R. W.: World Oil Mining Ltd., Hydraulic Mining of Oil Shales, Tar Sands, and Other Minerals, Canadian Patent 1,026,387, 1978.

3. *Fifth Annual Report and Five Year Review,* Alberta Oil Sands Technology and Research Authority, Government of Alberta, Mar. 31, 1980.

4. Stephenson H. G. (Mining Consultants) Ltd.: A Cost Estimate for Removal of Overburden, from Athabasca Oil Sands Using Hydraulic Dredges.

5. Ritter, R. A.: "Tailings Water Reclamation," *1981 AOSTRA/CP Conference on Advances in Petroleum Recovery and Upgrading Technology, Calgary, Alberta, May 24-26, 1981,* Alberta Oil Sands Information Centre, Edmonton, Alberta.

6. Lowry, T. S., and J. H. Cottrell: Cities Service Oil Company, Recovery of Bitumen from Bituminous Sand, U.S. Patent 3,117,922, Jan. 14, 1964.

7. Bowman, C. W.: "The Athabasca Oil Sands Development—50 Years in Preparation," *Origin and Refining of Petroleum, Ad. Chem. Ser.,* no. 103, ACS, pp. 81–93.

8. Taciuk, W.: "Development Status of the Taciuk Direct Thermal Processor for Tar Sands and Heavy Oils," *The 2d International Conference on Heavy Crude and Tar Sands, Caracas, Venezuela, Feb. 7-17, 1982.*

9. Meadus, F. W., B. D. Sparks, and I. E. Puddington: Canada National Research Council, Tar Sands Separation, Canadian Patent 1,031,712, May 23, 1978.

10. Hatfield, K., J. D. Miller, and A. G. Oblad: "Hot Water Processing of Oil Wet Tar Sand," *The 2d International Conference on Heavy Crude and Tar Sands, Caracas, Venezuela, Feb. 7-17, 1982.*

11. Kruyer, J.: "Oleophilic Separation of Tar Sands, Oil/Water Mixtures and Minerals," *The 2d International Conference on Heavy Crude and Tar Sands, Caracas, Venezuela, Feb. 7-17, 1982.*

12. Hallmark, F. O.: "The Unconventional Petroleum Resources of California," in Meyer, R. F., C. T. Steele, and J. C. Olsow (eds.): *The Future of Heavy Crude and Tar Sands,* (Proceedings of the First International Conference, June 4–12, 1979), McGraw-Hill, New York, 1981, pp. 69–82.

13. Farouk Ali, S. M.: "Current Steamflood Technology," *J. Pet. Technol.,* October 1979, pp. 1332–1342.

14. Harding, T. G.: "Marguerite Lake Project—Phase A.," *AOSTRA/CP Non-Conventional Oil Technology Seminar, Calgary, Alberta, May 29-30, 1980,* Alberta Oil Sands Information Centre, Edmonton, Alberta.

15. Imperial Oil Limited: "The Cold Lake Project, A Report to the Energy Resources Conservation Board, May 1978," Energy Resources Conservation Board, Calgary, Alberta.

16. Gorrill, R. G., P. Kitzan, and D. P. Komery: "The Design of the Peace River In-Situ Oil Sands Pilot Project," in Meyer, R. F., C. T. Steele, and J. C. Olson (eds.): *The Future of Heavy Crude and Tar Sands* (Proceedings of the First International Conference, June 4–12, 1979), McGraw Hill, New York, 1981, pp. 378–387.

17. Borregales, C. J.: "Production Characteristics and Oil Recovery in the Orinoco Oil Belt," in Meyer, R. F., C. T. Steele, and J. C. Olson (eds.): *The Future of Heavy Crude and Tar Sands,* (Proceedings of the First International Conference, June 4–12, 1979), McGraw-Hill, New York, 1981, pp. 498–509.

18. Bennion, D. W.: "Future of In-Situ Combustion in Canada," Report prepared for The Alberta Oil Sands Technology and Research Authority, May 22, 1981.

19. Wu, C. H.: "An Experimental Study on the Oil and Water Behavior in Porous Media Under Stepwise—Approximately Isothermal Conditions with Multi-Phase Fluid Flow," Ph.D. Dissertation, School of Engineering, University of Pittsburgh, Pittsburgh, 1968.

20. Chu, C.: "State-of-the-Art Review of Fireflood Field Projects," *SPE/DOE Second Joint Symposium on Enhanced Oil Recovery, Tulsa, Okla., Apr. 5-8, 1981.*

21. Jenkins, G. R., and J. W. Kirkpatrick: "Twenty Years Operation of an In-Situ Combustion Project," *J. Can. Pet. Technol.,* vol. 18, no. 1, 1979, pp. 60–66.

22. Towson, D. E.: "Electric Preheat Update," *The 2d International Conference on Heavy Crude and Tar Sands, Caracas, Venezuela, Feb. 7-17, 1982.*

23. Sresty, G., R. H. Snow, and J. E. Bridges: "The IITRI RF Process to Recover Bitumen from Tar Sand Deposits—A Progress Report," *The 2d International Conference on Heavy Crude and Tar Sands, Caracas, Venezuela, Feb. 7-17, 1982.*

24. Brown, A., C. H. Wu, and D. T. Konopnicki: Texaco Inc., Combined Multiple Solvent and Thermal Heavy Oil Recovery, U.S. Patent 4,004,636, Jan. 25, 1977; Wu, C. H., A. Brown, and W. L. Hall: Texaco Inc., Recovering Petroleum From Subterranean Formations, U.S. Patent 4,119,149, Oct. 10, 1978.

25. Redford, D. A., and M. R. Hanna: Alberta Oil Sands Technology and Research Authority, Gaseous and Solvent Additions for Steam Injection for Thermal Recovery of Bitumen from Tar Sands, U.S. Patent 4,271,905, June 9, 1981; Canadian Patent 1,102,234, June 2, 1981.

26. Marshall, B. W., H. R. Anderson, and J. J. Stosur: "A Review of the Operation and Performance of the Downhole Steam Generator in the Wilmington Field, Long Beach, Calif.," *The 2d International Conference on Heavy Crude and Tar Sands, Caracas, Venezuela, Feb. 7-17, 1982.*

27. Vandermeer, J. G., and T. C. Presber: "Heavy Oil Recovery from the Grosmont Carbonates of Alberta," *AOSTRA/CP Non-Conventional Oil Technology Seminar, Calgary, Alberta, Mar. 29-30, 1980,* Alberta Oil Sands Information Centre, Edmonton, Alberta.

28. Starr, J., J. M. Prats, and S. A. Messulam: "Chemical Properties and Reservoir Characteristics of Bitumen and Heavy Oil from Canada and Venezuela," in Meyer, R. F., C. T. Steele, and J. C.

Olson (eds.): *The Future of Heavy Crude and Tar Sands*, (Proceedings of the First International Conference, June 4–12, 1979), McGraw-Hill, New York, 1981, pp. 168–173.

29. Ondish, G. R., and A. J. Sunchanek: "Potential for Gulf Residual HDS in Upgrading Very Heavy Crudes," in Meyer, R. F., C. T. Steele, and J. C. Olson (eds.): *The Future of Heavy Crude and Tar Sands*, (Proceedings of the First International Conference, June 4–12, 1979), McGraw-Hill, New York, 1981, pp. 663–669.

30. Lunin, G., A. E. Silva, and J. M. Denis: "The Canmet Hydrocracking Process," *30th Canadian Chemical Engineering Conference, Edmonton, Alberta, October 1980.*

31. Van Zijll Langhout, W. C., C. Ouwerkerk, and K. M. A. Pronk: "Development of and Experience with the Shell Residue Hydroprocesses," *AIChE 88th National Meeting, Philadelphia, June 1980.*

32. Takeuchi, C., M. Nakamura, Y. Shiroto: "Upgrading of Athabasca Oil Sands Bitumen," *62d Canadian Chemical Conference and Exhibition, Vancouver, B.C., June 1979.*

33. Adams, F. H., J. G. Gatsis, and J. G. Sikonia: "The Aurabon Process: New Way to Upgrade Tars and Heavy Crudes: Process Description," in Meyer, R. F., C. T. Steele, and J. C. Olson (eds.): *The Future of Heavy Crude and Tar Sands*, (Proceedings of the First International Conference, June 4–12, 1979), McGraw-Hill, New York, 1981, pp. 632–635.

34. Urban, W.: "Mineral Oil Processing in the Scholven 300 Combined Hydrogenation Unit," *Erdöl Kolhe*, 1955, pp. 780–782.

35. Eccles, R. M.: "H-Oil Processing of Tar Sand Bitumen for Maximum Oil Yield and Cash Flow," in Meyer, R. F., C. T. Steele, and J. C. Olson (eds.): *The Future of Heavy Crude and Tar Sands* (Proceedings of the First International Conference, June 4–12, 1979), McGraw-Hill, New York, 1981, pp. 625–631.

36. Van Driesen, R. P., and J. Caspers: "Processing Heavier Crudes by Residual Hydrocracking," in Meyer, R. F., C. T. Steele, and J. C. Olson (eds.): *The Future of Heavy Crude and Tar Sands* (Proceedings of the First International Conference, June 4–12, 1979), McGraw-Hill, New York, 1981, pp. 618–624.

37. Aiba, T., H. Kaji, T. Suzuki, and T. Wakamatsu: "Residue Thermal Cracking by the Eureka Process," *Chem. Eng. Prog.*, vol. 77, no. 2, 1981.

38. Rakow, M. S., and M. Calderon: "The Dynacracking Process—An Update," *Chem. Eng. Prog.* vol. 77, no. 2, 1981.

39. Gearhart, J. A., and L. Garwin: "ROSE Process Improves Resid Feed," *Hydrocarbon Process.*, May 1976.

PART 6

SYNFUELS UPGRADING AND REFINING

Primary synthetic fuel liquids obtained from oil shale, coal, and oil sand contain mineral matter, sulfur, oxygen, and nitrogen in excess of that normally found in petroleum crude oil. Further, the molar ratio of hydrogen to carbon is significantly lower than that of normal refinery feed, as shown in the following table.

Liquid fuel	H/C molar ratio
Primary coal liquids	0.8 to 1.3
Bitumen	1.5
Shale oil	1.6
Petroleum crude oils	
Heavy asphaltic	1.6
Light paraffinic	2.0

The mineral matter and heteroatom impurities can be removed or greatly reduced, and the molar ratio of hydrogen to carbon can be elevated to that of petroleum crude oil by coking or hydrotreating or by combinations of the two.

The upgrading of shale oil and bitumen often involves a preliminary coking process followed by a hydrotreating unit operation, and hydrotreating can also be used as a first step to demetallize synthetic fuels prior to coking operations.

Coking is a thermal process which raises the hydrogen-to-carbon molar ratio in synfuels from oil shale or oil sands by rejecting carbon in the form of coke, while also producing some fuel gases. Coking of coal liquids produces mainly gases and carbon with a lower yield of liquids. The input mineral matter concentrates in the coke and is thus removed from the liquid products stream while the heteroatom content is largely unchanged.

Hydrotreating removes heteroatoms from primary synfuels by catalytic hydrogen addition to form ammonia, hydrogen sulfide, and water as by-products. The hydrogen-to-carbon molar ratio is also increased, giving a product very similar to petroleum crude oil or refined petroleum products.

The major coking and hydrotreating processes for liquid synfuel upgrading, together with major input feeds and products, are summarized below.

Process	Major synfuel feed(s)	Main products
Coking		
FLUID COKING	Shale oil, bitumen, liquefaction bottoms	Distillate oils, gases, and coke
FLEXICOKING		Distillate oil and gases
Hydrotreating		
H-Oil	Bitumen, coal-derived pyrolysis oil, solvent-refined coal	Upgraded fuel oil
Litol™	Bitumen, coal-derived pyrolysis oil, solvent-refined coal	Upgraded fuel oil
LC-Fining	Gasification oils, shale oil, coal liquids, bitumen	Upgraded fuel oil
Litol and Modified Litol™	Coal-derived light oils (C_6 to C_8) and gasification by-product oils	Benzene and toluene

The first five processes produce liquid fuel as the major product; the last process produces benzene and toluene chemicals. These six processes are presented in four chapters in this section.

EXXON FLUID COKING^{SM} AND FLEXICOKING^{SM} PROCESSES FOR SYNFUELS UPGRADING APPLICATIONS

S. F. MASSENZIO

Exxon Research and Engineering Company
Florham Park, New Jersey

INTRODUCTION

The need for utilization of lower-quality, hydrogen-deficient feedstocks such as bitumen from the Canadian oil sands, shale oil from Colorado, and coal liquids, is increasing in response to the diminishing reserves and increased cost of the more desirable crude oils. Such low-quality feeds present unique production, handling, and processing problems because of the high viscosity and bottoms content, as well as the high levels of feed sulfur, nitrogen, and ash. These facts, coupled with the growing demand for clean, light products, have created a high degree of conversion processing activity. FLUID COKING and FLEX-ICOKING are Exxon's proprietary coking processes which provide a means for converting low-quality feedstocks into clean, lighter liquids and fuel gas while complying with strict environmental regulations. This chapter will describe how the FLUID COKING and FLEXICOKING processes fit into the overall production and upgrading of these nonconventional feedstocks into high-quality, synthetic crude-oil substitutes.

PROCESS PERSPECTIVE

Coking Development Overview

The FLUID COKING and FLEXICOKING processes are continuous, noncatalytic, thermal cracking processes based upon the concept of *carbon rejection*. The heaviest, hydrogen-deficient portions of the feed, asphaltenes and resins, produce coke. Because the processes are thermal rather than catalytic, they are relatively insensitive to feed contaminants such as metals, sulfur, and nitrogen compounds. The heat required to crack the feed is provided by burning/gasifying a portion of the coke and transferring the heat by circulating coke in the fluidized state. During the early 1950s the FLUID COKING process was developed by utilizing the fluidized solids expertise gained from the Exxon Fluid Catalytic Cracking process. FLUID COKING technology was further extended during the early 1970s by adding high-temperature steam-air gasification of the coke, culminating in the FLEXICOKING process.

Commercial Facilities

The basic FLUID COKING process has been in commercial operation since 1954. There are presently ten operating units with a combined design capacity of 280,000 bbl/day. The most recent units to come on stream are the twin Syncrude Canada, Ltd. FLUID COKING units in 1978. These cokers are integrated at the mine site with downstream hydrotreating units to produce a high-quality synthetic crude oil from whole Athabasca oil-sands bitumen. This operation represents the first commercial synfuels application of FLUID COKING technology. Two similar oil-sands FLUID COKING units have been designed for the Alsands upgrading plant in Alberta, Canada. In addition, expansion or modernization studies are being carried out on several existing FLUID COKING units, and an unannounced European FLUID COKING unit is in the early stages of design. A list of FLUID COKING facilities and the status of each project is presented in Table 1-1.

The FLEXICOKING process technology was built upon this commercial FLUID COKING experience along with an extensive process development effort. The FLEXICOKING

TABLE 1-1   List of operating FLUID COKING units; current projects (as of IV:81)

Company	Location	Initial design fresh feed rate, bbl/sd	Initial start-up date	Comments
Exxon, U.S.A.	Billings, Mont.	3,800	Dec. 1954	Expanded to 7,000 bbl/sd
Tosco	Bakersfield, Calif.	4,000	Apr. 1957	Expanded to 7,000 bbl/sd
Tosco	Avon, Calif.	42,000	June 1957	
Getty	Delaware City, Del.	42,000	Aug. 1957	Currently running 44,000 bbl/sd
Hess	Purvis, Miss.	4,800	Dec. 1957	Currently running 8,000 bbl/sd
Pemex	Madero, Mexico	10,000	Feb. 1968	
Imperial	Sarnia, Ont.	14,000	Apr. 1968	
Exxon, U.S.A.	Benicia, Calif.	15,900	Apr. 1969	Currently running 24,000 bbl/sd
Syncrude	Mildred Lake, Alberta	72,900	July 1978	Athabasca oil-sands bitumen
Syncrude	Mildred Lake, Alberta	72,900	Oct. 1978	
Alsands	Fort McKay, Alberta	2 at 90,000/unit	——	Designed for Athabasca oil-sands bitumen
Unnannounced	Europe	55,000	——	In design

process was demonstrated during 1974–1975 in a 750-bbl/day prototype unit located at Exxon Company, U.S.A.'s Baytown, Texas, refinery. During 16 months of operation, the Baytown prototype FLEXICOKING unit successfully demonstrated the operability of the FLEXICOKING process on a wide range of feedstocks including high-metals Boscan residuum and athabasca oil-sands bitumen containing 0.75 percent by weight solids. The FLEXICOKING process was commercialized with the start-up in September 1976 of the Toa-Kawasaki FLEXICOKING unit. The Toa unit processes 21,000 bbl/day of high-sulfur Middle East vacuum residuum into clean fuel products, including a low-Btu gas. Performance of the unit has confirmed scale-up criteria developed from the operation of the Baytown prototype FLEXICOKING unit. A photograph of the Toa-Kawasaki facility showing the major process vessels is presented in Fig. 1-1.

Project Status

FLEXICOKING project activity is currently high; several projects are in various stages, as shown in Table 1-2. If the project progresses, the twin FLEXICOKING units for Esso Resources Canada, Ltd. for processing Cold Lake bitumen will represent the first commercial application of FLEXICOKING for the production of synfuels. These FLEXICOKING units, along with downstream hydrotreating and catalytic conversion, can produce a high-quality synthetic crude oil which can be processed in existing refineries. This project is currently on hold. Another potential FLEXICOKING application, although not under active study, is the upgrading of high-ash-containing streams from coal-liquefaction processes such as the Exxon Donor Solvent coal-liquefaction process. This potential application is covered in Chap. 1-1.

FIG. 1-1 The Toa-Kawasaki FLEXICOKING unit. Equipment spacing of the Toa unit is very tight—a common industry practice in Japan, where land is at a premium value. Toa vessels are large, but they do not represent the limit of Exxon's process and design technology.

PROCESS DESCRIPTION

FLUID COKING

A flow plan of a typical FLUID COKING unit is shown in Fig. 1-2. The FLUID COKING process employs two major fluid-solids vessels, a reactor and a burner, and coke circulates in a fluidized state between the vessels. Feed is injected through multiple nozzles into the reactor, where thermal cracking to lighter hydrocarbons and coke is accomplished in a thin film on the fluidized coke particles. The coke particles provide a large and efficient contacting area on which the cracking reactions can take place. The portion of the feed that is cracked to coke is deposited on the circulating coke from the burner. The reactor bed is fluidized by the product vapors produced from thermal cracking of the feed and by steam injected at the bottom of the reactor. The reactor is essentially at atmospheric pressure and at temperatures in the range of 480 to 540 °C.

Product vapors leave the fluidized reactor bed and pass through cyclones which remove most of the entrained coke particles and return them to the reactor. The cyclones discharge

TABLE 1-2 List of FLEXICOKING units; current projects (as of IV:81)

Company	Location	Initial design fresh feed rate, bbl/sd	Initial start-up date	Comments
Exxon, U.S.A.	Baytown, Texas	750	Feb. 1974	Prototype FLEXICOKING unit; shut down July 1975 and mothballed
Toa Oil	Kawasaki, Japan	21,300	Sept. 1976	
Esso Resources Canada	Cold Lake, Alberta	2 at 50,000/unit	———	Designed for Cold Lake bitumen; project on hold
Lagoven	Amuay, Venezuela	52,000	1982	Field construction under way
Shell	Martinez, Calif.	22,000	1982	Field construction under way
Esso Netherlands	Rotterdam, The Netherlands	———	———	In planning
Unannounced	———	50,000	———	In design

the vapor into the bottom of the scrubber, which is mounted directly above the reactor. In the scrubber, essentially all the remaining coke dust is scrubbed out and the products are cooled to condense the unconverted portion of the feed by a hydrocarbon pump-around stream. The resulting slurry is withdrawn from the scrubber and recycled to the reactor with the fresh feed. The lighter product vapors pass overhead to a conventional fractionator, where they are separated into wet gas, naphtha, and various gas oil fractions. The gas is compressed and further fractionated into the desired components.

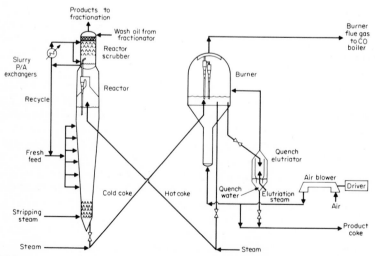

FIG. 1-2 Simplified process flow plan of a FLUID COKING unit.

The reactor coke passes through the stripping zone, where residual hydrocarbons are steam-stripped and ultimately recovered as reactor product. The stripped reactor coke flows down a standpipe and through a slide valve which controls the reactor bed level. A riser carries the reactor coke, which is fluidized by steam, to the burner. Air is introduced to the burner, where a portion of the coke is burned to supply the process heat requirements. The burner typically operates in the range of 595 to 650 °C. The burner bed coke flows down a standpipe and through a slide valve which controls coke flow and thus the reactor bed temperature. A riser carries the burner coke to the top of the reactor bed. This coke flow provides the heat required for the thermal cracking of the feed. Combustion products from the burner bed pass through two stages of cyclones, where entrained coke particles are recovered and returned to the burner bed. The combustion products pass through a stack valve (with a variable orifice) which controls the burner pressure. The burner flue gas is then typically burned in a CO boiler to generate high-pressure steam.

The portion of the reactor coke which is not burned is withdrawn from the burner to keep the coke inventory constant and control coke particle size. This is accomplished in the quench elutriator drum, where elutriation steam is injected and the finer coke particles pass overhead to the burner. The coarser coke particles are quenched with water in the elutriator drum and subsequently withdrawn as product. The product coke, which contains a substantial portion of the feed solids and ash, is normally pneumatically conveyed to storage. The coarser coke particles are replaced with small seed coke particles produced from steam attrition. These attriters, located at the bottom of the reactor dense bed, provide the energy required to reduce the size of the coke particles which have grown from deposition of a new coke layer.

FLEXICOKING

A flow plan of a typical FLEXICOKING unit is shown in Fig. 1-3. The reactor-side and the coke-transfer lines are identical to those just described for a FLUID COKING unit. The burner is replaced by a heater vessel, and a third fluid-solids vessel, the gasifier, is used to gasify the coke production.

The reactor coke is now carried to the heater, where the coke is partially devolatilized to methane and hydrogen. The devolatilized coke is circulated to the gasifier, where it is reacted with steam and air at an elevated temperature (760 to 980 °C) to produce H_2, CO, CO_2, N_2, H_2S, and a small quantity of COS. This coke gas, together with entrained coke fines, is returned to the heater bed, where some of its heat is picked up to provide a portion of the reactor heat requirements. The remaining reactor heat load is supplied by coke circulated from the gasifier to the heater. In this regard, the heater acts as a heat exchange vessel between the gasifier and the reactor.

The low-Btu gas passes through the heater bed and enters two stages of cyclones. The gas is subsequently cooled in a waste-heat boiler to generate high-pressure steam. Additional coke fines are recovered in external tertiary cyclones and a venturi scrubber. The essentially solids-free low-Btu gas can now be desulfurized by commercially available processes before being burned in boilers or process furnaces.

A purge of coke from the heater bed via a quench elutriator may be necessary. The quantity of heater bed purge is dependent upon the feed ash content. This purge, in addition to the tertiary cyclone and venturi scrubber coke fines, maintains the coke inventory at an acceptable ash level. The coke fines collected in the tertiary cyclones and venturi

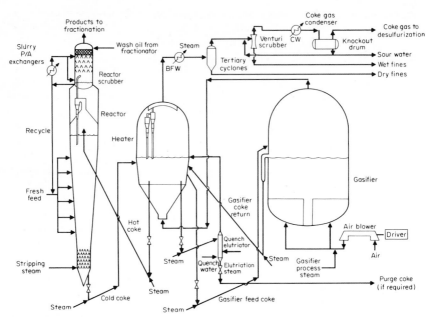

FIG. 1-3 Simplified process flow plan of a FLEXICOKING unit.

scrubber are rich in ash and generally represent a reasonable portion of the total purge necessary.

COKING-GASIFICATION MECHANISMS

As can be seen, the heart of the FLUID COKING and FLEXICOKING process is the formation of coke and its subsequent burning or gasification to supply the required process heat.

Coking

Coking is a thermal pyrolysis process in which the feed coats the bed coke particle, where the lighter portion of the feed vaporizes and subsequently cracks, leaving behind a liquid layer which forms coke. This liquid layer is a multicomponent mixture which for simplicity will be denoted as a "coke precursor." There are many reaction paths which contribute to the disappearance of the liquid layer. A general overall reaction which might be used to describe this process is

$$\text{Coke precursor (liquid)} \rightarrow \text{coke (solid)} + \text{oil (vapor)} + \text{gas}$$

It is assumed that the time required to flash off the lighter portions of the feed, leaving behind the liquid coke precursor film, is extremely short relative to the destruction of the coke precursors via the above reaction. The coke precursors can be broadly categorized as asphaltenes and resins. At coking temperatures, the asphaltenes and resins precipitate on the bed coke surface to form coke readily.

Gasification

Development of the FLEXICOKING process was a major engineering effort which involved the integration of carbon gasification with the commercially demonstrated FLUID COKING process. During the early 1970s this involved bench-scale laboratory studies, engineering studies, and development of a gasification model which incorporates the kinetics of the gasification reactions, bubble theory, and the physical parameters which describe the behavior of a fluid bed for a commercial-size unit.

These studies showed that a FLEXICOKING gasifier can be described in terms of three distinct sections: an oxidizing zone, a reducing zone, and a dilute phase, as shown in Fig. 1-4. Air and steam are fed into the bottom of the oxidizing zone, where O_2 reacts rapidly with coke to produce CO (reaction 1). The CO then reacts with additional O_2 in the voids between the coke particles to form CO_2 (reaction 2). The oxidizing zone is considered to be a very small portion of the gasifier bed immediately above the gas distributor. The remainder of the coke bed is considered to be the reducing zone, in which the slower reactions of carbon with CO_2 (reaction 3) and with steam (reaction 4) take place. These reactions are accompanied by the very fast water-gas shift reaction (reaction 5), which results in the interconversion of the gaseous reactants and products and fixes the composition of the gas leaving the gasifier. Only a small percentage of the gasification reactions takes place in the dilute phase above the fluidized bed.

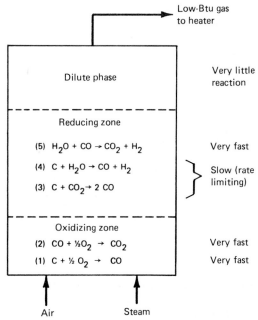

FIG. 1-4 Schematic of coke-gasification reactions occurring in a FLEXICOKING gasifier.

The basic concepts of gasification kinetics are founded on three types of interactions and reactions which influence the overall kinetics. The first type comprises the physical interactions in the gasifier, including the behavior of gas bubbles in the fluid bed and the gross movement of gas to and from the coke particles. The second type comprises the reactions which take place at the surface of the coke particles, such as reactions 1, 3, and 4. Finally, there are the reactions which occur in the gas phase, such as reactions 2 and 5. Reactions 1 and 2 are highly exothermic, while reactions 3 and 4 are highly endothermic and therefore set the bed depth and bed temperature. The air and steam rates must be set to consume the desired quantity of coke. The steam-to-air ratio, bed temperature, and bed depth must be set to balance the endothermic and exothermic reactions to provide the required process heat load.

FLUID COKING AND FLEXICOKING FOR SYNFUELS PRODUCTION

Heavy, hydrogen-deficient feedstocks such as bitumen from the Canadian oil sands, shale oil from Colorado, and coal liquids present unique processing problems. Conventional refineries are not equipped to handle these feeds because of the high content of bottoms and the high levels of nitrogen, sulfur, and ash. One approach to the problem is to upgrade these feeds into a usable crude-oil substitute at the production site. As mentioned previously, this route has been taken by Syncrude Canada, Ltd., Alsands, and Esso Resources Canada, Ltd.

Upgrading Athabasca Oil-Sands Bitumen

An example of how FLEXICOKING can be used as an upgrading process at the production site is shown in Fig. 1-5. The yields are based on whole Athabasca oil-sands bitumen fed directly to the reactor, but the processing scheme is generally applicable to other heavy feedstocks. The feed is thermally cracked to lighter hydrocarbon products and coke, which is gasified to produce a low-Btu gas. The reactor products are separated into C_3^-, C_4's, naphtha, and gas oil.

The C_3^- gas is scrubbed of H_2S and NH_3 and can be used either directly as fuel for the hydrogen plant, or, after hydrotreating, as feed for the hydrogen plant. The C_4's can be used similarly or blended into the motor gasoline pool. The coker naphtha and gas oil are treated in conventional hydrotreating equipment for olefin saturation, denitrogenation, and desulfurization and can then be recombined to produce a high-quality upgraded crude. The properties of the coker naphtha and gas oil prior to hydrotreating are shown in Table 1-3. The properties of the upgraded crude oil produced by FLEXICOKING plus hydrotreating are shown in Table 1-4. Also shown are the properties of the raw bitumen feed and a "premium" Middle East crude oil, Arabian Light. This comparison gives an indication of the quality improvement possible by coking and hydrotreating. Utilization of this processing route results in a low-viscosity, low-sulfur, bottomless crude oil which can be processed by a conventional refinery without vacuum distillation or residuum conversion. The sulfur and nitrogen content of the upgraded crude oil are low and can be varied by changes in hydrotreating severity. The low-Btu gas can be burned in boilers to generate steam and power after desulfurization by using conventional gas-treating technology. In addition, the low-Btu gas is suitable for firing in steam re-forming furnaces. Although not

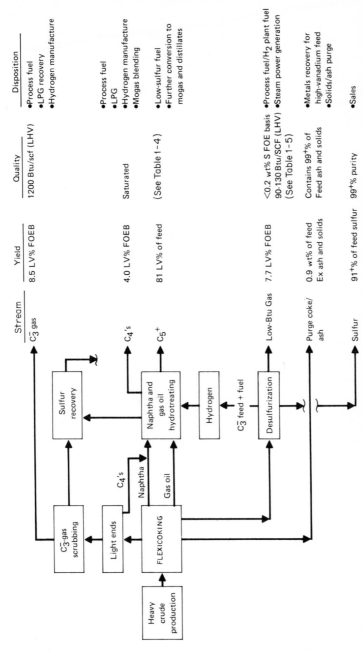

Stream	Yield	Quality	Disposition
C_3^- gas	8.5 LV% FOEB	1200 Btu/scf (LHV)	●Process fuel ●LPG recovery ●Hydrogen manufacture
C_4's	4.0 LV% FOEB	Saturated	●Process fuel ●LPG ●Hydrogen manufacture ●Mogas blending
C_5^+	81 LV% of feed	(See Table 1–4)	●Low-sulfur fuel ●Further conversion to mogas and distillates
Low-Btu Gas	7.7 LV% FOEB	<0.2 wt% S FOE basis 90-130 Btu/SCF (LHV) (See Table 1-5)	●Process fuel/H$_2$ plant fuel ●Steam power generation
Purge coke/ash	0.9 wt% of feed Ex ash and solids	Contains 99$^+$% of Feed ash and solids	●Metals recovery for high-vanadium feed ●Solids/ash purge
Sulfur	91$^+$% of feed sulfur	99$^+$% purity	●Sales

FIG. 1-5 Upgrading of Athabasca oil-sands bitumen via FLEXICOKING and hydrotreating. Shown are typical overall plant yields after FLEXICOKING and hydrotreating an Athabasca oil-sands bitumen with a 13 percent by weight Conradson carbon content. It should be noted that FOEB is defined as a barrel of 10° API fuel oil with a lower heating value of 6.05 MM Btu. LV% is the liquid volume of product divided by the liquid volume of feed, expressed as a percentage. Wt % is the weight of product divided by the weight of feed, expressed as a percentage. Mogas, or motor gasoline, refers to liquid products which can be used as transportation fuels after appropriate treating. Butanes (C_4's) are often blended into the motor gasoline pool to meet vapor pressure specifications for automobile fuels.

TABLE 1-3 Properties of FLEXICOKING liquid products prior to hydrotreating[*]

Fraction	Naphtha	Light gas oil	Heavy gas oil
Nominal cut range, °C	C_5 to 221	221 to 343	343 to 524
Gravity, °API	51.1	23.6	12.4
Sulfur, wt %	2.1	3.2	4.5
Nitrogen, wtppm	60	800	2600
Aniline point, °C	13	32	41
Bromine number, g / 100 g	96	35	21
Nickel, wtppm	——	——	0.2
Vanadium, wtppm	——	——	0.7
Conradson carbon, wt %	——	0.03	1.8

[*]Typical properties of liquid products which are produced after FLEXICOKING an Athabasca oil-sands bitumen with a 13 percent by weight Conradson carbon content.

TABLE 1-4 High-quality synthetic crude oil produced via FLEXICOKING plus hydrotreating

Properties of product	Conventional Arabian Light	Heavy crude oil[*]	
		Raw[†]	Upgraded[‡]
Gravity, °API	34.4	9.2	28.4
Conradson carbon, wt %	3.5	13.0	0.2
Sulfur, wt %	1.7	4.8	<0.3§
Nitrogen, wtppm	530	4,500	<1200§
V + Ni, wtppm	14	330	<1
Ash and solids, wtppm	30	7,500	<1
Viscosity, SSU at 38 °C	45	46,000	<180
C_5^+ distillation, vol %			
C_5 to 221 °C	35	5	23
221 to 343 °C	20	8	26
343 to 566 °C (343 to 524 °C)	32	37	(51)
566 °C+	13	50	——
Total	100	100	100

[*]Raw and upgraded heavy-crude-oil properties are shown for whole Athabasca oil-sands bitumen but can be considered representative of quality improvement attainable with any heavy crude oil.

[†]Feed to FLEXICOKING unit, as shown in Fig. 1-5.

[‡]Combined liquid product (C_5^+) after FLEXICOKING and hydrotreating, as shown in Fig. 1-5.

§Sulfur and nitrogen content will depend upon severity of gas oil and naphtha hydrotreating.

needed for combustion stability, the continuous firing of about 20 percent auxiliary fuel increases flexibility of operation.

Quality and Uses of Coking Products

A 6-month test program at the Baytown prototype FLEXICOKING unit demonstrated that the low-Btu gas is a stable, clean fuel which can burn without supplemental fuel. This gas burns to yield flue gas containing low contents of sulfur, NO_x, and particulates, thus eliminating the need for stack-gas cleanup. At the Toa-Kawasaki facilities, low-Btu gas is used at the refinery in the vacuum pipe still furnace and steam boilers, and a portion of it is sold to a nearby steel mill for use in boilers and rolling mill furnaces. A comparison of low-Btu gas with several conventional fuels is shown in Table 1-5.

Coke from FLEXICOKING and FLUID COKING has a number of possible uses which are dependent upon the ash and sulfur contents of the coke. Because of the higher consumption of coke in FLEXICOKING than in FLUID COKING, the purge coke from a FLEXICOKING unit contains a high concentration of the feed metals and mineral matter as oxides (ash). The gasification mechanism also effects significant desulfurization of the coke. In FLUID COKING, partially burning the coke concentrates the metals somewhat but does not reduce the sulfur content of the product coke. Typical coke properties for several feedstocks are shown in Table 1-6. Coke from FLUID COKING and FLEXICOKING can be burned as fuel. Fluid coke is burned for power generation, and combustion comparisons have shown FLEXICOKING purge coke to be similar to fluid coke. However, the high sulfur levels of some fluid cokes may require stack-gas desulfurization. Electrodes for the manufacture of aluminum can be made from fluid coke in admixture with delayed coke provided the sulfur and metals content of the fluid coke is low. Fluid cokes have also been used in ore-sintering operations and as fuel in cement manufacture; the low sulfur content of FLEXICOKING purge coke makes it an attractive fuel for this application also. Recovery of metals from FLEXICOKING purge coke may be economically attractive. This is especially true when a high-metals feed is processed. Conventional leaching technology can be applied to extract the metals, such as vanadium and nickel, from the FLEXICOKING purge coke.

It should be noted that the reactor-side yields and the qualities of the product from FLUID COKING the Athabasca oil sands bitumen would be very simular to those just discussed from FLEXICOKING. However, because the reactor coke make is partially burned instead of nearly completely gasified, burner off-gas will be produced and the quantity of product coke will increase. The yield of burner off-gas and product coke from an oil-sands operation would be about 0.6 percent by volume on an FOEB (fuel-oil equivalent barrel) basis and 9.5 percent by weight, respectively. The burner off-gas would typically be used as fuel for a CO boiler to raise high-pressure steam. The product coke could also be burned to generate steam or utilized as discussed above.

PROCESS ECONOMICS

The incentives for conversion processing via FLUID COKING or FLEXICOKING are dependent upon many factors such as price and availability of crude, prices of clean products, future product requirements, cost of hydrogen and purchased power, and the degree

TABLE 1-5 Comparison of low-Btu gas with conventional fuels

		Chemical and Physical Properties of Fuel					
Composition*		Typical low-Btu gas	Refinery gas	Natural gas	No. 6 fuel oil	Bituminous coal	Fluid coke
CO		19.9	—	—	—	—	—
CO₂		8.0	—	—	—	—	—
H₂		9.3	—	—	10.0	4.8	2.1
H₂O		5.2	—	—	—	3.5	—
N₂		55.3	—	—	2.0	1.5	—
O₂		—	—	—	—	6.2	—
CH₄		2.3	80.0	—	—	—	—
C₂H₆		—	20.0	—	—	—	—
Carbon		—	—	—	87.6	72.8	90.4
Sulfur		‡	—	—	0.3	2.2	7.3
Ash		Nil	Nil	—	0.1	9.0	0.2
Total		100.0	100.0	100.0	100.0	100.0	100.0
Molecular weight, lb/mol		26.1	18.9	—	—	—	—
Lower heating value (wet), Btu/scf		110.5	1050	—	—	—	—
Btu/lb		1607	21,100	—	17,300	12,600	14,500
Combustion Characteristics							
Adiabatic flame temperature at 20% excess air, °C		1390	1860	—	1845	1905	1815
Fuel requirements per 100 MM Btu	kscf	906	95.2	—	—	—	—
	klb	62.4	4.76	—	5.78	7.93	6.89
Air requirements at 20% excess air	lb air/lb fuel†	1.21	20.3	—	16.7	11.42	14.6
	klb/100 MM Btu†	75.5	96.6	—	96.5	90.6	100.6
Flue gas flow rates, klb/100 MM Btu†		137.9	101.4	—	102.3	98.5	107.5

*Compositions for gaseous fuels are given as vol %. Compositions for fuel oil, coal, and coke are given as wt %.
†Air flow rates are based on 20% excess air, which is a practical value for complete combustion of typical fuels.
‡Typically, less than 0.2 wt % fuel oil equivalent sulfur (after desulfurization).

TABLE 1-6 Coke properties for several feedstocks

Process parameters	Feedstock			
	Athabasca oil-sands bitumen	Boscan	Athabasca oil-sands bitumen	Arabian Heavy
Feedstock properties	4.8	5.2	4.8	4.6
Sulfur, wt %	250	1150	250	104
Vanadium, wtppm	FLEXICOKING		FLUID COKING	
Type of operation	Heater overhead		Burner bed	
Coke stream source				
Coke properties				
Density, g/cm³	1.9 to 2.1		1.4 to 1.6	
Average size, μm	< 10		170 to 220	
Surface area, m²/g	60 to 120		< 10	
Sulfur, wt %	2.6	2.3	7.5	5
Vanadium, wt %	4.7	15	0.3	0.1

of conversion desired, to mention a few. Some generalized economics for FLUID COK-ING and FLEXICOKING is presented in Table 1-7.

Capital Investment

The capital investment, which represents total erected costs for the on-site plant facilities, is based on 1980 U.S. dollars for a U.S. Gulf coast location. The investment reflects unit capacities ranging from 15,000 to 45,000 bbl per stream day (sd) with fresh feed Conradson carbons of 13 to 27 percent by weight. Included are the on-site coking facilities as well as the main coker product fractionator and an electric-motor-driven air blower. Invest-

TABLE 1-7 FLUID COKING and FLEXICOKING process economics

	FLUID COKING	FLEXICOKING
Investment, k$/(bbl·sd) capacity*		
on-site capital cost	1.6 to 2.4	3.4 to 4.8
Utility requirements (per bbl of feed)		
Electricity, kWh	5 to 7	8 to 14
Steam consumed, lb	75 to 130	45 to 175
Steam generated, lb	225 to 275	150 to 250
Cooling water, gal	50 to 60	200 to 450
Boiler feedwater, gal	30 to 35	20 to 35
Instrument and utility air, scf	20 to 25	25 to 30
Operating costs ($/bbl of feed)†		
Investment-related	2 to 3	4 to 6
Throughput-related	0 to 0.5	0.5 to 1.5
Total	2 to 3.5	4.5 to 7.5

*k$/(bbl·sd) capacity is defined as thousands of U.S. dollars (for a U.S. Gulf coast location) divided by the feed rate (bbl/sd) or throughput to the refinery process unit. In the petroleum industry, the standard barrel (bbl) has a capacity of 42 gal and a stream day (sd) is a 24-h period during which the refinery process unit is operating.

†Based on 1980 U.S. dollars for a U.S. Gulf coast location.

ment for a CO boiler on the FLUID COKING unit and for coke gas desulfurization on the FLEXICOKING unit are excluded. The capital investment includes direct material and labor, field labor overheads and erection fees, contractor's detailed engineering, basic engineering costs, and a project contingency of 30 percent. The investments are based on a reimbursable cost contract with a U.S. contractor.

Operating Costs

The operating costs are categorized as investment- and throughput-related. The investment-related operating costs include salaries, wages, benefits, repair material, supplies, taxes, administration, and a capital-recovery charge of 20 percent of the overall plant investment per year for a mature unit with an on-stream service factor of 0.9. This capital-recovery charge provides a 10 percent DCF return on investment. The throughput-related operating costs are based on the utility requirements shown, a fuel value of $35 per FOEB ($5.79 per MM Btu), and the following energy equivalents for various utilities:

Power:	15,000 Btu/kWh
Steam consumed:	1,300 Btu/lb
Steam generated:	1,000 Btu/lb
Cooling water:	70 kW/kgpm
Boiler feedwater:	8,150 kW/kgpm

Utility Requirements

As with the investments, the utility requirements reflect unit capacities ranging from 15,000 to 45,000 bbl/sd with fresh feed Conradson carbons of 13 to 27 percent by weight and are for the on-site coking facilities only. The throughput-related operating costs do reflect a credit for net steam generation.

OUTLOOK

FLUID COKING and FLEXICOKING are capable of upgrading nonconventional feedstocks to synthetic crude-oil substitutes as previously discussed for Athabasca oil-sands bitumen. There are, however, a number of technical and economic issues which must be addressed before selection of the optimum upgrading route for a particular location and feedstock. For example, the technical feasibility of upgrading via FLUID COKING or FLEXICOKING with perhaps subsequent hydrotreating must be evaluated against other possible processing routes including other thermal conversion processes, as well as hydroconversion processes. Here, the degree of development and engineering required before establishing a commercial design basis must be evaluated. Other issues which must be considered are the desired degree of conversion and quality of the upgraded products, the investment and operating costs, and potential environmental issues, as well as the degree of upgrading to be done at the production site. The ability of FLUID COKING and FLEXICOKING to handle a wide range of feedstocks and contaminant levels lends a favorable outlook for future applications of these conversion processes for upgrading nonconventional feedstocks.

THE H-OIL® PROCESS

JOSEPH E. PAPSO

Hydrocarbon Research, Inc. (Dynalectron Corporation)
Gibbsboro, New Jersey

PROCESSING PRINCIPLES

The H-Oil® process is unique in its ability to catalytically process high-metal, very-heavy residuum feeds. Commercial H-Oil® units have been designed and operated to hydrocrack residue for maximum distillate production and to produce low-sulfur residual fuel oils.

The H-Oil process can also be used to demetallize heavy residues so that they can be economically processed in delayed coking and Fluid Catalytic Cracking (FCC) units. It is

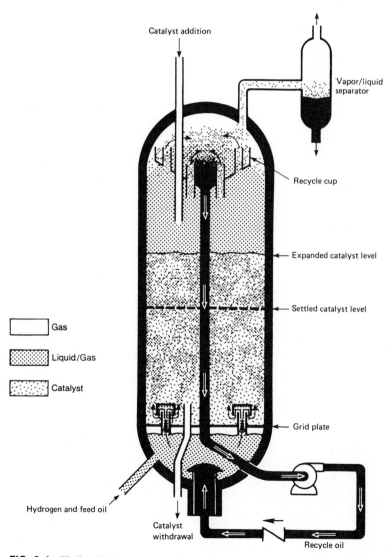

FIG. 2-1 Ebullated-bed reactor.

possible to upgrade feedstocks by reducing sulfur, nitrogen, and Conradson carbon levels to improve yields and product qualities in subsequent processing units.

In the H-Oil process, a mixture of oil and hydrogen is fed upward into an ebullated bed of catalyst. In this bed the catalyst is expanded somewhat in excess of its settled volume and is in a state of motion induced by the velocity of the oil and hydrogen. The ebullated bed avoids the limitations of analogous fixed-bed systems by permitting any fine solids in the feed to migrate through the bed and leave the reaction system. The motion of the catalyst, together with the high velocity of oil in the system, results in intimate contact between oil, hydrogen, and catalyst, effectively promoting the desired chemical reactions. Catalyst attrition is insignificant, since the particles in motion in the ebullated bed are cushioned by an envelope of oil. Furthermore, the catalyst is retained within the volume of the reactor so that it is completely unnecessary to provide means for separation of the catalyst from the product oil and recharging to the system.

Figure 2-1 is a diagram of the ebullated-bed reactor configuration for processing heavy oils. The feed consists of heavy oil and hydrogen, which, in the presence of the catalyst, react with the oil to produce the desired products. Prior to feed introduction, the catalyst is at the indicated lower level. The required catalyst expansion bed level is achieved by introducing the oil and hydrogen feed and circulating liquid from a point above the expanded catalyst level to the base of the reactor. The top of the ebullated bed is sharply defined from the liquid phase above it, unlike the condition in an analogous gas-solid fluidized bed. Catalyst loss by carryover is negligible.

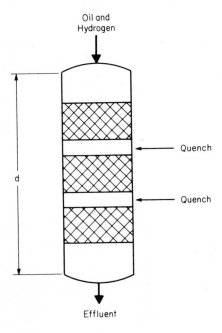

The ebullated-bed reactor offers many advantages over conventional fixed-bed designs. To compare the differences, see Fig. 2-2, which is a sketch of a fixed-bed reactor. Some of the advantages are:

1. Even though large amounts of heat are generated by the hydrogenation reactions, the temperature throughout the reactor is essentially isothermal because the internal and external recycle streams mix rapidly throughout the reactor.

2. Catalyst can be added to or withdrawn from the reactor on a regular basis, without interrupting the operation, thereby enabling a constant level of catalyst activity to be maintained. There is no need to continuously increase the temperature from the start of the run to the end of the run for quality control as is necessary in fixed-bed units.

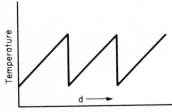

FIG. 2-2 Conventional temperature control in fixed-bed reactor.

3. Catalysts finer than those used in conventional fixed-bed systems can be employed to achieve a higher degree of catalyst activity.

4. Fuels containing some solids can be processed without danger of plugging the catalyst bed. Depending on their size, solids can be removed with the withdrawn spent catalyst or can stay in the product residuum as BS&W.

PROCESS DESCRIPTION

The inherent flexibility associated with an ebullated-bed reactor (the heart of the H-Oil® process) makes it possible to utilize an H-Oil unit in many different applications. For example, an H-Oil unit can be the prime operation of an entire project, such as crude-oil upgrading, where the plant is located in the producing fields. This plant could operate on simple hydrovisbreaking operating conditions to make heavy, tarlike oils pumpable, or the operation could be geared to upgrading the crude oil to make it a more valuable export commodity. The latter case has secondary options:

1. Either demetallize and/or desulfurize.

2. Alternatively, run at higher-severity operating conditions to hydrocrack the heavy oils to an all-distillate crude oil product. Any remaining residuum could be used for producing the hydrogen required for hydrocracking or the steam needed for field enhancement.

High severity has two distinct meanings in hydroprocessing. In conversion operations it refers to the percent by volume of disappearance of 524 °C+. In desulfurization operations, high severity is associated with the percent by weight removal of sulfur in the feedstock at minimum conversion of liquid products to gas. It is a controlled operation at

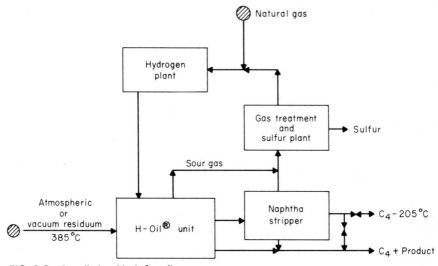

FIG. 2-3 Overall plant block flow diagram.

lower temperature and more reactor volume (space velocity) than high conversion, which is a combination of thermal and catalytic cracking.

Specific economics normally dictate the optimum location of the plant as well as the degree of severity needed to convert the residuum feed to lighter products.

Incorporated in this process description are six figures illustrating the overall plant block flow diagram and typical H-Oil flow sheets that encompass equipment facilities for most H-Oil applications. These flow sheets represent:

Figure 2-3: Overall block flow diagram.

Figure 2-4: Desulfurization. (For high-metals feedstocks, the first-stage reactor could be used for demetallization and the second for desulfurization.)

Figure 2-5: Once-through conversion.

Figure 2-6: Once-through conversion with distillate hydrotreating.

Figure 2-7: Conversion operation with vacuum gas oil recycled to extinction.

Figure 2-8: Conversion operation with vacuum bottoms recycle for high severity.

The basic flow pattern for all the H-Oil operating alternatives follows the same technique; the major differences among the applications are:

1. Series or parallel flow for a multiple-reactor system.

2. Number of pressure letdown levels and extent of liquid fractionation

3. Extent of gas-treatment facilities for recycle hydrogen cleanup

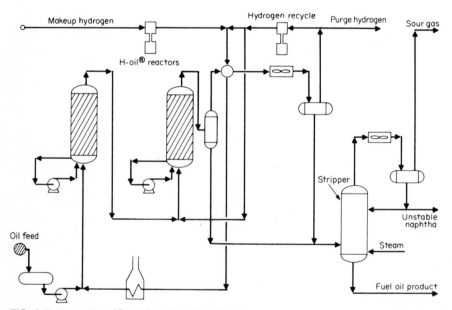

FIG. 2-4 Typical H-Oil® unit flow sheet, desulfurization.

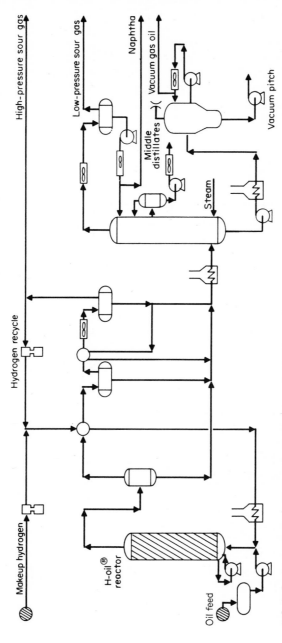

FIG. 2-5 Typical H-Oil® unit flow sheet, simple once-through conversion.

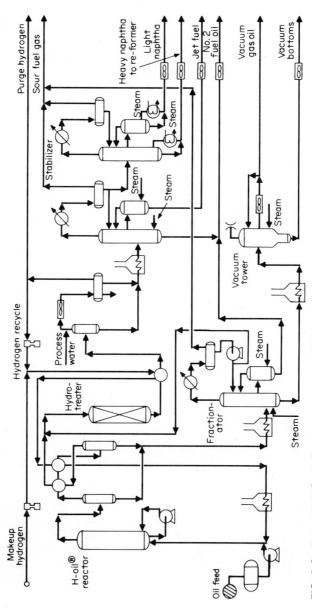

FIG. 2-6 Typical H-Oil® unit flow sheet, conversion with light distillate hydrotreating.

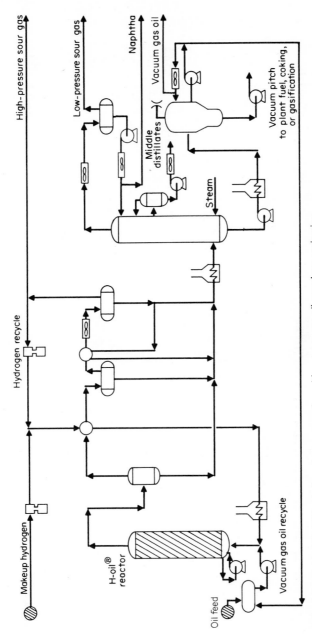

FIG. 2-7 Typical H-Oil® unit flow sheet, conversion with vacuum gas oil recycle to extinction.

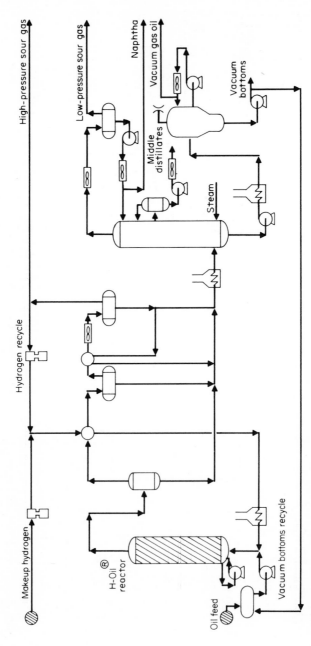

FIG. 2-8 Typical H-Oil® unit flow sheet, high conversion with vacuum bottoms recycle.

In each case the oil and hydrogen streams are heated separately (only the hydrogen heaters are shown in the figures), premixed in a transfer line, and fed to the bottom of the reactor. In a series operation the total fluid from the first reactor (oil plus gas) would go to the second reactor through the bottom, together with cold hydrogen as required to control the desired inlet temperature. Reactor fluid from the final reactor proceeds to a primary separator to split the liquid and gas streams. From this point on, the amount of equipment downstream of the primary separators will depend on the design basis for each particular application.

RESIDUUM-PROCESSING KINETICS

Chemical Reactions

Simultaneously hydrocracking and desulfurizing residues is probably the most commonly considered H-Oil application. This has been the principal application of each commercial H-Oil unit designed and built to date. The catalyst addition and withdrawal feature particularly favors such operation, since even feeds with extremely high metals contents can be processed at a constant catalyst activity without regeneration or shutdown. In addition to the commercial experience, extensive pilot plant background has been developed on stocks ranging from relatively clean mid-continent feedstocks through high-metals California and Venezuelan residues.

H-Oil processing of residues involves two general classes of chemical reactions. First are hydrogenation reactions, such as

1. Desulfurization
2. Hydrogenation of the products resulting from C-C bond cracking
3. Ring and olefin saturation

The rates of these reactions are favored by higher pressure, lower space velocity, higher catalyst activity, and higher temperature. Equilibrium considerations will have an effect on the extent of the saturation.

The second set of reactions involves cracking carbon-carbon bonds. These reactions are generally thermal rather than catalytic and are favored by higher temperature and lower space velocity. Pressure and catalyst activity are of lesser importance and, in the region of usual interest, equilibrium considerations have little effect.

To illustrate the relations between these two general classes of reactions, we have constructed a plot (Fig. 2-9) showing the relation between temperature and space velocity for each of these reaction types.[1] A sour West Texas vacuum residue has been chosen for this example, but a similar set of curves could be constructed for any feedstock. Percent desulfurization has been used to represent the effectiveness of hydrogenation, and percent conversion of material boiling over 524 °C has been used to indicate the extent of C-C cracking. This plot was prepared by using constant conditions of pressure, catalyst activity, and number of reactor stages.

Although the rates of both classes of reactions increase with temperature, the thermal hydrocracking reaction is more temperature-sensitive than the catalytic desulfurization reac-

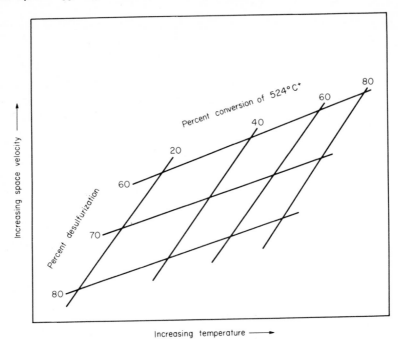

FIG. 2-9 H-Oil® processing of west Texas vacuum residue (conditions: constant pressure and catalyst activity).

tion. As noted earlier, pressure and catalyst activity will not have a great effect on cracking, but they are important for hydrogenation. Consequently, if either, or both, of these parameters are increased, the conversion lines will not change markedly, while the desulfurization lines will be shifted upward.

With a given feedstock, then, the designer or plant operator can choose to operate at whatever point within the grid of Fig. 2-9 best meets the needs of the situation. Further, other sets of constant conditions (pressure, catalyst activity, and number of stages) will result in somewhat different plots. Since the cracking reaction is more temperature-sensitive than desulfurization, and since desulfurization will respond to the other variables, there is a considerable degree of freedom in controlling these factors independently. For example, a plant can be designed for any one of several levels of desulfurization at a constant conversion. Similarly, a plant operator can change from a low-conversion and high-desulfurization operation to a high-conversion operation by raising reactor temperature at constant space velocity. This flexibility is subject, of course, to limitations imposed by hydrogen availability and the original design of the process facility. If, however, provision is made during the design stage for alternative operations, these can be added conveniently and economically.

Figures 2-10, 2-11, and 2-12 illustrate the effects of percent conversion of 524 °C+ in the feed on various qualities of the H-Oil bottoms product. Also included is a second parameter variable called relative reactor volume which can be translated to space velocity or feed rate.

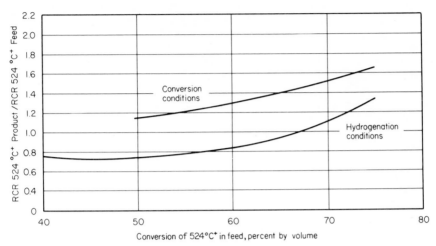

FIG. 2-10 RCR (Ramsbottom carbon residue) vs. conversion.

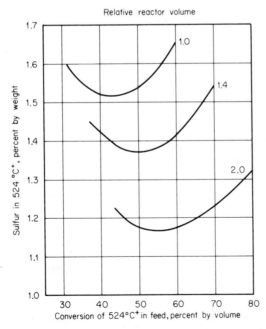

FIG. 2-11 Sulfur of H-Oil® bottoms vs. conversion and relative reactor volume. (Based on 3.3 percent by weight sulfur-H-Oil® feed.)

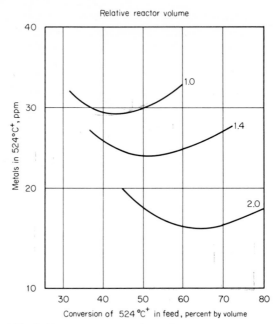

Relative reactor volume

FIG. 2-12   Metals content of H-Oil® bottoms vs. conversion and relative reactor volume. (Based on 69 ppm metals in H-Oil® feed.)

Desulfurization

In addition to the general reaction aspects noted above, there are certain factors of specific importance to the desulfurization of residues. These are discussed below.

Reaction mechanism

Various investigators[2,3] have shown that a first-order kinetic model adequately describes the desulfurization of the individual sulfur compounds contained in petroleum fractions. In the case of residual oils, however, we are not dealing with individual compounds or with a few similar species. Instead, we are dealing with a complex mixture of compounds having widely differing reaction rates. Since the compounds which are easier to react will tend to disappear first and those having a lower reaction rate constant will desulfurize last, desulfurization of these materials, viewed on an overall basis, does not follow a first-order kinetic model. Beuther and Schmid[4] found that desulfurization of residues can be represented adequately by use of a second-order model. These authors recognize that the individual reactions were probably first-order but that a second-order model would best represent the overall data. By using a pseudo-second-order approach, the fact of increasing difficulty of desulfurization is reflected in the concentration term of the rate equation, enabling one to use a fixed reaction rate constant.

HRI's work in this area has confirmed the observations of Beuther and Schmid. For the desulfurization of Kuwait atmospheric residue, average first- and second-order K values were calculated from data obtained under various operating conditions using fresh catalyst. Curves were then prepared and compared with the data: it was apparent that the second-order relation more accurately represented the situation. In using relationships, however, it should be borne in mind that this is an empiricism, and data extrapolations must be treated with caution.

The reaction rate in the final stages of a process will sequentially be reduced as market requirements dictate a lower-sulfur-content product.[5] With a second-order model the reaction rate in the final incremental portion of a reactor system at the 0.3 percent sulfur level will be only one-tenth what it would be at the 1 percent sulfur level.

Effect of H₂S on reaction rate

As the desulfurization reaction proceeds, H_2S is produced. This material, while mainly in the vapor phase, is in equilibrium with a concentration of dissolved H_2S in the liquid. Under certain conditions the mass action effect of this material can have a strong effect on the overall rate of the desulfurization reaction. When operations are directed to achieving very low levels of sulfur in the product, this effect can have important design consequences. The maximum desulfurization that can probably be economically justified would involve three reactors in series with catalyst counterflow and split hydrogen recycle to reduce H_2S concentration in the final stage.

Reactor staging

For practical purposes, the H-Oil reactor can be considered a completely back-mixed system. In such a reactor, the character of the reacting mixture is essentially identical to the material leaving the reactor. From the standpoint of the desulfurization reaction the second-order rate equation,

$$r = kc^2$$

shows that the reaction rate will be proportional to the square of the concentration of the effluent. As the process is required to produce a lower-sulfur-content product, the reaction rate will decrease rapidly. As noted earlier, the rate of removal at 0.3 percent sulfur concentration will be only one-tenth that at 1 percent sulfur. In theory, then, a single reactor to produce a 0.3 percent sulfur product would be well over ten times as large as one producing a 1 percent sulfur product, all other factors being equal.

In the H-Oil process this problem is solved by staging the reactor system, using two or more reactors in series. In this way the reaction is carried out at several decreasing sulfur contents. This permits higher reaction rates in those reactors in which the bulk of the desulfurization occurs.

Catalyst counterflow

A further area of design optimization relates to the catalyst-aging effects. If three H-Oil stages in series are used to achieve a very high degree of desulfurization, there is a distinct advantage to charging all of the makeup catalyst to the third reactor stage. This material would then be withdrawn and charged to the second stage; in like fashion, the second-stage catalyst would be charged to the first reactor. This accomplishes a countercurrent flow of oil and catalyst in which the freshest catalyst is exposed to the cleanest oil in the reactor, the oil which requires a high degree of catalyst activity in view of the low sulfur concentration present. At the same time, the faster reactions take place in a reactor in which the catalyst has reached its final equilibrium level prior to being discarded. Back-staging of catalyst in a three-stage system indicates that virtual equivalence is attained between the H-Oil system and a theoretical plug flow reactor.

DEVELOPMENT AND EXPERIENCE

Background

The need for effective hydrogen processing of residual oils had been apparent for some time. Heavy fuel oil continued to be a product of low value because of its sulfur, nitrogen, and metal content. Earlier attempts had been made to develop a residual hydrogenation process, but failure always resulted because of one of three main problems inherent in fixed-bed reactors:

1. The heat of hydrogen reacting with residuum produced an excessive temperature because of the highly exothermic reactions which caused coke to form in the catalyst bed.
2. Entrained solids accumulating in the catalyst bed resulted in high-pressure drop across the bed.
3. Frequent shutdown was needed to replace spent catalyst.

Following World War II, people in the petroleum and petrochemical industries realized they were rapidly approaching a time when processing residuum would not only become economically attractive but downright necessary. The major contributing factors were:

1. Projected exhaustion of light, sweet crude-oil reservoirs and the extreme unlikelihood of finding new fields to produce low-sulfur, high-gravity crude oil.
2. Increasing influence of the EPA in establishing emission standards for industrial plants, large residential complexes, and automobile exhausts. Each year the standards became more stringent and capital outlays were being made by all concerned to reduce SO_2 and NO_x emissions necessary to remain within the law.
3. Availability of the C_5-to-427 °C heart cut of the barrel of oil for which each phase of refining and petrochemical operations were vying as a feedstock. This material was in demand for catalytic re-formers, cat crackers, steam crackers, heating oils, and fuel-oil diluents. Soon it became apparent that any boiling range material within the C_5-to-427 °C range would have the same value because of its demand.

HRI began work on the H-Oil program in 1953. It went through a 2-year period of developing handling techniques and studying both catalyst and inert media. In 1955, a 25-bbl/day pilot plant was put into service in order to test the process on a commercial level. Five continuously operating pilot plants with feed-rate capacities ranging from 5 gal/day to 30 bbl/day were built. By 1960, HRI had confirmed that the ebullated-bed reactor principle had distinct advantages over fixed-bed reactors for processing high-sulfur, high-metal feedstocks.

In 1961, the first commercial H-Oil plant went into the design stage; it was a 2500-bbl/

day unit to be built in the Cities Service refinery at Lake Charles, Louisiana. Several engineering innovations were introduced in the design of the Lake Charles unit. Internal recycle pumps were selected to avoid the costs associated with external recycle lines and valves and to avoid potential difficulties with the high-pressure seals required for external pumps. Subsequent commercial experience with external recycle pumps proved that fears over high-pressure seals were unfounded. As developed, the external pump requires only one seal, the same as the internal pump. Only one valve is required; this is a very desirable feature of the recycle system. Also, the incremental cost of an external pump is relatively low and the benefits gained are well worthwhile. A hydraulic oil system was selected for the pump drive because of easy speed control and for its convenience as a coolant system for the drive motor. In other commercial experience, a submerged-type electric motor has been selected for the pump drive and has proved to be very successful. The unit was completed and placed on stream in 1963.

Experience

Laboratory experience

Since 1953, the work directly related to the H-Oil development has included over 150,000 hours of operating experience on a bench scale and with a process development unit (PDU). Feedstocks from all over the world have been processed. The feed materials range from atmospheric and vacuum residual oils to naphthenic and paraffinic crude oils, whole heavy crude oils, coal tars and coal extract, tar sands, and shale oil. Summaries of the work are given in Table 2-1, 2-2, and 2-3.

Commercial experience

For about the first 5 years, the Cities Service 2500-bbl/day unit was operated at the discretion of CSRD, the R&D division of Cities Service Company, to demonstrate that any new developments in equipment or catalyst types, shapes, and activities that were tested in the HRI laboratory could be duplicated in the larger unit. In 1967, the unit was taken over by the Cities Service Company, which expanded its capacity to process 6000 bbl/day of

TABLE 2-1 HRI's data base

Source of feedstocks tested
  Canada
  Egypt
  Indonesia
  Iran
  Iraq
  Kuwait
  Libya
  Malaysia
  Mexico
  Neutral Zone and Saudi Arabia
  Turkey
  United States
  Venezuela

TABLE 2-2   Accumulated hours of H-Oil® experience in bench-scale and pilot units

Feedstock	Hours
Petroleum source	
Athabasca bitumen	12,454
Alaskan	1,165
Bati Raman	140
Boscan crude oil	2,095
Brega	2,121
Buzurgan	405
California heavy crude oil	1,077
Cold Lake	1,093
Elwood vacuum bottoms	220
Eocene	2,770
Egyptian bottoms	239
Gach Saran	2,027
Iranian light	1,075
Iraq bottoms	489
Kirkuk	1,535
Khafji	5,612
Kuwait	51,350
Light Arabian	2,769
Light and Medium Arabian vacuum-bottom blend	1,911
Magrip	271
Medium Arabian vacuum bottoms	251
Mexican blends	2,837
Mid-continent blends	6,251
Ratawi	291
Safaniya	252
Texas (east and west)	27,819
Tia Juana	1,734
Umm Gudair	491
Venezuelan light	2,536
Venezuelan heavy	2,209
Miscellaneous U. S. blends	11,356
Coal source	
Coal tar and coal extract	2,862
Solvent-refined coal	761
Shale oil	1,197

high-sulfur West Texas residuum as a routine refinery operating unit. The objective of this expanded operation was to process feedstock for low-sulfur coke production in the Lake Charles refinery delayed-coking units.

The second commercial H-Oil unit was built at the Shuaiba Refinery of Kuwait National Petroleum Company (KNPC) in 1968. It represented the largest residual hydrocracker in the world, with two 13-ft-diameter reactors designed to run in parallel at throughputs of

TABLE 2-3   HRI's experience in two-stage H-Oil® bench-scale operations

Feedstocks	Hours
Buzurgan atmospheric residuum	405
Heavy Iranian atmospheric bottoms	582
Heavy Iranian vacuum bottoms	1467
Iraq atmospheric residuum	489
Kuwait atmospheric residuum	2951
Kuwait vacuum residuum	5597
Khafji atmospheric residuum	2349
Khafji vacuum residuum	1754
Light Arabian atmospheric residuum	352
Light Arabian vacuum residuum	781
Light and medium Arabian vacuum bottoms blend	1420
Lloydminster and Wainwright atmospheric residuum	863
Medium Arabian vacuum bottoms	251
Mixed blends (vacuum bottoms)	3222
Miscellaneous feeds	7605
Safaniya atmospheric residuum	382
Venezuelan atmospheric residuum	3400
Venezuelan vacuum residuum	764
West Texas atmospheric residuum	2207
West Texas vacuum residuum	9418

14,400 bbl/sd in each vessel. Operating severity was based on 75 percent conversion of 524 °C+ material present in a 482 °C+ vacuum residuum feed at the total throughput of 28,800 bbl/sd. The overall objective of the KNPC refinery was to maximize light and heavy distillates for the European market. Unfortunately, before the refinery was even completed, the middle-eastern Arab-Israeli conflicts resulted in the closing of the Suez Canal and thereby eliminated the marketing advantages Kuwait was to have enjoyed.

To compensate for KNPC's new markets, the product slate was revised to production of low-sulfur fuel oil instead of maximum distillates. The H-Oil unit operating conditions were switched from a basic conversion unit to a desulfurization process for the purpose of removing maximum pounds of sulfur from the refinery's overall fuel-oil pool. KNPC economics dictated to maximize throughput in the H-Oil unit rather than to desulfurize to a low level of sulfur at reduced capacity. This versatility of the H-Oil process is one of the unique advantages of the ebullated-bed reactors; the H-Oil unit provides a refiner with the capability to operate a processing plant to meet market demands. At KNPC, the throughput under the desulfurization mode of operation averages 53,000 bbl/sd, which is 85 percent above the design capacity of 28,800 bbl/sd at high conversion.

Figure 2-13 presents the chronology of H-Oil development through commercialization and projects through 1985. Figure 2-14 summarizes significant data related to each commercial H-Oil unit which has been built or has reached the final design stage. The data are naturally limited to information of a nonconfidential nature. Figure 2-15 presents data illustrating the success achieved by the KNPC with its H-Oil unit operating in the all-hydrogen Shuaiba refinery. This H-Oil unit has clearly illustrated its ability to meet design

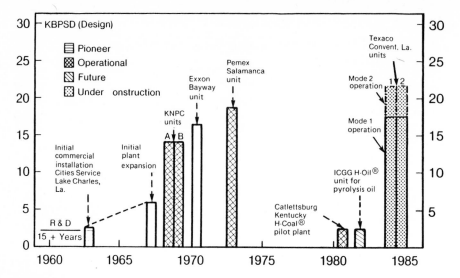

FIG. 2-13 Chronology of H-Oil®. Key: kB/sd, thousand barrels per stream day; CS, Cities Service; KNPC, Kuwait National Petroleum Company; BW, Bay Way; PEMEX, Petroleos Mexicans; LA, Louisiana; KY, Kentucky; ICCG, Illinois Coal Gasification Group.

HRI has participated in various design, engineering, and construction phases of the following H-Oil® projects:

A. Cities Service, Lake Charles, La.
    Capacity: 2500 bbl/sd
    Reactor
        Type   Hot wall: Wrapped
        Materials   Shell: 1146 steel
                  Liner: 3 Cr with 347 weld overlay
                  Heads: 3 Cr-Mo

B. Kuwait National Petroleum Co., Shuaiba, Kuwait
    Capacity: 28,800 bbl/sd
    Reactor
        Type   Cold wall: Wrapped
        Materials   Shell: Welcon 2HS steel
                  Liner: 347 SS
                  Heads: C–½ Mo

C. Humble, Bayway, N.J.
    Capacity: 16,500 bbl/sd
    Reactor
        Type   Hot wall: Wrapped
        Materials   Shell: 1147 steel
                  Liner: 347 SS clad
                  Heads: 3 Cr-1 Mo

FIG. 2-14 H-Oil® facilities. Key: Cr, chromium; Mo, molybdenum; HS, high strengh; SS, stainless steel.

D. Pemex, Salamanca, Mexico
   Capacity: 18,500 bbl/sd
   Reactor:
      Type   Hot wall: Monobloc
      Materials Shell: 2¼ Cr–1 Mo
                 Liner: 347 SS weld overlay
                 Heads: 2¼ Cr–1 Mo

E. Humble, Baton Rouge, La.
   Capacity: 17,600 bbl/sd
   Reactor
      Type   Hot wall: Monobloc
      Materials Shell: 3 Cr–1 Mo
                 Liner: 304L clad or 308L overlay
                 Heads: 3 Cr–1 Mo

F. Great Canadian Oil Sands (GCOS), Ft. MacMurray, Canada
   Capacity: 15,400 bbl/sd
   Reactor
      Type   Cold wall: Monobloc
      Materials Shell: 2¼ Cr–1 Mo
                 Liner: 347 weld overlay
                 Heads: 2¼ Cr–1 Mo

G. Illinois Coal Gasification Group (ICGG), Peoria, Ill.
   Capacity: 2500 bbl/sd
   Reactor
      Type   Cold wall: Monobloc
      Materials Shell: 2¼ Cr–1 Mo
                 Liner: 347 weld overlay
                 Heads: 2¼ Cr–1 Mo

H. Texaco, Convent, La.
   Capacity: 35,000 to 43,000 bbl/sd
   Reactor
      Type   Cold wall: Monobloc
      Materials Shell: 2¼ Cr–1 Mo
                 Liner: 347 weld overlay
                 Heads: 2¼ Cr–1 Mo

FIG. 2-14 (*Continued*)

conditions, flexibility in operating severity, and the plant reliability (service factor), the three criteria of a successful refinery operating unit.

The Pemex H-Oil unit located at the Salamanca, Mexico refinery was licensed by CSRD during a period of joint H-Oil ownership with HRI. This agreement has expired, and the unit's operation is now monitored by CSRD.

The Texaco unit, under construction at the time of writing, is located at Convent, Louisiana, and is expected to come on stream in late 1984 or early 1985. The unit will consist of two trains and is designed for dual modes of vacuum residuum processing. It will be capable of processing either 43,000 bbl/sd of fresh vacuum residuum feed, once through at 65 percent conversion of the 538 °C+ material, or of processing 35,000 bbl/sd of fresh feed at up to 90 percent conversion of the 538 °C+ material using vacuum-bottoms

- 14 years of commercial experience
- Operating well above design performance

	Design	Actual	% Over
Feed rate, bbl/sd	28,800	53,000	85
Vacuum residuum converted, bbl/sd	21,600	33,400	55

- Achieved 20-month run length
- Units available on average 95 percent of time between reactor turnarounds based on 10-year period.
- Operating data confirm laboratory correlations over wide range of conditions

FIG. 2-15 H-Oil® commercial operations at Kuwait.

recycle. (Conversion is normally referred to as the disappearance of the 524 °C+ material, but the actual cut point varies with refiners. Cut points most commonly used are 524 °C, 538 °C, and 565 °C).

CURRENT ACTIVITIES AND PROJECTED POTENTIAL

On a worldwide basis the availability of total gas and crude oil is rapidly diminishing. The known reserves of medium and light grades of crude oil are decreasing at a much faster pace than those of heavy crude oils. The so-called bottom of the barrel, which one can assume is either atmospheric or vacuum residuum, is getting to be a bigger share of the whole crude oil barrel. These residua are increasing in weight (lower API gravity) and have a higher content of sulfur, carbon, nitrogen, and metals; all these qualities make them undesirable feedstock material for normal refinery units—*except* H-Oil units.

H-Oil process capability and flexibility were described in detail in earlier sections of this chapter. Heavy crude oils, tar-sands bitumen, and kerogen from shale can be upgraded to high-quality refinery feedstocks via the H-Oil process. Alternatively, an H-Oil unit will be located within a refinery battery limits to process residual stocks at predetermined optimum severities to either desulfurize for low-sulfur fuel oil production or hydrocrack the 524 °C+ residue into naphtha and distillate products lighter than 524 °C+.

A very significant advantage of the H-Oil process has been the development of high-conversion technology. Residual oil conversion is generally kept below 75 liquid volume percent in single-pass operation. Attempts to increase conversion above a certain level, e.g., by increasing reactor temperature or residence time in an existing unit, will cause undesirable side reactions. These reactions produce higher gas yields and increase the hydrogen consumption.

To meet the challenge of increasing residual oil conversion, HRI has successfully extended the H-Oil process to high conversion, using its patented vacuum-bottoms recycle technique. This technique is an integral part of the analogous and related H-Coal® process, which uses the ebullated-bed reactor for liquid-phase hydroconversion of coal to distillate oil liquids. (The H-Coal process is covered in a separate chapter).

Beginning in the early 1970s, HRI carried out a large number of H-Oil runs using its vacuum-bottoms recycle technique in both bench-scale units and in a 30-bbl/sd PDU. Feedstocks including Canadian, Middle East, and U.S.-derived residua have been run, and conversion levels of 85 to 95 percent have been successfully demonstrated.

Vacuum-bottoms recycle increases the residual oil concentration in the H-Oil reactor liquid and thereby favorably increases the rate of the hydroconversion reactions. The presence of additional higher-boiling materials also suppresses vaporization of H-Oil products. These factors together account for the higher liquid products yield in recycle versus once-through operation.

Even at the same conversion level, hydrogen consumption is lower with vacuum-bottoms recycle than with the once-through mode because the yield selectivity is away from light gases in favor of gas oil products.

Facilities required to achieve vacuum-bottoms recycle in an existing single-pass, H-Oil unit represent a small portion of the total capital requirement. In any event, the additional cost can be readily justified in terms of the potential economic benefits. The exact benefits accruing to high-conversion operation are of course strongly influenced by the feedstock, the H-Oil yields, the relative pricing of the products, and by other site-specific economic factors.

According to various published sources,[6] a shortage of crude oil and natural gas will be felt by 1990 at the current rates of consumption and production. On the basis of estimates of construction periods (from contract signing to start-up), a minimum of five 50,000-bbl/day residuum-processing plants should be in the various stages of design and engineering requisite for final completion of all five plants by 1990 in order to prevent the world shortage of oil and gas.

Table 2-4 lists examples of processing applications that are being considered very seriously for near-term construction. Tables 2-5 to 2-9 illustrate various case studies completed recently for potential clients. Included are feedstock properties, nonconfidential operating conditions, process yields, and product properties. These estimates are based on actual laboratory data from processing each of the feedstocks shown. The following is a brief description of applications in which intense interest has been shown.

TABLE 2-4   Current H-Oil® applications under consideration

Feedstock				Operations Objectives		
Type	Crude source	Specific gravity, °API	S, %	Conversion, %	Desulfurization, %	End use
Coal tar	Kentucky coal	−6.2	1.61	78	96	Synthetic crude oil
Bitumen	Alberta tar sand	7.3	4.75	60	55	Coker feed
Atmospheric residuum	Lloydminster	8.9	4.60	90	70	Synthetic crude oil
Whole crude oil	Orinoco	8.6	4.09	44	60	Synthetic crude oil
Atmospheric residuum	Heavy Arabian	12.3	4.35	85	70	Distillate production

TABLE 2-5 H-Oil® processing of coal-derived oil from west Kentucky coal; estimated equilibrium yields

Feedstock properties		Operating data	
Specific gravity, °API	−6.2	485 °C+ conversion, vol %	78
Sulfur, wt %	1.61	Desulfurization, wt %	96
Nitrogen, wt%	1.10	Denitrogenation, wt %	46
Oxygen, wt %	7.49	Chemical hydrogen	
Ash, wt %	0.43	consumption, scf/bbl	2450
Quinoline insolubles, wt %	2.0	Catalyst replacement	
485 °C+ fraction, vol %	40.3	cost, ¢/bbl (1983 net $)	49

Yields and Product Properties

Fractions	wt %	vol %	°API	S, wt %	N, wt %	C, wt %	H, wt %	O, wt %
H_2S	1.7							
NH_3	0.7							
$CO + CO_2$	0.5							
H_2O	6.6							
C_1	1.1							
C_2	1.2							
C_3	1.3							
C_4 to 204 °C	11.4	16.8	53	——	0.13	86.80	12.9	0.1
204 to 345 °C	40.9	48.5	17	——	0.52	87.00	10.5	1.7
345 to 485 °C	28.4	31.2	6	<0.03	0.77	88.00	9.7	1.5
485 °C+	9.5	9.0	−12	0.5	1.00	88.60	7.7	2.0
Total (average C_4 +)	103.3	105.5	(15.1)	(0.06)	(0.60)	(87.45)	(10.25)	(1.47)

TABLE 2-6 H-Oil® processing of Athabascal Bitumen; 60% conversion H-Oil® followed by FLUID COKING (basis, 1000 tons bitumen)

	H-Oil® at 60% 524 °C+ conversion		Fluid Coking		Net Balances			
	tons	bbl	tons	bbl	tons	wt %	bbl	vol %
Bitumen feed	(1000)	(5675)			(1000)	(100.0)	(5675)	(100.0)
Hydrogen	(8)				(8)	(0.8)		
CO_x/H_2O	3				3	0.3		
H_2S/NH_3	23		5		28	2.8		
C_1,C_2,C_3	27		21		48	4.8		
C_4	9	85	6	56	15	1.5	141	2.5
C_5 to 225 °C	129	953	46	343	175	17.5	1296	22.8
225 to 345 °C	249	1600	8	53	257	25.7	1653	29.2
345 to 524 °C	346	2032	56	318	402	40.2	2350	41.4
524 °C+	222	1169	(222)	(1169)				
Burned coke			33		33	3.3		
Net coke			47		47	4.7		
Net liquid products		4670		770			5440	95.9

TABLE 2-7 H-Oil® processing of Lloydminster atmospheric bottoms; 90% conversion with vacuum-bottoms recycle

Feedstock properties		Operating data	
Specific gravity, °API	8.9	Conversion, vol %	90
Sulfur, wt %	4.6	Desulfurization, wt %	69.9
Oxygen, wt %	0.72	Chemical hydrogen consumption, scf/bbl	1180
Nitrogen, wt %	0.36	Catalyst replacement cost, ¢/bbl	54
Vanadium, wt ppm	144	(1983 net $)	
Nickel, wt ppm	76	No. of stages	1
Hydrogen, wt %			
(estimated)	10.69		
Carbon, wt %			
(estimated)	83.72		
524 °C+, vol %	58		

Yields and Product Properties							
Fractions	wt %	vol %	°API	S, wt %	N, wt ppm	(C + O), wt %	H, wt %
H₂S and NH₃	3.5						
H₂O	0.4						
C₁	1.1						
C₂	1.0						
C₃	1.6						
C₄ to 205 °C	16.3	22.5	62.0	0.14	345	85.60	14.47
205 to 345 °F	24.5	28.3	31.0	0.82	1,205	86.52	12.72
345 to 524 °C	47.1	49.9	17.1	1.97	3,600	86.49	11.18
524 °C+	6.3	5.8	−2.0	3.58	10,420	86.58	8.75
Total (average							
C₄+)	101.8	106.5	(27.2)	(1.47)	(2870)	(86.3)	(11.99)

Upgrading tar-sands bitumen[7]

Processing various heavy-oil and tar-sands reserves found around the world is a tailor-made operation for H-Oil because of some sand and other impurities that remain in the recovered bitumen. The sand in the bitumen passes harmlessly through the H-Oil reactors; some of the lighter sand may pass overhead with the effluent stream, and the balance will be removed together with spent catalyst that is withdrawn. The sand or particulate-matter content will determine the final disposition of H-Oil vacuum-tower bottoms. Options that have been considered are:

Fluid coking or delayed coking following a low-conversion H-Oil operation

Partial oxidation for production of hydrogen and/or low-Btu syngas for plant fuel

Fuel oil blending for export or plant fuel

Over 12,000 hours of experience in bench-scale and pilot units have been accumulated in processing tar-sands derivatives under a wide range of operating severities.

TABLE 2-8 H-Oil® processing of Orinoco crude oil, single-stage demetallization

Feedstock properties		Operating data	
Specific gravity, °API	8.6	524 °C+ conversion, vol %	44
Sulfur, wt %	4.09	Desulfurization, wt %	60
Nitrogen, wt %	0.40	Demetallization, wt %	83
Carbon, wt % (estimated)	84.84	Chemical hydrogen consumption, scf/bbl	530
Hydrogen, wt % (estimated)	10.62	Catalyst replacement cost, c/bbl (1983 net $)	25
RCR, wt %*	13.4		
Nickel, wt ppm	99		
Vanadium, wt ppm	417		
Vol %, 524 °C+	58		

Yields and Product Properties

Fractions	wt %	vol %	°API	S, wt %	N, wt ppm	C, wt %	H, wt %	V, wt ppm	Ni, wt ppm	RCR, wt %*
H₂S and NH₃	2.7									
C₁	0.6									
C₂	0.5									
C₃	0.7									
C₄ to 205 °C	8.0	11.0	61	0.10	120	85.45	14.43			
205 to 345 °C	20.00	23.0	30	0.60	650	86.52	12.81			
345 to 524 °C+	34.2	35.7	15	1.51	3000	86.84	11.35			
524 °C+	34.1	32.5	2	2.91	6000	86.89	9.58	178	77	0.8
Total										
(average C₄ +)	100.8	102.2	(16.9)	(1.70)	(3330)	(86.67)	(11.28)	(64)	(28)	28

*RCR = Ramsbottom carbon residue.

TABLE 2-9 H-Oil® processing of Arabian heavy atmospheric bottoms; 85% 524 °C+ conversion

Feedstock properties	
Boiling range	343 °C+
Specific gravity, °API	12.3
Sulfur, wt %	4.35
Carbon, wt % (estimated)	84.41
Hydrogen, wt % (estimated)	10.95
Nitrogen, wt %	0.29
CCR, wt %*	
Vanadium, wt ppm	85
Nickel, wt ppm	27
524 °C+, vol %	58.4

Operating data, vacuum bottoms recycle	
No. of stages	1
524 °C+ conversion, vol %	85
Desulfurization, wt %	70
Chemical hydrogen consumption, scf/B	1050
Catalyst replacement cost, ¢/B (1983 Net $)	43

Yields and Product Properties

Fractions	wt %	vol %	°API	S, wt %	N, wt ppm	H, wt %	C, wt %	CCR, wt %*	V, wt ppm	Ni, wt ppm
H$_2$S & NH$_3$	3.4									
C$_1$	1.4									
C$_2$	1.3									
C$_3$	1.5									
C$_4$ to 205 °C	17.8	24.0	62	0.15	210	14.48	85.35			
205 to 345 °C	26.9	31.0	34	0.74	690	13.04	86.15			
345 to 524 °C+	39.5	41.1	18	1.80	1970	11.42	86.59	0.5		
524 °C+	9.8	8.8	-3	2.95	8280	8.75	87.48	33.1	93	88
Total	101.6	104.9								
(average C$_4$ +)			(29.0)	(1.30)	(1930)	(12.19)	(86.32)	(3.7)	(9.2)	(8.7)

*CCR = Conradson carbon residue.

Upgrading coal-derived pyrolysis oil[8]

Pyrolysis oil derived from coal as a by-product of char manufacture is of little value because of its properties and high concentration of contaminants. It cannot be used as a fuel oil or even a petrochemical feedstock without hydroprocessing over a synthetic catalyst. The raw pyrolysis oil has a high specific gravity (-4 to $-6°$ API), a low hydrogen content, and high concentrations of oxygen, nitrogen, and sulfur. The char fines entrained in the raw oil range from 2 to 10 percent by weight and average about 5 μm in size. Application of H-Oil for upgrading pyrolysis naphtha provides processing flexibility to manufacture fuel oil or a wide-boiling-range distillate for combustion-turbine fuel. Almost 3000 hours of H-Oil experience have been accumulated in bench-scale and pilot units with pyrolysis oil, coal tars, and coal extracts.

Upgrading solvent-refined coal

Upgrading solvent-refined coal (SRC) represents one of the latest applications of the ebullating-bed H-Oil process. The objective is to economically modify a solid product from the SRC liquefaction process by catalytic hydrogenation in order to end up with a marketable synthetic crude oil. After over 750 hours of laboratory experience on bench-scale and pilot units, the results were extremely encouraging. Funding was arranged to install H-Oil facilities in conjunction with SRC operating plant facilities in Wilsonville, Alabama. The Wilsonville SRC plant is located within the power plant complex of the Southern Company Services, Inc.; it is an EPRI-sponsored program with Catalytic Inc. under a subcontract to be the operators of the plant. The H-Oil unit has been successfully run using a number of different H-Oil catalysts in both "boiler-fuel" and "all-distillate" modes of operation. Commercially acceptable levels of sulfur and nitrogen have been demonstrated.

The estimated current (1981) investments for a maximum-size single-train, H-Oil unit can vary anywhere from $2000 to $4000 per barrel per calendar day installed capacity in the U.S. Gulf coast. This variation, of course, is based on the severity of operation, which in turn determines throughput for a maximum reactor size that can be fabricated in terms of diameter, weight, or some shipping limitation of specific plant locations.

REFERENCES

1. Johnson, Axel R., Joseph E. Papso, Raymond F. Hippeli, and Govanon Nongbri: "H-Oil™ for Residue Elimination and Low Sulfur Coke Production," 35th Midyear Meeting API Division of Refining, May 1970.
2. Rosen, L: *Oil Gas J.*, vol. 57, no. 18, April 27, 1959, p. 153.
3. Banks, W. T., et al.: *Oil Gas J.*, vol. 58, no. 16, April 18, 1960, p. 131.
4. Beuther, H., and B. K. Schmid: section III, paper 20, PD7, World Petroleum Congress, Frankfurt, 1963.
5. Johnson, Axel R., Ronald H. Wolk, and Govanon Nongbri: "Production of Low Sulfur Fuels by the H-Oil Process," Japanese Petroleum Institute, November 30, 1972.
6. Evans, F.: 1979 HPI Market Data, Gulf Publishing Company, August 1978, pp. 2, 3.
7. Johnson, C. A., G. Nongbri, L. M. Lehman, and M. C. Chervenak: "Conversion of Bitumen and Heavy Crude Oils to More Valuable Products by the H-Oil Process," 28th Annual Meeting of the Petroleum Society of CIM, Edmonton, Alberta, June 1977.
8. Eccles, R. M., G. Nongbri, and R. A. Closius: "H-Oil Upgrading of Coal Derived Pyrolysis Oil," EPRI Sponsored Conference on Coal Pyrolysis, Palo Alto, Calif., February 1981.

LC-FINING PROCESS

RICHARD S. CHILLINGWORTH
JOHN D. POTTS
KENNETH E. HASTINGS
CHARLES E. SCOTT

Cities Service Research & Development Co.
Tulsa, Oklahoma

INTRODUCTION

LC-Fining is a catalytic hydrotreating process for upgrading heavy hydrocarbons including gas oils, residua, coal liquids, tars, and shale oil. Its outstanding feature is the reactor, which is of an upflow design that keeps the catalyst bed in an expanded state. Severity and catalyst selection may be varied for either hydrocracking or hydrodesulfurization. The process is licensed by C-E Lummus Company and Cities Service Company.

Research on the LC-Fining process began at Cities Service more than 20 years ago. The process was originally developed for upgrading bitumen from tar sands. Its first commercial applications were not in the synthetic fuels, but rather in upgrading petroleum residua.

The technology which led to the LC-Fining process was first demonstrated in 1963 at Cities Service's Lake Charles refinery with the construction of a 2500-bbl/day (16.6 m³/h) unit. The unit was later expanded, in 1971, to 6000 bbl/day (39.8 m³/h) to maximize sulfur removal from delayed coker feedstock.

AMOCO has recently licensed a 55,000-bbl/day LC-Fining unit for its Texas City, Texas, refinery. Feed from this unit will include residua from difficult-to-process heavy Mexican Maya crude oil.

Process description and details

The operation of an LC-Finer is best described by means of a process flow schematic (Fig. 3-1). Feed oil and hydrogen enter at the bottom and pass up through the catalyst bed. Effluent leaves at the top and goes to further processing. Since liquid feed does not supply velocity to expand catalyst above its settled level, recycle is taken from the top and is pumped back up through the reactor to generate sufficient velocity. The reactor contents very nearly approach isothermal conditions even though fresh feed and hydrogen enter the reactor at a temperature that is more than 55 °C below the operating temperature. Very high heats of reaction are dissipated in heating cold feed to reactor temperature.

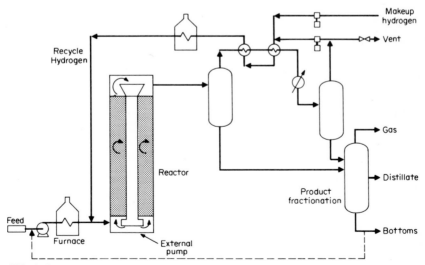

FIG. 3-1   LC-Fining process.

Reactor contents are much like a fluid which enables catalyst to be added and withdrawn during operation by use of lock-pots. Catalyst is generally added and withdrawn several times a week to maintain an equilibrium activity level which allows constant product quality to be produced.

Expanded-bed design allows very heavy feedstocks to be processed without ever shutting the unit down for catalyst replacement. There is ample free space between particles for entrained solids to pass through the catalyst bed without accumulation, plugging, or increased pressure drop. Since there is very little pressure drop and no increase in pressure drop with time on-stream, it is possible to utzilize an LC-Fining catalyst having a small diameter (½₂-in extrudates). Since the hydrogenation of heavy hydrocarbons is diffusion-controlled, use of the small particles results in considerable advantage in reaction rate.

OIL SHALE, TAR-SAND BITUMEN, AND HEAVY-OIL PROCESSING

Three exceedingly large hydrocarbon sources for synthetic crudes are oil shales, tar sands, and heavy-oil deposits. Several processes are commercial or in the process of being commercialized to exploit these resources to provide a supply of synthetic crude oil for refining. However, the initial product from the extraction step from these natural resources must be hydrotreated before being suitable for a refinery feedstock or as a fuel oil product.

The LC-Fining process has been applied to, and is well suited for, upgrading shale oil, tar-sand bitumen, and heavy oil. The process stabilizes the syncrude by reducing the oxygen and nitrogen levels. Pour point and viscosity, as well as the sulfur content, are reduced. Hydrocracking takes place, converting the high-boiling components into lighter fractions.

Table 3-1 presents the typical properties of raw shale oil, tar-sand bitumen, and a heavy oil. All have unacceptably high amounts of heteroatoms, although in different proportions.

TABLE 3-1 Properties of shale oil and tar-sand bitumen

	SO˙	TSB†	LHO‡
Specific gravity, °API	18.9	9.1	14.8
Elemental analysis, wt %			
Carbon	84.2	83.4	83.8
Hydrogen	11.3	10.5	10.9
Oxygen	1.8	1.1§	1.5
Nitrogen	1.9	0.4	0.3
Sulfur	0.8	4.6	3.6
Metals, ppm			
Vanadium	0.2	145	111
Nickel	2.0	66	59

˙SO = shale oil

†TSB = tar-sand bitumen

‡LHO = Lloydminster heavy oil

§By difference

Shale-Oil Upgrading

The processing of shale oil in the LC-Finer over a CoMo catalyst has been performed in two modes of operation: with and without recycling to extinction of 343 °C+ material. The product yields and qualities for the two operations are shown in Table 3-2. The heteroatom contents are reduced to comparable low levels for both cases although, as expected, the product yield structures are different. The 343 °C+ extinction case contains more material in each of the 343 °C− fractions. The C_1–C_3 gas make is much higher, as is the hydrogen consumption for the extinction case.

Tar-Sands Upgrading

Tar-sand bitumen is processed in the LC-Finer with two distinct modes of operation.[1] The first, and less costly, is the hydrovisbreaking mode, where the feed and hydrogen are processed through the reactor in the absence of a catalyst. Pressures as low as 1000 psig (6895 kPa) can be utilized, and conversion of the 524 °C+ material to lighter species can be as high as 75 percent by volume. Table 3-3 presents the product yield from the hydrovisbreaking operation.

Hydrotreating the whole bitumen in the presence of a ⅟₃₂-in extrudate catalyst provides increased upgrading over the hydrovisbreaking mode of operation. Desulfurization is increased from 23 to 91 percent with this operation. Severity of the process can be controlled by monitoring the throughput, the operating temperature, and the catalyst addition and withdrawal rate. The complete yield structure of catalytically hydrotreated bitumen is included in Table 3-3.

TABLE 3-2 Product yields and quality of hydrotreated shale oil

Parameter	Without 343 °C+ extinction	With 343 °C+ extinction
Specific gravity, °API	34.4	40.8
Yield, wt %		
C_1–C_3	2.89	5.81
C_4 to 204 °C	21.9	38.0
204 to 343 °C	53.8	70.5
343 °C+	30.3	——
Hydrogen consumption		
scf/bbl	1320	1930
(m^3/m^3)	(235)	(345)
Elemental analysis, wt %		
Carbon	86.29	85.75
Hydrogen	12.75	13.31
Oxygen	0.34	0.49
Nitrogen	0.60	0.43
Sulfur	0.02	0.02

TABLE 3-3 Product Yields of upgraded tar-sand bitumen

Yield	Hydrovisbreaking operation	Catalytic operation
Products, wt %		4.3
H_2S	1.1	
NH_3	0.1	0.1
C_1	0.7	1.2
C_2	0.6	1.0
C_3	0.9	1.9
C_4	0.5	1.3
C_5 to 204 °C	9.8	12.0
204 to 343 °C	29.1	37.6
343 to 524 °C	36.6	29.7
524 °C+	21.0	12.8
Total on fresh feed	100.4	101.9
Conversion, vol % 524 °C+	60	75
Desulfurization, wt %	23	91
Hydrogen consumption, scf/bbl	300	1250
(m^3/m^3)	(55)	(225)

Heavy-Oil Upgrading

LC-Fining can be used to upgrade the world's heavy crude oils to make them acceptable for transportation and refining in conventional facilities.[2] Normal heavy crude oils are too viscous for movement through existing pipelines and contain too much residuum and sulfur for economic processing in typical refineries. However, both conditions can be economically alleviated by processing the crude oils in an LC-Finer according to studies and pilot plant runs. Thus, use of this processing scheme could assist in making available hundreds of billions of barrels of crude oil in Canada, Venezuela, and the United States.

An example of a heavy oil is Lloydminster, from the Alberta-Saskatchewan border in Canada. Its properties are included in Table 3-1. Processing the 343 °C+ portion of the crude in the LC-Finer increased the API specific gravity by 14° and reduced the sulfur content by 86 percent by weight. Hydrogen consumption for the operation was 970 stdft³/bbl.

COAL LIQUEFACTION

At the present time the most active area for the LC-Fining process in synthetic fuels is in direct coal liquefaction. Continuous research in process development units (PDUs) on coal-extract hydroprocessing has been in progress since 1976. During that time several different coal-derived feeds have successfully been converted to higher-quality liquid products in a hydrogen-efficient manner.

The coal-liquefaction process using the LC-Finer is a two-stage liquefaction (TSL) process. The coal dissolution and the catalytic upgrading are performed in two separate steps. In this manner, the operating conditions for each step are optimized independently of each other to give the process maximum efficiency and flexibility.

The first stage in two-stage liquefaction is a noncatalytic coal dissolution step, such as the solvent-refined coal (SRC-I) process. Ground coal and a process-derived recycle solvent are heated with hydrogen, under pressure, and passed through a single-stage dissolver. Gases and light oil are separated from the effluent before being sent to filtration or solvent de-ashing to remove ash and insoluble organic material. The solvent–coal-extract solution is subjected to catalytic hydrotreating (LC-Fining) in stage two for hydrocracking and heteroatom removal. In an integrated mode of operation, a portion of the solvent-containing material is recycled back from LC-Fining to stage one to serve as slurry solvent.

Upgrading Coal Extracts

The majority of LC-Finer experience with coal is in catalytically upgrading SRC-I and short-contact-time (SCT) material produced at the 50-ton/day Fort Lewis, Washington, and the 6-ton/day Wilsonville, Alabama, facilities. The LC-Fining PDU work was performed separately from the coal-extract production and was therefore not integrated with the first stage with respect to the transfer of recycle solvent between the stages. These runs provided a large amount of data on optimizing operating conditions, determining yield structure and quality, rate of catalyst aging, and solvent concentration and characterization.

Feed components

The SRC-I for the upgrading studies was received as a solid material, having been filtered or solvent de-ashed and vacuum-distilled to 454 °C to remove solvent. Before SRC-I may be fed to the LC-Finer for hydroprocessing, it must be redissolved in a solvent to bring its viscosity down to pumpable limits and provide good operability within the unit. Most of the work did not employ product recycle around the LC-Finer, and so a foreign solvent (260 to 454 °C) was used. Of several solvents tested, the coal-derived Koppers heavy residue creosote oil (KC-Oil), hydrogenated to 7 percent by weight of hydrogen, was found to be the most satisfactory and representative of a native process solvent. Feed-blend ratios of 50:50 and 70:30 volume percent of SRC-I/solvent were run.

The other coal extract, SCT material, is similar to SRC-I in its production except that the coal slurry passes only through the preheater and bypasses the dissolver. Heated residence time is thus reduced, as is the hydrogen consumption. The SCT material received for LC-Fining was not de-ashed and contained its original process-derived solvent.

The SCT coal extract was topped before use to correspond to the 260 to 360 °C nominal initial boiling point (IBP) KC-Oils employed previously as solvents for SRC-I operation. The topped SCT simulated the recycle of a 260-to-454 °C or 316-to-454 °C solvent fraction.

Table 3-4 presents typical properties of SRC-I, KC-Oil, a nominal 50/50 volume percent SRC-I/KC-Oil feed blend, and SCT material. Where SRC-I is compared with raw shale oil and tar-sand bitumen (Table 3-1), it is immediately noted that the SRC-I is very deficient in hydrogen; it contains only half the amount present in the other two synfuels. The SRC-I sulfur content also is low, but nitrogen content is moderately high.

TABLE 3-4   Properties of SRC-1, KC-Oil, 50/50 vol % feed blend, and SCT

	SRC-I	KC-Oil	50/50 vol % feed blend	SCT
Specific gravity, °API	−15.7	−3.6	−11.5	−15.3
Elemental analysis, wt %				
Carbon	86.54	92.17	89.37	80.42
Hydrogen	5.88	6.70	6.32	6.82
Oxygen	4.65	0.59	2.52	7.56
Nitrogen	2.10	0.41	1.23	1.34
Sulfur	0.70	0.09	0.41	1.55
Ash, wt %	0.19	—	0.11	6.94
Metals, ppm				
Vanadium	7.3		4.5	25
Nickel	<30		1	23
Iron	321		200	12,000
Titanium	168		120	190
Sodium	521		19	200
Potassium	3.9		1	110
Calcium	<44		6	1,300

The emphasis on the upgrading work was to convert 454 °C+ material to 454 °C− liquids and to substantially reduce the nitrogen in the product. The production of a distillate product with no more than 0.30 percent by weight nitrogen was a specific goal.

Catalyst selection

Six commercially available catalysts were tested on a 50/50 volume percent SRC-I/KC-Oil feed blend.[3] The percent of 454 °C+ conversion, the denitrogenation level, and catalyst activity maintenance were monitored during each run. Included were these catalysts: three cobalt-molybdenum (CoMo), two nickel-molybdenum (NiMo), and one nickel-tungsten (NiW). The NiMo Shell 324 (⅟₃₂-in extrudate) catalyst gave the best results in all categories. Since then all LC-Finer work on coal has been done with NiMo Shell 324 as catalyst in the reactors. More extensive catalyst screening is planned in the future.

Product yields and analyses

Run 6 is representative of LC-Finer operation on a 50/50 volume percent SRC-I/KC-Oil feed blend over NiMo catalyst. Relative reactor pressure P_{rel} was 1.0; temperature ranged from 416 to 432 °C; and the relative liquid hourly space velocity SV_{rel} was 1.0. (Since total reactor pressure and space velocity are proprietary operating parameters, relative total reactor pressures P_{rel} between 1.0 and 1.35 together with relative volumetric space velocities SV_{rel} between 1.0 and 3.0 will be used.)

The 454 °C+ conversion level was 63.3 percent by weight with a hydrogen consumption of 3.61 percent by weight on feed, or 2860 stdft³/bbl (510 m³/m³). The distillate (C₅ to 454 °C) product nitrogen measured 0.17 percent by weight, well below the target maximum of 0.30 percent by weight, while the corresponding sulfur level was <0.06 percent by weight. These yields, presented in Table 3-5, are presented as a typical reactor output for the given feed on a single pass through the reactors.

Effects of feed and operational changes

Numerous feed and operating changes were tested against the base case to observe their effect on conversion and denitrogenation. Feed variations include changing the IBP of the foreign solvent, increasing the SRC-I/solvent feed ratio to 70/30 volume percent, and processing SCT material. The operational changes studied the effects of higher pressure and space velocity and increased conversion from recycling a 260 °C+ stream around the LC-Finer. The specific PDU runs are as follows:

Run 3:   41 days of operation with catalyst activity checkpoints; recycle 260 °C+ around LC-Finer.

Run 10:   31 days of operation with catalyst activity checkpoints; 50/50 percent by volume SRC-I/KC-Oil; cool-zone operation.

Run 12:   19 days of operation; SRC-I contained 1 percent by weight ash.

Run 13:   8 days of operation; non-de-ashed SCT (Wilsonville run 146).

Run 14:   11 days of operation; de-ashed SCT (Wilsonville run 146).

Run 16:   25 days of operation with catalyst activity checkpoints; non-de-ashed SCT (Wilsonville Run 145); higher pressure and space velocity.

Run 17:   32 days of operation with catalyst activity checkpoints; 70/30 percent by volume SRC-I/KC-Oil; higher pressure and space velocity

TABLE 3-5   Product yields and quality of hydrotreated SRC-I

Specific gravity, °API	2.9
Yield, wt % on feed blend	
C_1–C_4	6.69
C_5 to 199 °C	8.46
199 to 260 °C	11.05
260 to 343 °C	31.48
343 to 454 °C	22.98
454 °C+	19.21
Conversion of 454 °C+, wt %	63.5
Hydrogen consumption, scf/bbl	2860
(m^3/m^3)	(510)
Elemental analysis, wt %	
Carbon	91.73
Hydrogen	8.43
Oxygen	0.39
Nitrogen	0.38
Sulfur	<0.06
Ash, wt %	Trace

Catalyst activity Some loss of catalyst activity was observed in the base-case run, run 6.[3] The loss in catalytic activity was determined by operating the pilot unit under identical conditions at the beginning and end of a specific run (i.e., catalyst activity checkpoints). A similar technique was used for all the other runs which operated for a duration of 20 days or more. In order to remove the effect of catalyst deactivation on the 454 °C+ conversion or percent denitrification results of a run, the differences in data at the start of the run and at the end of the run were compared. The difference, if any, was divided by the length of the run to give the change in activity versus time. Each data point was then adjusted by this activity factor, with the later points receiving a larger adjustment than the earlier points. The data would then not show any catalyst-aging effect and would only reflect changes as a function of temperature or other operating variable differences.

However, several assumptions are inherent in this interpretation of the data. First, it is assumed that the change in the observed effect (such as conversion of 454 °C+ or percentage denitrification) is linear with respect to time. Second, it is assumed that the intermediate process parameter variations had no adverse effect on the catalyst deactivation function. For example, operation at constant temperature for a given interval of time would produce the same catalyst deactivation as varying temperatures (within limits) over the same interval of time. The assumptions appeared to hold true.

Space velocity and total reactor pressure In order to minimize process parameter perturbations, runs prior to run 16 were conducted at a nominally constant relative volu-

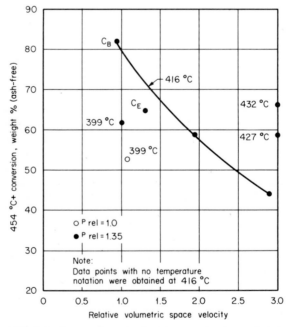

FIG. 3-2 Extract of non-de-ashed short-contact-time coal: 454 °C+ conversion vs. relative space velocity (run 16).

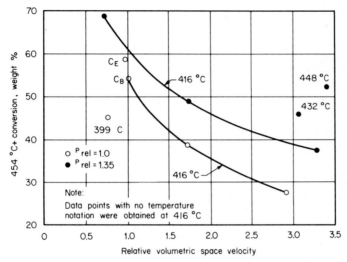

FIG. 3-3   Extract of solvent-refined coal (SRC-I): 454 °C+ conversion vs. relative space velocity (run 17).

metric space velocity (SV_{rel} = 1.0) and a constant relative total reactor pressure (P_{rel} = 1.0).[4] The main objective of runs 16 and 17 was to determine the effect of higher space velocity and higher total reactor pressure on SCT and SRC-I coal extracts, respectively.

Figures 3-2 and 3-3 show the percent 454 °C+ conversion relationship with liquid hourly space velocity for SCT material and SRC-I feeds, respectively. The data show a decrease in conversion with increased space velocity. Data points at temperatures other than the base temperature for the relationship (416 °C) are indicated to give a temperature perspective. In addition, it is observed that higher pressure results in higher conversion.

The higher-pressure operation also shows reduced catalyst deactivation. The following table summarizes the decrease with time in conversion observed in two low-pressure (P_{rel} = 1.0) and two high-pressure (P_{rel} = 1.35) runs. The conversion decrease was measured by using the catalyst activity checkpoints at the start and at the end of the run as described earlier.

Run number	Converstion decrease, wt % 454 °C+
6 (SRC-I)	9.3
10 (SRC-I)	8.3
16 (SCT), higher pressure	5.3
17 (SRC-I), higher pressure	Nil

Product distribution (comparison of SRC-I and SCT coal extracts)   Sufficient data have been accumulated and analyzed concerning the product distribution for LC-Fining of SRC-I and SCT coal extracts to allow a comparison to be made between the

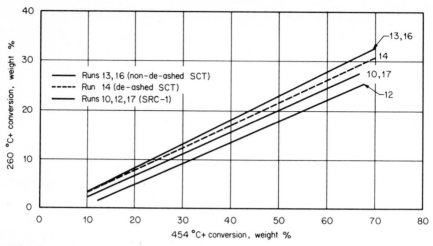

FIG. 3-4 Comparison of SRC-I and SCT coal extracts: 260 °C+ conversion vs. 454 °C+ conversion.

two types of coal extract. Figures 3-4 through 3-7 present the comparisons of product distribution for runs 10, 12, and 17 on SRC-I coal extract and runs 13, 14, and 16 on SCT coal extract.

Figures 3-4, 3-5, and 3-6 show that the non-de-ashed SCT coal extract (runs 13 and 16), when compared with de-ashed SCT coal extract (run 14), shows the greatest 260 °C+ and

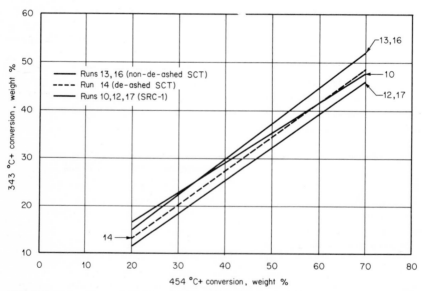

FIG. 3-5 Comparison of SRC-I and SCT coal extracts: 343 °C+ conversion vs. 454 °C+ conversion.

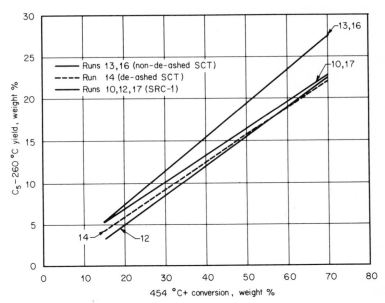

FIG. 3-6 Comparison of SRC-I and SCT coal extracts: C_5-to-260 °C yield vs. 454 °C+ conversion.

343 °C+ conversion together with the highest C_5-to-260 °C percent by weight yield for a given 454 °C+ conversion in the LC-Finer. The SRC-I run made with 1 percent by weight ash (run 12) was uniformly low in 260 °C+ and 343 °C+ conversion and also C_5-to-260 °C percent by weight yield.

Since relatively less conversion to distillates occurs in the first-stage SCT thermal step of the TSL process, it is not surprising that the SCT distillate produced in the LC-Finer is greater than that obtained from conventional SRC. However, these results do not imply that the total distillate yield for the overall TSL process has been optimized.

A comparison of the C_1–C_4 percent by weight yield from the LC-Finer (Fig. 3-7) shows that the SCT runs (13, 14, and 16) produce less C_1–C_4 gas yield than the SRC-I runs (10, 12, and 17) at a given 454 °C+ conversion.

Denitrification (comparison of SRC-I and SCT coal extracts) Figure 3-8 shows a comparison of the SRC-I and SCT runs in the LC-Finer with respect to percentage denitrification for the total liquid product. This plot also indicates the relative rate of loss in the denitrification activity when the normalized specific run data are plotted against time as equivalent periods of operation at SV_{rel} = 1.0. It will be recalled that runs 16 and 17 were made to evaluate the effect of increased space velocity.

In preparation for making a denitrification comparison, the nitrogen content of the total liquid product was normalized to 416 °C, SV_{rel} = 1.0, and P_{rel} = 1.0 by using proprietary LC-Fining correlations. The normalizing procedure also resulted in a change in the 454 °C+ conversion, which in turn resulted in a change in the amount of total liquid product produced. The normalized results incorporating the above noted procedure are plotted in Fig. 3-8.

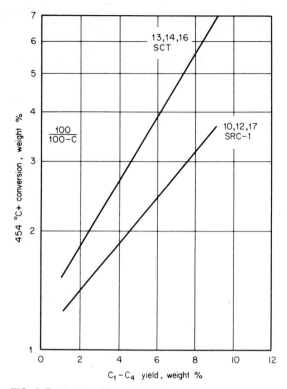

FIG. 3-7 C_1–C_4 yield vs. 454 °C+ conversion.

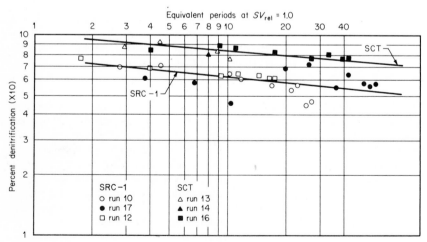

FIG. 3-8   Percentage denitrification vs. equivalent time. Data normalized to 416 °C; $P_{rel} = 1.0$, and $SV_{rel} = 1.0$.

The LC-Finer runs made with SCT feedstock (runs 13, 14, and 16) show a higher percentage feedstock denitrification than the runs made with SRC-I/KC-Oil solvent feed blends (runs 10, 12, and 17). It should be emphasized that the excellent percentage denitrification observed in the LC-Finer is representative of the total liquid product, which contains an unconverted 454 °C+ fraction.

These results are not meant to suggest that the percentage denitrification for the overall TSL process will necessarily be the same as obtained in the LC-Finer.

Recycle operation

Run 3 was made to determine the effectiveness of the LC-Fining process when operating in the recycle mode for improved conversion and denitrification of SRC-I extract using a process-derived recycle solvent (260 to 454 °C) and without an accumulation of unreacted 454 °C+ material.[5] Prior studies were always conducted with the SRC-I/solvent feed blend containing fresh prehydrogenated KC-Oil as a solvent. Consequently, the liquid products contained a large percentage of hydrotreated KC-Oil as well as product from the hydroprocessing of the SRC-I. Operations with a recycle solvent would eliminate this anomaly, and the liquid product would be derived entirely from the SRC-I feed.

The recycle operation was performed in the following manner. Once-through processing of fresh SRC-I/KC-Oil was undertaken for 11 periods (one period equals 24 h) to prepare sufficient internal recycle solvent for recycle pass I. The product was topped to 260 °C; the 260 °C+ material was analyzed for the 454 °C+ fraction; and sufficient fresh SRC-I was added to the 260 °C+ material to make a feed blend consisting of 50/50 fresh plus unconverted 454 °C+ internally generated solvent. This procedure was repeated with the product from any one recycle pass becoming the feed for the next recycle pass.

The behavior of the recycle run (run 3) with respect to 454 °C+ conversion is shown in Fig. 3-9. The data are normalized to 416 °C and $SV_{rel} = 1.0$ liquid hourly space velocity so that the effects of recycle and catalyst aging may be observed without bias. The five data points within the recycle operation regime represent the data from the five recycle

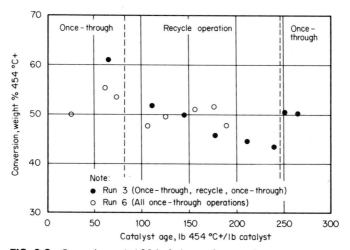

FIG. 3-9 Conversion at 454 °C+ during product recycle.

passes. Data points from an all once-through operation with a similar temperature program, run 6, are included on the figure for reference.

Comparing the once-through checkpoints at the start and at the end of run 3 defines the catalyst activity decline during the run. The amount of deactivation shown is similar to that found in the other life runs on NiMo catalysts. Therefore, it was concluded that the intervening recycle operation had caused no inherent damage to the catalyst.

During the recycle portion of the run the 454 °C+ conversion did, however, show a lower level than would be expected from a purely once-through operation as defined by the checkpoints. This might have been caused by the building up of unconvertible, or refractory, material within the system. It is recalled that the unconverted 454 °C+ material plus makeup fresh SRC was processed in the recycle passes. No effort was made to specifically draw off any 454 °C+ material to prevent refractory material buildup.

It is important to note that the 454 °C+ conversions cited for the recycle operation are based on the reactor feed (fresh plus recycled SRC) and the reactor effluent. This calculation is analogous to the once-through conversion calculation and allows for a per-pass evaluation of reactor efficiency in converting the 454 °C+ material fed to it. However, to arrive at a true conversion for the system under a continuous recycle operation, one must look at the unrecycled 454 °C+ leaving the system and the amount of fresh SRC entering the system. The calculated overall 454 °C+ conversion for this recycle run was 85 percent by volume based on fresh SRC. Denitrification was also excellent with a distillate product nitrogen level of 0.25 percent by weight.

Integrated Two-Stage Liquefaction

An LC-Finer is currently being used in a continuous Integrated Two-Stage Liquefaction (ITSL) pilot plant converting coal into liquid products. This two-stage process, shown schematically in Fig. 3-10, consists of SCT coal dissolution followed by Lummus Anti-Solvent De-ashing (ASDA) followed by catalytic hydrotreating (LC-Fining).

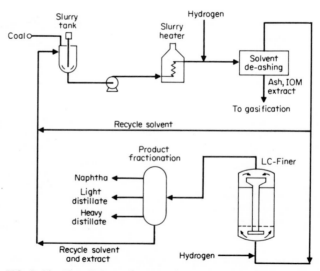

FIG. 3-10   Flow diagram of Integrated Two-Stage Liquefaction.

In the ITSL process, a 36 percent by weight coal feed slurry is fed to the first stage, where an SCT coal extract is prepared. The solvent for the coal feed slurry is process-derived recycle material taken from the ASDA overflow and the second-stage LC-Finer liquid product. The 260 °C+ SCT is processed through the ASDA unit, where some solvent is removed and recycled. The de-ashed 343 °C+ material is then fed to the LC-Finer for stage two of the two-stage process. Products are withdrawn, and a recycle stream containing 343 °C+ material is recycled to the first stage. In this manner, the 454 °C+ material is returned to the SCT unit for reprocessing. A drag stream of 454 °C+ material is removed as the ASDA underflow and is used for hydrogen generation. There is no net 454 °C+ product.

Yield and quality of products

A material balance calculated for the ITSL process shows an excellent yield of 2.9 bbl/ton (5.1×10^{-4} m^3/kg) of moisture-free (mf) coal with a hydrogen consumption of only 4.3 percent by weight. The complete yield analysis based on mf coal is presented in Table 3-6. Note the lower hydrogen consumption and lower gas yield in comparison to other liquefaction processes. This reduction is made possible by the lower-temperature operation of the LC-Finer. The high C_5-to-454 °C distillate yield compared with the low hydrogen consumption makes ITSL the most hydrogen-efficient of the coal-liquefaction processes. The 454 °C+ material produced is sufficient to keep the process in hydrogen balance via gasification.

The quality of the liquid products from ITSL also is excellent. The extremely low nitrogen and sulfur contents are shown in Table 3-6. Catalytic activity in the LC-Finer remains virtually unchanged in the ITSL process. With a catalyst age of over 1000 lb of 454 °C+ material per pound of catalyst, the conversion and denitrification levels were easily maintained. The stability of the process ensures constant product yield and quality.

TABLE 3-6   Product yields and quality from Integrated Two-Stage Liquefaction

	Net yield, wt % on MF coal	Nitrogen, wt %	Sulfur wt %
Product			
C_1–C_4	4.94		
C_5 to 199 °C	8.16	0.12	0.09
199 to 260 °C	8.99	0.13	0.08
260 to 343 °C	20.49	0.12	0.05
343 to 454 °C	11.15	0.23	0.08
Hydrogen consumption, wt % on mf Coal	4.27		
C_5 to 454 °C yield, bbl / ton mf coal	2.92		
(m^3/kg mf coal)	(5.1×10^{-4})		

All drag stream 454 °C+ material sent to gasification for hydrogen production.

Investment and operating costs

The presentation of an economic analysis as a basis for comparing synfuel processes is difficult because the results are so dependent on the methods by which and bases upon which the analyses are performed. For example, the debt/equity ratio, rate of return, plant life, and inflation are very influential factors and are not standardized among analysts. An analysis for a 30,000-ton/day (1.13×10^6-kg/h) ITSL plant has been performed, however, and some comparisons within the process are available.

The investment costs show the three major pieces of equipment to comprise 46 percent of the total cost as follows: SCT 17 percent, de-asher 9 percent, and LC-Finer 20 percent. The offsites, utilities, and storage account for 19 percent, and the hydrogen production accounts for 13 percent of the investment costs.

Of the operating costs, coal, when valued at $35 per ton, is the largest component at 51 percent of the total. Other significant contributors to the total operating cost are maintenance 16 percent, electric power 13 percent, and catalyst and chemicals 8 percent.

REFERENCES

1. Van Driesen, R. P., E. Wysocki, and R. A. Bell: "LC-Fining Applied to Bitumen from Oil Sand," C-E Lummus Brochure.
2. Van Driesen, R. P., J. Caspers, A. R. Campbell, and G. Lunin: "LC-Fining Upgrades Heavy Crudes," *Hydrocarbon Process*, vol. 58, no. 5, May 1979.
3. Potts, J. D., K. E. Hastings, and H. Unger: DOE FE-2038-25, Interim Report, August 1978.
4. Potts, J. D., K. E. Hastings, R. S. Chillingworth, and H. Unger: "LC-Fining of Solvent Refined Coal—SRC-1 and Short Contact Time Coal Extracts," *Upgrading Coal Liquids*, ACS Symposium Series 156, 1981, pp. 153–173.
5. Potts, J. D., K. E. Hastings, R. S. Chillingworth, and H. Unger: "LC-Fining of Solvent Refined Coal," *Coal Processing Technology*, CEP Technical Manual, vol. VI, 1980, p. 11–19.

THE MODIFIED LITOL™ PROCESS FOR BENZENE PRODUCTION

J. M. DUFALLO
W. A. SCHWARTZ
D. C. SPENCE

Air Products & Chemicals, Inc.
Allentown, Pennsylvania

INTRODUCTION

The Houdry LITOL™ process, in commercial use since 1964, has been producing reagent-quality benzene from secondary light oil, an aromatics-rich liquid by-product obtained during the coking of coal. As the synfuels industry develops, and fuels complexes based on solvent-refined coal (SRC) and coal-gasification technologies are designed and built, a large quantity of light-oil-type aromatic by-product liquids will become available. These streams will contain substantially more heavy aromatics and oxygen and nitrogen compounds than are present in coke oven light oil. The modified LITOL process provides a means of upgrading these difficult-to-handle, high-end-point streams.

LITOL PROCESS CHEMISTRY

The typical feedstock to a conventional LITOL unit is the C_6–C_8 fraction of liquid by-product resulting from the pyrolytic processing of coal to produce coke. This material, known as secondary light oil, represents 90 to 95 percent of the full-range liquid by-product and is rich in benzene, toluene, and C_8 aromatics (see Table 4-1). These potentially

TABLE 4-1   Typical raw light oil composition

	Full-range	Primary	Secondary
Density, kg/m³	887	950	883
ASTM distillation, °C			
Initial boiling point		165	80
10		185	85
50		203	89
90		250	115
End point		270	160
Composition, wt %			
CS_2	0.68	——	0.72
C_4-to-C_6 nonaromatics	0.59	——	0.62
Benzene	73.81	0.50	77.92
Thiophene	0.70	——	0.74
C_7-to-C_8 nonaromatics	0.75	——	0.79
Toluene	12.75	1.50	13.35
Methylthiophenes	0.15	——	0.16
C_8 aromatics	2.50	1.50	2.55
Styrene	1.35	0.50	1.40
C_9 aromatics	0.72	2.00	0.65
Vinyltoluene and methylstyrene	0.75	5.55	0.50
Indane and indene	2.45	37.50	0.60
Coumarone	0.15	2.50	——
Naphthalene	2.05	36.50	——
Alkylnaphthalenes	0.40	8.00	——
Heavies	0.20	3.95	——
Nitrogen, ppm	590	——	90
Chloride, ppm	2	——	5

valuable aromatics are contaminated by small amounts of paraffins, naphthenes, olefins, diolefins, aromatic olefins, and thiophene plus other sulfur-containing compounds.

In the LITOL process, these contaminants are converted to compounds that are easily separated from the valuable aromatics. All of the following reactions take place to some extent.

Hydrogenation of unsaturates

$$C_5H_{10} + H_2 \longrightarrow C_5H_{12} \tag{1}$$
$$\text{Pentene} + \text{hydrogen} \qquad \text{pentane}$$

Hydrocracking of nonaromatics

$$C_7H_{16} + 2H_2 \longrightarrow C_3H_8 + 2C_2H_6 \tag{2}$$
$$\text{Heptane} + \text{hydrogen} \qquad \text{propane} + \text{ethane}$$

Hydrodesulfurization

$$C_4H_4S + 4H_2 \longrightarrow C_4H_{10} + H_2S \tag{3}$$
$$\text{Thiophene} + \text{hydrogen} \qquad \text{butane} + \text{hydrogen sulfide}$$

Hydrodealkylation

$$C_6H_5CH_3 + H_2 \longrightarrow C_6H_6 + CH_4 \tag{4a}$$
$$\text{Toluene} + \text{hydrogen} \qquad \text{benzene} + \text{methane}$$
$$C_6H_4(CH_3)_2 + H_2 \longrightarrow C_6H_5CH_3 + CH_4 \tag{4b}$$
$$\text{Xylene} + \text{hydrogen} \qquad \text{toluene} + \text{methane}$$

Dehydrogenation of naphthenes

$$C_6H_{11}CH_3 \longrightarrow C_6H_5CH_3 + 3H_2 \tag{5}$$
$$\text{Methylcyclohexane} \qquad \text{toluene} + \text{hydrogen}$$

Nonselective reactions

$$\text{Aromatics} \longrightarrow \text{gas} + \text{coke} \tag{6}$$

Reaction 1 is of importance primarily in the pretreat section of the process where diolefins and styrenes are hydrogenated to levels which permit further processing at more severe conditions without fouling of equipment or excessive coking of the catalysts.

Operating conditions in the main LITOL reactors are so selected that reactions 2 and 3 proceed essentially to completion to permit production of very high purity aromatic products. The rate constant for the hydrodealkylation of toluene (4a) is smaller than that for the hydrocracking of nonaromatics by several orders of magnitude. Thus, it is possible to achieve high-purity products while retaining up to 50 percent of the toluene in the feed as a product (the remainder being dealkylated to benzene). The dealkylation of xylenes to toluene, however, proceeds at a rate approximately 3 times that of the toluene-to-benzene reaction. Therefore, C_8 aromatics are rapidly dealkylated to toluene and cannot usually be retained in quantities sufficient to justify their recovery. Reaction 5 does not proceed to a significant extent at LITOL conditions, and any aromatics produced by this reaction tend to be offset by the nonselective reactions of type 6.

TABLE 4-2 Typical aromatic by-product

	SRC* feedstock	Coal-gasification feedstock
Density, kg/m³	885	870
Distillation range, °C	C_5 to 280	C_5 to 245
Composition, wt %		
Aliphatics and alicyclics	22.46	7.95
Benzene	1.12	18.06
Toluene	4.24	15.60
C_8 aromatics	5.01	9.59
C_9+ aromatics	9.66	8.76
Indane and indenes	11.59	17.20
Coumarones and tetralins	10.65	——
Naphthalene	6.96	12.03
Alkylnaphthalenes	16.89	2.47
Biphenyl	4.47	——
Benzonitriles	2.07	——
Pyridines, anilines, and quinolines	0.49	3.45
Indoles	0.29	0.37
Phenols	——	3.71
Other nitrogen	2.82	——
Sulfur compounds	1.28	0.81
Total sulfur, ppm	3400	2800
Total nitrogen, ppm	5300	4600

*After caustic wash to remove tar acids.

Table 4-2 presents typical compositions for the liquid by-products which are expected to be derived from solvent-refined coal (SRC) and coal-gasification operations. In the SRC case, this stream, which will be caustic-washed for removal of tar acids, is actually a composite of the C_5-to-230 °C liquid piped directly from solvent refining plus the C_5 and heavier liquids produced in a hydrocracker and by delayed coker processing of the 230 °C and heavier SRC product.

In processing these feeds, additional reactions will become significant. Nitrogen and oxygen present in a variety of compounds of the pyridine, aniline, and phenol type must be hydrogenated to produce ammonia or water and a corresponding hydrocarbon. For example:

$$C_5H_5N + 6H_2 \longrightarrow NH_3 + C_2H_6 + C_3H_8$$
Pyridine + hydrogen → ammonia + ethane + propane

$$C_6H_5NH_2 + H_2 \longrightarrow NH_3 + C_6H_6$$
Aniline + hydrogen → ammonia + benzene

$$C_6H_5OH + H_2 \longrightarrow H_2O + C_6H_6$$
Phenol + hydrogen → water + benzene

Also, heavier monocyclic aromatics and alkylnaphthalenes will be dealkylated to benzene and naphthalene. As in conventional LITOL processing, reaction conditions and the relative inactivity of the catalysts for aromatic ring cracking permit the conversion of contaminants in these feeds to compounds easily separated from the products while a very high percentage of the valuable aromatics is retained.

CONVENTIONAL LITOL PROCESS

LITOL is a catalytic process with two principal reaction zones: a pretreat (hydrogenation) section and a hydrocracking–desulfurization–dealkylation section (Fig. 4-1). In conventional processing, the full-range light oil is prefractionated to reject the C_5 and lighter fraction which contains no aromatics and the C_9 and heavier fraction which is small in quantity and would consume a relatively large amount of hydrogen per unit of benzene produced. The C_6–C_8 heart cut is pumped to a vaporizer, where it is brought into contact with hot hydrogen-rich gas, and then to a pretreat reactor, where the highly polymerizable diolefins and styrene are hydrogenated. Some of the more reactive sulfur compounds are also converted at this point.

The partially hydrotreated stream is then heated to about 600 °C and passed through fixed-bed reactors where all remaining sulfur is converted to hydrogen sulfide, nonaromatics are hydrocracked, and alkyl aromatics are dealkylated. The reactor effluent is cooled and partially condensed by generating steam and by reboiling distillation columns. The liquid product is stabilized, and benzene product is recovered by conventional distillation. The benzene tower bottoms is either recycled to the reactors for ultimate conversion to benzene or further distilled for recovery of toluene product with recycle of only C_8 and heavier material.

The reactions taking place are net consumers of hydrogen and produce light hydrocarbons and hydrogen sulfide. The H_2S in the vapor from the effluent gas-liquid separator is normally rejected by absorption in amine solution. A portion of the hydrogen-rich recycle

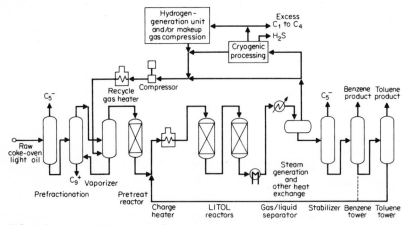

FIG. 4-1 Houdry LITOL process flow diagram.

gas is then processed cryogenically to selectively reject light hydrocarbons from the system. Since hydrogen is consumed, an external source of hydrogen is required. This may be available from other processing units, or it can be generated by steam re-forming a portion of the light hydrocarbons rejected by the cryogenic unit. Although the coke oven off-gases contain hydrogen, low pressure and high contaminant levels have made this an uneconomical source.

Conventional LITOL plants are in operation in Australia, Canada, Japan, the United Kingdom, and the United States with a combined capacity (in 1980) of over 500,000 tons/year of benzene. Yields for a typical LITOL operation producing benzene and toluene products are given in Table 4-3. Also shown are yields for processing the same feed with toluene recycled to extinction. Aromatic yield in both cases is greater than 99 percent of theoretical. Hydrogen consumption, including vent losses, is low at about 0.4 mole hydrogen per mole of aromatic product for the toluene-production case.

Benzene produced by the LITOL process is satisfactory for all commercial applications. It is unique in its extremely low content of thiophene and total sulfur, although it is produced from a high-sulfur feedstock. Table 4-4 shows properties of the total benzene product (not a heart cut) from two typical plants. The high purity is particularly evident in the 5.5 °C freeze point and the 99.97 percent benzene content.

Start-up and operation of commerical plants has generally been very smooth. Plant reliability is good, and high on-stream factors have been consistently demonstrated. *In situ* regeneration of the catalysts has been required at 6- to 12-month intervals with a LITOL catalyst total life of many years. Pretreat catalyst life is normally 3 years or more. The equipment design and operating techniques required are typical of any high-temperature, moderate-pressure hydrogen-processing plant.

LITOL has provided an economical method of producing benzene from secondary light

TABLE 4-3 Typical LITOL process yields from secondary light oil

| | Wt % of Hydrocarbon Feed | | | |
| | No toluene recycle | | Toluene recycle | |
Component	Feed*	Products	Feed*	Products
H_2	0.95	0.10	1.13	0.13
H_2S	——	0.41	——	0.41
C_1 to C_5	0.53	6.80	0.59	8.13
C_6-to-C_8 nonaromatics	0.89	0.02	0.89	0.02
Benzene	74.09	86.66	74.09	92.88
Toluene	19.22	7.36	19.22	0.02
Xylene and ethylbenzene	3.34	——	3.34	——
Styrene	0.62	——	0.62	——
C_9+ aromatics	0.70	0.01	0.70	0.01
Thiophene	1.02	——	1.02	——
Total	101.36	101.36	101.60	101.60

*After prefractionation. Includes makeup hydrogen (95 vol % H_2 and 5 vol % methane).

TABLE 4-4 Properties of LITOL-produced benzene

	Test	Plant A	Plant B
Benzene content, wt %	See note 1	—	99.97
Freeze point, °C	ASTM D-852	5.51 ± 0.02	5.50 (±0.0)
Boiling range, °C	—	0.3 (79.9 to 80.2)	—
Toluene, ppm	See note 1	—	86
Nonaromatics, ppm	See note 1		
n-Hexane		—	8
Methylcyclopentane		63	108
Cyclohexane		69	123
Thiophene, ppm	ASTM D-1685	0.3	0.1
Total sulfur, ppm	See note 2	0.6	0.4
Bromine index, mg / 100 g	ASTM D-1492	0.3	—
Basic nitrogen, ppm	See note 3	0.15 ± 0.05	—

1. Gas chromatograph (APCI) Analytical Method No. 30).

2. Granatelli method, *Anal. Chem.*, vol. 31, 1959, p. 434.

3. Moore method, *Anal. Chem.*, vol. 23, 1951, p. 1639.

oil. Based on a 1980 grass-roots installation, gross operating costs (including capital charges) for producing 56,935 tons/year of benzene from 77,162 ton/year of light oil were $72.6/ton. Taking a debit for feedstock cost at fuel value ($163/ton) and credit for fuel gas ($4.55/MM Btu) and toluene by-products ($385/ton), total production cost is $213/ ton of benzene. With benzene selling in 1980 for about $476/ton, the after-tax return on investment is greater than 50 percent.

In addition to processing secondary light oil, several existing plants supplement their feed with light aromatics–rich streams (150 °C end point) to produce additional benzene. Catalyst performance in these plants has been similar to that obtained in plants processing secondary light oil only.

Attempts to supplement LITOL unit feed with primary light oil have not been as successful. Both pilot plant and commercial tests indicated that the potential for fouling of equipment and coking of the catalysts were greatly increased by the heavier components present in the primary oil (Table 4-1). Pilot plant processing of 5 percent primary oil in secondary light oil showed the coking rate in the main LITOL reactors to be 2 to 4 times the rate for secondary oil alone. Coking in the pretreat section was higher by an even greater amount. These experiences led to the development of the modified LITOL process for use with heavier feedstocks containing significant amounts of unsaturated materials.

MODIFIED LITOL PROCESS

In the modified LITOL process (Fig. 4-2), a basic processing scheme similar to that of a conventional LITOL unit is used. The major differences are in the pretreat and product fractionation sections. A unique pretreat system was developed to partially hydrotreat heavy, difficult-to-handle aromatic streams. A pilot plant consisting of the new pretreat system and a conventional main LITOL reactor system was used to process secondary light

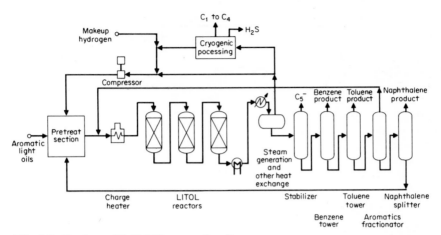

FIG. 4-2 Houdry modified LITOL process flow diagram.

oil containing various amounts of primary light oil. Product aromatic purities and cycle life were comparable with those obtained for conventional LITOL processing of secondary light oil alone.

The heaviest feed processed in the pretreater pilot plant was a 50/50 mixture of primary and secondary oil. The hydrotreating data, along with experimental dealkylation data for naphthalene boiling range feedstocks, indicate that the modified LITOL process can be used to produce high-purity benzene and naphthalene from high-end-point, difficult-to-handle feedstocks of the type which will be produced in SRC and coal-gasification operations.

Yields

Table 4-5 presents projected yields for processing the SRC and coal gasification by-product streams described earlier. The figures for the SRC-derived feed are based on recycle of aromatics heavier than toluene to maximize production of benzene and toluene. Naphthalene yield is based on once-through processing of the naphthalene and heavier fraction. Heavy ends recycle to increase naphthalene production is optional. The yields shown for the coal-gasification feedstock are based on complete recycle of toluene and heavy ends to maximize production of benzene and naphthalene. Hydrogen consumption and SNG (substitute natural gas) production are substantial for both feedstocks.

By-products

Production of SNG can amount to as much as 50 percent by weight of the feed (Table 4-5) and is the principal by-product of modified LITOL processing. This material consists primarily of C_1–C_4 hydrocarbons resulting from de-alkylation of aromatics and hydrocracking of nonaromatics. It is suitable for use directly as fuel gas, or it can be fed to a steam re-former to produce the hydrogen makeup gas required by the LITOL unit as well as other process units within the SRC or coal-gasification complex.

TABLE 4-5 Typical yields for modified LITOL process

| | Wt % of Hydrocarbon Feed | | | |
| | SRC | | Coal gasification | |
Component	Feed*	Products	Feed†	Products
H_2	4.55	0.41	4.26	0.39
N_2	——	——	0.36	0.36
H_2S	——	0.36	——	0.30
NH_3	——	0.65	——	0.56
C_1 to C_4	0.41	48.69	7.74	42.42
C_5+ nonaromatics	23.53	0.20	7.95	0.14
Benzene	1.12	15.45	18.06	55.17
Toluene	4.24	15.93	15.60	0.03
C_8 aromatics	5.01	——	9.59	0.07
C_9 aromatics	9.66	0.01	8.76	——
Indane, indenes, and tetralins	21.17	——	17.20	——
Naphthalene	6.96	16.58	12.03	12.31
Methylnaphthalenes	16.89	2.05	2.47	0.01
Biphenyl	4.47	4.47	——	——
Nitrogen compounds	5.67	——	3.82	——
Sulfur compounds	1.28	——	0.81	——
Phenols	——	——	3.71	——
Water	——	0.16	——	0.60
Total	104.96	104.96	112.36	112.36

*After caustic wash. Includes makeup hydrogen (98.9 vol % H_2, 1.1 vol % methane).

†Includes makeup hydrogen (81.0 vol % H_2, 0.5 vol % N_2, 18.5 vol % methane).

A vapor stream rich in H_2S and NH_3 will be rejected from the recycle-gas–purification section of the plant. This stream can be combined with other acid gases generated within the complex and fed to the gas-treatment plant for recovery of elemental sulfur and NH_3.

A very small stream of liquid hydrocarbons with about the same composition as fresh feed but containing some heavy polymers will be rejected from the pretreat section of the unit. This material will be blended with other liquid fuels produced elsewhere in the complex.

Waste Streams

During normal operation, the only waste effluent from a modified LITOL unit will be water, which is condensed at various points in the process. Although no plant data are yet available, it is estimated that this water will contain about 5000 wt ppm of H_2S, 4000 ppm of NH_3, and 2000 ppm of hydrocarbons. The H_2S and NH_3 can be stripped out and sent with the H_2S-rich vapor stream to the gas-treatment plant. The water can then be sent to the biological treatment facilities.

Solid wastes from the plant will consist of spent catalyst (discharged at intervals of several years) and spent treating clay (typical of clays used in refinery processing), which will be changed periodically as dictated by product-purity specifications. The catalyst can be sold to firms which recover metals from spent catalysts, and the clay will be steam-stripped of hydrocarbons in place before discharge and then used as landfill.

ECONOMICS OF MODIFIED PROCESS

Capital investment and utility requirements for a modified LITOL unit processing 330,695 tons/year of aromatics feed from SRC operations are given in Table 4-6. This feed rate corresponds to the liquids obtained from an SRC plant processing 6000 tons/day of coal, a figure which represents the projected capacity of one module of a commercial facility. A typical commercial installation might contain five of these modules and require three LITOL plants to process the available by-product liquids. The investment shown is for a grass-roots installation and is based on U.S. Gulf Coast construction in 1980.

Operating costs (1980 basis) are presented in Table 4-7 for the 330,695-ton/year plant. The utility prices used are consistent with a crude oil value of $26/bbl ($165/m^3). Makeup hydrogen is valued at cost plus return for steam re-forming of fuel gas.

Gross operating costs (excluding feedstock) for this operation, which yields 50,596 tons/year of benzene, are $517/ton of benzene product. Taking a debit of $163/ton (fuel value) for feedstock, and appropriate by-product credits for toluene ($385/ton), naphthalene ($240/ton), and fuel gas, results in a net operating cost of $240/ton of benzene. With benzene selling for about $476/ton, the after-tax return on investment (ROI) is 16.4 percent. This should be a conservative figure, since feedstock was priced at fuel value and, in fact, a portion of the feed will probably require hydrotreating to produce an acceptable fuel oil.

TABLE 4-6 Investment and utility requirements for modified LITOL processing

	SRC feedstock	Coal-gasification feedstock
Capacity, tons/year	330,695	220,465
Product yields, tons/year		
Benzene	50,596	121,255
Toluene	52,691	——
Naphthalene	55,060	27,117
Investment, 1980 U.S. Gulf Coast		
LITOL plant and auxiliaries	$29,000,000	$25,000,000
Offsites	7,500,000	6,000,000
Total	$36,500,000	$31,000,000
Utilities, units per ton benzene product		
Power, kWh	690	195
Steam, tons	0.58	−0.9*
Fuel, MM Btu	16.9	11.5
Cooling water, gal	3830	5750

*Net production

TABLE 4-7   1980 operating costs for processing 330,695 tons/year (300,000 Mg/year) of SRC plant aromatics

		$/ton of benzene
Fixed cost		
Depreciation, 15-year-straight line		48.1
Maintenance, 4% of investment		28.9
Insurance and taxes, 1½% of investment		10.8
Operating labor, $10/h, 4 men/shift		6.9
Overhead, 150% of operating labor		10.4
Supplies, including N_2		0.5
Subtotal		105.6 (116.4 $/Mg)
	1980 unit cost	
Variable cost		
Feedstock	$163/ton	1065.4
Utilities		
Power	$0.035/kWh	24.2
Steam	$12.00/ton	7.0
Fuel	$ 4.55/MM Btu	76.9
Cooling water	$0.10/(gal $\times$ 10^3)	0.4
Chemicals and catalyst		3.4
Hydrogen	$2.85/(scf $\times$ 10^3)	300.0
Subtotal		1477.3 (1628.4 $/Mg)
Credits		
Fuel	$4.55/MM Btu	682.0
Toluene	$385/ton	400.9
Naphthalene	$240/ton	261.2
Subtotal		1334.1 (1481.6 $/Mg)
Net processing cost		238.8 (263.2 $/Mg)
Benzene product value	$476/ton	476.0
Gross profit		237.2
NPAT(50% tax)		118.6 (130.7 $/Mg)
Return on investment, %		16.4

A commercial-size coal-gasification plant producing 7×10^6 m³(n)/day* of SNG would yield about 220,465 tons/year of C₅-to-245 °C by-product liquid as described in Table 4-2. Investment and utility requirements for processing this material to yield 121,255 tons/year of benzene and 27,117 tons/year of naphthalene are also given in Table 4-6. Economics is presented in Table 4-8 on the same basis as for the SRC-derived feedstock. The after-tax ROI for this case is over 37 percent.

*M³(n) symbolizes the normal cubic meter, i.e., a cubic meter of gas measured at 263 K and 101.6kPa.

TABLE 4-8 1980 operating costs for processing 220,465 tons/year (200,000 Mg/year) of coal-gasification aromatics

		$/ton of benzene
Fixed cost		
Depreciation, 15-year straight line		17.0
Maintenance, 4% of investment		10.2
Insurance and taxes, 1½% of investment		3.8
Operating labor, $10/h, 4 men/shift		2.9
Overhead, 150% of operating labor		4.3
Supplies, including N_2		0.6
Subtotal		38.8 (42.8 $/Mg)
	1980 unit cost	
Variable cost		
Feedstock	$163/ton	296.4
Utilities		
Power	$0.035/kWh	6.8
Steam	$12.00/ton	−10.8
Fuel	$ 4.55/MM Btu	52.3
Cooling water	$ 0.1/(gal × 10³)	0.6
Chemicals and catalyst		1.3
Hydrogen	$2.85/(scf × 10³)	78.1
Subtotal		424.7 (468.1 $/Mg)
Credits		
Fuel	$ 4.55/MM Btu	124.1
Naphthalene	$240/ton	53.7
Subtotal		177.8 (196.0 $/Mg)
Net processing cost		285.7 (314.9 $/Mg)
Benzene product value	$476/ton	476.0
Gross profit		190.3
NPAT (50% tax)		95.2 (104.9 $/Mg)
Return on investment, %		37.2

Since both cases produce a large quantity of naphthalene, its value has a significant impact on the profitability of the LITOL operations. In recent years, U.S. production of naphthalene has remained relatively constant at about 200,000 tons/year, two-thirds of which is obtained from coal tar and the remainder from petroleum. Petroleum-derived naphthalene is a higher-purity product and is normally sold at a higher price (up to $500/ton in 1980 vs. $240/ton for coal-tar naphthalene). Although naphthalene from the modified LITOL process would be comparable in purity to the petroleum grade, the lower price was used in the economic analyses presented in Tables 4-7 and 4-8. Since the naphthalene produced in one of these plants represents 10 to 25 percent of the 1980 U.S. market, the value of naphthalene would undoubtedly drop after several plants are built.

TABLE 4-9 Operating costs for naphthalene production from SRC hydrocrackate

Feed, tons/year		200,000 (181,436 Mg/year)
Product yields, tons/year		
Benzene		20,480
Toluene		22,180
Naphthalene		43,405
Investment, 1980 U.S. Gulf Coast		
LITOL plant and auxiliaries		$16,800,000
Offsites		4,200,000
Total		$21,000,000

Operating costs	1980 unit cost	$ per ton of naphthalene
Fixed cost		79.6
Variable cost		
Feedstock	$163/ton	751.1
Utilities		79.5
Chemicals and catalysts		3.2
Hydrogen	$2.85/(scf $\times$ 10^3)	100.0
Subtotal		1013.4 (1117.1 $/Mg)
Credits		
Fuel gas	$4.55/MM Btu	170.4
Fuel oil	$163/ton	297.7
Benzene	$476/ton	224.6
Toluene	$385/ton	196.7
Subtotal		889.4 (980.3 $/Mg)
Net processing cost		124.0 (136.8 $/Mg)
Naphthalene product value	$240/ton	240.0
Gross profit		116.0
NPAT (50% tax)		58.0 (63.9 $/Mg)
Return on investment, %		12.0

In 1980, about 60 percent of the naphthalene produced in the United States was used to make phthalic anhydride (PAN). o-Xylene has replaced naphthalene as the primary source of PAN with over 75 percent of U.S. production being based on o-xylene. At the right price, naphthalene should be able to displace o-xylene in PAN production and thereby expand the naphthalene market by a factor of two or more. Since PAN production costs are similar for either feedstock, and o-xylene was selling for about 20¢/lb ($440/Mg) in 1980, the maximum expected value for naphthalene is about $360/ton.

Conversely, if naphthalene were to be in long supply even with an expanded market, its value could drop almost to fuel value, or about $163/ton on the 1980 basis. With all other factors constant, this would result in reducing the ROI values for the operations described above to about 11 percent for SRC feed and 34 percent for coal-gasification feed. In this situation, plant economics will be improved by undercutting the LITOL feed to exclude the naphthalene fraction (about 210 °C end point).

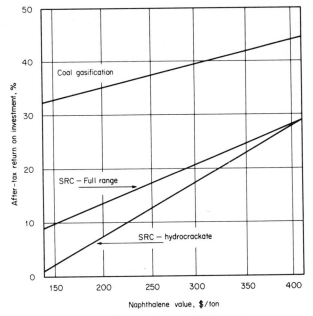

FIG. 4-3 Sensitivity of economics to naphthalene value.

If naphthalene value were high enough, it would be possible to design a LITOL plant to process a feedstock selected primarily for naphthalene production. The C_5-to-280 °C SRC hydrocrackate contains a large portion of the naphthalene and naphthalene precursors produced in the SRC plant. LITOL investment and operating costs (1980 base) for an operation of this type are given in Table 4-9. With naphthalene valued at $240/ton, the after-tax ROI is 12 percent. At a naphthalene value of $360/ton the ROI increases to almost 25 percent.

The sensitivity of the economics of the LITOL plant to naphthalene value is demonstrated by Fig. 4-3. The after-tax ROI for the three cases described is plotted as a function of naphthalene value with all other parameters constant. Since the coal-gasification feed produces mostly benzene, this case is least sensitive to naphthalene value and shows a very good profitability even with naphthalene at fuel value. The SRC full-range feed case is more sensitive, with ROI ranging from about 28 percent with naphthalene at *o*-xylene value to about 11 percent with naphthalene at fuel value. The SRC hydrocrackate-to-naphthalene case shows returns ranging from almost 30 percent to less than 5 percent as naphthalene value ranges from *o*-xylene to fuel equivalent.

SUMMARY

In over 15 years of commercial operations, the conventional LITOL process has proved to be an efficient method of producing high-quality benzene and toluene from light oil produced in steel plant coking operations. This experience, coupled with pilot plant tests on a new pretreat system specifically designed to deal with heavier feedstocks, indicates that the modified process will be equally successful in yielding high-purity aromatics from the liquid by-products of other coal-conversion processes. The economics of these operations is sensitive to by-product prices, but attractive returns on investment are projected over a wide range of values.

INDEX

1

APPENDIX

ACRONYMS AND ABBREVIATIONS USED IN THIS BOOK

ASWS	ammonium sulfide water stripper
B/D	barrels per day
BET	Brunauer, Emmett, and Teller surface area
BFW	boiler feedwater
BSD	barrels per stream day
BTX	benzene, toluene, and xylene
CCCG	combined-cycle coal gasification
CSD	critical solvent deashing
CV	calorific value
CW	cooling water
DAS	deashing solvent
DEA	diethanolamine
DOE	U.S. Department of Energy
DOS	days on stream
DPU	dust preparation unit
ED	expansion drum
EPA	U.S. Environmental Protection Agency
EPRI	Electric Power Research Institute
FOEB	fuel-oil equivalent barrel
HHV	higher heating value (of fuel)
HP	high-pressure (steam)
HTS	high-temperature shift (of synthesis gas)
IBP	initial boiling point
ID	inside diameter

IOM	including other minerals
K-MAC	Kerr-McGee ash concentrate
KO	knockout (pot)
LH	lock hopper
LHSV	liquid hourly space velocity
LPG	liquefied petroleum gas
LV	liquid volume
maf	moisture- and ash-free (analysis, usually refers to coal)
MHC	medium-heat content (gas)
MHV	medium heating value (of fuel)
MIS	modified *in situ* retorting (of oil shale, Occidental Oil Shale, Inc.)
MM	million
MP	medium-pressure (steam)
MTG	Methanol-to-Gasoline Process (Mobil)
NMHC	nonmethane hydrocarbons
PDU	process-development unit
PDWS	Primary Drinking Water Standards
POX	partial oxidation
QV	quench vessel
RCRA	Resource Conservation and Recovery Act
RON	research octane number
RVP	Reid vapor pressure (of gasoline)
SH	superheater
SNG	substitute (or synthetic) natural gas
SRC	solvent-refined coal
TDS	total dissolved solids
TEL	tetraethyllead
TPSD	tons per stream day
TSL	two-stage liquefaction
TSP	total suspended particulates
VGO	vacuum gas oil
VM	volatile matter
WHB	waste-heat boiler
WHSV	weight hourly space velocity